The Development of Mathematics Between the World Wars

Case Studies, Examples and Analyses

The Development of Mathematics Between the World Wars

Case Studies, Examples and Analyses

editor

Martina Bečvářová

Czech Technical University in Prague, Czech Republic

World Scientific

NEW JERSEY • LONDON • SINGAPORE • BEIJING • SHANGHAI • HONG KONG • TAIPEI • CHENNAI • TOKYO

Published by

World Scientific Publishing Europe Ltd.
57 Shelton Street, Covent Garden, London WC2H 9HE
Head office: 5 Toh Tuck Link, Singapore 596224
USA office: 27 Warren Street, Suite 401-402, Hackensack, NJ 07601

Library of Congress Cataloging-in-Publication Data
Names: Bečvářová, Martina, editor.
Title: The development of mathematics between the world wars : case studies, examples and analyses / editor, Martina Bečvářová, Czech Technical University in Prague, Czech Republic.
Description: New Jersey : World Scientific, [2021] | Includes bibliographical references and index.
Identifiers: LCCN 2020034549 | ISBN 9781786349309 (hardcover) | ISBN 9781786349316 (ebook for institutions) | ISBN 9781786349323 (ebook for individuals)
Subjects: LCSH: Mathematics--Europe, Eastern--History--20th century. | Mathematics--Study and teaching (Higher)--Europe, Eastern--History--20th century. | Europe, Eastern--Intellectual life. | Europe, Eastern--History--1918-1945.
Classification: LCC QA27.E85 D48 2021 | DDC 510.943/09042--dc23
LC record available at https://lccn.loc.gov/2020034549

British Library Cataloguing-in-Publication Data
A catalogue record for this book is available from the British Library.

For any available supplementary material, please visit
https://www.worldscientific.com/worldscibooks/10.1142/Q0273#t=suppl

Desk Editors: Balasubramanian Shanmugam/Michael Beale/Shi Ying Koe

Typeset by Stallion Press
Email: enquiries@stallionpress.com

https://doi.org/10.1142/9781786349316_fmatter

Preface

World War I, also known as the Great War, heavily affected academic life. In European countries, many talented researchers and students were killed in action; scientific activities were halted and they resumed only in the postwar years. However, this disruption turned out to be a catalyst for the birth of a new generation of mathematicians, the emergence of new ideas and theories, and the surprising creation of new and outstanding scientific schools (e.g., in Poland and Hungary). These phenomena are not sufficiently reflected in the existing literature.

This book presents an analysis of the transformation and development of scientific life of the mathematical communities during the interwar period in Germany, Russia, Poland, Hungary, and Czechoslovakia. It describes the impact of World War I in a broader historical context (social and political phenomena, roots of the revitalization of older mathematical centers, the foundation of new renowned and successful mathematical schools, analysis of the role of young, talented and motivated mathematicians in these processes, and description of the mathematical achievements, especially in Poland and Hungary). We pay special attention to the reconstruction and development of international cooperation, effects of religious persecution, rising anti-Semitism and political purges, destruction of mathematical research centers, attitude of universities and scientific societies in the struggle against terror, and life of the mathematical community in the dictatorship era, and so on.

The chapter *German Mathematics Between the Two World Wars* describes the roots of the high level of mathematics in Germany, especially in Berlin, Göttingen, and other centres. It documents the crucial role of Felix Klein and David Hilbert, who were behind the great advances in Göttingen, and shows how their leading position, not only in Germany, but also all of Europe, was achieved. A new generation of scientists, mostly educated in Göttingen, emerged after World War I. Some of them were not of German origin. Berlin, which had been a first-class mathematical center at least until the end of the 19th century, lost this position and recovered only slowly its former fame. The war led to inflation and financial crises, and the postwar era of new developments was interrupted by Nazis in 1933. New laws and discrimination against Jews led to the massive emigration of mathematicians, mostly to the United States. Of those who did not emigrate, only a few survived, while others were murdered in concentration camps or they committed suicide. Those who were able to start a new career abroad often helped to raise the level of mathematics in those countries that accepted them. These facts are illustrated in specific cases that are typical of the larger group.

The chapter *Russian Mathematics and Its Relation to German Mathematics (1914–1940)* describes how Soviet mathematics developed in the period between the two World Wars, and the roots of its fast expansion. Naturally, it mainly deals — with a few exceptions — with the two mathematical centers in Moscow and St. Petersburg. The Moscow school of the theory of real functions and the role of its founders Dimitri Fedorovich Egorov and Nikolai Nikolaevich Lusin are presented in detail. The mathematical contributions of members of the so-called Lusitania are briefly described, and it is shown how they influenced the further development of new branches of mathematics. Special attention is paid to individual fates and the role of international contacts. In the case of St. Petersburg, the influence of Vladimir Andreevich Steklov and Andrei Andreevich Markov is pointed out. The devastating influence of political events in the 1930s on mathematics is illustrated by the fates of selected mathematicians. The last part of the chapter deals with the mutual contacts of Russian and German mathematicians, which were very intensive, especially before the Nazi era. Despite their different political backgrounds, both states showed some common features in the development of mathematics.

The chapter *Roots of Modern Polish Mathematics (1815–1915)* presents a brief history of Polish mathematics and its cultural and historical background from 1815 to 1915. That century saw the foundations of the Polish school that emerged soon afterwards and blossomed between the two world wars.

The chapter *The Position of the Jagiellonian University and Its Mathematical Community Before and After WWI* describes the situation of the mathematical community of Kraków, which was historically influenced by Austro-Hungarian politics, culture and sciences. At the beginning of the 20th century, the Jagiellonian University in Kraków was particularly strong in differential geometry, differential equations, and mathematical logic. After World War I, when Kraków became a part of independent Poland, research on classical analysis and geometry continued, but new directions emerged. The mathematical community of Kraków was important also for its role in professional organization of Polish mathematics: in 1919, a mathematical society was established in Kraków; it was later transformed into the national Polish Mathematical Society. Thus, in the interwar period, the Jagiellonian University uniquely bridged traditional and modern mathematics.

The chapter *Transformation of the Warsaw Mathematical Community in the Aftermath of WWI* focuses on the birth and development of a new and strong mathematical community in Warsaw. In mathematics, the key people at the university were Stefan Mazurkiewicz, Zygmunt Janiszewski, and Wacław Sierpiński. Along with the logicians Jan Łukasiewicz and Stanisław Leśniewski, they became the leaders of a mathematical school following Janiszewski's vision and focusing on set theory and its applications. Another mathematical community centered around Samuel Dickstein efficiently functioned in Warsaw for several decades. In the "second Warsaw", there was an active group of researchers interested in mathematical analysis in a broad sense. Many outstanding results were obtained, but the Warsaw community was suffering from a lack of academic employment opportunities for creative mathematicians, as well as from anti-Semitic attitude, which caused a large number of mathematicians and logicians to emigrate from Poland.

The chapter *Formation and Activity of the Lwów School of Mathematics* is devoted to the creation of a unique and remarkable mathematical center. An outstanding role in its formation was played by Józef Puzyna, followed by Wacław Sierpiński, Zygmunt Janiszewski, and Stefan Mazurkiewicz. After the turmoil of World War I and the Polish-Bolshevik war, a key role was played by Hugo Steinhaus and Stefan Banach. Between the wars, they created a mathematical school of functional analysis. Other mathematical disciplines were also developed at this school, in particular, topology (mostly set-theoretical), algebra, probability theory, measure theory, etc. The unique style of doing mathematics in Lwów included long café sessions during which mathematical problems were formulated and discussed. Since the population of Lwów included Poles, Jews, Ukrainians, and Armenians, this chapter analyses the development of mathematics in this multicultural atmosphere. It also demonstrates diverse facets of mathematics in Lwów in the period between the two World Wars.

The fruitful period of Hungarian interwar mathematics and its famous mathematical achievements are analysed in the chapter *Impact of World War I on the Mathematical Community in Hungary*. The first Hungarian results in mathematics are associated with the name of Bolyai and appeared only after 1820. However, by the end of the 19th century, a small group of high school and university professors of mathematics emerged who were educated in foreign universities, especially German ones. These personalities then brought another generation of extremely good science and mathematics teachers. Poverty occurring in Hungary after World War I caused a great decline of this elite. Its members have profoundly influenced the history of the 20th century through their scientific work abroad. They were known as the Martians. Yet, despite the dismal situation in Hungary, many outstanding scientists and teachers continued to educate the next generation of mathematicians at home. This chapter provides a picture of the incredible qualitative development of Hungarian mathematics all the way to World War II, with a very brief outlook on the subsequent years.

The chapter *Mathematical Community in Czechoslovakia Between World Wars* focuses on the life, activities, and important results of the Central-European mathematical community from the end of the 19th century until the beginning of World War II. With the help of archival documents, primary literature, and secondary sources, the chapter analyzes and documents the following: a very interesting and rare phenomenon — the fruitful and peaceful coexistence of two different mathematical communities, Czech and German; their positions within the Czechoslovak Republic as well as Europe; their professional activities and achievements; as well as contacts and collaboration with other European mathematical communities. During the first half of the 20th century, mathematics in Czechoslovakia achieved the European level in research and teaching, thanks to the numerous contacts with foreign countries, construction of all types of new schools, and research institutions, and state support of publishing, social and traveling activities. However, both mathematical communities were too small to create important or famous mathematical schools as in Poland, Germany, Russia or Hungary. Mathematicians from Czechoslovakia carried out research in selected disciplines (e.g., mathematical analysis, modern algebra, topology, number theory, graph theory, theory of configurations, differential and integral geometry, and mathematical applications in economics) which corresponded to their own interests. Their research works followed the development of European mathematics, but contributed to only a few domains (topology, real functions, graph theory, and number theory). Further development was violently disrupted by the German occupation of 1939, which totally destroyed both mathematical communities.

This book also brings original insights into the history of selected mathematical disciplines; the corresponding chapters are devoted to the origin, significance and implications of some ground-breaking mathematical ideas and trends in mathematics, in particular, set theory, potential theory, combinatorics and discrete mathematics.

The chapter *The End of the 19th Century and the First Third of the 20th Century* is the first case study in this book devoted to the history of modern mathematics. It recalls a major skepticism toward the future development of certain branches of science, which was a consequence of fascinating results obtained during the 19th century. It illustrates the situation natural sciences and mathematics with numerous quotations from the works of great scientists and thinkers. The chapter discusses new trends in mathematics which influenced mathematical thinking at the turn of the 20th century. It focuses on the birth of the axiomatic approach and set theory and their role in the development of modern mathematics. The author pays particular attention to the third crisis of mathematics. The chapter explains why the crisis occurred, how and who attempted to solve it, and why it could not be resolved. The crisis in the foundations of mathematics bothered mainly those researchers dealing with logic and the rigorous development of mathematics. The chapter describes why some mathematicians embarked on the difficult and challenging journey of mathematical accuracy and struggle for exactness, and shows that their attempts influenced the axiomatic development of various mathematical disciplines.

The chapter *Set Theory — A Vehicle to Modern Mathematics in Poland* shows some important moments in set theory, it initiated by Georg Cantor in the last decades of the 19th century, and before World War I it already gained wide recognition, aspiring to become an independent field and even a basis for all mathematics. Among the first developers of the theory were the Polish mathematicians of the time who were introducing it in their textbooks, university lectures, or research. After the war, Wacław Sierpiński became the leader of a group in Warsaw that soon rose in importance. The group was concentrated around the journal *Fundamenta Mathematicae*, founded in 1920 and devoted specifically "to the theory of sets and its applications". The foundation of the journal was also influential in the start of the Polish school of mathematics. The journal became (and still is) one of the most important ones in the world, and Polish mathematicians contributed more than 20% of all the papers, in particular, largely promoting such areas as the validity and role of the Axiom of Choice and the Continuum Hypothesis, the development of descriptive set theory, and the theory of large cardinals. The chapter tells the story up to World War II, emphasizing the role of the journal in advancing Polish mathematics to world status.

The chapter *Transformation of Mathematics Between World Wars: The Case of Potential Theory* offers a walk through the history of mathematical analysis as well as the corresponding mathematical community in the first half of the 20th century. During the last three centuries, potential theory has attracted the most famous mathematicians. It is shown to what extent the evolution of this field moved in directions determined in advance by leading scientists and how it inspired the introduction of new methods and mathematical tools of a large scope. The transformation of ideas towards abstract concepts gradually led to ultimate solutions of problems that had been open for many decades. In particular, attention is focused on the development of knowledge concerning the Dirichlet problem, the basic boundary value problem of mathematical physics. The counterexample of Henri Lebesgue, showing that this problem is not always solvable in the classical sense, marked a decisive step in broadening the horizons in the development of potential theory. New tools, such as measure and integration theory and the notion of capacity, as well as other techniques, turned out to be substantial for progress in this discipline. The historical development is described not only from the mathematical point of view but also regarding the background of relationships within the mathematical community. For instance, the confrontation relations between the American mathematicians Norbert Wiener and Oliver Kellogg as well as the competition between the French and American mathematicians in the 1920s are mentioned. This contribution may lead to a better understanding of the viability of mathematics during the first half of the 20th century and, in the case of potential theory, provides clues to the explanation of its later development. It also illustrates the search for unity in mathematics, which is typical of the period discussed.

The chapter *Transformation of Mathematics Between World Wars: The Case of Combinatorics* provides an overview of the major discoveries in combinatorics between the Wars. In comparison with standard references on the history of combinatorics, the text focuses on a shorter historical period and provides additional details as well as some interesting excerpts which illustrate the spirit of the original works. It describes the renewal of interest in Latin squares, the birth of Ramsey theory and combinatorial set theory, the development of graph theory, and more. It shows that new results in combinatorics were often motivated by problems from other disciplines (set theory, group theory, logic, number theory), and also by practical problems (design of experiments, and construction of electricity networks). Graph theory initially had a bad reputation of being a science of trivial problems arising from recreational mathematics, but the publication of the first textbook helped to change this attitude. The final section is of a biographical character, and describes how the lives of mathematicians mentioned in previous sections were affected by the World Wars and by the rise of anti-Semitism.

The book presents the history of the interwar mathematical communities in selected European countries and several case studies on the development of some mathematical discipline, in a way that is accessible to a wide audience of mathematicians and the historians of mathematics and natural sciences. The text is supplemented with numerous original quotations as well as figures, which facilitate a deeper and better understanding of the subject.

We gratefully acknowledge the support of the Czech Science Foundation, project no. 18-00449S entitled *The impact of WWI on the formation and transformation of the scientific life of the mathematical community.*

Prague 2020

Martina Bečvářová

https://doi.org/10.1142/9781786349316_fmatter

About the Editor

Martina Bečvářová studied mathematics and physics at Charles University in Prague, where she received her PhD in the history of mathematics (1997). In 2016, she became a professor at Charles University in Prague. Since 1998, she has been teaching mathematics at the Faculty of Transportation Sciences of the Czech Technical University in Prague. Her research interests include the history of mathematics, the history of scientific societies and schools and the history of education in mathematics. She has published, as an author or co-author, 17 monographs and several research papers (in Czech, English, German, Italian, Polish, Croatian and Bulgarian).

https://doi.org/10.1142/9781786349316_fmatter

About the Authors

Vojtech Bálint graduated from the Faculty of Science of Pavol Jozef Šafárik University in Košice. Since 1971 till 2016, he taught at the University of Transport in Žilina, later renamed to University of Žilina, where he held positions of associate professor and extraordinary professor. His main interests are in discrete geometry, teaching of mathematics, and history of mathematics. He is the author of more than 50 original research publications in discrete geometry and more than 50 publications on teaching of mathematics. For almost half a century, he has been involved in mathematical olympiads, and served as the leader of the Slovak national team in the International Mathematical Olympiad 11 times during the period 1996–2013.

Jindřich Bečvář graduated from the Faculty of Mathematics and Physics of Charles University in Prague in 1970, where he has been working since 1969. He is the author of more than a hundred scientific, professional and educational works and articles, lecture notes, textbooks, monographs or book chapters (linear algebra, algebra, history of mathematics). Since the end of the 1980s, he has been the organizer of national summer schools in the history of mathematics and seminars for university and high school teachers. From 1992 to 2016, he supervised 10 successful doctoral students. In 1994, he co-founded the book series *Dějiny Matematiky* (History of Mathematics), in which 65 volumes have already been published. He is a member of the program committee of the international conference History of Mathematics.

Stanisław Domoradzki studied mathematics at the Jagiellonian University in Kraków. He received PhD in the scientific discipline of mathematics in the field of history of mathematics (1995, Pedagogical University in Kraków). He obtained habilitation in historical sciences in the field of history of science/history of mathematics (2012, L.&A. Birkenmajer Institute for the History of Science, Polish Academy of Sciences, Warsaw). Since 2012, he has been a professor at the University of Rzeszów, first at the Faculty of Mathematics and Natural Sciences, and since

2019 at the College of Humanities. He is the author of over 140 publications, several books on the history of mathematics, history of mathematical education, mathematical culture throughout history, as well as on mathematical education and the concept of pedagogical inheritance. He collaborates with research centers in Austria, Czech Republic, Greece, Slovakia, Italy, Ukraine, and Russia. He is editor-in-chief of *Antiquitates Mathemaicae*, a journal on the history of mathematics published by the Polish Mathematical Society.

Roman Duda studied mathematics at the University of Wrocław. After obtaining PhD in topology at the Institute of Mathematics of the Polish Academy of Sciences (1961), he spent a year in Cambridge and became a professor in 1976. His early research interests in point-set topology led to the two-volume monograph *An Introduction to Topology* (1986, in Polish). After broadening his interests to history and philosophy of mathematics, he wrote other monographs including *Mathematicians of the XIX and XX Century Related to Poland* (2012, in Polish), *Pearls from a Lost City: The Lvov School in Mathematics* (two Polish editions in 2007 and 2014, American translation 2014), and *History of Mathematics in Poland* (2019, in Polish). He published over 200 papers (in Polish, English, French, and German) and several books. He is an honorary member of the Polish Mathematical Society and of the International Academy of the History of Science.

Ivan Netuka graduated in mathematics from Charles University in Prague, Faculty of Mathematics and Physics, where he taught since 1968 till 2020. He was appointed full professor in 1986, and served as the dean of the faculty in the period 1999–2005. An internationally recognized expert in mathematical analysis, Ivan Netuka was a leading representative of the Prague school of potential theory, an author or co-author of several monographs, textbooks, and more than hundred journal articles. Besides mathematical analysis, he had a strong interest in the history of mathematics. He received a number of domestic and foreign awards, including Chevalier dans l'Ordre des Palmes Académiques, Österreichisches Ehrenkreuz für Wissenschaft und Kunst I. Klasse, the Golden Commemorative Medal of Charles University, and the Bernard Bolzano Honorary Medal for Merit in Mathematical Sciences awarded by the Academy of Sciences of the Czech Republic.

Antonín Slavík studied computer science and mathematics at Charles University in Prague, Faculty of Mathematics and Physics, where he is currently an associate professor. His primary interests are in differential equations, integration theory, and history of various mathematical disciplines. He is an author or co-author of several books, book chapters, and numerous research papers in these areas. He is involved in the popularization of mathematics, and has written a number of expository articles in English and Czech.

Margaret Stawiska-Friedland studied mathematics and philosophy at Jagiellonian University in Kraków (Poland). She received her PhD in mathematics from

Northwestern University (Evanston, Illinois, USA) in 2001. She held several academic appointments in Poland and the USA. Since 2011, she has been working full time for the American Mathematical Society as an associate editor of *Mathematical Reviews*/MathSciNet in Ann Arbor (Michigan, USA). Her research interests include complex analysis, dynamical systems and potential theory as well as history of mathematics and logic in Central and Eastern Europe in the 19th and 20th centuries. She has authored or co-authored over 30 research publications in the related areas in English and Polish.

Jiří Veselý graduated in mathematics from Charles University in Prague in 1962. For many years, he taught at the same university, Faculty of Mathematics and Physics, mainly real and complex analysis. His interests are connected with potential theory and history of analysis. He is an author and co-author of 17 textbooks and more than 60 papers. He spent longer periods in the USSR, Denmark, and the United States. He was among the main organizers of several major conferences and founded the Czechoslovak TEX User Group. After retirement from the position of associate professor in 2005, he also lectured at the Technical University in Liberec. As an active member of the Union of Czech Mathematicians and Physicists, he is involved in the popularization of mathematics and its history.

Mykhailo Zarichnyi studied mathematics at the Ivan Franko University of Lviv. He obtained his PhD in 1983 and habilitation in 1992, both at the Moscow State University. In 1982, he began working at the Ivan Franko University of Lviv. And since 2003, he has been working at the University of Rzeszów, where he is currently a professor of the College of Natural Sciences. His scientific interests include topology of infinite-dimensional manifolds, categorical topology, asymptotic topology, and history of mathematics. He is the author of three monographs and about 150 research papers.

https://doi.org/10.1142/9781786349316_fmatter

Contents

https://doi.org/10.1142/9781786349316_0001

Chapter 1

German Mathematics Between the Two World Wars

Jiří Veselý

1. Introduction

At the beginning of the 19th century, Germany had 17 universities and three technical universities.[1] Education of mathematics had a high standard and its centers were visited by many foreign guests. With a few exceptions, we will mainly concentrate on two places, Berlin and Göttingen, as the most important traditional mathematical centers. The names of states are mentioned in a slightly vague sense since they were changing at that time.

2. Berlin Until 1918

In Berlin, where the university was founded in 1810,[2] there taught Leopold Kronecker (1823–1891), Ernst Eduard Kummer (1810–1893), and Karl Theodor Wilhelm Weierstrass (1815–1897). The first of them was born in a wealthy Jewish family and obtained broad classical education. At the "gymnasium", his teacher was Kummer, who later became his lifelong friend. Kronecker studied in Berlin and received his doctorate in 1845 for a thesis, *De unitatibus complexis* [On Complex Units][3], on algebraic number theory. One of his two advisors was Johann Gustav Peter Lejeune Dirichlet (1805–1859), who highly regarded the thesis. After

---

[1]Here is a list of the corresponding towns, including the years of foundation of the university: Heidelberg (1386), Leipzig (1409), Rostock (1419), Greifswald (1456), Freiburg (1457), München (1472), Tübingen (1477), Halle-Wittenberg (1502), Marburg (1527), Jena (1558), Würzburg (1582), Giessen (1607), Kiel (1665), [University of Arts in] Berlin (1695), Göttingen (1734), Erlangen-Nürenburg (1743), and Münster (1780). Technical universities were in Braunschweig (1745), Freiberg (1765), and Clausthal (1775). We use for both Technical University and Technische Hochschule the same abbreviation: TU.

[2]Now known as *Humboldt-Universität zu Berlin.*

[3]Published much later in *Journal für Mathematik* **93**(1882), 1–52.

Figure 1. The main building of Berlin University.[4]

Kronecker received his doctorate, he followed his father's wish and returned home to manage his family estates. He was a talented mathematician and a very good businessman, too. He was doing mathematics regardless of the fact that between 1845 and 1853 he did not publish.

In 1855, Kronecker returned to Berlin where he remained for the rest of his life.[5] From 1861 onwards he was a member of the Berlin Academy. This honour was accompanied by the right to lecture at the university, which Kronecker did. In 1866 he was offered the Riemann's chair in Göttingen but decided to stay in Berlin, where he became a university professor in 1883 (the first employment in his life). He held this position until his death in 1891.

Kronecker is one of the founders of group theory and, for sure, one of the most influential algebraists of the 19th century. When Georg Ferdinand Ludwig Philipp Cantor (1845–1918) started to publish his "Mengenlehre" he almost immediately opposed his ideas (and he was not the only one among famous mathematicians who did; see Kline (1972), p. 1003). Even today, his statement *Die ganzen Zahlen hat der liebe Gott gemacht, alles andere ist Menschenwerk* [God created the integers, all else is the work of men] is quoted. This comes from Kronecker's obituary, where his whole mathematical profile is described together with a list of his works.[6]

---

[4]Christian Wolf, Wikimedia Commons, https://commons.wikimedia.org/wiki/File:Frontansicht_des_Hauptgeb%C3%A4udes_der_Humboldt-Universit%C3%A4t_in_Berlin.jpg [14.4.2020].

[5]Biographical data are often taken from *MacTutor History of Mathematics Archive* by O'Connor and Robertson (2020). Almost all data were verified from independent sources.

[6]Kronecker's statement is from a lecture delivered in Berlin, 1886; see H. Weber: Leopold Kronecker [Obituary], *Jahresbericht der Deutschen Mathematiker-Vereinigung* **2**(1893), pp. 5–31, where it is quoted on p. 19.

Kummer was Professor of Berlin University since 1855 and Weierstrass joined him in 1864.[7] In 1861 they together founded the first German mathematical seminar for advanced mathematics (see Begehr *et al.* (1998a) Volume 1, p. 129). In contrast to Kronecker, who had only 12 doctoral students, they had many more: Kummer 58 and Weierstrass 46.[8] One of them, Adolf Kneser (1862–1930), student of Kummer and later on professor in Breslau, expressed his opinion as follows: *Weierstrass was unquestioned ruler of the whole enterprise* (see Begehr *et al.*, 1998b, p. 74).

Kummer's main mathematical contributions lie in number theory. His interest in optics led him to some discoveries in geometry. He also tried to prove Fermat's Last Theorem and was able to prove the assertion for all positive integers up to 100 and for others of a special form.

Weierstrass is one of those mathematicians who made mathematics a real Queen of Sciences. He contributed many beautiful results which form the basics for both real and complex function theory. Most of them are connected forever with his name. Here are just a few examples of his results: he proved fundamental statements on continuous real functions on a compact interval, the existence of a continuous nowhere differentiable function on an interval, studied complex functions via power series (and more). A great part of his impact on mathematics and the way in how it is done now is contained in his lectures. His collected works are contained in seven volumes the first two of which appeared before his death — he realized that publishing of the materials of his lectures would be a difficult task for anybody else. Besides, the first of those books appeared in 1894 and the seventh in 1927.

All three "giants" cooperated quite well, but after 1870 their personal relationships started to deteriorate. Kronecker's and Weierstrass' ideas about infinity were quite different. As a result, Kronecker not only disallowed basics of set theory of Cantor but also he even denied the existence of transcendental numbers. Because of mutual quarrels with Kronecker, Weierstrass wanted around his 70s to leave Berlin for Switzerland (see Begehr *et al.*, 1998b, p. 75). The lectures of Weierstrass were very influential and helped to increase the precision of mathematical thought. His results on uniform approximation or fundamentals of complex analysis are now classical. But at the end of the 19th century, the golden years of Berlin mathematics were over.

---

[7]Both Kummer and Weierstrass started their careers as "gymnasium" teachers.

[8]The number of students depends on the way how it is defined. Here it means that the person was among the advisors of the student's doctoral dissertation (an equivalent of PhD thesis).

3. Göttingen Until 1918

In the case of Göttingen we will mention the most important persons who formed mathematics at the university. It was founded in 1734 by king George II of England and Elector of Hanover. Note that Göttingen mathematics goes back to Johann Carl Friedrich Gauss (1777–1855), Georg Friedrich Bernhard Riemann (1826–1866), and Dirichlet. Rudolf Friedrich Alfred Clebsch,[9] (1833–1872), a specialist in algebraic geometry, came to Göttingen in 1868. Previously he also taught at secondary schools and at various German universities.

Clebsch suggested in 1867 to found a society for *all* German mathematicians and also a new mathematical journal.[10] Clebsch was not able to work out the idea of the society since he soon after that suddenly died.

The idea was adapted by the architect of Göttingen mathematics Felix Christian Klein (1849–1925). He came there at first as a private docent in 1871. A year later (at the age of 23!) he was appointed professor at Erlangen and after stays at TH München[11] and Leipzig he returned back to Göttingen in 1886. As a geometer and the author of the *Erlangen program*,[12] he was a well-known scientist and so he could devote strength to his other love — organizing work for mathematics. In 1895 he was joined by David Hilbert (1862–1943) and in 1902 by Hilbert's colleague from Königsberg and Hilbert's good friend Hermann Minkowski (1864–1909). While the golden era of Berlin mathematics at the end of the 19th century was over, the fame of Göttingen attracted mathematicians from all over the world like a magnet.

How this fame accrued needs an explanation. Klein worked always very hard. In 1882, after competing with Jules Henri Poincaré (1854–1912) in completion of the theory of automorphic functions, Klein's health collapsed. The aim of both was to formulate and prove the so-called uniformization theorem, which claims that every simply connected Riemann surface is conformally equivalent to one of the following Riemann surfaces: the open unit disk, the complex plane, or the Riemann sphere. It is the basic theorem of this theory. Klein succeeded in formulating such a theorem

---

[9] Known frequently only as Alfred Clebsch.

[10] At that time, there existed only the *Journal für die reine und angewandte Mathematik*, published since 1826 in Berlin (also known as Crelles journal). A year later, Clebsch and Carl Gottfried Neumann (1832–1925) started to publish *Mathematische Annalen*. For details of *Mathematische Annalen* see Behnke (1973). Both these journals exist up to now.

[11] Now Technische Universität München, but the original name was different. Later for some time it was the Königlich Bayerische Technische Hochschule München.

[12] On the occasion of Klein's admission to the faculty of the University of Erlangen he delivered in 1872 a speech *Vergleichende Betrachtungen über neuere geometrische Forschungen* in which he explained his views on metric, projective and non-Euclidean geometries. These views are shortly designated as the Erlangen program. See also Birkhoff and Bennett (1988) and Gray (2005).

and described the strategy by which it can be proved. So, when he came back to Göttingen in 1885, his scientific career was almost over and he could concentrate on other sides of his personality. He became an excellent teacher and organizer of professional and scientific mathematical life at Göttingen and later in the whole Germany.

Hilbert was born near Königsberg, the town with a famous German university founded in 1544. In that town, he studied and spent more than one third of his life. There he got to know his lifelong friend Minkowski and reached first success in mathematics. At the end of his study, he went to Leipzig where he visited Klein's lectures and had a talk in his seminar. Klein made careful notes of the whole of Hilbert's talk and carried those notes carefully. Later he added to Hilbert's talk a comment: *When I heard his* "Vortrag", *I knew immediately that he was the coming man in mathematics.*[13] In Leipzig, Hilbert met Christian Hugo Eduard Study (1862–1930).[14] On the recommendation of Klein, both Hilbert and Study travelled to Paris in 1886. They both, but independently, reported to Klein their impressions from meetings with French mathematicians. On the way back from Paris, Hilbert visited Göttingen for the first time.

In 1888, he found time for travelling. He planned to visit 21 prominent mathematicians (see Reid, 1970, p. 29), among them Paul Albert Gordan (1837–1912). From him he picked up a difficult, now so-called Gordan's problem, concerning the structure of algebraic invariants: is there a kind of basis for the system of all such invariants?[15] Hilbert solved the problem in full generality. Publishing the result was like a bolt from the blue: quite a new method leading to an unexpected result. The result led to discussions and by no means was accepted by all mathematicians, but Klein seemed to recognize the importance of the work — *wholly simple and, therefore, logically compelling.* It was at this time that he decided he must get Hilbert to Göttingen at the first opportunity.[16] And, as for Hilbert, he decided to change the field of interests. Soon after that, in 1893, he found new proofs of the transcendence

[13]Quoted from Reid (1970, p. 19).

[14]More frequently referred to as only Eduard Study.

[15]Gordan worked with Clebsch on invariant theory and algebraic geometry. He is also known for giving simplified proofs of the transcendence of e and π.

[16]For details see again Reid (1970, p. 33, and pp. 247–251), written by Hermann Weyl. Hilbert concluded the series of related articles on the problem of invariants by a long one with the title Über die vollen Invariantensysteme, *Mathematische Annalen* **42**(1893), pp. 313–373. The final sentence of it says: *Hiermit sind, glaube ich, die wichtigsten allgemeinen Ziele einer Theorie der durch die Invarianten gebildeten Functionenkörper erreicht.* [Thus I believe the most important goals of the theory of the function fields generated by invariants have been attained.]

of e and π, the result of Charles Hermite (1822–1901) and Klein's student Carl Louis Ferdinand von Lindemann (1852–1939).[17] See Reid (1970, p. 41).

In 1892, there was a chance to get a position for Hilbert in Göttingen, but Hermann Amandus Schwarz (1843–1921) who moved to Berlin was replaced by Heinrich Weber (1842–1913). Then, three years later, Weber moved to Strasbourg and this time Klein already reached his goal: the chair was offered to Hilbert and so he came to Göttingen. In 1902, Minkowski accepted a chair at the University of Göttingen — in this case it was Hilbert who arranged that the chair be created especially for Minkowski. He held it until his unexpected death in 1909.

Together with the world exposition (Chicago Columbian Exposition) and fair (Chicago World's Fair) in connection with the spectacular celebration of the 400 years anniversary of the discovery of America, a mathematical congress was organized in 1893.[18] Klein was one of its key speakers; see Tobies (1981, p. 61). After the congress Klein delivered a series of lectures at Evanston.[19]

The invitation and appreciation of these lectures increased Klein's popularity and, together with the relation to Friedrich Theodor Althoff (1839–1908), departmental director of the Royal Prussian Ministry for Educational Affairs, made Klein a dominant person in the German mathematical community.[20]

Klein together with Cantor[21] is credited for the idea of international congresses of mathematicians in the 1890s. The first International Congress of Mathematicians (ICM) was held in Zürich in August 1897. At the turn of the century, there was the ICM 1900 in Paris. Hilbert was asked to deliver one of the main lectures and a relatively long time was spent thinking about the subject of the talk. The decision which led to the famous talk on 23 problems (now known as Hilbert's problems) was partly due to his colleague Minkowski. This talk contributed to the popularity

---

[17]While Lindemann was a professor at Königsberg, his first two doctoral students in 1885 were Hilbert and Minkowski.

[18]Klein was invited by his former students Henry Seely White (1861–1943), Oscar Bolza (1857–1942) and Heinrich Maschke (1853–1908). See Rowe (2018, p. 142).

[19]These lectures were published a year later as Klein (1894).

[20]In Reid (1970), we can read about Althoff on p. 43: *His great goal was to build up mathematics in Germany. He was a good friend of Klein's — the two had served in the army together during the Franco-Prussian War — and he thought very highly of Klein's opinion.* He worked at the Ministry since 1882 and was a proponent of establishing centers for different sciences according to already existing traditions of excellence, and this he did. For mathematics it was Göttingen. See also vom Brocke (1991), Bečvářová (2019, pp. 154–157), Tobies (1981, pp. 60–63), and Tobies (2012, p. 43).

[21]Cantor lived till 1856 in St. Petersburg and then in Germany. He studied in Berlin, and for a short time also in Zürich and Göttingen. Schwarz was a colleague of him during their studies in Berlin. Cantor obtained doctorate in 1867 for the dissertation *De aequationibus secundi gradus indeterminatis* [On the Indeterminate Equations of the Second Degree]. Almost all the rest of his life he worked in Halle where he was a full professor since 1879. He worked on his set theory and related questions. Unfortunately, he had a mental illness and so his work was interrupted by stays at a sanatorium. He proved e.g., that the set of transcendental numbers is uncountable and thus "bigger" than the set of algebraic numbers.

of Hilbert as well as Göttingen. When Minkowski read the page proofs of "Mathematische Probleme" before Hilbert delivered the lecture in Paris, he wrote to him on July 28, 1903 in a letter: *Your power of attraction for young mathematicians will be even stronger, if that is still possible, after this speech, which will probably be read by every mathematician without exception. ... You've taken out a lease on the mathematics of the twentieth century, and most will be happy to acknowledge that you are its general manager.*[22] Although Hilbert spoke in Paris only about some problems, all 23 were published later.[23]

It is interesting to quote from Rowe (2018, p. 408), about the context of Hilbert's talk in Paris: ...*Hilbert himself saw mathematics in very broad terms, a vision sustained by strong views about the nature of mathematical thoughts. The same can be said of his leading rival, Henri Poincaré, whose ideas had a lasting impact on philosophers of science. ... Hilbert originally conceived his famous lecture "Mathematical Problems" as a reply to Poincaré's lecture at the inaugural ICM in Zurich. Stressing foundations, axiomatics, and number theory, he set forth a vision of mathematics that was at once universal and purist. Aside from the sixth of his twenty-three Paris problems he gave only faint hints of links with other fields. The central theme of Hilbert's address, in fact, was based on the claim that mathematics, as a purely rigorous science, was fundamentally different from astronomy, physics, and all other exact sciences.*

In 1904, two famous applied mathematicians came to Göttingen: Carl David Tolmé Runge (1856–1927) and Ludwig Prandtl (1875–1953). In that year, Klein persuaded Göttingen University to offer Runge a chair of applied mathematics. In that post Runge remained until his retirement in 1925.

Runge was thus the first professor of applied mathematics in Germany. He came to Göttingen from Hannover where he was a professor since 1886 and where he worked on problems of wavelengths of the spectral lines together with the physicist Heinrich Gustav Johannes Kayser (1853–1940). They published important results in several quite long articles in the *Abhandlungen der Königlich Preußischen Akademie der Wissenschaften in Berlin* (1888–1893) [now *Proceedings of the Berlin Academy of Science*], e.g., *Spektren der Elemente*. Runge is also known for results on numerical methods[24] and on the approximation theorem in complex function theory.

---

[22]See Rowe (1989, p. 213), and Hazewinkel (2005, pp. 732–747).

[23]They were initially published in D. Hilbert, Mathematische Probleme, *Nachrichten von der Königliche Gesellschaft der Wissenschaften zu Göttingen* (1900), pp. 253–297 and later in *Archiv der Mathematik und Physik* **1**(1901), pp. 44–63, 213–237. An English translation appeared in D. Hilbert, Mathematical Problems, *Bulletin of the American Mathematical Society* **8**(10), pp. 437–479.

[24]See the article C. Runge, Über die numerische Auflösung von Differentialgleichungen, *Mathematische Annalen* **46**(1895), pp. 167–178.

Prandtl studied mechanical engineering at TH München since 1894 and from 1901 he was a professor (at the age of 26!) at TH Hannover. In 1904 he lectured on the ICM in Heidelberg about boundary layers and thus he in fact laid the foundation for German fluid dynamics and aeronautics. The same year, he accepted Klein's offer and, together with his friend Runge, came to Göttingen. It became, slightly later, a world renowned center of aeronautics research. Prandtl made numerous contributions to boundary layer theory, aerofoil theory, and gas dynamics, and he is the father of modern fluid dynamics. In 1925 he became the first director of the *Kaiser-Wilhelm-Institut für Strömungsforschung*.[25]

The growth of the applied mathematics group in Göttingen was connected very much with aeronautics. In 1906 the Hungarian engineer Theodor von Kármán (1881–1963) joined the group. Kármán got a two years fellowship to stay in Göttingen and Prandtl soon offered him a chance to work on the construction of a wind tunnel for tests on models for aviation. It was finished in 1908.[26] Kármán obtained doctorate in 1908 and became a private docent in 1909. He spent some time in London and Paris and in 1912 he left Göttingen for a while to Hungary. Then he went to Aachen with which he remained connected, with some interruptions, up to 1930. In Aachen he worked mainly on the frictional resistance of fluids, boundary-layer calculations, and problems of turbulence.

Another man who was connected with Göttingen and Aachen was Hilbert's first doctoral student Ludwig Otto Blumenthal (1876–1944). He came to Göttingen to study medicine, but soon changed his mind. He was very much influenced by theoretical physicist Arnold Johannes Wilhelm Sommerfeld (1868–1951), who was since 1894 an assistant of Klein. From 1900, Klein arranged for Sommerfeld a position at Königliche Technische Hochschule Aachen (now RWTH Aachen).[27] In 1906 Sommerfeld left for München, but it was he who recommended Blumenthal for the chair of mathematics at RWTH.[28]

Blumenthal was, since 1905, Managing Editor of *Mathematische Annalen* and he worked at this position *de facto* for 33 years. He was also a Dean, member

[25]For details on Prandtl's life and work see Eckert (2019).

[26]A remark: Some sources like the history of *Max Planck Institute for Dynamics and Self-Organization* say that it was in 1907. Prandtl founded in Göttingen *two* research institutions: in 1915, the *Modellversuchsanstalt aerodynamic* [Aerodynamic Experimental Station], and later, after the war, in 1924 Kaiser-Wilhelm-Institut für Strömungsforschung [Kaiser Wilhelm Institute for Fluid Dynamics]. Both are considered to be predecessors of the *Max Planck Institute for Dynamics and Self-Organization*.

[27]Abbreviation is created from **R**heinisch-**W**estfälische **T**echnische **H**ochschule Aachen.

[28]In Sommerfeld's handwritten notes from the last years of his life we can read: *Blumenthal war mein Lieblingsschüler* [Blumenthal was my favorite student]. It is mentioned in Behnke (1958, p. 387). More information about Sommerfeld can be found in Eckert (2013).

Mathematische Probleme.

Vortrag, gehalten auf dem internationalen Mathematiker-Kongreß zu Paris 1900.

Von

D. Hilbert.

Wer von uns würde nicht gern den Schleier lüften, unter dem die Zukunft verborgen liegt, um einen Blick zu werfen auf die bevorstehenden Fortschritte unsrer Wissenschaft und in die Geheimnisse ihrer Entwickelung während der künftigen Jahrhunderte! Welche besonderen Ziele werden es sein, denen die führenden mathematischen Geister der kommenden Geschlechter nachstreben? welche neuen Methoden und neuen Thatsachen werden die neuen Jahrhunderte entdecken — auf dem weiten und reichen Felde mathematischen Denkens?

Die Geschichte lehrt die Stetigkeit der Entwickelung der Wissenschaft. Wir wissen, daß jedes Zeitalter eigene Probleme hat, die das kommende Zeitalter löst oder als unfruchtbar zur Seite schiebt und durch neue Probleme ersetzt. Wollen wir eine Vorstellung gewinnen von der muthmaßlichen Entwickelung mathematischen Wissens in der nächsten Zukunft, so müssen wir die offenen Fragen vor unserem Geiste passiren lassen und die Probleme überschauen, welche die gegenwärtige Wissenschaft stellt, und deren Lösung wir von der Zukunft erwarten. Zu einer solchen Musterung der Probleme scheint mir der heutige Tag, der an der Jahrhundertwende liegt, wohl geeignet; denn die großen Zeitabschnitte fordern uns nicht blos auf zu Rückblicken in die Vergangenheit, sondern sie lenken unsere Gedanken auch auf das unbekannte Bevorstehende.

Die hohe Bedeutung bestimmter Probleme für den Fortschritt

Figure 2. Part of the first page of Hilbert's article.[29]

of the Senate, and President of the local Aachener Mathematischen Gesellschaft. In 1924, he became a member of the directorate of Deutsche Mathematiker-Vereinigung (DMV)[30] and the society's executive editor of the *Jahresbericht der DMV*. Because of these and teaching duties, he was not as scientifically active as in his young years when he worked on complex function theory and also on some problems of aeronautics and applied mathematics.

[29] *Nachrichten von der Gesellschaft der Wissenschaften zu Göttingen, Mathematisch-Physikalische Klasse* (1900), p. 253.

[30] DMV was founded in 1890 in Bremen and its first meeting took place in 1891 in Halle.

Figure 3. Felix Klein (1849–1925), architect of Göttingen mathematics.[31]

Let us make some remarks on Klein's other activities. He influenced also secondary school education,[32] mathematics teaching at technical high schools and universities (see Bečvářová, 2010; Schubring, 2010; Weigand *et al.*, 2016; Rowe, 2018). Klein practically all his life took care of secondary school curricula. Already in his dissertation, he expressed the following opinion: *Es ist wünschenwert, daß neben der Euklidischen Methode neuere Methoden der Geometrie in den Unterricht der Gymnasien eingeführt werden* [It would be desirable that besides Euclidean methods, also new geometric methods would be presented at grammar schools, too]. Problems of the curricula of mathematics at schools of all levels were among Klein's main interests during all his life.[33]

Klein also initiated the project the Encyklopädie der mathematischen Wissenschaften mit Einschluß ihrer Anwendungen.[34] During the ICM 1908 in Rome

[31]Klein in his Leipzig days. From Wikimedia Commons, https://commons.wikimedia.org/wiki/File:Felix_Klein,_Leipzig_years.png [19.4.2020].
[32]Not only in its content, but also from the point of legislative: in Prussia there had existed a qualification of an applied mathematics teacher since 1898.
[33]For details see Tobies (1981, pp. 75–85).
[34]It was published from 1898 to 1935 by B. G. Teubner Verlag and altogether it had 20,000 pages in 23 books. The successful project had international character and was announced at ICM 1908 in Rome. In a slightly modified form, the Encyklopädie later appeared also in French. A well-known historian of mathematics, Ivor Grattan-Guinness (1941–2014) wrote: *Many of the articles were the first of their kind on their topic, and several are still the last or the best. Some of them have excellent*

it was decided to establish the International Commission on Mathematical Instruction (ICMI) and Klein as a reformer of mathematics instruction was elected as its president. It gave him a chance to expand his opinion to the international level. In fact, Klein succeeded in linking the task of the ICMI with the agenda of curricular change. The next ICM 1912 in Cambridge decided that the activity of the Commission should be prolonged until 1916.[35]

Klein was honored in many different ways. Unfortunately, in connection with the WWI Klein's signature appeared at the end of 1914 on the document *An die Kulturwelt! Ein Aufruf* (known mostly as Manifesto of the Ninety-Three), which was expressing support for German military actions. Klein signed it as the only mathematician, together with 92 other German scholars and intellectuals. For the foreigners, the text was a document of German chauvinism. Klein was excluded from the Paris Academy and the Western allies inflicted a boycott on German scientists.[36]

Figure 4. David Hilbert (1862–1943).[37]

information on the deeper historical background. This is especially true of articles on applied mathematics, including engineering, which was stressed in its title. See Ivor Grattan-Guinness, *Routes of Learning: Highways, Pathways, Byways in the History of Mathematics*, Johns Hopkins University Press, Baltimore, 2009, xiv + 372 pages. ISBN 978-0-8018-9248-6. See pp. 44–45.

[35]See Bečvářová and Netuka (2010) for details.

[36]The document denied any war crimes committed by the Germans in Belgium, spoke on a German cultural mission in Europe, and backed German militarism. As was revealed much later, probably none of the signatories except the initiators had seen the text – they had just been asked telegraphically whether they would agree to sign. See Schubring (1925). For the text and all signatories see https://wwi.lib.byu.edu/index.php/Manifesto_of_the_Ninety-Three_German_Intellectuals [20.8.2019].

[37]Reproduced from *Acta Mathematica* (1882–1912). Table générale des tomes 1–35.

All the duties and functions in which he always tried to be active had a negative impact on his health and so he spent a year in a sanatorium. He retired in 1913, stopped his teaching but all his privileges and rights were preserved. So he still had all the affairs of mathematics and physics at Göttingen University in his hands.[38]

Minkowski suddenly died at the beginning of the year 1909 and Edmund Georg Hermann Landau (1877–1938) came to Göttingen as his follower. When Klein retired in 1913 it was not easy to replace him: Constantin Carathéodory (1873–1950) occupied his chair from 1913 and from 1914 he was also the editor of the *Mathematische Annalen*, but in 1918 he left for Berlin. He was followed by Erich Hecke (1887–1947) and when he left after a year for Hamburg, Landau's favourite for the chair was Issai Schur (1875–1941). Instead of him, Hilbert's PhD student Richard Courant (1888–1972) was chosen.

When Klein started to become too old for demanding activities, Hilbert was slowly undertaking his role of making Göttingen an attractive place for mathematicians. Both Klein and Hilbert had many doctoral students.[39]

4. Back to Berlin

When in 1891 Kronecker died, his position was obtained by Ferdinand Georg Frobenius (1849–1917) and when Weierstrass retired his chair was in 1892 given to Schwarz. Later, in 1902, when Lazarus Immanuel Fuchs (1883–1902) died, Frobenius and Schwarz wanted to replace him by Hilbert, but he turned down the offer preferring to continue as a professor in Göttingen. Fuchs' chair was then filled by Friedrich Hermann Schottky (1851–1935). Both Schwarz and Schottky came to Berlin after their best days of mathematical research. While Klein's dynamic leadership of Göttingen moved it to the place of the best mathematical center in Germany, Berlin was slowly losing this position.

Schwarz was a student of Weierstrass. He contributed significantly to complex function theory, calculus of variations and potential theory, but only before his appointment in Berlin in 1892; he taught there till 1917 when he was succeeded by Erhard Schmidt (1876–1959). Frobenius, who had jointly led the department with

[38]In Weigand *et al.* (2016, p. 5), there is a quote of the Bohr brothers which describes Klein's position: *Klein commanded not only mathematics as a whole but also all of the natural sciences. Through his powerful personality, which was complemented by his handsome appearance, he became a leading figure in the faculty and at the entire university. ... Over the years, Klein became more and more of a Zeus, enthroned above the other Olympians. He was known among us as "the Great Felix", and he controlled our destinies.*

[39]Sometimes students had more than one advisor. Klein had 68 students (the first in 1871 and the last one in 1920) while Hilbert had 76 students (the first in 1898 and the last one in 1934). See the Mathematics Genealogy Project at https://genealogy.math.ndsu.nodak.edu/index.php [1.7.2019].

Schwarz, was a pure mathematician who held the opinion that applied mathematics should belong to schools like TU's. His relations to Göttingen mathematics and to Klein were strictly negative. He contributed to the theory of elliptic functions, group theory, especially Abelian groups and representation theory. Schottky is probably best known for his contribution to the generalization of Picard's Great Theorem, which says the following: *If an analytic function f has an essential singularity at a point z_0, then on any punctured neighborhood $U(z_0)$ of z_0, the function f takes on all complex values, with at most a single exception, infinitely often.*

All Berlin professors nominated between 1892 and 1917 also studied in Berlin and so their thinking and opinions were much more uniform than the thinking of the teachers in Göttingen; Klein appreciated people who *discussed and quarreled* on problems.[40]

5. World War I and Its Impact

The natural development of sciences was interrupted by World War I (WWI) in which Germany lost more than 7 million people.[41] It affected very much also universities: the number of students diminished to one half (in the case of foreign students even much more, down to 15–35%),[42] young employees in a great number enlisted in the military, and universities suffered from a lack of finances. Standard lectures and curricula were not changed, but new departments and some new courses appeared. Nevertheless, universities were much more different than before the war: there were many women among students and the relations among students and professors were less formal.

Secondary schools had to face more difficulties than universities including not only the falling number of teachers but also the lack of space (schools were used as offices, barracks, and hospitals), coal for heating, etc. Some universities including Göttingen began to organize supplementary courses for veterans and for those who were not able to follow the standard program of study.[43]

[40]The whole Berlin mathematics was based on algebraic rather than analytic tradition; see Begehr *et al.* (1998b, pp. 72–96). For a much more detailed description of this era in Berlin together with an extensive collection of documents and tables see the two-volume monograph Begehr *et al.* (1998a).
[41]Reported causalities differ very much, e.g., the United States War Department figures it as follows: total mobilized German force 11,000,000; total casualties 7,142,558 (including killed and died 1,773,700; wounded 4,216,058; prisoners and missing 1,152,800).
[42]In Jarausch (1984, p. 318), there are the following numbers: in summer term of 1914 from 74,225 students 39,585 went to a front while in summer semester of 1918 there were 57,382 soldier students of 84,636 enrolled, i.e., more than two thirds.
[43]For much more detailed information about that time in Germany see Jarausch (1984), or the comprehensive internet source *International Encyclopedia of the First World War 1914–1918 online — 964 authors*, Editorial Board of 110 members, available at https://encyclopedia.1914-1918-online.net/home.html [10.7.2019] and Snyder (2015).

During the war, life in Göttingen was not easy; even to get food was a problem, but the lack of international contacts was for Hilbert much worse. Let us give an example of the social atmosphere in Göttingen: In 1917 Hilbert learned that the famous French mathematician Jean Gaston Darboux (1842–1917), the man he admired, had died in Paris. He immediately started to prepare for publication of his obituary. When it appeared in the *Nachrichten*, a mob of students gathered in front of his house and insisted on the destruction of all copies of the obituary. But Hilbert refused.

Quite often during the war Hilbert travelled to Zürich to visit his old friend and Klein's PhD student Adolf Hurwitz (1859–1919). Hurwitz became a private docent at Göttingen in 1882 and two years later he received professorship in Königsberg, where he became a friend of Hilbert and Minkowski. He spent there nine years before leaving for Zürich. He had really a big dilemma since shortly after accepting the offer from Zürich, Schwarz left Göttingen and there was a chance to go there, but Hurwitz turned it down.[44]

Hurwitz was a very complex person: mathematician, philosopher, musician, and very good organizer; see Pólya (1969). His obituary in Hilbert (1935) written by Hilbert in 1919 contains some details from his life and a description of his mathematical works. It is interesting to note that his result on polynomials with positive coefficients and roots having negative real parts[45] is still quite often used in technical applications.

6. Between 1918 and 1933

The time after the war was also very difficult. It formally ended by the Treaty of Versailles signed in 1919 in which Germany was declared responsible for all the losses and damages (so-called War Guilt Clause). Its predecessor was the two-side treaty between Soviet Russia and Germany signed in Brest-Litovsk in 1918. The Treaty of Versailles was considered in Germany as injustice and together with high reparations it formed the basis of Germany's future political development. We do not want to go into details. Note that negotiations were long and difficult and led later, in 1922, to the Rapallo Treaty between the Union of Soviet Socialist Republics (USSR) and the Weimar Republic (Germany), two states which were for different reasons the most isolated.

There existed also a boycott of German scientists. Germans were excluded from two post-war congresses of ICM in Strasbourg (1920) and Toronto (1924). Application of the collective responsibility principle thus deprived German mathematicians from important international contacts for years. Despite the fact that some mathematicians did not agree with such a solution, a majority of them accepted

---

[44] See also Bečvářová (2019) and Reid (1970, pp. 143–145).

[45] Such polynomials are called Hurwitz polynomials.

it and with respect to the cruelty and losses during WWI it is understandable. These facts describing the relations between states must be mentioned for a better understanding of the further development of mathematics.[46]

Arthur Moritz Schoenflies (1853–1928) came from a wealthy Jewish family. He studied in Berlin from 1870 to 1875 where he was taught by Kummer and Weierstrass. He was also very much influenced by Siegfried Heinrich Aronhold (1819–1884), who belonged to the staff of the Königliche Bauakademie zu Berlin. After obtaining the qualification for teaching he was a probation teacher at the Friedrich-Wilhelms-Gymnasium in Berlin. In 1877 he obtained a PhD in Berlin under Kummer and Weierstrass. Schoenflies continued to teach and worked on his habilitation which took place in Göttingen in 1884. From 1892 to 1899 he was a professor there, and after a short stay in Königsberg, he went to Frankfurt am Main in 1911 where he helped with the foundation of the university. He was its first full professor of mathematics and played a significant role in establishing its first department of mathematics.[47]

A. Schoenflies was also responsible for the appointment of two other colleagues, Ludwig Georg Elias Moses Bieberbach (1886–1982) in 1915 and later, in 1921, Max Wilhelm Dehn (1878–1952). A. Schoenflies' previous interests (geometry, set theory, topology, together with applications of the group theory in crystallography) formed an important background for mathematics in Frankfurt. Except those two mentioned, A. Schoenflies had at the institute other well-known mathematicians Paul Epstein (1871–1939), Ernst David Hellinger (1883–1950) and Otto Szász (1884–1952). We will return to Frankfurt once more below.

As many other mentioned mathematicians, Otto Toeplitz (1881–1940) went to Göttingen after finishing his study in Breslau where he obtained a PhD in 1905. At that time, Hilbert worked on integral equations and Toeplitz under Hilbert's influence started to build up a theory of forms in infinite-dimension spaces.

One of Felix Klein's last assistants, who also edited the first volume of his *Gesammelte Mathematische Abhandlungen I* (1921), was Alexander Markowich Ostrowski (1893–1986). He was born in Kiev and his talent for mathematics was discovered by Dmitry Aleksandrovich Grave (1863–1939). When he was 15, Ostrowski took part in Grave's seminar. For formal reasons he was denied to enter Kiev University. Grave found for him a chance to study in Germany in Marburg. Ostrowski came there just at the beginning of the war and so he became a civil prisoner. He spent the war studying and in 1918 he moved to Göttingen. There he gained in 1920 a PhD for a thesis supervised by Hilbert and Landau. The thesis partially answered Hilbert's

[46]For details see Bečvářová (2019).

[47]A. Schoenflies had an older brother Samuel Martin Schoenflies (1840–1879) who was also a mathematician and whose career was very similar. Because of the big difference of their ages Samuel had a strong influence on his brother. Both obtained their PhD in Göttingen but Samuel became a Gewerbeschul-Director [Director of an Industrial School] in Lübeck. In 1876 he became a professor at Polytechnikum in Riga, where he also died.

18th Problem. In 1920 he went to Hamburg, where he worked on his habilitation. This took place in 1922 and in the next year he came back to Göttingen.[48]

It is very difficult to characterize his mathematics since his interests were quite broad. He worked on different parts of algebra, but also on number theory, geometry, topology, on various questions of real and complex function theory. He was also interested in applications which were for him an important source of stimuli. As an example let us mention his influence on the valuation theory. He contributed to it in 1913 for the first time[49] and then again in 1934; in both cases it was of crucial importance.[50] Ostrowski was the last doctoral student of Klein.

7. Göttingen After the War

Let us turn to another member of the new generation. Recall that Courant came to the Göttingen University in 1907. At that time he met there the Hungarian mathematician Alfred Haar (1885–1933), who defended his doctoral thesis in 1909 and became Hilbert's assistant. A year later, Courant became Haar's colleague as an assistant of Hilbert and served this position for two years.

In 1912, Courant married the mathematician Nelly Neumann (1886–1942), a bit older colleague from his study at Breslau University. Later on, in the beginning of 1916, they divorced. Nelly resumed her maiden name and never remarried.[51] *During the four semesters that Courant served as Hilbert's assistant, Hilbert was devoting himself almost exclusively to subjects in analysis, the area of mathematics which developed in closest connection with problems of physics and mechanics. Analysis had not been emphasized as a field of advanced endeavor in Breslau. Although the students there studied some of the higher developments of the calculus, they tended to write their dissertations on subjects in algebra and geometry. Nelly confessed to Richard that analysis "frightened" her.*

"Don't think that analysis is so much harder than other things," he wrote back. "Perhaps completely the opposite. I am getting into it now, and I find that it is not actually so dangerous."

---

[48]Later on, Ostrowski spent a year as a Rockefeller Research Fellow in England and then he was offered a position in Basel, where he remained from 1927 for the rest of his active life. He retired in 1958 but produced new results up to his late 80s. His collected papers form six volumes published in 1983–1985 and he took part in their production. The rest of his life he spent with his wife in their villa Casa Almarost (ALexander MARgret OSTrowski) close to the Lake of Lugano; see Gautschi (2002).

[49]Basic principles of the theory were published by the Hungarian mathematician József Kürschák (1864–1963) in 1912. It is based on four axioms somewhat similar to the axioms imposed on a norm in a normed linear space.

[50]For details see Roquette (2002). Note that Ostrowski together with his wife Margret (née Sachs) established in the 1980s a prize for the best mathematical achievements; see Gautschi (1998).

[51]During World War II (WWII) she was deported as a Jew to the East in 1941 and murdered in Minsk in 1942.

In fact, Courant took to analysis as if it were his natural element. All his future mathematical work was to be done in subjects to which he was introduced during his student days by Hilbert.[52]

Then he took part in WWI, where he was wounded in 1915.[53] When Courant returned to Göttingen he married Runge's daughter Nina in January 1919. A few months later he became there a private docent. In 1920 he got a chair in Münster but Klein and Hilbert were able to bring him back to Göttingen in the same year.[54]

In the first two decades of the 20th century, Klein and Hilbert shifted their interests slightly more towards theoretical physics, namely to general relativity theory. When Minkowski passed away in 1909, Hilbert started to regularly invite best theoreticians to lecture within so-called Wolfskehl lectures.[55] The first who gave these lectures in Göttingen was Poincaré in 1909 followed by Hendrik Antoon Lorentz (1853–1928) in the next year. In 1912 these lectures were delivered by Sommerfeld. Leo Corry (born 1956) wrote that *by the year 1913 Hilbert's interest in a wide variety of physical disciplines had become a truly central feature of his current research and teaching concerns.* This quote appears in Roselló (2019, p. 151).

In 1915, Hilbert, Klein, Runge and several physicists suggested the theoretical physicist Albert Einstein (1879–1955) for a corresponding membership in the Königliche Gesellschaft der Wissenschaften zu Göttingen and the suggestion was approved.

Einstein studied in Zürich at the Eidgenössische Technische Hochschule (ETH) where he obtained a teacher qualification in 1900. After teaching at several places, he worked for the Swiss patent office in Bern from 1902 to 1909. In 1905, he received doctorate from the University of Zürich and in the same year he published his special theory of relativity.[56] Since 1912 he had worked on the general theory of relativity. He had changed many places and obviously liked travelling.

When Einstein came to know the news, he wrote to Hilbert within a few days: *Ich danke Ihnen für die freundliche Mitteilung, die Wahl zum korr. Mitgliede betreffend* [I want to thank you for the friendly letter informing me about the election as

[52]See Reid (1970, pp. 26–27).

[53]His brother Ernst was killed in 1916. Runge's oldest son Bernhard was killed, too, while the younger son Wilhelm (1895–1987) was selected by Courant to help him in developing a so-called earth telegraph. Victims were also on the other side, e.g., René Eugène Gateaux (1889–1914) was killed at the beginning of the war while in 1916 two sons of Jacques Hadamard (1865–1963) were killed at Verdun. Four professors of Italian universities died in the war, too.

[54]Approximately at the same time also the physicists James Franck (1882–1964) and Max Born (1882–1970) came to Göttingen. Under Born, who was a student of Runge, Göttingen became one of the world's leading physics centers.

[55]It was an invitation to deliver a series of lectures financed by Wolfskehl-Stiftung, which also provided a substantial honorarium of 2500 Marks. Paul Wolfskehl (1856–1906) in his last will established a prize for the solution of Fermat's Last Problem and the lectures were financed from the interest from that prize money. See Rowe (2018, p. 195).

[56]Einstein spent more than a year as a full time professor of theoretical physics at the German University of Prague.

corresponding member]. Then he officially thanked the society, and in his letter he wrote *Es gereicht mir zur Freude, einer Gesellschaft anzuhören, der die Wissenschaft so viel verdankt.* [It makes me happy to be a member of a society whom science has so much to thank for.]

8. The Case of Noethers

The next personality we have to mention is a woman, but before that we have to make a general remark. Women in different parts of Germany were accepted for study at universities (or technical universities) in different times. In Prussia it was since 1909. Before that, studying abroad was possible and some less formal approaches to education were available, too. Max Noether (1844–1921), professor of mathematics in Erlangen, friend and co-author of Clebsch, had four children. Two of them, Fritz Alexander Ernst Noether (1884–1941) and Amalie Emmy Noether (1882–1935)[57] were mathematicians. E. Noether is very often described as the most important woman in the history of mathematics of the 20th century or even at all. She studied in Erlangen and Göttingen. Her dissertation *Über die Bildung des Formensystems der ternären biquadratischen Form* was completed under Gordan in 1907 and since then she taught in Erlangen (without being paid!) till 1915 when she was invited by Klein and Hilbert to join them in Göttingen. In between, she wrote some articles inspired by Hilbert's work and improved Hilbert's results on mathematical physics.

In 1915 Hilbert offered in Göttingen a lecture, which under his name was delivered by E. Noether. Since 1919 she started to teach under her name as a private docent. She laid the foundation of a famous school of algebraists, contributed to mathematics and physics, and had at least 22 doctoral students. Some of her collaborators, like Richard Brauer (1901–1977), Helmut Hasse (1898–1979), and Bartel Leendert van der Waerden (1903–1996), also became leading algebraists. From the viewpoint of physics her most important accomplishment (known as Noether's theorem) describes the relationship between the symmetries of the laws of nature and conservation principles.

She also studied a special type of rings, which are now called Noetherian rings, but it is just a small fraction of her scientific merits. Note that the famous van der Waerden's two-volume book *Moderne Algebra*, issued several times and also translated into other languages, was originally based on the lectures of Emil Artin (1898–1962) and E. Noether.

Being a man, F. Noether had a better chance for study. After his "Abitur" [school-leaving exam], he entered Erlangen University where he spent five semesters. Among his teachers were Gordan and his father M. Noether. Then he continued in München, where his dissertation was supervised by Aurel Edmund Voss (1845–1931). He also

---

[57]Almost exclusively referred to as Emmy Noether. For details of her importance for the development of mathematics see Kimberling (1972).

gave credit to Sommerfeld for valuable inspiration.[58] Then he spent two years in Göttingen working on his habilitation thesis. He submitted it at TH Karlsruhe and obtained the right to lecture. He took active part in the war and after being wounded he was sent to do research in ballistics. After the war he returned to TH Karlsruhe.

In 1922 natural sciences and mathematics in Göttingen were separated from other disciplines studied at the Philosophical Faculty, and the name Mathematisches Institut der Universität came into use. At the beginning of the same year Hilbert was 60 years old and it was a chance for the journal *Naturwissenschaften* to organize for him a celebration. The last January issue was devoted to the anniversary.

Figure 5. Hilbert's 60th birthday party. Courtesy of the Archives, California Institute of Technology.[59]

[58]Klein wrote with Sommerfeld a voluminous book Über Die Theorie des Kreisels I–IV (1910) [The Theory of the Top I–IV] about rigid body dynamics. Sommerfeld refers in its fourth volume to Fritz Noether as a collaborator on the part of the work dealing with gyroscopes.

[59]Hilbert's 60th birthday party; the description is from Reid (1970, p. 238). Front row, left to right: Richard Courant, Franz Hilbert, Mrs. Courant (Nina Runge), Hertha Sponer (later Mrs. Franck), Mrs. Grotrian; second row: Mrs. Esslen (later Mrs. Springer), Mrs. Landau, Mrs. Hilbert, David Hilbert, Mrs. Hofmann, Mrs. Minkowski; third row: Ferdinand Springer, Felix Bernstein (behind Mrs. Landau), Mrs. Prandtl, Edmund Landau, Mrs. Franck, Fanny Minkowski (at end of row); fourth row: Ernst Hellinger, Erich Hecke (behind Landau), Walter Grotrian (behind Mrs. Hofmann); fifth row: Peter Debye, Theodore von Kármán (behind Hecke), Mrs. Rüdenberg (Lily Minkowski), Paul Bernays, Leonard Nelson, "Klarchen" (second from end of row).

Blumenthal, Hilbert's "oldest student", sketched Hilbert's career and contributions. This introduction was followed by articles by his former pupils on Hilbert's work in algebra, geometry, analysis, mathematical physics and the philosophy of mathematics.[60] During the birthday banquet Hilbert's older colleague Klein (then on the wheel chair) gave him a copy of the notes on "Vortrag" which Hilbert delivered in Klein's seminar at Leipzig in 1885 (see above).

As we have seen so far, Göttingen played a decisive role in the development of mathematics, not just within Germany. After the war the man who created its shape was approaching the end of his life. Klein was alive and worked at some committees, but he was moving only on the wheelchair and so he lectured at home for a small audience since 1916. In certain activities he was helped by his daughter Elisabeth Steiger (1888–1968),[61] but Klein was still active despite his age.

Klein could be proud looking back at the world science center which he created. Half of his success was due to the talent for choosing the right people regardless of their nationality, race or sex. He loved both pure and applied mathematics and what he was doing was always led by the feeling that it is good for mathematics. As a result he was not always beloved, especially in Berlin.

At the celebration of his 75th birthday, he received many honors, even from Berlin — he obtained a honorary membership of the DMV and GAMM (see below) as well, and the Faculty of Philosophy and Laws of Berlin University granted him the title Doctor rerum politicarum honoris causa. After a misstep in 1924, he was confined to his bed and he died in the middle of 1925.[62] Next day, Hilbert remembered him in a small circle of colleagues.

At the meeting after his funeral (during which Runge delivered a farewell speech) Courant started his talk in the aula of the Göttingen University: ... *eine Epoche in der Geschichte der Mathematik ist abgeschlossen* [... an Epoch of the History of Mathematics is Over.][63] At his grave it was written "Sincere et Constanter". The same year, Runge retired and was replaced by Gustav Herglotz (1881–1953), but his chair has not been connected with applied mathematics any more.[64]

9. Göttingen After Klein's Death

Hilbert's health was steadily worsening. Shortly after Klein's death, doctors recognized the reason: their diagnosis was *pernicious anemia* and they gave him at best

[60]The authors of all contributions were Jewish. A part promised by Courant was not finished and so it was not published.

[61]Known also by her nickname "Putti". She married in 1914 and her husband Dr. Robert Steiger, conductor of Academic Orchestra, Göttingen, died in the first days of the war.

[62]His wife who took care of him died two years later.

[63]See Tobies (1981, p. 94).

[64]For details see Tobies (1981, pp. 92–94).

just a few months. Hilbert decided to ignore it — he disagreed with the diagnosis. Fortunately, just at that time in the U.S., George Richard Minot (1885–1950) applied in 1926 a method discovered a year ago, on Hilbert's disease. It seemed to work. The whole story in which some "maths friends" helped to obtain the drug for Hilbert's treatment is described in Reid (1970, pp. 179–180).

Some new young faces appeared in Göttingen. Artin was a student of mathematics and chemistry in Leipzig, where he was taught by Herglotz. After receiving a PhD in 1921 he spent more than a year in Göttingen before taking a position at University of Hamburg in 1923; the University was founded in 1914. Two years later, he was appointed extraordinary professor and after another year, he became a full professor. For another 11 years, Artin together with Hecke and Wilhelm Johann Eugen Blaschke (1885–1962) directed the activities of the Mathematical Seminar of Hamburg University. Blaschke came to Hamburg after short stays in Prague, Leipzig, Königsberg, and Tübingen and remained there from 1919 until 1953.[65]

Artin was well known as an excellent teacher. His main field was algebra but contributed to other parts of mathematics: his little booklet *Einführung in die Theorie der Gammafunktion* (1931) became a classic; together with his *Galois Theory* (1942) and *Theory of Algebraic Numbers* (1959)[66], it was included in Rosen (2007), where other facts on Artin are available.[67] Artin had many interests outside mathematics; he loved chemistry, astronomy, biology, and music. He was an accomplished musician playing the flute, harpsichord, and clavichord.

Carl Ludwig Siegel (1896–1970) started his university studies in Berlin in 1915 but had to interrupt them since he was called to the army. From the army he was discharged as an unusable soldier. Since 1919 he studied in Göttingen with Landau. There he received a PhD in 1920. On his dissertation *Approximation algebraischer Zahlen*, Jean Alexandre Eugène Dieudonné (1906–1992) wrote: ... [*it*] *was a landmark in the history of Diophantine approximations.* After a short stay in Hamburg, Siegel returned to Göttingen, where he obtained habilitation in 1921. Next year, he received a professorship at the Goethe University of Frankfurt established in 1914, where he was a successor of A. Schoenflies.

One of the many students who came to study in Göttingen was Otto Eduard Neugebauer (1899–1990). Before that he served in the army, then he studied in Graz,

---

[65]Some details of his career, especially his stay in Prague, are available in M. Bečvářová, *Matematika na Německé univerzitě v Praze v letech 1882–1845* [Mathematics at the German University in Prague between 1882 and 1945], Karolinum, Praha, 2016, 446 pages. ISBN 978-80-246-3182-0.

[66]Notes by Gerhard Wurges from the lectures held at the Mathematisches Institut in Göttingen in the winter semester of the year 1956/1957.

[67]Artin had to leave with the whole family to the U.S in 1937 since his wife Natasha was of Jewish origin. After WWII they got American citizenship in 1946. At that time Artin got a permanent position in Princeton. Nevertheless he returned back to Hamburg in 1958 and a year later his marriage with Natasha was divorced. Artin died in Hamburg at the end of 1962. He was one of those German mathematicians who returned back from emigration.

Munich and from 1922 he listened to lectures of Courant, Landau and E. Noether. Later, he became an assistant of Courant who was at the time the head of the Mathematical Institute. In 1924, Neugebauer went to the University of Copenhagen where he worked with Harald August Bohr (1887–1951)[68] and with whom he studied almost periodic functions. They published their results in Bohr and Neugebauer (1926). Neugebauer's interests switched to the history of mathematics and his thesis *Die Grundlagen der ägyptischen Bruchrechnung* was written under Courant and Hilbert.[69]

William Srećko Feller (1906–1970)[70] studied at the University of Zagreb and he moved to Göttingen at the age of 19. In 1927 he received PhD when his thesis *Über algebraisch rektifizierbare transzendente Kurven* had been accepted for publication. In 1928 he moved to Kiel where he became a private docent in 1929. The same year, there came Erhard Tornier (1894–1982) with whom he wrote several papers. They were friends.[71] Feller became an expert in probability theory with close ties to applications.

After treatment, Hilbert overcame his illness and he retired in 1930. For Göttingen and the whole Germany, only three years of relatively difficult but stable life remained. When in 1933 Nazis started to change Weimarer Republik[72] into a dictatorship, Hilbert still lectured. In the winter semester of 1933/1934 he gave the last lecture and after it he stopped to come to the Institute of Mathematics. The reason will be explained in detail below but before that we will mention two other mathematicians who were able to enjoy a few years before the change.

Willy (William) Prager (1903–1980) studied at the Institute of Technology in Darmstadt where he received diploma in engineering in 1925 and a year later, he became doctor of engineering. Since 1929, he was for three years an acting director of the Institute of Applied Mechanics in Göttingen. From it he went to Karlsruhe in 1932, where he was appointed professor of technical mechanics. He learned many things from Prandtl and became an expert in applied mathematics. As an indication of just how rapid his success had been, we note that his appointment in Karlsruhe made him the youngest professor in Germany. This appointment was certainly due to the leading international reputation he had already achieved, with over 30 papers and a book already showing the depth of his contributions to applied

[68]Harald was the older brother of Niels Henrik David Bohr (1885–1962), who was known as a nuclear physicist and Nobel Prize winner in 1922.

[69]Very good characteristics of the main Göttingen "pillars" Klein, Hilbert and E. Noether including their work are contained in Kramer (1972, pp. 401–433, 627–632, 656–660).

[70]Srećko is a Croatian equivalent of Felix. Feller used several names; depending on the land where he lived he was Willi, Willy, William or Will.

[71]See Schilling and Woycznski (2015), where it is possible to find many details of Feller, including the list of his works.

[72]Officially Deutsches Reich, since this name was not changed from 1871 until 1945.

mathematics. However, life in Germany was about to change when Hitler came to power on 30 January 1933.

Another interesting guest of Göttingen was the French Bourbakist André Weil (1906–1998). He visited Göttingen when he studied at the École Normale Supérieure, but he took part also in a seminar at the Collège de France held by Jacques Salomon Hadamard (1865–1963).[73] He finished his studies in 1925, spent a half year in Rome and a bit later in 1927 he used the support of the Rockefeller Foundation for an almost one year stay at Göttingen. His contacts with E. Noether, Courant and others influenced his "maths taste". Then he returned to Paris to finish his PhD, which he received in 1928.[74]

10. Frankfurt Mathematics Seminar

After Siegel's appointment in Frankfurt, A. Schoenflies was not mathematically active but Siegel found there several young mathematicians with whom he created an excellent mathematical center. Together with Dehn, Hellinger, Epstein and Szász they organized the well-known Mathematisch-historisches Seminar. The seminar was devoted to the history of mathematics and took place from 1925 until at least 1931.

Dehn and Hellinger were both PhD students of Hilbert. Dehn studied in Freiburg and Göttingen. He was the first who solved one of Hilbert's problems from his talk at the ICM 1900. It was the third Hilbert problem and it consisted in the following: Any two polygons in the plane of the same area can be divided into a finite number of polygonal pieces in such a way that from the pieces of the first one it is possible to compose the second one. Hilbert posed the question whether an analogous problem for polyhedrals also has a positive solution and Dehn proved in 1900 that the answer was *no*.

Hellinger studied in Heidelberg and Breslau. Once he received a PhD in 1907, he taught for two years in Göttingen and then in Marburg. Since 1914 he worked as a professor in Frankfurt. Together with Toeplitz, he wrote for Klein's *Encyclopedia der Mathematischen Wissenschaften* the article *Integralgleichungen und Gleichungen mit unendlich vielen Unbekannten*, which became classic. Hellinger studied integral equations, infinite systems of equations, real functions, and continued fractions. His dissertation *Die Orthogonalinvarianten quadratischer Formen von unendlichvielen Variablen* (1907) written under Hilbert was inspired by Hilbert's intensive study of integral equations. In it, he introduced integrals of a special type, now known as the *Hellinger integral*.

Szász studied in Budapest, but during his studies he went to Göttingen, Munich and Paris. He got his PhD in Budapest in 1911 under Lipót Fejér (1880–1959) and

[73]For the life and work of Hadamard, see Maz'ya and Shaposhnikova (1999).

[74]It is very interesting to take a look at the mathematical history of Göttingen in Neuenschwander and Burmann (1999).

became a private docent there. In 1914, he started in Frankfurt, also as a private docent but later became there a full professor. Besides, in Frankfurt he was for six years a colleague of Bieberbach. In the Szász obituary,[75] we can read: *His life and energy were dedicated to the promotion of simple and beautiful problems of mathematics, in particular of the classical analysis.*

Epstein studied at the oldest Jewish secondary school in Germany called Philanthropin. His father was a professor there and he was also the head of the Frankfurt Observatory. Epstein continued his study at the University of Strasbourg (at that time "Straßburg" was a part of Germany) and finished it in 1895. He remained there and taught at the technical school and also at the university as a private docent. During the war he served in the German army. After the war the situation in Strasbourg was not stable and so it was easily occupied by the French troops at the end of 1918. In 1919, following the Treaty of Versailles, the city was annexed by France.

When Strasbourg was given to France, Epstein went back to Frankfurt where he lectured at the university since 1919. In the school year 1923/1924, he obtained a paid position of extraordinary professor of didactics and the history of mathematics. His results have found application in crystallography and in mathematical physics.

The abovementioned historical seminar in Frankfurt was popular among students and colleagues. Its uniqueness is described in Bergmann *et al.* (2012, pp. 119–132).[76] The seminar was quite often visited by Otto Toeplitz from Bonn or by the French mathematician Weil. Toeplitz became a friend with Hellinger with whom he wrote several papers. In 1913 Toeplitz got an extraordinary professorship at the University of Kiel, where he became an ordinary professor in 1920. In 1928 he accepted a chair at the University of Bonn, where he joined Felix Hausdorff (1868–1942).

At the time when the seminar was active, Weil was a quite young student. He wrote in his autobiography about the seminar: *It was only later that I attended it, on subsequent visits to Frankfurt, a place I made a point of visiting as often as I could. I am not sure whether it was already in the summer semester of 1926 that, during a seminar session devoted to Cavalieri, Dehn showed how this text had to be read from the viewpoint of the author, taking into account both what was commonly accepted in his lifetime and the new ideas that Cavalieri was trying to the best of his ability to implement. Everyone participated in the discussion, contributing what he could to the group effort.*[77]

---

[75]It was written by another famous Hungarian mathematician Gábor Szegő (1895–1985) and published in the *Bulletin of the American Mathematical Society* **60**(1954), pp. 261–263.

[76]Siegel devoted to the abovementioned colleagues an address given in 1964 on the occasion of the 50th Anniversary of the foundation of Frankfurt University. See Siegel (1979).

[77]Cited from Bergmann *et al.* (2012, pp. 125–126). As was already mentioned, Weil used the support from the Rockefeller Foundation for a visit to Göttingen in 1927 and produced there his first mathematical results. He profited a lot from discussions with Göttingen colleagues.

Being equipped with a knowledge of several languages (when he was about 16 he read Sanskrit), Weil taught at several universities. He substantially influenced modern parts of mathematics, like topology, differential geometry and complex analytic geometry. He successfully combined an intimate masterful understanding of these parts of mathematics, e.g., in the study of harmonic analysis of topological groups. He was always hard working, creative and open to new ideas and activities. He had a sense of humour as shown by a story: He was arrested twice in his life, in Finland and also in France.[78] From the prison at Rouen he wrote: ... *I am even a bit worried — if it's only in prison that I work so well, will I have to arrange to spend two or three months locked up every year?*[79]

11. ICM 1928 in Bologna

Tension was growing (partly because of the political, social and other problems caused by WWI) among mathematicians of different nationality, but also in Germany itself. The ICM 1928 was organized by the Italians in Bologna. It was the first ICM after the war from which German mathematicians were not excluded but it turned out that some of them were against the participation. Hilbert was a leading person of those who wanted to participate while Bieberbach actively fought on the opposite side. Courant and other Göttingen mathematicians took the same view as Hilbert but almost all Berlin mathematicians strongly opposed it. Not all of them, e.g., Hilbert's student and professor in Berlin E. Schmidt, had the same standpoint as his former teacher.[80]

The whole affair started when the Dutch intuitionistic mathematician Luitzen Egbertus Jan Brouwer (1881–1966), a very German-oriented person, wrote at the beginning of 1928 an open letter to DMV in which he was against the attendance of German mathematicians at the ICM in Italy. After some quarrels, Hilbert, who was not entirely healthy at that time, led the German delegation of 67 people to Bologna. In addition to his invited lecture he planned to give a short "political" speech, a manuscript of which (written probably by his wife Käthe) is now in the manuscript division of the State and University Library in Göttingen. The speech was probably not delivered. It was supposed to contain the sentences: *Die Mathematik kennt keine Rassen ... Für die Mathematik ist die gesamte Kulturwelt ein einziges Land.* [Mathematics knows no races ... For mathematics, the whole cultural

---

[78]In Finland he was considered a spy since he had with him some papers in Russian. He returned to France rather late and so he was sentenced as a deserter. For details see A. Weil: *The Apprenticeship of a Mathematician*, Birkhäuser, Basel, 1992, 197 pages. ISBN 978-3-7643-2650-0.

[79]Another quote cited in H. Eves, *Mathematical Circles Adieu*, Boston, 1977, 181 pages, ISBN 978-0-87150-240-7: *God exists since mathematics is consistent, and the Devil exists since we cannot prove it.*

[80]For details see, e.g., Bečvářová (2019, p. 96).

world is a single country].[81] The ICM 1928 looks like the removal of the last part of boycott of German mathematicians and mathematics after WWI. But later on, when Roland George Dwight Richardson (1878–1949) commented on the first wave of German emigrants to the U.S. in 1934, he said: *Since the war, we have been constantly compelled to think of colleagues as nationals and not as citizens in the international domain.*[82]

Through contacts with N. Bohr and H. Bohr, Courant was able to arrange money from the Rockefeller Foundation[83] for the construction of a new building of the Mathematical Institute and Neugebauer helped a lot with its construction plans. The whole action started already in 1923. A representative of the Foundation Wickliffe Rose (1862–1931) travelled to Europe in 1923–1924. He informed European representatives about the Foundation. Another aim of his travel was to get an overview of the centers of European science. In Germany, Rose visited Göttingen and Berlin. Weyl, with whom he consulted on the situation of German mathematics, explicitly directed him to Göttingen, and so Rose at the end of his trip had conversations with Hilbert and Runge. He also met Born and Franck.

The next important negotiations took place in 1925 when Augustus Trowbridge (1870–1934), experimental physicist of Princeton University and a new face of one of the Rockefeller institutions, visited Göttingen. Among other things he discussed with Courant an old idea of Klein, who had planned the creation of a new mathematical institute. Later on, Courant discussed it with the Bohr brothers in 1926 and wrote: *...following a proposal by the Bohr brothers to turn to the International Educational Board with a request to enable the completion of Klein's great endeavour by founding a Mathematical Institute.* It should work probably even without their help but of course *Niels Bohr in particular, had plenty of experience with the Rockefeller people and knew their inclinations from his own institute for theoretical physics in Copenhagen.*[84]

[81] The text is quoted from page 56 of R. Siegmund-Schultze, "Mathematics knows no races". A Political Speech that David Hilbert Planned for the ICM in Bologna in 1928, *Mathematical Intelligencer* **38**(2016), pp. 56–66. ISSN 0343-6993.

[82] See the contribution of R. Siegmund-Schultze in Parshall and Rice (2002, p. 336).

[83] Rockefellers were able to collect a tremendous amount of money within a few generations which were later partly used for philanthropy. At the beginning of the 20th century, John Davison Rockefeller, junior (1874–1960), started to prepare a project of The Rockefeller Foundation which was created in 1913. From promoting individuals mentioned several times above, it moved to such big projects in Europe like the support of the Institute for Theoretical Physics of Copenhagen's University (1921), Institut Henri Poincaré in Paris (1928), and Mathematical Institute of the University of Göttingen (1929). For details see *The Rockefeller Foundation History* at http://www.fundinguniverse.com/company-histories/the-rockefeller-foundation-history/ [15.7.2019].

[84] For details see Siegmund-Schultze (2001, especially p. 147), from where we took the last two quotes.

And so, after all, the dream came true. A very expensive and well-designed building containing a large reading hall and reference library was completed in 1929. The library was a continuation of the reading room known as *Lesezimmer*, one of the first (mathematical) libraries with open shelves, which Klein created in Göttingen after his arrival in 1886. [85] This innovation of Klein's (which is now quite common) promoted informal contact within the Göttingen mathematical community. Lesezimmer is mentioned quite often in different memoirs of people studying or visiting Göttingen.

The construction of the building required a big effort. *Very few people were aware of the key role played by Courant in the difficult negotiations with the Rockefeller Foundation.* Hilbert was aware of it and he found a way how to speak poetically on it and praise Courant's work behind-the-scene at the opening ceremony. He said: *The idea of this institute — cherished and nurtured by Felix Klein — lived once upon a time and was charming and lovely, like Sleeping Beauty, and we who have been here a while took proud pleasure from it; but the Wicked Witch of inflation put this Sleeping Beauty into a sleep so deep that all forgot about her until Prince Charming Courant awakened her to new life.*[86]

Before shifting round to Berlin and some other places, let us mention that Göttingen in its golden years attracted mathematicians from all over the world. Several presidents (at least six — see Rowe, 2018, p. 116) of the American Mathematical Society (AMS) studied there under Klein, and also many other American and other foreign mathematicians spent at least some time there. The list of all of them looks almost like "who is who in mathematics/physics". Some of them we have already mentioned and some others will be presented below. Our choice of those who were introduced above was not subordinated to a special task and it is somewhat unfair that we did not choose, e.g., Vojtěch Jarník (1897–1970), Hugo Dyonizy Steinhaus (1887–1972), Teiji Takagi (1875–1960), Eugene Paul Wigner (1902–1995), Norbert Wiener (1894–1964), and others (contacts of Russian mathematicians with an emphasis on their relations with Göttingen will be introduced in another chapter).

It was already said that the last member of the "big trinity" Weierstrass passed away in 1897 and that their followers represented by Frobenius and Schwarz came to Berlin when their scientific career was almost over. Frobenius dominated Berlin mathematics in the period from 1892 to 1917. In Begehr *et al.*, 1998, pp. 83–84, he is characterized as follows: *He considered himself to be a scholar whose duty it was to contribute to the knowledge of pure mathematics. ... occasionally choleric, quarrelsome, and given to invectives. He felt a particular aversion to Göttingen*

[85]Some details can be found also in Rowe (1989, p. 202).

[86]Both quotes are from Bergmann *et al.* (2012, pp. 82 and 83). The second one shows another facet of Hilbert's personality.

Figure 6. Mathematical Institute in Göttingen constructed in 1929.[87]

mathematics.... It was to a large extent Frobenius who removed Berlin mathematics from the development in Germany and who, as a consequence, led Berlin into comparative isolation. On the other hand, he was an excellent (pure) mathematician. He contributed substantially to the theory of group characters, but the complete list of his important results would be much longer.[88]

Schur, who was already mentioned in connection with Landau, was a PhD student of Frobenius. He obtained habilitation in Berlin in 1903 but was not appointed to a chair there until 1921 when he succeeded Schottky. Meanwhile, he spent three years in Bonn, and Frobenius considered him as his successor and suggested him unsuccessfully several times for a position in Berlin. Of course, Frobenius knew a possible reason and in 1910 he wrote: *... as a Jew and a Russian he unfortunately has few prospects of being accepted by a Prussian university.*[89] Schur continued the study of representation theory of his teacher and contributed also to number theory. In many articles he treated problems of analysis, mostly with algebraic methods. His relationship to applications was indifferent, but he was not against them.

E. Schmidt replaced Schwarz when he retired in 1917 and Carathéodory replaced Frobenius a year later. Together they led the department but Carathéodory left

---

[87]A picture of the Faculty of Mathematics, University of Göttingen. Author Daniel Schwen, Wikimedia Commons, https://commons.wikimedia.org/wiki/File:Mathematik_G%C3%B6ttingen.jpg [20.4.2020].

[88]For a more detailed characterization of Frobenius' further mathematical merits, see Begehr *et al.* (1998, pp. 84–87).

[89]This is quoted from Bečvářová (2019, p. 87).

after a year[90] and so E. Schmidt had to fill the vacant chair. His list of candidates contained Brouwer, Hermann Klaus Hugo Weyl (1885–1955), and Herglotz in this order. The professorship was offered to each of them but all turned it down. Then Hecke did the same and so the position was left vacant until 1921 when it was offered to Bieberbach, who accepted it.

He is well known in connection with the so-called Bieberbach conjecture which he formulated in 1916: If a function f expressed by a power series

$$f(z) = z + a_2z^2 + a_3z^3 + \cdots + a_nz^n + \cdots$$

maps injectively the unit disk $\{z : |z| < 1\}$ on a domain in the complex plane, then $|a_n| \leq n$ for all positive integers n. He proved that $|a_2| \leq 2$.[91]

There was no chair for applied mathematics in Berlin but E. Schmidt planned to change it. He was the main person who pushed for the founding of an Institut für Angewandte Mathematik. When it was created he had to fill both the new chair of applied mathematics and the post of the director of the Institute of Applied Mathematics. It was in 1920 that Richard von Mises (1883–1953) agreed to take both these positions. It was a great luck for Berlin since after many years in 1965 Ostrowski wrote about it: *Only with the appointment of Richard von Mises to the University of Berlin did the first mathematically serious German school of applied mathematics with a broad sphere of influence come into existence.*[92] His opinion on omitting Göttingen was probably due to the conception of the study of applied mathematics at both places. Note that E. Schmidt had a lion's share in introducing and promoting applied mathematics in Berlin.

The career of von Mises started in Vienna, where he studied at Technische Hochschule. In 1908 he received his doctor's degree, obtained habilitation in Brno (now Czech Republic) and in the next year he became associate professor of applied mathematics at the University of Strasbourg. During the WWI he served in the Austro-Hungarian army. He was a pilot and he also lectured about the airplane construction. The lectures are covered in von Mises (1957), published first in 1918.

Von Mises was very active in the 1920s. In 1921 he founded the *Zeitschrift für Angewandte Mathematik und Mechanik* (ZAMM) and until 1933 he was its editor and frequent contributor. A year later, he founded together with Prandtl

[90] In Begehr *et al.* (1998) the reason and circumstances are described in the following way: *It was his fate and that of mathematics and of mathematicians in Berlin that he had to resign from his positions at the University of Berlin and the Academy only after 15 months, when his patriotism forced him to follow the call of his country, Greece.* See also Georgiadou (2004).

[91] Through the effort of other mathematicians the corresponding inequality was proved for several first positive integers but the conjecture itself was proved by Louis de Branges de Bourcia (born 1932), an American mathematician of French origin, in 1985.

[92] See Siegmund-Schultze (1993, p. 365).

Gesellschaft für Angewandte Mathematik und Mechanik (GAMM). [93] Both exist up to the present. He worked arduously on the elements of probability theory and despite the fact that his basic notion "Kollektiv" of the theory appeared to be a dead end, he can be considered one of the fathers of probability theory. [94]

Another applied mathematician was Hilda Geiringer-Pollaczek (1893–1973). After her study in Vienna from 1913 to 1917 and work as an associate editor for a mathematical journal, she got the research position as an assistant to Richard von Mises in 1921. The same year she married Leo Félix Pollaczek (1892–1981) and a year later their daughter Magda was born.[95] Geiringer's life in Berlin was somewhat difficult: her move from pure to applied mathematics was demanding, her teaching load combined with the care of her daughter took a lot of time, and to be employed at von Mises' Institute was not easy.

In 1927, Geiringer became a private docent in applied mathematics at the University of Berlin and continued to conduct research on probability and mathematical theory of plasticity. It was not a smooth process. Her work *Über starre Gliederungen von Fachwerken* submitted in 1925 had two reviewers, von Mises and Bieberbach. There were some mistakes in the work which were overlooked by von Mises and Bieberbach's criticism was not pleasant. So she withdrew the work, and a half year later in 1926 she submitted a new one *Über die Poissonsche Verteilung und die Entwicklung willkürlicher Verteilungen.* The whole situation of her habilitation is complicated and maybe some other (political) factors played a role.[96] In fact, Geiringer was the second woman in Germany who obtained habilitation in mathematics[97] (see also the story of Ruth Moufang later in this chapter).

12. From 1933 to the Beginning of WWII

From a more general view, German mathematicians had several possibilities in the coming Nazi era. We will report on almost all who decided to collaborate with the Nazi regime and we will also make a note of some of those who stayed in Germany; their fates were different; some of them were sent to Konzentrationslager (KZ) where they were killed, and only a few survived. Some others committed suicide or survived

---

[93]Prandtl was chairman of GAMM from 1922 until 1933.

[94]Andrey Nikolaevich Kolmogorov (1903–1987) wrote in his *Grundbegriffe der Wahrscheinlichkeitsrechnung* (1933) [Foundations of the Theory of Probability] ...*In laying out the assumptions needed to make probability theory applicable to the world of real events, the author has followed in large measure the model provided by Mr. von Mises*... The quote is from page 91 of the paper G. Shafer, V. Vovk, The Sources of Kolmogorov's Grundbegriffe, *Statistical Science* **21**(2006), pp. 70–98. ISSN 0883-4237.

[95]The marriage broke up in 1922. She later used the hyphenated name Pollaczek-Geiringer and since 1934 her maiden name Geiringer. She married von Mises in the U.S. in 1943.

[96]In Siegmund-Schultze (1993), it is described in detail. Both the mentioned works were published later on, at least substantial parts of them.

[97]More precisely, she was habilitated for *Applied Mathematics.*

in different ways. We will try to mention especially those who resisted. Anyway, many more were persecuted and our description of individual fates cannot be complete. The other big group is formed by those who tried to emigrate — not all of them found the chance to do it and hence some of them also died. With a few exceptions, these emigrants did not come back and their relation to the country was mostly very negative. We will briefly supply information on the fates of the abovementioned mathematicians, again not complete. Altogether there were 145 emigrants (mostly to the U.S.); 17 were killed or committed suicide and 72 were persecuted.[98]

The whole process of political changes started much sooner but without substantial consequences for mathematics. In 1919, Adolf Hitler (1889–1945) joined the Deutsche Arbeiterpartei, which later changed its name to Deutsche Nationalsozialistische Arbeiterpartei (NSDAP). In 1923, after an attempt called "Bierkeller-Putsch" he was put into jail for five years, but after nine months he was released. While imprisoned, he wrote the first part of his book *Mein Kampf*. There he blamed the parliament of the Weimar Republic, Jews, Social Democrats as well as Marxists for the misery of Germany.

NSDAP gained more and more votes from 1928 to 1932 and at the beginning of 1933 president Paul von Hindenburg (1847–1934) named Hitler chancellor. Shortly after it Hitler gained unlimited power through the so-called "Ermächtigungsgesetz" [Enabling Act], which began the process of transforming the Weimar Republic into Nazi Germany. One of the first laws issued by the Nazis was the Gesetz zur Wiederherstellung des Berufsbeamtentums. As a consequence of it many mathematicians (including almost all Jews) in Germany lost their positions. Exceptions of those who fought in the war were "solved" a bit later by application of the Nürnberger Gesetze. These laws got rid Jews of all human rights.[99]

Let us briefly describe some individual fates. From those who emigrated just a few did it before 1933. It was, e.g., Albert Einstein (1879–1955), Emil Julius Gumbel (1891–1966) and von Kármán.

Einstein received a doctorate from the University of Zürich for his thesis *Eine neue Bestimming der Moleküldimensionen* [On a New Determination of Molecular Dimensions] in 1905 and in the same year he published his special theory of relativity.[100] At the end of 1932 he left for the U.S. where he applied for permanent residency in 1935 and became a citizen of the U.S. since 1940.[101]

---

[98]These numbers are cited from Siegmund-Schultze (2009) and are probably almost precise.

[99]For details see Pinl (1969, 1974), Segal (2003) or Siegmund-Schultze (2009).

[100]Einstein lived in Prague with his wife Mileva and two children in 1911.

[101]In 1896 Einstein renounced German citizenship (because of a danger to be considered a deserter) and lived stateless for three years. In 1899 he applied for Swiss citizenship. Later, he also applied for the Austrian citizenship, but the process was not finished in time and so he remained Swiss. Later on, he became in a very strange way a German citizen, but the whole problem of his nationality is much more complicated. Those who are interested in it should take a look at https://artsandculture.google.com/exhibit/QQrk18h6 [20.8.2019].

Just a few words about John (János) von Neumann (1903–1957): As it was said by Paul Richard Halmos (1916–2006), legends about him *started accumulating in his childhood*[102] — he was for sure a prodigy. He studied chemistry in Berlin till 1923 and then he moved to Zürich, where he obtained diploma in chemical engineering from the ETH in 1926. He was very much interested in mathematics and was in contact with Weyl and George Pólya (1887–1985). Also in 1926, with a thesis on set theory entitled *Az általános halmazelmélet axiomatikus felépítése* [Axiomatisation of the Set Theory], he received doctorate in mathematics in Budapest. Then he lectured in Berlin from 1926 to 1929 and in Hamburg from 1929 to 1930, of course with the interruption caused by his stay in Götttingen.

He was invited to the U.S. and arrived there in 1930. In the next year he became a visiting lecturer at Princeton University. In fact, von Neumann was not a political refugee, he just used a chance to work at a top place. Later on, in 1931, he was appointed professor there. He kept his position in Germany, which he regularly visited until 1933. He contributed, e.g., to set theory (axiomatics), measure theory, Hilbert spaces and operator theory, mathematical logic, group theory, and also influenced probability theory (a proof of ergodic theorem). The applications he did concerned game theory, hydrodynamics, theory of automata, and also geometry and physics. Moreover, he would have been a good candidate for a patron of computer sciences.[103]

Gumbel was a well-known pacifist and together with the Noethers one of the rather leftist mathematicians in Germany. He graduated from the University of München and taught statistics as a professor of the University of Heidelberg. He was also a political writer who wrote about the crimes of "Brownshirts" (SA). In 1914 he obtained a doctorate in political economy for the thesis *Die Berechnung des Bevölkerungsstandes durch Interpolation*, which was published in 1916.

In Berlin, he worked in the Civil Service and later for Telefunken. He was suspended several times for his political activities. In 1926 he left Germany for Moscow to work on the mathematical archives of Marx. He also took part in the development of extreme value theory.[104]

The case of von Kármán was similar to Einstein's one — for both, the chance for emigration appeared due to their research qualities and personal contacts. While he

[102]See P.R. Halmos, The Legend of John von Neumann, *American Mathematical Monthly* **80**(1973), pp. 382–394.

[103]In private life he married twice (1930, 1938), always with a Hungarian. He was known for organising parties — he liked to socialize a lot and entertained regularly at home. For further information on von Neumann see Ulam (1958).

[104]In 1932 he was forced by the Nazis to leave Germany for France from where he emigrated to the U.S. in 1940. After WWII, Gumbel published a key book on the topic of statistics of extremes. For more detailed information on Gumbel see Siegmund-Schultze (2009), or the paper L. Fernández and M. Scherer, Emil J. Gumbels last course on the "Statistical theory of extreme values": A conversation with Tuncel M. Yegulalp, *Extremes* **21**(2018), pp. 97–113.

was in Aachen he was invited to the U.S. to help with the design of a wind tunnel. His next visit to the U.S. took half a year and in 1930 he left Germany with his mother and sister for the U.S., feeling the danger of rising anti-Semitism. Altogether they moved to Pasadena, their new home. In 1932 he was one of the founders of the Institute of Aeronautical Sciences, one of the world's top institutions for aeronautics and related fields.[105]

Weyl left Göttingen still in time but almost too late. In fact, he was leaving the town for the second time in his life, which needs an explanation. His PhD dissertation *Singuläre Integralgleichungen mit besonder Berücksichtigung des Fourierschen Integraltheorems* (1908) was supervised by Hilbert and submitted in Göttingen. Then he taught there until his habilitation in 1913. In the same year he married Helene Joseph (1893–1948) who had been a student of Edmund Husserl (1859–1938).[106] They went to Switzerland where Weyl had a position at ETH Zürich. He left it in 1930 when, after Hilbert's retirement, he was appointed a chair in Göttingen.

Unfortunately, he was able to stay there only for three years since his wife Helene was of Jewish descent. This was the main reason why they decided to leave for the US. Weyl, by a lucky chance, was offered a position at the Institute for Advanced Studies (IAS) in Princeton which he held since the end of 1933. His colleagues there were, e.g., Einstein or Oswald Veblen (1880–1960). He is an author of the first unified field theory for which the Maxwell electromagnetic field and the gravitational field appear as geometrical properties of space-time. One of his most influential books is *Gruppentheorie und Quantenmechanik* (1928, English translation 1931). In 2016, as a subtitle of a paper Sury (2016) on Weyl, B. Sury used *Mathematician, Physicist, Philosopher and Poet*, which gives a clear idea of his personality. Another good source of information on Weyl is Atiyah (2003).

13. Some Successful Emigrants...

For various reasons, like ties to place, country, family and friends, only a small number of mathematicians left Germany before 1933. One of the difficulties later on was that too many people needed help in a very short time. Emigration was a difficult task and especially for older scientists it was almost out of the question.

Eighteen mathematicians left Göttingen in 1933.[107] Let us recall a many times quoted answer of Hilbert; when he was asked by the Nazi Reich Minister Bernhard

[105] For many more details see the article S. Goldstein, Theodore von Kármán, 1881–1963, *Biographical Memoirs of Fellows of the Royal Society of London* **12**(1966), pp. 335–365. Note that in 1956 *The von Kármán Institute for Fluid Dynamics* was created in Sint-Genesius-Rode, Belgium. See https://www.vki.ac.be/ [27.07.2019] and other Internet sources.

[106] Helene Joseph-Weyl studied philosophy and mathematics in Göttingen; she was known as a translator from Spanish and English. After her death in 1948, Weyl married Ellen Bär from Zürich in 1950.

[107] Others included, e.g., Franck or Born to name those we mentioned above.

Rust (1883–1945) in 1934: *Is it really true, Herr Professor, that [the Mathematical Institute] suffered so much from the departure of the Jews and their friends?*, his answer was *Suffered? No, it hasn't suffered, Herr Minister. It simply doesn't exist anymore*! It can be found in Fraenkel and Cohen-Mansfield (2016, p. 135).[108] It is not clear whether it really happened but it precisely describes the situation.[109]

Courant left for the U.S. where he settled in New York. At the beginning he had to face difficulties but in 1935 he got a permanent position and since 1936 he was a professor of the New York University. He built there a division which bears his name since 1964: *Courant Institute of Mathematical Science.* During his life he wrote several books. His co-authors were Hilbert, Hurwitz, Fritz John (1910–1994), and Herbert Robbins (1915–2001).[110] Note that the introduction of the finite elements method goes back to his article from 1943. He had some other offers for a position but due to contacts in the U.S. his decision was a simple one. Needless to say, thanks to this position, he could help many other mathematicians.

His assistant Kurt Otto Friedrichs (1901–1982), who came to study in Göttingen in 1922 and during the stay there helped Courant with the "Courant–Hilbert book", went in 1929 to Aachen to visit von Kármán. He presented his habilitation thesis in Göttingen in the same year. He was very much influenced by von Neumann, who came to Göttingen for postdoctoral studies under Hilbert during 1926–1927.

Friedrichs obtained the position of a professor at the TH Braunschweig in 1931 and in 1933, shortly after Hitler came to power, he met a Jewish girl there, Nellie Bruell (1908–1994). She was dismissed from her position as an assistant at the TH. To be able to marry her, which was against Nuremberg Laws, they emigrated with Courant's help to the U.S. He worked there with Courant and was also for a year the director of the Courant Institute. He studied analytic functions, contributed to the moment problem, and so on.[111]

Sometimes personal political orientation played a role. E. Noether found a place also in the U.S. at Bryn Mawr College where she got a one-year visiting professorship.

---

[108]The author of the book Adolf Abraham Halevi Fraenkel (1891–1965) worked mainly in set theory. He could not be a direct witness of the event, since in 1934 he was already in Palestine.

[109]The last years of Hilbert's life were not happy. He was always optimistic about the future of human culture and science. His pupil Hermann Weyl wrote about him: *No mathematician of equal stature has risen from our generation ... Hilbert was singularly free from national and racial prejudices; in all public questions, be they political, social or spiritual, he stood forever on the side of freedom.* Hilbert died in 1943 and only a few people accompanied him on his last journey. He is buried in Göttingen, not far from Klein's grave. On his gravestone we can read *Wir müssen wissen, wir werden wissen!* [We must know, we will know!]. Hilbert said it in his retirement speech to the Society of German Scientist and Physicians in 1930.

[110]C. Reid's book Reid (1976) on Courant contains in some places something like "critical description" of Courant's features. He did not always give proper credit to co-authors. See the whole story at pp. 223–226 and pp. 229–232 of Reid (1976).

[111]Details of his life and career are available in C.S. Morawetz, Kurt Otto Friedrichs, *Biographical Memoirs of the National Academy of Sciences* **40**(1969), pp. 69–90.

It was not a proper place for somebody who was an invited speaker at the ICM 1928 and the ICM 1932 but it was close to Princeton and no better offer was available at the time when she probably felt a danger. In a year she started to lecture once a week at the IAS, Princeton. In 1934 she visited for a while her brother and some friends in Germany. In 1935 she died in the U.S. — her death was completely unexpected. Doctors found that she had a tumour and so she had to undergo an operation which was successful, but a few days after that she got probably a viral infection, collapsed and died.[112]

When F. Noether lost his position in Breslau, he decided to go to Russia where he was appointed as professor at the Institute of Mathematics and Mechanics at the University of Tomsk in 1933. His wife did not feel well in Russia and so she returned to Germany where she committed suicide. F. Noether remained with their two sons in Tomsk. While he was there he visited ICM 1936 in Oslo as the only participant coming from USSR. His further fate was unclear for a long time, but his sons left Russia in time and survived.[113]

F. Noether was not the only mathematician who emigrated to Russia. Stefan Bergman (1895–1977) studied in Vienna where he became Diplomingenieur in 1920. Since 1921, he was employed at von Mises' Institute in Berlin and worked for a doctorate under von Mises' supervision. He submitted a thesis *Über die Entwicklung der harmonischen Funktionen der Ebene und des Raumes nach Orthogonalfunktionen.* He was well educated in both pure and applied mathematics and developed a powerful method in potential theory which he applied to electrical engineering, elasticity and fluid flow. In 1930 he became a private docent of the Institute for Mathematics and of the Institute for Applied Mathematics as well.

When he lost as a Jew his position, he went for two years to Tomsk in Russia (1934–1936) and then to Tbilisi in Georgia (1936–1937). Due to Stalin's purges of foreign scientists he had to escape to Paris where he was a member of the Institut Henri Poincaré and in 1939 he had to flee for the third time, this time to the U.S. He was able to apply his broad mathematical knowledge on demographical problems and, in collaboration with the Curies, on the study of radioactivity. Especially his applications in the fields of hydrology and meteorology helped considerably in France and later in the U.S. in the forecasting of river floods.

During his stay in France he wrote a monograph on the kernel function and its applications in complex analysis. Now the term "Bergman kernel" commemorates his studies.[114]

---

[112]The monograph Dick (1981), provides details of her life.

[113]In 1989 they were informed about the death of their father in 1941 (see below).

[114]Bergman kernel is a reproducing kernel for the Hilbert space of all square integrable holomorphic functions on a domain D in $\mathbb{C}^n$. See, e.g., https://www.encyclopediaofmath.org/index.php/Bergman_kernel_function [5.12.2019].

Neugebauer is now known mainly as a historian of mathematics, but in spite of the fact that he switched the field of interest to history of mathematics, almost all mathematicians owe him many thanks. In cooperation with Springer, he founded in 1931 the reviewing journal *Zentralblatt für Mathematik und ihre Grenzgebiete* (*Zbl*); he was also its first editor-in-chief. The initiative leading to its foundation came from Courant, Neugebauer and H. Bohr. It is another good point for German interwar mathematics.

When Neugebauer refused as a new acting director of the Göttingen Mathematical Institute to sign an oath of loyalty to the new government in 1934, he was promptly suspended as "*untragbar*" and was denied access to the Institute building; in fact, he had the position of director for just one day.[115] For some time of uncertainty, he went to Copenhagen, where H. Bohr arranged for him a three-year appointment as a professor. His research there was partly supported by the Rockefeller Foundation. From Copenhagen he also continued his editorial work for *Zbl*.[116]

In 1938 Blaschke criticized the growing proportion of English reviews in *Zbl*, but Neugebauer sharply replied that *Zbl* is an international journal. After all, further development led to resignation not only of Neugebauer but also of many members of the editorial board of *Zbl* and of many reviewers of *Zbl* as well. Neugebauer went to the U.S. where in collaboration with the American Mathematical Society he started preparation for issuing a new reviewing journal, *Mathematical Reviews* (*MR*). He came, together with his wife Grette, two children and his student and assistant Olaf Henrik Schmidt (1913–1996)[117], to the U.S. in 1939, and they succeeded to publish the first issue of *MR* in January 1940. It was a great success, but it is another story. At that time *Zbl* lost its unique position.[118]

F. John, a student of Courant and one of his collaborators and co-authors, was not of Aryan descent and so he also decided for emigration. Thanks to the support of St. John's College, Cambridge, he could go to England and later in 1935 to the U.S., where University of Kentucky in Lexington offered him a position of an assistant professor. Since the beginning of his mathematical career, John studied the Radon transform. He used it for the study of the general properties of linear partial differential equations, convex geometry and the mathematical theory of water waves.

[115]This information is contained in Segal (1980, p. 48). Neugebauer was Courant's choice. He knew him for years and they were quite close. And Neugebauer was not a Jew.

[116]The role of Neugebauer in the development of Göttingen mathematics is more decisive than one might expect. This is described in the book Jones *et al.* (2016) by Rowe in his contribution *From Graz to Göttingen: Neugebauer's Early Intellectual Journey.* It gives also a look at the personal relations in Göttingen's mathematical community based on available documents rather than on oral communications as in the case of the book Reid (1976).

[117]O.H. Schmidt became a Doctor of Philosophy in 1943 with a thesis *On the Relation Between Ancient Mathematics and Spherical Astronomy.* In 1945 he returned to Europe. He became a full professor of the Mathematical Institute of the University of Copenhagen in 1965.

[118]For details see Swerdlow (1993).

It is worth mentioning that some states did many things for German mathematicians and scientists (it was not quite easy), e.g., the U.S., England, Turkey and Palestine.

14. From Turkey to the U.S.

The fate of a group of scientists who had to leave Germany in 1933 is closely related to the modern history of Turkey. It started in 1923, when the Republic of Turkey was founded with Mustafa Kemal Atatürk (1881–1938) as its first president. Atatürk modernized Turkey into a secular, industrial state. It included a reform of the university system. Darülfünün [House of Multiple Sciences] was closed in 1933 and the new Istanbul University was created. Its Institute of Mathematics was established upon the recommendations of Courant, Franck and Born who were officially invited to Turkey. Courant worked out the final version of the recommendations and its ideas were based on his Göttingen experience. He essentially suggested the creation of an institute of applied mathematics based on close ties with other applied sciences, but did not accept the offer to lead it. He also suggested several names of possible candidates for a leading position of the institute, among which was the name of Prager, who came to Turkey slightly later. In Berlin, von Mises and Geiringer accepted offers from Istanbul University.[119]

Von Mises was offered a chair there in 1933 but he tried to ensure his German pension rights before his departure to Turkey at the end of the year. It was the reason why he supported Karl Theodor Vahlen (1869–1945) as a successor for his former position since he promised him help in this case.[120] Geiringer went to Brussels in 1933 where she was appointed to the Institute of Mechanics. After a year, she left for Istanbul to work with von Mises and they taught there together with Prager who joined them also in 1934.[121] All three, von Mises, Geiringer and Prager adjusted quite well in Turkey to the new life, and they mastered the Turkish language at a good level to be able to write some texts for students. When Atatürk died in 1938, all three independently went to the U.S., where von Mises became a professor at Harvard University. Geiringer followed him to the U.S. where she found a position at Bryn Mawr College.[122]

[119]For details of German mathematicians in Istanbul see Eden and Irzik (2012).

[120]Much later, von Mises wrote about it: *Dr. Vahlen emphatically assured me that my departure abroad to accept the offered chair would not change anything with regard to my pension rights accrued during my previous years of service.* In Siegmund-Schultze (2009, pp. 329–330), some details are given: *Compensation for von Mises was approved only after his death. In the process of making their decision the ministry even asked von Mises's former colleague and prominent Nazi-mathematician, Ludwig Bieberbach, for advice.* See again Siegmund-Schultze (2009, p. 330).

[121]Some sources mention that Prager fought his dismissal through the German courts and even won his case. I was not able to verify the information from independent sources. See, e.g., http://shellbuckling.com/presentations/deceased/pages/page_66.html [20.8.2019].

[122]She did not find a better position which should correspond to her qualities. Later, she married von Mises in 1943. Prager left Turkey in 1941 and joined Brown University, Providence.

We saw that until the 1930s only two women, E. Noether and Geiringer, were able to obtain habilitation in mathematics. The third one was Ruth Moufang (1905–1977) who gained her PhD under Dehn in 1931. After a year spent in Rome and another one in Königsberg, she returned back to Frankfurt. At that time, Szász was banned from teaching because he was a Jew. Dehn and Hellinger were forced to retire for the same reason in 1935 and Epstein resigned voluntarily to avoid being dismissed. Moufang taught in Frankurt and worked on her thesis dealing with geometry. She submitted it to the Goethe University in Frankfurt in 1936. But it did not fit the Nazi policy. Hitler's minister of education refused Moufang permission to teach because she was a woman, and so she became an industrial mathematician working for Krupps in 1937.[123]

Many more such fates could be described but we have included only some of them for illustration. Their choice was more or less random. For further reading we recommend, for example, the books and monographs Bergmann *et al.* (2012), Segal (2003) and Siegmund-Schultze (2009) where many more fates of mathematicians are discussed.

15. Some Tragic Stories

There were other fates of those mathematicians who were less happy. The majority of them were Jews. Let us start with Landau, but his case needs a small introduction. Landau was an excellent mathematician; nevertheless his attitude to other colleagues was not simple. He used to criticize their results sometimes in connection with his ability to simplify proofs — he was always devoted to finding the simplest one. He did it also with proofs of some theorems of Bieberbach and Blaschke and so their relations to him were not good. He was known as a Zionist and he also spent some time with his family in Jerusalem.[124] Landau underestimated Nazis for a long time. In Siegmund-Schultze (2009, p. 83), we can read about his reaction on learning the news about plans to build a concentration camp: *In that case I had better reserve a room with a balcony, south view, as fast as possible.*

As a mathematician Landau is known for his special style of writing. He wrote more than 250 papers and 8 books, mostly from number theory. We mentioned his ability to simplify expositions and proofs. For example, he found a new approach to

[123]To have an idea of her abilities, we mention that in 1942 she became the Head of the Department of Applied Mathematics and Mechanics at Krupps. At that time she studied elasticity. She continued her academic career, but after the war. It is an example of the general attitude of the Nazis to the "women question". See also Srinivasan (1984).

[124]Details may be found in Corry and Schappacher (2010), in which his talk at the opening ceremony of the Hebrew University in Jerusalem in 1925 is reproduced and commented. He and his family found that they were not able to live there and so they soon returned back to Germany.

Picard's theorem (see above) or a new proof of Weierstrass' approximation theorem that is now used in many textbooks.[125]

After a student boycott of his lectures in 1933, he retired in 1934. Then, he moved from Göttingen to Berlin. He also spent some time abroad in England, and in Holland.[126] Shortly after the stay in Holland, he died in Berlin in 1938 from a heart attack. Thus he was spared from the cruelties which came later.

Anti-Semitism gained gradually a much more brutal form. In 1933 the Nazis organized an action of boycott of Jewish businesses under the slogan *Kauf nicht bei Juden*! Anti-Semitism took place before too, but this time the Nazis carried out the first nationwide, planned action against Jews. Other actions took place later. A certain peak was reached by the "Kristallnacht", which was a kind of medieval program against Jews, carried out in the 20th century by the SA forces and civilians. It occurred at the night from the 9th to 10th November 1938. Hundreds of synagogues were burned down, many shops were completely destroyed and many Jews were killed. In this way, the Nazis showed their genuine face...

It happened just a day after Hausdorff's 70th birthday. He was born in Breslau and studied at the University of Leipzig where he graduated in 1891. He worked under Heinrich Bruns (1848–1919) on the applications of mathematics in astronomy, but his main interest at that time was in literature. His friends were writers and artists and he himself published literary works under the name Paul Mongré. In Segal (2003, p. 456), Segal has written: *As the scion of a wealthy family, he did not have to worry about making a career as a mathematician; for him, mathematics, both as research and as a subject to teach, was more an avocation than anything else.* He taught in Leipzig, Bonn and from 1913 onwards as a professor at the University of Greifswald. A year later, his famous book *Grundzüge der Mengenlehre* appeared, in which he presented a theory of topological and metric spaces and enriched both with many results; see Hausdorff (1914). This book with some improvements was republished in 1927.

Hausdorff was too old for emigration and so he did not plan to do so but after the Kristallnacht he lost all illusions. Nevertheless it was too late. Conditions became worse and worse and so, when he received, at the beginning of 1942, summons to come with his wife and his sister-in-law to the monastery Endenich in the vicinity of Bonn, which was a place from where Jews were sent to concentration camps, they committed suicide. In his farewell letter to his lawyer Dr. Wollstein, he remarked:

[125]Landau's proof was, among almost a hundred others existing today, probably the first one in which the constructed approximating functions are *obviously* polynomials.

[126]These stays abroad are mentioned in N. Schappacher, Edmund Landau's Göttingen: From the Life and Death of a Great Mathematical Center, *Mathematical Intelligencer* **13**(1991), pp. 12–18.

Auch Endenich ist noch vielleicht das ende nich! [Also Endenich is still perhaps not the End!].[127]

We have already mentioned Hausdorff's Mengenlehre but to list all his merits would be difficult and somewhat useless: every mathematician knows at least some notions connected with his name and some of his contributions to set theory, topology, probability and other parts of mathematics. Note that he was not the only Nazi victim of this kind. Similarly the voluntary departure from life was chosen, e.g., by Epstein in 1939, Friedrich Moritz Hartogs (1874–1943),[128] and some others.

Some fates were more complicated: already in 1933, Blumenthal was jailed for two weeks as a suspected communist. Then he was released but was suspended from the university, officially not for being a Jew but for his membership in the Deutsche Liga für Menschenrechte and in the Gesellschaft der Freunde des neuen Russlands. Later on, he had to finish his editorial work for the *Mathematische Annalen* and according to the new by-laws, his membership in the DMV was cancelled. The Blumenthals successfully sent their children Ernst and Margarete to England in 1936. In 1939, they sold their house in Aachen and left for the Netherlands.

During their stay in a house very close to Utrecht, they were able to visit their children in England but unfortunately they returned, just a few days before the beginning of the war, back. Then they lived in a flat which they rented in Delft. In 1943 they were sent to the concentration camp at Vught and later to another one at Westerbork, where Mrs. Blumenthal soon died.

Blumenthal was sent on his own request to the ghetto in Theresienstadt to meet his sister, but when he arrived in January 1944, there, she was already half a year dead. In November 1944, Blumenthal died there, too. He was a good man, and a good mathematician and he has done a lot for his beloved mathematics. He knew many languages and the Czech language which he started to learn in Theresienstadt was his tenth one.[129]

In a way, a very similar fate is that of Robert Erich Remak (1888–1942), who studied mathematics and physics partly also in Göttingen. He was a good algebraist who contributed also to algebraic number theory and mathematical economy, but according to some witnesses he was, like Landau, a somewhat "difficult person". He received his doctoral degree in 1911 and after several attempts, he passed his

[127]See Bergmann *et al.* (2012, pp. 94–104), where the facsimile of the letter is reproduced on p. 103.

[128]Known mostly as Fritz Hartogs. He was a specialist in the theory of several complex variables and set theory. As a Jew he was sent after the Kristallnacht for a short time to KZ Dachau. After coming back, he and his Aryan wife divorced but they continued to live together. In 1943 he committed suicide.

[129]It is worth mentioning that Blumenthal's mathematical work became highly interesting at the end of 20th century. *His work on orthogonal polynomials whose zeros are dense in intervals, ... led to over two dozen recent papers in the field.* This is a quote from Butzer and Volkmann (2006) where more detailed information can be found.

habilitation in 1929. As a private docent he lectured from 1929 to 1933 in Berlin. He introduced independently the generalized solution of the Dirichlet problem[130] and is also credited for the prediction of the importance of computers. During the Kristallnacht he was arrested and sent to KZ Sachsenhausen. When he was released he tried to emigrate to the U.S., but in vain.[131]

16. Some Nazi Mathematicians

We have already mentioned several Nazi mathematicians. One of them was Bieberbach, who together with Vahlen founded the journal *Deutsche Mathematik* (*DM*) in 1936. Bieberbach was its editor-in-chief.[132] The mathematicians who published in *DM* included Bieberbach, Vahlen, Tornier, Paul Julius Oswald Teichmüller (1913–1943), Gustav Doetsch (1892–1977), Wilhelm Süss (1895–1958) and others.

It is somewhat surprising that Bieberbach, who had almost all that a scientist may wish to have, became suddenly an active Nazi promoting racial views on mathematical thinking. In 1938 he found out in connection with the publishing of Weierstrass' collected works that Schur was still a member of the Weierstrass Commission of the Prussian Academy of Sciences which took care of the publication. He — as another member of the Commission — expressed his astonishment that it is still possible and Max Planck (1858–1947), another member of the Commission, promised "to solve it". Within a week Schur had to resign from the Commission and also from his membership in the Academy.[133] In 1939 Schur left for Palestine where he died two years later as a poor man.[134]

When the Nazi students led by Teichmüller protested against Landau in Göttingen, Teichmüller wrote to him an explanatory letter: *You spoke of your belief that what happened yesterday was an anti-Semitic demonstration. My standpoint was, and continues to be, that an anti-Jewish individual action might rather be directed against everyone else than against you. I am not concerned with making difficulties for you as a Jew, but only with protecting — above all — German students of the second semester from being taught differential and integral calculus by a teacher of a race quite foreign to them. I, like everyone else, do not doubt your*

---

[130]See the Chapter written by Ivan Netuka where the story of the so-called PWB method of the Dirichlet problem is explained.

[131]In 1942 he was again arrested in Amsterdam and sent to KZ Auschwitz [Oświęcim] where he was murdered in the same year.

[132]The last 1942 issue of DM appeared with a two-year delay. Except for mathematical articles the journal published also ideological articles advertising "Deutsche" mathematics (in contrast to the "Jewish" one).

[133]See Siegmund-Schultze (2009, pp. 75–76), where the text and the facsimile of the hand-written document are reproduced.

[134]He was forced to sell his library for financial reasons to the Institute for Advanced Study in Princeton so he could afford to leave Germany for Palestine. This information is mentioned also in Siegmund-Schultze (2009, p. 153).

ability to instruct suitable students of whatever origin in the purely abstract aspects of mathematics. But I know that many academic courses, especially the differential and integral calculus, have at the same time educative value, inducting the pupil not only to a conceptual world but also to a different frame of mind. But since the latter depends very substantially on the racial composition of the individual, it follows that an Aryan student should not be allowed to be trained by a Jewish teacher.[135]

Teichmüller was an ardent Nazi but a good mathematician. He started to work with Hasse, who was appointed to fill Weyl's chair in Göttingen in 1934. The appointment of Hasse was not approved by certain Nazi hard-liners since, though he was a strong supporter of Hitler, he did not believe in politicizing mathematics.[136]

However, despite the fact that Hasse worked in a different area of mathematics than Teichmüller, it was Hasse who was chosen by Teichmüller as his thesis supervisor. Teichmüller's main mathematical contributions lie in the area of geometric function theory and in the theory of quasi-conformal mappings. In April 1934, Bieberbach commented on the boycott of students in Göttingen against Landau (the protest was almost surely led by Teichmüller): *A few months ago differences with the Göttingen student body put an end to the teaching activities of Herr Landau. ... This should be seen as a prime example of the fact that representatives of overly different races do not mix as students and teachers. ... The instinct of the Göttingen students was that Landau was a type who handled things in an un-German manner.*[137]

In 1935 Teichmüller graduated from Göttingen University and remained there for another year as an assistant of Hasse. During that time, Rolf Herman Nevanlinna (1895–1980) was a visiting professor in Göttingen and probably influenced Teichmüller's future study of quasiconformal mappings. This could be a possible reason of his move to Berlin in which Bieberbach's circle was close to him, both politically and mathematically. Of his 34 papers, 21 appeared in the *DM. In 1937 in Berlin he earned his qualification to lecture, and in January 1939 he became a lecturer. He passed each examination with honors. He won a position at the University of Berlin in the summer of 1939, when he was already a soldier, for he had been drafted for an 8 week enlistment. Before the 8 weeks were up, the unholy war began. Teichmüller served in the Norwegian campaign, and after that received a position with the High Command of the Wehrmacht.*[138]

---

[135] This long quote can be found in Siegmund-Schultze (2009, p. 72). For the whole of Teichmüller's original letter, see N. Schappacher, E. Scholz (eds.), Oswald Teichmüller–Leben und Werk, *Jahresbericht der Deutsche Mathematik-Vereinigung* **94**(1992), pp. 1–39.

[136] The politically complicated situation in Göttingen at the time is described in Segal (1980), or in Segal (2003, p. 147).

[137] See Siegmund-Schultze (2009, p. 73).

[138] See Abikoff (1998, p. 12). In the article, more details on Teichmüller are given. Note that he could stay in Berlin but returned voluntarily to active duty on the Russian front. *He very probably arrived at the front in the fall of 1943. Nothing has been heard from him since September 11, 1943. It*

Vahlen belonged to the old generation of mathematicians of the first half of the 20th century and he was also an "old Nazi". He was born in Vienna but studied in Berlin, where he obtained a doctorate in 1893. His thesis *Beiträge zu einer additiven Zahlentheorie* was written under Fuchs and Schwarz. Later, he slowly turned to applied mathematics and published on ballistics, aerodynamics, the magnetic compass, and celestial mechanics. From 1911 on, he was a full professor in Greifswald. In 1922 he joined the NSDAP and became its "Gauleiter" [district leader] for Pomeranian. In 1924 he was responsible as a rector of Greifswald University for protest against the Weimar Republic and taking down its flags. The university placed him on leave for political abuse of his function, and in 1927 he was dismissed without the right for a pension but Nazis rehabilitated him later.

In 1933 he joined "Sturmabteilung" (SA) and after three years he switched to "Schutzstaffel"(SS) and so he became "SS-Brigadeführer". As a full professor of mathematics at the Humboldt University of Berlin since 1933, he replaced Mises but he left, probably much sooner, mathematics for politics. In the next year, he became the head of the Science Office at the *Reichserziehungsministerium.* He had some other politically important functions.[139]

Tornier was another ardent Nazi connected with Bieberbach's and Vahlen's Deutsche Mathematik. He studied in Breslau, Berlin and Marburg. He specialised in measure theory in probability. He profited very much from a chance to be in Kiel since 1931 together with Feller, with whom he wrote at least one article. Another colleague of Feller was Fraenkel, and Tornier was his temporary replacement while Fraenkel was in Jerusalem from 1929 to 1931. Tornier hoped to get Fraenkel's chair but he obtained a much better position in Göttingen. In 1934 he appeared at the same place with Hasse, a colleague from his study days in Marburg.

The position of Hasse, who was a candidate for Courant's position as a director of the Mathematical Institute in Göttingen, was difficult since Tornier, for a long time a secret Nazi, had been appointed to fill another chair there and to be interim Institute director until Hasse's arrival to Göttingen. Besides, in July, when Hasse came to Göttingen in connection with his future appointment, he was greeted by a demonstration of pro-Nazi students. Teichmüller and Tornier were Nazi exponents in Göttingen. H. Bohr characterized Tornier in a letter to Veblen as *"Vertrauensmann" in mathematical questions for the Nazi party and the government.*[140]

Hasse and Tornier were far from being good partners, which would be beneficial for two heads of departments of the same university; nevertheless they were able to report:

---

was reported to his mother that her son had been missing since the heavy fighting along the Dnieper. See again Abikoff (1998, p. 12).

[139]For Vahlen's role in the Nazi regime see Begehr *et al.* (1998, pp. 132–133), Segal (2003, p. 34, p. 144), and further numerous remarks in the text.

[140]See, e.g., Segal (1980, p. 52), and also Campbell and Higgins (1984).

Tornier (30 June, 1934): *By order of the Ministerial director Prof. Dr. Vahlen, Professor Hasse will begin his work here at the beginning of July. Since all old difficulties have been removed, both Vahlen and I expect that no new artificial ones will be created.*

A few days later:

Hasse (2 July, 1934): *Today I have taken over the leadership of the institute together with Dr. Tornier.*[141]

Note that the time coincides with the Röhm-Putsch or Nacht der langen Messer [Night of Long Knives] in which Hitler's political opponents and some members of SA (including Röhm) were killed. These quotes also illustrate the atmosphere of the time.

They had not much time to cooperate since Tornier left for Berlin in 1936 where he became a professor.[142]

Doetsch and Süss were almost of the same age and both had to take part in the WWI. Both of them survived and continued their studies as it was common at that time — at several different universities. Doetsch submitted his doctoral dissertation *Eine neue Verallgemeinerung der Borelschen Summabilitätstheorie der divergenten Reihen* (1920), which was supervised by Landau in Göttingen, while Süss' dissertation *Begründung der Inhaltslehre im Raum ohne Benutzung von Stetigkeitsaxiomen* (1920) was written under Bieberbach in Frankfurt.

Doetsch then worked with Bernstein on a problem involving convexity. In 1915 they proved an elementary but interesting theorem which says that *every real mid-convex (Jensen convex) function on an open convex subset G of the Euclidean m-space $\mathbb{R}^m$ which is upper bounded at least in a neighborhood of a point $x \in G$ is already continuous at G.*[143]

Slightly later, they studied problems connected with the Laplace transform, which became Doetsch's main theme for his whole life. In 1937 he published a monograph on the theory and application of the Laplace transform[144] and it became a popular standard text on this subject.

Doetsch was a pacifist and this could be one of the reasons why he started to support Nazis to avoid difficulties in the coming political system. His active support of the new regime in 1933 stemmed probably from the fact that he was afraid about his professorship in Freiburg where he had a position since 1931. On the other hand he did not join the NSDAP.

[141] For both quotes and a detailed description of these events see Segal (2003, pp. 149–151).

[142] Tornier was later pensioned because of his eccentric life and affairs in Berlin.

[143] They published it as Zur Theorie der konvexen Funktionen, *Mathematische Annalen* **76**(1915), pp. 514–526. The result was later generalised, e.g., in W. Sierpiński, Sur les fonctions convexes mesurables, *Fundamenta Mathematicae* **1**(1920), pp. 125–128.

[144] *Theorie und Anwendung der Laplace-Transformation*, Grundlehren der mathematischen Wissenschaften, Volume 67, Springer, Berlin–Heidelberg, 1937, 438 pages. ISBN 978-3-642-98721-2.

Süss finished his dissertation after the pause caused by the WWI but having no chance for a proper place in Germany he went to Japan, where he taught German at the University of Kagoshima. In 1928 he returned and went to Greifswald. There he worked with Helmuth Kneser (1862–1930), who wrote his doctoral thesis *Untersuchungen zur Quantentheorie* [145] in Göttingen under Hilbert. It seems quite natural that Süss also started to become interested in Hilbert's works, especially in *Grundlagen der Geometrie.* In Greifswald and later in Freiburg, where he replaced Alfred Loewy (1873–1935), as a full professor in 1934,[146] he studied convex bodies and relative differential geometry. He was offered this professorship since he was recommended by Doetsch.

Süss' approach to his career was different from Doetsch's one. He started later, but he joined NSDAP in 1937 and a year later also the "NS-Dozentenbund". Whereas up to 1937, Doetsch and Süss were on friendly terms, this year brought a break, which was definitely finished in 1940. Their previous cooperation influenced the struggles for power in the DMV and some others which were in the German mathematical community at that time. We will try to describe, at least partially, some of them.

17. Gleichschaltung of the Mathematical Community

It is interesting to follow the consequences of a struggle for the implementation of the *Führerprinzip* in mathematics. That principle copied the functionality of a military organization, where a similar authority principle (The commander is always right) is used even today and even in some democratic countries. Besides the DMV and GAMM, examples of German mathematical organizations included the Mathematische Reichsverband (MR) for secondary school teachers, founded by Georg Karl Wilhelm Hamel (1877–1954)[147] in 1921, and the much older Deutsche Verein zur Förderung des mathematischen und naturwissenschaftlichen Unterrichts, founded in 1891.

These societies behaved in a different way in 1933 — naturally, the DMV and GAMM were organizations of highly qualified scientists while MR was a political interest organization. MR's former chair and new "Führer" Hamel made the following statement: *We want to cooperate sincerely and loyally in accordance with the total state. Like all Germans, we place ourselves unconditionally and happily in the service of the National Socialist movement, behind its Führer, our Chancellor Adolf Hitler.*[148] DMV and MR were mutually tied together and MR can be seen from the first point of view as a pro-Nazi political wing of DMV.

[145] Published in *Mathematische Annalen* **84**(1921), pp. 277–302.
[146] Being a Jew, Lewy was forced to retire in 1934 under the Civil Service Law.
[147] Known for the so-called Hamel basis of $\mathbb{R}^1$ considered as a vector space over the rationals.
[148] This translation of Hamel's words can be found in Mehrtens (1989, p. 48).

The situation around DMV is described in the comprehensive and nicely written papers Remmert (1999) and Remmert (2004), which have together more than 90 pages (see also Schubring, 2010; and Tobies, 1982). We will try to describe briefly a part of their content with some comments. We restrict ourselves to DMV, but it is still complicated since it was under the pressure of "Gleichschaltung" and the whole case was interlocked with other important events as well. On top of it, changes followed quite rapidly and it was difficult for the actors to predict what will happen tomorrow.

The president of DMV was always elected for one year. Oskar Perron (1880–1975) was the president in 1933; he was elected in the same year in which Hasse, the successor of Weyl, came to Göttingen. The other top three positions in DMV (members of its board) were practically permanent: Blumenthal (editor of Jahresbericht der DMV) just had to retire and was replaced by Konrad Knopp (1882–1957), and the other two were Bieberbach (secretary) and Hasse (treasurer).

Bieberbach's goal was to gain an absolute power in DMV (and with it in hands, to govern German mathematics). He was, at least in the beginning, supported by Doetsch and Süss, and he was not such a good politician and diplomat as Süss, which will be obvious a bit later. A report on Bieberbach's lecture on races in mathematics appeared in *Deutsche Zukunft* under the title *Neue Mathematik* in April 1934. Shortly after that, Bieberbach wrote several articles on the same topic, one of them in the *Proceedings of the Berlin Academy.* He mentioned the SS-organized boycott of lectures of Landau in Göttingen during the preceding fall, which was mentioned above. His considerations, which were based on the racial-typological theories of Nazi psychologist Erich Rudolf Ferdinand Jaensch (1883–1940), were registered abroad. Distinguished foreign mathematicians such as H. Bohr, and Godfrey Harold Hardy (1877–1947) openly criticized Bieberbach.

H. Bohr wrote a newspaper article about Bieberbach's thoughts and the conflict was growing. At the meeting of the DMV at Bad Pyrmont in September 1934, Bieberbach proposed a change of the statutes: One year term for the president of DMV should be changed and the Führer principle (based on the slogan *Ein Volk, ein Reich, ein Führer*) applied. As a candidate for the role of president, he suggested Tornier, who in return nominated Bieberbach. What Bieberbach really wanted became clear when he used a quote of Hitler *Alle Macht oder keine* in a reply to Doetsch's doubts on his approach to the case of DMV and related things.

Neither Bieberbach nor Tornier was approved in Pyrmont by the DMV, and Blaschke was elected by acclamation for two years, which was understood as a partial Führer principle. Bieberbach as a secretary was responsible for applying for the corresponding change of society by-laws in Leipzig but he did not do it. In 1934, he also published in the *Jahresbericht der DMV* an "Open Letter" to H. Bohr. The letter was highly critical of H. Bohr for his comments on Bieberbach's racist views. This was done *against the wishes of his co-editors, who had been afraid that its publication in the official DMV journal by DMV secretary would give the impression that his views represented those of the DMV. This was the beginning of Bieberbach's conflict with the DMV.*[149] The DMV was critical of his actions and Bieberbach was forced to resign from his editorial position. Blaschke resigned as well.

The original plans of Bieberbach for the Bad Pyrmont meeting were supported by other players in the game for power, Doetsch and Süss. They wanted to demand the resignation of Blaschke, Hasse and Knopp, but not of Bieberbach, from all offices in the DMV. The situation again changed and as it was said, Bieberbach and Blaschke resigned. The whole case led to the resignation of some members of the DMV, like Bohr, Courant, von Neumann and Weyl. In 1935, Hamel was elected president of the DMV, followed by E. Schmidt from Berlin and two others, always for a year. In this situation, which was full of animosities of those taking part in this game for power, a new president was elected in 1938. It was Süss who remained in that position until 1945.

Today, more than 80 years after those events, it is not easy to evaluate what happened. Many documents were destroyed but even all those remaining have probably still not been examined. It is difficult to believe that these things really took place. But they had important consequences since they were a part of the developments which led to WWII and which caused the death of many millions of people.

References

W. Abikoff, Oswald Teichmüller, *Mathematical Intelligencer* **8**(1998), pp. 8–16, 33. ISSN 0343-6993.

M. Atiyah, Hermann Weyl (November 9, 1885–December 9, 1955), *Biographical Memoirs of the National Academy of Sciences* **82**(2003), pp. 320–335. ISSN 0077-2933.

M. Bečvářová and I. Netuka, *Jarník's Notes of the Lecture Course Punktmengen und Reelle Funktionen by P.S. Aleksandrov* (*Göttingen 1928*), Matfyzpress, Prague, 2010, 148 pages. ISBN 978-80-7378-113-2.

M. Bečvářová and I. Netuka, *Dopady první světové války na mezinárodní spolupráci matematiků* [Impact of the World War I on the International Collaboration of Mathematicians]. Česká technika – nakladatelství ČVUT Praha, 2019, 240 pages. ISBN 978-8001-06540-2.

H.G.W. Begehr *et al.* (eds.), *Mathematik in Berlin. Geschichte und Dokumentation* [1. und 2. Halbband]. Shaker Verlag, Aachen, 1998, xii + 670 + xvi + 670 pages. ISBN 3-8265-4225-8.

H.G.W. Begehr *et al.* (eds.), *Mathematics in Berlin.* Springer Basel AG, Basel, 1998, xii + 208 pages. ISBN 978-3-7643-5943-0.

H. Behnke, Otto Blumenthal zum Gedächtnis, *Mathematische Annalen* **136**(1958), pp. 387–392. ISSN 0025-5831.

H. Behnke, Rückblick auf die Geschichte der Mathematischen Annalen, *Mathematische Annalen* **200**(1973), pp. i–vii. ISSN 0025-5831.

[149] This is a quote from Remmert's paper Remmert (1999, p. 17).

B. Bergmann, M. Epple, and R. Ungar (eds.), *Transcending Tradition. Jewish Mathematicians in German-Speaking Academic Culture.* Springer, Heidelberg, 2012, 12 + 294 pages. ISBN 978-3-642-22463-5.

B. vom Brocke, Friedrich Althoff: A Great Figure in Higher Education Policy in Germany, *Minerva* **29**(1991), pp. 269–293. ISSN 0026-4695.

G. Birkhoff and M. Bennett, Felix Klein and his Erlanger Programm, pp. 145–176, in W. Aspray and P. Kitcher (eds.), *History and Philosophy of Modern Mathematics*, Volume XI. University of Minnesota Press, Minnesota, 1988, viii + 386 pages. ISBN 978-0-8166-5305-8.

H. Bohr and O. Neugebauer, Über lineare Differentialgleichungen mit konstanten Koeffizienten und fastperiodischer rechter Seite, *Nachrichten von der Gesselschaft der Wissenschaften zu Göttingen. Mathematisch-Physikalische Klasse*, 1926, pp. 8–22.

P. Butzer and L. Volkmann, Otto Blumenthal (1876–1944) in Retrospect, *Journal of Approximation Theory* **138**(2006), pp. 1–36. ISSN 0021-9045.

D.M. Campbell and J.C. Higgins (eds.), *Mathematics. People, Problems, Results.* Wadsworth, Belmont, CA, 1984, 971 pages in 3 volumes. ISBN 978-0-534-03202-9.

L. Corry and N. Schappacher, Zionist Internationalism through Number Theory: Edmund Landau at the Opening of the Hebrew University in 1925, *Science in Context* **23**(2010), pp. 427–471. ISSN 0269-8897.

A. Dick, *Emmy Noether, 1882–1935.* Birkhäuser, Boston 1981, xiv + 184 pages. ISBN 978-1-4684-0535-4.

M. Eckert, *Arnold Sommerfeld. Science, Life and Turbulent Times 1868–1951.* Springer, Berlin, 2013, 488 pages. ISBN 978-1-4614-7460-9.

M. Eckert, *Ludwig Prandtl. A Life for Fluid Mechanics and Aeronautical Research*, Springer Biographies. Springer International Publishing Berlin–Heidelberg, 2019, xvi + 360 pages. ISBN 978-3-030-05663-6.

A. Eden and G. Irzik, German Mathematicians in Exile in Turkey: Richard von Mises, William Prager, Hilda Geiringer, and their Impact on Turkish Mathematics, *Historia Mathematica* **39**(2012), pp. 432–459. ISSN 0315-0860.

A.A. Fraenkel and J. Cohen-Mansfield (eds.), *Recollections of a Jewish Mathematician in Germany.* Birkhäuser (Springer International Publishing), Switzerland, 2016, xx + 234 pages. ISBN 978-3-319-30845-6 [the German version was published in 1967].

W. Gautschi, Alessandro M. Ostrowski (1893–1986): la sua vita e le opere [Alexander M. Ostrowski (1893–1986): His Life, Works and Students], Expanded Text of the Lecture Presented at a Meeting of Ostrowski Foundation, Bellinzona, Switzerland, May 24–25, 2002, published in *Bollettino dei docenti di matematica* **45**(2002), pp. 9–19. Available also at https://www.cs.purdue.edu/homes/wxg/AMOengl.pdf [10.7.2019].

W. Gautschi, Ostrowski and the Ostrowski Prize, *Mathematical Intelligencer* **20**(1998), pp. 32–34. ISSN 0343-6993.

I. Grattan-Guinness (ed.), *Landmark Writings in Western Mathematics 1640–1940.* Elsevier, Amsterdam, 2005, 1040 pages. ISBN 978-0-444-50871-3.

J. Gray, *Felix Klein's Erlangen Program, 'Comparative Considerations of Recent Geometrical Researches'*, in Grattan-Guinness (2005), pp. 544–552.

M. Georgiadou, *Constantin Caratheodory — Mathematics and Politics in Turbulent Times.* Springer, Berlin–Heidelberg, 2004, xxviii + 651 pages. ISBN 978-3-540-20352-0.

F. Hausdorff, *Grundzüge der Mengenlehre.* Verlag von Veit & Comp., Leipzig, 1914, 476 pages.

M. Hazewinkel, *David Hilbert, Paper on 'Mathematical Problems'*, in Grattan-Guinness (2005), pp. 732–747.

D. Hilbert, *Gesammelte Abhandlungen III. Analysis, Grundlagen der Mathematik, Physik, Verschiedenes. Nebst einer Lebensgeschichte.* Springer, Berlin, 1935, xii + 436 pages.

K.H. Jarausch, German Students in the First World War, *Central European History* **17**(1984), pp. 310–329. ISSN 1569-1616.

A. Jones, Ch. Proust, and J.M. Steele (eds.), *A Mathematician's Journeys: Otto Neugebauer and Modern Transformations of Ancient Science*, Archimedes 45, Springer, Cham, Heidelberg, London: Springer, 2016, xi + 342 pages. ISBN 978-3-319-25863-8.

C.H. Kimberling, Emmy Noether, *American Mathematical Monthly* **79**(1972), pp. 136–149. ISSN 0002-9890.

E.E. Kramer, *The Nature and Growth of Modern Mathematics.* Princeton University Press, Princeton, New Jersey, 1972, xxvi + 758 pages. ISBN 978-0-691-02372-4.

F. Klein, *Lectures on Mathematics* (*Evanston Colloquium*). Macmillan & Co., New York and London, 1894, ix + 109 pages.

M. Kline, *Mathematical Thought from Ancient to Modern Times 1–3.* Oxford University Press, New York and Oxford, 1972, xviii + 1212 + xxiv pages in 3 volumes. ISBN 978-0-19-506135-2.

J.J. O'Connor and E.F. Robertson, *MacTutor History of Mathematics*, University of St. Andrews, Scotland, UK, available at https://www-history.mcs.st-andrews.ac.uk/index.html [20.8.2019].

V.G. Maz'ya and T.O. Shaposhnikova, *Jacques Hadamard, a Universal Mathematician*, History of Mathematics 14, London Mathematical Society, London, 1999, 574 pages. ISBN 978-0-8218-1923-4.

H. Mehrtens, The 'Gleichschaltung' of Mathematical Societies in Nazi Germany, *Mathematical Intelligencer* **11**(1989), pp. 48–60. ISSN 0343-6993. Translation of the article Die 'Gleichschaltung' der mathematischen Gesellschaften im nationalsozialistischen Deutschland, pp. 83–103, in S.D. Chatterji, I. Fenyö, U. Kulisch (eds.), *Jahrbuch Überblicke Mathematik*, Volume 18 (1985), Bibliographisches Institut, 195 pages. ISBN 978-3-411-03103-0.

R. von Mises, *Fluglehre: Theorie und Berechnung der Flugzeuge in Elementarer Darstellung*, 6th edition. Springer, Berlin, 1957, xii + 402 pages.

E. Neuenschwander and H-W. Burmann, *Die Entwicklung der Mathematik an der Universität Göttingen*, Available at https://www.uni-math.gwdg.de/entwicklung.xhtml [10.7.2019].

K.H. Parshall and A.C. Rice (eds.), *Mathematics Unbound: The Evolution of an International Mathematical Community, 1800–1945*, History of Mathematics, Volume 23, American Mathematical Society and London Mathematical Society Providence, 2002, 406 pages. ISBN 978-0-8218-2124-4.

F. Paschen, Carl Runge [Obituary], *The Astrophysical Journal* **69**(1929), pp. 317–321.

M. Pinl, Kollegen in einer dunklen Zeit, *Jahresbericht der Deutschen Mathematiker-Vereinigung* **71**(1969), pp. 167–228; **72**(1971), pp. 165–189; **73**(1971/72), pp. 153–208. ISSN 0012-0456.

M. Pinl and A. Dick, Kollegen in einer dunklen Zeit. Schluß, *Jahresbericht der Deutschen Mathematiker-Vereinigung* **75**(1974), pp. 166–208. ISSN 0012-0456.

G. Pólya, Some Mathematicians I Have Known, *The American Mathematical Monthly* **76**(1969), pp. 746–753. ISSN 0002-9890.

C. Reid, *Hilbert. With an Appreciation of Hilbert's Mathematical Work by Hermann Weyl*, Springer, New York–Heidelberg–Berlin, 1970, xii + 292 pages. ISBN 978-0-387-04999-1.

C. Reid, *Courant in Göttingen and New York. The Story of an Improbable Mathematician.* Springer, Berlin-Heidelberg, 1976, iv + 314 +xviii pages. ISBN 978-3-540-90194-9.

V.R. Remmert, Mathematicians at War. Power Struggles in Nazi Germany's Mathematical Community: Gustav Doetsch and Wilhelm Süss, *Revue d'Histoire des mathématiques* **5**(1999), pp. 7–59. ISSN 1262-022X.

V.R. Remmert, Die Deutsche Mathematiker-Vereinigung im 'Dritten Reich' I, II: Krisenjahre und Konsolidierung (1st part), Fach- und Parteipolitik (2nd part), *Mitteilungen der Deutschen Mathematiker-Vereinigung* **12**(2004), pp. 159–177, pp. 223–245. ISSN 0942-5977.

J. Roselló, *Hilbert, Göttingen and the Development of Modern Mathematics.* Cambridge Scholars Publishing, Cambridge, 2019, xii + 272 pages. ISBN 978-1-5275-2331-9.

M.I. Rosen (ed.), *Exposition by Emil Artin: A Selection*, History of Mathematics, Volume 30. American Mathematical Society, Providence, 2007, 343 pages. ISBN 978-0-8218-4172-3.

P. Roquette, History of Valuation Theory (Part I), *Fields Institute Communications* **32**(2002), pp. 291–335. ISSN 1069-5265.

D.E. Rowe, Klein, Hilbert, and the Göttingen Mathematical Tradition. *Osiris* **5**(1989), pp. 186–213. ISSN 0369-7827.

D.E. Rowe, *A Richer Picture of Mathematics. The Göttingen Tradition and Beyond.* Springer International Publishing AG, 2018, xx + 468 pages. ISBN 978-3-319-67818-4.

R.L. Schilling and W.A. Woyczyński, William Feller: A Biography, pp. 17–41 in Volume I, and pp. 1–25 in Volume II, in R.L. Schilling, Z. Vondraček, W.A. Woyczyński (eds.), *Selected Papers I, II* [of W. Feller]. Springer Verlag International, Cham, 2015, xxxvi + 820 and xxxiv + 785 pages. ISBN 978-3-319-16858-6 and ISBN 978-3-319-16858-5.

G. Schubring, 120 Jahre Deutsche Mathematiker-Vereinigung. Neue Ergebnisse zu ihrer Geschichte, *Mitteilungen der Deutschen Mathematiker-Vereinigung* **18**(2010), pp. 103–108. ISSN 0942-5977.

G. Schubring, *Felix Klein. Düsseldorf 1849–Göttingen 1925*, Available at http://www.icmi history.unito.it/portrait/klein.php [17.07.2019].

S.L. Segal, Helmuth Hasse in 1934, *Historia Mathematica* **7**(1980), pp. 46–56. ISSN 0315-0860.

S.L. Segal, *Mathematicians Under the Nazis.* Princeton University Press, Princeton and Oxford, 2003, xxiv + 530 pages. ISBN 978-0-691-00451-8.

C.L. Siegel, On the History of the Frankfurt Mathematics Seminar, *Mathematical Intelligencer* **1**, pp. 223–230. ISSN 0343-6993.

R. Siegmund-Schultze, Hilda Geiringer-von Mises, Charlier Series, Ideology, and the Human Side of the Emancipation of Applied Mathematics at the University of Berlin During the 1920s, *Historia Mathematica* **20**(1993), pp. 364–381. ISSN 0315-0860.

R. Siegmund-Schultze, *Mathematicians Fleeing from Nazi Germany. Individual Fates and Global Impact.* Princeton University Press, Princeton and Oxford, 2009, xxviii + 474 pages. ISBN 978-0-69114041-4.

R. Siegmund-Schultze, *Rockefeller and the Internationalization of Mathematics Between the Two World Wars.* (Documents and Studies for the Social History of Mathematics in the 20th Century). Birkhäuser Verlag Basel–Boston–Berlin, 2001, xiii + 341 pages. ISBN 978-3-7643-6468-7.

T. Snyder, *Bloodlands: Europe Between Hitler and Stalin*. Vintage, London, 2015, xx + 524 pages. ISBN 978-0-09-955179-9.

B. Srinivasan, Ruth Moufang, 1905–1977, *Mathematical Intelligencer* **6**(1984), pp. 51–55. ISSN 0343-6993.

B. Sury, Hermann Weyl (1885–1955). Mathematician, Physicist, Philosopher and Poet, *Resonance — Journal of Science Education* **21**(2016), pp. 1065–1068. ISSN 0971-8044.

N.M. Swerdlow, Otto E. Neugebauer (26 May 1899–19 February 1990), *Proceedings of American Philosophical Society* **137**(1993), pp. 137–165. ISSN 0003-049X.

R. Tobies, *Felix Klein*, Biographien hervorragender Naturwissenschaftler, Techniker und Mediziner, Band 50. BSB B.G. Teubner Verlagsgesellschaft, Leipzig, 1981, 104 pages.

R. Tobies, Die Stellung deutscher mathematischer Fachgesellschaften zur angewandten Mathematik (bis 1917), *Mitteilungen der Mathematische Gesellschaft der Deutschen Demokratischen Republik* **1,2**(1982), pp. 133–154.

R. Tobies, *Iris Runge — A Life at the Crossroads of Mathematics, Science and Industry*. Birkhäuser (Springer), Basel, 2012, xxii + 432 pages. ISBN 978-3-0348-0229-1.

S. Ulam, John von Neumann (1903–1957), *Bulletin of the American Mathematical Society* **64**(1958), pp. 1–49. ISSN 0273-097.

H.-G. Weigand *et al.*, The Legacy of Felix Klein, *Proceedings of the 13th International Congress on Mathematical Education ICME-13*. Springer Open, Hamburg, 2016, vi + 226 pages. ISBN 978-3-319-62597-3.

https://doi.org/10.1142/9781786349316_0002

Chapter 2

Russian Mathematics and Its Relation to German Mathematics (1914–1940)

Jiří Veselý

1. Introduction

European traditions, culture and history were very different from Russian ones. The same was true of the educational systems. For example, when the first Russian university in Moscow was founded in 1755, Germany had already 12 universities. Russia only reached that number just before the October revolution in 1917.

Let us quote from Vucinich (1960, p. 161): *The great mathematical tradition in the Soviet Union is a legacy of the* ancient régime. *Its development has been comparatively short: at the time when the West was working on the complex machinery of the calculus, Russia did not have even an elementary textbook in mathematics. During the XVIIth century only one mathematical book was published in Russian providing elementary knowledge useful to the "sellers and buyers". The first textbook, prepared by L.F. Magnitskii,*[1] *appeared in 1703 and for several decades was the most widely used source of mathematical knowledge in Russia.*

In what follows we will take a closer look at the field of mathematics. Note that mathematics was not part of the curriculum in the first Russian schools. The choice of mathematicians is somewhat random, and it covers by no means everyone, but we hope that it will give the reader some basic information. Biographies of many Russian and Soviet mathematicians can be found on the website (O'Connor and Robertson, 2020).

[1]Leonty Filippovich Magnitskii (1669–1739) taught in the *School of Mathematics and Navigation*, which was founded in Moscow by Tsar Peter the Great in 1701.

2. Moscow Until 1918

The history of the Moscow mathematical school is closely related to two well-known mathematicians: Dmitri Fedorovich Egorov (1869–1931) and Nikolai Nikolaevich Lusin (1883–1950). These names are known to all students of introductory courses in analysis, where they are introduced to Egorov's and Lusin's theorems. Both Egorov and Lusin studied at the university in Moscow and, among others, were taught by Nikolai Vasilievich Bugaev (1837–1903), one of the founders of the Moscow Mathematical Society, in 1864, and its president from 1891 to 1903. He took part in the first International Congress of Mathematicians (ICM), which was organized in 1897 in Zürich, where he delivered a lecture, *Les mathématiques et la conception du monde au point de vue de la philosophie scientifique.*

Bugaev was a teacher of two other well-known mathematicians: Boleslav Kornelievich Mlodzeevskii (1858–1923)[2] and Pavel Alekseevich Nekrasov (1853–1924).[3]

Figure 1. Pictures of Egorov (1869–1931).[4]

Egorov entered the university in 1887 and started to publish while still a student. His thesis, *Second-order Confocal Surfaces in a Space of Constant Curvature* (1891, Russian), was highly praised. One reviewer wrote: *In this paper, in addition to the*

[2]Mlodzeevskii spent about one and a half year abroad in Zürich, Paris and Göttingen, where he worked with Hermann Amandus Schwarz (1843–1921). Further details of his career can be found in Egorov (1925). It is mentioned on p. 450 that Mlodzeevskii was the first to give lectures on the theory of functions of a real variable and set theory at the Moscow University. For details, see Medvedev (1986).

[3]More about Nekrasov and his critics from a Marxist's position, Mikhail Khrisanfovich Orlov (1900–1936) and (indirectly) Ernst (Yaromirovich) Kolman (1892–1979), may be found in Seneta (2004).

[4]The first picture is from Wikimedia Commons, see https://commons.wikimedia.org/wiki/File:Dmitri_Egorov.jpg [20.4.2020], and the second from http://kondakov.ws/blog/Esli-zvezdy-zazhigayut,-znachit,-et [4.5.2020].

independent, very elegant and simple solution of the problem proposed, the originality and logical rigor of the exposition of the basic general geometrical principles deserve special mention, as does also the very successful working out of many details.[5] In 1899 he received his master's degree and two years later, after completing his dissertation *On a Class of Orthogonal Systems* (1901, Russian), he obtained the degree of Doctor of Pure Mathematics. For the academic year 1902/1903, he was sent abroad. He visited Berlin, Paris and Göttingen and when he returned to Moscow he was appointed an extraordinary professor in 1903 and an ordinary professor in 1904.

He published the result that is now known as Egorov's theorem in Egorov (1911): Théorème. — *Si l'on a une suite de fonctions mesurables convergente pour tout les points d'un intervalle AB sauf, peut-être, les points d'un ensemble de mesure nulle, on pourra toujours enlever de l'intervalle AB un ensemble de mesure η aussi petite qu'on voudra et tel que pour l'ensemble complémentaire* [de mesure $= m(AB) - \eta$] *la suite est uniformément convergente.*

[If we have a sequence of measurable functions convergent for all points of an interval AB except, perhaps, the points of a set of zero measure, we can always remove from the interval AB a set of measure η as small as we like and such that on the complementary set [of measure $= m(\mathrm{AB}) - \eta$] the sequence is uniformly convergent.][6]

Within a year Lusin published in Lusin (1912) a theorem which now bears his name. We quote again the original version of the theorem:

Si $f(x)$ est une fonction mesurable définie dans l'intervalle $0 \leq x \leq 1$, il existe dans cet intervalle un ensemble P parfait et non dense jouissant des propriétés

1° *La fonction $f(x)$ est continue dans P (relativement à P)*
2° *La mesure de P est supérieure à $1 - s$, $s > 0$ aussi petit qu'on veut.*

[If $f(x)$ is a measurable function defined on interval $0 \leq x \leq 1$, then there exists in this interval a perfect and nowhere dense set P having properties

1° The function $f(x)$ is continuous on P (with respect to P)
2° Measure of P is bigger than $1 - s$, where $s > 0$ is as small as we want.][7]

---

[5]The quote is from Kuznetsov (1971, p. 126).

[6]The theorem was generalized many times. Sometimes it is called the Severini–Egorov theorem since Carlo Severini (1872–1951) proved it independently in 1910. He published it under the title *Sulle successioni di funzioni ortogonali* [*On Sequences of Orthogonal Functions*] in the Italian journal *Atti dell'Accademia Gioenia*. Severini did not claim the priority and Leonida Tonelli (1885–1946) credited him the priority for the first time, in 1924.

[7]In both cases (Egorov's theorem and Lusin's theorem) we tried to preserve the original formulations. Lusin's theorem was later generalized. Lusin also remarked that every measurable function has this *"C-property"* and that any function having C-property is measurable.

The above theorems are the first important results of the Moscow school of the theory of functions. Sometimes, the year 1911 is considered as the year of its foundation, e.g., in Zdravkovska and Duren (2007, p. 35), or Demidov (1985, p. 78). It is possible that its roots are connected with the philosophical background of Bugaev, since both Egorov and Lusin were close to it.[8] The journal *Comptes Rendus Hebdomadaires des Séances de l'Académie des Sciences*, where both results were published, will be referred to several times and abbreviated as *C.R.*

Egorov, as a teacher, was able to recognize the talents of his students. It was also the case of Lusin, who wrote his dissertation *The Integral and Trigonometric Series* (1915, Russian) under Egorov's supervision. Lusin studied at the Moscow university from 1901 to 1906 and for some time was hesitating to devote his life entirely to mathematics. It was Egorov who helped Lusin to work abroad in Göttingen and Paris, while he was preparing his dissertation. Access to literature and contact with already-famous mathematicians was invaluable to him, especially for preparatory work on the dissertation.[9]

Lusin's fellow students were Boris Nikolaevich Bugaev (1880–1934), son of Professor N.V. Bugaev, better known as the poet Andrei Bely or Biely, and Pavel Aleksandrovich Florensky (1882–1937), a Russian Orthodox theologian, priest, philosopher and mathematician. Let us quote from Graham and Kantor (2009, p. 79): *At Moscow University Luzin was influenced not just by his teachers but also by his fellow students, who debated politics, philosophy, and mathematics with him. Among these mathematics students were Pavel Florensky and Boris Bugaev... Both of these men were, like Luzin, destined for fame. Pavel Florensky was deeply interested in religion and would eventually become a priest and a renowned polymath. Several years after they all graduated from the university (Bugaev in 1903, Florensky in 1904 and Luzin in 1905), Florensky had an important influence on Luzin by turning him away from secularism and to religious faith.* Later, when Lusin was struggling with depressions, Florensky helped him a lot.

In Graham and Kantor (2009), the authors describe Lusin's disillusion with the revolution in 1905. Traumatized by these events he was unable to continue his mathematical studies, and more importantly — he completely lost interest in mathematics and contemplated suicide. It was at that time that Florensky helped him to return to his path. Later, in 1908, Lusin wrote to Florensky: *I felt as if I had leaned on a pillar — I owe my interest in life to you.*[10]

Let us mention another student of Egorov, Vyacheslav Vassilievich Stepanov (1889–1950). He enrolled in the Faculty of Mechanics and Mathematics at the

[8]There is an article by Demidov (1985) where one can find details.

[9]See Phillips (1978), we will refer back to this paper several times since it contains translations of important quotes from Russian sources.

[10]See Graham and Kantor (2009), motto on p. 66 and the text on p. 83.

Moscow University in 1908. After studying under Egorov, he was sent to Göttingen where he attended lectures of Edmund Georg Hermann Landau (1877–1938) and David Hilbert (1862–1943). He spent there a period from 1912 to 1915 and then he returned to Moscow where he started to work as a docent at the Moscow University. His first lecture was *On the Dirichlet Problem* (1915, Russian). He collaborated closely with Egorov and was also much influenced by Lusin. In 1928, Stepanov became an ordinary professor of mathematics at the Moscow University and he occupied this position until his death.

The Moscow school of the theory of functions is connected with Lusin's seminar.[11] *In the fall of 1914 Luzin took up his teaching duties as a privat-dozent at Moscow University. ...* [His] *lectures were concentrated on the theory of functions of a real variable and the analysis seminar which he (and for a time, Egorov) conducted for more than a decade.* As a teacher, Lusin was inspiring. Adolph-Andrei Pavlovich Yushkevich (1906–1993), one of those who attended Lusin's lectures, wrote about him in 1968: *Those who heard him never forgot the remarkable enthusiasm which they experienced at his lectures. It seemed as if he were creating the theory anew before the eyes of the audience — as indeed was the case in the special course on the theory of functions, where he told of newly obtained, even incomplete or still unsettled, developments.*

Figure 2. Pictures of Lusin (1869–1931).[12]

[11]All the four following italicized parts are from Phillips (1978, pp. 288–289). Note that the older transcription "Luzin" differs from "Lusin", which is used in more recent works.

[12]The first picture is reproduced from Demidov and Lëvshin (2016), published by AMS, and the second from the Free Social Encyclopedia Alchetron, https://alchetron.com/Nikolai-Luzin [10.4.2020].

Significantly, Pavel Sergeevich Aleksandrov (1896–1982) did not spare praise: *In this activity* [teaching] *his creative individuality was manifested. ... The richness of N.N. Luzin's creative ideas was so great that almost every meeting with him was, for his students, a source of new problems and unexpected ideas. N.N. Luzin's lectures, devoid of outward brilliance and oratorical devices, seized his audiences with creative intensity. Luzin possessed the extraordinary art of presenting a mathematical result so that the listener was forced to participate in the process of obtaining it, thus transforming the lecture into a kind of self-made laboratory of thought* (1955).

The first year at the Moscow University brought disappointment to Aleksandrov. In Aleksandrov (1979), he wrote: *My first semester at the university during the autumn of 1913 was rather a disappointment, after my happy and very full life at school. I already knew everything we were taught in mathematics in the first year, and it seemed to me that I knew it in a much better form than that in which it was presented to us in the first-year mathematics course. I do not think that I was far wrong. But I found myself an outlet. In the comfortable round reading room of the university library I found Cantor's memoirs on set theory and I began to read them with delight.* His disappointment quickly disappeared.

3. The Birth of Lusitania (Approx. 1915–Approx. 1927)[13]

Aleksandrov was among Lusin's first students attending his seminar. Others were Aleksandr Yakovlevich Khinchin (1894–1959), Dmitrii Evgen'evich Men'shov (1892–1988) and Mikhail Yakovlevich Suslin (1894–1919).[14] Later all of them became professors of the Moscow University except Suslin who died soon after World War I (WWI). Let us take a closer look at their careers.

Aleksandrov started his studies at the Moscow University in 1913. He was ahead of his fellow students. He was advised by his older colleague Stepanov to take part in Egorov's seminar, which he did from 1914.

About a year later he solved the first difficult problem concerning the power of any uncountable Borel set. He proved that any such set contains a perfect set (i.e., closed set without isolated points) and so its cardinality is the continuum.[15] Note

[13]The time of the existence of the Moscow mathematical school is sometimes divided into subperiods, see, e.g., Lyusternik (1967), where the author gave them names like "Luzitania" or "post-Luzitania". Instead, in Phillips (1978), "generations" of Lusitanians are used; see below.

[14]More about Suslin and his results can be found in Igoshin (1996).

[15]Lusin gave the task of solving this problem to Aleksandrov. Later Aleksandrov described the whole story in Aleksandrov (1979). To solve it he introduced an operation which is now called the A-operation. He described how Lusin was skeptical of his proposed method of the solution and how, despite this skepticism, he obtained the solution. He lectured about this result in October 1915 and on that occasion he met for the first time Suslin, Pavel Samuilovich Urysohn (1898–1924) and Wacław Sierpiński (1882–1969).

that the problem was independently, and virtually at the same time, solved by Felix Hausdorff (1868–1942).[16]

The result appeared under the title Sur la puissance des ensembles mesurables B in *C.R.* **162**(1916), pp. 323–325. For Aleksandrov's view on its discovery see his memories Aleksandrov (1979). Other results of Lusin's pupils appeared soon afterwards.

Men'shov started his studies at the Moscow College of Engineering, but after six months he left the school as he had decided to study higher mathematics. In 1912 he entered the Faculty of Physics and Mathematics at the University of Moscow, where he attended the lectures of Egorov and others.[17] *In the autumn of 1914 Luzin, who had just come back from scientific visits to Göttingen and Paris, began to give a course on function theory. Men'shov's first research paper was connected with this course. Luzin presented it so that the theory of functions of a real variable was expounded as a living, developing branch of mathematics, in which there was still much to be done. He shared some outstanding problems with his students.*

One of these concerned the clarification of the mutual relationship between two generalizations of the Lebesgue integral: the Borel integral and the Denjoy integral. Men'shov, as a student in his third year, gave a solution to this problem by proving that the Denjoy integral is wider than the Borel integral. This result, which he obtained in the autumn of 1914, was the topic of Men'shov's first research paper, which was later (1916) published in Matematicheskii Sbornik.

In the note *Sur l'unicité du développement trigonométrique*,[18] Men'shov proved the existence of a trigonometric series with infinitely many nonzero coefficients, which is convergent to the zero function, and in this way solved the uniqueness problem.

Khinchin's father was an engineer in a paper factory. In 1906, he sent his son to a private school in Zürich for two years to complete his education. Then Khinchin attended a good secondary school in Moscow, where he graduated in 1911, and in the autumn of that year he entered the Moscow University. Thereafter he started to attend Lusin's seminar and contributed to its fame.

Looking into the *C.R.* we can find two of his papers from that time. In Sur une extension de l'integrale de M. Denjoy, *C.R.* **162**(1917), pp. 287–291, he introduced a generalization of the Denjoy integral.[19] Within a short time it was followed by Sur la derivation asymptotique, *C.R.* **164**(1917), pp. 142–144, in which an asymptotic derivative of a measurable function is studied. From 1922 he worked at the

[16]See F. Hausdorff, Die Mächtigkeit der Borelschen Mengen, *Mathematische Annalen* **77**(1916), pp. 430–437.

[17]The following text is from the article Kolmogorov *et al.* (1982, p. 203), published on the occasion of Men'shov's 90th birthday.

[18]Published in *C.R.* **163**(1916), pp. 433–436.

[19]Now the integral is called the Denjoy–Khinchin integral and is different from the Denjoy integral.

Moscow University where, from 1932, he was also the head of the Department of Mathematical Analysis.[20]

Yet another student had a good start. Suslin published in *C.R.* **164**(1917), pp. 88–91, the paper *Sur une définition des ensembles mesurables B sans nombres transfinis.* Unfortunately, his life was short and so his first article was also the last one published during his life.[21] Nevertheless, he recognized that Lebesgue had made a mistake: One of the lemmas (stated by Lebesgue without proof) was false. Suslin was close to Aleksandrov, who wrote in his memories Aleksandrov (1979): *Suslin proposed that my new set-theoretical operation should be called the A-operation and that the sets obtained by applying it to closed sets should be called A-sets. He emphasized that he was proposing this terminology in my honor, by analogy with Borel sets, which by then were usually called B-sets.*

Lusin later studied these sets and gave some other possible definitions (or characteristics). These sets are now usually called *analytic sets.* From the beginning of the 1920s, Lusin's seminar, and the people around it, formed a community known as "Lusitania".[22]

Sometimes Lusitania is divided into "generations" like in Phillips (1978), where on p. 295 we can read: *Among the Luzitaniyans during this period were some of his* [i.e., Lusin's] *colleagues and older students: I.I. Privalov, V.V. Stepanov, P.S. Aleksandrov, D.E. Menshov and A.Ya. Khinchin. In the second (post-revolutionary) generation of students were P.S. Uryson, V.N. Veniaminov, A.N. Kolmogorov, V.V. Nemytskii, N.K. Bari, S.S. Kovner, V.I. Glivenko, L.A. Ly[u]sternik, M.A. Lavrentev and I.G. Schnirelman. P.S. Novikov, L.V. Keldysh and E.A. Selivanovskii, who arrived in 1923, made up the third and final generation of Luzitaniyans.* [*This list was compiled mainly from Lavrentev* (1974), *and supplemented by Bari and Golubev* (1959). *It agrees substantially with the lists given by other authors.*]

Some comments are needed: Ivan Ivanovich Privalov (1891–1941) started his study of mathematics at the Moscow university in 1909. Already as an

[20]He also held some other positions but this one he occupied until 1957. In the second volume of the *Proceedings of the Fourth Berkeley Symposium on Mathematical Statistics and Probability* which was held in 1960 and published in 1961 by the University of California Press, two of the first articles are devoted to Khinchin's contributions to the field.

[21]His other two short articles were published after his death. Note that Suslin had health problems which were made considerably worse by the severe food shortage at that time. He had to stop his work and return to his home in the countryside where he lived with his parents. There he caught typhus and died.

[22]A part of this community/order/society game were some special terms, nicknames, etc. More about them can be found in Lyusternik (1967, Part 2). Another example: members of Lusitania had some "grades" of the type "$\aleph_x$" where the index x was a positive integer connected with the number of publications, rewards and similar things. In this chapter we mention a majority of the members of Lusitania, but by no means all of them.

undergraduate student he was a member of the Student Mathematics Circle and served as its president. In 1911 he went for the summer semester to Göttingen, where he attended lectures of Hilbert, Felix Christian Klein (1849–1925), and Landau.[23] Privalov graduated in 1913 being examined on his paper *The Reducibility Problem in the Theory of Linear Differential Equations* (Russian). For master's degree he was examined in 1915 and in the next year he became private docent at the Moscow university.

In 1918 Privalov became a professor of pure mathematics at the Saratov university but in 1921 he returned to Moscow where he was appointed professor, and the head of the Department of Function Theory, at the Institute of Mathematics and Mechanics of the University of Moscow.[24] He was the head of the Institute for the rest of his life. With regard to his lectures, we read in Lyusternik (1967, Part 3, p. 69): *At MGU* [Moscow University] *Privalov began to give, with great enthusiasm, a course on the theory of functions of a complex variable (before him this course had been given by B.K. Mlodzeevskii). On the basis of this course he wrote a famous university textbook on complex function theory. Then (in his own words) he got tired of this course and began to give (also with great enthusiasm) a course of analysis.* From his numerous works let us mention the important monograph *Subharmonic Functions* (1937, Russian).

Privalov had almost no students, which Aleksandrov explained as follows: *A universal character in both mathematical and general cultural interests gave exceptional scope and rare breadth to Hausdorff's university lectures. But probably because of this he did not have his own students: he demanded too much of them and he himself knew too much. This characteristic, incidentally, was also possessed by our great Soviet mathematician I.I. Privalov, who for this reason did not have many students: not everyone could endure his extraordinary exactitude. The scientist's knowledge apparently subdued many just-beginning young people. ... It seemed that it was impossible to imitate him, and students are not able to avoid imitating their teacher to some extent.*[25]

We already wrote about Stepanov above. He liked, e.g., the topic of functions of several variables, and Aleksandrov mentioned this in the obituary published in Uspekhi Matematicheskich Nauk: *In works published in 1923 and 1925 Stepanov established the necessary and sufficient conditions under which a function of two variables, defined on a measurable plane set of finite measure greater than zero, possesses a total differential almost everywhere on that set.*[26]

[23]He was there at the same time as Lusin who was seven years older than Privalov, but they were postgraduate students at the same time. They became close friends.

[24]The Institute was founded in 1922. For more details of Privalov's life and for the description of his work see Kuznetsov and Solomentsev (1967).

[25]See Kuznetsov and Solomentsev (1967, p. 170), where Aleksandrov's opinion is reproduced.

[26]See *Uspechi* **5**(1950), no. 5, pp. 3–10. Translations called *Russian Mathematical Surveys* of the *Uspechi* journal started with volume **15**(1960), this quote is a translation from Aleksandrov (1950).

One of the themes which played a key role in Stepanov's life was the theory of almost periodic functions. He defined a generalization of the notion and several times he returned to this topic. Similarly, he was coming back to functions of several variables. In the article Myshkis and Oleinik (1990) we can read about his stays in Göttingen: *There he made the acquaintance of Luzin, who proved to be a big influence on the later development of Vyacheslav Vasil'evich's mathematical interests, although, as he affirmed himself later, his "first love" was differential equations. ... In terms of the merit of his work, Stepanov's greatest achievements were in the development of the theory and applications of differential equations. He became a kind of connecting link between specialists in the abstract fields of mathematics on the one hand, and applied mathematicians and physicists on the other.*

And one more quote from Myshkis and Oleinik (1990): *Stepanov's survey articles played a large part in the development of mathematics in our country, among them his survey of the qualitative theory of differential equations at the Second All-Union Mathematical Congress (1934), and likewise two books which are of particular interest.* The second of those books which he wrote together with his student Viktor Vladimirovich Nemytskii (1900–1967) is *Qualitative Theory of Differential Equations* (1947, Russian).[27]

It is worth mentioning that Stepanov's seminar about trigonometric series was attended, among others, by Nemytskii, Andrei Nikolaevich Kolmogorov (1903–1987) and Gleb Aleksandrovich Seliverstov (1905–1944), who died in WWII. In the article Bogolyubov *et al.* (1983), the importance of the seminar for Kolmogorov is described in the following way: *His formation as a mathematician was greatly influenced by Stepanov's seminar on trigonometric series. In this seminar unsolved problems were raised before the participants, answers to which seemed to be essential.*[28]

In 1922, after A.N. Kolmogorov wrote his first independent work (on the values of Fourier coefficients), he became the pupil of N.N. Luzin. See below.

4. Daily Life in Lusitania

It is quite natural that several members of Lusitania were good friends and that they shared other interests besides mathematics. Nemytskii's close friend, and later his wife, was Nina Karlovna Bari (1901–1961). She enrolled at the Moscow University in 1918, when the university was reopened after political changes in the country. It was the first post-revolutionary enrollment, and women were admitted to the university for the first time. Bari graduated in 1921 (after only three years of study) and was

[27]Under this title the English translation of the second Russian edition appeared in the Princeton Mathematical Series 22, Princeton University Press 1960, viii + 523 pages. It also contains Nemytskii's survey of the results obtained by the Russians, especially those connected with his and Stepanov's seminar. The last edition of the book (ISBN 978-0-691-65228-3) appeared in 2016.
[28]See Bogolyubov *et al.* (1983, p. 10).

registered as Lusin's research student at the Research Institute for Mathematics and Mechanics of the Moscow University, where she was preparing for an academic career. From 1928 until her premature tragic death, she worked at the Moscow University where she became a professor in 1932.[29]

The first work of Bari *Sur l'unicité du développement trigonométrique* appeared in *C.R.* **177**(1923), pp. 1195–1197. A set $E \subset [0, 2\pi]$ is called a U-set if the convergence of the series

$$\frac{a_0}{2} + \sum_{k=1}^{\infty} (a_k \cos kx + b_k \sin kx),$$

to zero for all $x \in [0, 2\pi] \setminus E$ implies that $a_0 = 0$, and $a_k = b_k = 0$ for all positive integers k. Otherwise, the set E is called an M-set. In this work Bari showed that *the union of countably many closed U-sets is again a U-set.*

Bari continued her research on trigonometric series and formulated the characterization problem: Find a necessary and sufficient condition for a perfect set to be a U-set (1927). This question still fascinates researchers. In addition to this major problem, Bari studied problems concerning superpositions of absolutely continuous functions. An interesting result which she proved[30] is the following: *Any continuous function $F(x)$ can be expressed in the form*

$$F(x) = f_1(\varphi_1(x)) + f_2(\varphi_2(x)) + f_3(\varphi_3(x)),$$

where functions $f_k(t)$ and $\varphi_k(x)$, $k = 1, 2, 3$, are absolutely continuous.[31]

Bari was the first woman who presented, two years after graduation, her results to the Moscow Mathematical Society. She had many other interests such as art, music, and ballet. She also enjoyed hiking. With her husband Nemytskii she shared two great passions: mathematics and hiking in the mountains. Twice, she spent a long period abroad; six months in Paris in 1927, and during the academic year 1928/1929, she obtained support from the Rockefeller Foundation which she used for another stay in Paris. She gave a lecture at the ICM 1928 (Bologna). She wrote several popular textbooks and was active in drawing young people into research. She had at least six PhD students who later became leading mathematicians.

---

[29]She taught at other institutions. For details, see her obituary in *Russian Mathematical Surveys* **17**(1962), no. 1, pp. 119–127. The article contains a description of her main results and the list of her works. The circumstances of her death are not quite clear. She fell in front of a train of the Moscow underground shortly after finishing the monograph on trigonometric series. She was, without a doubt, one of the best female mathematicians of the 20th century.

[30]Together with Men'shov.

[31]The notation used in the formulation of the theorem is reproduced from the above-mentioned obituary. Bari also proved that sums of two such compositions are not sufficient. Bari's major work is the monograph *Trigonometric Series* (1961, Russian). Soon after its publication it was translated as Bary (1964).

Another active member of Lusitania was Lev Genrikhovich Schnirel'mann (1905–1938).[32] He was a prodigy with a special gift for mathematics. In 1921 he enrolled at the Moscow University at the age of 16, without a formal high school leaving certificate.[33] He contributed to the calculus of variations, algebra, topology, and number theory.

In 1929, Schnirel'mann was appointed to the Chair of Mathematics at the Novocherkassk Industrial Institute. After a year, he returned to Moscow, where he worked at the Steklov Mathematical Institute of the Academy of Sciences from 1934 onwards. Modest and critical of his own results, he started to publish in 1929. The same year he defended his dissertation *On Some Nonlinear Problems of Geometry and Analysis* (Russian). His first co-author was a colleague from Lusitania, Lazar Aronovich Lyusternik (1899–1981). Together they began to apply topological methods to the study of closed geodesics — a problem going back to the French mathematician Henri Poincaré.

In 1930, Schnirel'mann switched to additive number theory, attacking the Goldbach problem of 1742.[34] Schnirel'mann made the first breakthrough in this problem when he proved that every integer is the sum of a *bounded number* of primes. His result was based on the notion of density, which he introduced. Schnirel'mann's ideas completely revolutionized additive number theory, and *Schnirel'mann's density* remains a fundamental notion. Some of his other results are easy to explain, e.g., he proved that *for any closed plane curve there are four points on it that form the vertices of a square.*

Lyusternik was also a student of Lusin. He entered the Mathematical Department of the Faculty of Physics and Mathematics in 1918, and he graduated in the academic year 1921/1922. Later he defended his dissertation (similar to present-day PhD dissertation) on the subject of *Direct Methods of Analysis* (1926, Russian).[35] In 1928 he was a member of the Russian delegation at the ICM 1928 and in 1929–1930, and later in 1934–1935, he worked with Schnirel'mann; together they wrote at least seven papers.[36] In 1931 Lyusternik was awarded a professorship at the Moscow University. He was also the first editor-in-chief of the journal *Uspekhi Matematicheskikh Nauk*

[32]The transcription of his name changes, it also appears as Schnirelman, Shnirlman, etc.

[33]It was Lusin's intervention which helped him. He recognized Schnirel'mann's great talent for mathematics.

[34]Christian Goldbach (1690–1764) conjectured in a letter written in 1742 to Leonhard Euler (1707–1783) that every even integer greater than 2 can be represented as a sum of two primes and that every odd number is the sum of three primes. In the era of computers the conjecture has been checked for numbers up to at least 4×10^{17}. See *Goldbach Conjecture Verification*, http://sweet.ua.pt/tos/goldbach.html [5.11.2019].

[35]The facts about Lyusternik are from Aleksandrov *et al.* (1960). Another source cites it as *Direct Methods of Calculus of Variations*, which corresponds to the title of his second publication in *Matematicheskii Sbornik*, which appeared in the same year.

[36]Schnirel'mann worked intensively with Pontryagin in the years 1931–1934; see below.

[Russian Mathematical Surveys] and was, until the end of his life, a member of its editorial board. He also wrote an article, which was translated as Lyusternik (1967), providing a lot of information on Lusitania at the time of its existence. It gives, especially in its second part, a picture of Lyusternik's personality, his views and political orientation.

Just a few words about V.N. Veniaminov (1895–1932), about whom it is not easy to find detailed information. He is remembered by his colleagues — here is a short note from Aleksandrov (1979, p. 282): *Veniaminov was a nice young man, of open and benevolent character, who was liked by everybody. After graduating from the university he did postgraduate work under I.I. Privalov, specializing in the theory of functions of a complex variable. Then he became a prominent figure in higher technical education; and, if I am not mistaken, a professor at the Air Force Academy. But there were complications in his later life, and he committed suicide long before reaching old age.*[37]

Figure 3. Pictures of Aleksandrov (1896–1982).[38]

Now we will describe how Aleksandrov met Uryson who was two years younger. Uryson entered the Moscow University in 1915 with a plan to study physics, but soon afterwards he changed his mind, probably having been influenced by lectures of Egorov and Lusin, and mathematics became his main interest. We can read about Uryson and his teaching: *His lectures were distinguished by their clarity. They usually concluded with a completed result. But, in the period 1918–1920 in Moscow,*

[37]He was really a professor of the Air Force Academy, which is confirmed by Lyusternik (1967, Part 3, p. 69).

[38]The first picture is from the Wikimedia Commons, see https://commons.wikimedia.org/wiki/File: Paul_S_Alexandroff_2.jpg [10.4.2020], and the second from the World Biographical Encyclopedia, Inc., https://prabook.com/web/pavel.alexandrov/949189 [10.4.2020].

P.S. Uryson developed as a mathematician and became one of the central figures of Moscow mathematics of that time.[39]

On the same page Lyusternik mentioned that: *In the autumn of 1921 Pavel Samuilovich began his cycle of lectures on dimension theory. We at once understood that Uryson had "found his theme", that the period of searching had changed for him into a period of scientific maturity.*[40] In the same year he became a private docent, and in 1923, a full professor of the Moscow University.

Uryson and Aleksandrov spent a great deal of time together. They worked on problems which were of interest to both of them. In the summer of 1921, Lusitanians rented a "dacha" [cottage] on the bank of the Klyaz'ma River, and the following year Uryson and Aleksandrov rented another cottage at the same place, in order to work together on the theory of compact topological spaces.[41]

Their first visit to Göttingen was the realization of Uryson's plan. As a part of this plan, Uryson and Aleksandrov prepared a series of lectures on the theory of relativity. The lectures were a success and earned them money for the journey. Other formalities were not as easy to overcome, but in May 1923, Aleksandrov, Kovner and Uryson reached Göttingen. They were equipped with a letter of recommendation written by Lusin to Klein. Their lectures and results were much appreciated. The results in the form of preliminary communications were submitted for publication in the *Mathematische Annalen.*[42]

Aleksandrov and Uryson went to Göttingen also in 1924, and again it was a success. The only disappointment was that their communications were not yet published, but this happened shortly after their second visit. For the whole story, in which Alexander Markowich Ostrowski (1893–1986) played a negative role, see Aleksandrov (1979, p. 300).[43] On the way from Göttingen to France they visited Hausdorff and discussed mathematics together. Aleksandrov wrote in Aleksandrov (1979, p. 286): *When Uryson and I visited Hausdorff in the summer of 1924 and talked a lot with him, particularly about descriptive set theory, Hausdorff put the*

---

[39]See Lyusternik (1967, Part 3, p. 77).

[40]Complete results appeared after Uryson's death, under the title Mémoire sur les multiplicités Cantoriennes I, II, *Fundamenta Mathematicae* **7**(1925), pp. 30–137 and **8**(1925), pp. 225–351. Together with the first part of the paper, Aleksandrov's announcement of the death of Uryson was published.

[41]It is described as an idyllic time in Aleksandrov (1979). It was preceded by their joint visit to a concert on March 30, 1921, after which they walked the whole night around Moscow, arriving back to their lodging at 5 a.m. They decided *to count as the beginning of our friendship our walk of the night of 30–31 March 1921.* The next day, Aleksandrov married and he described it as follows: *The next day another event in my life occurred: I got married to Ekaterina Romanovna Eiges. This marriage was not a success but a mistake. Ekaterina was a woman made for domestic life and I was totally unfit for such a life, and any marriage would have been a mistake for me.* Both quotes can be found on p. 295 of Aleksandrov (1979).

[42]They almost "filled" *Mathematische Annalen* **92**(1924), see pp. 267–274, 294–301 (Aleksandrov), pp. 275–293, 301–304 (Uryson), pp. 258–266 (Aleksandrov and Uryson).

[43]For Ostrowski, see Gautschi (2002). The article of Walter Gautschi about Ostrowski is available in English at https://www.cs.purdue.edu/homes/wxg/AMOengl.pdf [20.4.2020].

question to us directly: what should one call the new sets that Lusin was popularizing everywhere under the name of "analytic". I firmly replied to Hausdorff that Suslin (who had died nearly five years before) was the first mathematician to prove that it was really a matter of new sets, whereas I had spent almost two years trying to prove that there are no such new sets and had only come up with a new definition of the old class of Borel sets. Therefore, I declared, these sets should be called Suslin sets. After some hesitation Uryson backed me up, Hausdorff agreed to our joint proposal and in the new edition of his "Mengenlehre" they were called Suslin sets.

After a week's stay in Bonn where they were guests of the Hausdorffs, and where they went every morning to swim across the Rhine and back, Aleksandrov and Uryson continued on their travels to Blaricum (close to Amsterdam). They visited Luitzen Egbertus Jan Brouwer (1881–1966) and at the end of July they went to Paris. Then they decided to spend some time in Bretagne, in the village of Batz-sur-Mer, and do some work there. The end is known to almost every mathematician: while bathing in a wild sea, Uryson was thrown on the rocks and died, on Sunday, August 17.

Andrei Nikolaevich Tikhonov (1906–1993) was quite close to Aleksandrov and Uryson. He enrolled at the Faculty of Physics and Mathematics of the Moscow University in 1922. He attended lectures of Mlodzeevskii, who soon became ill and died. The course was continued by Egorov. In Lapko and Lyusternik (1982) on p. 132 the authors quote Tikhonov: *In Autumn 1922 I wanted to enroll with N.N. Luzin who was then the most popular Professor of Mathematics. But the enrollment was already closed. I began to go to the topology seminar, and in 1923/24 I attended the lectures of P.S. Aleksandrov on topology. Three of the audience, N.B. Vedenisov, V.V. Nemytskii and I, systematically wrote down these lectures and I made a fair copy of all three of these rough notes. After 40 odd years, at the jubilee celebrations for P.S. Aleksandrov, I produced this fair copy for him.*

For a moment, we will leave the members of the second generation of Lusitania and turn to the next, and the last one. Lyudmila Vsevolodovna Keldysh (1904–1976) joined Lusitania in 1923. Later she graduated under the supervision of Lusin, as did her future husband, Petr Sergeevich Novikov (1901–1975). *Among the members of Lusitania, they were the closest to Luzin, the head of that school, in spirit and in their research interests. Together they wrote a remarkable commentary to Luzin's book on analytic sets. Their first papers were directed towards the solution of problems of interest to Luzin. A portrait of Luzin always hung above Keldysh's desk in the study she shared with P.S. Novikov in their apartment.*[44]

Novikov started his studies in 1919 but, due to the civil war, he began later in 1922 again, since in 1920 he joined the Red Army. He finished his studies in 1925

[44]See Adian (2005) or Chernavsky (2006), where other details of L.V. Keldysh and her family can be found.

and remained at the university. He continued studying and was supervised by Lusin. In 1929 he obtained the equivalent of PhD. In 1934 he was invited to join the newly organized Steklov Mathematical Institute. At first, he worked in the Department of Real Function Theory, and from 1957 he was the head of the Department of Mathematical Logic. In 1935 Novikov received a doctorate in mathematics, and in the same year he married L.V. Keldysh.[45] Four years later he became a full professor of the Moscow University.

Both L.V. Keldysh and P.S. Novikov worked in different fields of mathematics.[46] Keldysh's first known result was published in one of Lusin's works, and her first papers appeared in 1936. She became a specialist in set theory and theory of functions. Novikov worked in mathematical logic and theory of algorithms.[47]

Now we will say a few words about Mikhail Alekseevich Lavrentev (1900–1980) who, together with M.V. Keldysh, published more than 15 papers by 1939.[48] In fact, Lavrentev supervised M.V. Keldysh's PhD dissertation *Complex Variable and Harmonic Function Representation by Polynomial Series* (Russian), which was defended in 1937. The dissertation has its origin in a topic on which they worked together.[49]

While his father was in Göttingen, M.A. Lavrentev attended secondary school there. In the 1920s the family moved to Moscow and so it was altogether natural that M.A. Lavrentev's postgraduate studies took place at the Moscow University, under the supervision of Lusin. During that time he worked in descriptive set theory and obtained a series of important results. He defended his PhD thesis in 1927, and the university then sent him, for six months, to Paris. After this stay, Lavrentev's interests began to extend to other branches of analysis. In 1929 he was appointed a professor, and the head of the Department of Mathematics at the Moscow Institute of Chemical Technology. M.A. Lavrentev obtained fundamental results in set theory, theory of functions and theory of approximation of functions of a complex variable. He also contributed to the theory of conformal and quasiconformal mappings, and the theory of differential equations. From 1931 he was a professor of the Moscow University, where he taught from 1931 to 1941.

---

[45] Her brother was the well-known Mstislav Vsevolodovich Keldysh (1911–1978). In 1927 he was refused entrance to the Institute of Civil Engineers because of his social class. His sister, L.V. Keldysh, persuaded him to study mathematics. See also Adian (2005) and the text below.

[46] She had five siblings and also five children. One of them was Sergei Petrovich Novikov (*1938). He was a president of the Moscow Mathematical Society from 1985 till 1996. He worked in geometry, topology and mathematical physics and was honored in many ways; in 1970 he received the Fields Medal.

[47] In 1952 he proved that the so-called word problem for groups, which was formulated in 1912, is insolvable.

[48] M.V. Keldysh was the president of the USSR Academy of Sciences (1961–1975) and a key researcher in the Soviet space program.

[49] Lavrentev's father Aleksei Lavrentevich Lavrentev (1876–1952) was a mathematician. He finished University at Kazan, in 1897, and spent two years (1910–1912) in Paris and in Göttingen. There he met Lusin, and they became friends.

Figure 4. Pictures of Kolmogorov (1903–1987).[50]

As an example of his early results from 1925 we mention the construction of a differential equation $y' = f(x, y)$ with the right hand side continuous in a certain domain D, such that at least two locally different integral curves pass through each point of D.[51]

Kolmogorov is the last member of Lusitania we want to deal with in detail. Among Lusitanians he was probably the most famous one and many people consider him the best mathematician of the 20th century. From 1920 he was a student of the Moscow University, where he also attended some non-mathematical courses. For quite a long time he was interested in Russian history. He researched old manuscripts concerning agrarian relations in the ancient Novgorod, and also conjectured how some areas were settled, which was later confirmed.

Soon he became a colleague of people such as Aleksandrov, Lusin, Suslin, and Uryson, and all of them helped him in various ways with his early mathematical research. His most influential teacher at that time was probably Stepanov. He attended Stepanov's seminar on trigonometric series. Therefore, it seems quite natural that in 1922 he completed his first independent paper dealing with the order of magnitude of the Fourier coefficients. The same year he became a pupil of Lusin.[52]

[50]Both pictures are from the electronic journal "Man without Border" (Russian), from the interview of E. Belega with V.M. Tikhomirov, see http://www.manwb.ru/articles/science/natural_science/Kolmogorov/ [4.5.2020]. Courtesy of the journal.

[51]More about Lavrentev can be found in *Russian Mathematical Surveys* **16**(1961), no. 4, pp. 143–153, where his publications until 1960 are also listed.

[52]Information (sometimes incomplete) on pupils of Lusin and other mathematicians can be found in Keller *et al.*, 2020.

Influenced by Lusin and Suslin, he worked on the theory of operations on sets, and he finished that work in 1922.[53] Another of Kolmogorov's results from 1922 is well known: He constructed a Fourier–Lebesgue series which was divergent almost everywhere, and a bit later improved the example and constructed everywhere divergent Fourier–Lebesgue series.[54] For his whole life Kolmogorov was interested in the theory of Fourier series and orthogonal functions. His ideas were a source of inspiration for young researchers and helped them in their investigations.

Kolmogorov is the author of over 300 research papers and monographs, textbooks, and historical notes, and they span almost every area of mathematics except number theory — this characterization replaces a long list of areas which he influenced. He is best known for his contributions to probability theory. In 1924 he started to be interested in probability and quite soon became an authority. He is considered the "father" of modern probability theory, following the publication of his *Grundbegriffe der Wahrscheinlichkeitsrechnung* [Fundamental Concepts of Probability Theory].[55] According to Keller *et al.* (2020), Kolmogorov had 82 students, among whom were, e.g., Israil Moiseevich Gelfand (1913–2009), Yakov Grigorevich Sinai (*1935) and Vladimir Igorevich Arnold (1937–2010).[56]

Kolmogorov graduated in 1925, but continued his studies at the Moscow University for four more years as a "research student". He was forced to finish this type of study, as new rules were introduced in 1929. Aleksandrov helped him to get a position in the Institute of Mathematics and Mechanics at the Moscow University. In the years from 1929 to 1933, Kolmogorov worked on measure theory, which culminated in his publication of the aforementioned book. In a way, it was a solution for one of the famous Hilbert's problems and it established the foundation for the creation of the theory of random processes.[57]

There is an interesting moment concerning the relationship of Kolmogorov and Lusin. In the series *Istoriko-Matematicheskie Issledovaniya* [Historico-Mathematical Investigations] in the Volume **28**(1985) [ed., A.P. Yuskevich, Nauka, Moscow 1985, 352 pages] a letter of Kolmogorov was published with comments about the article Men'shov (1983) by Men'shov, which appeared in the previous volume.

[53]The work, which was published much later in 1928, influenced some other mathematicians like Leonid Vitaliyevich Kantorovich (1912–1986) and Alexey Andreevich Lyapunov (1911–1973).

[54]These results were published in *Fundamenta Mathematicae* **4**(1924), pp. 324–328 and in *C.R.* **183**(1926), pp. 1327–1328.

[55]It appeared in the series *Ergebnisse der Mathematik und ihrer Grenzgebiete*, Springer, Berlin, 1933, iv + 62 pages.

[56]Kolmogorov and his three mentioned students are laureates of Wolf's Prize, which is the third most prestigious international academic award in mathematics, after the Abel Prize and the Fields Medal. They received it the following years: Gelfand (1978), Kolmogorov (1980), Sinai (1996/1997) and Arnold (2001).

[57]There is an interesting article about the origins of *Grundbegriffe*: G. Shafer, V. Vovk, The Sources of Kolmogorov's Grundbegriffe, *Statistical Science* **21**(2006), pp. 70–98. ISSN 0883-4237. It shows the role played by Kolmogorov's predecessors.

In that article it was stated that Kolmogorov had no need to be supervised. Kolmogorov asked for publication of the following: *I would like to add a specification. In fact I am a direct student of N.N. Lusin. In 1921 he learned about my first result in the theory of trigonometric series (the existence of the Fourier–Lebesgue series coefficients of which goes arbitrarily slowly to 0). He invited me to come regularly to his home At Arbat, 25, together with a group of my peers (L.V. Keldysh, P.S. Novikov, I.N. Khlodovskii). These discussions were very useful for me and lasted until my enrolling to PhD studies in 1925, when N.N. Luzin became my official scientific supervisor. They continued during my further studies (with the exceptions of N.N. Lusin's travels abroad) and these discussions with him were stimulating and of huge importance for me.* From 1936, Kolmogorov did much work on *Bol'shaya Sovetskaya Entsiklopediya* [The Great Soviet Encyclopedia]. For more information on Kolmogorov, see Andrews *et al.* (2000).[58]

Now we return to more general aspects of Moscow mathematics. We will describe "the case of Egorov". His career, which started in the first decade of the 20th century, seemed very successful. He was a professor from 1904, an internationally well-known mathematician and a co-founder of the Moscow Mathematical School. In 1917 he became the Secretary of the Moscow Mathematical Society, and he was elected its vice-president in 1921. A year later he became president of the Society and he held this position until 1930. He was also the editor of the journal *Matematicheskii Sbornik* (Russian).

In 1923 Egorov became the director of the Institute for Mechanics and Mathematics at the Moscow State University. In 1924 he was elected a corresponding member of the Academy of Sciences of the USSR, and from 1929 he was an honorary member. During his time at the university, he was, for some time, a popular Dean of the Faculty.

Let us quote an opinion of Aleksandrov, from Kuznetsov (1971, p. 152): *Exceptionally restrained, slightly dry, Egorov was to a high degree master of his art both as regards the internal, logical and purely mathematical structure of his lectures and as regards their external, verbal structure, and, what is more important, these lectures could serve as a matchless example of mathematical rigor, but this really brilliant mastery of composition, free from any external effects, was aimed at the trained listener, which explains why Egorov was primarily a professor for advanced courses. However, the formal merits, refinements of calculation, always the shortest path of logical deduction, possibly sometimes even overwhelmed the listener, who in Egorov's lectures received every mathematical theory in its definitive crystallized form.*

... in Kharkiv was the All-Union Congress of Mathematicians, held June 24–29, 1930. The leading mathematician in Kharkiv at the time, no doubt involved with

[58]Very detailed information on Kolmogorov may also be found in the (Russian) book Shiryaev (2006). See also Bogolyubov *et al.* (1983).

the organization of the conference, was S.N. Bernstein, who, like Kolmogorov, had little tolerance for Marxist interference with science, and eventually left Kharkiv, probably on account of ideological differences with Orlov... There was a motion to send greetings from this mathematical Congress to the Communist Party Congress; the motion was not supported by Bernstein and Egorov, in particular. The telegram was, nevertheless, sent.[59] In September 1930, Egorov was arrested.[60]

5. Conflict Due to Faith

Why was he arrested? His conflict with the Soviet government lay in the sphere of religion. Egorov was a deeply religious man.[61] When the Church was repressed after the revolution, Egorov defended it. In 1922–1923, and again in 1928, there was a mass execution of clergy. Egorov tried to shelter academics who had been dismissed from their posts. Generally, he was critical of the Soviet regime. In 1929 he was dismissed from the position of director of the Institute for Mechanics and Mathematics. Some time later, he was arrested and imprisoned. The Moscow Mathematical Society refused to expel him. Then the so-called "Initiative Group" published in the journal *Nauchnyi Rabotnik* [Scientist] 11–12 (1930), pp. 67–71, a declaration denouncing Egorov as "a reactionary and a churchman". The declaration is too long to be reproduced here. Its signatories were L.A. Lyusternik, L.G. Schnirel'mann, A. Gel'fond, L. Pontryagin and Nekrasov. I have not seen anywhere a public expression of the regret of any of them. Later, in November 1930, Kolman became president of the Society. As to the identity of those who were not mentioned up to now: Aleksandr Osipovich Gel'fond (1906–1968) was a member of the Academy of Sciences of USSR and is known for proving the transcendency of numbers a^b with algebraic $a \neq 0, 1$, and algebraic irrational b (thus he solved Hilbert's 7th Problem in 1934). Nekrasov was probably K.P. Nekrasov, a son of Pavel Alekseevich Nekrasov (1853–1924), former rector of the Moscow University. At that time K.P. Nekrasov was a PhD student. The initials K.P. are taken from Demidov and Lëvshin (2016, pp. 45 and 223). See also Seneta (2004, p. 343).

By that time, Egorov was already in jail. He was first imprisoned in Moscow and later in Kazan, where he went on a hunger strike. The wife of Nikolai Grigorievich Chebotaryov (1894–1947) tricked everyone by declaring that he was dead, and she took him from the prison hospital to her own home. Egorov died a day later in Chebotaryov's arms. Only a few people attended Egorov's funeral.

[59]See Phillips (1978, p. 340). As for the mentioned interferences they concerned Kolman.
[60]For the following topic see also Graham and Kantor (2009).
[61]See also the paper Ford (1991) where more details are given.

6. A Mathematical Friendship

At the end of the 1920s a very important life-long friendship between two mathematicians began. In the article Gurzadyan (2004) we read: *In the summer of 1929, Andrei N. Kolmogorov, 26, and Pavel S. Aleksandrov, 33, two mathematicians just starting their life-long friendship, visited Sevan. For about a month they enjoyed the beauties of the lake, living in a cell in the monastery on the island of Sevan. They visited other sites of Armenia, but still actively continued their research. Both mathematicians mentioned Sevan in their memoirs ...*

In 1935 Aleksandrov and Kolmogorov bought part of an old manor-house in the village of Komarovka. Kolmogorov wrote about it: *On returning from a trip along the Volga and to the Caucasus in the fall of 1929 Pavel Sergeevich and I decided to live together somewhere near Moscow. The same fall we leased a three-room half-house in the village of Klyazma, where we settled together with my aunt Vera Yakovlevna. She was then already 66 years old, but she was still able to manage our household ...*

But in 1935 the need clearly arose to have our own comfortable home. The problem unexpectedly resolved itself. While skiing around the area in the early spring of that year, we saw a fairly large white house on the edge of a field and we said in unison, "How nice it would be if that house were for sale." Miraculously, it turned out that the house, which belonged to Anna Sergeevna Shtekker, a sister of Konstantin Sergeevich Stanislavskii, was indeed for sale.

7. The Case of Lusin

As to the case of the academician Lusin, which took place later in 1936, we are able to present just a very small fragment of what is now known.[62] Importantly, and contrary to the hope of all the participants interrogating Lusin on the platform of a special "Emergency Commission" of the Academy of Sciences of USSR, records of it remain preserved. They are available in Russian and English and were published in Demidov and Lëvshin (2016). The commission consisted of G.M. Krzhizhanovsky (Chairman), and its members were A.E. Fersman, S.N. Bernstein, O.Yu. Schmidt, I.M. Vinogradov, A.N. Bakh, N.P. Gorbunov, L.G. Schnirel'mann, S.L. Sobolev, P.S. Aleksandrov, and A.Ya. Khinchin.

Lusin returned to Russia from his stay in France, where he had been living as a Rockefeller Fellow from 1928 to 1930.[63] After his return he was shocked by what had happened to Egorov. To become less visible he left the university and started

[62]For details, see the book Demidov and Lëvshin (2016). Very condensed information about Lusin's case is presented in Kutateladze (2007).

[63]Receiving this support was exceptional due to his being of the age of 45. For further details, see Siegmund-Schultze (2001).

to work in the Central Aero-hydrodynamics Institute. However, he still worked at the Steklov Mathematical Institute as the head of the Division of Function Theory. The Institute was in Leningrad (now St. Petersburg) until 1934 when it moved to Moscow.

A trap for Lusin was set by Lev Zakharovich Mekhlis (1889–1953), editor-in-chief of the main Russian newspaper "Pravda", in cooperation with Kolman[64], who seems to be the initiator of the whole action. It is difficult to present a proof of this fact, but it is known that Kolman denunciated Lusin in a secret letter (see Demidov and Lëvshin (2016, p. 12)) already in 1931. It is also highly probable that he was behind the declaration in Egorov's case (Demidov and Lëvshin (2016, p. 355)).

Lusin was invited to a secondary school and after the visit he was asked to share his impressions with the readers of the newspaper *Izvestiya* from examinations he followed there. It appeared under the title *A Pleasant Disillusionment* and Lusin expressed his exaggerated praise.[65] After five days the director and political leader of the school published in *Pravda* an answer in which he expressed the necessity of comradely criticism and refused hypocritical praise. He posed a question: *Was it not your purpose to gloss over our defects and thereby harm our school?* It was the beginning of a campaign against Lusin.

When the case developed from some articles published in *Pravda*, the Emergency commission investigated Lusin's accusation. Just to give the reader a feeling of the way Lusin was publicly attacked, we present the last item of the article *Enemies Wearing a Soviet Mask* (1936, Russian, in *Pravda*, 3rd of July): *It won't work, Mister Luzin! The Soviet scholarly community has torn off your mask of a conscientious scholar, and you now stand before the world naked and insignificant. You, preaching "pure science" and all the while betraying the interests of science, selling it out in the interests of your former bosses and the present bosses of fascist science. The Soviet community will perceive the history of Academician Luzin as yet another object lesson showing that the enemy is not laying down his arms, that he is masquerading in ever more skillful ways, that his methods of mimicry become more and more varied, and that vigilance remains the most important trait of every Bolshevik and every Soviet citizen.*

The Emergency Commission had to answer some questions concerning Lusin. When the Commission formulated these questions, they were presented by Fersman, as follows:

(1) *Is Luzin an eminent scholar?*
(2) *What did he have to do with the foundation of the Moscow Mathematical School?*

[64]The transcription of his name varies; it could also be Kol'man due to the form of the name he used in Russia.
[65]Under the title it is included as one of the *Appendices* in Demidov and Lëvshin (2016).

(3) *What is his connection with the old academic Black-Hundred group that used to be at Moscow University?*
(4) *What did he have to do with the writing of recommendations?*
(5) *What did he have to do with publishing abroad?*
(6) *What is his relation to the work of his colleagues and the younger staff? And finally, what is his social profile?*

More than 160 printed pages in the book Demidov and Lëvshin (2016) are minutes from the discussions, questioning, accusations, etc. The person who most actively wanted to prove that Lusin was guilty was Aleksandrov. Among the other things he said to Lusin was that he should ... *found the courage and modesty necessary to admit the accusations against you. Had you done that, you would have resurrected the respect you once enjoyed, which you once deserved. And in fact, only from an inner impulse could you have achieved your academic and social resurrection, which all of us, naturally, would have welcomed.*

What was the conclusion of the negotiations of the commission? It was formulated in numbered items, and the last one, the ninth, says: *9. All that is stated above, summarizing copious factual material at the Academy of Sciences, when carefully studied, fully confirms the characterization of Luzin given in* "Pravda".

Considering the epoch when this event took place, and the situation in the USSR at the time, one could have expected only one result, i.e., that Lusin's punishment would be a clear warning to others. But it did not happen. Lusin was neither executed nor arrested, and he remained a member of the Academy. But, for some time he was unemployed, and obtained a position at the Institute of Automation and Telemechanics of the Academy of Sciences of the USSR only in 1939. Even the secret denunciation of Kolman, mentioned above, had some consequences. Lusin, at the time of the ICM 1932 in Zürich the president of the Moscow Mathematical Society, was not allowed to attend it.[66]

A natural question is: Who saved Lusin? For example, a Nobel Laureate, known for his work in low-temperature physics, Pyotr Leonidovich Kapitsa (1894–1984), wrote a letter to Vyacheslav Mikhailovich Molotov (1890–1986) in which he supported Lusin. He wrote: *I was puzzled, amazed, and shocked by the article in* "Pravda". It was returned, with Molotov's remark *Return to citizen Kapitsa as irrelevant.* Even now it is not quite clear why Lusin was spared from the fatal end of his case, but it is believed that it was because of instructions "from above".

[66]Also, publication of the results in foreign journals started to be more difficult and even during Lusin's case, was considered to be something wrong. In Vucinich (2000) we can read: *Among many other grim consequences, the Luzin affair brought an end to the publication of Soviet mathematicians' scholarly papers in foreign journals, thereby ending valuable contact with the international community of mathematicians.*

Potentially, it did not fit into Stalin's global strategy of the processes which then followed...

Among all the members of the Emergency Commission, only Sergei Natanovich Bernstein (1880–1968), Aleksei Nikolaevich Krylov (1863–1945), and in a small part Ivan Matveevich Vinogradov (1891–1983), tried to defend Lusin.[67] Others, such as Bari, L.V. Keldysh, Men'shov and P.S. Novikov, refused to come to a session of the Commission, due to the fact that it could be dangerous for them. The key attacking role was played by Aleksandrov, the head of the Moscow topological school. Also active at the meetings of the Commission were Kolmogorov, Lyusternik, Khinchin, and Schnirel'mann. The political attacks on Lusin were vigorously supported by S.L. Sobolev (1907–1989) and O.Yu. Schmidt (1891–1956).[68] Many people believe that behind the campaign against Lusin was Kolman, sometimes called the "Red Professor". For example, in his article *Sabotage in Science* (1931, Russian) which he wrote for the Party's theoretical journal *Bolshevik*[69] he attacked not only Egorov, but also Veniamin Fedorovich Kagan (1869–1953), Nikolai Vasilievich Bogomolov (1915–2011) and Nikolai Ivanovich Vavilov (1887–1943). Note that the fate of the academician Vavilov, a botanist and geneticist, was similar to Egorov's — he died in jail.[70]

A few words on Otto Yulyevich Schmidt (1891–1956), one of the members of the Emergency Commission. He graduated from the classical gymnasium in Kiev in 1909 with the Gold Medal, and entered the Faculty of Physics and Mathematics of Kiev University. Among his teachers was Dmitry Aleksandrovich Grave (1863–1939). Because of Grave, Kiev had a good school of algebra, and under his influence Schmidt published three algebraic papers in the years 1911–1913. Schmidt graduated from Kiev and married in 1913. Three years later, he published the monograph *The Abstract Theory of Groups* (1915, Russian), and received the title of a private docent. During WWI, Kiev University was evacuated to Saratov. Schmidt lived in Saratov in the period 1915–1916, and after the university returned to Kiev, he began his career as a university teacher. From the beginning of his career, the administrative work for the university, and the town of Kiev, took over.

In 1917 Schmidt was appointed a clerk of the Ministry of Food, and after the October Revolution, he had several political positions. In 1921 Schmidt became the

[67]In Graham and Kantor (2009, p. 153), it is confirmed, but the authors claim: *Only one member of the commission, the older mathematician S.N. Bernshtein, openly tried to defend him.*

[68]See Demidov and Lëvshin (2016, pp. 355–357). Note that only Schmidt was an old Party member and Sobolev was at the time one of the most active members of the *KomSoMol* in the Academy. It should be noted that the Commission contained only three members implacably hostile to Luzin: Schnirel'mann, Aleksandrov, and Khinchin. See Demidov and Lëvshin (2016, p. 19).

[69]It appeared in *Bolshevik* **2**(1931), pp. 71–81.

[70]All I know about Kolman, who was my lecturer of Marxism and who wrote the book *We Should Not Have Lived That Way* (Russian), Chalidze Press, New York, 1982, 375 pages, supports an opinion expressed in Lorentz (2002), where Kolman is described as a somber "black angel".

director of the State Publishing House and under his leadership the publishing of scientific journals and research monographs was resumed. He was also appointed the editor-in-chief of the Great Soviet Encyclopedia. In 1923 he became a professor of mathematics at the Moscow State University. At the same time, he became the head of the Department of Algebra and set up an active school of group theory. Later on, he became better known as a polar investigator and mountaineer.

Sof'ja Aleksandrovna Yanovskaya (1896–1966) became, as a young student in Odessa, a member of the illegal Red Cross. In 1918 she joined the Bolshevik party, where in 1920–1923 she worked as a member of a regional committee. In 1923 she was sent to the Moscow Institute of Red Professorate. From 1925 she began to conduct a seminar on the methodology of mathematics and natural sciences for students and research students of the Moscow University. In 1929 she graduated from the Institute of Red Professorate. Her work *Categories of Hegel's Quantity and the Essence of Mathematics* (1928, Russian) was followed by a long series of works of historico-methodological character.

She worked on the publication of the mathematical manuscripts of Marx, which were accompanied by her long article *On the Mathematical Manuscripts of Marx* (1933, Russian). In connection with Lusin's case, there was a meeting at the Moscow University where Yanovskaya was the main speaker. She said about Lusin: *He has acted arrogantly, unethically, and disruptively, relying on his complete untouchability.* ...She was supported by Aleksandrov, Pontryagin, [Felix Ruvimovich] Gantmacher [(1908–1964)], Lyusternik, and Kolmogorov.[71] The resolution from that meeting is also available in Demidov and Lëvshin (2016), Appendix XII. Yanovskaya also criticized Schmidt and Kagan, and described parts of the Great Soviet Encyclopedia, which they had written, as antimarxist and idealistic. She especially pointed out the omission of Engels and Lenin.[72]

Up to now we have covered members of Lusitania and a few others who took part in Lusin's case. But, by 1923, Lusitania had started to change. Meanwhile, Moscow became an acknowledged European mathematical center and its members started to establish important mathematical contacts. Already in 1922, Aleksandrov and Uryson created a nucleus of the Moscow topological school and they were soon joined by Lev Semenovich Pontryagin (1908–1988).

Pontryagin, due to an accident, was blind from the age of 14. Nevertheless, he was able, with the help of his mother who read him everything including mathematical

[71]Soviet mathematics during the Stalin Era, and the role of Yanovskaya, is described in Vucinich (2000). It includes materials not only on the role of Yanovskaya, but also of other, sometimes well-known mathematicians.

[72]Lusin's case is one of the biggest tragedies of Russian mathematics. There were later some other strange cases such as the article in Pravda (1974) (see Kushner, 2008) directed against Solzhenitsyn and signed by Kolmogorov and Aleksandrov. Note that Solzhenitsyn studied mathematics and physics at the Rostov State University.

texts,[73] to study mathematics at the Moscow University. He entered the university in 1925 and graduated in 1929. Of all the advanced courses, Pontryagin preferred those given by Aleksandrov, and he was strongly influenced by him for many years.

From 1929, Pontryagin worked at the Mechanics and Mathematics Faculty of the Moscow University. From 1934, he worked at the Steklov Mathematics Institute of the USSR Academy of Sciences. There, he led the Department of Differential Equations until his death. He contributed mainly to the fields of topology, geometry, algebra, and the theory of dynamical systems. He always paid attention to possible applications. Pontryagin can be considered the founder of two Russian schools of mathematics, namely of algebraic topology and optimal control theory. Pontryagin was able to solve Hilbert's 5th Problem for Abelian groups, using the characters on locally compact Abelian groups that he had introduced, in 1934.

S.N. Bernstein finished high school in Odessa in 1898. He then went with his elder sister to Paris, where he studied mathematics at the Sorbonne. After a year he changed his mind and entered the École d'Electrotechnique Supérieure. Nevertheless, he was still interested in mathematics and so he went to Göttingen where he spent two semesters, from the autumn of 1902.[74] He was supervised by Hilbert. He then returned to the Sorbonne where he submitted his thesis in which he solved Hilbert's 9th Problem. He received his doctorate in 1904 and went to Heidelberg where he attended the ICM 1904, and spent some more time returning to St. Petersburg in 1905. Russia did not recognize foreign qualifications for university posts and so he had to start again. After more exams he taught for one year at the Women's Polytechnic College and when he could not find a position at the university, he moved to Kharkiv in 1908.

In Kharkiv he submitted another thesis, and after being examined, he became a docent. His most famous work, providing a constructive proof of the Weierstrass approximation theorem from 1885, by approximating polynomials, dates back to 1911.[75] Altogether, S.N. Bernstein taught at the Kharkiv University from 1907, for 25 years. He received many honors, also in the USSR, but when he protested against the politicization of teaching and science in 1930, he was removed from the position of the director of the Kharkiv Mathematical Institute, which he had held from 1928.

In 1932 S.N. Bernstein left Kharkiv for Leningrad where he became the head of the Department of Probability Theory and Mathematical Statistics of the

[73]She had her own special terminology, e.g., the symbol $\bigcap$ she read "tails down", etc.

[74]More about Göttingen can be found in Neuenschwander and Burmann (2020).

[75]For this result he received a prize from the Belgium Academy of Science. Here is his comment on it: *The example of the problem of the best approximation of the function* $|x|$*, posed by* [Charles-Jean Étienne Gustave Nicolas] *de la Vallée-Poussin* [(1866–1962)] *reaffirms, once again, the fact that a well-posed specific question leads to theories of a much more general significance.* It appeared as Sur l'approximation des fonctions continues par des polynomes in *C.R.* **152**(1911), pp. 502–504 and then as Démonstration du théorème de Weierstraß, fondée sur le calcul des probabilités, *Soobschenya Charkovskovo Matematicheskogo Obchestva* **13**(1912–1913), no. 2, pp. 1–2.

Figure 5. Pictures of Bernstein (1903–1987).[76]

Mathematical Institute of the USSR Academy of Sciences. In 1934 the Institute was divided into two parts, and the mathematical one was known as the *Steklov Institute of Mathematics* (in talks usually shortened to *Steklovka*). In the same year, the Steklov Institute moved from Leningrad to Moscow. From 1933 S.N. Bernstein lectured at the Leningrad University. Probably his most important work was in probability theory including its applications, in particular to genetics.[77]

Another member of the Committee from Leningrad was Aleksei Nikolaevich Krylov (1863–1945). From 1916 he was a full member of the Russian Academy of Sciences. He was an applied mathematician and his main results were in shipbuilding and the oscillating motion of a ship and related problems (known from Euler's time as "naval science"). He also worked on the theory of gyroscopes and other topics in mathematical physics, approximation theory, and history of mathematics.[78]

The last Leningrad member of the Committee was Vinogradov.[79] He was probably one of the greatest number theorists in the world. As a young boy he was good at

[76]Pictures from the Archives of the Russian Academy of Sciences, http://isaran.ru/?q=ru/person&guid=06FA90F5-FD82-F3EA-5E89-F91794BA99CD [10.4.2020], and the Oberwolfach Photo Collection, https://opc.mfo.de/detail?photo_id=320 [20.4.2020].

[77]See also Sinai (2003) and S.N. Bernstein's obituary in *Russian Mathematical Surveys* **24**(1969), no. 3, pp. 169–176, written by P.S. Aleksandrov *et al.*

[78]His daughter Anna Alekseevna Krylova (1903–1996) later became Anna Kapitsa after marrying the famous physicist Kapitsa, mentioned above.

[79]Sobolev was, at that time, already in Moscow and was much younger than the other participants. He later became one of the most famous representatives of the Siberian Branch of the USSR Academy of Sciences. He worked with Smirnov and studied functionally invariant solutions of the wave equation. In 1939 he was elected a full member of the USSR Academy of Sciences. At the time he was only 31 years old, which was a remarkable achievement. It made him the youngest full member of the Academy of Sciences. At the ICM 1958 in Edinburgh, he was a member of the delegation led by Vinogradov and there gave an invited address on partial differential equations. From 1960 to 1978 Sobolev worked at the Institute of Mathematics of the Siberian Branch of Academy, and was also a professor at Novosibirsk University.

drawing but aspects of more practical education won. After graduation he enrolled at the Faculty of Mathematics and Physics of the University of St. Petersburg. From 1915 he was trained for an academic career. Already at a young age he solved some very difficult problems, and in the years 1914–1918 he produced top level works. For a short time he moved to Perm but from 1920 he was back in Petrograd and became a professor at the Petrograd State University and also at the Polytechnic.

He developed a so-called method of trigonometric sums, which became a fundamental tool in number theory. He also contributed important results on Waring's Problem and solved questions connected with the Goldbach Conjecture. This conjecture has its origin in an exchange of letters between Euler and Goldbach, in the year 1742, and claims that every even integer greater than two *is the sum of two prime numbers*; see above. Thanks to Vinogradov we know the partial solution of the so-called weak Goldbach conjecture, i.e., that *every odd number greater than 7 can be expressed as the sum of three odd prime numbers.* He proved that *it is true for all sufficiently large n.* When the Steklov Institute was established, Vinogradov was appointed its first director. He remained in that position, with the exception of the years 1961–1964, for the rest of his life.

Egorov was not the only victim of repressions from the circle of Lusin's colleagues. Lusin's friend Florensky was arrested in 1928 and the warrant on him was signed by Genrikh Grigoryevich Yagoda (1891–1938), the director of the People's Commissariat for Internal Affairs (NKVD). He was given a relatively light sentence: three-year exile to the city of Nizhny Novgorod, where he was to report regularly to the police. Soon he was allowed to return home. In 1933 he was again arrested, and now NKVD was better prepared. Being tortured, he signed his confession.

He was sent to one of the Gulags, and to cut a long story short, the result is contained in a document from Lubianka archives: *The death sentence passed on Florensky Pavel Aleksandrovich by the Leningrad Region NKVD Troika was executed on December 8, 1937. Commandant of the Leningrad Region NKVD Senior Lieutenant K. Polikarpov.*[80]

Note that Schnirel'mann committed suicide in 1938. As for the reason, there are some doubts. Pontryagin claimed in his memoirs that Schnirel'mann gassed himself, because he was probably depressed. In an interview by Eugene Borisovich Dynkin (1924–2014) with the Soviet and Jewish physicist Akiva Moiseevich Yaglom (1921–2007), another reason is given: Yanovskaya reported that Schnirel'mann had been recruited as an informer by the NKVD, and that this was his reaction. According to Yanovskaya he left the message *I die as honest to comrades and Soviet order*

[80]See Graham and Kantor (2009, p. 145). In the book it is possible to find many interesting details of Florensky, Egorov and Lusin.

(Russian).[81] As in many other cases, it is difficult to find the truth. Even Gel'fond, with whom Schnirel'mann had led the Seminar on Number Theory, was afraid to tell students after the event that Schnirel'mann had committed suicide.

In Graham and Kantor (2009) there is another version on p. 167: *He* [Schnirel'mann] *told his close friend and mathematical associate, Lazar Lyusternik, that he had done "something very bad under pressure". Lyusternik could guess what had happened. At that time the secret police frequently demanded such signed denunciations and then later showed these documents to other detainees in subsequent interrogations, "proving" that a person's friends and colleagues had accused them of awful crimes.*

8. Mathematics in St. Petersburg/Petrograd/Leningrad/ St. Petersburg

We already touched upon Leningrad mathematics in connection with members of the Emergency Commission. Before exploring the relevant part of its mathematical history we will make some more general remarks. The town was founded by Tsar Peter the Great (1682–1725) as St. Petersburg, in 1703, and from 1712 it was the capital of Russia; later on its name was changed: in 1914 to Petrograd, in 1924 to Leningrad and in 1991 again to St. Petersburg.[82]

Note that a year after the revolution, in 1917, Lenin moved his government to Moscow, which became the capital of the new country. Peter the Great created the Russian Academy of Sciences (1724)[83] and the first Russian University (1724).[84] He also invited foreign scholars to come and teach. Thus, in the St. Petersburg Academy, the first academicians were Leonhard Euler (1707–1783), the brothers Nicolaus Bernoulli (1695–1726) and Daniel Bernoulli (1700–1782), Joseph-Nicolas Delisle (1688–1768) and Anders Johan Lexell (1740–1784).[85]

---

[81]See https://ecommons.cornell.edu/bitstream/handle/1813/17258/Yaglom_12-2-1988_transcript.pdf;jsessionid=5EC963B102CD53E1505E73E3F4AFC413?sequence=5 [11.10.2019].

[82]For Russians, especially for local residents, the city is referred to by its nickname *Piter*.

[83]In fact, the Academy was created in accordance with Peter's decree in 1925; see Nazarov and Sinkevich (2019, p. 1).

[84]Moscow University, now Lomonosov's University, was founded in 1755. St. Petersburg University has been in operation continuously since 1819.

[85]Both Euler and N. Bernoulli died in St. Petersburg. Euler's almost encyclopedic knowledge of contemporary mathematics is known to every mathematician. From his results recall, e.g., the solution of the Basel problem (summation of the squares of reciprocals of all positive integers to $\pi^2/6$), the problem of Königsberg bridges, and disproof of the conjecture that all Fermat's numbers $F_n = 2^{2^n} + 1$ are prime — by no means the most important, but well-known for their simple formulations. N. Bernoulli contributed to the theory of differential equations, probability and fluid dynamics. His younger brother D. Bernoulli was not happy in St. Petersburg and returned to Basel. He also obtained valuable results in mathematical physics and theory of the tides, and was the author of a method for measuring blood pressure. Delisle is known as the main architect of the so-called *Atlas Rossicus*, a collection of maps of Russia. Lexell was a Finnish–Swedish astronomer,

What followed we will try to describe more systematically, giving some details. The development of St. Petersburg's mathematics is closely connected with Pafnuty Lvovich Chebyshev (1821–1894). For the interwar period, two important personalities of St. Petersburg mathematics, Andrei Andreevich Markov senior (1856–1922), and Vladimir Andreevich Steklov (1864–1926), were both his pupils.

Chebyshev studied in Moscow where he was most affected by Nikolai Dmitrievich Brashman (1796–1866), a co-founder of the Moscow Mathematical Society in 1864. Chebyshev's numerous contributions to mathematics, mechanics and other fields are well known (e.g., in approximation theory), as is his role of teacher at the St. Petersburg University. He taught there from 1847 until 1882[86] and among those he influenced were, e.g., Aleksandr Nikolayevich Korkin (1837–1908), Egon Ivanovich Zolotarev (1847–1878), A.A. Markov Sr, Lyapunov, Grave, Nikolay Yakovlevich Sonin (1849–1915), Georgy Fedoseevich Voronoy (1868–1908), and Vladimir Andreyevich Markov (1871–1897).[87]

A.A. Markov Sr contributed to probability theory.[88] He showed his mathematical talent rather early — he wrote his first mathematical work while he was a secondary school pupil. The result was not new but it caught the attention of professors of the St. Petersburg University, Korkin and Chebyshev.

When Markov Sr entered the St. Petersburg University, Chebyshev became his teacher. Markov obtained a bachelor's degree in 1878, a master's degree in 1880 and a doctorate in 1884. From Chebyshev he took over his main field of interest, probability theory. In 1887 Chebyshev sketched a proof of the central limit theorem. Later on, Markov Sr succeeded in proving the general result using Chebyshev's method in a rigorous way. He also contributed to the law of large numbers, which led to the introduction of the so-called Markov chains. Markov became an extraordinary professor of the university in 1886 and an ordinary professor in 1894 (see Steffens, 2006, p. 99). In 1890 he was elected an extraordinary member of the Academy, and — as a successor of Chebyshev — an ordinary member of the Academy, in 1896.[89]

In 1905 he became a merited professor with the right to retire, which he did immediately. From 1905 till 1910, he lectured at the university on the calculus of finite differences. In Steffens (2006, p. 99), it is stated that *Protesting against a decree of the ministry of education Markov rejected a further teaching activity at the St. Petersburg University. In connection with students' riots in 1908, professors*

mathematician, and physicist who spent most of his life in Russia and contributed to spherical trigonometry.

[86]See Steffens (2006, p. 77).

[87]For more details see again Steffens (2006, pp. 77–78) (data) and throughout the book.

[88]His, also very talented, younger brother V.A. Markov died young of tuberculosis. Similarly Zolotarev, who defended his PhD thesis written under Chebyshev in 1867, was run over by a train and soon afterwards died from blood poisoning.

[89]For a more detailed text on Markov Sr see Basharin *et al.* (2004).

Figure 6. Old buildings of the Saint Petersburg University.[90]

and lecturers of St. Petersburg University were ordered to maintain observations over their students. Firstly Markov rejected this decree and after that he wrote an explanation where he expressed his refusal to be an "agent of the governance". Finally he saw no recourse other than full retirement from the university.

In 1913 the council of St. Petersburg elected nine scientists honorary members of the university. Markov was among them, but his election alone was not affirmed by the minister of education. The affirmation was done only four years later, after the February revolution in 1917.

After the October Revolution, he asked to be sent to a small place where he taught at a secondary school. Later, he had a chance to return to Petrograd, where he soon died.[91]

Now we will describe Steklov's life story. In 1882 Steklov started his studies at the Faculty of Physics and Mathematics of the Moscow University. After a year, he

---

[90]Author NSmirnova, https://ru.m.wikipedia.org/wiki/%D0%A4%D0%B0%D0%B9%D0%BB:%D0%92%D0%B8%D0%B4_%D0%BD%D0%B0_%D0%B7%D0%B4%D0%B0%D0%BD%D0%B8%D0%B5.jpg [15.2.2021]. The building on the right-hand side is sometimes called *Twelve Collegia* and it was built between 1722 and 1744.

[91]Markov Jr had the same given names and was also a well-known mathematician. He was born in 1903 and died in 1979.

failed to pass an exam in geography[92] and so he left Moscow, and in the same year 1883 he entered a similar faculty in Kharkiv. He met there Aleksandr Mikhailovich Lyapunov (1857–1918) and became one of his two most well-known students and also a friend. Under Lyapunov's supervision he submitted (an equivalent of) a PhD thesis and later also a doctoral dissertation. He worked on problems connected with potential theory, electrostatics and hydromechanics. In 1902, when his former supervisor Lyapunov moved to St. Petersburg, Steklov was appointed to the Chair of Applied Mathematics at the Kharkiv University.[93]

By 1896 he had started to publish abroad in the *Mathematische Annalen*, and soon afterwards, regularly (in French) in the *C.R.* He also corresponded with some top European mathematicians.[94] In 1906 Steklov moved to St. Petersburg, practically to the same position as in Kharkiv. In 1908 both couples, the Lyapunovs and the Steklovs, traveled to Italy to take part in the ICM 1908, in Rome.

Steklov also visited the ICM 1912 in Cambridge, where he was elected one of its vice-presidents, together with Jacques Salomon Hadamard (1865–1963) and Samuel Giuseppe Vito Volterra (1860–1940). The last congress he visited was the ICM 1924, in Toronto.[95]

---

[92]Here is a quote from Smirnov's unpublished manuscript *Excerpts from my diaries* (see Kuznetsov (2014, p. 32)): *The last oral exam was in physical geography taught by the stern professor Stoletov. Rather quickly, I have managed to study this easy discipline within the lecture course. My reply to the questions formulated on the card was excellent. Suddenly, Stoletov asks: "What date is the longest day in Moscow" I was completely taken aback by this question. My silence lasted several minutes. Stoletov was glassy staring at me and, at last, he said sluggishly: "Complete ignorance". He writes unsatisfactory in my record-book, and I am ruined because my marks for all other difficult exams were excellent...*

[93]Lyapunov studied in St. Petersburg, where he defended his thesis *On Stability of Elliptic Equlibrium Forms of a Rotating Liquid* (1985, Russian). He became a private docent in Kharkiv, and after receiving a doctorate from the Moscow University in 1892, he was, from 1893 to 1902, a professor of the Kharkiv University. In 1901 he was elected to the Russian Academy in St. Petersburg, and the following year, he became an academician in applied mathematics. He taught as a professor in St. Petersburg from 1902 to 1917 and then he moved to Odessa, where he got a post at the university. In 1918 his wife died from tuberculosis and he shot himself the same day, and in another three days he died.

[94]See Kuznetsov *et al.* (2014), where some other information on Steklov is available, including a long list of sources. Other facts on Steklov, and the quotes, are from this paper. See also Kuznetsov (2014).

[95]Steklov was a Russian and Soviet mathematician, mechanician and physicist with many interests. Note that N. Kuznetsov in Kuznetsov *et al.* (2014) states: ... *After Steklov's death the Institute was named after him. In 1934, simultaneously with the relocation of the Academy from Leningrad to Moscow, the Physical-Mathematical Institute was divided into the following two parts: the P.N. Lebedev Physical Institute and the V.A. Steklov Mathematical Institute (even before that the division of seismology became a separate institute). The Leningrad (now St. Petersburg) Department of the latter was founded in 1940, and it has been an independent institute since 1995. To summarize, it must be said that the role of V.A. Steklov in Petrograd was similar to that of R. Courant who organized mathematical institutes, first in Göttingen and then in New York.*

Figure 7. Pictures of Steklov (1864–1926).[96]

In 1906 Steklov moved to St. Petersburg and the same year a group of very talented students enrolled at the university. Steklov later commented on it as follows: *I should note that the class of 1910 is exceptional. In the class of 1911 and among the fourth-year students who are about to graduate there is no one equal in knowledge and abilities to Messrs. Tamarkin, Friedmann, Bulygin, Petelin, Smirnov, Shohat, and others. There was no such case during the fifteen years of teaching at Kharkiv University either.*[97] Steklov was an excellent mathematician, a very good organizer and teacher and was quite close to the new regime. This also probably explains why Steklov's name was chosen for the institute and how "Steklovka" was created.[98]

Steklov's works were usually very close to applications. In the beginning, they were devoted mostly to mechanics, especially to hydromechanics. His principal field of interest was mathematical physics and corresponding problems of analysis, now known as the Dirichlet and the Neumann problems. Let us quote an evaluation: *Although Neumann, Hermann Schwarz, Poincaré, G. Robin, E. Le Roy, and others had suggested methods of solving such problems, they did not elaborate their rigorous grounding, and their methods were applied to relatively restricted classes of surfaces. The precision of analysis in the general investigation of the problem was*

[96]Both pictures are from Wikimedia Commons, https://commons.wikimedia.org/wiki/File:Vladimir_Andreevich_Steklov.jpeg [20.4.2020], and https://commons.wikimedia.org/wiki/File:Steklov.jpg [20.4.2020].

[97]The quote is from Kuznetsov *et al.* (2014, p. 11).

[98]To illustrate Steklov's opinions, we quote from his diaries: *St. Petersburg has been renamed Petrograd by Imperial Order. Such trifles are all our tyrants can do — religious processions and extermination of the Russian people by all possible means. Bastards! Well, just you wait. They will get it hot one day!* See Kuznetsov *et al.* (2014, p. 13).

first achieved by Lyapunov and Steklov.[99] Note that Steklov's monograph (originally planned in three volumes, but only two of them appeared) is in the list of the most influential works published in the first half of the 20th century.[100]

There was a similarity between Moscow and St. Petersburg: both had a group of talented mathematicians, and so St. Petersburg had also its "Lusitania". We have to choose from its members — the best known among them were Steklov's students, Alexander Alexandrovich Friedmann (1888–1925) and Vladimir Ivanovich Smirnov (1887–1970), as well as Markov's students, Abram Samoilovitch Besicovitch (1891–1970) and Yakov (Jacob) Davydovich Tamarkin (1888–1945).

Younger members of the aforementioned group had different fates. Probably the most famous one was Friedmann. His biography[101] describes his best known result: his solution to Einstein's field equations provided early evidence of the expanding universe, and the theoretical underpinnings for both the Big Bang and steady state models of the universe in 1922. Einstein thought at first that the solution was erroneous but soon changed his mind.

Figure 8. Pictures of Markov senior (1856–1922).[102]

[99]See https://www.encyclopedia.com/science/dictionaries-thesauruses-pictures-and-press-releases/steklov-vladimir-andreevich [20.9.2019].
[100]See p. 13 in J.-P. Pier (ed.), *Developments of Mathematics 1900–1950*, Birkhäuser, Basel, 1994, xvii + 729 pages. ISBN 978-3-0348-9913-0. Both Steklov and Markov are buried in St. Petersburg.
[101]E.A. Tropp, V.Ya. Frenkel, A.D. Chernin, *Alexander A. Friedmann: The Man Who Made the Universe Expand*, Cambridge, 1993, 267 pages. ISBN 978-0-521-38470-4.
[102]Both pictures are from Wikimedia Commons, https://commons.wikimedia.org/wiki/File:AAMarkov.jpg [20.4.2020] and https://commons.wikimedia.org/wiki/File:Andrei_Markov.jpg [20.4.2020].

During WWI, Friedmann volunteered to serve with the Russian Air Force, at first as a technical expert and later as a bomber pilot. After the war, he worked as a professor of theoretical mechanics at the University of Perm. In 1920 he returned to St. Petersburg where he became familiar with Einstein's general theory of relativity, and entered this field. In the following years he traveled through Europe. He visited Germany, Norway and the Netherlands. Unfortunately, probably during his vacation in the Crimea, he contracted typhoid fever and died at the young age of 37.

Another member of the above-mentioned student group, created in 1908 in St. Petersburg, was Smirnov. At this time lectures were rather irregular, the study was demanding, and the students had to be very independent. Smirnov was an excellent student and graduated from the university with a First Class Diploma in 1910. He then taught mathematics at a very good secondary school until 1912, when he became a graduate student of Steklov. At this time Smirnov was interested in the latest results in analysis, such as the theory of integral equations. Since Steklov was skeptical about these new directions, Smirnov studied the works of René Maurice Fréchet (1878–1973) and the Riesz-Fischer theorem, on his own. When WWI emerged, Smirnov's perspectives started to become less favorable.

In 1918 Smirnov defended his thesis *The Inversion Problem of a Linear Second-Order Differential Equation with Four Singular Points* (Russian). After the October Revolution and WWI, life in Petrograd was difficult and so Smirnov was happy when he got a job in Simferopol, Crimea. He went there with his wife and her two daughters from her first marriage. Both Smirnov and his wife were considered enemies of the regime. They received a summon to the so-called "court". A "judge" allowed him to go home, but his wife was executed.[103] Smirnov supported his step-daughters for many years.

With the help of Steklov, Smirnov was able to return to Petrograd in 1921. *Smirnov returned to his* alma mater *and never left the University again. At this moment, the long-term activity of [him] as an organizer of mathematical life in Petrograd-Leningrad began.* Another important fact has to be mentioned: *Already in the early 1920s, Smirnov and Tamarkin began to write a textbook for technicians and physicists. The first two volumes were published in 1924 and 1926, respectively. However, the joint work did not continue for long because Tamarkin soon fled from the country. For further editions Smirnov extended and revised the material alone. Since 1930, "A Course of Higher Mathematics" began to appear only under the name of Smirnov.*

[103]See D. Apushkinskaya, A.I. Nazarov and Vladimir Ivanovich Smirnov (1887–1974), *Complex Variables and Elliptic Equations* **63**(2018), pp. 897–906, ISSN 1747-6933. Quotations from the part on Smirnov, below, are from this article.

9. Emigrants

Tamarkin graduated in 1906 from the Second Gymnasium of St. Petersburg, where Friedmann was his schoolmate.[104] Together they wrote their first mathematical paper[105] and their school awarded them gold medals.

Both Tamarkin and Friedmann graduated in 1910 and got their first academic positions. They continued working for their master's degrees in applied mathematics. Tamarkin's book on boundary value problems appeared in 1917. He was a professor at several institutions, the fourth of which was the Naval Academy, where he started to teach in 1922. He was hard-working and produced several books: *Approximate Computations* (Russian, with Friedmann), Leningrad, 1925, 150 pages, *A Course in Higher Mathematics for Engineers and Physicists* (Russian, with Smirnov), Leningrad 1924, Volume 1, 420 pages, and Leningrad, 1925, Volume 2, 408 pages. He also collaborated with Grigorii Mikhailovich Fichtenholz (1888–1959), with whom he published *A Course in Mathematics for Engineers* (Russian), Leningrad, 1926, 300 pages.

Einar Carl Hille (1894–1980), with whom Tamarkin collaborated for a long time, wrote a comprehensive article about him.[106] Among other things, he described Tamarkin's life: [*Tamarkin*] *had been very well-to-do before the revolution razed all values. That his mathematical library was first rate, is not surprising, but he was also intensely fond of music and he possessed a music library of some 6000 items including a complete collection of Russian music. He used to engage a string quartet to play at his home weekly; the concert lasted from eight in the evening until five or six in the morning with an intermission for a grand supper at midnight and good breakfast in the morning. Shostakovich, at the age of sixteen, played part of his first symphony at one of these concerts, in 1924 ... J.D. was a menshevik* (*social democrat*) *and, as such, he was fair game both to Okhrana of the czar and the Cheka, after the October revolution.*

In 1925 Tamarkin left for the USA, and was joined by his family in 1926. It seems quite understandable that he was probably afraid for his family, knowing what had happened to Smirnov. At first, he lectured for two years at Dartmouth College, and he was called to Brown University in 1927. Altogether he wrote more than 125 publications (more than 50 with Hille) including six books. He collaborated with 20, mostly well-known, mathematicians and a list of his merits would be very long. The happy time for him ended in 1934 when his wife suddenly died. He published only four papers after 1935 but he devoted a lot of time to his lecture notes and

[104] Smirnov had studied at the same gymnasium and graduated this a year earlier.

[105] J. Tamarkine, A. Friedmann: Sur les congruences du second degré et les nombres de Bernoulli, *Mathematische Annalen* **62**(1906), pp. 409–412.

[106] Jacob David Tamarkin — His Life and Work, *Bulletin of the American Mathematical Society* **53**(1947), pp. 440–457.

care for his students. The previously mentioned paper of Hille contains a detailed description of his works.[107]

Besicovitch's life story is in some way similar to Tamarkin's one. He was born to a family of jewelers, by no means a rich one. They lived frugally with six children, two daughters and four sons, in St. Petersburg. All the children were talented and all studied at the university. Their mother was probably a highly intelligent person, but without a higher education, since she married at the age of 15; her husband was 25 years older. The children agreed that she had shown a conspicuous ability for mathematics. During their studies, the older children supported the younger ones.

In the 1920s, the duties of a university teacher under the regime were demanding, and Besicovitch was not satisfied. He was offered a Rockefeller Fellowship to work abroad, but he needed approval in order to accept the offer. However, it was repeatedly refused.[108] He decided to leave illegally, which he did in 1924, together with Tamarkin, via Latvia. His wife stayed in Russia, and in 1926 their childless marriage was dissolved. He then spent a year in Copenhagen with Harald Bohr (1887–1951), who worked on the theory of almost periodic functions. He was invited by Godfrey Harold Hardy (1877–1947) to Oxford. Contrary to the usual strategy[109] his Rockefeller Fellowship was prolonged for a year. Then, with the help of colleagues, he received a fellowship from Trinity College, Cambridge.

While he was in Perm,[110] he had become friendly with Maria Ivanovna Denisova, the widow of another mathematician, and her two daughters. He brought them to England, and in 1928 he married the elder daughter, then aged 16.

Besicovitch was an excellent mathematician. He contributed to the theory of almost periodic functions, to measure theory, especially to fractional Hausdorff measures and — more generally — to real and also complex analysis. An interesting problem which he solved was the so-called *Kakeya problem.*[111] Not quite precise but an understandable form of it *asks for a set* $S \subset \mathbb{R}^2$ *of a minimal area in which a unit segment can be turned through* 360° (*of course, translations of the segment are also allowed*). If we require that S must be *convex* then the minimal area is achieved

[107] Note that many mathematicians who emigrated from Europe found their place at Brown University, Providence, RI, USA. This approach to immigrants was strongly supported by the president of the university and Tamarkin began working there in 1927. He worked there until his retirement. See Rowe (2018, p. 424).

[108] It could be one of the reasons why he decided to emigrate.

[109] For Rockefeller support of the sciences, see Siegmund-Schultze (2001).

[110] The university in Perm was founded in 1916 as a branch of St. Petersburg University. The town was later (1940–1957) renamed to Molotov to honor Vyacheslav Michailovich Molotov (1890–1986) known for the Molotov-Ribbentrop Pact (1939).

[111] Named to honor Sōichi Kakeya (1886–1947).

by an equilateral triangle of height 1 and area $1/\sqrt{3}$.[112] Besicovitch proved in 1919 that such a set without the requirement of convexity could have an arbitrarily small positive area.[113]

Moscow and St. Petersburg were the most important centers of mathematics, but there were others like Kharkiv, Kiev, Perm, Kazan. Many other important mathematicians could be added to our preceding text, but due to limited space we have only had the chance to deal with some of them.[114]

Jakov Viktorovich Uspensky (1883–1947)[115] finished his studies at the St. Petersburg University in 1906. As an excellent student he was chosen for a university career. In 1911 he defended his dissertation *Some Applications of Continuous Parameters in the Theory of Numbers* (1910, Russian). From 1912 he was a private docent and taught, among others, Friedmann and Tamarkin. From 1915 he was an extraordinary professor and from 1917 a full professor of the university. In 1921, recommended by Krylov, Markov Sr and Steklov, he became an ordinary member of the Russian Academy of Sciences (he replaced Lyapunov who died in 1918). He attended the ICM 1924, where he delivered a lecture on results he had obtained together with his student, Boris Alekseevich Venkov (1900–1962), and then he delivered three lectures in Chicago. He returned to the U.S. and in the period 1926–1927 he visited several universities to deliver lectures. In 1929 he again visited the U.S. and did not return to the USSR for several reasons. During his previous stay in the U.S. he had married, and his wife refused to live in the USSR. Also, conditions for work in mathematics in the USSR were worse, since Marxism penetrated mathematics. From 1929 to the end of his life, he was a professor at Stanford University.[116]

10. German and Russian Mathematics and Mathematicians

Let us start with a more general view. For several centuries Germans took part in the process of education in Russia, which probably created some kind of tradition. German teachers were popular in the families of Tsars, but more interesting is that there were a number of German schools throughout the country. For example, from 1726 there was a German gymnasium in St. Petersburg, whose teachers were

[112]It was proved by Gyula (Julius) Pál in 1920. See J. Pál, Ein Minimumproblem für Ovale, *Mathematische Annalen* **83**(1921), pp. 311–319.

[113]In the work Sur deux questions d'intégrabilite des fonctions, *Journal of Physico-Mathematical Society* (Perm) **2**(1919/1920), pp. 105–123.

[114]It is possible that some mathematicians emigrated from the USSR because of the fact that they were Jews and so their potential careers were somewhat limited.

[115]Known also as James Victor Uspensky.

[116]When Kapitsa was in England, where he studied and then worked with Ernest Rutherford (1871–1937) in the *Cavendish Laboratory*, Uspensky wrote to him about Krylov's warning not to come back to the USSR. In 1934 Kapitsa returned to Russia to visit his parents and then he was not allowed to travel back. Kapitsa was Krylov's son-in-law. See Ermolaeva (1997) for details.

Germans and all the subjects were taught in German for about 30 years. Later, in 1869, approximately 25% of the secondary school teachers and almost 10% of the lecturers at higher or university type schools in St. Petersburg were Germans.[117]

Another part of the relationship of Russians to the German educational system is connected with women's studies. Several Russian women completed their studies at German universities, including the study of mathematics. Probably the best known is the case of Sofia Vasilyevna Kovalevskaya (1850–1891). At that time, women were not allowed to attend universities, so she left Russia, and she started to study at Heidelberg University in 1869. After two years she moved to Berlin, and since she could not officially attend classes at the university, Karl Theodor Wilhelm Weierstrass (1815–1897) became her "private teacher".

She published several articles, and one of them on differential equations contains the Cauchy–Kovalevskaya theorem (1874).[118] Based on these articles she earned her doctorate in mathematics *summa cum laude* in Göttingen. Thus, with the support of Weierstrass, she could skip the required lectures and examinations. In 1884 she became an extraordinary professor of mathematics in Sweden, and for a time, an editor of the *Acta Mathematica*. She also wrote some non-mathematical books. Unfortunately, at the age of 41, she contracted influenza and died.

Kovalevskaya was followed by other Russian women. As presented in Maurer (2016), later other German universities allowed women students, e.g., Gießen in 1900 and Jena in 1902. Of the overall number of female students, foreign women formed a large part. For example, in the 1895/1896 school year in Berlin, they made up more than a half. Some of them were very successful.

In the period 1890–1894, Zapol'skaya, Lubov Nikolaevna (1871–1943) studied mathematics at the Bestuzhev Courses in St. Petersburg.[119] Then she decided to continue her studies in Göttingen and left there in 1895. Her dissertation, *Über die Theorie der Relativ-Abel'schen kubischen Zahlkörper*, was supervised by Hilbert, who was also the examiner of her final exam in 1900. She received the title in 1902, when her dissertation was printed and published. The same year she returned to Russia.[120] She had to rewrite her Göttingen dissertation for the Russian master's dissertation; she was the first woman to obtain the master's degree at the Moscow University or even on any Russian university. At that time she had already thought

[117]See *Dvoryanskii rod Rogge* [Nobility Rogge], https://genrogge.ru/gr.htm [30.8.2019].

[118]It is the main local existence and uniqueness theorem for analytic partial differential equations associated with the Cauchy initial value problems.

[119]These courses were the largest and most prominent women's higher education institution in Imperial Russia.

[120]The year 1902 is from Maurer (2016, p. 80). In the *Mathematics Genealogy Project* on the address https://genealogy.math.ndsu.nodak.edu/id.php?id=7342 [11.10.2019] she is listed as Ljubowj Sapolsky, with the year 1900. Her printed dissertation had vii + 481 + vi pages.

of a Russian doctorate and wrote to Hilbert for advice on finding a topic. In the letter she had also written:

Während meiner Studien in Göttingen habe ich, abgesehen von dem direkten Zwecke der Erwerbung der wissenschaftlichen mathematischen Kenntnisse in einer der besten deutschen Universitäten, noch ein anderes rein praktisches Ziel in's Auge gefasst: mit Hilfe des erlangten Göttinger Doctorgrades auch in Russland den russischen Frauen die Bahn zum Erwerben der russischen wissenschaftlichen Grade zu brechen, — Sie werden wohl gehört haben, dass in Russland die Damen in die Universitäten nicht zugelassen werden können. [In addition to the immediate purpose of acquiring mathematical knowledge in one of the best universities of Germany, I also envisaged another practical aim while studying in Göttingen: to pave the way for Russian women to acquire Russian academic degrees by means of the Göttingen doctorate — you probably know that in Russia women cannot be admitted to the universities.][121]

The last woman we want to mention is Vera Myller (born Lebedeva) (1880–1970), who was also supervised by Hilbert. She wrote her dissertation *Die Theorie der Integralgleichungen in Anwendungen auf einige Reihenentwicklungen* under Hilbert, and received her doctorate in 1906. In 1907, she changed her name and followed her husband[122] to Romania. In 1918 she became the first female full professor of mathematics in Iaşi.[123]

Professors at Russian universities quite often studied in Germany, or spent some time there. After the defense of his master's dissertation, Bugaev spent two and a half years abroad, in order to prepare for his further career at the Moscow University. In Berlin he attended the lectures of Weierstrass and Ernst Eduard Kummer (1810–1893). He presented his habilitation thesis *Number Identities Connected with Properties of the Symbol e* (1866, Russian), which was, according to Keller *et al.* (2020) supervised by Kummer, Weierstrass and Joseph Liouville (1809–1882), and he was appointed a professor of the Moscow University in 1867.

Both Mlodzeevskii and Egorov were taught by Bugaev, and we have already mentioned their studies abroad. Both spent some time in Paris and Göttingen. Egorov was also in Berlin, while Mlodzeevskii's choice was Zürich.[124]

Chebyshev loved to travel to foreign countries and, especially, loved Paris above all. His French was excellent and so he felt there like at home. He had good relations

[121]See Maurer (2016, p. 81).

[122]Her husband was the Romanian mathematician Alexander Myller (1879–1965) who studied in Göttingen from 1902 to 1906 and was supervised by Hilbert.

[123]Note that many American women also studied in Germany at the time.

[124]We know from Kuznetsov (1971) the names of those mathematicians whose lectures Egorov attended: *In Berlin, he attended lectures of Schwarz, Frobenius and Hensel, in Paris of Poincaré, Darboux, Goursat, Hadamard and Lebesgue, and in Göttingen of Hilbert, Klein and Minkowski.* See Kuznetsov (1971, p. 127). From Egorov (1925) we know that Mlodzeevskii's most influential foreign colleague was H.A. Schwarz.

with many French mathematicians like Hadamard, Joseph Alfred Serret (1819–1885) and Liouville. Chebyshev considered himself a world citizen. *Once, much later in his career, Chebyshev objected to being described as a "splendid Russian mathematician" and said that surely he was a "world-wide mathematician" rather than a Russian mathematician.*[125] He visited England, France, Belgium and Germany in 1852 and the journey was primarily oriented on mechanisms. His mathematics had a computational character.

Chebyshev's results found their way to Göttingen despite the limited availability of Russian papers. Klein knew this problem. He was aware of Chebyshev's works since he was informed about them by letters of Markov Sr and others. Klein gave a talk before the Göttingen Mathematical Society on some Chebyshev's results on aproximation in 1895.[126] Chebyshev's collected works were published by Markov Sr and Sonin in 1899. In 1902 Paul Kirchberger (1878–?) defended the thesis *Über Tschebyschefsche Annäherungsmethoden* [About Chebyshev's Approximations Methods], which was supervised by Hilbert.

The connection between Russian mathematicians and German mathematics is extensively documented in the papers of Tobies (1985) and of Demidov (2000), so we will concentrate only on the time between the wars, and on the most important facts. We will not provide extensive information on the *Deutsche Mathematische Vereinigung* or the *Moscow Mathematical Society.*[127] In general, mathematicians from St. Petersburg were less attracted to the Berlin school than Moscow mathematicians. After WWI, Muscovites changed the intensity of their relationship with French mathematics and started a more intensive collaboration with Germans, particularly those from Göttingen.

One more general remark: From a global point of view, smaller mathematical centers like Kharkiv, Kazan, Kiev, Odessa and others, having shorter traditions, were in a different position. In Demidov (2000) we can read about the Berlin school: *Of course, the influence of Weierstrass on young people was especially strong, and above all on provincial scientists, since in St. Petersburg and Moscow they were much more influenced by their own leaders. For example, negative relation of Chebyshev to the complex variable paralyzed initiative of his pupils and descendants. Young universities (of the Kiev type) were more open to the influences from outside. There new directions of research were more easily planted — new ideas of the theory of functions of a complex variable or analytic theory of differential equations* (Russian).[128]

WWI broke the connection between German scientists and scientists from other countries. After the war, the two large countries were isolated: Russia, with its new

---

[125] This is a quote from O'Connor and Robertson (2020).

[126] See Steffens (2006, pp. 131–141).

[127] For information, see, e.g., Remmert (2004) and Demidov *et al.* (2016).

[128] Quotes from the series *Istoriko-Matematicheskie Issledovaniya* had to be translated by the author of this chapter since this quite important source is not available in English, in contrast to, e.g., *Uspekhi Matematicheskich Nauk* [Russian Mathematical Surveys].

Bolshevik government, and Germany.[129] According to the treaty which ended the war between Russia and the Central powers, Russia lost approximately a third of its population, a quarter of its territory, around three-quarters of its coal and iron, a third of its factories and a quarter of its railroads. Thus, the new Soviet government had delivered peace but had lost swathes of territory and, in the process, established a one-party state.[130]

After WWI, peace negotiations took place, among which the most important was the Treaty of Versailles. It changed the Brest–Litovsk treaty and new independent countries were founded in the lost Russian territory. It contained clauses on reparations and on military restrictions. The German people did not accept the conditions and that fact was later used by the Nazis to come to power. The Treaty of Rapallo was an agreement signed in April 1922, between Germany and the Russian Soviet Federative Socialist Republic (RSFSR), under which each country renounced all territorial and financial claims against the other. The treaty did not contain military provisions, but they soon followed.[131]

Due to mutual official connections, it was, at least for some scientists, easy to develop former or quite new contacts. We have already mentioned the visit of Aleksandrov, Uryson and Kovner to Göttingen in 1923, which was one of the important places. Mathematicians from 15 countries (excluding Germany) lectured in Göttingen between 1919 and 1932, and the number of 17 lectures by Russian mathematicians is the second biggest one.[132] These lectures were mostly appreciated, and consequently, they sometimes led to cooperation. Thus, Aleksandrov wrote a book Alexandroff and Hopf (1935) with Heinz Hopf (1894–1971).[133]

German mathematicians also lectured, for even longer, in Moscow. Amalie Emmy Noether (1882–1935) delivered a course in Moscow in the winter of 1928/1929. Her lectures greatly influenced the career of Aleksandr Gennadievich Kurosh (1908–1971). Recollections about the beginnings of his scientific foundations are reproduced in the obituary, written by Aleksandrov: *In 1926 Pavel Sergeevich Aleksandrov began to give lectures at the university. In that time he had studied in one of the Smolensk secondary schools. I attended his lectures on the theory of sets, the theory of functions, and topology. In 1928 I was kept on at the university by Aleksandrov as a postgraduate student. At that time Emmy Noether was in Moscow and gave a course*

[129] After they were connected by the Brest–Litovsk treaty, ratified in March 1918 by the Fourth All-Russian Congress of Soviets, their relations started to improve.

[130] The Left Communist Bolshevik fraction and the Left Socialist Revolutionaries did not agree with the treaty and withdrew from the Sovnarkom [Council of People's Commissars].

[131] Both sides established elaborate military cooperation, which was publicly denied; more detailed studies on it appeared only in the 1990s. Some details are available, e.g., in the thesis of Ian Johnson, *The Secret School of War: The Soviet-German Tank Academy at Kama*, The Ohio State University, 2012, http://rave.ohiolink.edu/etdc/view?acc_num=osu1338500708 [11.10.2019].

[132] As we know from Tobies (1985), 18 lectures were delivered by Dutch visitors.

[133] The project of further volumes was not realized.

at the Moscow State University on abstract algebra, which Aleksandrov attended. Under the fresh impression of these lectures he gave a course on modern algebra at Smolensk. It was then that my scientific interests were formed.[134] Thus, Kurosh's "mathematical life" started in Smolensk where he also privately studied papers by Schmidt on group theory.

With Aleksandrov's help he had the chance to be in direct contact with Schmidt. At that time Schmidt headed the Institute of Mathematics and Mechanics of the Moscow State University. The Institute was created by him, and he held the chair of higher algebra. From 1930 he held a seminar on the theory of groups, and the seminar played a significant role in the creation of the Soviet school of general algebra.

Schmidt soon left the post of director of the Institute. He was more and more absorbed by his Arctic expeditions, which followed one after another, and so most of the teaching and the everyday practical organizational work fell on Kurosh, who had been an assistant in the Department of Higher Algebra since 1930. It was during this time when everything was changing and the old system of subdividing the academic process into lectures and seminars was stopped. However, in the academic year 1932/1933, the traditional system of giving lectures and conducting exercises separately was again established. Some new things remained, practical studies were introduced into the teaching process. Kurosh soon defended his doctoral dissertation *Research on Infinite Groups* (1936, Russian) and became a professor a year later.

One more quote from the Kurosh obituary (Aleksandrov *et al.*, 1972): *Kurosh was also charged with giving two parallel courses of lectures in higher algebra .... It became immediately clear that Kurosh was a brilliant spirited lecturer who combined in himself an infectious enthusiasm for the subject of his lectures with a clarity carried to the limit in the construction of a course, beginning from a general plan for the whole course and ending with a selection of notation and the arrangement of material on the blackboard.*[135]

Let us now turn to some other German guests in Russia. Ludwig Prandtl (1875–1953) delivered three lectures in Moscow in 1929. In 1930, on the occasion of the first Mathematical Congress of Russian Mathematicians, mentioned above, Germany was represented by Wilhelm Johann Eugen Blaschke (1885–1962), Ludwig Otto Blumenthal (1876–1944) and Leon Lichtenstein (1878–1933). In Tobies (1985), it is documented that some works of Aleksandrov, Khinchin, Kolmogorov and Uryson were written in Göttingen. Soviet mathematicians quite often published their results in German journals, at least until the 1930s.

Another quite well-known mathematician had a fate connecting Germany and the USSR. Chaim Herman Müntz (1884–1956) graduated in Berlin in 1906. His

[134] See Aleksandrov *et al.* (1972, p. 174).
[135] See Aleksandrov *et al.* (1972, p. 179).

dissertation *Zum Randwertproblem der Partiellen Differentialgleichung der Minimalfläche* (1910) was led by Schwarz and Friedrich Hermann Schottky (1851–1935) and he defended it in the same year. Later in his life Müntz was engaged in pedagogy, partly in private schools or as a private tutor, but he did not abandon research.

He published papers on problems from modern projective geometry and the axiomatics of geometry. Some of them appeared in the *Mathematische Annalen.* He spent the years 1920–1929 working on other mathematical problems, mostly in Göttingen and Berlin. He sought a position, and tried to obtain habilitation and to get the permission to lecture, but in vain. In 1929 he accepted a position in Leningrad, where he spent part of his career, until 1938.

What is of interest for us is the fact that he proved a generalization of the Weierstrass approximation theorem, which was presented by S.N. Bernstein as a conjecture during the ICM 1912. Two years later Müntz proved the following theorem: *Any continuous function on a closed bounded interval can be uniformly approximated by polynomials in functions* $x^{\alpha_0}, x^{\alpha_1}, x^{\alpha_2}, x^{\alpha_3}, \ldots$ *if and only if*

$$0 = \alpha_0 < \alpha_1 < \alpha_2 < \cdots \quad \textit{and} \quad \sum_{k=1}^{\infty} \frac{1}{\alpha_k} = +\infty.$$

For the theorem (without proof) and details of Müntz's life, including his non-mathematical activities and his bibliography, see Ortiz and Pinkus (2005). Obviously, for $\alpha_k = k$, $k = 0, 1, 2, \ldots$, we obtain Weierstrass' approximation theorem. In 1916 the theorem was independently proved by the Hungarian mathematician Otto Szász (1884–1952).

Müntz was a member of the official Russian delegation to the ICM 1932 in Zürich,[136] and he gave a talk on boundary value problems in mathematical physics. In 1937, the Müntzs were expelled from the USSR, and later, in February 1938, they moved to Sweden where they requested political asylum.

The list of mutual contacts could be longer — there were some mathematicians who traveled quite a lot. As already mentioned, Aleksandrov and Uryson visited Hausdorff, and later on, after Uryson's death, Aleksandrov visited Göttingen several times. In 1931 he and Kolmogorov made a long scientific trip[137] during which they visited Berlin, Göttingen, Munich and France.[138]

---

[136] Other members were Aleksandrov, who delivered a contribution on dimension theory, Chebotaryov, who gave a plenary lecture on Galois theory, and Kolman, an ideologist in this delegation. He gave two talks, the first about quaternions and the second about the foundations of differential calculus in the works of Karl Marx. See Ortiz and Pinkus (2005, p. 10).

[137] It was a 10-month trip to Germany and France. See Andrews *et al.* (2000, p. 27).

[138] Note that the Russian translation of Hausdorff's *Mengenlehre* appeared in 1937. The translation was based on the second edition (1927), but it differs substantially from the translated text. Comments on it are contained in the "Preface" by Aleksandrov and Kolmogorov, but in short, it is a different book, not a translation. It was translated by N.B. Vedenisov, while Aleksandrov and Kolmogorov are responsible for the additions/changes. See the comments in Hausdorff (2002). In the

Fritz Alexander Ernst Noether (1884–1941), and likewise his sister Emmy Amalie Noether (1882–1935), were both politically left-oriented intellectuals.[139] F. Noether was an applied mathematician. After his postdoctoral studies in Göttingen, he became an assistant at the Technische Hochschule in Karlsruhe. In the summer of 1911 he submitted his work, *Über den Gültigkeitsbereich der Stokesschen Widerstandsformel*, and obtained the permission to lecture and was appointed private docent for theoretical mechanics and mathematics. During WWI he was wounded on the German–French front. After the war he returned to teaching at the Technische Hochschule in Karlsruhe as an associate professor, and during the period 1922–1934 he was a full professor at the Technische Hochschule in Breslau.

When the Nazis started to apply their politics in Germany in 1933, both E. and F. Noether decided to emigrate. E. Noether found a place at Bryn Mawr College in the USA, where she unexpectedly died after a surgery. F. Noether emigrated to Russia with his wife and two sons. He found a place at the University of Tomsk where he was offered a professorship at the Institute of Mathematics and Mechanics. In 1935 he took part as a honorary guest at the Emmy Noether Memorial Session in Moscow.[140] F. Noether's wife, Regina, suffered from severe depression and returned to Germany where she committed suicide in 1935. F. Noether took part in the ICM 1936 in Oslo as the only participant who was allowed to come from Russia/USSR. It was probably due to the fact that he also had a German passport. Gel'fond and Khinchin did not come, despite the fact that they were supposed to deliver invited lectures.[141]

In November 1937, at the height of the Stalinist purges, F. Noether was arrested by the NKVD at his home in Tomsk. His two sons survived by being deported from Russia to Sweden, and from there to the USA. Gottfried Emanuel Noether (1915–1991) became a statistician there.[142]

For a long time the sons of F. Noether tried to get further information about the fate of their father. After many years they received a letter. The whole letter, from

Preface to the Russian edition, Aleksandrov and Kolmogorov wrote about the changes: *All that work* [changes] *which was far behind the frames of usual work of translator, was done by N.B. Vedenisov with our permanent editorial instructions and, naturally, in our and only our total responsibility.* (The last part of the sentence, i.e., "with ... responsibility" is emphasized.)

[139]For more about E. Noether, see also Kimberling (1972).

[140]It was organized by Aleksandrov, who was at that time president of the Moscow Mathematical Society. As a key speaker, he delivered the eulogy.

[141]The same happened with two other speakers, Edmund Georg Hermann Landau (1877–1938) and Francesco Severi (1879–1961). See Bečvářova and Netuka (2019, p. 111). At the ICM 1920 in Rome there was also only one participant from Russia, whom I was not able to identify. At the ICM 1924 in Toronto were Steklov, Nikolai Günther (1871–1941), Vladimir Alexandrovitch Kostitzin (1883–1963) and Aleksandr Vasil'evich Vasil'ev (1953–1929). At the ICM 1928 in Bologna were Bernstein, Krylov, V.V. Demchenko and Nikolai Nikolaevich Parfentiev. On the ICM 1932 see above. The lists are not complete, e.g., emigrants from Russia are not listed.

[142]His older brother, Hermann Dietrich Noether (1912–2007), was a chemist.

the Embassy of the USSR in Washington, signed by Andrei Parastaev, the First Secretary of the Embassy, was published[143] and we quote from it:

> *... On October 23, 1938 Professor Noether was found guilty of allegedly spying for Germany and committing acts of sabotage and was sentenced in Novosibirsk to 25 years of imprisonment. He served time in different prisons.*
>
> *On September 8, 1941 the Military Collegium of the USSR Supreme Court sentenced Professor F. Noether to death, on the accusation of engaging in anti-Soviet agitation. He was shot in Orel on September 10, 1941. His burial place is unknown.*

Needless to say that the letter also contained an announcement that the USSR Supreme Court had passed a decree, No. 308-88, by which F. Noether was fully rehabilitated. Both the fates of Müntz and F. Noether are sad and the second one had a tragic end.

The next story is that of George Günter Lorentz (1910–2006), who spent time in Russia and Germany until 1949. He was born in Petrograd but his father had to leave for political reasons and moved to Tiflis (since 1936 Tbilisi), the capital of Georgia. Lorentz had a German father and a Russian mother, but his first language was Russian. At Tiflis he studied at the Institute of Technology (1926–1928) and then moved to Leningrad where he continued his study at the School of Mathematics and Mechanics of the Leningrad State University. He graduated in 1931. There he attended a seminar on Banach spaces, and Markov's seminar on topology and worked on the so-called candidate's dissertation (an equivalent of a PhD) without a supervisor. This was, of course, not easy.[144]

In 1937 Lorentz's father lived in Tbilisi (Tiflis). He was arrested and sentenced to eight years in prison. He died in a concentration camp in 1938. In the meantime, Lorentz had started to prepare a book on functional analysis, but he was not able to finish it before WWII.[145]

11. Stalin's Era

The leader of the October Revolution, Vladimir Ilyich Lenin (1870–1924)[146] served as the Head of Government of Soviet Russia from 1917 to 1924. In 1924 he died,

[143]See G.E. Noether, Letter to the Editor, *Integral Equations and Operator Theory* **13**(1990), pp. 303–305. ISSN 0378-620X.

[144]See de Boor and Nevai (2010), where details of the life and work of Lorentz can be found.

[145]After the German attack on the USSR the Lorentzs made several attempts to leave Leningrad across the Ladoga Lake along the so-called "Road of Life", and in the end they were successful. They lived in Kislovodsk. Lorentz wrote about this time: *In August 1942 the city was abandoned by the Russian forces without a fight. Tanny* [Mrs. Lorentz] *and I were registered as ethnic Germans by the occupation authorities. The German advance stalled far away from the oil fields of Baku, one of its targets. Because of the Russian victories at Stalingrad, it was forced to hastily retreat. In the first days of January 1943 we left Kislovodsk with numerous other refugees. In this way we came to the refugee camps of Kalush and Toruń in Poland.*

[146]He was born as Vladimir Ulyanov.

probably as a consequence of the wounds he had sustained in 1918, during an assassination attempt. Upon Lenin's death, Joseph Vissarionovich Stalin (1878–1953)[147] was officially hailed as his successor. A vast majority of the events described above happened during the time of Stalin's reign.

Figure 9. Nonaggression Pact signed by Ribbentrop and Molotov in 1939.[148]

We have mentioned the tragic fates of some mathematicians and their relatives. They must be understood as a part of events on a much larger scale. Mass repressions started after the assassination of the Petrograd Cheka leader Moisei Uritsky (1873–1918). These repressions are sometimes called the "Red Terror". It was a time when all opponents of the Bolsheviks were exterminated.

There were some similarities in the development of Germany and of the Soviet Union — both states were, for some time, in the hands of one party and were governed by the dictators Stalin and Adolf Hitler (1889–1945). Antisemitism, which permeated in many countries, but with different intensities, was also apparent in the USSR. In some texts by Sergei Petrovich Novikov (*1938) we can read, e.g., about special, more difficult entrance exams for Jewish children. He also described the

[147]He was born as Joseph Dzhugashvili. Note that Stalin was of Georgian origin.
[148]Wikimedia Commons, https://commons.wikimedia.org/wiki/File:MolotovRibbentropStalin.jpg [20.4.2020].

behavior of Vinogradov, Pontryagin and some other mathematicians. Pontryagin, to whom people sometimes referred by his nickname "Pont", was used by Vinogradov, Aleksandrov and other "big players" to do things which they openly did not want to do themselves.[149]

Many thousands of people died in the USSR at the time of collectivization in the 1930s, and later, at the time of the so-called Moscow processes. Some German mathematicians died in concentration camps or committed suicide.[150] For a long time there was no reliable information on the gulags. We do not know how many mathematicians found the end of their life there, we only know the total number of victims. The book Snyder (2015) gives a picture of what happened there. We now know that, e.g., Sergei Pavlovich Korolev (1907–1966), the man who developed the USSR space program, spent a long time in a gulag. This final part is too short, but it could be a longer and sadder story, in which mathematics would play only a small role...

References

S.I. Adian, Lyudmila Vsevolodovna Keldysh (On the Centenary of her Birth), *Russian Mathematical Surveys* **60**(2005), pp. 581–588. ISSN 0036-0279.

P.S. Aleksandrov, Vyacheslav Vasil'evich Stepanov [Obituary], *Uspekhi Matematicheskich Nauk* **5**(1950), no. 5, pp. 3–10 (Russian).

P.S. Aleksandrov *et al.*, Lazar' Aronovich Lyusternik (On the Occasion of his 60th Birthday), *Russian Mathematical Surveys* **15**(1960), no. 2, pp. 153–168.

P.S. Aleksandrov *et al.*, Aleksandr Gennadievich Kurosh [Obituary], *Russian Mathematical Surveys* **27**(1972), pp. 176–193.

P.S. Aleksandrov, Pages from an Autobiography, *Russian Mathematical Surveys* **34**(1979), no. 1, pp. 267–302. ISSN 0036-0279.

P. Alexandroff and H. Hopf, *Topologie, Erster Band*, Springer, Berlin, 1935, xiii + 638 pages.

G.E. Andrews *et al.*, *Kolmogorov in Perspective*, History of Mathematics 20, American Mathematical Society and London Mathematical Society, Providence, RI, 2000, x + 230 pages. ISBN 978-0-8218-0872-6.

N.K. Bary, *A Treatise on Trigonometric Series. Volumes I and II.* A Pergamon Press Book, The Macmillan Co., New York, 1964, xxiii + 553 + xix + 508 pages. ISBN 978-1-4832-1076-6.

G.P. Basharin, A.N. Langville and V.A. Naumov, The Life and Work of A.A. Markov, *Linear Algebra and Its Applications* **386**(2004), pp. 3–26. ISSN 0024-3795.

M. Bečvářová and I. Netuka, *Dopady první světové války na mezinárodní spolupráci matematiků* [Impact of the World War I on the International Collaboration of Mathematicians], Česká technika — nakladatelství ČVUT, Praha, 2019, 240 pages. ISBN 978-8001-06540-2.

[149]For the above-mentioned texts the reader can see S.P. Novikov's pages at the http://www.mi-ras.ru/~snovikov/Content.html [11.10.2019]. S.P. Novikov was the son of P.S. Novikov, husband of L.V. Keldysh. S.P. Novikov received the Fields medal in 1970 for solving some difficult problems. From 1985, he was the president of the Moscow Mathematical Society, and as the successor of Kolmogorov, held this position until 1996.

[150]See the documentation in Siegmund-Schultze (2009).

N.N. Bogolyubov, B.V. Gnedenko and S.L. Sobolev, Andrei Nikolaevich Kolmogorov (On his 80th Birthday), *Russian Mathematical Surveys* **38**(1983), no. 4, pp. 9–27. ISSN 0036-0279.

C. de Boor and P. Nevai, In Memoriam George G. Lorentz (1910–2006), *Journal of Approximation Theory* **162**(2010), pp. 465–491. ISSN 0021-9045.

A.V. Chernavsky, Ljudmila Vsevolodovna Keldysh (To her Centenary, Part I, II), *Newsletter of the European Mathematical Society* **58**(2005), pp. 25–29, **59**(2006), pp. 17–21. ISSN 1027-488X.

S.S. Demidov, N.V. Bugaev and the Origin of the Moscow School of the Theory of Functions of a Real Variable, pp. 113–124, in A.P. Jushkevich *et al.*, *Istoriko-Matematicheskie Issledovaniya*, Volume 29, Nauka, Moscow, 1985, 352 pages (Russian).

S.S. Demidov, Russian Mathematicians in Berlin in the Second Half of 19th Century and the Beginning of 20th Century, pp. 71–83, in S.S. Demidov *et al.*, *Istoriko-Matematicheskie Issledovaniya* (Vtoraya Seria), Volume 5(40), Janus-K, Moscow, 2000, 392 pages. ISBN 978-5-8037-0052-4 (Russian).

S.S. Demidov and B.V. Lëvshin, *The Case of Academician Nikolai Nikolaevich Luzin*, History of Mathematics 43, American Mathematical Society, Providence, RI, 2016, xxxi + 376 pages. ISBN 978-1-4704-2608-8.

S.S. Demidov, V.M. Tikhomirov and T.A. Tokareva, The Moscow Mathematical Society and the Development of Mathematics in Russia (On the 150th Anniversary of the Society's Creation), *Transactions of the Moscow Mathematical Society* **77**(2016), pp. 127–148. ISSN 0077-1554.

D.F. Egorov, Sur les suites de fonctions mesurables, *Comptes Rendus Hebdomadaires des Séances de l'Académie des Sciences* **152**(1911), pp. 244–246.

D.F. Egorov, Boleslav Kornelievich Mlodzeevskii (Obituary), *Matematicheskii Sbornik* **32** (1925), pp. 449–452 (Russian).

N. Ermolaeva, Uspenskii Yakov Viktorovich, in V.V. Schelochaev and N.I. Kaniskheva (eds.), *Golden Book of Emigration. The First Third of 20th Century. Encyclopaedic-biographical Dictionary*, Russian Political Encyclopaedia (ROSSPEN), Moscow, 1997, 745 pages. ISBN 978-5-86004-038-0 (Russian).

C.E. Ford, Dmitrii Egorov, Mathematics and Religion in Moscow, *The Mathematical Intelligencer* **13**(1991), no. 2, pp. 24–30. ISSN 0343-6993.

W. Gautschi, Alessandro M. Ostrowski (1893–1986): la sua vita e le opere [Alexander M. Ostrowski (1893–1986): His Life, Works and Students], Expanded text of the lecture presented at a meeting of Ostrowski Foundation, Bellinzona, Switzerland, May 24–25, 2002, published in *Bollettino dei docenti di matematica* **45**(2002), pp. 9–19, https://www.cs.purdue.edu/homes/wxg/AMOengl.pdf [10.7.2019].

L. Graham and J.-M. Kantor, *Naming Infinity. A True Story of Religious Mysticism and Mathematical Creativity*, The Belknap Press of Harvard University Press, Cambridge, Massachusetts–London, England, 2009, 239 pages. ISBN 978-0-674-03293-4.

V.G. Gurzadyan, Kolmogorov and Aleksandrov in Sevan Monastery, Armenia, 1929, *The Mathematical Intelligencer* **26**(2004), no. 2, pp. 40–43.

F. Hausdorff, *Gesammelte Werke. Volume II, Grundzüge der Mengenlehre*, Springer, Berlin, 2002, xvi + 884 pages. ISBN 978-3-540-42224-2.

V.I. Igoshin, A Short Biography of Mikhail Yakovlevich Suslin, *Russian Mathematical Surveys* **51**(1996), pp. 371–383. ISSN 0036-0279.

A. Jones, Ch. Proust and J.M. Steele (eds.), *A Mathematician's Journeys. Otto Neugebauer and Modern Transformations of Ancient Science*, Archimedes 45, Springer,

Cham–Heidelberg–New York–Dordrecht–London, 2016, xii + 342 pages. ISBN 978-3-319-25863-8.

C.H. Kimberling, Emmy Noether, *American Mathematical Monthly* **79**(1972), pp. 136–149. ISSN 0002-9890.

A.N. Kolmogorov *et al.*, Dmitrii Evgen'evich Men'shov (On his 90th Birthday), *Russian Mathematical Surveys* **37**(1982), no. 5, pp. 203–215. ISSN 0036-0279.

B.A. Kushner, Kolmogorov in Reminiscences of his Pupils (Review), *The Mathematical Intelligencer* **30**(2008), no. 2, pp. 60–63. ISSN 0343-6993.

S.S. Kutateladze, Roots of Luzin's Case, *Journal of Applied and Industrial Mathematics* **1**(2007), pp. 261–267. ISSN 1990-4789.

N.G. Kuznetsov, The Legacy of Vladimir Andreevich Steklov in Physics: Work and School, *Newsletter of the European Mathematical Society* **91**(2014), no. 1, pp. 31–38. ISSN 1027-488X.

N. Kuznetsov *et al.*, The Legacy of Vladimir Andreevich Steklov, *Notices of the American Mathematical Society* **61**(2014), pp. 9–22. ISSN 0002-9920.

P.I. Kuznetsov, Dimitri Fedorovich Egorov (On the Centenary of His Birth), *Russian Mathematical Surveys* **26**(1971), no. 5, pp. 125–164. ISSN 0036-0279.

P.I. Kuznetsov and E.D. Solomentsev, Ivan Ivanovich Privalov (On the 90th Anniversary of his Birth), *Russian Mathematical Surveys* **37**(1967) no. 4, pp. 152–174.

A.F. Lapko and L.A. Lyusternik, From the History of Soviet Mathematics, *Russian Mathematical Surveys* **22**(1982), no. 6, pp. 11–138. ISSN 0036-0279.

G.G. Lorentz, Mathematics and Politics in the Soviet Union from 1928 to 1953, *Journal of Approximation Theory* **116**(2002), pp. 169–223. ISSN 0021-9045.

N. Lusin, Sur les propriétés des fonctions mesurables, *Comptes Rendus de l'Académie des Sciences de Paris* **154**(1912), pp. 1688–1690.

L. Lyusternik, The Early Years of the Moscow Mathematics School, *Russian Mathematical Surveys* **22**(1967), no. 1, pp. 133–157, no. 2, pp. 171–211, no. 4, pp. 55–91.

T. Maurer, Russian Women in German Universities — Pioneers of Female Higher Education? *Vestnik Sankt-Peterburskogo Universiteta. Istorya* **3**(2016), pp. 68–84, https://dspace.spbu.ru/handle/11701/5616 [5.5.2020]. ISSN 1812-9323.

J.J. O'Connor and E.F. Robertson, *MacTutor History of Mathematics*, University of St. Andrews, Scotland, https://www-history.mcs.st-andrews.ac.uk/index.html [5.5.2020].

F.A. Medvedev, On Lectures of B.K. Mlodzeevski on the Theory of Functions of Real Variable, given in Autumn 1902 at Moscow University, pp. 130–147 in A.P. Jushkevich *et al.*, *Istoriko-Matematicheskie Issledovaniya*, Volume 30, Nauka, Moscow, 1986, 368 pages. ISBN 978-5-8037-0383-9 (Russian).

M.T. Keller *et al.*, *Mathematics Genealogy Project*, North Dakota State University, Fargo, North Dakota, available at https://genealogy.math.ndsu.nodak.edu/ [20.8.2019].

D.E. Men'shov, Reminiscences of the Early Years and of the Beginning of the Moscow School of the Theory of Functions, pp. 312–333, in A.P. Jushkevich *et al.*, *Istoriko-Matematicheskie Issledovaniya*, Volume 27, Nauka, Moscow, 1983, 344 pages (Russian).

A.D. Myshkis and O.A. Oleinik, Vyacheslav Vasil'evich Stepanov (On the Centenary of his Birth), *Russian Mathematical Surveys* **46**(1990), no. 6, pp. 179–182. ISSN 0036-0279.

A.I. Nazarov, G.I. Sinkevich, History of Leningrad Mathematics in the First Half of the 20th Century, *Notices of the International Congress of Chinese Mathematicians* **7**(2019), pp. 47–63, https://arxiv.org/pdf/1812.03231.pdf [20.8.2019].

E. Neuenschwander and H.-W. Burmann, *Die Entwicklung der Mathematik an der Universität Göttingen*, https://www.uni-math.gwdg.de/entwicklung.html [10.7.2019].

E.L. Ortiz and A. Pinkus, Herman Müntz: A Mathematician's Odyssey, *The Mathematical Intelligencer* **27**(2005), pp. 22–31. ISSN 0343-6993.

E.R. Phillips, Nicolai Nicolaevich Luzin and the Moscow School of the Theory of Functions, *Historia Mathematica* **5**(1978), pp. 275–305. ISSN 0315-0860.

V.R. Remmert, Die Deutsche Mathematiker-Vereinigung im 'Dritten Reich' I, II: Krisenjahre und Konsolidierung (1st part), Fach- und Parteipolitik (2nd part), *Mitteilungen der Deutschen Mathematiker-Vereinigung* **12**(2004), pp. 159–177, pp. 223–245. ISSN 0942-5977.

D.E. Rowe, *A Richer Picture of Mathematics. The Göttingen Tradition and Beyond*, Springer International Publishing AG, 2018, xx + 468 pages. ISBN 978-3-319-67818-4.

E. Seneta, Mathematics, Religion, and Marxism in the Soviet Union in the 1930s, *Historia Mathematica* **31**(2004), pp. 337–367. ISSN 0315-0860.

A.N. Shiryaev (ed.), *Kolmogorov in Reminiscences of his Pupils*, Moskovskii Tsenter Nepreryvnogo Matematicheskogo Obrazovaniya (MTsNMO Publishers) Moscow, 2006, 472 pages. ISBN 978-5-94057-198-8 (Russian).

R. Siegmund-Schultze, *Rockefeller and the Internationalization of Mathematics Between the Two World Wars.* (Documents and Studies for the Social History of Mathematics in the 20th Century), Birkhäuser Verlag, Basel–Boston–Berlin, 2001, xiii + 341 pages. ISBN 978-3-7643-6468-7.

R. Siegmund-Schultze, *Mathematicians Fleeing from Nazi Germany. Individual Fates and Global Impact*, Princeton University Press, Princeton and Oxford, 2009, xxviii + 474 pages. ISBN 978-0-69114041-4.

Y. Sinai, *Russian Mathematicians in the 20th Century*, World Scientific Publishing Co., New Jersey, 2003, xi + 700 pages. ISBN 978-981-02-4390-6.

T. Snyder, *Bloodlands: Europe Between Hitler and Stalin*, Vintage, London, 2015, 524 pages. ISBN 978-0-09-955179-9.

K.-G. Steffens, *The History of Approximation Theory. From Euler to Bernstein*, Birkhäuser, Boston–Basel–Berlin, 2006, xix + 219 pages. ISBN 978-0-8176-4353-9.

R. Tobies, Zu den Beziehungen Deutscher und Sowjetischer Mathematiker während der Zeit der Weimarer Republik (Gastvortrag in Greifswald, April 1985), *Mitteilungen der Mathematische Gesellschaft der Deutschen Demokratischen Republik* **1**(1985), pp. 66–80.

A. Vucinich, Mathematics in Russian Culture, *Journal of the History of Ideas* **21**(1960), pp. 161–179. ISSN 0022-5037.

A. Vucinich, Soviet Mathematics and Dialectics in the Stalin Era, *Historia Mathematica* **27**(2000), pp. 54–76. ISSN 0315-0860.

S. Zdravkovska and P.L. Duren (eds.), *Golden Years of Moscow Mathematics*, History of Mathematics 6, 2nd edition, American Mathematical Society and London Mathematical Society, Providence, RI, 2007, xii + 306 pages. ISBN 978-0-8218-4261-4.

https://doi.org/10.1142/9781786349316_0003

Chapter 3

Roots of Modern Polish Mathematics (1815–1915)

Roman Duda

1. General Background

Vast territories of Central-Eastern Europe are inhabited by several nations, many with an interesting historical past, but in the 19th century most of them were subjected to a foreign domination. The whole area was politically controlled by a few great powers: Turkey in the south, Germany and Austria in the middle, and Russia to the east. With the exception of Turkey, which was then compelled to drive somewhat back, thus allowing liberation of Greece, Bulgaria, Romania, and Serbia, other powers kept their grip.

The 19th century was also a time of fundamental changes in nearly all spheres of public life. Great progress has been achieved in economy, means of communication, and education. In particular, educational reforms resulted in the establishment of complete state systems consisting of three levels: basic schools (teaching lasted usually four years), secondary schools with the final exam called matura (main type: gymnasia, where teaching lasted eight years), universities and polytechnics at the top. The system had a positive feedback: matura qualified for university studies while teaching in gymnasia was allowed for university graduates only. Moreover, universities were becoming not only centers of high education but also of research, thus combining the two. In consequence, the level of both education and research has grown considerably and the 19th century was a time when mathematics (and other sciences as well) have blossomed.

Together with the rising level of education there grew social and political ambitions of common people. National consciousness, awakened in the process, has found its expression also in forming national scientific communities.[1] The principal goal of those communities was to establish scientific terminology in a native language, to disseminate widely that terminology through valuable textbooks (translated or original ones), to found local scientific societies, and in some cases to found native scientific journals. And if such a community grew sufficiently well to have scientists within its ranks working on a national soil and possessing productive contacts abroad, one could say that it has attained some maturity and thus has become a part of a common scientific community. In this respect mathematics was a specific area since in places where there was neither high level mathematics before nor mathematical research, its germs had to be transplanted from afar. There were different ways to do that (see Duda (2014)) but in most regions of Central-Eastern Europe it has been done successfully up to the outbreak of World War I (cf. Bečvářová and Binder, 2011; Duda, 2015).

Poland underwent the same evolution but her initial situation was considerably different from that of most nations in the area. She was a state with a long history behind (cf. Davies, 1982; 1984), in the past enjoying periods of good mathematics at the old universities of Kraków and Vilnius, and already possessing, at the end of the 18th century, a well established native terminology in elementary mathematics. However, the 18th century saw a period of her decline which ended in partitions 1772–1793–1795. After the last partition in 1795 the whole territory of Poland was divided between three powers: Austria, Prussia, and Russia. The powers have introduced the policy of integration of those newly acquired territories with their empires and that policy was continued until 1918 (with the exemption of Austria after 1867, see below), thus preventing free national development. Consequently, for a period 123 years long (1795–1918), Poland has disappeared from political maps of Europe. In the Russian part each generation was raising its arms to fight for the political freedom and recovery of the state, but in other parts (and eventually in the Russian one as well) there was also an advancement of scientific life analogous to that in other countries of the area.

2. Polish Background

To understand the case of Polish mathematics in that period some additional historical facts should be recalled.

The 18th century saw a decline of the Polish-Lithuanian Commonwealth. The state became so weak that in 1772 there came the First Partition in which Austria, Prussia and Russia took off large parts of it. However, in the remaining part of the country, just about that time, there begun a movement towards recovery in which particular significance must be attributed to the Commission on National Education which was

[1] The problem of national communities in mathematics has found already some interest, cf. Parshall (1995).

appointed by the Parliament in 1773 with the aim to work out a uniform school system (on a state scale), to prepare curricula and textbooks in Polish facilitating their realisation, and to educate teachers for reformed schools.[2] Thus the Commission became in fact a ministry of national education, the first in the history, and for the two decades (1773–1794) it worked very hard doing its best to carry out the task. At the end, it left behind a complete school system comprising elementary parochial schools, secondary ones, up to universities in Kraków and Vilnius (Lwów was then in Austria and thus beyond its range), renamed the universities Main Schools and gave them the task to supervise the system in Poland or Lithuania, respectively, worked up curricula and textbooks for all types of schools (mainly secondary ones), widely distributed those textbooks, and prepared teachers in large numbers. The language of instruction was Polish (including Main Schools, where it replaced Latin), and in mathematics the set of textbooks completed a two centuries long work[3] upon elaborating Polish mathematical terminology in elementary mathematics (first mathematical textbooks in Polish appeared as early as in the 16th century).

Figure 1. Map of Central-Eastern Europe in 1880.[4]

[2]In Polish *Komisja Edukacji Narodowej Korony Polskiej i Wielkiego Księstwa Litewskiego*. There is a vast literature on the activities of that Commission including an excellent monograph Jobert (1941), translated into Polish.

[3]Więsław (2002) has found over 40 school textbooks from the period 1764–1795. Initially some of them were still in Latin, but Polish soon prevailed. They were published mostly in Vilnius, Warsaw, and Kraków, some in Lwów, Poznań and other cities.

[4]The map is taken from the book Davies (1984, p. 176).

The country has been reviving and in 1791 the Parliament has passed a new Constitution which had to become a basis for further reforms and thus intended to strengthen the state. The neighbours, however, understood the threat and decided to destroy the recovering state till it was militarily weak. So in 1793 there came the Second Partition and in 1795 the final Third Partition, thus completely rubbing the state from political maps.

During Napoleon's wars with Austria and Prussia, there was established a small and half-independent Duchy of Warsaw (allied to France), comprising some lands around the former capital of the state. It was a time of great hopes for patriotically minded people that the Polish state will recover but those hopes were shattered soon after Napoleon's defeat in 1813. In 1815 Vienna Congress took place. Victorious powers established a new political order in Europe and founded Saint Alliance as its guard. In particular, the Congress confirmed the suppression of Poland and set up new borders. Those borders, which existed until World War I, have divided Poland into four parts (see the map): Great Duchy of Poznań (with Poznań) and Pomerania (with Gdańsk) became a part of Kingdom of Prussia (after 1871 a part of German Empire), Galicia (with Lwów and, after 1846, also with Kraków) was incorporated into the Austrian Empire, and by far the largest share of the former state was captured by Russians and divided into two parts: Annexed Territories (with Vilnius) were directly incorporated into the Russian Empire, while Congress Kingdom (with Warsaw as its capital) has enjoyed a limited and temporal self-independence. Those were borders dividing one nation for a century and so no wonder that the nation, subdued to a hostile policy of Germanisation or Russification, respectively, undertook incessant efforts towards political and spiritual recovery.

The situation was different in each of the four parts of the partitioned country. In the Prussian part there was an almost continuous anti-Polish pressure and so was the Polish anti-German resistance, steady but without great strokes, succeeding in ushering a steady growth of social self-organisation and economic strength of the Polish population. In the Austrian part the first period (1772–1872) was equally ignominious, but after the defeat of Austria in 1867 in a war with Prussia, there came a radical change: Austria offered some freedom to its provinces. In consequence, Galicia got a rather wide autonomy within the Austrian Empire and Lwów, which became the provincial capital, has entered the period of rapid growth both in size and meaning. Lwów University (founded 1661) has become a vivid scientific center with the rapidly growing level of mathematics. Russian parts went through harder times but the beginnings were rather promising. After 1815 the Congress Kingdom enjoyed some liberties which allowed founding the Royal University in Warsaw in 1816, and the Annexed Territories also had then a good time: the old Vilnius university (founded 1578) was promoted to become the Main School in the whole area with Polish as the language of instruction and Vilnius became then "a cultural capital of Poland" (see Beauvois, 2010). But in 1831 there came an unsuccessful November Rising and the next three decades were the worst ones. First came severe reprisals.

All officers loyal to the National Government were cashiered and deprived of their property and all the remaining Polish soldiers were drafted into the Russian Army and sent to Caucasus. Some 80,000 Poles walked in irons to Siberia. Universities in Warsaw and Vilnius were both closed, Polish language forbidden in public, number of schools cut by half and many Russified. Some recovery came after 1862 when there existed in 1862–1869 the Main School (a Polish college in Warsaw of a university type). However, the collapse of January Rising (1863–1864) was followed by another wave of bloody repressions and mass deportations to Siberia. In particular, the Main School was closed and re-established as a Russian university, which acted in 1869–1915. And still two decades later, in the 1880s, there begun another and this time more durable recovery which had a decisive influence on further development. The movement was called Warsaw Positivism and its adherents argued that "to take place among the modern nations of Europe, the Poles must first improve the trade and industry of the Polish provinces, build towns and railways, and raise the literacy and national consciousness of the population"(Davies (1984, p. 170)).

After drawing a general picture, we come now to a more detailed story in each part separately.

3. Prussian Part

The Prussian part included Poznań which in the period of independent Poland was a centre of excellent education on a college level. There were efforts to raise the college to the university rank but all failed (one of the obstacles was a resistance on the part of Kraków University in defense of its monopoly in the Kingdom of Poland). Nevertheless, the province offered a friendly atmosphere which invited also some scholars from abroad, e.g., Jan Ámos Komenský (1592–1670) who had spent many years in Leszno, and Antoni Hoene, the father of Józef Maria Hoene-Wroński (1776–1853), who had settled in Poznań (both came from Bohemia).

After partitions and Vienna Congress, the Prussian part comprised two provinces of former Poland: Great Duchy of Poznań and Pommerania (renamed as Western Prussia) with Gdańsk. The provinces became a part of the Prussian Kingdom (since 1871 a part of the German Empire) and were subjected to a constant Germanisation pressure, particularly severe under the Prussian Colonization Commission (1886–1913), which aimed to strengthen the German element in the East. The use of Polish language was restricted, Polish schools on a secondary level were not allowed; as a rule Polish initiatives were treated unwillingly. One example: the Fourth Congress of Polish Physicians and Naturalists[5] took place in Poznań in 1884; when there appeared in 1898 an analogous initiative to organise in Poznań another such

[5]In Polish *IV Zjazd Lekarzy i Przyrodników Polskich*, see Dydyński and Ławrynowicz (1933). Since 1891 Polish mathematicians were forming a section in those congresses. See Maligranda (2019).

congress, the authorities first agreed but at the very last moment retracted their agreement and the Congress was not held.

In such an atmosphere all efforts to establish a Polish university in Poznań under Prussian rule were doomed to fail (the university was founded only in 1919 within the frame of the Polish state). German authorities agreed only to Poznań Society of Friends of Sciences,[6] which was founded in 1857 and earned much gratitude of posterity, but until 1919 had no mathematician within its ranks. Only in 1906 a German polytechnic was founded in Gdańsk and it was the only institution of a university type in the Prussian part. In such an atmosphere Poles from those two provinces who intended to study at a university were compelled to go to nearby German universities in Berlin, Breslau, Königsberg, Göttingen, etc. or to study abroad (the most popular lands were Switzerland and France). Many went and those who returned have contributed to the national cause. In consequence, there was a steady growth of national self-consciousness and self-organisation of the Polish population living in the German Empire, but until the end there was no mathematics at research level.

4. Congress Kingdom

At the beginning of the 19th century the best situation among parts of the divided country was in the Congress Kingdom. The Kingdom had enjoyed then some liberties and in 1802 there has been founded the Society of Friends of Sciences[7] and in 1816 there was established the Royal University,[8] both institutions in Warsaw. The Society had been founded by influential aristocrat dilettantes and had later supported several Polish scientific endeavours including founding libraries and publishing its "Memoirs" (in Polish). The university, patterned after a recently (in 1811) founded one in Berlin, was equally ambitious but facing a great need of the country for lawyers, medicines, teachers, etc., it felt compelled to concentrate efforts on education, postponing research to better times (which never came under Russian rule).

After founding the university, a problem of professors had arisen. In medicine and law they were chosen from among outstanding practitioners but in mathematics (and other sciences within the faculty of philosophy as well) it was hardly possible. Mathematics was first taught by some college teachers but the government decided to offer to best graduates a two-year scholarship abroad, most often in Paris, as a time to prepare for a professorship. The system worked and in the 20s the department of mathematics comprised already several mathematicians of good rank, including Augustyn Frączkiewicz (1798–1883) and Kajetan Garbiński (1798–1847).

[6]In Polish *Poznańskie Towarzystwo Przyjaciół Nauk*. See Wojtkowski (1928) and Grot (1957).
[7]In Polish *Towarzystwo Przyjaciół Nauk*. See Kraushar (1911).
[8]In Polish *Uniwersytet Królewski*. See Bieliński (1907–1912).

The former was a graduate of Kraków University but both have studied also in Paris. Frączkiewicz restricted himself to teaching but it was a modern teaching, his lectures included Pfaff's theory of differential equations and other new topics, while Garbiński had in addition some original publications in French and German journals which were the first Polish contributions to common mathematics since Adam Adamandy Kochański (1631–1700)[9] and an émigré, Józef Maria Hoene-Wroński.[10]

The two institutions, Society and University, acted until 1832. When Russians had crushed the unsuccessful November Rising (1831–1832), both were closed and their properties, including rich library collections, were plundered by the victors and partly taken to St. Petersburg. The number of elementary schools was cut by half. Only Russian state gymnasiums remained. They were rare and there was not a single higher school at the university level.

The recovery has begun in 1862 when there was established the so-called Main School in Warsaw[11] (the School was of the university rank, but Russians did not agree with that name) and some other liberties were allowed. Among professors of mathematics at the Main School were the old Augustyn Frączkiewicz and then a younger mathematician, Władysław Zajączkowski (1837–1898), who was educated in Kraków and had original results in differential equations. He also wrote a good extensive textbook on the subject (in Polish). Later, he also played a distinguished role in Galician Lwów as a professor of a Polytechnic School. To victimise Poles after the unsuccessful January Rising (1863–1864), the Main School was transformed in 1869 into a Russian university but with limited access for Poles, and other liberties were also withdrawn.

After the downfall of the January Rising, repressions were severe. In addition to some thousands, who fell in battles, hundreds were hanged during and after it, tens of thousands deported to Siberia, and hundreds of landowners expropriated. The country fell in decline. It lost a large share of its elite, the population became impoverished, and the Russification policy strengthened. The Kingdom lost remnants of any autonomy, became directly incorporated into Russian Empire and renamed Vistulaland.[12]

Among personal victims was a talented young mathematician, Roman Żuliński (1834–1864), a gymnasium teacher and the author of a good textbook on differential calculus. In the January Rising he became an effective minister of communications in the underground National Government and was hanged together with the last leader of the Rising (Romuald Traugutt, 1826–1864) and some other ministers. In addition, a dozen or so active mathematicians emigrated, mostly to Paris.

---

[9]The name is well known, cf. Taton (1978) and Heinekamp (1978).

[10]Short biographies of all the other Polish mathematicians mentioned below can be found in the book Duda (2007).

[11]In Polish *Szkoła Główna*. See Kieniewicz (1981).

[12]In Russian *Privislanskij Kraj*.

A more durable recovery had begun in Warsaw only in the 1880s, after two decades since the January Rising. In 1882 Russians agreed to the foundation of the Bank of Support for People Working in Sciences,[13] called shortly Mianowski Bank after its patron Józef Mianowski (1804–1879), rector of the closed Main School and a person of great merits. Although confined in its activities (e.g., independent initiatives of the Bank were forbidden, applicants could only be Russian citizens, no scholarships could be taken abroad), the Bank had enjoyed a wide support on the part of Polish society and managed to subsidise many important initiatives, including scholarships for talented young people, financing scientific enterprises, writing or translating textbooks at the university level and their further publication, supporting scientific journals, etc.

In 1888 there appeared the first volume of *Transactions in Mathematics and Physics*[14] (the journal existed until 1952) and in 1897 — the first volume of *Mathematical News.*[15] They were not the first Polish scientific journals (of older provenance were *Annals of the Scientific Society in Kraków* and its successors, see Section 7 or Paris *Memoirs*, see Section 6 below), but they were the first ones devoted specifically to mathematics. Appearing regularly, they invited mathematicians from all parts of the country and from abroad. Supporting original research and publishing results, they played an invaluable role in integrating the Polish mathematical community at that time still rather modest.

The Bank had been highly esteemed as an unofficial Polish "Ministry of National Education" and its "minister" was considered to be Samuel Dickstein (1851–1939), a graduate of the Main School in mathematics, one of the founders of the Bank, and its most prominent activist. Dickstein had some original papers in algebra but later turned to the history of mathematics, to editing journals like *Transactions* ... or *Mathematical News*, and to all other kinds of services to the rising Polish scientific community.

This rising tendency in the Kingdom during the two last decades of the 19th century was strengthened after Revolution 1905 when Russians felt compelled to weaken somewhat their anti-Polish policy. On Dickstein's initiatives, they agreed to the foundation of the *Mathematical-Physical Circle*,[16] the *Society of Scientific*

---

[13]In Polish *Kasa pomocy dla osób pracujących na polu naukowym imienia Józefa Mianowskiego.* See Zasztowt (2011).

[14]In Polish *Prace Matematyczno-Fizyczne.* The journal was edited initially by two mathematicians S. Dickstein and W. Gosiewski and by two physicists E. Natanson and W. Natanson. Later the board was changing, but S. Dickstein remained there until 1939.

[15]In Polish *Wiadomości Matematyczne.* Dickstein was its founder, editor and publisher (using his own funds). The journal has appeared in 1897–1939 until the outbreak of World War II and the death of Dickstein. Since 1955 it has been edited as the second series of *Annals of the Polish Mathematical Society.* See Duda (2015).

[16]In Polish *Koło Matematyczno-Fizyczne.* See Wuczyńska (2007).

Courses[17] and the *Warsaw Scientific Society*[18] (in all three cases Dickstein was a charter member). The "Circle" was the first Polish professional organisation comprising mathematicians, physicists, and astronomers, both researchers and teachers. The "Society of Scientific Courses" had organised legal university courses in Polish (again Russians did not agree to founding a Polish university); these legal courses had replaced formerly illegal ones organised by the so-called "Flying University" (see Miąso (1990)). The "Warsaw Scientific Society" was a revitalisation of the old "Society of Friends of Science" but was more ambitious, intending to be an Academy of Sciences (upon which Russians neither agreed). All three institutions played a major role in awakening and supporting scientific ambitions in the Congress Kingdom and in integrating a scientific community in Warsaw.

5. Annexed Territories

Comprising more than a half of the previous territory of the fallen Polish state, the Territories were turned in 1802 into one huge region called "educational district" with the Vilnius university as the institution to oversee the whole educational system in the region. The liberally disposed tsar Alexander I (1777–1825) has allowed using Polish as the only official language in the Vilnius University and as the language of instruction in most schools of the "district". Under these circumstances, in spite of the Russian framing, the Polish culture in the region continued to grow.

In 1803 there was founded Lyceum in Wolhynian Krzemieniec as a college with ambitions to become eventually a university. One of its founders and the first director was mathematician Józef Czech (1762–1810), formerly professor of Kraków University. In a short time the Lyceum succeeded in collecting a good staff and in attaining a high teaching level.

That good period ended after three decades. In November 1831 there broke out a rising in Warsaw, directed against Russians. It spread out rapidly throughout both Russian parts, Congress Kingdom and Annexed Territories. After its failure, there came reprisals even more severe than those in the Congress Kingdom. Hundreds of people were hanged; tens of thousands deported to Siberia, and properties of those somehow connected with the rising were taken from their owners and given to Russians. Vilnius University was closed and its belongings were taken to St. Petersburg, a great many Polish schools in the "district" were closed while some turned into Russian ones. Lyceum in Krzemieniec was equally closed and its staff and resources were taken to Kiev where they have served to found in 1834 a new Russian university. In particular, the first professors of mathematics in that Kiev university were

[17]In Polish *Towarzystwo Kursów Naukowych*. See Skubała-Tokarska (1965).
[18]In Polish *Towarzystwo Naukowe Warszawskie*. Closed by communists in 1952, it had recovered in 1991. See Nawroczyński (1950).

Stefan Wyżewski (1783–1850), Franciszek Miechowicz (1783–1852), and Grzegorz Hreczyna (1796–1840) — all three from Krzemieniec Lyceum.[19]

Three decades later, in 1863 there came the January Rising (again both in Congress Kingdom, where it broke out, and in Annexed Territories to which it soon spread out). Its fall next year still worsened the general situation and strengthened the anti-Polish policy.

In the miserable and oppressive conditions, several millions of Poles were still living in the Annexed Territories. Polish students from the Territories were storming Russian universities in St. Petersburg, Moscow, Dorpat (nowadays Estonian Tartu, until 1912 with German as the language of instruction), Kiev, Odessa, Kharkov and some other, forming there large communities. For instance, Polish community in St. Petersburg consisted, at the end of the 19th century, of around 60,000 people, including 3,000 students.[20] Students of mathematics founded there in 1880 the Circle of Polish Mathematicians, which edited four volumes of papers written by its members. Some students of mathematics from Russian universities made brilliant careers, including Julian Sochocki (1842–1927) and Jan Ptaszycki (1854–1917) in St. Petersburg, Bolesław Młodziejewski (1858–1923) in Moscow, Cezary Russyan (1867–1935) in Kharkov, Jan Sleszynski (1854–1931) in Odessa, Antoni Żbikowski (1827–1900) in Kazan — all became professors.[21]

Slow recovery in Vilnius began only in 1907 with the foundation there of the Society of Friends of Sciences (see Ilgiewicz (2008)). However, until refounding there the university in 1919, the Society had no mathematician within its ranks.

6. Emigration

Throughout the 19th century until World War I, Poland was not an independent country and its population was subjected to different forms of oppression. No wonder then that a great many Poles went abroad either for the sake of individual safety or simply to look for better conditions to live. After the two risings, November and January alike, there were even waves of émigrés several thousand people strong.

One of the first Polish émigrés was Józef Maria Hoene-Wroński (1776–1853), who settled down in Paris and made there his name as a mathematician and philosopher. Another one was Jakub Filip Kulik (1793–1863) from Lwów who went to Czech lands, eventually settling down in Prague. A promising mathematician was Ignacy Domeyko (1802–1889), a graduate of the Vilnius University; compelled to go abroad, he went via Paris to Chile where he rendered great services to that country and became a national hero. They were individual forerunners but after

[19]Vilnius University and Krzemieniec Lyceum did not recover until 1919 and 1920, respectively (when they were refounded by Piłsudski).

[20]A detailed account of that situation is provided in Róziewicz (1984). See also Polskiye profesora i studenty v universitetach Rossiyi (XIX-Naczalo XX v.) (1995).

[21]More details see Duda (2013).

1831 the emigration took on a much larger scale. After the defeat of November Rising in 1832 around 9,000 people left the country. Many of them were well educated, including later mathematicians: Antoni Bukaty (1808–1876), Józef Godebski (1803–1868), Hieronim Martynowski (1807–1861), Henryk Grach Niewęgłowski (1807–1881), Adolf Sągajło (1806–1877), and others.

Another large wave of émigrés came after the defeat of the January Rising in 1864, including the mathematicians Władysław Folkierski (1842–1904), Władysław Gosiewski (1844–1911), Edward Habich (1835–1909), and Karol Loewenhart (1840–1917). The former two eventually returned to Warsaw, but Habich went to Chile and Loewenhart settled in Argentina.

One of the few Polish emigration centers was Paris. In the second half of the 19th century there were so many Poles that they founded some schools including a preparatory one to enter Paris Polytechnic (see Rederowa (1972)). Not confining themselves to teaching only, they founded also a Society of Exact Sciences which acted in 1880–1882 until the death of its patron Jan Działyński (1829–1880). The Society published 12 volumes of its *Memoirs*,[22] including articles on mathematics (a dominating part), physics, chemistry, astronomy, medicine, natural sciences, technology, and adding information on the current scientific life. Papers were in Polish, while authors were coming from all parts of the partitioned Poland and from abroad. There were 40 mathematical papers by 17 authors[23] and although they were rather contributory, they nevertheless witnessed a rising level of reception of modern mathematics, growing research ambitions, and high position of mathematics in the Polish scientific community. The integrating role of *Memoirs* can be hardly overestimated. Another important field, for which the Society also deserves much credit, was the *Kórnik Library* which published in Polish several dozens of academic textbooks in different branches of mathematics, thus facilitating contact with modern mathematics for Polish youth in the country and spreading up the Polish terminology in higher mathematics.

The dominating direction of Polish emigration in the 19th century was West, not only France but also other European countries, the United States and South America (see Duda (2012)).

7. Galicia

The story of the Austrian part of Poland was different. For more than a century, Galicia was maltreated. A peripheral province in the Austrian Empire, it has been ruthlessly exploited and subjected to an imperial policy of imposing the German language and strengthening Austrian elements.

[22]In Polish *Pamiętnik Towarzystwa Nauk Ścisłych w Paryżu*, 12 volumes (1870–1882).
[23]See Koroński (2014c). For a more detailed description, see Więsław (2013).

Until 1846 Kraków was a capital of a small half-independent Republic (under the protection of Austria). The university, reformed yet by the Commission of National Education, tried to maintain a high level but its peripheral position, small budget and uncertain general situation pre-empted it from playing any major role. Nevertheless, in 1815 there was founded a Scientific Society related to the university, devoted to the cultivation of Polish science with a stress on literature and history but admitting also exact sciences. In the years 1817–1872 the Society edited 44 volumes of its *Annals* including 28 mathematical papers by 11 Polish mathematicians from Kraków, Lwów and Chile (Habich).[24] For nearly four decades in 1832–1870, it was the only Polish journal publishing mathematical papers. Their small number shows how modest Polish mathematics then was.

The situation changed in the 1870s when Galicia obtained a wide autonomy with a local parliament ruling education and schools in the province.

First to use the new chance was the Kraków Scientific Society which in 1872 transformed itself into the Academy of Arts and Sciences and widely extended its activities.[25] In the next 20 years 1874–1894 it edited 18 volumes of its *Memoirs* with 43 mathematical papers from Kraków and Lwów (see Koroński (2014a,b)). In addition to that, the Academy had been editing also *Dissertations and Reports of Meetings* of its faculties and in one of those series there appeared 90 mathematical papers (see Koroński (2014d)). And since 1885 there began to appear *Bulletin* of the Academy[26] which was the first Polish journal publishing papers in foreign languages, thus making achievements of Polish mathematicians admissible to a wide mathematical community. Editorial activity of the Society and the Academy was an inestimable factor for the growth of Polish science, in particular of mathematics, as shown by the rapidly growing number of publications.

The old universities in Kraków (founded 1364) and Lwów (founded 1661), previously Germanised, now became repolonised and entered a period of rapid growth in both magnitude and the value of research, and soon became centers of Polish cultural revival, particularly in mathematics (see Domoradzki (2012)).

The first Polish professor of mathematics in Lwów University, after its repolonisation, was Wawrzyniec Żmurko (1824–1889) who got some appreciation for constructing mathematical instruments. Among his students was Józef Puzyna (1856–1919), a talented mathematician (the author of a good monograph on analytic functions) who devoted himself deliberately to a wide teaching with the intention to raise highly the level of teaching mathematics in the Lwów University. Upon his invitation there came to Lwów in 1908 Wacław Sierpiński (1882–1969) who assembled around him

[24]For more information, see Koroński (2014).

[25]In Polish *Akademia Umiejętności*. See Hulewicz (1958).

[26]The original name was *Anzeiger der Akademie der Wissenschaften in Krakau*, but in 1910 it was changed to *Bulletin International de l'Académie de Sciences de Cracovie* and has remained since then.

a group of talented young mathematicians including Stefan Mazurkiewicz (1888–1945), Zygmunt Janiszewski (1888–1920), Stanisław Ruziewicz (1889–1941), and some others. Members of the group have concentrated their efforts on set theory, which then included topology and was a new branch of mathematics, and began to receive original results (see the chapter *Set Theory — A Vehicle to Modern Mathematics in Poland* in this volume). Although dissolved during World War I, the group gave rise to the Warsaw school of mathematics in the interwar period.

The first mathematician of high rank in the 19th century Kraków University was Franciszek Mertens (1840–1927) who was there a professor for 20 years from 1865 to 1884, when he moved to proper Austria, first to Graz and then to Vienna. He excelled in number theory but had no students of his own. The chair after him was taken by Marian Aleksander Baraniecki (1848–1895) who did his best to keep good mathematics but died early. And at the turn of the 19th and 20th centuries Kazimierz Żorawski (1866–1953) came to Kraków, followed a few years later by Stanisław Zaremba (1863–1942). Both were already known for their contributions to world mathematics, Żorawski to the theory of Lie groups and Zaremba to the theory of differential equations and its applications in physics.

The two universities, in Lwów and Kraków alike, have thus attained a rather high level in mathematics in the two decades preceding World War I. They contributed original results in a few areas to common mathematics and became known to the world.

8. Raising Mathematics

During the long period of partitions, in spite of all political upheavals and difficulties, there was incessant work on raising the general level of education including mathematics with a goal to reach the status of a nation on a par with the leading ones. An impetus, provided by the Commission of National Education in the last decades of an independent state, was still perceptible at the beginning of the 19th century. In particular, textbooks prepared under the Commission were still used, the work on mathematical terminology in mathematics was continued, and general impulses towards raising up remained strong.

To raise science, however, some conditions should be satisfied, including established scientific institutions (universities, academies) offering some posts, scientific journals and printing shops in which articles and books could be published, and most important talented people. All of that reached some maturity only in the last two decades of the 19th century in Warsaw, Kraków and Lwów.

In Warsaw there was a Russian University (with a limited access for Poles), but after 1882 there have been founded some Polish institutions like Mianowski Bank and scientific journals partly supported by it. There were people well educated in mathematics (mostly abroad) and there was a general interest in mathematics supported by books in Polish. Young people wanted to study in Polish and so "Flying

University", which offered courses on a university level in Polish, has enjoyed great popularity (courses in Polish were illegal until 1907 but somehow tolerated). See Miąso (1990).

Recovery of good mathematics on the Polish soil, however, has begun in Kraków. As noticed in the previous section, the first mathematician of eminence there was Franciszek Mertens, a professor of Jagiellonian University, widely known for his contributions to number theory. After his moving to Austria proper, at the end of the 19th century the two mathematical chairs were taken by Stanisław Zaremba and Kazimierz Żorawski. Zaremba came from France where he excelled in differential equations, while Żorawski completed his Warsaw studies in Germany and wrote highly evaluated papers in the theory of Lie groups and differential geometry.[27] All three were known to the world, and the latter two also had students with research ambitions.

The situation in Lwów has developed more slowly but in 1908 there came Wacław Sierpiński who took a chair and started a seminar on the then new set theory. He was joined soon by Zygmunt Janiszewski who obtained his PhD in Paris for a thesis in topology, Stefan Mazurkiewicz, Stanisław Ruziewicz, and some others. Sierpiński's seminar on set theory had later seminal consequences.

9. Prospects in 1915

In 1915 a Polish university was established in Warsaw. Mathematical chairs were offered to people from Lwów, Janiszewski and Mazurkiewicz (Sierpiński has been then interned in Russia). Upon the invitation of Mianowski Bank, Janiszewski proposed a program to initiate "an independent Polish mathematics".[28] The essence of his program consisted of gathering all active mathematicians upon one field, possibly new, to found a special journal devoted particularly to the chosen field, and to publish exclusively in foreign languages (used in congresses). The proposal seemed to be hazardous: if the choice were improper, the whole generation of young mathematicians would be lost; at that time there was no single journal restricted to only one specific field of mathematics, and neglecting Polish language seemed abusive. However, Janiszewski was supported by Mazurkiewicz and soon thereafter by Sierpiński (who joined the two in 1918 after his return from Russia), and all three decided upon the choice of "set theory and its applications" as their research field, continued their hitherto research in it, assembled talented students around themselves, founded *Fundamenta Mathematicae* (the first volume appeared in 1920) and were successful.[29]

---

[27]K. Żorawski is the only Polish mathematician mentioned in the book Klein (1979). See Volume 2, p. 202.

[28]Janiszewski's article Janiszewski (1918) was translated into English in the book Kuzawa (1968).

[29]There are some monographs and many papers on the Polish school of mathematics, cf. Kuzawa (1968) and Duda (2012).

10. Concluding Remarks

The century 1815–1915 was a period of rigid borders dividing the former Polish-Lithuanian state and of continuous efforts in all parts of the partitioned country towards recovery. In several consecutive generations, in Poland and Lithuania alike, those efforts took on a character of military conspiracy but also a steady work on raising the general level of public life. The former led to a few armed steps out, including the two great national risings, November 1831 and January 1863, comprising both parts within the Russian Empire, while the latter, although less visible, was eventually even more important. A part of it was a steady tendency towards increasing the range and raising the level of education.

Important factors for the cultural status of mathematics seem to be: educational system from elementary schools up to universities, embracing wide parts of youth, editorial activity including journals and textbooks, well furnished libraries, vivid centers of research, professional scientific organisations. Great educational reforms in the 19th century led to a rapid growth of the number of schools and their range, there have been appearing new journals and many new textbooks, schools were often supplied with libraries, revived universities supported research, and here and there appeared professional societies.

Political conditions, however, ruling in all the four parts of divided Poland hampered that general growth. Of great importance was the problem of language, in the Prussian part and two Russian ones the official language of instruction in all types of schools (exempting a few elementary ones) was either German or Russian, respectively, while in the Austrian Galicia the autonomy has allowed repolonisation of the universities in Kraków and Lwów and a gradual repolonisation of gymnasiums. Thus in the last decades of the period 1815–1915 by far the best conditions for the growth of Polish mathematical culture have reigned in Galicia but it took place also in other parts albeit at a much slower speed. In particular, textbooks in Polish witnessed the fact that the Polish terminology in elementary mathematics was evolving well and that in higher mathematics it had achieved maturity.

In the 19th century there appeared many people well educated in mathematics and living on that knowledge (university professors, gymnasium teachers, etc.). They began to meet in conferences (e.g., congresses of Polish physicians and naturalists) and tended to found professional organisations. There appeared Polish scientific societies comprising mathematicians (Kraków and Warsaw) but not the general mathematical society (founded only in 1919).

To meet growing mathematical needs, there appeared Polish journals, first edited by general scientific societies (Warsaw and Kraków), which published papers from many areas of science, including mathematics, but at the end of the 19th century there appeared also two independent journals denoted specifically to mathematics (both in Warsaw). Mathematical books were published not only in Warsaw, Kraków, and Lwów, but also in other cities (Poznań and Vilnius) and even abroad (Paris).

If mathematics has to enjoy a wide regard, it must be supported by research. Research can be done by people enjoying a privilege of social safety (stable and reasonably well-paid positions) and not living in isolation; in practice they were mostly university professors (Lwów and Kraków), but there were also gymnasium teachers (e.g., Tarnów and Płock) and white-collar workers (Warsaw). Results of research were published mainly in Polish journals and in Polish, but some also abroad (Paris and Vienna).

One can thus say that in spite of the heavy conditions prevailing in Poland in the period 1815–1915, Polish mathematics has recovered from the decline of the previous century, achieved some maturity, here and there even attaining an international level, and formed an editorial and organisational net. In short, generations up to 1915 have succeeded in building a solid basis to be used by great talents of the next generation.

References

D. Beauvois, *Wilno — Polska Stolica Kulturalna Zaboru Rosyjskiego 1803–1832* [Vilnius — A Cultural Capital of Poland under Russian Rule 1803–1832], Wydawnictwo Uniwersytetu Wrocławskiego, Wrocław, 2010, 824 pages. ISBN 978-83-229-3085-4.

M. Bečvářová and Ch. Binder (eds.), *Mathematics in the Austrian-Hungarian Empire*, Proceedings of a Symposium held in Budapest on August 1, 2009 during the XXIII ICHST, Prague, 2011, 176 pages. ISBN 978-80-7378-190-3.

J. Bieliński, *Królewski Uniwersytet Warszawski 1816–1831* [The Royal Warsaw University 1816–1831], 3 volumes, Warszawa, 1907–1912, 767 + 755 + 877 pages.

N. Davies, *God's Playground. A History of Poland*, Columbia University Press, New York, 1982, 2 volumes, 812 + 905 pages (Several reprints). ISBN 83-7006-052-8 (Polish translation).

N. Davies, *Heart of Europe. Short History of Poland*, Oxford University Press, Oxford, 1984, 511 pages (Several reprints). ISBN 0-19-285152-7.

S. Domoradzki, *The Growth of Mathematical Culture in the Lvov Area in the Autonomy Period (1870–1920)*, History of Mathematics, Volume 47, Matfyzpress, Prague, 2012, 331 pages. ISBN 978-80-7378-178-1.

R. Duda, Stulecie "Wiadomości Matematycznych" [One Hundred Years of "Wiadomości Matematyczne"], *Wiadomości Matematyczne* **33**(1997), pp. 111–135. ISSN 2080-5519.

R. Duda, *Matematycy XIX i XX Wieku Związani z Polską* [Mathematicians of the XIX and XX Century Related to Poland], Wrocław, 2007, 587 pages. ISBN 978-83-229-3316-9.

R. Duda, Matematyka Polska w Międzywojennym Dwudziestoleciu [Polish Mathematics in the Interwar Twenty Years], *Nauka Polska* **21**(46) (2012), pp. 121–155. ISSN 1230–5480.

R. Duda, Matematycy Polscy i Polskiego Pochodzenia w Europie Zachodniej i Obu Amerykach — Ludzie i Losy [Polish Mathematicians and of Polish Origin in Western Europe and both Americas — People and Their Fates], *Analecta* **21.1–2**(40–41) (2012), pp. 227–251. ISSN 1230-1159.

R. Duda, Matematycy Polscy i Polskiego Pochodzenia w Rosji Carskiej i Związku Sowieckim — Ludie i Losy [Polish Mathematicians and of Polish Origin in Tsarist Russia and Soviet Union — People and Their Fates], *Nauka Polska* **22**(47) (2013), pp. 113–147. ISSN 1230–5480.

R. Duda, Some Remarks Concerning Reception of Mathematics in Central-Eastern Europe in the Years 1850–1920, *Technical Transactions — Fundamental Sciences* 1-NP **7**(2014), pp. 99–106. ISSN 0011-4561.

R. Duda, The Emergence of National Mathematical Research Communities, *Technical Transactions — Fundamental Sciences* 2-NP **20**(2015), pp. 153–163. ISSN 0011-4561.

J. Dydyński and A. Ławrynowicz, Zjazdy Lekarzy i Przyrodników Polskich od 1869 roku [Congresses of Polish Physicians and Naturalists Since 1869], *Medycyna* **7.17**(1933), pp. 517–538.

Z. Grot, Poznańskie Towarzystwo Przyjaciół Nauk 1857–1957 [Poznań Society of Friends of Sciences 1857–1957], *Kwartalnik Historii Nauki i Techniki* **2**(1957), no. 4, pp. 651–661.

A. Heinekamp, Kochański als Leibniz-Korrespondent, *Organon* **14**(1978), pp. 72–106. ISSN 0078-6500.

J. Hulewicz, *Akademia Umiejętności w Krakowie 1873–1918. Zarys dziejów* [Academy of Arts and Sciences in Kraków 1873–1918. An Outline of Its History], Wrocław, 1958, 218 pages.

H. Ilgiewicz, *Societas Academiae Vilnensis. Towarzystwo Przyjaciół Nauk i jego poprzednicy (1907–1989)* [Societas Academiae Vilnensis. The Society of Friends of Sciences and Its Forerunners (1907–1989)], Warszawa, 2008, 668 pages. ISBN 978-83-924311-4-5.

Z. Janiszewski, Stan i Potrzeby Matematyki w Polsce [The State and Needs of Mathematics in Poland], *Nauka Polska* **1**(1918), pp. 11–18 (Translated into English in the book Kuzawa (1968)).

A. Jobert, *La Commission d'Education Nationale en Pologne 1773–1794*, Dijon, 1941, 491 pages (Translated into Polish).

S. Kieniewicz (ed.), *Dzieje Uniwersytetu Warszawskiego, 1807–1915* [A History of the Warsaw University, 1807–1915], Volume I, Warszawa, 1981, 606 pages. ISBN 978-83-01018321.

F. Klein, *Vorlesungen über die Entwicklung der Mathematik in 19. Jahrhundert*, Berlin, 1926–1927, 2 volumes, 385 + 208 pages (Reprint 1979). ISBN 3-540-09235-8, ISBN 0-387-09235-8.

J. Koroński, Mathematical Publications in the Annals of the Cracow Scientific Society (1817–1872), *Technical Transactions — Fundamental Sciences* 1-NP **7**(2014a), pp. 119–124. ISSN 0011-4561.

J. Koroński, Mathematical Publications in the Memoirs of the Academy of Arts and Sciences, *Technical Transactions — Fundamental Sciences* 1-NP **7**(2014b), pp. 125–129. ISSN 0011-4561.

J. Koroński, A Note on Mathematical Publications of Polish Mathematicians in the Memoirs of the Society of Exact Sciences in Paris (1870–1882), *Technical Transactions — Fundamental Sciences* 1-NP **7**(2014c), pp. 131–135. ISSN 0011-4561.

J. Koroński, A Note on the Mathematical Publications in the Dissertations and Reports of Meetings of the Academy of Arts and Sciences in Cracow in the Years 1874–1951, *Technical Transactions — Fundamental Sciences* 1-NP **7**(2014d), pp. 191–200. ISSN 0011-4561.

A. Kraushar, *Towarzystwo Warszawskie Przyjaciół Nauk 1800–1832. Monografia historyczna osnuta na źródłach archiwalnych* [The Warsaw Society of Friends of Sciences 1800–1832. A Historic Monograph Based Upon Archival Sources], 9 volumes, Kraków–Warszawa, 1900–1911, 407 + 318 + 338 + 408 + 476 + 498 + 527 + 510 + 158 pages.

S.M. Kuzawa, *Modern Mathematics. The Genesis of a School in Poland*, New Haven, 1968, 143 pages.

L. Maligranda, Matematyka na Zjazdach Lekarzy i Przyrodników Polskich 1869–1937 [Mathematics at the Congresses of Polish Physicians and Naturalists 1869–1937], *Antiquitates Mathematicae* **13**(2019), pp. 145–176. ISSN 2353-8813.

J. Miąso, Tajne Nauczanie w Królestwie Polskim w Świetle Dokumentów Wadz Rosyjskich [Secret Teaching in the Polish Kingdom in the Light of Documents of Russian Authorities], *Rozprawy z Dziejów Oświaty* **33**(1990), pp. 47–85. ISSN 0080-4754.

B. Nawroczyński, *Towarzystwo Naukowe Warszawskie. Materiały do Jego Dziejów w Latach 1907–1950* [Warsaw Scientific Society. Materials to Its History in the Years 1907–1950], Warszawa, 1950, 153 pages.

K.H. Parshall, Mathematics in National Contexts (1875–1900). An International Overview, *Proceedings of International Congress of Mathematicians* (Zürich, 1994), Basel, 1995, Volume 2, pp. 1581–1591. ISBN 3-7643-5153-5.

Polskiye profesora i studenty v universitetach Rossiyi (*XIX — Naczało XX v.*) [Polish Professors and Students in the Universities of Russia (XIX — Beginning of XX c.)], Conference in Kazan 13–15 October 1993, Warszawa, 1995, 205 pages. ISBN 83-85618-16-3.

D. Rederowa, *Polski emigracyjny ośrodek naukowy we Francji w latach 1831–1872* [Polish Scientific Emigration Centre in France in the Years 1831–1872], Wrocław, 1972, 248 pages.

J. Róziewicz, *Polsko-rosyjskie powiązania naukowe 1795–1918* [Polish-Russian Scientific Ties 1795–1918], Wrocław, 1984, 348 pages. ISBN 978-83-040-17221.

Z. Skubała-Tokarska, Towarzystwo Kursów Naukowych i jego znaczenie w życiu naukowym Królestwa Polskiego [The Society of Scientific Courses and Its Meaning in the Scientific Life of the Polish Kingdom], *Studia i Materiały z Dziejów Nauki Polskiej, Series A*, **8**(1965), pp. 41–96.

R. Taton, Adam Kochański dans le monde scientifique de son époque, *Organon* **14**(1978), pp. 43–50. ISSN 0078-6500.

W. Więsław, *Matematyka polska epoki Oświecenia* [Polish Mathematics in the Enlightment Era], Warszawa, 2002, 359 pages. ISBN 978-83-88839-29-0.

W. Więsław, Publikacje Towarzystwa Nauk Ścisłych w Paryżu [Publications of the Society of Exact Sciences in Paris], in W. Więsław (ed.), *Dzieje matematyki polskiej II* [History of Polish Mathematics II], Wrocław, 2013, pp. 215–230. ISBN 978-83-910055-8-3.

A. Wojtkowski, *Historia Towarzystwa Przyjaciół Nauk w Poznaniu* [A History of the Society of Friends of Sciences in Poznań], Warszawa, 1928, 450 pages.

K. Wuczyńska, Setna rocznica zawiązania koła matematyczno-fizycznego w Warszawie [One Hundred Anniversary of Setting up the Mathematical-Physical Circle in Warsaw], *Antiquitates Mathematicae* **1**(2007) pp. 213–218. ISSN 2353-8813.

L. Zasztowt, *Kasa imienia Mianowskiego 1881–2011* [The Mianowski Bank 1881–2011] Warszawa, 2011, 467 pages. ISBN 978-83-7545-268-6.

https://doi.org/10.1142/9781786349316_0004

Chapter 4

The Position of the Jagiellonian University and Its Mathematical Community Before and After WWI

Stanisław Domoradzki and Margaret Stawiska-Friedland

1. Intellectual Atmosphere in Kraków at the Turn of the 19th and 20th Centuries

Mathematics has had a venerable tradition in the ancient Polish city of Kraków, which from 1038 to 1596 was the capital of Poland. The University of Kraków (called the *Jagiellonian University* since 1817) was founded in 1364 and got its first chair of mathematics and astronomy in 1405.[1]

After the partitions of Poland, in 1796, Kraków fell under Austrian occupation. In the years 1815–1846 (by a decision of the Vienna Congress) it functioned as a Free City of Kraków, or Kraków Republic, under the control of Prussia, Austria and Russia. In 1846, after an unsuccessful Polish uprising, it was incorporated into the Austrian Empire, where it remained until 1918. Starting in the 1850s, the Austrians converted the city into a military fortress, which stifled its economic growth. However, when the Empire lost a war with Prussia, it started making concessions to occupied nations. Granting autonomy to the province of Galicia in 1866, with a Polish as the official language, reinforced the role of Kraków as a center of Polish

[1]Nicolaus Copernicus (1473–1543) studied there. Another notable mathematician and astronomer from the University of Kraków was Jan Brożek (or Brzozek), known as Joannes Broscius (1585–1652). For more information, see Ciesielski and Pogoda (2012); Gołąb (1964); Waltoś (Date unknown).

Figure 1. Jan Brożek (1585–1652), a mathematician, astronomer, poet, writer, and musician, and rector of the Kraków Academy.[2]

culture and learning,[3] even though Lwów was the official capital. Still, the two cities were often perceived in contrast to each other, Lwów as supposedly open and advanced, and Kraków as closed and backward.

Kraków (as well as its image) started to change in the last decade of the 19th century, which brought intellectual and cultural changes in connection with modern developments of art and science. Tadeusz Boy-Żeleński (1874–1941), a physician, satirist, essayist, literary and theater critic, translator and social activist, dates the beginning of these changes precisely to the year 1893, the opening of a new theater under the direction of Tadeusz Pawlikowski (1861–1915).[4] Gabriela Zapolska[5] debuted there as a playwright. About the same time, Julian Fałat[6] renewed

[2]From the Museum of the Jagiellonian University. Photograph by Janusz Kozina.

[3]In particular, Kraków School of History, represented by Józef Szujski (1835–1883), Michał Hieronim Bobrzyński (1849–1935) and Stanisław Smolka (1854–1924), while generally critical about the Polish past, significantly shaped Polish national consciousness.

[4]Under Pawlikowski's direction the dramas of Henrik Ibsen, Maurice Maeterlinck, Tadeusz Miciński, Stanisław Przybyszewski and Stanisław Wyspiański were staged.

[5]Maria Gabriela Stefania Janowska neé Korwin-Piotrowska, by her first marriage Śnieżko-Błocka (1857–1921), pseudonym Zapolska, an actress, playwright, novelist and publicist. Her works represented naturalism while addressing important social issues, including the woman question.

[6]Julian Fałat (1853–1929), a painter, one of the most outstanding Polish watercolorists, a representative of realism and impressionist landscape. In 1895–1909 he was the rector of the Academy of Fine

the Academy of Fine Arts, Zenon Przesmycki[7] and Ludwik Szczepański,[8] edited art- and literature-oriented periodicals, *Chimera* and *Życie*, respectively. Feliks Jasieński[9] propagated Japanese art, a collection of which he brought to Kraków. Kazimierz Tetmajer[10] in poetry and Stanisław Przybyszewski[11] in prose represented the Decadent Movement with its "art for art's sake" motto. Finally, there was Kraków-born (and partially Paris-educated) Stanisław Wyspiański (1869–1907), a painter, playwright and poet, theater theorist and designer (stage sets, costumes, interiors, furniture and stained glass, among others). All these personalities were part of the *Młoda Polska* [Young Poland] movement, exhibiting modernist tendencies.[12]

The city transformed itself into the "Great Kraków" from 1907 to 1915 under the mayor Juliusz Leo,[13] by increasing nearly seven-fold its territory (through incorporating the neighboring communities), regulating the rivers and extending the water, sewage and transportation systems. The university experienced changes, too. New buildings were constructed and the number of professorial chairs tripled. A major change was admission of women to academic education — first as auditors only, and then, after a central decision made in Vienna in 1897, also as regular students.[14] The number of male and female students coming from the Polish Kingdom (under the Russian rule) grew, particularly after the school strike in 1904–1905.[15]

Arts in Kraków. During his term he appointed many great painters and sculptors as professors: Leon Wyczółkowski (1852–1936), Jacek Malczewski (1854–1929), Józef Pankiewicz (1866–1940), Teodor Axentowicz (1859–1938), Konstanty Laszczka (1865–1956), Jan Stanisławski (1860–1907) and Ferdynand Ruszczyc (1870–1936).

[7]Zenon Przesmycki (1861–1944), pseudonym Miriam, a poet and translator, representative of Parnassianism. A literary and art critic, discoverer and editor of the writings of Cyprian Kamil Norwid. He edited *Chimera* in 1901–1907.

[8]Ludwik Szczepański (1872–1954), a writer, publicist, journalist and poet.

[9]Feliks Jasieński (1861–1929), pseudonym Manggha, an art critic and collector. He donated his collection, including 6,500 Japanese exhibits, to the National Museum in Kraków.

[10]Kazimierz Przerwa-Tetmajer (1865–1940), a poet, writer and mountaineer, and promoter of Tatra Mountains folklore.

[11]Stanisław Przybyszewski (1868–1927), a novelist, dramatist, and poet of the decadent naturalistic school. His drama is associated with the Symbolist movement. He spent his formative years in Berlin and wrote both in German and in Polish.

[12]For more information see Boy-Żeleński (1932).

[13]Juliusz Franciszek Leo (1861–1918), a politician, economist and lawyer, and professor of the Jagiellonian University, working on public finance and fiscal policy. Mayor from 1904 for three terms until his death, and also a parliamentary representative.

[14]The first female students enrolled in 1894/1895 at the Jagiellonian University were Jadwiga Sikorska (1871–1963), Stanisława Dowgiałłówna (1865–1933) and Janina Kosmowska. All graduated in pharmacy: Sikorska and Kosmowska in Kraków, and Dowgiałłówna in Moscow.

[15]For more information see Suchmiel (2004).

At the end of the 19th century some groundbreaking scientific achievements took place in Kraków.[16] One of them happened somewhat earlier, in 1883 — it was the liquefaction of oxygen, nitrogen and carbon monoxide by Zygmunt Wróblewski (1846–1888), a professor of experimental physics, and Karol Olszewski (1846–1915), a professor of chemistry at the Jagiellonian University. In 1894 Olszewski also liquefied a newly discovered noble gas, argon. These achievements made Kraków one of the world centers of low-temperature physics for a period. Another achievement was the discovery and description of the adrenaline hormone in 1895 by Napoleon Cybulski (1854–1919), a professor at the Faculty of Medicine. Cybulski was also an early pioneer of electroencephalography, which he used in his study of brain activity.[17] Tadeusz Browicz (1847–1928), a professor of pathological anatomy (and the rector of the Jagiellonian University in 1894–1895), discovered the bacillus that causes typhoid fever (in 1874) as well as macrophage cells present in the liver sinusoid (in 1899, later than but independently of Karl Wilhelm Ritter von Kupffer (1829–1902), who is commonly credited with the discovery) and described them precisely.[18] Leon Marchlewski (1869–1946), a professor of chemistry, described the molecular structure of chlorophyll (in his habilitation, published 1895) and pioneered the methods of spectrophotometry in chemical structural analysis.[19] Władysław Natanson (1864–1937), a professor of theoretical physics since 1894, came up with statistics of indistinguishable particles (now known as Bose–Einstein statistics) in his study of radiation in 1911.[20] In 1913, Marian Smoluchowski (1872–1917) was appointed to a chair of experimental physics. This outstanding physicist made also some lasting contributions to mathematics, in particular the theory of Brownian motion [concurrently with Albert Einstein (1879–1955)] and a model of statistical mechanics based on it, as well as the Smoluchowski equation, a limit case of the Fokker–Planck equation.[21]

It was also at the turn of the 19th/20th centuries that Kraków mathematics started a systematic rise to a world-class level, thanks to three newly appointed professors. They were legitimate and worthy members of Kraków scientific community with achievements recognized abroad. Paulin Kazimierz Stefan Żorawski (1866–1953) was born in the Polish Kingdom. In 1884–1888 he studied at the Imperial University in Warsaw and in 1891 got his doctorate in Leipzig under Sophus Lie (1842–1899). His PhD thesis *Über Biegungsinvarianten. Eine Anwendung der Lieschen Gruppentheorie* was published in *Acta Mathematica*. It was cited by Sophus

[16]For general information see Waltoś (2008).

[17]For more information see Śródka (1990); Wróblewski (1920).

[18]For more information see Śródka (2006).

[19]For more information see Sarnecka-Keller (1974).

[20]For more information see Średniawa (2007).

[21]For more information see Fuliński (1998).

Lie and Felix Klein (1849–1925). Żorawski took a chair of mathematics at the Jagiellonian University in 1893 and became an ordinary professor in 1898. Since 1900, he was a member of the (Polish) Academy of Arts and Sciences in Kraków; since 1910, of the Royal Bohemian Society of Sciences in Prague. His research (altogether about 76 works) concerned mainly differential geometry and mechanics; early in his career he also wrote some pioneering works in iteration theory.[22]

Figure 2. Kazimierz Żorawski (1866–1953).[23]

Stanisław Zaremba (1863–1942) was born in the Russian Empire. He got a diploma in engineering at St. Petersburg in 1886 and a doctorate in mathematics in Paris in 1889 (it was the scientific degree given to French nationals, not the one for foreigners). His PhD thesis [praised by Émile Picard (1856–1941) and Gaston Darboux (1842–1917)] solved a problem concerning isothermal surfaces in a uniform solid, posed in 1858 by the Academy of Sciences in Paris. His solution implied previous partial results obtained by Bernhard Riemann (1826–1866) and Wilhelm Weber

[22]For more information see Ślebodziński (1956, 1969); Stawiska (2016); and also Suszczyński (date unknown); Wójcik (2015).

[23]Photograph from the Archives of the Jagiellonian University, Album: Rectors of the Jagiellonian University 1861/1862–1921/1922.

(1804–1891). Zaremba was appointed a professor of mathematics at the Jagiellonian University in 1900. He was a member of the Academy of Arts and Sciences since 1903. His research was mainly in differential equations, potential theory and mathematical physics; he was also interested in foundations of mathematics. He gave the first example of a domain for which the Dirichlet boundary problem has no solution, made contributions to the direct calculus of variations and initiated the theory of reproducing kernels.[24]

Figure 3. Stanisław Zaremba (1863–1942).[25]

The third professor (although initially in title only) was Jan Sleszyński (1854–1931). Also known as Ivan Vladislavovich Sleshinsky, he was born in the Russian Empire. He studied in Odessa and Berlin, getting a doctorate in 1882 under Karl Weierstrass (1815–1897). He was a professor of the Odessa University. After his retirement he lectured in Kraków (at the invitation of Zaremba) since 1911 and was formally appointed a professor in 1919. His research concerned number theory, logic and probability. Sleszyński independently obtained a criterion of convergence of continued fractions, better known as Pringsheim's criterion. He also gave the first rigorous proof of a version of the Central Limit Theorem.[26] In 1903–1904 some

[24]See Ciesielska (2016); Ciesielska and Ciesielski (2016); Domoradzki and Pelczar (2009); Duda (2009); Koroński (2015); Szafraniec (2016, 2017). In this and many other cases we consulted Duda (2007) for biographical and professional data. Stanislaw Zaremba's son, Stanislaw Krystyn Zaremba is mentioned below; we will use the initial S. to refer to the father and S.K. to refer to the son.
[25]Photograph from the Archives of the Jagiellonian University (Signature SII 619).
[26]See Jadacki (1998); Obojska (2015); Sierpiński (1959).

lectures were given by Cezary Russjan (1867–1935; doctorate from the Odessa University in 1900), who ultimately took the chair of mathematical analysis at the University of Kharkiv in 1907.[27]

For a long time there were no posts for assistants in the mathematical chairs. Professors lectured on course topics as well as conducted recitation classes. In the second decade of the 20th century Żorawski, Zaremba and Sleszyński were aided by private docents Antoni Maria Emilian Hoborski (1879–1940) and Alfred Rosenblatt (1880–1947).

Hoborski studied in Kraków (1897–1901), Paris (licentiate in 1909) and Göttingen. He taught high school in Kraków, Tarnów, Jasło, Brzeżany and Nowy Sącz. He got a doctorate at the Jagiellonian University in 1908 (under Zaremba) and habilitation in 1911. His research interests were differential geometry and differential equations.

Figure 4. Antoni Hoborski (1879–1940).[28]

Rosenblatt (the son of Józef Rosenblatt, a law professor without chair) studied in Vienna, Kraków and Göttingen (mathematics, its applications and physics). He got a doctorate at the Jagiellonian University in 1908 and *venia legendi* in 1913.

[27]See Domoradzki and Stawiska (2018, 2019) and the references cited therein.
[28]Photograph from the Archives of the Jagiellonian University (Signature FII 5583).

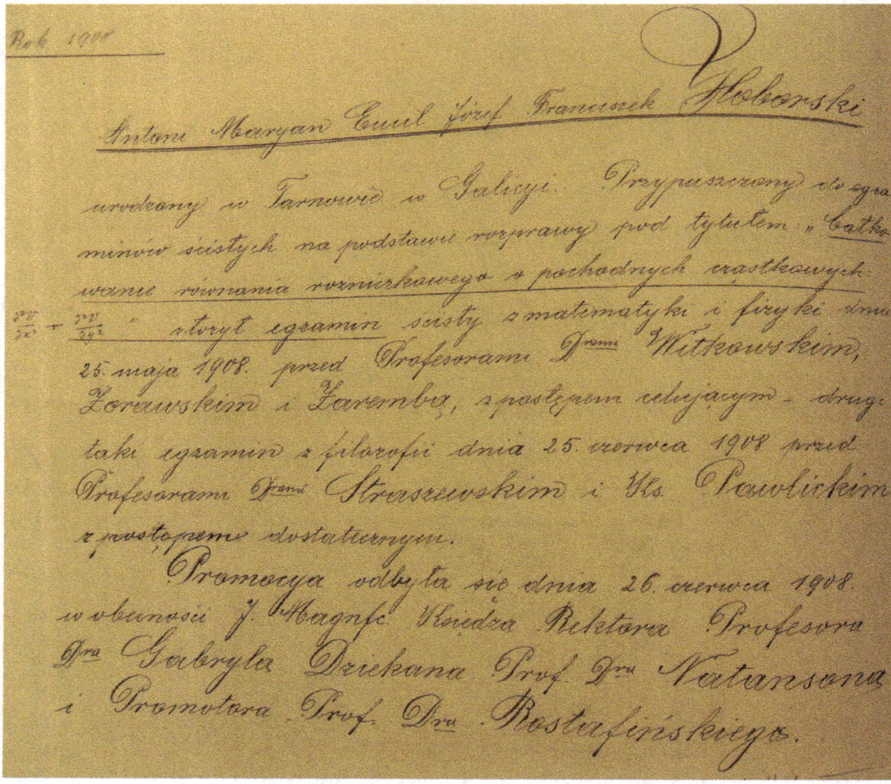

Rok 1908

Antoni Maryan Emil Józef Franciszek Hoborski

urodzony w Tarnowie w Galicyi. Przypuszczony do egzaminów ścisłych na podstawie rozprawy pod tytułem „Całkowanie równania różniczkowego o pochodnych cząstkowych $\frac{\partial^2 V}{\partial x^2} + \frac{\partial^2 V}{\partial y^2}$" złożył egzamin ścisły z matematyki i fizyki dnia 25 maja 1908 przed Profesorami Drami Witkowskim, Żorawskim i Zarembą, z postępem celującym – drugi taki egzamin z filozofii dnia 25 czerwca 1908 przed Profesorami Drami Straszewskim i Ks. Pawlickim z postępem dostatecznym.

Promocya odbyła się dnia 26 czerwca 1908 w obecności J. Magnif. Księdza Rektora Profesora Dra Gabryla Dziekana Prof. Dra Natansona i Promotora Prof. Dra Rostafińskiego.

Figure 5. Hoborski's doctoral exams and degree award.[29]

His research concerned algebraic geometry, analytic functions and many other areas. He introduced the notions of set theory early in his lectures.[30]

The standard curriculum in mathematics was established before 1918 and comprised mathematical analysis (differential and integral calculus), analytic geometry, introduction to higher mathematics, differential equations, differential geometry, and theory of analytic functions. Less frequently, lectures were given in number theory, power series, synthetic geometry, algebraic equations, higher algebra and projective geometry and occasionally, also in theory of elliptic functions, theory of transformations, theory of conformal mappings, spherical trigonometry, calculus of variations, analytical mechanics, algebraic geometry, elementary geometry, kinematics of continuous media, equations of physics, probability, theory of determinants, mathematical logic, integral equations and introduction to the methodology of mathematics. Modern topics related to transformations made their first appearance in a Polish

[29]Photograph from the Archives of the Jagiellonian University (Signature WF II 508).
[30]See Ślebodziński (1969).

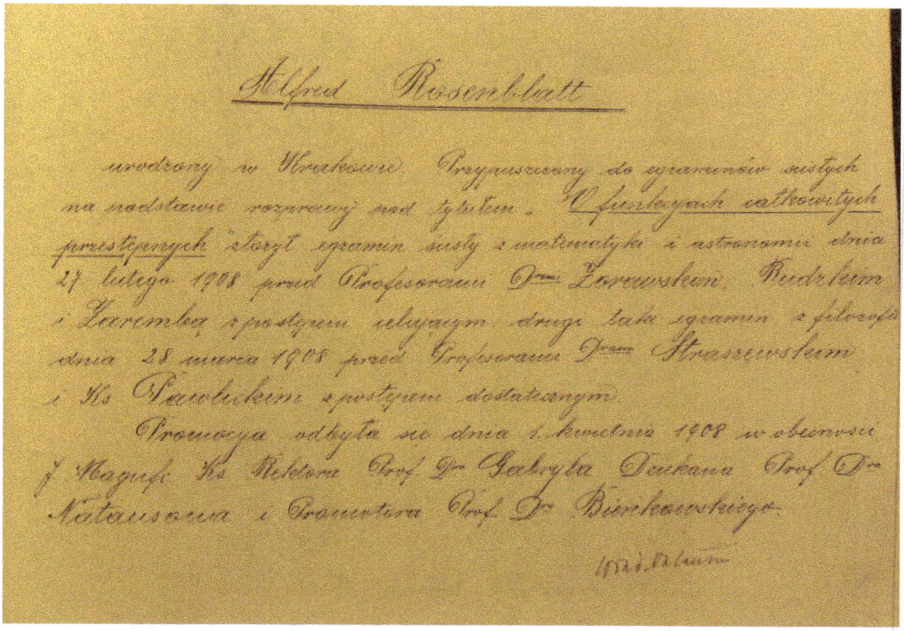
Alfred Rosenblatt

urodzony w Krakowie. Przypuszczony do egzaminów ścisłych na podstawie rozprawy pod tytułem „O funkcyach całkowitych przestępnych" złożył egzamin ścisły z matematyki i astronomii dnia 27 lutego 1908 przed Profesorami Drami Żorawskim, Rudzkim i Zarembą z postępem celującym drugi taki egzamin z filozofii dnia 28 marca 1908 przed Profesorami Drami Straszewskim i Ks. Pawlickim z postępem dostatecznym.

Promocya odbyła się dnia 1. kwietnia 1908 w obecności J. Magnif. Ks. Rektora Prof. Dra Gabryla Dziekana Prof. Dra Natansona i Promotora Prof. Dra Bieńkowskiego.

Figure 6. Rosenblatt's doctoral exams and degree award.[31]

academic curriculum, most likely thanks to Żorawski.[32] A course in mathematics for naturalists was introduced in 1912/1913, taught by Włodzimierz Stożek (1883–1941). He studied in Kraków (finished 1907) and Göttingen (finished 1909), taught in the St. Hyacinthus gymnasium (Gimnazjum Św. Jacka) and simultaneously served as an assistant to Żorawski.[33] According to Władysław Ślebodziński (1884–1972), who was then a student,[34] the reason that in some years advanced topic courses were not offered at all was professors' heavy teaching loads.

2. World War I and New Mathematicians in Kraków

The status of Kraków as a military fortress affected the university. When the Great War broke out, the military commander Karl Kuk (1853–1935) ordered closing the gates of Kraków on 17 October, 1914. The Austrians evacuated over 63,000 people. The Russian army advanced fast, coming near Kraków and surrounding it. Two battles were fought. Ultimately the Russians were stopped near Łapanów (about 42 km southeast of Kraków) on 12 December, 1914, and forced to retreat after heavy

[31]Photograph from the Archives of the Jagiellonian University (Signature WF II 508).
[32]A description of some interesting advanced courses (including a course in set theory given by Zaremba in 1910/1911) can be found in Ciesielska and Domoradzki (2014).
[33]See Gołąb (1964).
[34]See Ślebodziński (1969).

Figure 7. Włodzimierz Stożek (1883–1941).[35]

losses. The Jagiellonian University was closed for the winter semester in the academic year 1914/1915. Many of its buildings were requisitioned for military purposes. The Academy of Letters, which managed to avoid requisitions, came to help the university in making its buildings available. On 1 May, 1915, the university restarted teaching activities. The classes were held during the shortened summer semester, among them two lectures in mathematics by Sleszyński. Regular academic activities continued after 1915, even though some faculty and students were enlisted in the military. During the war some future mathematicians had pursued their studies at the Jagiellonian University, among them Stefan Kaczmarz (1895–1939), Władysław Nikliborc (1889–1948) and Tadeusz Ważewski (1896–1972).

In the early years of the war many Poles served in *Legiony Polskie* [Polish Legions], a separate formation within the Austro-Hungarian army which had originated from former Polish paramilitary forces. The Legions were reorganized in 1915. In 1917 most soldiers and officers refused to swear an oath of loyalty to the German emperor (an ally of Austro-Hungary). This situation was referred to as the Oath Crisis. As a result, the former Russian subjects were interned in the camps in Beniaminów and Szczypiorna, and the Austro-Hungarian subjects were enlisted in the Austro-Hungarian army.

The events of the war brought to Kraków some established mathematicians. Franciszek Leja (1885–1979) came from a peasant background. Supported by various

[35]Photograph from the collection of Stanisław Domoradzki.

jobs and scholarships he studied at the University of Lwów from 1904 to 1909.[36] Then he taught high school in Kraków and Bochnia and did some mathematical research. One of his publications caught the attention of Żorawski. Financed by a scholarship from the Academy of Letters, Leja continued his studies in Paris and London from 1912 to 1913. He fought in the war from 1914 to 1915. After dissolution of the Eastern Legion in 1915 he taught in Kraków and worked part time at the Jagiellonian University. He got his doctorate in 1916 under Żorawski.

Figure 8. Franciszek Leja (1885–1979).[37]

Witold Wilkosz (1891–1941) grew up in Kraków and was Banach's classmate in the Fourth Gymnasium. Even though he showed early an interest in mathematics, he decided to study Oriental languages in Beirut when his paper on semitology brought him a scholarship. In 1912 he enrolled at the Royal University in Turin to study mathematics and obtained a doctoral degree in 1914 under Giuseppe Peano (1858–1932). The degree was not formally granted because of the outbreak of the war. Wilkosz was summoned to return to the Austro-Hungarian empire and fought in Polish Legions. After the Oath Crisis he continued his studies in mathematics at the Jagiellonian University. In 1918 he obtained another PhD degree (under Zaremba) and habilitation in 1920.

There were other people trained in mathematics present in Kraków during the war. Leon Chwistek (1884–1944), a logician, philosopher, painter and art theorist, wounded in combat; Otto Nikodym (1887–1974), a graduate of the Lwów University

[36]A detailed account of Leja's studies can be found in Domoradzki (2012). See also Duda (2014).
[37]Photograph from the Archives of the Jagiellonian University (Signature FI 8458).

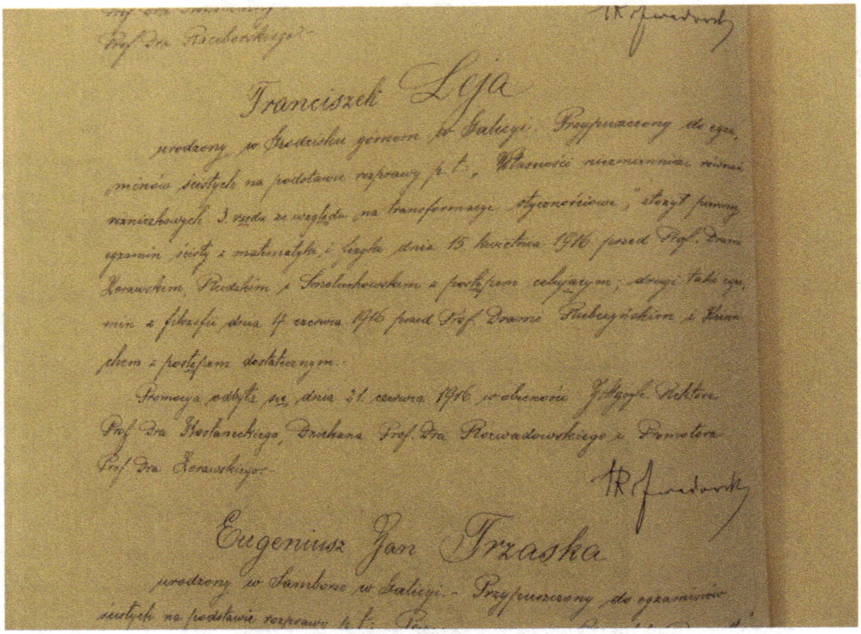

Prof. Dra Raciborskiego

Franciszek Leja

urodzony w Grodzisku górnem w Galicyi. Przypuszczony do egzaminów ścisłych na podstawie rozprawy p.t.: „Własności niezmiennicze równań różniczkowych 3. rzędu ze względu na transformacye stycznościowe,“ złożył egzamin ścisły z matematyki i fizyki dnia 15 kwietnia 1916 przed Prof. Drami Żorawskim, [illegible] i Smoluchowskim z postępem celującym; drugi taki egzamin z filozofii dnia 4 czerwca 1916 przed Prof. Drami [illegible] i [illegible] z postępem dostatecznym.

Promocya odbyła się dnia 21. czerwca 1916 w obecności J. Magnif. Rektora Prof. Dra [illegible], Dziekana Prof. Dra [illegible] i Promotora Prof. Dra Żorawskiego.

Eugeniusz Jan Trzaska

urodzony w Samborze w Galicyi. – Przypuszczony do egzaminów ścisłych na podstawie rozprawy p.t. [illegible]

Figure 9. Leja's doctoral exams and degree award.[38]

Figure 10. Witold Wilkosz (1891–1941).[39]

[38]Photograph from the Archives of the Jagiellonian University (Signature WFII 508).
[39]Photograph from the Archives of the Jagiellonian University (Signature F II 5691).

teaching in the Real Fourth Gymnasium at that time; Ślebodziński, a graduate of the Jagiellonian University who had to interrupt his further studies in Göttingen; Jan Norbert Kroo (1886–1942?), a physicist with a PhD from Göttingen; Stefan Banach (1892–1945), returning home after the closure of the Lwów Polytechnical School (where he completed half of the engineering course) and leading a team of roadworkers for living; Hugo Steinhaus (1887–1972), a graduate of Göttingen with a PhD supervised by David Hilbert (1862–1943) and an employee of the Center for the Reconstruction of the Country. Various groups met informally for discussing mathematics and finally a mathematical society was established on April 2, 1919, later giving rise to the Polish Mathematical Society.[40]

As the Austro-Hungarian empire was collapsing, Kraków became the seat of the Polish Commission for Liquidation, which held temporary authority over Galicia and Cieszyn Silesia. It was also the first Polish city to be liberated, on October 31, 1918. Lieutnant Edward Stamm (1886–1940), a mathematician, philosopher and historian of science, raised the Polish white-and-red flag on the radiotelegraph station in Dębniki, which he commanded.[41]

3. Mathematics in Kraków Between the Two Wars

3.1. *Chairs of Mathematics and Their Staffing*

There were no war casualties among Kraków mathematicians, but Smoluchowski died in an epidemic of dysentery in 1917. Still, the mathematical community of Kraków experienced quite big shifts right after the war, as many of those who spent the war years in Kraków moved to different academic centers (old and new ones) in independent Poland.

Steinhaus got habilitation at the Lwów University and moved there once classes resumed after the Polish-Ukrainian war in 1920 (spending some time in his hometown Jasło as a private scholar). Banach followed him, initially to work as an assistant to Antoni Łomnicki (1881–1941) and then to embark on the amazing scientific path as a professor of the Lwów University.[42]

Żorawski (who served as rector in the years 1917 and 1918, stepping into this function after Smoluchowski's death) accepted a professorship at the Warsaw Polytechnic, established in 1915 under German and Austrian auspices. In 1926 he became a professor of mathematics at the University of Warsaw. Ślebodziński took a position

[40]See Domoradzki (2009); Duda (2009, 2018); Iwiński (1975); Leja (1969). Regarding Poles studying in Göttingen, see Ciesielska *et al.* (2019); Ciesielska (2020).

[41]See Domoradzki and Stawiska (2018) and references cited therein and Kantor (2008).

[42]Recall that Lwów mathematics about that time experienced the death of Józef Puzyna (1856–1919) and the move of Wacław Sierpiński (1882–1969), Stefan Mazurkiewicz (1888–1945) and Zygmunt Janiszewski (1888–1920) (as professors) and Kazimierz Bartel (1882–1941) (as a politician) to Warsaw and of Zdzisław Krygowski (1872–1955) to Poznań.

at the newly established University of Poznań, where he worked until 1939. He got his PhD under Żorawski's supervision in 1928.

Hoborski became a professor of mathematics and the first rector of the Academy of Mining, which was established in Kraków in 1919.[43] Kroo for some time lectured on mechanics at the Free Polish University in Warsaw (a private institution with full academic credentials), but was unsuccessful in getting a permanent position anywhere and left academia.

Nikodym moved to the University of Warsaw, where he got his PhD in 1924[44] and habilitation in 1927. His habilitation was transferred to the Jagiellonian University, but ultimately he chose the University of Warsaw. He lectured there from 1930 until 1939.[45] Stożek, after finalizing his PhD in 1922, moved to the Lwów Polytechnic. Kaczmarz and Nikliborc also moved to Lwów after their graduation,[46] where they got their doctorates and started successful careers: Kaczmarz at the Lwów University, Nikliborc at the Lwów Polytechnic (later at the Warsaw Polytechnic). Leja left Kraków in 1924 to become an extraordinary professor at the Warsaw Polytechnic (he also had an offer from Poznań).

These personal moves were sometimes misinterpreted as caused by the dominance of old-school mathematics represented by Zaremba and lack of opportunities in Kraków for junior mathematicians wanting to follow new trends. Such a point of view was expressed by Nikolai Nikolaevich Lusin (1823–1950) in his letter of 1926 to Arnaud Denjoy (1884–1974).[47] But it did not reflect the reality,[48] since it ignored the emergence of several new academic centers with positions for mathematicians as well as the fact that new directions in mathematics (including topology and set theory) were represented in the research of the Kraków mathematicians. Because of the "exodus" of talented people Lusin predicted quick dissolution of the Kraków mathematical community, but it did not happen. Instead, during the two decades between the two world wars, one could observe remarkable continuity of mathematics research done in Kraków by a coherent group of researchers.[49]

[43]The Academy was scheduled to open in 1914/1915, but the outbreak of the war prevented its activities.

[44]Otto Nikodym's wife Stanisława Nikodym (1897–1988) neé Lilienthal was the first woman recipient of PhD in mathematics in Poland, also in 1924. She worked first as a librarian at the Mathematical Cabinet, and then as an assistant in the Chair of Mathematics at the Warsaw Polytechnic.

[45]For more information on Otto and Stanisława Nikodym see Piotrowski (2014).

[46]Before moving, Kaczmarz spent a brief time teaching at the Academy of Mining. For more information see Maligranda (2007).

[47]See Lusin (1983); Ślebodziński (1948).

[48]Lusin acknowledged in the letter that he did not know firsthand the situation of the Kraków mathematics, not having visited Kraków during his trip to Poland.

[49]At the time of Lusin's visit, his host Sierpiński was dismissive of mathematical achievements of Kraków scholars. Later, writing a summary of Polish mathematics in 1959, Sierpiński acknowledged that in Kraków *Prof. Zaremba and his students devoted themselves to classical analysis.* See Sierpiński (1959, p. 54).

After Żorawski's leaving, Zaremba took over the first chair of mathematics, while Sleszyński formally became a professor in the second chair of mathematics. In 1921 the third chair was created. After some controversy engaging many in the Polish mathematical community,[50] it was given in 1922 to Wilkosz. His mathematical research concerned measure theory, real analysis, foundations of mathematics and many other areas. In addition to mathematics and Semitic languages, he was also interested in new developments in physics (especially quantum mechanics) and radio technology and broadcasting, among other things. In his PhD thesis in 1921 Banach gave credit to Wilkosz for their mathematical discussions and announced some joint results, which however were never published.[51] Wilkosz introduced many new subjects to the university curriculum, such as foundations of mathematics, set theory, theory of quadratic forms, theory of functions of a real variable, geometric topology and group theory. He wrote textbooks in set theory, topology of the plane, algebra and theoretical arithmetic of integers (through an axiomatic approach).[52]

Sleszyński retired in 1924 and the university decided to discontinue the chair he occupied, the first chair of mathematical logic in Poland and possibly in the whole world. Chwistek, who lectured at the university while also teaching in Kraków gymnasia, ultimately took a chair of logic in Lwów in 1930.[53] He was followed there by his students: Jan Skarżeński, Jan Hercberg (1908–1941?) and Władysław Benedykt Hetper (1909–1941?).

Leja returned to the Jagiellonian University in 1935 to take over the first chair after Zaremba's retirement. In a poll among mathematicians from various Polish academic centers many candidates were named as possible successors: Aleksander Rajchman, Otto Nikodym, Alfred Rosenblatt, Stefan Kaczmarz, Juliusz Schauder, Stanisław Ruziewicz, Stanisław Saks, Władysław Nikliborc, Władysław Ślebodziński, Alfred Tarski, Arnold Walfisz, Adolf Lindenbaum, Karol Borsuk, Stanisław Mazur, Stefan Banach, Antoni Zygmund, Bronisław Knaster, Juliusz Rudnicki, Witold Pogorzelski, Włodzimierz Stożek, Wacław Sierpiński, Antoni Hoborski, Stefan Mazurkiewicz, Antoni Plamitzer, Kazimierz Bartel, Antoni Łomnicki, Antoni Przeborski, Stefan Straszewicz, Stefan Kempisty, Kazimierz Kuratowski, and Hugo Steinhaus. Zaremba himself proposed Banach, Leja, Nikliborc and Rosenblatt (in this order) and ultimately Leja won. Recall that Leja's early interests, clearly influenced by Żorawski, were differential equations and topological groups. He introduced

[50]For more information see Przeniosło (2011).

[51]In the paper Sur les opérations dans les ensembles abstraits et leur application aux équations intégrales, *Fundamenta Mathematicae* **3**(1922), pp. 133–181, based on his thesis, Banach wrote: *Mr. Wilkosz and I have some results (which we propose to publish later) on operations whose domains are sets of Duhamelian functions* ... (pp. 133–134, translated by the second author).

[52]For more information see Średniawa (2019).

[53]The staffing of this chair was also controversial, as Chwistek competed for it with Alfred Tarski. Their rivalry caused a lasting rift between mathematicians and philosophers in Lwów. For more information see Woleński (1995; 2015; 2018).

Figure 11. Leon Chwistek (1884–1944).[54]

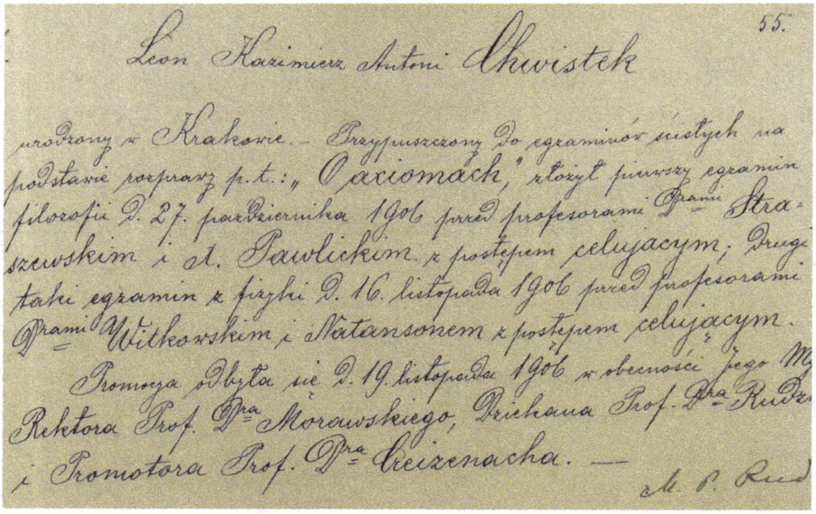

55.

Leon Kazimierz Antoni Chwistek

urodzony w Krakowie. – Przypuszczony do egzaminów ścisłych na
podstawie rozprawy p.t.: „O axiomach," złożył pierwszy egzamin
filozofii d. 27. października 1906 przed profesorami Drami Stra-
szewskim i St. Pawlickim z postępem celującym; drugi
taki egzamin z fizyki d. 16. listopada 1906 przed profesorami
Drami Witkowskim i Natansonem z postępem celującym.
Promocja odbyła się d. 19. listopada 1906 w obecności Jego M
Rektora Prof. Dra Morawskiego, Dziekana Prof. Dra Kudz
i Promotora Prof. Dra Creizenacha. —

Figure 12. Chwistek's doctoral exams and degree award.[55]

[54]Photograph from the collection of Stanisław Domoradzki.
[55]Photograph from the Archives of the Jagiellonian University (Signature WF II 508).

the abstract notion of a topological group, independently of Otto Schreier (1901–1929). Later Leja concentrated his research on complex analysis, potential theory and interpolation (more aligned with Zaremba's interests), which ultimately brought him worldwide fame (Leja's extremal function, Leja interpolation sequences, Leja's polynomial lemma).[56]

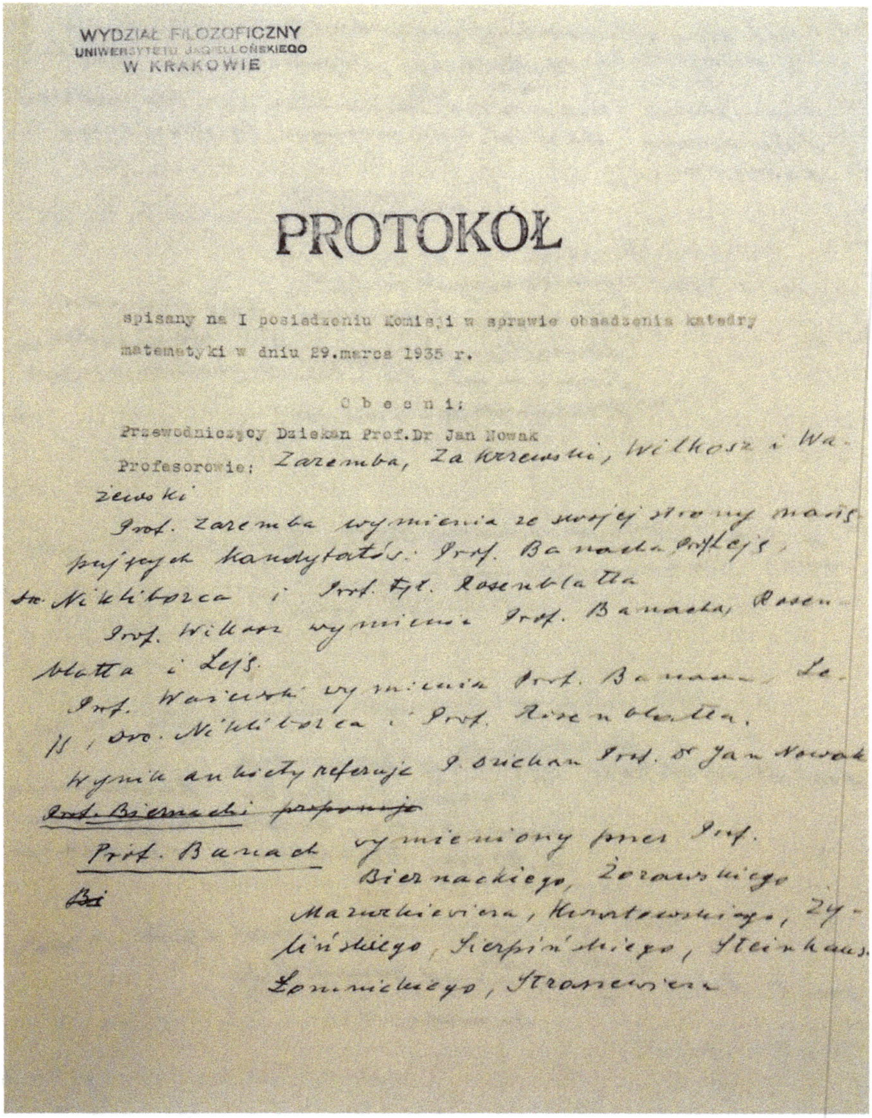
WYDZIAŁ FILOZOFICZNY
UNIWERSYTETU JAGIELLOŃSKIEGO
W KRAKOWIE

PROTOKÓŁ

spisany na I posiedzeniu Komisji w sprawie obsadzenia katedry matematyki w dniu 29. marca 1935 r.

Obecni:

Przewodniczący Dziekan Prof. Dr Jan Nowak

Profesorowie: Zaremba, Zakrzewski, Wilkosz i Ważewski

Prof. Zaremba wymienia ze swojej strony najs[illegible] kandydatów: Prof. Banacha, Prof. Leja, Dr. Nikliborca i Prof. Tyt. Rosenblatta

Prof. Wilkosz wymienia Prof. Banacha, Rosenblatta i Leja.

Prof. Ważewski wymienia Prof. Banacha, Leja, Doc. Nikliborca i Prof. Rosenblatta.

Wyniki ankiety referuje P. Dziekan Prof. Dr Jan Nowak

~~Prof. Biernacki proponuje~~

Prof. Banach wymieniony przez Prof. Biernackiego, Żorawskiego, Mazurkiewicza, Kuratowskiego, Żylińskiego, Sierpińskiego, Steinhaus, Łomnickiego, Straszewicza

Figure 13. Minutes of the committee on staffing of the first chair of mathematics.[57]

[56]See Kleiner (1985); Siciak (1982).

[57]Photograph from the Archives of the Jagiellonian University (Signature WF II 163).

Rosenblatt was unsuccessful in getting a university chair in Poland, partially because of rising anti-Semitic tendencies and partially because his rich and diverse scholarly output was underappreciated. In 1936 Rosenblatt accepted a position at the San Marcos University in Lima, Peru.[58]

We should mention another gain for Kraków mathematics due to World War I. Tadeusz Banachiewicz (1882–1954), a native of Warsaw, worked during the war at the Astronomical Observatory of the Yuryev University (Dorpat), won the competition for a professor position there in 1918 and was also named the director of the observatory. Ultimately he returned to Poland and in 1919 took a chair of astronomy at the Jagiellonian University. Even though he was primarily an astronomer, he made a lasting contribution to mathematics by inventing the (non-associative) algebra of Cracovians, which he used to simplify astronomical calculations.[59]

3.2. *The New Generation of Mathematicians and Their Research Interests*

Several mathematicians of the new generation educated between the two wars joined the faculty of the Jagiellonian University.[60] Tadeusz Ważewski started his studies already during the war. Influenced by Zaremba's lectures, he switched his interests from physics to mathematics. He subsequently went to Paris, where he studied from 1921 to 1923 and was awarded a doctorate in 1924.[61] In his thesis *Sur les courbes de Jordan ne renfermant aucune courbe simple fermeé de Jordan* he constructed a universal dendrite.[62] For a few years he worked on problems in point-set topology. One of his results on the hyperspace of a locally connected metric continuum was obtained independently and published almost simultaneously by the prominent Austrian topologist Leopold Vietoris (1891–2002).[63]

Ważewski became a deputy professor in 1926, got his habilitation in 1927, and was nominated an extraordinary professor in 1933 (proposed already in 1929, his candidacy was initially ignored by the ministry).[64] Starting in the 1930s, Ważewski devoted more and more interest and time to differential equations. He successfully

---

[58]For more information see Ciesielska and Maligranda (2014a,b).

[59]Banachiewicz is also credited with a proof of Schur's determinant formula. See Domoradzki and Stawiska (2019) for references and more information.

[60]For an account of other distinguished graduates, see Domoradzki and Stawiska (2015a,b).

[61]Four eminent French mathematicians have signed on Ważewski's doctoral diploma: Émile Borel, Arnaud Denjoy, Paul Montel, the members of the doctoral committee, and Paul Émile Appell, the rector of the University of Paris. See Domoradzki (2013).

[62]His methods were not purely "constructivist", since he used countable choice.

[63]See Domoradzki and Stawiska (2015a).

[64]See Dybiec (2000); Przeniosło (2011).

applied the newest achievements of topology to the study of properties of solutions of differential equations: his groundbreaking result was Ważewski's retract principle, published in 1947.[65]

UNIVERSITÉ DE PARIS

NOUS, Doyen & Professeurs de la Faculte des Sciences

Vu l'article 15 du décret du 21 juillet 1897;

Vu la délibération, en date du 28 mars 1898, du Conseil de l'Université de Paris & l'arrêté ministériel du 1er avril 1898 approuvant cette délibération;

Après avoir constaté que M Ważewski ... né à ... (Pologne) le ... septembre 1896, a rempli toutes les conditions exigées par les règlements, l'avons admis à présenter & soutenir une thèse sur les courbes de Jordan ne renfermant aucune courbe simple fermée de Jordan.

Et, attendu qu'il a subi cette épreuve avec succès, l'avons déclaré digne du titre de Docteur de l'Université de Paris.

Paris, le ... Avril 1924

Le Doyen, Les Membres du Jury,

Nous, Président du Conseil de l'Université de Paris, vu la déclaration ci-dessus, conferons à M Ważewski le titre de Docteur de l'Université de Paris & lui délivrons le présent diplôme.

Paris, en Sorbonne, le ... Avril 1924

Figure 14. Ważewski's doctoral diploma from Paris.[66]

Stanisław Gołąb (1902–1980) studied mathematics at the Jagiellonian University in the years from 1920 to 1924, but his talent was discovered earlier, while he was still in gymnasium, taught by Hoborski. In 1923 Gołąb started working as a deputy assistant to Hoborski, and after his graduation was promoted to an assistant in the Chair of Mathematics at the Academy of Mining. He worked in geometry, inspired by Hoborski (who wished to establish a scientific school in geometry in Kraków). Since 1930, he conducted contract lectures at the Jagiellonian University

[65]For more information see Pelczar (1991; 2000).

[66]From the Archives of the Jagiellonian University (Signature WF II 515).

Figure 15. Ważewski's university identificator from 1935.[67]

and in 1931 he obtained a PhD degree at the Jagiellonian University. The thesis *O uogólnionej geometrii rzutowej* [On Generalized Projective Geometry][68] was related to his earlier joint work with Jan Arnoldus Schouten (1883–1971). We note that his supervisor was Zaremba; the referee was Wilkosz. In 1932 Gołąb obtained habilitation at the Jagiellonian University. It was later extended to the Faculty of Mining at the Academy of Mining, where he received the title of professor in 1939.[69]

In addition to classical differential geometry, he published papers on tensor calculus (then on the rise because of its importance for the theory of relativity), spaces with linear or projective connection, Riemann, Minkowski and Finsler spaces and general metric spaces.[70] In metric geometry, he proved, e.g., that, under certain mild assumptions, approximation of a point x of a Minkowski space by points in a given closed subset C is unique if the distance function to C is differentiable at x,[71] which was one of the first results of this type. Another memorable result obtained by Gołąb says that the perimeter of the unit disk in the plane with a norm (a Minkowski

[67]Photograph from Archives of the Jagiellonian University. Courtesy of A. Pelczar (1937–2010), a former PhD student of Ważewski.

[68]Published as Gołąb (1930).

[69]See Pawlikowska-Brożek (2003).

[70]See Gancarzewicz and Pogoda (2000); Kucharzewski (1982).

[71]See Gołąb (1938).

plane) can take any value between 6 and 8 and the extremal values are taken if and only if the disk is a regular hexagon or a parallelogram.[72] He also continued the traditions of Żorawski, Leja and Hoborski by introducing and studying the notion of a pseudogroup of transformations.[73]

Figure 16. Stanisław Gołąb (1902–1980).[74]

Stanisław Krystyn Zaremba (1903–1990) was the son of Stanisław Zaremba and his Provençal wife Henryka (Henriette Leontine) neé Cauvin. S.K. Zaremba started his studies at the Jagiellonian University and continued at the Sorbonne in Paris from 1924 to 1927. Ultimately he got a master's degree in mathematics from the Jagiellonian University in 1929. Not wanting to be perceived only as his father's son,[75] he moved to Wilno (Vilnius), where since 1929 he was an assistant at the Stefan Batory University. There he got a PhD degree on the basis of the thesis *Sur l'allure des intégrales d'une équation différentielle ordinaire du premier ordre*

---

[72]See Gołąb (1932).
[73]See Gołąb (1939).
[74]Photograph from the collection of Stanisław Domoradzki.
[75]Zbigniew Łomnicki (1904–1994), a mathematician from Lwów and S.K. Zaremba's collaborator in emigration after World War II, pointed out that the pre-war work of Zaremba had a purely theoretical character. In a letter to Zygmunt Wilhelm (William) Birnbaum (1903–2000; a statistician from Lwów, Łomnicki's co-worker at an insurance company and a professor of mathematics at the University of Washington, Seattle, since 1939) dated November 12, 1951, Łomnicki wrote: *The greatest sensation for us is that* [S.K.] *Zaremba, who dealt with differential equations "im Grossen" and detested applications, is writing a paper with engineers and even a book on applying topology to investigating singular points of solutions etc.* (*illegible*) (p. 2). On the other hand, it is known that S. Zaremba Sr.'s work was often motivated by applications.

dans le voisinage de l'intégrale singulière (supervised by Juliusz Rudnicki). In 1936 Zaremba got his habilitation at the Jagiellonian University on the basis of the thesis *On paratingent equations*,[76] in which he introduced paratingent equations, a generalization of differential equations nowadays known as differential inclusions. In 1937 he came back to Kraków, first as an adjunct, and later as a docent at the Jagiellonian University.[77]

Stanisław Turski (1906–1986) studied physics and astronomy at the Jagiellonian University from 1924 to 1928 while also working as a schoolteacher. Since 1927 he worked as a mathematician, starting at the level of an assistant [78] at the Jagiellonian University and at the Academy of Mining. He obtained his PhD under the supervision of Wilkosz, presenting a thesis *On a Generalization of Theorems on Uniformity of Integrals of a Hyperbolic Equation*. He wrote several papers on partial differential equations, number theory and functions of a complex variable.

Adam Bielecki (1910–2003) finished his studies in 1931, obtaining a master's degree in philosophy in the field of mathematics. In 1935 he obtained a doctorate on the basis of the thesis *On Integral Representation of m-Dimensional Surfaces Contained in the n-Dimensional Euclidean Space by Implicit Functions* (see Bielecki (1935)). The problem — of equivalence of the implicit and parametric representations — was posed by Wilkosz, who supervised the thesis. Bielecki learned about it from Ważewski and solved it using the technique of C^{∞}-partition of unity, later reintroduced and refined by Laurent Schwartz (1915–2002). From 1935 to 1936 Bielecki worked in the Theoretical Physics Seminar at the Jagiellonian University, becoming a senior assistant in the Chair of Theoretical Physics in 1936.[79]

Overall, from 1918 to 1939, lectures in mathematics were given by: Sleszyński (until 1924), Zaremba, Hoborski (a professor at the Academy of Mining in Kraków), Rosenblatt, Wilkosz, Nikodym, Chwistek (in 1919–1931), Stożek, Ważewski, Leśniak, Gołąb, Adam Marian Różański[80] (descriptive geometry; at the Department of Agriculture), Leja (from 1936), S.K. Zaremba (from 1937), Jerzy Spława-Neyman (1927–1929, commuting from Warsaw). There were also lectures in mathematics for natural sciences and lectures in elementary mathematics from the higher standpoint, oriented towards future teachers.[81] From 1928 to 1939 about 135

[76]See Zaremba (1935; 1936).

[77]For more information see Pawlikowska-Brożek (2003) and Zaremba (Date unknown).

[78]In the independent Republic of Poland academic institutions could employ assistants. Wilkosz was credited with being active in hiring them, unlike Zaremba (see Turowicz, 1989). The first assistants in mathematics were Jan Leśniak and Irena Wilkosz.

[79]See Kaczor *et al.* (2004).

[80]Adam Marian Różański (1874–1940) — an engineer, specialist in water resource management. Since 1924 a professor in the Laboratory of Agricultural Engineering, later a dean of the department at the Jagiellonian University. Arrested by the Germans in 1939 in Sonderaktion Krakau, died in the Sachsenhausen concentration camp.

[81]See Dybiec (2000) and Przeniosło (2011).

students graduated from the Jagiellonian University with master's degree in mathematics.[82]

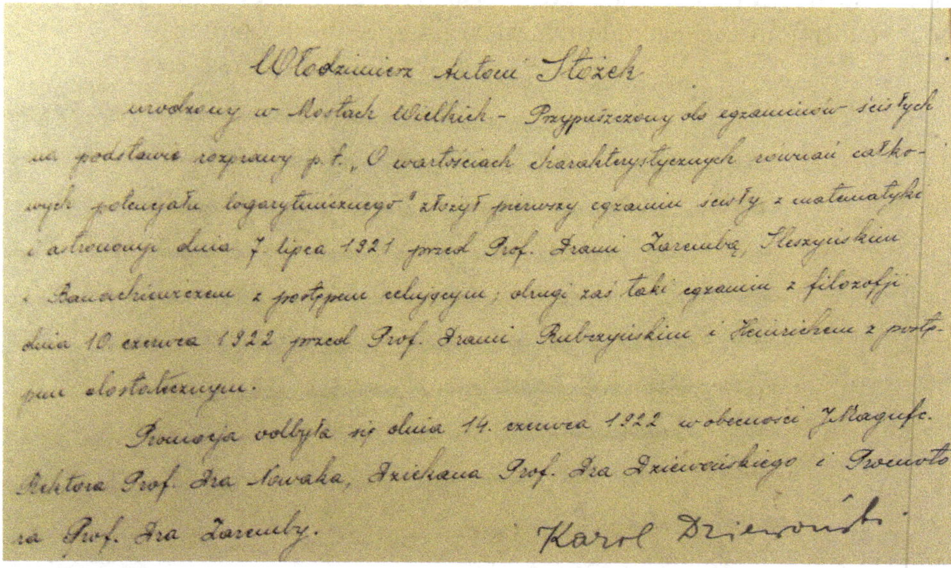
Włodzimierz Antoni Stożek

urodzony w Mostach Wielkich – Przypuszczony do egzaminów ścisłych na podstawie rozprawy p.t. „O wartościach charakterystycznych równań całkowych potencjału logarytmicznego" złożył pierwszy egzamin ścisły z matematyki i astronomji dnia 7. lipca 1921 przed Prof. Drami Zarembą, Śleszyńskim i Banachiewiczem z postępem celującym; drugi zaś taki egzamin z filozofji dnia 10. czerwca 1922 przed Prof. Drami Rubczyńskim i Heinrichem z postępem dostatecznym.

Promocja odbyła się dnia 14. czerwca 1922 w obecności J. Magnif. Rektora Prof. Dra Nowaka, Dziekana Prof. Dra Dziewulskiego i Promotora Prof. Dra Zaremby.

Karol Dziewoński

Figure 17. Stożek's doctoral exams and degree award.[83]

Some graduates of the Jagiellonian University had positions in other academic centers. They successfully either continued working in classical areas (Nikliborc at the Lwów Polytechnic on mechanics and functions of complex variables, Ślebodziński at the University of Poznań on differential geometry, Hetper at the Lwów University on logic, and S.K. Zaremba at Vilnius on differential relations) or made a transition to research topics considered more modern (Kaczmarz at the Lwów University on orthogonal series, and Andrzej Turowicz (1904–1989) at the Lwów University on functional algebras). Finally, mathematicians from other centers interested in classical disciplines such as (broadly understood) mathematical analysis and mathematical physics sought contacts with Kraków to obtain their PhDs and habilitations at the Jagiellonian University. The full list of recipients of these degrees is given in Appendix A; let us just comment briefly that one of them, Witold Pogorzelski (1895–1963), a graduate of the universities at Nancy and Paris, was a professor of mathematics at the Warsaw Polytechnic since 1922 for 40 years (not counting, the period of the Nazi German occupation) and at the Free Polish University from 1920 to 1939. Among his assistants at the Polytechnic there were Stanisław Saks

[82]Available data are incomplete. For more information see Domoradzki and Stawiska (2015b) and references cited therein.

[83]Photograph from the Archives of the Jagiellonian University (Signature WF II 508).

(1897–1942) and Antoni Zygmund (1900–1992). Pogorzelski gave lectures from 1921 to 1951 as a docent at the University of Warsaw in theoretical physics, differential equations and integral equations. His research concerned integral and differential equations, analytic functions, mathematical physics and probability.[84]

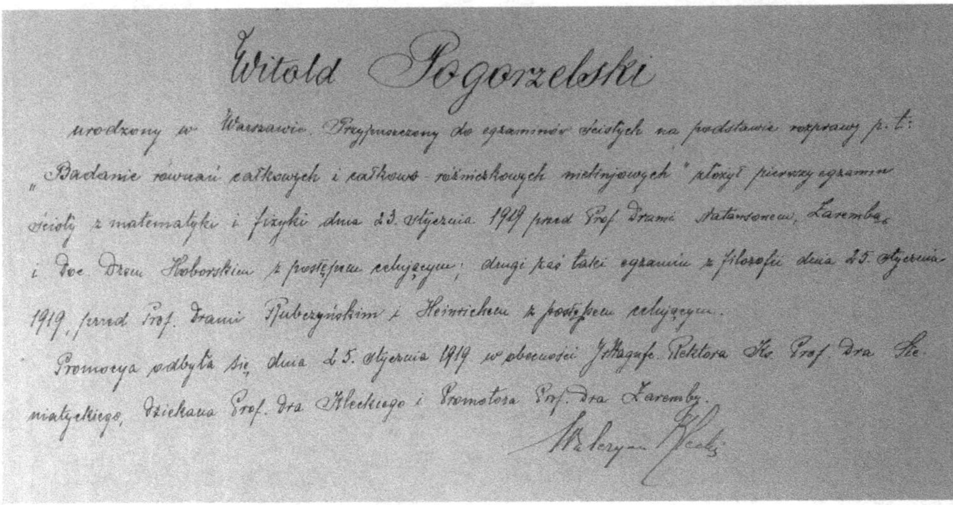
Witold Pogorzelski

urodzony w Warszawie. Przypuszczony do egzaminów ścisłych na podstawie rozprawy p.t.: „Badanie równań całkowych i całkowo-różniczkowych nielinjowych" złożył pierwszy egzamin ścisły z matematyki i fizyki dnia 23. stycznia 1919 przed Prof. Drami Natansonem, Zarembą i Doc. Drem Hoborskim z postępem celującym; drugi zaś taki egzamin z filozofii dnia 25. stycznia 1919, przed Prof. Drami Rubczyńskim i Heinrichem z postępem celującym.

Promocya odbyła się dnia 25. stycznia 1919 w obecności JMagnif. Rektora Ks. Prof. Dra Ste-niatyckiego, Dziekana Prof. dra Kleckiego i Promotora Prof. dra Zaremby.

Waleryan Kłecki

Figure 18. Minutes on Pogorzelski's doctoral exams and degree award.[85]

The scientific interaction of Kraków mathematicians with their colleagues from other places in Poland did not result in joint publications, and generally was not intense, but it happened. An important instance is the interest of Tadeusz Ważewski in the theory of retracts, which was developed in the 1930s by mathematicians from the Warsaw school, Karol Borsuk (1905–1982) and his PhD supervisor Mazurkiewicz. Ważewski used this theory to prove his famous Retract Theorem (published 1947), thus creating a breakthrough topological approach to differential equations. Also, Menachem Wojdysławski (1919–1942?), a graduate of the University of Warsaw in 1939 (who later perished in the Holocaust), generalized in his work some results of Ważewski on the properties of hyperspace of a locally connected metric continuum, using the notion of retract.

3.3. *Foreign Contacts, Relations with Physicists and Philosophers*

Kraków was not an isolated outpost of mathematics and its mathematicians traveled widely.[86] The representatives of the older generation took advantage of their previous foreign contacts and made several trips with scientific purposes: S. Zaremba went

[84]See Przeworska-Rolewicz (2008) for more information.
[85]From the Archives of the Jagiellonian University (Signature WF II 508).
[86]For more information see Przeniosło (2010).

to Paris in 1931, 1934 and 1935 (the 1934 trip was to give a series of lectures at Henri Poincaré Institute on mechanics of viscous liquids) and Wilkosz went to Italy in 1925. Younger mathematicians started with study trips: Ważewski in 1921/1922 and 1922/1923 to Paris, S.K. Zaremba from 1924 till 1927, and also to Paris. Gołąb traveled a lot: before his PhD, he went in 1928 to Delft to work with Schouten and in 1929–1930 he visited several prominent differential geometers: Schouten in Delft, Tullio Levi-Civita (1873–1941) and Enrico Bompiani (1889–1975) in Rome, Ludwig Berwald (1883–1942) and Václav Hlavatý (1894–1969) in Prague. Later, in 1937, he gave a series of lectures in Hamburg.

Kraków mathematicians participated in international conferences. For example, Chwistek and Gołąb went to Prague for the Second Congress of Mathematicians from Slavic Countries (1934); Leja, Wilkosz, Rosenblatt and a historian of science Aleksander Birkenmajer (1890–1967) went to Turnu Severin for the Congress of Romanian Mathematicians (the second one, held in 1932); Rosenblatt went to the Third International Congress for Applied Mechanics in Stockholm (1930); Hoborski and Gołąb went to a conference in differential geometry at the Moscow University (1934) organized by Veniamin Fedorovich Kagan (1869–1953) and his school. A few of them (counting Leja, who was for some time affiliated with Warsaw) gave talks at the International Congresses of Mathematicians.

Figure 19. A group of mathematicians in the courtyard of Collegium Maius: Turski, NN, Gołąb, NN, Hoborski, NN, S. Zaremba, Schouten, Wilkosz, NN, and Ważewski.[87]

[87]Photograph from the collection of Stanisław Domoradzki.

In 1928 (Bologna) there were: F. Leja: *Sur quelques propriétés frontières des séries entières doubles* (Session I-D); A. Rosenblatt: *Varietà algebriche a tre e più dimensioni*; *Sopra le varietà algebriche a tre dimensioni fra i cui caratteri intercedono certe disuguaglianze* (Session II-A); L. Chwistek: *Nouvelles recherches sur les fondements de la mathématique* (Session VI); S. Gołąb: *Sur un théorème rentrant dans le calcul fonctionnel et son application géométrique* (Session I-C).

In 1932 (Zürich) there were: S. Zaremba: *Sur la notion de la force dans la mécanique* (Session VIa); W. Wilkosz: (a) *La propriété de Darboux du Jacobien généralisé*, (b) *Sur le théorème fondamental de la théorie des déformations continues* (Session IIb); A. Rosenblatt: *Sur les ondes de gravité* (Session VIb); S. Gołąb: *Einige Bemerkungen über die Winkelmetrik in den Finslerschen Räumen* (Session IIIa).

In 1936 (Oslo) there were: S. Zaremba: *Un théorème général relatif aux équations aux dérivées partielles du second ordre linéaires et du type hyperbolique*; F. Leja: *Sur les séries des polynômes homogènes de deux variable*; T. Ważewski: *Quelques propriétés de caractère intégral de l'equation* $P(x,y)(\partial z/\partial x)+Q(x,y)(\partial z/\partial y)=0$; S. Gołąb: *Über das Anholonomitätsobjekt von Schouten und van Dantzig*. S.K. Zaremba and W. Wilkosz also participated in the Oslo Congress, without giving talks.[88]

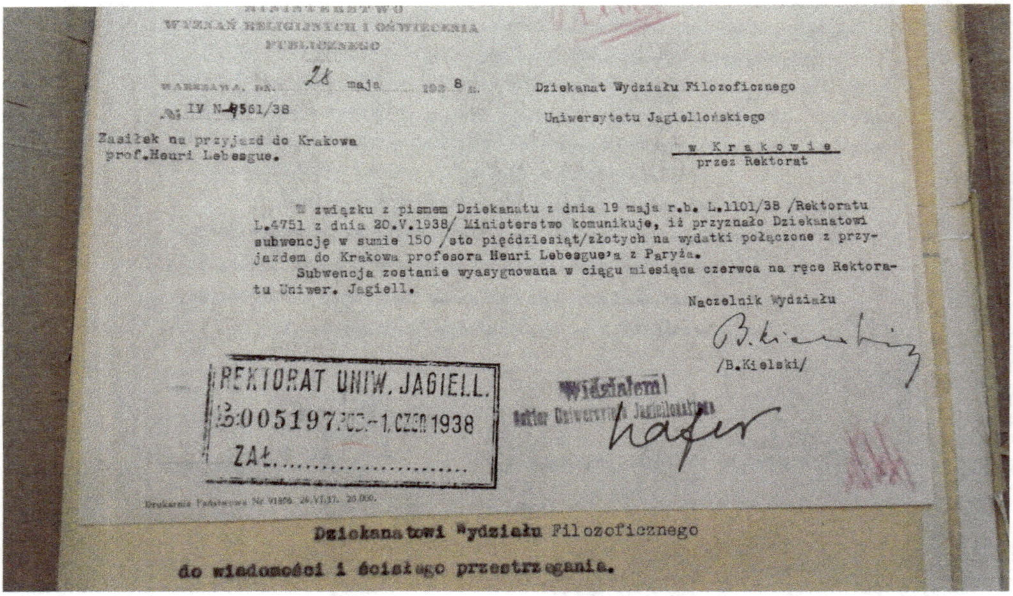

Warszawa, dn. 28 maja 1938 r.

IV N-4561/38

Zasiłek na przyjazd do Krakowa prof. Henri Lebesgue.

Dziekanat Wydziału Filozoficznego
Uniwersytetu Jagiellońskiego
w Krakowie
przez Rektorat

W związku z pismem Dziekanatu z dnia 19 maja r.b. L.1101/38 /Rektoratu L.4751 z dnia 20.V.1938/ Ministerstwo komunikuje, iż przyznało Dziekanatowi subwencję w sumie 150 /sto pięćdziesiąt/ złotych na wydatki połączone z przyjazdem do Krakowa profesora Henri Lebesgue'a z Paryża.

Subwencja zostanie wyasygnowana w ciągu miesiąca czerwca na ręce Rektoratu Uniwer. Jagiell.

Naczelnik Wydziału
/B.Kielski/

REKTORAT UNIW. JAGIELL.
1938
ZAŁ. ....

Dziekanatowi Wydziału Filozoficznego
do wiadomości i ścisłego przestrzegania.

Figure 20. Subsidy for Lebesgue's visit to Kraków in 1938.[89]

---

[88]See ICM Proceedings (1893–2014) and Ciesielski (2018).

[89]Photograph from the Archives of the Jagiellonian University (Signature WF II 528).

Foreign mathematicians also visited Kraków, sometimes for a longer period. Recall that Sleszyński's chair, financed from Władysław Kretkowski's fund, became vacant after Sleszyński's retirement in 1924. In 1925/1926, the same fund financed guest lectures by three French mathematicians, one trimester each: Georges Bouligand (1889–1979) from Poitiers, on potential theory;[90] Joseph Kampé de Fériet (1893–1982) from Lille, on hypergeometric functions; Maurice Janet (1888–1983) from Caen, on systems of partial differential equations.[91] Visitors who came for a shorter time were giving talks at the meetings of the Kraków branch of the Polish Mathematical Society, e.g., Schouten, Wilhelm Blaschke (1885–1962) and Paul Montel (1876–1975) (all in 1935). Henri Lebesgue (1875–1941) also visited the Jagiellonian University, in 1938.[92]

With the view of facilitating international scientific communication, some modern mathematicians took interest in developing an international language, among them Giuseppe Peano, creator of *Latino sine Flexione* (also called Interlingua).[93] This language had a strong advocate in the person of Edward Stamm, a mathematician, philosopher and historian of mathematics, from 1918 until 1921 stationed in Kraków as a military commander of the radiotelegraphic station in Dębniki and an early member of the Polish Mathematical Society. Stamm corresponded with Peano and published a paper in logic in the first issue of Annales de la Société Polonaise de Mathématique (1922) written in Latino sine Flexione.

Relations with physics were important for mathematics throughout its history. In the modern period (roughly 1890–1940) they took on new aspects because of radical developments in physics, such as quantum mechanics, special and general relativity theory and statistical physics, which required new mathematical approaches.

In Kraków there was a substantial interest in physics among mathematicians. Among other things, S. Zaremba dealt with hydrodynamics of viscous liquids and theoretical mechanics, collaborated with mineralogist Stefan Kreutz (1883–1941) on topics in crystallography and polemized with general relativity theory (which he eventually accepted).[94] Żorawski worked on the theory of motion of a continuous medium and on hydrodynamics of perfect liquids. Rosenblatt worked on many problems inspired by physics and wrote a valuable monograph on hydrodynamics of viscous liquids. Wilkosz lectured on elasticity theory, mechanics of atoms and applications of analysis to theoretical physics. The work of his PhD

[90]The contents of these lectures were published as Bouligand (1926).

[91]Even though no state funds were involved, Warsaw mathematicians did not like their Kraków colleagues issuing invitations on their own; they tried to prevent such actions in the future as they would rather have visitors closer to the interests of the Warsaw school. See Lusin (1983).

[92]For more information see Dybiec (2000); Gołąb (1964).

[93]See Gray (2008).

[94]See Pogoda (2016); Polak (2016).

student Duwid Wajnsztejn (1916–1940) related to quaternions, Clifford algebras and hypercomplex numbers was done with applications to quantum mechanics in mind.[95] Nikodym lectured on Hilbert spaces in 1929/1930 and on spectral theory of infinite matrices in 1930/1931. Gołąb's work on tensors was related to general relativity and in 1933/1934 he gave a course on the introduction to projective relativity theory. Ważewski taught a course on theoretical mechanics, for which the recitation classes were taught by his assistant Roman Leitner (1914–2008). Bielecki conducted seminars in hydrodynamics, elasticity theory, principles of quantum mechanics and electromagnetism; he collaborated with physicists Jan Weyssenhoff (1884–1972)[96] and Myron Mathisson (1897–1940).[97] Pogorzelski's habilitation in Kraków was in mathematical physics, on radiation theory and energy quanta.[98]

Graduates of mathematics, among them Karol Kozieł (1910–1996) and Lidia Stankiewicz-Piegza (1910–1948), were employed as assistants in the Astronomical Observatory by Banachiewicz. They further developed computational methods for astronomy, especially those dealing with the fast solving of systems of linear equations, inverting power series, spherical polygonometry and the least-squares method. There was a chance for the development of statistical physics in Kraków, but the premature death of Smoluchowski prevented it.[99]

Since the late 19th century mathematicians and philosophers [David Hilbert, Bertrand Russell (1872–1970), Rudolf Carnap (1891–1970), etc.] turned to examination of the axiomatic method and its applicability to whole mathematics or other sciences, especially physics. The character of axioms themselves was debated [Henri Poincaré (1854–1912) treated them as definitions].[100]

Kraków mathematicians also had a vivid interest in foundations and axiomatic systems. Sleszyński represented the tendency to give mathematical reasonings in full. He understood the need for logic in mathematics as a theory of proofs and he believed that it also offered tools to shorten proofs. His lectures on proof theory were written up by S.K. Zaremba and published as an imprint of the Mathematical and Physical Circle of the Students of the Jagiellonian University (Kółko Matematyczno-Fizyczne Uczniów Uniwersytetu Jagiellońskiego).

---

[95] Results of Wajnsztejn's PhD thesis were published in Wajnsztejn (1938).

[96] Weyssenhoff wrote a few mathematical articles, on multi-vector densities and calculus of variations.

[97] Mathisson got his PhD in physics in 1930 from the University of Warsaw with Czesław Białobrzeski (1878–1953). He corresponded with Albert Einstein on general relativity and solved a problem in hyperbolic differential equations posed by Hadamard. He spent the year 1938 in Kraków. See Sauer and Trautman (2008).

[98] For more information see Średniawa (1986).

[99] For more information see Mietelski (2011); Pogoda (2007); Ślebodziński (1969); Średniawa (2007).

[100] See Gray (2008).

Chwistek studied in detail *Principia Mathematica* of Bertrand Russell and Alfred North Whitehead (1861–1947)[101] and developed his pure type theory before moving from Kraków to Lwów.[102]

Wilkosz, as a former student of Giuseppe Peano in Turin, saw logic as a tool for developing deductive systems, associated with the axiomatic method. He wrote on set-theoretic possibility of defining abstracts.

S. Zaremba, influenced by the French views, claimed that logic was auxiliary to mathematics. He defended this claim in published polemics with other mathematicians and logicians [Jan Leopold Łukasiewicz (1878–1956), Kazimierz Kuratowski (1896–1980) and Tadeusz Hipolit Czeżowski (1889–1981)]. At the same time, he admitted that considerations of axiomatics of geometry helped to create the notion of space and the notion of logical independence of statements. He regarded set theory as a precise account of infinity. He also made an attempt at axiomatization of mechanics. He even supervised a PhD degree in philosophy of Henryk Greniewski (1903–1972), who wrote a thesis entitled *A Contribution to Logical Foundations of Mechanics* (actually under the direction of Chwistek, who in 1926 did not then have a habilitation and could not yet supervise PhD candidates).[103]

Several other mathematicians from Kraków were interested in logical problems (Hoborski, Gołąb, S.K. Zaremba) or in axiomatic systems of geometry (Bielecki, Turowicz and Anna Zofia Krygowska (1904–1988)).[104] Another doctorate related to mathematical logic was awarded in 1930 to Jan Hilary Kon (thesis *O warunkach stosowalności definiowania przez indukcję matematyczną* [On Conditions of Applicability of Defining by Mathematical Induction]), supervised by the physicist Konstanty Zakrzewski (1876–1948). It is also worth mentioning that S. Zaremba and Natanson were awarded honorary doctorates in philosophy by the Jagiellonian University, in 1929 and 1930, respectively.

Modern philosophy of mathematics also paid attention to cognitive foundations of mathematics.[105] This direction of research was present in the works of students of Władysław Heinrich (1869–1957), a professor of psychology and methodology of natural sciences, the holder of the Chair of Philosophy at the Jagiellonian University from 1911 until 1939. In the interwar period he supervised two PhD theses

[101] Chwistek maintained correspondence with Russell and Russell's recommendation helped him win the chair of logic in Lwów. Another scientific correspondent of Russell in Kraków was Michał Dziewicki (1851–1928), a philosopher, translator, editor and a lector of English at the Jagiellonian University. See Bremer (2016).

[102] See Chwistek (2012).

[103] Greniewski also prepared another thesis, *An Attempt of Deductive Theory of Causality* under the direction of Wilkosz. For more information see Trzęsicki (2016).

[104] For more information see Domoradzki and Stawiska (2015a,b), Murawski (2004, 2014), Pawlikowska-Brożek (2003); Urbaniak and Godziszewski (2018).

[105] See Gray (2008).

dealing with mathematical knowledge: of Stanisław Bilski (1893–1934; thesis *Poznanie aprioryczne w epistemologii B. Russella* [A Priori Knowledge in B. Russell's Epistemology]), in 1920, and of Stanisław Skrzeszewski (1901–1978; thesis *Uwagi o podstawach logiki* [Remarks on Foundations of Logic]), in 1931. After World War II, Danuta Gierulanka (1909–1995), a graduate of mathematics with a master's degree from 1932, did experimental research on acquiring mathematical knowledge in Heinrich's laboratory under his direction. For this work she was awarded a PhD in psychology in 1947.[106]

3.4. *Kraków and the Polish Mathematical Society*

In the 19th century, well into the beginning of the 20th century, dissemination of mathematics happened at the meetings of numerous scientific societies, Congresses of Naturalists and Physicians as well as in publications. Kraków mathematicians published in the proceedings of the Academy of Arts and Sciences in Kraków, scientific Polish-language journals in Warsaw (*Prace Matematyczno-Fizyczne* and *Wiadomości Matematyczne*) and in international mathematical journals. The scientific activities of Polish scholars had not only professional, but also cultural and patriotic aspects, manifesting the existence of the Polish scientific community transcending the borders of the three occupying powers.[107] With the increasing professionalization of mathematics[108] it became necessary to create a specialized organization for mathematicians.

The first such organization in Polish territories was *Towarzystwo Matematyczne we Lwowie* [Mathematical Society in Lwów],[109] established in 1917 under the leadership of Józef Puzyna. On 2 April, 1919, the Mathematical Society in Kraków was founded, by (signed) Stefan Banach, Dr. Leon Chwistek, Dr. Antoni Hoborski, Dr. UJ, Dr. Franciszek Leja, Dr. Alfred Rosenblatt, Dr. UJ, Dr. Jan Sleszyński, Prof. UJ, Dr. Stanisław Zaremba, Prof. UJ, Dr. Kazimierz Żorawski, Prof. UJ, and gymnasium teachers: Julian Chmiel, Mirosław Gibas, Dr. Ludwik Hordyński, Ludwik Kaszycki, Otton Nikodym, Antoni Rozmus, Xawery Stankiewicz, and Dr. Antoni Wilk.[110] The purpose of the Society was the universal cultivation

[106]For more information see Woleński (1995; 2015).

[107]At the First Congress of Naturalists and Physicians in Kraków in 1869 Józef Dietl (1804–1878), a Polish physician, pioneer of balneology, politician, professor and rector of the Jagiellonian University, and mayor of Kraków in 1866–1874, said:

> *It matters only that, dispersed in different parts and corners of the country, we should find this place and ourselves in it, regain our feeling of strength, vitality and connection, of which we were deprived by age-old oppression of relations hostile toward us.*

[108]See Parshall (2005) for an overview of this phenomenon in various places over the Western world.

[109]For more information see Domoradzki (2012).

[110]The abbreviation UJ means the Jagiellonian University.

of pure and applied mathematics through talks, publication of mathematical works, scientific collaboration with foreign mathematicians and organization of scientific congresses.[111]

In 1920 this Society was transformed into the Polish Mathematical Society. New branches of the Society were formed: Lwów (1921), Poznań, Warszawa, Wilno (1923). Since 1922 the Society published (annually) its journal *Annales de la Société Polonaise de Mathématique*, based in Kraków and edited by S. Zaremba. It was meant primarily for the works of members of the Society. Many mathematicians from other centers in Poland [e.g., Sierpiński, Stefan Kempisty (1892–1940), Kazimierz Abramowicz (1888–1936), Mieczysław Biernacki (1891–1959)] and from abroad [e.g., Élie Cartan (1869–1951), Maurice Fréchet (1878–1973), Hlavatý] published there. All papers were in international languages, mostly French and other "Congress" ones (one exception was the already mentioned paper in Interlingua by Stamm). Besides original works, *Annales* also published detailed records of meetings of the branches. Thanks to these records we know titles and abstracts of many talks given at those meetings; not all of them resulted in subsequent publications, so these records preserve an important historical evidence of mathematical interests and work (altogether over 1,100 talks were given in Warsaw, Lwów and Kraków in the interwar period; published records are incomplete).

From 1919 to 1936 the Polish Mathematical Society was the only central institution representing Polish mathematics. The seat of the governing body of the Society was Kraków until 1935.[112]

It is worth adding here that another mathematical periodical, *Opuscula Mathematica*, was published in Kraków since 1937. It was established at the initiative of Hoborski and was affiliated with the Academy of Mining.

4. Conclusion

The presence of modern topics and significant results in the research of Kraków mathematicians in the interwar period cannot be denied. As far back as 1919, Mazurkiewicz (writing in support of Janiszewski's program of accelerating the development of Polish mathematics and boosting its status by concentrating research efforts in very few selected disciplines) admitted that Kraków had the potential to become an important center in geometry. Other mathematicians (Sierpiński, Kuratowski) were at times critical of S. Zaremba's leadership, but they also acknowledged the achievements of their Kraków colleagues.

A question that is still debated is: Did Kraków mathematicians form a school between the world wars? Historians tend to answer it in the negative, writing that:

[111]The idea to make popularization of mathematics one of the purposes was not accepted.

[112]For more information see Domoradzki (2009); Duda (2018); Gołąb (1961); Iwiński (1975); Kuratowski (1980); Pawlikowska-Brożek (1988); Przeniosło (2010).

A mathematical school in the mold of the Warsaw-Lwów school did not emerge in Kraków. Bringing about collective work on selected issues was unsuccessful. A feature of the Kraków university mathematics was individual work of separate scholars.[113]

or

Of no less importance for the development of Warsaw and Lwów communities was also the conviction of necessity of creation of scientific schools, while in Kraków the old idea of only individual scholarly work was upheld.[114]

Such statements seem to imply that the Warsaw school based on collaboration on selected topics was the only possible model of a mathematical school. But mathematical schools existed even though collaboration of scientific peers on joint publications was still an exception rather than a rule at that time.[115]

Anyway, collaboration did happen in Kraków, although on a smaller scale than in Warsaw. A list of collaborative papers written by Kraków mathematicians in the interwar period is given in Appendix B (it could be extended by a few more items where the author solved a problem posed by another mathematician).

Further, a closer analysis of the activities of the Kraków mathematicians in the interwar period shows that they shared many interests and often influenced each other. Gołąb (1964, p. 78) admitted that

> *As much as lectures of Żorawski and Zaremba brought in a new spirit to Kraków mathematics, one must say that they did not form their own schools in the strict sense.*

But from his subsequent analysis of Żorawski and Zaremba's influences he concluded that beginnings of schools appeared. Commenting elsewhere on his teacher Hoborski's desire to establish a school of geometry in Kraków and his conscious effort in this direction, Gołąb wrote (1969, p. 39):

> *... he managed to fulfill it to a large extent, or make a significant start anyway.*

These words can be applied to all Kraków-related scientific schools that blossomed later: the Kraków school of differential equations (concentrated around Ważewski), Kraków school of potential theory and complex analysis (around Leja), Kraków school of differential geometry (around Gołąb), Wrocław school of differential geometry (around Ślebodziński) and Kraków school of control theory, developed mostly thanks to Ważewski and based on the theory of differential inclusions developed by S.K. Zaremba [simultaneously with but independently of André Marchaud (1887–1973)].[116] The ground for these schools was prepared

---

[113]See (Dybiec 2000, p. 406).

[114]See (Przeniosło 2011, p. 58). A similar view was implicitly present in Kuratowski (1980).

[115]For comparison in the international setting, see, e.g., Parshall (1996).

[116]For more information see Górniewicz (2011).

through modern developments taking place in Kraków mathematics in the interwar period.

Appendix A: Doctorates and Habilitations in Mathematics at the Jagiellonian University in 1918–1939[117]

Doctorates (with full titles, and indicating multiple works submitted where applicable):

- 1918
  Witold Wilkosz (S. Zaremba); thesis *Z teorii funkcji absolutnie ciągłych i całek Lebesgue'a* [On Theory of Absolutely Continuous Functions and Lebesgue Integrals]
- 1919
  Stefan Kempisty (K. Żorawski); thesis *O funkcjach nawpół ciągłych* [On Semi-Continuous Functions]
  Witold Pogorzelski (S. Zaremba); thesis *Badanie równań całkowych i całkowo-różniczkowych nieliniowych* [A Study of Integral and Integro-Differential Non-linear Equations]
- 1920
  Juliusz Rudnicki (S. Zaremba); thesis 1. *Badanie pewnego typu wzrastania funkcji* [A Study of a Certain Type of Growth of Functions], 2. *Określenie pewnej funkcji całkowitej przy pomocy granicy* [A Definition of a Certain Entire Function by Means of a Limit]
- 1922
  Włodzimierz Stożek (S. Zaremba); *O wartościach charakterystycznych równań całkowych potencjału logarytmicznego* [On Characteristic Values of Integral Equations of the Logarithmic Potential]
- 1926
  Henryk Greniewski (S. Zaremba/W. Wilkosz; logic); thesis 1. *Przyczynek do podstaw logicznych mechaniki* [A Contribution to Logical Foundations of Mechanics] (manuscript); 2. *Próba dedukcyjnej teorji przyczynowości* [An Attempt at a Deductive Theory of Causality]
- 1931
  Stanisław Gołąb (S. Zaremba); thesis *O uogólnionej geometrii rzutowej* [On Generalized Projective Geometry]
- 1935
  Adam Bielecki (W. Wilkosz); thesis *O integralnym przedstawianiu m-wymiarowych powierzchni zawartych w n-wymiarowej przestrzeni euklidesowej*

[117]Based on the chronicles of the Jagiellonian University, author unknown (1924, 1936, 1938).

za pomoc funkcji uwikłanych [On Integral Representation of m-dimensional Surfaces Contained in an n-dimensional Euclidean Space through Implicit Functions]
Stanisław Turski (W. Wilkosz); thesis *O pewnym uogólnieniu twierdzeń o istnieniu i jednolistności całek równań hiperbolicznych typu* $z_{xy} = f(x, y, z, z_x, z_y)$ [On a Generalization of Theorems on Existence and Univalence of Integrals of Hyperbolic Equations of the Type $z_{xy} = f(x, y, z, z_x, z_y)$]

- 1939
  Duwid Wajnsztejn (W. Wilkosz); thesis *Interpretacja macierzowa liczb Clifforda* [A Matrix Representation of Clifford Numbers]

  Habilitations: S. Mazurkiewicz (1919), W. Wilkosz (1920), W. Pogorzelski (1921), J. Rudnicki (1921), F. Leja (1924), T. Ważewski (1927), L. Chwistek (1928), S. Gołąb (1932).

Appendix B: Collaborative Papers by Kraków Mathematicians in the Interwar Period

An alphabetical list by authors (*Source*: zbMATH database)

A. Bielecki and J.W. Weyssenhoff: *Quaternions, 4-dimensional rotations and Cayley's formula*, Bulletin de l'Académie Polonaise des Sciences et des Lettres, Séries A, 1936, pp. 216–227. Zbl 0015.12002.

A. Bielecki and S.K. Zaremba: Sur les points singuliers des systèmes de deux équations différentielles ordinaires, *Annales de la Société Polonaise de Mathématique* **15**(1936), pp. 135–139. Zbl 0018.12302.

A. Bielecki, M. Mathisson and J.W. Weyssenhoff: Sur un théorème concernant une transformation d'intégrales quadruples et intégrales curvilignes dans l'espace de Riemann, *Bulletin de l'Académie Polonaise des Sciences et des Lettres, Séries A*, 1939, pp. 22–28. Zbl 0015.12002.

V. Hlavatý and S. Gołąb: Zur Theorie der Vektor- und Punktkonnexion, *Prace Matematyczno-Fizyczne* **39**(1932), pp. 119–130. JFM 58.0762.01.

F. Leja and W. Wilkosz: Sur une propriété des domaines concaves, *Annales de la Société Polonaise de Mathématique* **2**(1924), pp. 222–224. JFM 50.0490.01.

A. Rosenblatt and S. Turski: Sur les coefficients des séries de puissances univalentes dans le cercle unite, *Comptes Rendus Hebdomadaires des Séances de l'Académie des Sciences, Paris* **200**(1935), pp. 1270–1272. Zbl 0011.26101.

A. Rosenblatt and S. Turski: Sur la représentation conforme de domaines plans, *Bulletin des Sciences Mathématiques, Deuxième Série*, **60**(1936), pp. 309–320. Zbl 0015.11602.

A. Rosenblatt and S. Turski: Sur la représentation conforme de domaines plans, *Comptes Rendus Hebdomadaires des Séances de l'Académie des Sciences, Paris* **202**(1936), pp. 899–901. Zbl 0013.31203.

J.A. Schouten and S. Gołąb: Über projektive Übertragungen und Ableitungen, *Mathematische Zeitschrift* **32**(1930), pp. 192–214. JFM 56.0627.02.

J.A. Schouten and S. Gołąb: Über projektive Übertragungen und Ableitungen. II., *Annali di Matematica Pura ed Applicata, Serie Quarta* **8**(1930), pp. 141–157. JFM 56.0628.01.

T. Ważewski and S.K. Zaremba: Les ensembles de condensation des caractéristiques d'un système d' équations différentielles ordinaries, *Annales de la Société Polonaise de Mathématique* **15**(1936), pp. 24–33. Zbl 0017.39901.

References

A. Bielecki, O integralnem przedstawianiu m-wymiarowych powierzchni zawartych w n-wymiarowej przestrzeni euklidesowej zapomocą funkcyj uwikłanych [On Integral Representation of m-dimensional Surfaces Contained in the n-dimensional Space by Implicit Functions] (PhD thesis) *Dodatek do Rocznika Polskiego Towarzystwa Matematycznego, Kraków* **7**(1935), pp. 1–38.

G. Bouligand, Sur le problème de Dirichlet, *Annales de la Société Polonaise de Mathématique* **4**(1926), pp. 59–112.

T. Boy-Żeleński, *Znasz-li ten kraj?* (*Cyganeria krakowska*) [You Know that Land Perhaps? (The Kraków Boheme)] Biblioteka Polska, series Biblioteka Boya, Warszawa, 1932, 216 pages.

J. Bremer, Michał Dziewicki — edytor, logik, lektor, tłumacz, powieściopisarz [Michał Dziewicki — An Editor, Logician, Instructor, Translator, Novelist], *Przegląd Filozoficzny — Nowa Seria* **25**(2016), no. 3, pp. 209–226. ISSN 1230-1493.

L. Chwistek, The Principles of the Pure Type Theory (1922), Translated by Adam Trybus with an Introductory Note by Bernard Linsky, *History and Philosophy of Logic* **33**(2012), no. 4, pp. 329–352. ISSN 0144-5340.

D. Ciesielska, Fotografia: Grupa polskich stypendiów i stypendystów. Getynga, lato 1907. Trudne identyfikacje [A Photograph. A Group of Polish Scholarships and Scholarship Recipients, Göttingen Summer of 1907. Difficult Identifications], *Kwartalnik Historii Nauki i Techniki* **65**(2020), no. 4, p. 19 (In print).

D. Ciesielska, Stanisław Zaremba (1863–1942). Publikacje, odczyty i wykłady [Stanisław Zaremba (1863–1942) — Publications, Talks and Lectures], *Kwartalnik Historii Nauki i Techniki* **60**(2016), no. 4, pp. 71–99. ISSN 0023-589X.

D. Ciesielska and K. Ciesielski, Stanisław Zaremba (1863–1942) i jego działalność na rzecz matematyki [Stanisław Zaremba (1863–1942) and His Activity for the Benefit of Mathematics], *Kwartalnik Historii Nauki i Techniki* **60**(2016), no. 4, pp. 37–71. ISSN 0023-589X.

D. Ciesielska and S. Domoradzki, On Mathematical Lectures at the Jagiellonian University in the Years 1860–1918. An Essay Based on Manuscripts, *Technical Transactions — Fundamental Sciences 1-NP* **111**(2014), pp. 59–71. ISSN 0011-4561.

D. Ciesielska and L. Maligranda, Alfred Rosenblatt (1880–1947), *Wiadomości Matematyczne* **50**(2014), pp. 221–259. ISSN 2080-5519 (Polish).

D. Ciesielska and L. Maligranda, Alfred Rosenblatt (1880–1947). Publikacje, odczyty, wykłady [Alfred Rosenblatt (1880–1947) — Publications, Talks, Lectures], *Antiquitates Mathematicae* **8**(2014), pp. 3–45. ISSN 1898-5203.

D. Ciesielska, L. Maligranda and J. Zwierzyńska, Doktoraty Polaków w Getyndze. Matematyka. [Doctorates of Poles in Göttingen. Mathematics.], *Analecta* **28**(2019), no. 2, pp. 73–116.

K. Ciesielski, Wykłady polskich matematyków na Międzynarodowych Kongresach Matematyków [Lectures of Polish Mathematicians at the International Congresses of Mathematics], *Wiadomości Matematyczne* **54**(2018), no. 1, pp. 55–65. ISSN 2080-5519.

K. Ciesielski and Z. Pogoda, On Mathematics in Kraków throughout Centuries, *Wiadomości Matematyczne* **48**(2012), pp. 313–324. ISSN 2080-5519.

Dodatek do kronik Uniwersytetu Jagiellońskiego za lata akademickie 1923/24-1933/34. Spis prac naukowych [Addendum to the Chronicles of the Jagiellonian University for the Academic Years 1923/24–1933/34. The List of Research Works], W Krakowie w drukarni Uniwersytetu Jagiellońskiego pod zarządem Józefa Filipowskiego, Krakow, 1936, 446 pages.

S. Domoradzki, *The Growth of Mathematical Culture in the Lvov Area in the Autonomy Period (1870–1920)*, History of Mathematics, Volume 47, Matfyzpress, Prague, 2012, 331 pages. ISBN 978-80-7378-178-1.

S. Domoradzki, Cztery doktoraty z matematyki uzyskane przez Polaków we Francji przed 1939 r. [Four Doctorates in Mathematics Obtained by Poles in France Before 1939], pp. 93–100, in J. Bečvář, M. Bečvářová (eds.), *34. Mezinárodní Konference Historie Matematiky* [34th International Conference on History of Mathematics], Matfyzpress, Praha, 2013, 202 pages. ISBN 978-80-7378-234-4.

S. Domoradzki and A. Pelczar, O założycielach Polskiego Towarzystwa Matematycznego [On the Founders of the Polish Mathematical Society], *Wiadomości Matematyczne* **45**(2009), pp. 217–240. ISSN 2080-5519.

S. Domoradzki and M. Stawiska, Distinguished Graduates in Mathematics of Jagiellonian University in the Interwar Period, Part I: 1918–1926, *Technical Transactions — Fundamental Sciences* 2-NP **112**(2015), pp. 98–115. ISSN 0011-4561.

S. Domoradzki and M. Stawiska, Distinguished Graduates in Mathematics of Jagiellonian University in the Interwar Period, Part II: 1926–1939, *Technical Transactions — Fundamental Sciences* 2-NP **112**(2015), pp. 117–141. ISSN 0011-4561.

S. Domoradzki and M. Stawiska, Polish Mathematics and Mathematicians in World War I, Part I, *Studia Historiae Scientiarum* **17**(2018), pp. 23–49. ISSN 1731-6715.

S. Domoradzki and M. Stawiska, Polish Mathematics and Mathematicians in World War I, Part II, *Studia Historiae Scientiarum* **18**(2019), pp. 55–92. ISSN 2451-3202.

R. Duda, Polskie Towarzystwo Matematyczne na tle dziejów [Polish Mathematical Society on the Background of History], *Wiadomości Matematyczne* **45**(2009), pp. 241–289. ISSN 2080-5519.

R. Duda, *Matematycy XIX i XX wieku związani z Polską* [Mathematicians of the 19th and 20th Centuries with Links to Poland] Wydawnictwo Uniwersytetu Wrocławskiego, Wrocław, 2007, 587 pages. ISBN 978-83-229-3316-9.

R. Duda, *Pearls from a Lost City: The Lvov School of Mathematics.* Translated from the 2007 Polish original by Daniel Davies, History of Mathematics, Volume 40, American Mathematical Society, Providence, RI, 2014, xii + 231 pages. ISBN 978-1-4704-1076-6.

R. Duda, One Hundred Years of the Polish Mathematical Society, *Antiquitates Mathematicae* **12**(2018), pp. 3–30. ISSN 1898-5203.

J. Dybiec, *Uniwersytet Jagielloński 1918–1939* [Jagiellonian University 1918–1939] Nakładem Polskiej Akademii Umiejętności, Kraków, 2000, 759 pages. ISBN 978-83-86-95689-0.

A. Fuliński, On Marian Smoluchowski's Life and Contribution to Physics, *Acta Physica Polonica B* **29**(1998), no. 6, pp. 1523–1537. ISSN 0587-4254.

J. Gancarzewicz and Z. Pogoda, Stanisław Gołąb (1902–1980), pp. 357–362, in B. Szafirski (ed.), *Złota Księga. Wydział Matematyki i Fizyki UJ* [The Golden Book of the Faculty of Mathematics and Physics of JU] Księgarnia Akademicka, Kraków, 2000, 551 pages. ISBN 978-83-7188-326-2.

S. Gołąb, Über eine verallgemeinerte projektive Geometrie, *Prace Matematyczno-Fizyczne* **37**(1929/1930), pp. 1–63.

S. Gołąb, Quelques problèmes métriques de la géometrie de Minkowski, *Travaux de l'Academie des Mines Cracovie* **6**(1932), 179 pages.

S. Gołąb, Sur la fonction représentant la distance d'un point variable à un ensemble fixe, *Comptes Rendus Hebdomadaires des Séances de l'Académie des Sciences, Paris* **206**(1938), pp. 406–408.

S. Gołąb, Über den Begriff der "Pseudogruppe von Transformationen, *Mathematische Annalen* **116**(1939), pp. 768–780.

S. Gołąb, Z dziejów założenia Polskiego Towarzystwa Matematycznego [On the History of Founding of the Polish Mathematical Society], *Wiadomości Matematyczne* **4**(1961), pp. 155–159. ISSN 0373-8302.

S. Gołąb, Zarys dziejów matematyki w UJ w XX w. [An Outline of the History of Mathematics in JU in the 20th Century], pp. 75–86, in S. Gołąb (ed.), *Studia z dziejów katedr Wydziału Matematyki, Fizyki i Chemii* [Studies in the History of Chairs of the Faculty of Mathematics, Physics and Chemistry], Wydawnictwa Jubileuszowe Uniwersytetu Jagiellońskiego, Volume 15, Kraków, 1964, 280 pages.

S. Gołąb, Antoni Hoborski organizator polskiej szkoły geometrycznej [Antoni Hoborski the Organizer of the Polish Geometrical School], *Wiadomości Matematyczne* **12**(1969), pp. 33–48. ISBN 0373-8302.

L. Górniewicz, Differential Inclusions — The Theory Initiated by Cracow Mathematical School, *Annales Mathematicae Silesianae* **25**(2011), pp. 7–25. ISSN 2391-4238.

J. Gray, *Plato's Ghost. The Modernist Transformation of Mathematics*, Princeton University Press, Princeton, NJ, 2008, x + 515 pages. ISBN 978-0-691-13610-3.

J. Gray, Plato's Ghost. *ICM Proceedings 1893–2014*, International Mathematical Union. Online document. https://www.mathunion.org/icm/proceedings [3.10.2019].

T. Iwiński, *Ponad pół wieku działalności matematyków polskich. Zarys historii Polskiego Towarzystwa Matematycznego 1919–1973* [Over a Half of the Century of Activity of Polish Mathematicians. An Outline of the History of the Polish Mathematical Society], PWN, Warszawa, 1975, 220 pages.

J.J. Jadacki, Jan Śleszyński, *Wiadomości Matematyczne* **34**(1998), pp. 83–97. ISSN 0373-8302.

M. Kantor, Before the Liberty Oak Was Planted. Jagiellonian University and World War I, *International Relations Office of Jagiellonian University Kraków, UJ Newsletter* **36**(2008), pp. 11–13. Online document. https://www.uj.edu.pl/documents/1333512/66 4d6b01-e7b6-46d1-a7ab-dfe75cd44f43 [12.4.2019].

W. Kaczor, T. Kuczumow and W. Zygmunt, Adam Bielecki (1910–2003), *Wiadomości Matematyczne* **40**(2004), pp. 213–228. ISSN 0373-8302.

W. Kleiner, Franciszek Leja, *Annales Polonici Mathematici* **46**(1985), pp. 1–11. ISSN 0066-2216.

J. Koroński, Stanisław Zaremba (1863–1942) and His Results in the Field of Differential Equations, *Technical Transactions — Fundamental Sciences* 2-NP **112**(2015), pp. 207–217. ISSN 0011-4561.

Kronika Uniwersytetu Jagiellońskiego za lata szkolne 1921/22 i 1922/23 [The Chronicle of the Jagiellonian University for the School Years 1921/22 and 1922/23], Jagiellonian University, Kraków, 1924, 110 pages.

Kronika Uniwersytetu Jagiellonńskiego za lata akademickie 1934/35 i 1935/36 [The Chronicle of the Jagiellonian University for the Academic Years 1934/35 and 1935/36] W Krakowie w drukarni Uniwersytetu Jagiellońskiego pod zarządem Józefa Filipowskiego, Krakow, 1938, 391 pages.

M. Kucharzewski, Stanislaw Gołąb — Life and Work, *Aequationes Mathematicae* **24**(1982), pp. 1–18. ISSN 0001-9054.

K. Kuratowski, *Pół wieku matematyki polskiej 1920–1970*, Wiedza Powszechna, Warszawa, 1973, 191 pages. English translation: *A Half Century of Polish Mathematics. Remembrances and Reflections*, Translated from the Polish by Andrzej Kirkor. With a preface by S. Ulam, International Series in Pure and Applied Mathematics, Volume 108, Pergamon Press, Inc., Elmsford, NY, PWN — Polish Scientific Publishers, Warsaw, 1980, viii + 200 pages. ISBN 0-08-023046-6.

F. Leja, Powstanie Polskiego Towarzystwa Matematycznego [The Emergence of the Polish Mathematical Society], *Wiadomości Matematyczne* **12**(1969), pp. 3–8. ISSN 0373-8302.

Z. Łomnicki, *A Letter to Zygmunt Wilhelm (William) Birnbaum, 11/12/1951*, Zygmunt William Birnbaum Collection, University of Washington, University Libraries, Accession Number 5266-001. Online document. https://www.lib.washington.edu/specialcollections/collections/exhibits/spotlight-on-zygmunt-william-birnbaum-papers [14.02.2021].

N.N. Lusin, List do Arnauda Denjoy z 1926 r. [A Letter to Arnaud Denjoy from 1926], *Wiadomości Matematyczne* **25**(1983), pp. 65–68. ISSN 0373-8302 (Polish, translated from the French by Andrzej Mąkowski).

L. Maligranda, Stefan Kaczmarz (1895–1939), *Antiquitates Mathematicae* **1**(2007), pp. 15–61. ISSN 1898-5203 (Polish).

J. Mietelski, Tematyka prac astronomicznych w pierwszej siedzibie Obserwatorium krakowskiego [The Topics of the Astronomical Works in the First Location of the Kraków Observatory], *Prace Komisji Historii Nauki PAU* **10**(2010/2011), pp. 191–204. ISSN 1731-6715.

R. Murawski, Philosophical Reflection on Mathematics in Poland in the Interwar Period. Provinces of Logic Determined, *Annals of Pure and Applied Logic* **127**(2004), pp. 325–337. ISSN 0168-0072.

R. Murawski, *Filozofia matematyki i logiki w Polsce międzywojennej*, Monografie Fundacji na Rzecz Nauki Polskiej, Wydawnictwo Naukowe Uniwersytetu Mikołaja Kopernika, Toruń, 2011, 268 pages. ISBN 978-83-231-2670-6. English translation: *The Philosophy of Mathematics and Logic in the 1920s and 1930s in Poland*, Translated from the 2011 Polish original by Maria Kantor, Science Networks. Historical Studies, Volume 48, Birkhäuser, Springer, Basel, 2014, xii + 228 pages. ISBN 978-3-0348-0830-9.

L. Obojska, Some remarks of jan śleszyński regarding foundations of mathematics of stanisław leśniewski [sic], *Technical Transactions — Fundamental Sciences 2-NP* **112**(2015), pp. 235–245. ISSN 0011-4561.

K.H. Parshall, How We Got Where We Are: An International Overview of Mathematics in National Contexts (1875–1900), *Notices of the American Mathematical Society* **43**(1996), pp. 287–296. ISSN 0002-9920.

K.H. Parshall, The Emergence of the American Mathematical Research Community, pp. 183–202, in G. Van Brummelen, M. Kinyon (eds.), *Mathematics and the Historian's Craft*, The Kenneth O. May Lectures, CMS Books in Mathematics/Ouvrages de

Mathématiques de la SMC, Volume 21, Springer, New York, 2005, xvi + 357 pages. ISBN 978-0387-25284-1.

Z. Pawlikowska-Brożek, Matematyka polska w latach 1918–1951 [Polish Mathematics in the Years 1918–1951], *Studia i Materiały z Dziejów Nauki Polskiej, seria II*, **2**(1988), pp. 13–59. ISSN 0860-1011.

Z. Pawlikowska-Brożek, Stanisław Gołąb, pp. 62–63, in S. Domoradzki, Z. Pawlikowska-Brożek, D. Węglowska (eds.), *Słownik Biograficzny Matematyków Polskich* [A Biographical Dictionary of Polish Mathematicians], Państwowa Wyższa Szkoła Zawodowa, Tarnobrzeg 2003, 286 pages. ISBN 9788391729335.

Z. Pawlikowska-Brożek, Andrzej Turowicz, pp. 246–247, in S. Domoradzki, Z. Pawlikowska-Brożek, D. Węglowska (eds.), *Słownik Biograficzny Matematyków Polskich* [A Biographical Dictionary of Polish Mathematicians], Państwowa Wyższa Szkoła Zawodowa, Tarnobrzeg, 2003, 286 pages. ISBN 9788391729335.

Z. Pawlikowska-Brożek, Stanisław K. Zaremba, pp. 268–269, in S. Domoradzki, Z. Pawlikowska-Brożek, D. Węglowska (eds.), *Słownik Biograficzny Matematyków Polskich* [A Biographical Dictionary of Polish Mathematicians], Państwowa Wyższa Szkoła Zawodowa, Tarnobrzeg, 2003, 286 pages. ISBN 9788391729335.

A. Pelczar, Równania różniczkowe w Polsce. Zarys historii do połowy lat siedemdziesiątych XX wieku [Differential Equations in Poland. An Outline of the History Until the Mid-1970s], *Wiadomości Matematyczne* **37**(1991), pp. 63–118. ISSN 0373-8302.

A. Pelczar, O matematyce i matematykach w Uniwersytecie Jagiellońskim [On Mathematics and Mathematicians at the Jagiellonian University], pp. 214–237, in B. Szafirski (ed.), *Złota Księga wydziału matematyki i fizyki UJ* [The Golden Book of the Faculty of Mathematics and Physics of JU], Księgarnia Akademicka, Kraków, 2000, 551 pages. ISBN 83-7188-326-9.

A. Pelczar, Tadeusz Ważewski (1896–1972) uczony i nauczyciel [Tadeusz Ważewski — A Teacher and a Scholar], pp. 341–356, in B. Szafirski (ed.), *Złota Księga wydziału matematyki i fizyki UJ* [The Golden Book of the Faculty of Mathematics and Physics of JU], Księgarnia Akademicka, Kraków, 2000, 551 pages. ISBN 83-7188-326-9.

W. Piotrowski, Jeszcze w sprawie biografii Ottona i Stanisławy Nikodymów [More on the Matter of the Biography of Otto and Stanisława Nikodym], *Wiadomości Matematyczne* **50**(2014), pp. 69–74. ISSN 2080-5519.

Z. Pogoda, Początki geometrii różniczkowej w Polsce [Beginnings of Differential Geometry in Poland], *Antiquitates Mathematicae* **1**(2007), pp. 115–129. ISSN 1898-5203.

Z. Pogoda, Stanisław Zaremba i krystalografia [Stanisław Zaremba and Crystallography], *Kwartalnik Historii Nauki i Techniki* **60**(2016) no. 4, pp. 131–139. ISSN 0023-589X.

P. Polak, Stanisława Zaremby filozoficzna koncepcja nauki [Stanisław Zaremba's Philosophical Conception of Science], *Kwartalnik Historii Nauki i Techniki* **60**(2016), no. 4, pp. 99–131. ISSN 0023-589X.

M. Przeniosło, Polskie Towarzystwo Matematyczne w okresie międzywojennym [Polish Mathematical Society in the Interwar Period], *Rocznik Lubelski* **36**(2010), pp. 65–76. ISSN 0080-3510.

M. Przeniosło, Matematycy krakowscy w dwudziestoleciu międzywojennym [Kraków Mathematicians in the Interwar Period], *Res Historica* **32**(2011), pp. 57–71. ISSN 2082-6060.

D. Przeworska-Rolewicz, *Mój mistrz i promotor Witold Pogorzelski (1895–1963)* [My Master and Supervisor Witolf Pogorzelski (1895–1963)], Institute of Mathematics, Polish Academy of Sciences, IM PAN, Print 688, Warszawa, 2008, pp. 1–19.

M. Sarnecka-Keller (dz. naukowa) [Scholarly Activity], J. Hulewicz (dz. polityczna) [Political Activity], Marchlewski Leon Paweł, pp. 542–545, in E. Rostworowski (ed.), *Polski Słownik Biograficzny* [Polish Biographical Dictionary], Volume 19 (Wawrzyniec Machowski — Maria Kazimiera), Zakład Narodowy imienia Ossolińskich — Wydawwnictwo Polskiej Akademii Nauk, Wrocław, 1974, 664 pages.

T. Sauer and A. Trautman, Myron Mathisson: What Little We Know of His Life, *Acta Physica Polonica B, Proceedings Supplement* **1**(2008), pp. 7–26. ISSN 1899-2358.

E. Seneta, Jan Śleszyński (1854–1931) jako probabilista [Jan Śleszyński (1854–1931) as a Probabilist], *Wiadomości Matematyczne* **34**(1998), pp. 99–104. ISSN 0373-8302.

J. Siciak, Franciszek Leja (1885–1979), *Wiadomości Matematyczne* **24**(1982), pp. 65–90. ISSN 0373-8302.

W. Sierpiński, The Warsaw School of Mathematics and the Present State of Mathematics in Poland, *The Polish Review* **4**(1959), no. 1/2, pp. 51–63. ISSN 0032-2970.

M. Stawiska, Prace Kazimierza Żorawskiego z teorii iteracji [Works of Kazimierz Żorawski in Iteration Theory], *Antiquitates Mathematicae* **10**(2016), pp. 121–132. ISSN 1898-5203.

J. Suchmiel, Emancypacja naukowa kobiet w uniwersytetach w Krakowie i we Lwowie do roku 1939 [Scholarly Emancipation of Women at the Universities in Kraków and Lwów Before 1939], *Prace Naukowe Akademii im. Jana Długosza w Częstochowie, Seria Pedagogika* **13**(2004), pp. 115–123. ISSN 1734-185X.

S. Suszczyński, Kazimierz Żorawski (tekst polski, angielski i francuski) [Kazimierz Żorawski (Polish, English and French Texts)], Online document. Webpage *Wybitni polscy matematycy*, Instytut Matematyczny Polskiej Akademii Nauk, https://www.impan.pl/Great/Zorawski/Zorawski-pl.html [14.02.2021].

F.H. Szafraniec, Początki własności reprodukowania: Bergman, Szegő, Bochner — a może jednak Zaremba [Beginnings of the Reproducing Property: Bergman, Szegő, Bochner — and Maybe Zaremba After All], *Wiadomości Matematyczne* **52**(2016), pp. 53–67. ISSN 2080-5519.

F.H. Szafraniec, Przypadek Stanisława Zaremby — oportunizm czy nonszalancja [Stanisław Zaremba's Case — Opportunism or Nonchalance], *Kwartalnik Historii Nauki i Techniki* **61**(2017), no. 1, pp. 117–129. ISSN 0023-589X.

W. Ślebodziński, Władysław Nikliborc et son œuvre scientifique, *Colloquium Mathematicum* **1**(1948), pp. 322–330. ISSN 0010-1354.

W. Ślebodziński, L'œuvre scientifique de Kazimierz Żorawski, *Colloquium Mathematicum* **4**(1956), pp. 74–88. ISSN 0010-1354.

W. Ślebodziński, Wspomnienia matematyka z lat 1903–1968 [Reminiscences of a Mathematician from the Years 1903–1968], *Wiadomości Matematyczne* **12**(1969), pp. 17–31. ISSN 0373-8302.

W. Ślebodziński, Kazimierz Żorawski, *Wiadomości Matematyczne* **12**(1969) pp. 49–64. ISSN 0373-8302 (Polish).

B. Średniawa, W dwudziestą rocznicę śmierci prof. Witolda Wilkosza [On the 20th Anniversary of the Death of Prof. Witold Wilkosz]. Online document. Institute of Physics of the Polish Academy of Sciences, http://www.ifpan.edu.pl/ON-1/Historia/art/28wil.pdf [5.10.2019].

B. Średniawa, Współpraca matematyków, fizyków i astronomów w Uniwersytecie Jagiellońskim w XIX i pierwszej połowie XX wieku. Studia z historii astronomii, fizyki i matematyki w Uniwersytecie Jagiellońskim [Collaboration of Mathematicians, Physicists and Astronomers at Jagiellonian University in the 19th Century and the First Half of the 20th Century. Studies in History of Astronomy, Physics and Mathematics at

Jagiellonian University], *Zeszyty Naukowe Uniwersytetu Jagiellońskiego, Prace Fizyczne* **25**(1986), pp. 53–82. ISSN 0083-4335.

B. Średniawa, Autobiografia. Wspomnienie fizyka teoretyka w 90-lecie urodzin i w 60-lecie promocji doktorskiej [An Autobiography. A Reminiscence of a Theoretical Physicist on the 90th Birthday and the 60th Anniversary of the Doctoral Promotion], *Kwartalnik Historii Nauki i Techniki* **52**(2007), no. 3/4, pp. 7–19. ISSN 0023-589X.

B. Średniawa, Władysław Natanson (1864–1937). For the 110th Anniversary of Formulation of the Thermodynamics of Irreversible Processes and the 95th Anniversary of Formulation of Quantum Statistics, *Concepts of Physics* **4**(2007), pp. 705–723. ISSN 2331-6535.

A. Śródka (ed.), *Biogramy uczonych polskich* [Biographical Entries of Polish Scholars], Ośrodek Informacji Naukowej, Polska Akademia Nauk, Część VI: Nauki medyczne, zeszyt 1: A–Ł [Part VI: Medical Sciences, Issue 1: A–Ł] Ossolineum, Wrocław, 1990, 414 pages. ISBN 9788304016828.

A. Śródka, R.W. Gryglewski and W. Szczepański, Browicz or Kupffer Cells?, *Journal of Physiology and Pharmacology* **57**(2006), pp. 183–185. ISSN 1233-9687.

K. Trzęsicki, Henryk Feliks Greniewski 1903–1972, pp. 358–367 in P. Salwa and A. K. Wróblewski (eds.), *Portrery Uczonych Profesorowie Uniwersytetu Warszawskiego* [Portraits of Scholars Professors of Warsaw University] 1915–1945 (A-Ł), Monumenta Universitatis Varsoviensis, Wydawnictwa Uniwersytetu Warszawskiego, 2016. ISBN 978-83-235-1784-9.

A. Turowicz, unpublished tape recording, before 1989, courtesy of Z. Pawlikowska-Brożek.

R. Urbaniak and M.T. Godziszewski, Material Implication and Conversational Implicature in Lvov-Warsaw School, pp. 117–132, in Á. Garrido and U. Wybraniec-Skardowska (eds.), *The Lvov-Warsaw School. Past and Present*, Studies in Universal Logic, Birkhäuser, Springer, Cham, 2018, xviii + 815 pages. ISBN 978-3-319-65429-4.

S. Waltoś, Uniwersytet Jagielloński w Krakowie. Tradycja i współczesność [Jagiellonian University in Kraków. Tradition and Contemporary Period] (2008). Online document. https://web.archive.org/web/20080201102745/http://www.uj.edu.pl/dispatch.jsp?item=uniwersytet%2Fhistoria%2Fhistoriatxt.jsp [14.02.2021].

D. Wajnsztejn, α-Matrizen und Clifford-Zahlen, *Annales de la Société Polonaise de Mathématique* **16**(1938), pp. 162–175.

J. Woleński, Mathematical Logic in Poland 1900–1939: People, Circles, Institutions, Ideas, *Modern Logic* **5**(1995), pp. 363–405. ISSN 1047-5982.

J. Woleński, Philosophy of Exact Sciences (Logic and Mathematics) in Poland in 1918–1939, *Technical Transactions — Fundamental Sciences 2-NP* **112**(2015), pp. 255–265. ISSN 0011-4561.

J. Woleński, Mathematical Logic in Poland 1918–1939, *Antiquitates Mathematicae* **12**(2018), pp. 99–109. ISSN 1898-5203.

W. Wójcik, Kazimierz Żorawski, pp. 521–523, in B. Orłowski: *Polski wkład w przyrodoznawstwo i technikę. Słownik polskich i związanych z Polską odkrywców, wynalazców oraz pionierów nauk matematyczno-przyrodniczych i techniki* [Polish Contributions in Natural Sciences and Technology. Dictionary of Polish and Related-to-Poland Discoverers, Inventors and Pioneers of Natural and Mathematical Sciences and Technology], Redakcja naukowa Bolesław Orłowski, Tom IV S–Z, Publisher ASPRA-JR (in cooperation with Instytut Historii Nauki im. L. i A. Birkenmajerów Polskiej Akademii Nauk. Instytut Pamięci Narodowej, Komisja Ścigania Zbrodni przeciwko Narodowi Polskiemu), Warszawa, 2015, 544 pages. ISBN 978-83-7545-572-4.

A.K. Wróblewski, Polish Physicists and the Progress in Physics (1870–1920), *Technical Transactions — Fundamental Sciences* 1-NP **111**(2014), pp. 255–273. ISSN 0011-4561.

S.K. Zaremba, O równaniach paratyngensowych [On Paratingent Equations], *Dodatek do Rocznika Polskiego Towarzystwa Matematycznego* **9**(1935), pp. 1–22 [rozprawa doktorska (PhD thesis)].

S.K. Zaremba, Sur les équations au paratingent, *Bulletin des Sciences Mathématiques* **60**(1936), pp. 139–160.

S.K. Zaremba, *Personally written CV*, unpublished copy courtesy of Dr. Z. Pawlikowska-Brożek, Kraków, no date, 2 pages.

https://doi.org/10.1142/9781786349316_0005

Chapter 5

Transformation of the Warsaw Mathematical Community in the Aftermath of WWI

Stanisław Domoradzki and Margaret Stawiska-Friedland

1. Transformation of the University in Warsaw and the Role of Young, Talented and Spirited Mathematicians in This Process

1.1. *Higher Education and Mathematics in Warsaw at the End of the 19th Century*

Between 1795 and 1918 there was no Polish state. The territories of the Commonwealth of Two Nations (Poland and Lithuania) were divided among three neighboring powers: Prussia, Austria and Russia. Warsaw (Warszawa) — the former seat of the king and *Sejm* [Parliament] — changed hands several times. In 1815, by the decision of the Congress of Vienna, it became the capital of the new Polish Kingdom, ruled by the Tsar of Russia and united with the Russian Empire. A cycle of Polish struggles for freedom and increased Russian repressions followed.[1]

The *Królewski Uniwersytet Warszawski* [Royal University of Warsaw] established in Warsaw in 1816 was subsequently closed down, replaced by the *Szkoła Główna* [Main School], and then closed down again. In 1869 a Russian-language *Uniwersytet Cesarski* [Imperial University] was opened. Poles constituted about 70% of the student body, but the university was treated more as an instrument of Russification than as an educational opportunity.[2] While not highly ranking and isolated from academic centers abroad, it nevertheless represented a decent level in some disciplines, among them mathematics. Some prominent Russian mathematicians were professors at the Imperial University or at the Polytechnic Institute, which opened in Warsaw in 1898. For example, Vasili Afanasievich Anisimov (1860–1907, differential equations, calculus of variations, functions of a complex variable),

[1]See e.g., Duda (2019); Kieniewicz and Grochulska (1981); Marczewski (1948) for more detailed chronology and description.
[2]See Bartnicka (2014).

Nikolai Yakovlevich Sonin (1849–1915, special functions, numerical integration) and Vsevolod Ivanovich Romanovskii (1879–1954, differential equations, probability, statistics) taught at the university. Dmitri Dmitrievich Mordukhai-Boltovskoi (1876–1952, abelian integrals, ordinary differential equations, differential and algebraic geometry, transcendental numbers and functions, history of mathematics) and Georgy Feodosevich Voronoi (1868–1908, number theory, geometry) were affiliated with both institutions.[3]

Some former professors of the Main School of Polish nationality continued to work at the Imperial University: Nikodem Pęczarski (1814–1877, analytic and descriptive geometry), Władysław Zajączkowski (1837–1898, differential calculus, theory of integrals, difference calculus, variational calculus, probability) and Tytus Babczyński (1830–1910, in the chair of mechanics, advanced algebra, differential calculus, mathematical physics). In turn, some future prominent Polish mathematicians started their academic education at the Imperial University.

Marian Aleksander Baraniecki (1848–1891) started his studies still at the Main School but finished them at the Imperial University in 1870. He continued his studies at the Jagiellonian University in Kraków [with Franciszek Mertens (1840–1927)] and Leipzig, where he got his PhD. He returned to the Imperial University to lecture in the years 1876–1885 and then took a chair in mathematics at the Jagiellonian University.

Kazimierz Żorawski (1866–1953) studied in Warsaw in 1884–1888. Due to his work in astronomy, he was selected to prepare for his future academic career. He continued his studies in mathematics and mechanics in Leipzig and got a doctorate in 1891 under Sophus Lie.[4]

Wacław Sierpiński (1882–1969) studied mathematics in 1889–1903, drawing the attention of Voronoi and winning a gold medal for the best student's work.[5]

Michał Feldblum (1875–1926) got his candidate degree in 1897 (thesis *Ueber die Riccatische Gleichung*) and went on to get his PhD in 1899 with David Hilbert in Göttingen [thesis *Ueber elementar-geometrische constructionen*, published in Warsaw in the printing house of Saturnin Józef Sikorski (1862–1922)].

Lucjan Emil Böttcher (1872–1937) spent only a year as a student at the Imperial University (1893/1894), expelled for participation in a patriotic demonstration, but

[3]See for example Duda (2016); Nalbandjan (1998); Syta and van de Weygaert (2009).

[4]Later on, Żorawski worked at the Polytechnic School in Lwów. In 1898 he became a professor at the Jagiellonian University and in 1919 at the Warsaw Polytechnic. He wrote 76 papers, mostly in differential geometry and mechanics, which met with recognition from abroad. He pioneered the topic of iteration of functions in Polish mathematics. He was a member of many learned societies. See, e.g., Ślebodziński (1969).

[5]This work was published in 1906 in Polish, with an abstract in French, in the 17th volume of Prace Matematyczno-Fizyczne [Mathematical and Physical Works], under the title *O pewnem zagadnieniu z rachunku funkcyj asymptotycznych* [*Sur un problème du calcul des fonctions asymptotiques*]. Sierpiński did not want to publish it in Russian at the time of intense Polish protests.

he got a solid background in analytic function theory and mechanics for his future studies at the Lwów Polytechnic and the University in Leipzig (where he got a doctorate in 1898 under Sophus Lie), as well as his own scholarly work.[6]

Samuel Dickstein (1851–1939) also graduated from the Imperial University (in 1870), and in 1876 obtained a master's degree there (equivalent to a doctorate). He was active as a teacher of mathematics, organizer of scientific life (*Koło Matematyczno-Fizyczne* [Mathematical-Physical Circle] for teachers, *Towarzystwo Naukowe Warszawskie* [Warsaw Learned Society], *Towarzystwo Kursów Naukowych* [Society for Scientific Courses]) and an editor and publisher. He authored over 260 papers and books on algebra, number theory, set theory and history of mathematics (including a monograph *Hoene-Wroński. Jego życie i prace* [Hoene-Wroński. His Life and Works], published by the Academy of Arts and Sciences in Kraków in 1896), and translated foreign works, not only contributing to their dissemination, but also to the creation of Polish mathematical terminology. He was a member of many learned societies, a vice-president of the International Academy of History of Exact Sciences, wrote reviews of Polish mathematical works for the review journal *Jahrbuch über die Fortschritte der Mathematik* and gave a talk at the International Congress of Mathematicians in Heidelberg in 1904.[7] For many years he was an informal leader of the Warsaw mathematical community.[8]

Figure 1. Samuel Dickstein.[9]

[6]Böttcher's work concerned iteration of complex functions and contained many foundational ideas of holomorphic dynamics, subsequently developed independently by Pierre Fatou (1878–1929); Gaston Julia (1893–1978); Samuel Lattès (1873–1918). See Domoradzki and Stawiska (2014) for more information.

[7]See Ciesielski (2018).

[8]For more information see Domoradzki (2015); Kolankowski and Pawlikowska-Brożek (2003).

[9]Photograph from the collection of the Library of the Mathematical Institute of the Polish Academy of Sciences.

No Polish academic structures existed in Warsaw (or anywhere in the Polish Kingdom) at the turn of centuries, but there were some substitutes, both legal and illegal ones.

In 1888 Dickstein joined the Natanson brothers,[10] Władysław (1864–1937), a physicist,[11] and Edward (1861–1940), a chemist and industrialist, along with the mathematician and physicist Władysław Gosiewski (1844–1911),[12] as an editor of a mathematics and physics journal, *Prace Matematyczno-Fizyczne* [Mathematical and Physical Works].[13]

In 1897 he established a new journal, *Wiadomości Matematyczne*, which he ran with his own funds. *Wiadomości Matematyczne* had its own unique character and published articles concerning mathematics (including original research works), its role in the sciences and in the world's development; it also presented people who did mathematics.

Since 1884, along with Aleksander Czajewicz (1843–1926), Dickstein also oversaw the publishing of *Biblioteka Matematyczno-Fizyczna* [Mathematical and Physical Library], a series of academic textbooks and monographs, continued later as *Dzieła i Rozprawy Matematyczne* [Mathematical Works and Dissertations].

He was also involved in another publishing enterprise: in 1898 the first series of *Poradnik dla Samouków* [Guide for the Self-Study] appeared, published by Aleksander Heflich (1866–1936) and Stanisław Michalski (1865–1949).[14] It was a collection of surveys in various disciplines, addressed to those who wished to study at the advanced level but could not do so within official structures (lacking means or formal certificates to enroll at a university, or being excluded on the basis of gender, nationality, religion, class status or political views). Each article was written at the elementary, intermediate or advanced level. Dickstein wrote the parts devoted to mathematics and history of mathematics. Articles in history of mathematics by Feldblum, Władysław Smosarski (1876–1960, candidate degree in mathematics at the Imperial University in 1899 with a gold medal) and Władysław Kwietniewski

---

[10]Dickstein was related to the Natansons through his marriage to their cousin Paulina Natanson (1861–1928), a talented pianist and a student of Ignacy Jan Paderewski (1860–1941). Both families were representatives of the progressive Polish-Jewish community of entrepreneurs, scholars and social activists. They either adhered to the Reform Judaism [Władysław and Edward Natanson's father, Ludwik Natanson (1822–1896), a physician, co-founded the Reform synagogue in Warszaw in Tłomackie street; Samuel Dickstein served on the executive committee of the Warsaw Jewish Community, concerning himself with Jewish schools] or converted to Catholicism (Władysław and Edward Natanson). For more information see Cała (1989).

[11]Władysław Natanson was one of the most distinguished Polish physicists and a professor at the Jagiellonian University. For more information, see Chapter 4 and the references therein.

[12]Gosiewski studied at the Warsaw Main School and in Paris. He worked as a high school teacher and as a clerk in the Credit Society. He published over 67 works in probability, theory of elasticity and theoretical physics.

[13]For more information see Duda (1980, 1997).

[14]Additional copies were printed in 1901. In 1915 a new series started.

(1837–1902, a former professor of the Main School), were also included.[15] The series was financed by *Kasa im. Józefa Mianowskiego* [The Mianowski Fund],[16] established in 1881 by a group of alumni of the Warsaw Main School. The purpose of the Mianowski Fund was to support research and publication of scholarly works by Poles and in time, it became the largest organization of this type.[17]

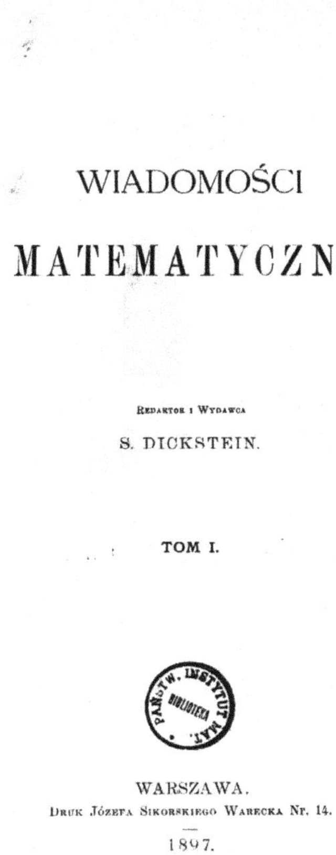

P.202

WIADOMOŚCI

MATEMATYCZNE.

Redaktor i Wydawca

S. DICKSTEIN.

TOM I.

WARSZAWA.

Druk Józefa Sikorskiego Warecka Nr. 14.

1897.

Figure 2. Cover of *Wiadomości Matematyczne.*[18]

[15]See Pawlikowska-Brożek (1992).

[16]Józef Mianowski (1804–1879) was a physician and a rector of the Warsaw Main School. He shielded students from repressions for their participation in the January Uprising in 1863–1864.

[17]Among many others, the Mianowski Fund financed the books *O związkach między zasadą najmniejszego działania, a najprawdopodobniejszym układem* [On Connections Between the Least Action Principle and the Most Probable Configuration] by Władysław Gosiewski (1888) and *Wstęp do fizyki teoretycznej* [An Introduction to Theoretical Physics] (1890) by Władysław Natanson. See Domoradzki (1992, 2015).

[18]From the collection of the Library of the Mathematical Institute of the Polish Academy of Sciences.

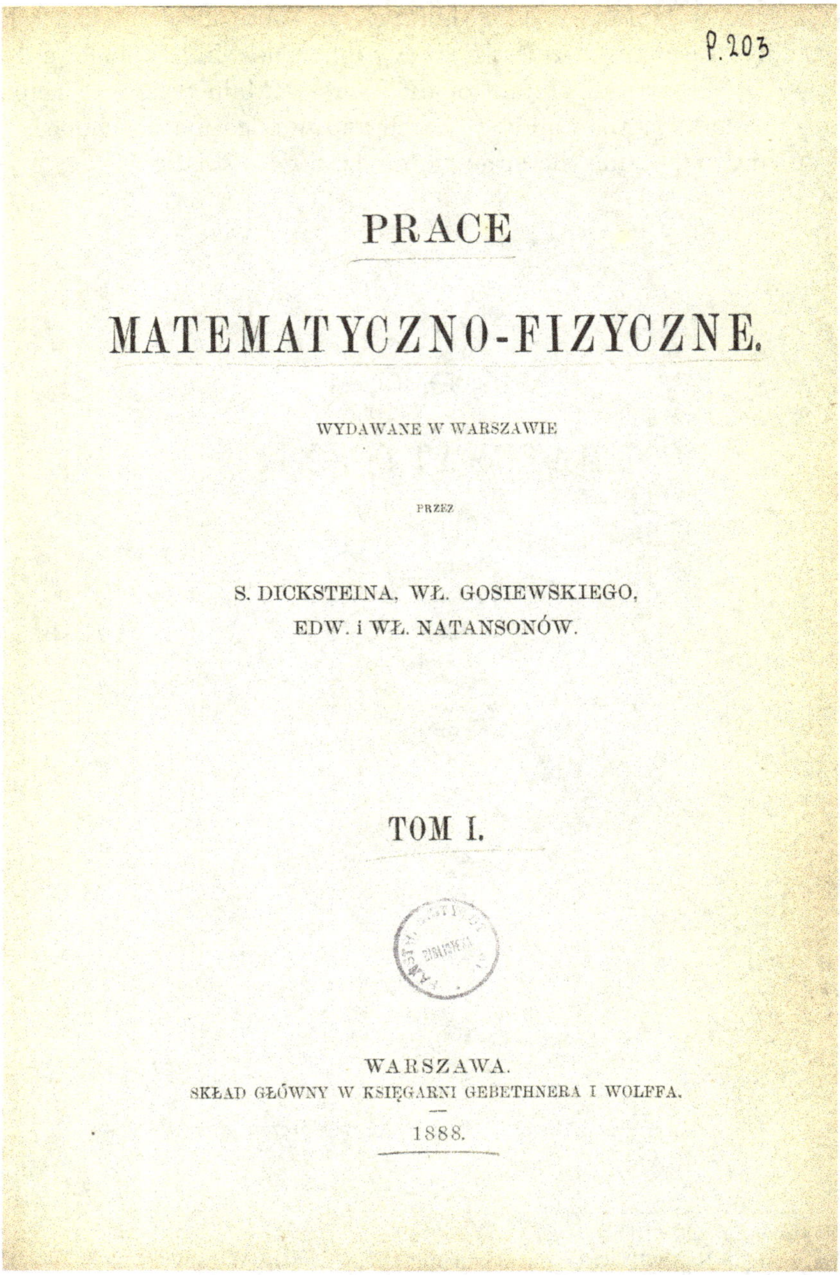

PRACE

MATEMATYCZNO-FIZYCZNE.

WYDAWANE W WARSZAWIE

PRZEZ

S. DICKSTEINA, WŁ. GOSIEWSKIEGO,
EDW. i WŁ. NATANSONÓW.

TOM I.

WARSZAWA.
SKŁAD GŁÓWNY W KSIĘGARNI GEBETHNERA I WOLFFA.

1888.

Figure 3. Cover of *Prace Matematyczno-Fizyczne.*[19]

[19]From the collection of the Library of the Mathematical Institute of the Polish Academy of Sciences.

An important initiative of the late 19th century Warsaw, illegal from the viewpoint of Russian authorities, was the *Uniwersytet Latający* [Flying University], organized by Jadwiga Szczawińska-Dawidowa (1864–1910)[20] and operating in 1884–1905. The instruction was in Polish and included, among others, courses in Polish history and literature, all of which violated official policies. The classes were financed by the students' dues and met in private apartments, often changing location for fear of the police (hence the name). Initially this clandestine university served the educational needs and ambitions of women,[21] who were not allowed to study at academic institutions in the Russian Empire (making for another violation), and thus it was also known as the *Uniwersytet Babski* [Petticoat University], but then expanded also to men. Dickstein and Władysław Kwietniewski gave lectures in mathematics.[22] The authorities mostly turned a blind eye to the activities of the Flying University, despite their illegal character.

On January 28, 1905, the Polish students of the Warsaw academic schools initiated a strike, which soon spread to high schools and to other cities in the Kingdom. The strikers postulated re-introduction of Polish as the language of instruction, abolition of police supervision of students, abolition of admission quota on the basis of nationality, religion and class status, and granting women access to higher education. The Polish youth boycotted Russian schools. As a result, the university suspended operations in the years 1904–1908 and some of the Polytechnic faculty were relocated to Novocherkassk in Southern Russia.

On the other hand, the authorities allowed the establishment of more Polish-language private schools, which enjoyed more liberties but did not carry full accreditation, and legalized the Society for Scientific Courses, which replaced the Flying University. The lineup of instructors in mathematics at the Society was impressive.

Besides Dickstein, there were Stefan Kwietniewski (1874–1940, PhD from the University of Zürich), Juliusz Rudnicki (1881–1948, a graduate of Sorbonne, PhD from the Jagiellonian University in Kraków), Zygmunt Straszewicz (1860–1927, a graduate in mechanical engineering of Eidgenössische Technische Hochschule in Zürich and a former student of mathematics at the Imperial University), Lucjan Zarzecki (1873–1925, a graduate of the St. Petersburg University), Tadeusz Banachiewicz (1882–1954, a graduate of the Imperial University with a gold medal, PhD in astronomy from the University of Kazan) and Sierpiński (additional studies in Göttingen, PhD from the Jagiellonian University, Kraków), Zygmunt Janiszewski (1888–1920) and Stefan Mazurkiewicz (1888–1945), the future founders of the

[20]Jadwiga Szczawińska-Dawidowa was an educator, a writer and a social activist. She was married to an educator and a psychologist, Jan Władysław Dawid (1859–1914), who also lectured at the Flying University.

[21]The future Nobel Prize winner in physics and chemistry, Maria Skłodowska-Curie (1867–1934), was a student at the Flying University.

[22]For more information see Bartnicka (2014); Cywiński (1971).

Warsaw Mathematical School. Among students, there was Stefan Straszewicz (1889–1983, PhD from the University of Zürich in 1914). Later, he also taught at the Society. But the creation of a full-fledged Polish university in Warsaw had to wait for a different occasion.

1.2. *A Prelude to the Warsaw Mathematical School: Sierpiński, Mazurkiewicz and Janiszewski at the Lwów University*

At the turn of the 19th and 20th centuries, new directions and methods emerged in mathematics, thanks to the works of Georg Cantor (1845–1918), Felix Klein (1849–1925), Felix Hausdorff (1868–1942), Sophus Lie (1842–1899), David Hilbert (1862–1943), Ernst Zermelo (1871–1953), Giuseppe Peano (1858–1932), Henri Poincaré (1854–1912), Émile Borel (1871–1956), Henri Lebesgue (1875–1941), Maurice Fréchet (1878–1973), Bertrand Russell (1872–1970), Alfred North Whitehead (1861–1947), Luitzen Egbertus Jan Brouwer (1881–1966) and others.

Polish mathematicians educated in Leipzig, Göttingen, Paris and other foreign academic centers became acquainted with theory of groups of transformations, set theory, topology, and measure theory, and developed them in their own research. At that time the only Polish-language academic institutions were located under the Austro-Hungarian occupation, in Kraków and Lwów, and this is where modern mathematical ideas were first transplanted and allowed to flourish.

Żorawski and Stanisław Zaremba (1863–1942) were passing their interests to a new generation of mathematicians at the Jagiellonian University in Kraków.[23] At the Lwów University, Józef Puzyna (1865–1919), a former student of Karl Weierstrass and a professor in the first chair of mathematics, actively worked on creating a modern mathematical center.[24] He understood the importance of set theory and pioneered its applications to complex function theory in his monograph;[25] he wanted to bring in mathematicians familiar with the modern approaches. He offered a position to Sierpiński in 1908, who promptly got his habilitation and in 1910 became an extraordinary professor in the newly created second chair of mathematics.[26]

Sierpiński's research in the Lwów period partially concerned number theory (his early interest), but he also took up new problems in set theory. Up to 1914, he published 75 works in mathematics, including three books in Polish: *Teoria liczb niewymiernych* [The Theory of Irrational Numbers] (1910), *Zarys teorii mnogości* [An Outline of Set Theory] (1912), and *Teoria liczb* [Number Theory] (1912). Some

[23]The received view is that Żorawski and Zaremba "did not create a school". See Borsuk (1978); Kuratowski (1980); Przeniosło (2010). We present evidence against it in Chapter 4.

[24]For more information see Domoradzki (2017).

[25]*Teorya funkcyj analitycznych* [Theory of Analytic Functions], Volume I, 1898, and Volume II, 1900.

[26]Sierpiński taught advanced analysis and set theory, advanced algebra and applications of set theory to analysis, and ran the lower-level mathematical seminar.

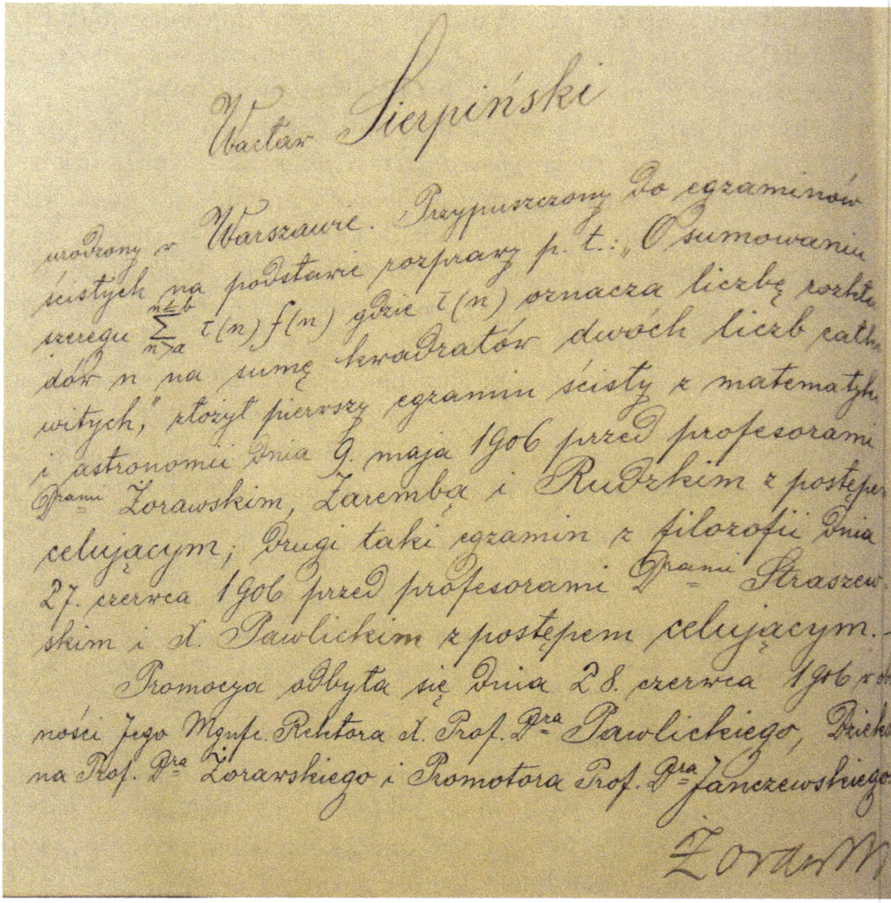

Wacław Sierpiński

urodzony w Warszawie. Przypuszczony do egzaminów ścisłych na podstawie rozprawy p. t.: „O sumowaniu szeregu $\sum_{n>a}^{n\le b} \tau(n) f(n)$ gdzie $\tau(n)$ oznacza liczbę rozkładów n na sumę kwadratów dwóch liczb całkowitych," złożył pierwszy egzamin ścisły z matematyki i astronomii dnia 9. maja 1906 przed profesorami Drami Żorawskim, Zarembą i Rudzkim z postępem celującym; drugi taki egzamin z filozofii dnia 27. czerwca 1906 przed profesorami Drami Straszewskim i St. Pawlickim z postępem celującym.—

Promocya odbyła się dnia 28. czerwca 1906 r. w obecności Jego Magnif. Rektora St. Prof. Dra Pawlickiego, Dziekana Prof. Dra Żorawskiego i Promotora Prof. Dra Janczewskiego.

Żorawski

Figure 4. Minutes of Sierpiński's doctoral exams and degree award at the Jagiellonian University, Kraków.[27]

of his achievements of that period became mathematical classics; for example in 1912 he published his construction of a square-filling curve in the article *Sur une nouvelle courbe continue qui remplit toute une aire plane.*[28]

Sierpiński gathered around himself a group of young mathematicians. First there were Janiszewski and Mazurkiewicz, educated at leading European universities.[29]

[27]Photograph from the Archives of the Jagiellonian University.

[28]Bulletin International de l'Academie Polonaise des Sciences et des Letters. *Classe de Sciences Mathematiques et Naturelles. Série A* **8**(1912), pp. 462–478. For more information see Ciesielska (2012); Domoradzki (2012); Domoradzki and Zaricznyi (2014); Domoradzki *et al.* (2018); Schinzel (1971); Sierpiński (1959).

[29]See, e.g., Duda (2019); Kuratowski (1969).

Janiszewski studied in Zürich, Munich, Göttingen, Marburg and Graz. He obtained his PhD in Paris (thesis *Sur les continus irreductibles entre deux points*) in 1911 under Lebesgue,[30] in 1913 he got his habilitation in Lwów.

Mazurkiewicz studied in Kraków, Munich and Göttingen and got his PhD in Lwów in 1913. In his thesis *Przyczynki do teorii mnogości* [Contributions to Set Theory] he proved some results concerning planar continua and dealt with some aspects of the definition of dimension.[31] Soon others followed in their footsteps.

In 1913 Stanisław Ruziewicz (1889–1941)[32] defended his PhD thesis *O funkcyi ciągłej monotonicznej nie posiadającej pochodnej w nieprzeliczalnej mnogości punktów* [On Continuous Monotone Function without Derivative at Uncountable Set of Points] (under the supervision of Puzyna, as Sierpiński was not an ordinary professor yet).

Another future mathematician who came under Sierpiński's substantial influence in Lwów was Otto Nikodym (1887–1974), then a student in mathematics and physics (graduation 1911).

The group worked on set theory, topology and applications of set theory to real analysis, making rapid advances in these relatively new disciplines. In 1914 a joint paper by Sierpiński and Mazurkiewicz appeared in *Comptes Rendus de l'Academie de Paris*. As observed by Sierpiński's later student, it was the first joint publication of two Polish mathematicians.[33]

The young Lwów mathematicians were also in close contact with a group of Lwów philosophers gathered around Kazimierz Twardowski (1866–1938) and constituting the first generation of the philosophical Lwów–Warsaw school. Among them, Jan Łukasiewicz (1878–1956, professor of philosophy in Lwów in 1911–1915)[34] and Stanisław Leśniewski (1886–1939, doctorate with Twardowski in 1912) were developing interest in mathematical logic. Sierpiński's and Janiszewski's habilitation lectures, given before the Council of the Faculty of Philosophy (where mathematics belonged), had a philosophical character: Sierpiński's title was *Pojęcie odpowiedniości w matematyce* [The Notion of Correspondence in Mathematics], Janiszewski's was *O realizmie i idealizmie w matematyce* [On Realism and Idealism in Mathematics] (both talks were published in Przegląd Filozoficzny [Philosophical Survey], in 1909 and 1916, respectively).[35]

---

[30]He was able to nostrify it in Lwów only in 1917.

[31]In his review Sierpiński considered Mazurkiewicz's thesis outstanding.

[32]Ruziewicz became a professor of mathematics in Lwów. After falling out with the authorities and losing his chair at the Lwów University, he took a chair at the Academy of Commerce. Killed by the Germans along with other Lwów professors soon after the Nazi invasion of Lwów.

[33]See Schinzel (1984).

[34]As a student at the Lwów University, Łukasiewicz took courses in mathematics from Puzyna.

[35]See Murawski (2014).

World War I cut this fruitful collaboration. Sierpiński, who traveled to the Russian Empire for his vacation, was interned as an enemy alien (first in Vyatka, and then in Moscow). Janiszewski enlisted in the Polish Legions and fought on the side of the Central Powers. Mazurkiewicz found himself in Warsaw, where he engaged in committees working on establishing a Polish university and polytechnic.

2. Establishment of Polish Academic Institutions in Warsaw and the Beginnings of the Warsaw Mathematical School

At the outbreak of World War I the occupying powers sought the support of Poles for their sides and were willing to make some concessions. The Russians promised "self-government" for Polish territories unified under the rule of tsars, but soon had to retreat as the Central Powers advanced. In the spring and summer of 1915 the Imperial University was evacuated to Moscow (and later to Rostov-on-Don) and the Polytechnic was evacuated to Nizhny Novgorod. All files, libraries and equipment were taken away. The Germans captured Warsaw on 5 August, 1915, and on 4 September the German general Hans von Beseler (1850–1921) became Governor-General of the Warsaw general governorate. In line with the concept of *Mitteleuropa*[36] preparations started for creating a buffer Polish state. An important role was to be played by Polish-language academic institutions. In the fall of 1915 Beseler gave a statute to the University of Warsaw and the Warsaw Polytechnic. Zygmunt Straszewicz became the first rector of the Warsaw Polytechnic. At the university lectures in mathematics were given by Dickstein (algebra), Rudnicki (analytic geometry) and S. Kwietniewski (projective geometry). The first two chairs of mathematics were given to Janiszewski and Mazurkiewicz.[37] Starting in 1916, they ran a seminar in topology, probably the first one in the world in this discipline. Łukasiewicz became a professor of philosophy.

In 1917 the new journal *Nauka Polska. Jej potrzeby, organizacja i rozwój* [Polish Science. Its Needs, Organization and Development], published by the Mianowski Fund, announced a survey on mathematics. Janiszewski's short response *O potrzebach matematyki w Polsce* [On the Needs of Mathematics in Poland] not only addressed organizational and professional aspects of scientific life, but also declared the direction of the future development of Polish mathematics.[38] The postulates

---

[36] *Mitteleuropa* meant unity of East Central and South East Europe under political and economical control of Germany, possibly along with Austro-Hungary. Outlined by the German politician Friedrich Naumann (1860–1919) in a book of eponymous title published in 1915 as a liberal idea emphasizing common interests and culture, it later took a nationalist and expansionist character. See Górny (2015).

[37] Because of his military service, Janiszewski was available only part-time. Meanwhile, the University of Lwów applied for his release from the army to resume teaching alongside Puzyna, in the absence of Sierpiński.

[38] See Janiszewski (1963). For the English translation, see Kuzawa (1968).

became later known as the "Janiszewski Program". Among other things, Janiszewski advocated attracting talented youth to mathematics, concentrating research in one discipline and creating a specialized journal devoted to this discipline, as this would let Polish mathematics advance and raise its status. He did not name any particular discipline, but with certainty he meant topology.[39] It was still a relatively new discipline, in which he worked himself and collaborated with others, and in which he saw a chance for Polish mathematics to get ahead and meet with recognition from abroad. His response was grounded in what he already did in Lwów with his colleagues Sierpiński and Mazurkiewicz (at the initiative of Puzyna, whose idea was to establish a modern mathematical center) and continued to do with Mazurkiewicz (and soon with Sierpiński again) in Warsaw.[40]

The independent Polish state was proclaimed on November 11, 1918. Janiszewski returned from his hiding in the Radom region, where he lived after the Oath Crisis[41] and ran a shelter and an elementary school for orphaned children. At the university he taught analytic geometry (with recitations) and continued to run an advanced seminar in topology. Some other faculty members who were teaching at the University at its inception continued to do so. Dickstein was made an honorary professor in mathematics and history of mathematics.[42] Kwietniewski taught projective geometry as a *docent tymczasowy* [temporary docent]. Rudnicki obtained his habilitation in Kraków in 1921 and in 1921–1923 was an extraordinary professor of mathematics at the Faculty of Mechanics of the Warsaw Polytechnic.[43] In 1923 he moved to the newly opened University of Wilno (Vilnius) upon being named an ordinary professor of mathematics there. But overall the University of Warsaw and the Warsaw Polytechnic expanded. Sierpiński came back from the internment in Russia, where he managed to collaborate in 1918 with Moscow mathematicians, in particular Nikolai Nikolaevich Lusin (1883–1950). He spent a short time in Lwów to finalize the proceedings of Ruziewicz's habilitation, but ultimately went to Warsaw, where he accepted a chair of mathematics at the university.

Leśniewski obtained habilitation in 1918 and became a professor of philosophy of mathematics (holding what was then probably the only such chair in the

[39] Janiszewski's program did not exclude other mathematical disciplines. Writing in 1919 in support of the program, Mazurkiewicz acknowledged Kraków's chance to become an important center in geometry.

[40] For more information see Duda (2014); Garrido and Wybraniec-Skardowska (2018); Kauffman (2015); Pawlikowska-Brożek (1992).

[41] The Polish Legions were supposed to become a part of the *Polnische Wehrmacht* [*Polska Siła Zbrojna*], dependent on the German and Austro-Hungarian authorities. Most soldiers and officers refused to swear an oath of loyalty to the German emperor. Depending on their pre-war citizenship, they were either interned or enlisted into the Austro-Hungarian army. Some left the military forces and went into hiding.

[42] See Gleichgewicht (2002).

[43] See Garlicki (1982).

Figure 5. Wacław Sierpiński.[44]

world). Kazimierz Ajdukiewicz (1890–1963), a former student of philosophy and mathematics in Lwów (doctorate with Twardowski in 1912), got his habilitation at the University of Warsaw for the dissertation *Z metodologii nauk dedukcyjnych* [On Methodology of Deductive Sciences] and was an extraordinary professor of philosophy in the years 1925–1928.[45] Antoni Przeborski (1871–1941),[46] a repatriate from Kharkiv,[47] took a chair of theoretical mechanics at the University of Warsaw.

Żorawski left his chair at the Jagiellonian University in Kraków to become a professor at the Warsaw Polytechnic in 1919, also giving lectures in differential geometry at the University. Witold Pogorzelski (1895–1963), who got his PhD in 1919 at the Jagiellonian University in Kraków and his habilitation in 1921 also there, replaced Rudnicki in the chairs of mathematics at the Mechanical and Electrical Faculties of the Polytechnic. He gave commissioned lectures at the University in integral and differential equations and in theoretical physics. S. Straszewicz was a replacement of Żorawski, who transferred to the University in 1926. He lectured as a *docent* at

[44]Photograph from the collection of the Library of the Mathematical Institute of the Polish Academy of Sciences.
[45]Later Ajdukiewicz was a professor at the universities of Lwów, Poznań and Warsaw.
[46]Przeborski graduated from the University of Kiev, got his master's degree in Moscow and supplemented his studies in Göttingen. He held a chair of mathematics at the Kharkiv University since 1902, becoming a full professor in 1908.
[47]On the basis of the Treaty of Riga in 1921, ending the Polish–Bolshevik war.

the University, the Polytechnic, Officers' Schools of Topographers and at the *Szkoła Główna Gospodarstwa Wiejskiego* [Major School of Rural Economy]. After getting his habilitation in 1926 at the University of Warsaw (thesis *Über die Zerschneidung der Ebene durch Abgeschlossene Menge*) he became an extraordinary contract professor in 1927 at the Warsaw Polytechnic (promoted from a deputy professor) and in 1928 he took up management of the chair of mathematics in the faculties of civil engineering and hydraulic engineering (later combined).

Figure 6. Franciszek Leja (1885–1979) and Otakar Borůvka (1899–1995), Eighth Congress of Polish Mathematicians, 1953.[48]

In 1925 Franciszek Leja (1885–1979, habilitation 1922) came from Kraków to become an extraordinary professor of mathematics at the Polytechnic at the Faculty of Chemistry. He also lectured at the University as a *docent*. Classes in mathematics at the Polytechnic (including descriptive and projective geometry) were also taught by Przeborski, Ludomir Wolfke (1882–1937), Paweł Węziagolski (1883–1929, a graduate of mathematics and architecture at the St. Petersburg University, an architect), Stanisław Garlicki (1875–1935, a graduate of the University of Paris), Wacław Gniazdowski (1864–1938, a textile engineer), Stefan Bóbr (1886–1958), Henryk Czopowski (1863–1935, a professor of theoretical mechanics), Leon Marek Suzin (1901–1976, an architect, a lecturer in the Chair of Descriptive Geometry and Perspective), Julian Bonder (1900–1975, a graduate of the Warsaw Polytechnic and Sorbonne, specializing in fluid dynamics, PhD 1931, habilitation 1935) and Kazimierz Zarankiewicz (1902–1959, see below). Stanisława Nikodymowa (1897–1988, see below) worked as an assistant in Leja's chair. Stefan Kulczycki (1893–1960), a former student of Stanisław Zaremba, worked in Pogorzelski's chair.[49]

[48]Photograph from the collection of the Library of the Mathematical Institute of the Polish Academy of Sciences.
[49]See Pawlikowska-Brożek (1988).

Janiszewski died during the Spanish flu epidemic before he could see fulfillment of some of his postulates. But the preparations for the specialized mathematical journal *Fundamenta Mathematicae* were set in motion in his lifetime. The journal was devoted to set theory and its applications, mainly to topology and the theory of real functions, and published in "international" languages. The first issue came out in 1920, shortly after Janiszewski's death. His joint work with Kazimierz Kuratowski (1896–1980, until 1921 Kuratow, see below) appeared posthumously in this issue, along with articles by Mazurkiewicz, Sierpiński, Ruziewicz, Stefan Banach (1892–1945), Hugo Steinhaus (1887–1972) and Witold Wilkosz (1891–1941). The choice of papers only by Polish authors was not merely pragmatic, but purposeful: in accordance with Janiszewski's wish it showed that there was a community of mathematicians in now-independent Poland with common research interests. In 1962 Kuratowski wrote that the publication of the first volume of *Fundamenta Mathematicae* was the first indicator of the Polish mathematical school. The journal got noticed abroad and was praised, for example, by Lebesgue. Despite some concerns, there was no shortage of suitable material: 32 volumes appeared before 1939.[50]

Figure 7. Polish mathematicians at the International Congress of Mathematics in Oslo, 1936: Banach, Knaster, Zarankiewicz, Kuratowski.[51]

Janiszewski emphasized that any mathematician needed collaborators and a research community in general. The initial years of the University of Warsaw were a period of intense growth for the mathematical community. In addition to

[50]See Duda (2016); Kuratowski (1963); Kuzawa (1968, 1970); Przenioslo (2009); Świątkowski (1988).
[51]Photograph from the collection of the Library of the Mathematical Institute of the Polish Academy of Sciences.

those already mentioned in connection with the Warsaw Polytechnic, there were many others. The first person to obtain a PhD in mathematics from the new University of Warsaw was Aleksander Rajchman (1895–1940), a graduate of the University of Paris, who was employed as an assistant in the Mathematical Seminar run by Sierpiński and Mazurkiewicz. In 1919 he submitted a PhD thesis *O szeregach trygonometrycznych sumowalnych metodą Poissona* [On Trigonometric Series Summable by Poisson's Method] (published in *Prace Matematyczno-Fizyczne* **30**(1919), pp. 19–88).[52] However, the award of his degree was delayed due to a political argument with Mazurkiewicz until 1921.[53]

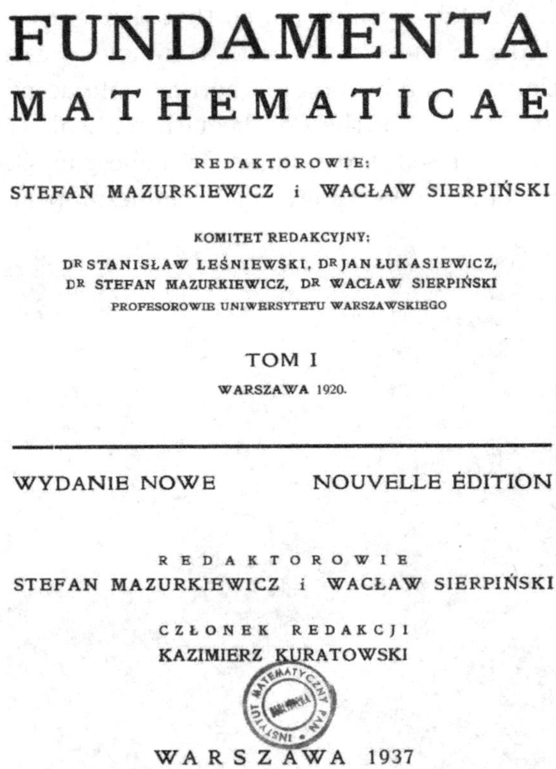
FUNDAMENTA
MATHEMATICAE

REDAKTOROWIE:
STEFAN MAZURKIEWICZ i WACŁAW SIERPIŃSKI

KOMITET REDAKCYJNY:
DR STANISŁAW LEŚNIEWSKI, DR JAN ŁUKASIEWICZ,
DR STEFAN MAZURKIEWICZ, DR WACŁAW SIERPIŃSKI
PROFESOROWIE UNIWERSYTETU WARSZAWSKIEGO

TOM I
WARSZAWA 1920.

WYDANIE NOWE NOUVELLE ÉDITION

REDAKTOROWIE
STEFAN MAZURKIEWICZ i WACŁAW SIERPIŃSKI

CZŁONEK REDAKCJI
KAZIMIERZ KURATOWSKI

WARSZAWA 1937
SEMINARIUM MATEMATYCZNE, OCZKI 3
PARIS, LIBRAIRIE GAUTHIER-VILLARS ET Cie
QUAI DES GRANDS AUGUSTINS, 55

Figure 8. Cover of *Fundamenta Mathematicae*.[54]

[52]In 1914 and 1915 Rajchman applied for starting the PhD proceedings at the Jagiellonian University, but was rejected because he lacked certification of attending a sufficient number of lectures. In 1914/1915 he visited the University of Vienna supported by Władysław Kretkowski Fund.

[53]See Maligranda and Piotrowski (2017).

[54]From the collection of the Library of the Mathematical Institute of the Polish Academy of Sciences.

Some students in mathematics who were graduating in the 1920s got interested in topology and quickly obtained doctorates: Kazimierz Kuratowski (PhD 1921, the University of Warsaw, thesis published under the title *Sur l'operation A de l'Analysis Situs* in *Fundamenta Mathematicae* **3**(1922), pp. 182–199, supervisor Sierpiński, replacing deceased Janiszewski), Stanisław Saks (1897–1942, PhD 1922, the University of Warsaw, thesis *Przyczynek do topologii powierzchni i obszarów płaskich* [A Contribution to Topology of Surfaces and Planar Domains], published in part as *Sur l'homéomorphie des varietes à deux dimensions*, *Fundamenta Mathematicae* **5**(1924), pp. 288–320, supervisor Mazurkiewicz), Bronisław Knaster (1893–1980, PhD 1923, thesis *Un continua dont tout sous-continus est indécomposable*, *Fundamenta Mathematicae* **3**(1922), pp. 247–286, supervisor Mazurkiewicz), Zarankiewicz (PhD 1923, thesis *Sur les points de division dans les ensembles connexes*, *Fundamenta Mathematicae* **9**(1927), pp. 124–171, supervisor Sierpiński).[55] Antoni Zygmund (1900–1992) developed interest in trigonometric series under the influence of Rajchman (PhD 1923, thesis *O metodzie Riemanna w teorii szeregów trygonometrycznych* [On Riemann's Method in the Theory of Trigonometric Series], published as *Sur la théorie riemanienne des séries trigonométriques*, *Comptes Rendus Hebdomadaires des Séance de l'Academie des Sciences*, Paris, 1923, pp. 521–523, supervisor Mazurkiewicz).

Figure 9. Antoni Zygmund (1900–1992), after World War II.[56]

[55] Between the two wars Zarankiewicz worked on connected sets and division of the plane. He also studied the conformal representation of a doubly connected set in the plane onto an annulus. Later he switched his interests to graph theory. After World War II a professor of mechanics at the Warsaw Polytechnic. Active in the International Astronautical Federation. See Bergman *et al.* (1964).

[56] Photograph from the collection of the Library of the Mathematical Institute of the Polish Academy of Sciences.

A few mathematicians whose work was interrupted by war or delayed by material difficulties also took the opportunity to complete their degrees.

Władysław Ślebodziński (1884–1972), a graduate of the Jagiellonian University (1903–1908), took a position at the *Państwowa Wyższa Szkoła Budowy Maszyn* [Higher State School of Machine Construction] in Poznań[57] in 1921 and got his PhD with Żorawski in Warsaw in 1928 (thesis *O pewnej klasie przestrzeni Riemanna* [On a Class of Riemann Spaces], Sierpiński signed as supervisor).[58]

Nikodymowa returned to her studies in Warsaw after having taught basic classes in mathematics to army recruits and got her PhD in 1925 (thesis *Sur les coupures du plain faites par les ensembles connexes et les continus*, *Fundamenta Mathematicae* **7**(1925), pp. 15–23, supervisor unknown). Her husband, who taught high school in Kraków since 1911, got his PhD on the same day as she did (thesis based on the works *Sur une propriété de l'opération A*, *Fundamenta Mathematicae* **7**(1925), pp. 149–154 and *Sur les points linéairement acessibles des ensembles plans*, *ibid.*, pp. 250–258, supervisor Sierpiński).[59]

Franciszek Sieczka (1890–1962), a Roman Catholic priest ordained in 1916, interrupted his studies in mathematics to serve as a chaplain in the Polish–Bolshevik war in 1919–1921. He obtained his PhD in 1924 (thesis Naturalna definicja krzywizny [A Natural Definition of Curvature], supervisor Sierpiński).[60]

Tadeusz Felsztyn (until 1920 Feldstein, 1897–1963), a native of Lwów, studied at the Lwów Polytechnic and the University of Göttingen before World War I. During the war he served in the Polish Legions and in 1918 took part in the Defense of Lwów. He continued serving in the Polish military and fought in the Polish–Bolshevik war. He defended his PhD in 1925 as an active officer in the rank of Major and the director of the *Centralna Szkoła Strzelecka* [Central Shooting School] in Toruń (thesis *Topologiczna definicja czaszy* [On the Topological Definition of a Cap], supervisor Mazurkiewicz).[61]

[57]The city of Poznań was mostly under Prussian (German) occupation after 1795. It returned to independent Poland as a result of the Greater Poland Uprising, December 1918–February 1919.

[58]Ślebodziński taught in Poznań until 1939. During the German occupation he got involved in clandestine teaching, for which he was arrested and imprisoned in the Auschwitz concentration camp. After World War II a professor of mathematics at the University of Wrocław. Worked mainly in differential geometry. See Ślebodziński (1969).

[59]See Ciesielska (2017a,b); Derkowska (1983); Piotrowski (2014); Szymański (1990).

[60]After obtaining his PhD Sieczka taught mathematics in the Lower Seminary in Płock. He published a few more scholarly papers and problem sets in mathematics for high schools. Later he was entirely engaged in pastoral service in the village of Radzanów (Mazovia). See Piotrowski (2015).

[61]In the 1930s Felsztyn, promoted to Lieutenant Colonel (*Podpułkownik*), was the director of the Centralne Biuro Balistyczne [Central Ballistic Bureau] in Warsaw and a lecturer of ballistics at the Warsaw Polytechnic. Taken a prisoner of war by the Soviets in 1939 and released in 1941, he joined the army of Gen. Władysław Anders (1892–1970) and commanded the 19th Regiment of Infantry. After World War II he lived and taught high school mathematics in Great Britain.

Kazimierz Abramowicz (1888–1936), a repatriate from Kiev, nostrified his Russian degree in 1922 (thesis *O funkcjach hypergeometrycznych z dowolną ilością punktów pozornie osobliwych* [On Hypergeometric Functions with an Arbitrary Number of Removable Singularities], supervisor Mazurkiewicz). A former extraordinary professor at the University of Perm and a lecturer at the Kiev Polytechnic on his way to start an academic appointment at the new University of Poznań, he took advantage of the more favorable rules for degree nostrification at the University of Warsaw (no residency requirement, thesis could have been published prior to submission).

Another repatriate getting his PhD in mathematics at the University of Warsaw in the early 1920s was Jerzy Spława-Neyman (1894–1981, later Neyman). A former student and lecturer at the University of Kharkiv interested in measure theory and mathematical statistics, he took a job as a senior statistical assistant at the State Scientific Agricultural Institute in Bydgoszcz after his arrival in Poland.[62] Soon he traveled to Warsaw in search of better opportunities and became an assistant at the university, thanks to Przeborski. He also taught at the *Szkoła Główna Gospodarstwa Wiejskiego* (*SGGW*) [Major School of Rural Economics] (an academic-level institution) and worked at the *Państwowy Instytut Meteorologiczny* [State Meteorological Institute]. His PhD thesis was related to one of his publications on the application of statistics to agriculture (*Próby uzasadnienia rachunku prawdopodobieństwa do doświadczeń polowych* [Attempts at Justifying Calculus of Probability in Field Experiments], *Roczniki Nauk Rolniczych* [Annals of Agricultural Sciences] **10**(1923), pp. 1–51, partially translated as Splawa-Neyman (1990)). The examination committee in mathematics consisted of Sierpiński, Mazurkiewicz and Przeborski (there was nobody able to examine in statistics at the academic level, as Mazurkiewicz was only developing an interest in probability about that time). Meanwhile, Sierpiński chose one of the theoretical works of Neyman for publication in *Fundamenta Mathematicae*; it appeared as *Sur un théorème métrique concernant les ensembles fermés*, *Fundamenta Mathematicae* **5**(1924), pp. 328–330.[63]

3. An Unprecedented Collaboration Between Mathematicians and Logicians

Mathematical logic was important in the Warsaw School of Mathematics since its inception. In the 1915 edition of *Poradnik dla samouków* [The Guide for Self-Study] Janiszewski presented "logistics" (as it was then sometimes called), namely, logic

[62]The institute was established in 1902 by the Prussian government as the first research institute in Polish territories under the German occupation. Named *Kaiser Wilhelms Institut für Landwirtschaft zu Bromberg*, it started operating in 1906. After Bromberg/Bydgoszcz was incorporated into the Polish state in 1920, it was renamed *Państwowy Instytut Naukowo-Rolniczy* [State Scientific Agricultural Institute]. Kazimierz Bassalik (1897–1960), a botanist and microbiologist, was its first director in Poland.

[63]For more information see Reid (1998).

done by methods specific to mathematics. He stressed autonomous character of logic as a mathematical discipline, not a mere tool of mathematics, although he acknowledged the usefulness of logic in the analysis of mathematical reasoning. In 1918 the atmosphere at the University was favorable for appointing Leśniewski as professor of philosophy of mathematics. Sierpiński evaluated his habilitation thesis, consisting of the works *Podstawy ogólnej teorii mnogości. I.* [Foundations of the General Set Theory, I] and *Krytyka logicznej zasady wyłączonego środka* [A Criticism of the Logical Principle of the Excluded Middle]. During the Polish–Bolshevik war, Sierpiński, Leśniewski and Mazurkiewicz worked together as codebreakers for the Cipher Bureau of the General Staff of the Polish Army. This work could have motivated Leśniewski's two publications outside of logic, in group theory (*Über Funktionen, deren Felder Gruppen mit Rücksicht auf diese Funktionen sind, Fundamenta Mathematicae* **13**(1929), pp. 319–332, and *Über Funktionen, deren Felder Abelsche Gruppen in bezug auf diese Funktionen sind, ibid.*, **14**(1929), pp. 242–251).

When the journal *Fundamenta Mathematicae* was launched, its issues were supposed to alternate between mathematical and logical problematics. In practice, mathematical submissions dominated. Further, because of Sierpiński's criticism of Leśniewski's article on the foundations of mathematics, Leśniewski and Łukasiewicz resigned from the editorial board in 1928. Despite this fallout, Warsaw mathematicians and logicians collaborated with each other in many aspects. Since Warsaw logicians also had ties with the strong philosophical community of Warsaw, it is fit to quote here Edward Poznański (1901–1974), a graduate of mathematics, who described Warsaw mathematics, logic and philosophy of the interwar period as the *three sides of a triangle*.[64]

The University of Warsaw was reorganized in 1927 in line with ministerial educational reforms. Mathematics was assigned to the *Wydział Matematyczno-Przyrodniczy* [Faculty of Mathematics and Natural Sciences], while philosophy became part of the *Wydział Humanistyczny* [Faculty of Humanities]. A Philosophical Seminar led by Łukasiewicz was part of mathematics. In 1934 a Seminar in Philosophy of Mathematics run by Leśniewski was created.

Łukasiewicz's and Leśniewski's works during their Warsaw period had mostly mathematical character and their results were milestones in logic. Among other things, Łukasiewicz initiated many-valued logic and introduced a parenthesis-free notation, later called the Polish notation. Leśniewski developed his own axiomatic systems (in his own notation); his mereology, dealing with the relation "part/whole" was meant as an alternative to set theory and reflected his nominalist views. As an alternative to type theory, he proposed a concept of semantic categories, later developed further by Ajdukiewicz in the foundations of categorial grammar.

[64]See Duda (2004); McFarland (2014); Murawski (2004a,b); Poznański (2010); Simons (2018); Woleński (1995, 2004, 2018).

Ajdukiewicz left Warsaw for a chair at the Jan Kazimierz University in Lwów in 1928, but his work on methodology of deductive sciences in his Warsaw years produced definitions of consequence, theorem and proof, which were adopted in applications of logic to mathematics.

On the other hand, Warsaw mathematicians studied problems related to logic, in particular to the axiomatization of theories. Sierpiński continued his work on the Axiom of Choice, in which he developed serious interest during his Moscow period. He did not advocate either its rejection or acceptance; he just studied its consequences in set theory, measure theory and theory of real functions as well as its possible replacements and/or elimination from proofs. In the same way he dealt with the Continuum Hypothesis, on which he published the monograph *Hypothèse du continu* (*Monografie Matematyczne* [Mathematical Monographs], no. 4, Seminarium Matematyczne Universytetu Warszawskiego, Warszawa, 1934, 5 + 192 pages). Kuratowski came up with axioms for topology in terms of the closure operation. Later, extending his joint work with Tarski, he showed a relation between logical operations and descriptive complexity of sets. Mazurkiewicz proposed axiomatics for probability.[65]

Figure 10. Jan Łukasiewicz (Award of honorary doctorate at the University of Dublin).[66]

Students in mathematics attended the lectures and seminars of Łukasiewicz, Leśniewski, Ajdukiewicz and Tadeusz Kotarbiński (1886–1981) and sometimes made

[65]See Betti (2008); Borowik and Kurtowski (1997); Jadacki (2018); Kuratowski (1947, 1979, 1981); Simons (2018); Urbaniak (2015); Woleński (2004, 1995); Woleński (2018).

[66]Photograph from the collection of the Library of the Mathematical Institute of the Polish Academy of Sciences.

contributions outside their main interests. Kuratowski owed his first publication as a student (*O definicji wielkości*. (*Sur la définition de la grandeur*) [On the definition of magnitude], *Przegląd Filozoficzny* **20**(1917), pp. 288–306) to a discussion in Łukasiewicz's seminar. Knaster translated *L'Algèbre de la logique* by Louis Couturat (1868–1914) into Polish. The book was used as an introductory text on logic. Zygmunt Kobrzyński (1893–1944?), who went on to obtain his PhD with Łukasiewicz (1930, thesis *O wyznacznikach terminów logicznych* [On Determinants of Logical Terms]) edited the lectures of Janiszewski on analytic geometry for publication. Russell's *Principia Mathematica* were proposed as reading on the guide for incoming students, with the caveat that they were an advanced text.[67]

Alfred Tarski (1901–1983, until 1924 Tajtelbaum), considered one of the greatest logicians of all time, was the only PhD student of Leśniewski. Enrolled at the University of Warsaw to study biology, he soon switched to mathematics impressed by Leśniewski's lectures. His PhD was awarded in 1924 for the thesis *O wyrazie pierwotnym logistyki* [On the Primitive Term of Logistics] (published in part in *Przegląd Filozoficzny* **26**(1923), pp. 85–94, and in *Fundamenta Mathematicae* **4**(1923), pp. 196–200, as *Sur le terme primitif de la Logistique*). Tarski got his habilitation in 1925, in 1929 was made a senior assistant and in 1930 an adjunct. To supplement his meager income, since 1925 he also taught mathematics at Stefan Żeromski Gimnazjum in Warsaw. In his research he departed from Leśniewski's systems (although, besides an axiomatic system for protothetics from his PhD thesis which eliminated the connective of conjunction, he also gave a formalization of geometry of solids in mereology).

Tarski's most important work was in metamathematics, namely a general mathematical theory of deductive systems, which makes a distinction between theorems about objects of a given system (inside it) and theorems about the system itself (outside of it). Tarski formulated a semantic theory of truth (in the publication *Pojęcie prawdy w językach nauk dedukcyjnych* [The Notion of Truth in the Languages of Deductive Sciences], Towarzystwo Naukowe Warszawskie, Warszawa, 1933, vii + 116 pages) and proved a theorem of undefinability of the set of true sentences in a formal language with negation and Gödel's numbering. He proved the decidability of many mathematical and logical theories, including the first-order theory of real numbers, perfecting the method of quantifier elimination in the process. He also proposed a topological interpretation of connectives of classical, intuitionistic and modal (S4) logic, and topological semantics for intuitionistic logic. He wrote joint

[67]See the personal file of Kobrzyński at the University of Warsaw and also Bartnicka (2014); Kuratowski (1979); Woleński (2015).

Figure 11. International Symposium on Foundations of Mathematics, 2–8 September 1959, Alfred Tarski (second from right).[68]

papers with Sierpiński (on arithmetic of cardinal numbers),[69] Kuratowski (on the connection between logical operations, in particular quantifiers, and projective sets), Łukasiewicz (on propositional calculus), Banach (on a paradoxical decomposition of a ball, now known as Banach–Tarski paradox) and Adolf Lindenbaum (1904–1941) (see below).

Not being a professor, Tarski could not formally supervise PhDs, but he and Kuratowski co-supervised the PhD of Andrzej Mostowski (1913–1975, PhD 1939,

[68]Photograph from the collection of the Library of the Mathematical Institute of the Polish Academy of Sciences.

[69]In the third edition of his *Zarys teorii mnogości. Część pierwsza: Liczby pozaskończone* [An Outline of Set Theory. Part I: Transfinite Numbers], 3rd edition, Wydawnictwo Kasy im. Mianowskiego, Warszawa, 1928, viii + 267 pages, Sierpiński credited Tarski with many ideas and results on set theory.

O niezależności definicji skończoności w systemie logiki [On the Independence of the Definition of Finiteness in a System of Logic]). Mostowski and Tarski published one joint paper before World War II, on Boolean rings. Tarski taught university courses on set theory, methodology of mathematics, foundations of elementary geometry and arithmetic. He wrote two high-school textbooks: one in logic and one in geometry, the latter jointly with Wacław Schayer (1905–1959) and Zygmunt Chwiałkowski (1884–1952).[70] Prolific and profound, he remained so in his later career, working at the intersections of logic, algebra and geometry.[71]

From Tarski's seminar came another important result. Mojżesz Presburger (1904–1943?) in his master thesis *Über die Vollstandigkeit eines gewissen Systems der Arithmetik ganzer Zahlen* in 1929 proved the consistency, completeness and decidability of the first-order theory of the natural numbers in the signature with equality and addition.[72] Presburger also helped with editing Łukasiewicz's textbook *Elementy logiki matematycznej* [Elements of Mathematical Logic], offering many original ideas, among them a decision procedure for a propositional calculus.[73]

It is worth noting that other mathematicians from the University of Warsaw also developed a theory of truth about the same time as Tarski, but from a very different viewpoint. *Pojęcie prawdy na terenie fizyki* [The Notion of Truth in the Domain of Physics][74] by Poznański and Aleksander Wundheiler (1902–1957) presented an operational theory of truth, closer to neo-positivist views in the philosophy of science. Wundheiler got his PhD in 1930 at the University of Warsaw with Przeborski (thesis *Geometria reonomiczna jako podstawa mechaniki absolutnej* [Rheonomic Geometry as a Foundation of Absolute Mechanics], published as *Rheonome Geometrie. Absolute Mechanik*, *Prace Matematyczno-Fizyczne* **40**(1933), pp. 97–142) and later introduced the notion of a geometric object, independently but almost simultaneously with David van Dantzig (1900–1959).

Poznański did not pursue further research. He was active in the Warsaw society *Towarzystwo Przyjaciół Uniwersytetu Hebrajskiego w Jerozolimie* [Society of Friends of the Hebrew University in Jerusalem] and dealt with the acquisition of books for its mathematical library.[75]

---

[70]See Betti (2008); Feferman and Feferman (2008); Gruber (2016); Mancosu *et al.* (2009); McFarland *et al.* (2014); Woleński (1901–1883).

[71]Tarski left Poland in summer 1939 for a conference in the USA and stayed there because of World War II. Since 1942 he was a professor of mathematics at the University of California, Berkeley.

[72]This system is now referred to as Presburger arithmetic.

[73]Presburger did not continue his research; later he worked as an actuary and published occasionally on actuarial mathematics. He perished in the Holocaust. See Zygmunt and Purdy (2014).

[74]Bibliographic data: *Odbitka z księgi pamiątkowej ku uczczeniu 15-lecia pracy nauczycielskiej w Uniwersytecie Warszawskim Prof. Tadeusza Kotarbińskiego* [An Offprint from a Memoir in Honor of the 15th Anniversary of Prof. Tadeusz Kotarbiński's Teaching at the University of Warsaw], Drukarnia im. Kasy Mianowskiego, Warszawa, 1934, 50 pages.

[75]Poznański emigrated to Palestine in 1939 and became the Registrar of the Hebrew University in Jerusalem. Wundheiler left Warsaw in 1939 for a conference in the USA, where he remained because

Tarski's collaborator Lindenbaum obtained his PhD in 1928 with Sierpiński (thesis *O własnościach metrycznych mnogości punktowych* [On Metric Properties of Point Sets], not published in full, but quoted by Sierpiński) and found employment in Łukasiewicz's Philosophical Seminar. He taught measure theory, algebra, actuarial mathematics, foundations of mathematics and mathematical foundations of sciences. He and Tarski introduced the notion of a Lindenbaum–Tarski algebra for a logical theory and studied such algebras. Lindenbaum's lemma, akin to Kuratowski's lemma (see below), states that any consistent theory of predicate logic can be extended to a completely consistent theory. Lindenbaum also wrote joint papers with Mostowski (on Axiom of Choice), Edward Szpilrajn (1907–1976, from 1940 he used the name Marczewski) and Andrzej Koźniewski (1907–1939). From 1903 until 1904, he also credited his high school and university classmate Mojżesz Dawid Kirszbraun (1903–1942) for many discussions on mathematical problems. Lindenbaum gave a talk at the International Congress of Scientific Philosophy in Paris in 1935 on simplicity.[76] The topics of talks he gave at the meetings of the Polish Mathematical Society and other organizations show that he was also interested in geometry. He and Tarski announced a monograph on one-to-one mappings, to be published in the series *Monografie Matematyczne* [Mathematical Monographs], but it did not get written. He also served as a mentor to new students in logic.[77]

Janina Hosiasson-Lindenbaum (1899–1942, Lindenbaum's wife) got her PhD at the Faculty of Philosophy in 1926 (thesis *Uprawnienie rozumowania indukcyjnego* [Justification of Inductive Reasoning], supervisor Kotarbiński, Part I was published as *Definicje rozumowania indukcyjnego* [Definitions of Inductive Reasoning], *Przegląd Filozoficzny* **31.4**(1928), pp. 110–128). She worked on paradoxes of probability and gave a talk at the International Congress of Scientific Philosophy in Paris in 1935 on the issues of probability. She taught recitation classes for Neyman's lectures in probability and translated Russell's books into Polish.[78]

There were a few more doctorates concerning formal logic awarded at the University of Warsaw in the interwar period. In the Faculty of Mathematics and Natural Sciences, the recipients were: Mordchaj Wajsberg (1902–1941?), in 1931 (thesis *Aksjomatyzacja trójwartościowego rachunku zdań* [Axiomatization of Multi-Valued Propositional Calculus], supervisor Łukasiewicz),[79] Stanisław Jaśkowski (1906–1965), in 1932 (thesis *Reguły wnioskowania za pomocą założenia* [Rules of

---

of World War II. He taught at several academic institutions and in 1948 became a professor of mechanics at the Illinois Institute of Technology in Chicago. See Koterski and Uebel (2017).

[76]The issue of "simplicity" of formal systems was interesting also for David Hilbert. Following Hilbert's notes it is coming to be regarded as "Hilbert's 24th Problem".

[77]See Purdy and Zygmunt (2018); Zygmunt and Purdy (2014).

[78]Adolf Lindenbaum and Janina Hosiasson-Lindenbaum managed to escape in 1939 to Eastern Poland occupied by USSR and then to Lithuania, but ultimately perished in the Holocaust. See Purdy and Zygmunt (2018); Zygmunt and Purdy (2014).

[79]Apart from the axiomatization of multi-valued, intuitionistic and intermediate logic, Wajsberg worked on semantics of modal logic, in particular C.I. Lewis's system. He continued his research

Inference by Supposition], published as *On the Rules of Suppositions in Formal Logic*, *Studia Logica* **1**(1934), pp. 1–32, supervisor Łukasiewicz),[80] Bolesław Sobociński (1906–1980), in 1937 (thesis *Aksjomatyzacja pewnych wielowartościowych systemów teorii dedukcji* [Axiomatization of Certain Multi-Valued Systems of Deduction Theory], Roczniki Prac Naukowych Asystentów Uniwersytetu imienia Józefa Piłsudskiego w Warszawie [Annals of Research Works of the Assistants of the Józef Piłsudski University in Warsaw] **1**(1936), pp. 399–417, supervisor Łukasiewicz),[81] Jerzy Słupecki (1904–1987), in 1938 (thesis *Dowód aksjomatyzowalności pełnych systemów wielowartościowych rachunków zdań* [A Proof of Axiomatizability of Complete Systems of Multi-Valued Propositional Calculi], published in *Comptes Rendus des Séances de la Sociétè des Sciences et des Lettres de Varsovie, Classe III*, **32**(1939), pp. 110–128, supervisor Łukasiewicz).[82] In the Faculty of Humanities, the PhD recipients in logic were: Jan Franciszek Drewnowski (1886–1978), PhD 1927 (thesis *Podstawy logiki Bernarda Bolzano* [Foundations of Logic of Bernard Bolzano], supervisor Kotarbiński),[83] Kazimierz Pasenkiewicz (1897–1995), PhD 1933 (thesis *Opis aktów za pomocą których może być zbudowany system formalny teorji dedukcji* [On Cognitive Acts Constructing Deductive Theories], supervisor Kotarbiński).[84]

---

and publications while working as a teacher in Kovel and Łomża. He perished in the Holocaust. The exact date and place of his death are unknown. See Piotrowski (2013); Surma (1973).

[80]Jaśkowski developed systems of natural deduction almost simultaneously with Gerhard Gentzen (1909–1945) and gave such systems, e.g., for the classical as well as intuitionistic propositional calculus. He also proposed matrix semantics for intuitionistic logic. After World War II, as a professor at the University of Toruń, he introduced paraconsistent logic. See Indrzejczak (2018); Kotas and Pieczkowski (1967).

[81]Sobociński graduated in philosophy. After his PhD he worked in Leśniewski's Seminar in Philosophy of Mathematics and got his habilitation in 1939. During World War II he was active in the nationalistic underground. He emigrated in 1946 and was a professor of philosophy at the Notre Dame University (South Bend, Indiana, USA) since 1961. In his Warsaw period he was interested in the axiomatization of propositional calculi and Leśniewski's systems. See Świętorzecka (2018); Tatarkiewicz (1998).

[82]Słupecki was born in Manchuria, then part of the Russian Empire, a son of a Polish military engineer. The family repatriated in 1921. In his thesis Słupecki solved the problem of completeness of multi-valued logic. After World War II he was a professor at the University of Wrocław and the Pedagogical College in Opole. He had many students and established his own logical school, studying mainly the operation of consequence in various logical systems. See Woleński (2018); Woleński and Zygmunt (1989).

[83]Drewnowski served in the Russian, and then Polish army. Together with Sobociński, Fr. Jan Salamucha (1903–1944) and Fr. Józef Maria Innocent Bocheński (1902–1995) he formed Koło Krakowskie [the Kraków Circle], a philosophical group interested in the application of logic to Roman Catholic theology. Taken as a prisoner in the defense of Warsaw in 1939 against the Germans, he spent World War II in the prisoner's camp in Murnau. After the liberation he spent some time in Italy and Great Britain, returning to Poland in 1947. He did not pursue further research. See Piotrowski (2014); Wolak (2005).

[84]Pasenkiewicz was born to a Polish family in Kiev. He fought in the Russian, and then Polish army. Wounded in the Polish-Bolshevik war, he lost a leg. During World War II he ran a transportation and consumer-goods cooperative in Warsaw and in 1944 fought in the Warsaw Uprising. He belonged

Figure 12. Stanisław Jaśkowski (1906–1965) after World War II.[85]

One PhD related to logic was also awarded in the Faculty of Catholic Theology, to Salamucha in 1927 [thesis *Teoria wynikania modalnego Arystotelesa* [Aristotle's Theory of Modal Implication], supervisors Łukasiewicz and Stanisław Kobyłecki (1864–1939)].[86]

4. The Emergence of "Two Warsaws"[87]

4.1. *"Sierpiński's Warsaw"*

Sierpiński was an unquestionable leader of the emerging Warsaw school of mathematics. He worked and published papers and books in set theory (descriptive set

---

to Polska Partia Socjalistyczna [Polish Socialist Party], later merged with the communist party into Polska Zjednoczona Partia Robotnicza [Polish United Workers' Party]. After the war he was a professor at the Jagiellonian University. He supervised a few PhD theses and wrote several textbooks on logic. See Piotrowski (2014); Woleński (2018).

[85]Photograph from the collection of the Library of the Mathematical Institute of the Polish Academy of Sciences.

[86]Salamucha was ordained a priest in 1925. He became a professor of Christian theology at the Faculty of Theology of the Jagiellonian University. Arrested in Sonderaktion Krakau in 1939 and taken to the Sachsenhausen concentration camp, he was released in 1941. He served as a chaplain of *Narodowe Siły Zbrojne* [National Armed Forces], a nationalistic armed group of resistance, and was killed in the Warsaw Uprising. See Piotrowski (2014); Wolak (2005).

[87]We use the term "two Warsaws" to describe two parts of the Warsaw mathematical community following the fictionalized account of the interwar period in Mioduszewski (2014). The term is credited there to Borsuk.

theory, Axiom of Choice and Continuum Hypothesis, arithmetic of cardinal numbers, infinite combinatorics) and related areas, such as topology, theory of real functions, measure theory and Baire categories. It is difficult to comment on his achievements, not only because of a very large number of his publications (over 300 in the interwar period).[88] The difficulty lies in the fact that Sierpiński's results laid the foundations for the disciplines mentioned above, shaped their development and revealed the connections between them. As one of the numerous equivalents of Continuum Hypothesis, he proved the existence of two subsets E_1, E_2 of the plane such that E_1 is countable in every horizontal line, E_2 is countable in every vertical line, and their union is the whole plane. Among other things, he introduced the notions and initiated the study of totally disconnected spaces, of almost disjoint sets, of inaccessible cardinals and of iterations of transfinite order.[89] He also promoted collaboration among mathematicians. Jointly with Mazurkiewicz he classified countable compact topological spaces and characterized them in terms of ordinal numbers. Jointly with Kuratowski he characterized locally closed sets. Jointly with Ruziewicz he gave an example of a totally disconnected subset of the real line which has at most one point in common with each of its translations. Jointly with Zygmund he gave an example of a function (of a real variable) which is discontinuous on every subset of cardinality continuum. Last but not least, jointly with Banach and Włodzimierz Stożek (1883–1941) he wrote school textbooks on arithmetic and geometry.[90]

Mazurkiewicz served in various positions in the university administration, but this did not curb his activity as a researcher and mentor. He continued to work on topology and, among other things, proved that almost all subcontinua of the square are hereditarily indecomposable. Jointly with Banach he showed that the set of continuous real-valued functions on the interval $[0, 1]$ which has a derivative at least at one point is a set of the first category in the space of all continuous real-valued functions on $[0, 1]$. His problems were taken up by others: for example, in his work *Un Théorème sur les lignes de Jordan*, *Fundamenta Mathematicae* **2**(1921), pp. 119–130, he asked whether every dendrite is homeomorphic with a subset of a plane. This was answered positively in 1923 by Tadeusz Ważewski (1896–1972) from the Jagiellonian University (Kraków), who constructed a universal dendrite in his

---

[88] See Sierpiński (1974).

[89] See also Sinkevich (2011); Wilczyński (2012).

[90] During World War II, Sierpiński worked as a municipal clerk and ran clandestine university courses in his private apartment. After the liberation he lectured briefly at the Jagiellonian University but returned to Warsaw in 1946, where he continued his research and teaching as a professor. In later years he returned to number theory, one of his initial interests. He re-launched *Fundamenta Mathematicae* in 1945 and became editor-in-chief of *Acta Arithmetica*. See Adamowicz (1984); Kuratowski (1972); Marczewski (1996); Schinzel (1971, 1984).

PhD (obtained in Paris). Later Mazurkiewicz expanded his mathematical interests and supervised (at least formally) a number of PhDs on analysis.[91]

The founders of the Warsaw mathematical school were soon joined by the first generation of their students. In particular, the career of Kazimierz Kuratowski progressed in a dazzling way. His PhD was quickly followed by habilitation, also in 1921, and in 1923 he was made a deputy professor.

In his early period Kuratowski formulated a definition of an ordered pair, now commonly taught to all students of advanced mathematics. In his PhD, as a consequence of his closure axioms, he proved the (closure-complement) 14-set theorem (saying that the largest number of distinct sets obtainable by repeatedly applying the set operations of closure and complement to a given starting subset of a topological space is 14). While studying the possibility of eliminating transfinite induction from some proofs in topology, he discovered a now-famous lemma about the existence of maximal elements in certain partially ordered sets.[92] Together with Knaster, he gave an important example of a 2-connected set, called the Knaster–Kuratowski fan. He also introduced a notion of convergence for nets of compact subsets of metric spaces (Kuratowski limits).

In 1928 Kuratowski became a professor of mathematics at the Wydział Ogólny [General Department] of the Lwów Polytechnic. In Lwów he collaborated with Banach, with whom he proved non-existence of a non-trivial non-atomic translation-invariant measure on the family of all subsets of the real line as a consequence of the Continuum Hypothesis. He also worked on graph theory and proved a theorem characterizing planar graphs. He published papers with his students in Lwów: Stanisław Ulam (1909–1984, PhD 1933 with Kuratowski and Stożek), Edward Otto (1908–1986) and Tadeusz Posament (1905–1941). One of the joint results was the Kuratowski–Ulam theorem, a Baire-categorical analog of Fubini's theorem for arbitrary second countable Baire spaces.

After the liquidation of the General Department in 1934,[93] Kuratowski returned to Warsaw and was appointed a professor at the University. He continued to work, among other things, on theory of dimension, projective hierarchies and theory of Borel sets in general metric spaces, where he made significant contributions.

---

[91]Mazurkiewicz remained in Warsaw during World War II. The manuscript of his treatise on probability burned in the Warsaw Uprising. After the liberation he managed to reconstruct a part of it soon before his death in an operation for a gastric ulcer. See Kuratowski (1947).

[92]This lemma is commonly but erroneously referred to as "Zorn lemma", sometimes more correctly as "Kuratowski–Zorn lemma". Kuratowski proved it along with many equivalent statements in 1922 and Max Zorn (1906–1993) rediscovered it in 1935. It is equivalent to the Axiom of Choice. More detailed view is offered by P.J. Campbell: *The Origins of "Zorn's Lemma"*, *Historia Mathematica* **5**(1978), pp. 77–89. See also Marczewski (1960).

[93]The Department was liquidated in response to professors' protest against imprisoning a group of parliamentary representatives of the center-left in the Brest Fortress.

He showed that every metric space can be isometrically embedded into a Banach space (through Kuratowski embedding). Together with Knaster and Mazurkiewicz he proved a theorem on certain coverings of a simplex by closed sets, which is equivalent to Brouwer's fixed point theorem (proved in 1910) for a continuous map of a compact convex set into itself as well as to Sperner's lemma in combinatorics [Emanuel Sperner (1905–1980), a German mathematician, proved it in 1928]. This theorem is now known as KKM theorem and has been generalized in several ways. He also published joint results with foreign mathematicians: Karl Menger (1902–1985), Gordon Thomas Whyburn (1904–1969) and John von Neumann (1903–1957), as well as with a representative of "the other Warsaw", Stefan Straszewicz (on a generalization of a theorem by Janiszewski).

Since 1928 Kuratowski was a member of the editorial board of *Fundamenta Mathematicae*, which by then became a leading journal in set theory, topology and related areas, publishing the works of many distinguished mathematicians. The journal was so important that its Volume 1 was reprinted in 1937. Kuratowski assisted Szpilrajn and Stefania Braunówna (1908–1943) with comments on its contents in the *Annèxe*. At the initiative of Kuratowski a series of publications *Monografie Matematyczne* [Mathematical Monographs] was started in 1932. The first volume of his own major work *Topologie* appeared as number 3 in the series.[94]

Knaster was another very successful alumnus of the Warsaw mathematical school. Together with Kuratowski in 1921 he formulated the first general theory of connected sets. In 1922 he constructed the first example of a hereditarily indecomposable continuum (a pseudo-arc). He suffered some setbacks from malaria he contracted during his service in the Polish–Bolshevik war,[95] but ultimately picked up his research and teaching at the University of Warsaw. He got his habilitation in 1929. Among his other achievements in the interwar period, there are: construction (jointly with Kuratowski) of a connected, locally connected and hereditarily non-perfect set (Knaster–Kuratowski fan, in Polish *miotełka*, "a little broom"), an example (jointly with Mazurkiewicz) of λ-connected sets whose continuous images are not λ-connected, Knaster–Tarski theorem on sets of fixed points of an order-preserving function from a complete lattice to itself. Knaster also studied problems of decomposition of the plane, the operation of removal of points from a connected space (and topological characterization of resulting sets) and the properties of indecomposable and irreducible continua, of which he constructed many

[94]During World War II Kuratowski went into hiding because of his Jewish origin and was active in clandestine teaching. After the war he continued his research and teaching as a professor at the University of Warsaw. He was an initiator and the first director of the State Institute of Mathematics, launched in 1949, later the Mathematical Institute of the Polish Academy of Sciences. See Borsuk (1960); Engelking (1998); Kuratowski (1979, 1981); Marczewski (1972).

[95]Knaster served in the medical corps because of his prior medical studies.

examples. He ran an advanced seminar in topology and was an editor of *Monografie Matematyczne*.[96]

Set theory, topology and theory of functions of real variable kept attracting a new generation of students. PhD degrees related to these areas were awarded to Aleksander Grużewski (1895–1968), in 1926 (thesis *O zagadnieniu alfa P. Urysohna* [On the Alpha Problem of P. Urysohn], supervisor Mazurkiewicz),[97] Piotr Szymański (1900–1965), in 1926 (thesis *O sumie dwóch continuów nieprzywiedlnych* [On the Union of Two Irreducible Continua], published in parts as 1. *Sur les constituants d'ensembles situés sur des continus arbitraires*, *Fundamenta Mathematicae* **10**(1927), pp. 363–364, 2. La somme de deux continus irréductibles, *ibid.*, **11**(1928), pp. 1–15, supervisor Mazurkiewicz),[98] Jerzy Popruženko (1893–1973, also known in the transcription Poprougenko), in 1930 (thesis *O pewnych własnościach funkcji rzeczywistych zbioru* [On Some Properties of Real Set Functions], supervisor Sierpiński),[99] Stefan Sokołowski (1904–1940), in 1936 (thesis *O własnościach funkcji* $d_n(x, h)$ [On Properties of the Function $d_n(x, h)$], supervisor Sierpiński).[100]

Some other PhD recipients of that period made their own mark in mathematics. Nikodym transferred his habilitation from the Jagiellonian University to the University of Warsaw and continued to work mainly on measure theory. His main achievement is the Radon–Nikodym theorem (which he proved in 1930), establishing the existence of the density (or Radon–Nikodym derivative) of a measure which is absolutely continuous with respect to a given measure. Among other things, he gave an example of a subset of the unit square in $\mathbb{R}^2$ with the complement of Lebesgue measure zero, such that, for any point in this set, there is a straight line that only

[96]With the outbreak of World War II Knaster escaped to Soviet-occupied Lwów, where he became a professor of the Ukrainian university. At that time he worked with Banach and Steinhaus on the fair division problem. After the Nazi invasion he remained in Lwów and fed lice on his body for the production of typhus vaccine in the institute of the Polish microbiologist Rudolf Weigl (1883–1957). This work saved him from persecution against Jews and Polish intellectuals. Since 1945 he continued his research and teaching as a professor at the University of Wrocław. See Duda (1987).

[97]Grużewski lectured at the *Szkoła Główna Handlowa* [Main School of Commerce]. After World War II he headed the Actuarial Section of the Państwowy Instytut Matematyczny [State Mathematical Institute], later *Instytut Matematyczny Polskiej Akademii Nauk* [Mathematical Institute of the Polish Academy of Sciences].

[98]Later Szymański switched his interests to fluid dynamics. He worked at the Warsaw Polytechnic.

[99]Born in Odessa, Popruženko completed a university course in philosophy there. Living in Warsaw since 1920, he enlisted in the Polish army and fought in the Polish–Bolshevik war. Afterwards he worked at the Polish State Printing Office and studied mathematics. Since 1946 a professor at the University of Łódź, since 1962 an ordinary professor. He worked on set theory in relation with topology and functions of a real variable.

[100]As a teenager, Sokołowski fought in the Defense of Lwów. After finishing his studies at the University of Warsaw he worked at the Center for Ballistic Research in Rembertów. In 1933 he graduated from the *Szkoła Podchorążych Rezerwy Artylerii* [School for Reserves of Officer Cadets of Artillery] in Włodzimierz. Promoted in 1935 to first lieutenant, he fought in the 1939 defensive war. Killed by NKVD in Katyń in the spring of 1940. See Piotrowski (2013).

intersects the set at that point (a Nikodym set) as well as introduced the Nikodym distance between two measurable sets in $\mathbb{R}^d$ (as the volume of their symmetric difference). During his Warsaw period Nikodym also gave an axiomatic definition of a Riemann surface of topological character.[101]

Figure 13. Otto Nikodym (1887–1974).[102]

Zygmunt Zalcwasser (1898–1943, PhD 1928, thesis *O osobliwościach funkcji ciągłych zmiennej rzeczywistej* [On Singularities of Continuous Functions of a Real Variable], supervisor Sierpiński) introduced in 1930 (an ordinal) rank for the set A of (everywhere) pointwise convergent sequences of real-valued continuous functions on $[0,1]$ — that is, a "map" from $A \subset C[0,1]$ to the class of all ordinals, which measures how uniform the convergence is. Nowadays there exist ranks of various kinds, suitable for the notions of convergence, differentiability, integrability and others, which sustain mathematicians' interest in ordinal numbers.

Nachman Aronszajn (1907–1980, PhD 1929, thesis *On Metrics and Metrization* [*O metryce i metryzacji*], supervisor Mazurkiewicz) constructed in 1934 a tree (in the sense of descriptive set theory) which is uncountable and has no uncountable branches and no uncountable levels.[103] He also worked with Karol Borsuk (1905–1982) on products of retracts. He went on to get another PhD, from the

[101]After World War II Nikodym taught briefly at the Academy of Mining in Kraków. In 1946 he and Stanisława Nikodym emigrated from Poland. Since 1948 he taught at Kenyon College, Ohio, USA. See Ciesielska (2017); Derkowska (1983); Piotrowski (2013); Szymański (1990).

[102]Photograph from the collection of Stanisław Domoradzki.

[103]The existence (or non-existence) of certain Aronszajn trees is related to the Continuum Hypothesis and Generalized Continuum Hypothesis.

University of Paris with Fréchet, in 1935 (thesis *Sur les decompositions des fonctions analytiques uniformes et sur leurs applications*) and switched his interests to analysis.[104]

Figure 14. Karol Borsuk (1905–1982).[105]

Borsuk got his PhD in 1931 with Mazurkiewicz. His thesis *O retraktach i zbiorach związanych* [On Retracts and Related Sets] (publications: (1) *Sur les rétractes, Fundamenta Mathematicae* **17**(1931), pp. 152–170, (2) *Quelques théorèmes sur les ensembles unicohèrents, ibid.*, **17**(1931), pp. 171–209, (3) *Einige Sätze über stetige Streckenbilder, ibid.*, **18**(1932), pp. 198–213) introduced new topological concepts of retract and absolute neighborhood retract. Solving a problem posed by Ulam, he proved in 1933 that every continuous function from an n-sphere into Euclidean n-space maps some pair of antipodal points to the same point. The result (which can be re-stated in terms of retractions) is known as Borsuk–Ulam theorem and implies, among others, the KKM theorem. Borsuk further studied the topological properties of spaces through their mappings into spheres. His habilitation in 1934 was titlebreak *O zagadnieniu topologicznego scharakteryzowania sfer euklidesowych* [On the Problem of Topological Characterization of Euclidean Spheres]. From 1934 till 1938 he was a docent at the University of Warsaw and in 1938 was appointed an extraordinary professor, succeeding Żorawski upon the latter's retirement. He pioneered research in algebraic topology in Poland, proving a homotopy extension theorem,

[104]Aronszajn emigrated to the USA and later worked mainly on theory of reproducing kernels. He was a professor of mathematics at the University of Kansas from 1951 until retirement in 1977.

[105]Photograph from the collection of the Library of the Mathematical Institute of the Polish Academy of Sciences.

studying homological aspects of the fixed point theorem on compact sets and introducing cohomotopy groups, among other things.[106]

Figure 15. Edward Szpilrajn (Marczewski), left, and Jaroslav Kurzweil (born 1926), Eighth Congress of Polish Mathematicians, 1953.[107]

Szpilrajn[108] distinguished himself already as a student, getting noticed by Kuratowski (with whom he published a paper). He got his PhD in 1932 with Sierpiński, the thesis title was *O funkcjach miarowych i warunku Baire'a* [On Measure Functions and Baire Condition] (published as *O mierzalności i warunku Baire'a*, in *Comptes Rendus du I Congrès des Mathématiciens des Pays Slaves, Warszawa*, 1929, pp. 297–303). In the years 1933–1934 he worked as an assistant, and later senior assistant, at the University of Warsaw. Since the academic year 1936/1937 he gave commissioned lectures on the theory of measure and integral.

[106]During World War II Borsuk ran a stationery store in Warsaw, which served as a contact point for the underground Home Army [Armia Krajowa]. To supplement his income, he invented and produced a board game *Hodowla Zwierzątek* [Animal Husbandry], now also known as *Superfarmer* or *Have you Herd?* After the war he continued his research and teaching as an ordinary professor at the University of Warsaw. He introduced shape theory and constructed many examples, among them a curve known as the Warsaw Circle. See Borsuk (1978); Dydak (2012); Eilenberg (1993).

[107]Photograph from the collection of the Library of the Mathematical Institute of the Polish Academy of Sciences.

[108]The outbreak of World War II found Szpilrajn in Lwów, but he decided to return to Warsaw. In order to hide from the Nazis he assumed the name Marczewski, which he kept using after the war. Imprisoned in a labor camp in Breslau (Wrocław) after the Warsaw Uprising, he remained in that city after the liberation and became a professor at the University of Wrocław. He got his habilitation in 1945 in Warsaw. He was forced into early retirement during the anti-Semitic actions by the ruling communist party in the aftermath of the 1968 student protests.

The requirement of habilitation was waived in his case, because he had many publications on related themes. His sufficient condition for a σ-algebra of sets to be invariant under the (Suslin) A-operation revealed certain connections between measure and Baire categories. Such connections were a topic of high interest for Sierpiński at that time. Szpilrajn also investigated a problem of equivalence of classes of sets and proved that in $\mathbb{R}^n$ there exists a sequence of non-dense sets non-equivalent to any sequence of projective sets, answering a question of Ulam. As a tool he introduced a characteristic function of a sequence of sets (now called Marczewski function). In 1930 he proved yet another maximal principle (besides those of Kuratowski and Lindenbaum), which states that every strict partial order is contained in a total order. Regarding dimension theory, he proved in 1937 that, for arbitrary metrizable separable space X, topological dimension coincides with the Hausdorff dimension under one of the metrics in X which induce the given topology of X. With his friend Stanisław Kierst (1909–1945),[109] Szpilrajn published a paper on values of entire or meromorphic functions in the complex plane, which used methods of Baire categories in an essential way. He also worked on subharmonic functions.[110]

Zenon Waraszkiewicz (1909–1946, also known as Zacharski) got his PhD in 1932 with Mazurkiewicz (thesis *O odwzorowaniach ciągłych kontynuów* [On Continuous Mappings of Continua]). He constructed a continuous family of many planar continua which are not continuous images of each other (Waraszkiewicz spirals).[111]

Samuel Eilenberg (1913–1998, in the faculty reports spelled Ejlenberg), who got his PhD in 1936 with Kuratowski, was in fact a student of Borsuk (who could not yet supervise PhDs). The thesis *O zastosowaniach topologicznych odwzorowań na okrąg koła* [On Topological Applications of Mappings onto the Circumference of a Disk] was published under the same title in *Wiadomości Matematyczne* **41**(1935), pp. 1–32, a related publication was *Transformations continues en circonférence et la topologie du plan*, *Fundamenta Mathematicae* **26**(1936), pp. 61–112. Under the influence of Borsuk Eilenberg got interested in algebraic topology and in 1938 published a paper on the action of the fundamental group on the higher homotopy groups

[109] A graduate of mathematics and an accomplished chess player, Kierst died in the labor camp Neuengamme, where he was imprisoned by the Germans after the Warsaw Uprising.

[110] See Hartman *et al.* (1979).

[111] Waraszkiewicz taught mathematics in high schools and worked as an assistant at the Warsaw Polytechnic. He obtained habilitation in 1937. During World War II he was involved in clandestine teaching in Warsaw. From 1945 until his death a professor at the University of Łódź and the *Wyższa Szkoła Gospodarstwa Wiejskiego* [Higher School of Rural Economy] in Łódź. See Derkowska (1983); Mioduszewski (2001).

Figure 16. Atuo Komatu (1909–1995) and Samuel Eilenberg (1913–1998) (right).[112]

of a space. After emigrating to the USA he came up with axiomatics of homology theory (together with Norman Steenrod, 1910–1971) and laid the foundation for homological algebra and category theory [together with Saunders Mac Lane (1909–2005)].[113]

A few other graduates of the University of Warsaw took up research in topics related to set theory and topology. Witold Wolibner (1902–1961, see the next subsection) gave a topological characterization of non-accessible points of a closed subset of the plane. Chaim Chojnacki (1912–1991, later Haim Hanani) completed his studies in Vienna and Warsaw and wrote a master's thesis *O krzywych istotnie niespłaszczalnych w przestrzeni trójwymiarowej* [On Essentially Non-Planar Curves in a Three-Dimensional Space]. In 1935 he emigrated to Mandatory Palestine (controlled by the British) and obtained the first doctorate in mathematics awarded

[112]Photograph from the collection of the Library of the Mathematical Institute of the Polish Academy of Sciences.

[113]Since 1947 Eilenberg was a professor of mathematics at the Columbia University in New York City.

by the Hebrew University of Jerusalem in 1938 (thesis *A Contribution to the Four Color Problem*, supervisor Abraham Fraenkel, 1891–1965).[114]

Julian Perkal (1913–1965) got his master's degree in mathematics in 1937 and worked as a technician in a surveying office. He answered positively in dimension $n = 3$ the question of Borsuk whether every set of diameter 1 in n-dimensional Euclidean space can be decomposed into $n + 1$ sets with diameters less than 1. He published the proof in 1947.[115] He got his PhD with Steinhaus in 1950 (thesis *Uwagi o oznaczaniu objętości pni drzewnych* [Remarks on Determining Volumes of Tree Trunks]) and habilitation in 1957. He worked mainly in applications of mathematics, in particular biometrics and biostatistics. He introduced Perkal indicators, used in factor analysis.[116]

Zygmunt Charzyński (1914–2001) studied at the University of Warsaw from 1932 until 1939 and published his first work in 1933 (*Sur les fonctions dont la dérivée symétrique est partout finie*, *Fundamenta Mathematicae* **21**(1933), pp. 214–225). Since 1938 he also worked as an assistant to Sierpiński.[117]

Stefania Braun (1908–1943, also known in the feminine form Braunówna) got her master's degree on the basis of the thesis *O rzutowości operacji Hausdorffa* [On Projectivity of Hausdorff's Operation]. She published a few papers in set theory, among them *Sur quelques propositions équivalentes à l'hypothèse du continu*, *Fundamenta Mathematicae* **19**(1932) pp. 17–21), jointly with Sierpiński.

Menachem Wojdysławski (1919–1942?) graduated from the University of Warsaw in 1939. In his work he generalized some results of Ważewski on the properties of hyperspace of a locally connected metric continuum using the notion of retract (*Retractes absolus et hyperspaces des continu*, *Fundamenta Mathematicae* **32**(1939), pp. 23–25). He also strengthened Kuratowski's result on embedding: Kuratowski–Wojdysławski theorem states that every bounded metric space is isometric to a closed subset of a convex subset of some Banach space. In 1940–1941 he worked in Lwów, then under Soviet occupation, as an assistant at the Ukrainian Ivan Franko University, which replaced the Jan Kazimierz University. Both Braun and Wojdysławski perished in Holocaust.

[114]Hanani fought for the establishment of the Jewish state of Israel and was imprisoned by the British. In 1955 he was appointed a professor of mathematics at Technion — Israel Institute of Technology in Haifa. He switched his interests to combinatorial design theory. See Duda (2000).

[115]Taken prisoner by Germans in the defensive war of 1939, Perkal escaped to the Soviet occupational zone and was later resettled in Uzbekistan. He worked as a teacher and surveyor. In 1946 he returned to Poland and started working at the University of Wrocław.

[116]See Krzyśko *et al.* (2018).

[117]In 1948 Charzyński got his PhD at the University of Warsaw for the thesis Sur les fonctions univalentes bornees under Władysław Nikliborc (1889–1948). For most of his post-war career he worked at the University of Łódź, since 1966 as an ordinary professor. He was the founder of the Łódź school of analytic functions. See Jóźwik and Terepeta (2015).

4.2. *The "Other Warsaw"*

One could get the impression that the Warsaw school of mathematics managed to concentrate all its efforts on set theory, measure theory and topology, according to Janiszewski's proposal. Indeed, in the initial years it was hard for a student to be introduced to research outside of the topological and set-theoretical current.[118] But soon alongside the prevailing disciplines a strong current of ideas related to (broadly understood) mathematical analysis developed.

The key figure in its development was Aleksander Rajchman. A docent at the University of Warsaw since 1925 (when he got his habilitation) and a professor since 1926 at the *Free Polish University* in Warsaw[119] (Chair of Geometry), he influenced mathematical interests of many students, especially through his seminars in the theory of real functions and in probability. He was known for his open criticism of what he saw as the dominance of set-theoretic topics in Warsaw, apparently being of the opinion that "set theory is not mathematics." His own work was related to trigonometric series, probability and statistics. He characterized uniqueness sets for trigonometric series and studied their structure; he investigated the asymptotic behavior of Fourier transforms of measures on a circle, among which he distinguished a class now called Rajchman measures; he defined a formal multiplication of trigonometric series and considered orthogonal series. Jointly with Saks, he gave a new proof of Lebesgue's theorem on differentiability almost everywhere of monotone functions. He proved a strong law of large numbers for uncorrelated random variables. He compiled statistical tables for use in insurance mathematics. In his polemics with the economist Michał Kalecki (1899–1970), he pointed out certain solutions of some differential equations with delayed argument describing business cycles.[120] He represented leftist political and economical views. He publicly protested against anti-Semitic actions at Polish academic schools, which resulted in (failed)

---

[118] Szolem Mandelbrojt (1899–1983), who before World War I took private lessons from Rajchman, enrolled in 1917 at the university and got fascinated by Janiszewski. However, wishing to pursue interests in analysis rather than topology, he moved first to Kharkiv, and then to Paris, where in 1923 he got his PhD under Jacques Hadamard (1865–1963). Mandelbrojt worked on harmonic and complex analysis. Since 1945 he was a professor at Collège de France. See Mandelbrojt (1985).

[119] *Wolna Wszechnica Polska* [Free Polish University] was established in 1918 as a continuation of the Society for Scientific Courses. It was a private enterprise, but its diplomas and degrees were equivalent to those awarded by state schools. It was friendly towards women, Jews and people with leftist political views. Since 1928 it also had a branch in Łódź. For more information see Błędowski *et al.* (1916); Domoradzki and Stawiska (2019).

[120] On this matter Rajchman collaborated with Samuel Fogelson (1902–after 1941, also spelled Fogielson), a graduate in mathematics at the University of Warsaw and an assistant at the Free Polish University working in descriptive statistics. A communist and a Soviet spy, Fogelson crossed to the Soviet occupational zone after World War II broke out. The circumstances of his death are unknown. See Maligranda and Piotrowski (2016).

disciplinary action and ultimately in the cancellation of his lecture in probability in 1936.[121]

Rajchman and Zalcwasser were professors of mathematics at the Free Polish University. Classes in mathematics were also taught by Stanisław Landau (1881–1956, later Ziemecki, the chair of the Laboratory of Physics in the Faculty of Horticulture at SGGW, taught mathematics for naturalists) and Jan Krassowski (1883–1947, a professor of astronomy at the University of Warsaw, taught and ran recitations in general mathematics as well as astronomy). Stefan Aleksander Szulc (1881–1956) taught general statistics, [122] Edward Szturm de Sztrem (1885–1962) taught statistics for economics.[123] A mathematical seminar was run, at different times, by Stefan Straszewicz with Aleksander Rybarczyk, as a senior assistant, Straszewicz and Antoni Wakulicz (1902–1988)[124] and by Rajchman, with Jakub Rywkind as a junior assistant.

Rajchman's influence was strongest on the following mathematicians: Zygmund, Saks, Zalcwasser, Halina Milicer-Grużewska (1901–1981, PhD 1927 with Mazurkiewicz, thesis *O różniczkowaniu całek Jacksona* [On Differentiation of Jackson Integrals])[125] and Wacław Kozakiewicz (1911–1959 PhD 1935, promotion 1936, thesis title unknown, supervisor Mazurkiewicz).[126] The work of the last two concerned probability and statistics.

Zalcwasser got his habilitation in 1929 at the Free Polish University, where he became a professor. He mainly studied notions of analysis by set-theoretic methods (see the preceding subsection), but also worked on the summability of trigonometric series and polynomials related to modular functions.[127]

---

[121]In 1940 the Germans arrested Rajchman and took him to the Sachsenhausen-Oranienburg concentration camp, where he died. See Borsuk (1978); Maligranda and Piotrowski (2017).

[122]Szulc was a professor at the Free Polish University. He also worked in the Główny Urząd Statystyczny [Central Office of Statistics, GUS].

[123]Szturm de Sztrem, a director of GUS, organized the second census in 1931.

[124]Wakulicz studied at the University of Warsaw from 1920 until 1924 and worked as a high school teacher. During World War II he was involved in clandestine teaching. Since 1946 he worked at the Silesian Polytechnic in Gliwice. In 1949 he got his PhD with Sierpiński (thesis *O sumie skończonej liczby liczb porządkowych* [On the sum of a finite number of ordinal numbers]). An extraordinary professor since 1963, he published a few papers on number theory. See Duda (2016).

[125]Before World War II Milicer-Grużewska worked as a teacher and a clerk in Warsaw. After the war she worked at the University of Warsaw, SGGW and the Mathematical Institute of Polish Academy of Sciences. She got her habilitation in 1957. See Duda (2016).

[126]In 1933–1938 Kozakiewicz worked as a senior assistant in the *Zakład Statystyki Matematycznej* [Group of Mathematical Statistics] at SGGW in Warsaw. In 1938 he went on a scholarship to France. His further studies were interrupted by World War II. He voluntarily enlisted in Polish armed forces organized in France. In 1944 he crossed to Spain, from where he went to Canada. He was a professor at the University of Saskatchewan and the Université de Montréal.

[127]In 1939 Zalcwasser was forced by the Germans to live in the Warsaw Ghetto. He perished in the Treblinka extermination camp.

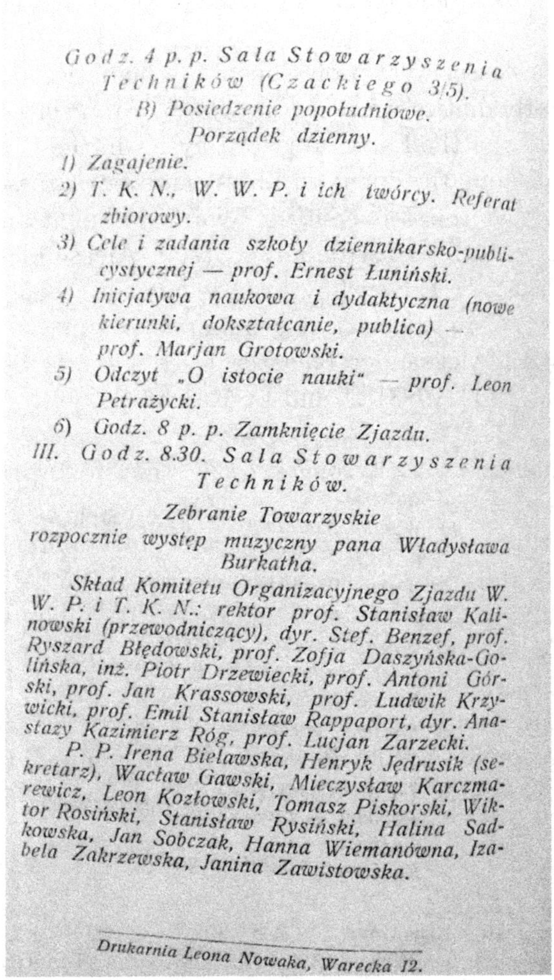

Godz. 4 p. p. Sala Stowarzyszenia Techników (Czackiego 3/5).

B) Posiedzenie popołudniowe.

Porządek dzienny.

1) Zagajenie.
2) T. K. N., W. W. P. i ich twórcy. Referat zbiorowy.
3) Cele i zadania szkoły dziennikarsko-publicystycznej — prof. Ernest Łuniński.
4) Inicjatywa naukowa i dydaktyczna (nowe kierunki, dokształcanie, publica) — prof. Marjan Grotowski.
5) Odczyt „O istocie nauki" — prof. Leon Petrażycki.
6) Godz. 8 p. p. Zamknięcie Zjazdu.

III. Godz. 8.30. Sala Stowarzyszenia Techników.

Zebranie Towarzyskie

rozpocznie występ muzyczny pana Władysława Burkatha.

Skład Komitetu Organizacyjnego Zjazdu W. W. P. i T. K. N.: rektor prof. Stanisław Kalinowski (przewodniczący), dyr. Stef. Benzef, prof. Ryszard Błędowski, prof. Zofja Daszyńska-Golińska, inż. Piotr Drzewiecki, prof. Antoni Górski, prof. Jan Krassowski, prof. Ludwik Krzywicki, prof. Emil Stanisław Rappaport, dyr. Anastazy Kazimierz Róg, prof. Lucjan Zarzecki.

P. P. Irena Bielawska, Henryk Jędrusik (sekretarz), Wacław Gawski, Mieczysław Karczmarewicz, Leon Kozłowski, Tomasz Piskorski, Wiktor Rosiński, Stanisław Rysiński, Halina Sadkowska, Jan Sobczak, Hanna Wiemanówna, Izabela Zakrzewska, Janina Zawistowska.

Drukarnia Leona Nowaka, Warecka 12.

Figure 17. Program of the anniversary congress of the Free Polish University.[128]

Zygmund and Saks made lasting contributions to analysis. Zygmund worked on the sets of uniqueness, Fourier series with random coefficients, summability, absolute convergence and other issues related to trigonometric series. He also studied generalizations of the notion of derivative and smoothness, and in the process introduced function spaces now called Zygmund classes. He got his habilitation in 1926 at the University of Warsaw and taught there as well as at the Warsaw Polytechnic. From 1929–1930 he was on a scholarship (from Rockefeller Foundation) in England, where he started collaborating with Raymond Paley (1907–1933). In 1930 he was appointed a professor of mathematics at the Stefan Batory University in

[128]From the collection of Stanisław Domoradzki.

Wilno (Vilnius). There he supervised a brilliant student, later his colleague, Józef Marcinkiewicz (1910–1940), with whom he made further advances in harmonic analysis and interpolation operators. He published a monograph, *Trigonometric Series*, as number 5 of *Monografie Matematyczne* [Mathematical Monographs] in 1935 and *Funkcje Analityczne* [Analytic Functions], jointly with Saks, as number 10 in 1938 (vi + 431 pages).[129]

Saks got his habilitation in 1926 at the University of Warsaw. He was employed in 1921–1939 as an assistant in the Chair of Mechanics at the Warsaw Polytechnic and lectured as a docent at the university. He studied differentiability of integrals (including the surface area), tangents to curves and surfaces, generalized differentiability of subharmonic functions and other topics. He wrote a joint paper with Banach on bounded sequences in the spaces L^p. In the series *Monografie Matematyczne* he published *Théorie de l'intégrale* (number 2, 1933, xii + 290 pages), rewritten and reissued as *Theory of the Integral* (number 7, 1937, vi + 347 pages) and, jointly with Zygmund, *Funkcje Analityczne* [Analytical Functions].[130]

Other doctorates related to analysis were awarded to: Stanisław Dobrowolski (1898–?), in 1925 (thesis *O wartościach granicznych uogólnionych średnich arytmetycznych* [On Limit Values of Generalized Arithmetic Means], supervisor Mazurkiewicz),[131] Michał Kerner (also spelled Koerner, 1902–1943), in 1926 (thesis *Extremum między punktami sprzężonymi w przestrzeni* [An Extremum between Conjugate Points in the Space], supervisors Przeborski and Mazurkiewicz), Wolibner, in 1930 (promotion 1931, thesis *Przyczynek do teorii funkcji analitycznych* [A Contribution to the Theory of Analytic Functions], supervisor Mazurkiewicz).[132]

Despite significant interests and some achievements of Warsaw mathematicians in probability, the potential "Warsaw school of probability and statistics" did not materialize. It looked like Neyman could join for good the community already including Hosiasson-Lindenbaum, Rajchman, Mazurkiewicz and their students and collaborators mentioned above, plus a few more master's degree students from Rajchman's

[129]Zygmund managed to emigrate to the USA in 1940. Since 1947 he was a professor of mathematics at the University of Chicago. Together with his student Alberto Calderón (1920–1998) he created the theory of singular integrals. See Lorentz (1993); Stein (1999); Anonymous (1992).

[130]Saks was mobilized in the 1939 defensive war. From 1940–1941 he was in Lwów. After returning to Warsaw he was arrested and killed by Gestapo in 1942. See Zygmund (1987); Wojtaszczyk (1982).

[131]Further details of Dobrowolski's life are unknown.

[132]Wolibner worked as a senior assistant at chairs of mathematics as well as at the Aerodynamical Institute of the Warsaw Polytechnic. During World War II he was involved in clandestine teaching in the vicinity of Staszów. Since 1947 a professor of theoretical mechanics at the Wrocław Polytechnic, and since 1953 also in IMPAN. He worked on analytic functions, special functions, aerodynamics and hydrodynamics. See Charzyński *et al.* (1963).

seminar as well as several descriptive statisticians working in various state institutions.[133] At that time Neyman did pioneering work on randomized experiments and sampling and their applications to agriculture. He got his habilitation in 1926 and lectured at the University of Warsaw and SGGW; from 1924 till 1925 he also commuted to Kraków every week to lecture at the Jagiellonian University.[134] For a while he also worked for the seed company K. Buszczyński & Sons.[135] He organized a Group of Mathematical Statistics at the Faculty of Horticulture of SGGW and Biometrics Laboratory at the *Instytut Biologii Doświadczalnej im. M. Nenckiego* [M. Nencki Institute of Experimental Biology].[136] He directed the research of several young people (whom he called his "pups") and promoted their results for publication in international journals. Meanwhile, he traveled (on scholarships from the Fundusz Kultury Narodowej [National Cultural Fund][137] and from the Rockefeller Foundation) to Paris and London. In London he started collaborating with Egon Pearson (1895–1980) on the theory of hypothesis testing, which became a milestone in inferential statistics. Neyman became known internationally and in 1928 he was an Invited Speaker of the International Congress of Mathematicians in Bologna.[138] But, disheartened by the lack of a permanent academic position for himself and by the cutting of funds for the laboratories where his students worked, he left Poland in 1934, first taking a position at the University College London and in 1938 moved to the USA for a professorship at the University of California, Berkeley.[139]

Among the people whom Neyman influenced in Warsaw, two were active in developing mathematical methods of statistics. One was Stanisław Marian Kołodziejczyk (1907–1940), who transferred from the Jagiellonian University and in 1930 finished his master's degree in mathematics at the University of Warsaw. From 1929 until 1930 he worked in the Biometrics Laboratory at the Nencki Institute. In 1931 he received a year-long scholarship from the National Cultural Fund for special studies on probability and mathematical statistics. In the same year he worked

---

[133]Łukasiewicz wrote in 1913 a paper on logical issues of probability, but in his Warsaw period he did not deal with this subject.

[134]Among Neyman's students in Kraków there were Stanisław Kołodziejczyk (1907–1940, see below) and Oskar Lange (1904–1965, later a neo-Marxian economist).

[135]Konstanty Buszczyński (1856–1921) was a landowner and graduate of the Riga Polytechnic. In 1892 he became a co-owner and the head of the company, which specialized in the seeds of sugar beet, a major crop in Poland. Neyman assessed the quality and commercial value of their seeds and published his results in trade journals and pamphlets, as well as in *Wiadomości Matematyczne*.

[136]The Institute was founded in 1918, based on three pre-existing laboratories belonging to the Warsaw Scientific Society. It was named after Marceli Nencki (1847–1901), an organic chemist noted for his work on the structure of hemoglobin and the synthesis of phenyl salicylate.

[137]The National Cultural Fund was a continuation of the Mianowski Fund.

[138]See Ciesielski (2018).

[139]Neyman remained in Berkeley for the rest of his life, where he originated a strong school in mathematical statistics. He received many honors and was a plenary speaker of the ICM in 1954 in Amsterdam. See Reid (1998).

as an instructor for the *Drugi Powszechny Spis Ludności* [Second General Census] of Poland. He became a volunteer assistant in the Group of Mathematical Statistics at the Faculty of Horticulture of SGGW and in 1934 he started giving lectures at SGGW in advanced mathematics and in the theory and applications of statistics. He got his PhD at the University of Warsaw at the Faculty of Mathematics and Natural Sciences in 1936 (promotion 1939, thesis *O pewnej klasie hipotez statystycznych związanych z metodą najmniejszych kwadratów* [On a Class of Statistical Hypotheses Related to the Least Squares Method], published in *Kwartalnik Statystyczny* [Statistical Quarterly] **11**(1936), issue 3–4, pp. 357–427, supervisor unknown).[140]

Karolina Iwaszkiewicz (1902–1999, later Gintowt) taught in a crafts school in her native Wilno[141] even before completing high school in 1920. Since 1921 she worked as a junior assistant in the Astronomy Group at the Mathematics and Natural Science Faculty of the Stefan Batory University under Władysław Dziewulski (1878–1962).[142] This resulted in her publishing a paper in astronomy. In 1926 she started her studies at SGGW in Warsaw and graduated with honors in horticultural engineering in 1931. Her thesis, supervised by Neyman, was the basis of their joint publication *Counting Virulent Bacteria and Particles of Virus*, *Acta Biologiae Experimentalis* **6**(1931), pp. 110–142. In the years 1929–1938 she worked at SGGW, and in 1928–1937 also in Biometrics Laboratory at M. Nencki Institute. In 1934 she got a doctorate in horticultural sciences at SGGW for the thesis *Uogólnienie metody korelacji wielorakiej na przypadek gdy eliminowana zmienna jest niemierzalna* [A Generalization of the Method of Multiple Correlation to the Case when the Eliminated Variable is not Measurable] (supervisor Michał Korczewski (1889–1954),[143] Mazurkiewicz on the examining committee; promotion 1938). She gave lectures and ran recitations in theory of statistics and advanced mathematics for the students in the Faculties of Horticulture and Agriculture of SGGW. She published several

---

[140]Mobilized in 1939 as a reserve second lieutnant of cavalry, Kołodziejczyk was taken prisoner by the Soviets and killed in Katyń. See Krzyśko *et al.* (2018).

[141]Wilno (Vilnius) was the capital of Lithuania united with Poland in the Commonwealth of Two Nations. It was occupied by Russians since 1795 and by Germans in 1915–1918. Later it changed hands between Poles, Bolsheviks and Lithuanians in armed conflicts. Taken by General Lucjan Żeligowski (1865–1947) in 1920, it was made the capital of a puppet state of Central Lithuania and incorporated with it into Poland in 1922. The independent Republic of Lithuania did not recognize this decision and kept in its constitution an article claiming Vilnius as the capital of Lithuania.

[142]Władysław Dziewulski graduated from the Imperial University in 1901 with a silver medal. He got his PhD and habilitation at the Jagiellonian University, where he worked in 1903–1919, first in the Chair of Astronomy, and then in the Astronomical Observatory. Since 1920 a professor at the Stefan Batory University in Wilno. After World War II he co-organized the Nicolaus Copernicus University in Toruń and its Astronomical Observatory. He worked on celestial mechanics, stellar astronomy and astrophysics.

[143]Korczewski was a botanist, biochemist and phytophysiologist. A graduate of the Jagiellonian University, he got his PhD and habilitation there. Since 1922 a professor at SGGW in Warsaw.

papers, some with Neyman, and one jointly with Neyman and Kołodziejczyk: *Statistical Problems in Agricultural Experimentation*, Supplement to the *Journal of the Royal Statistical Society* **2**(1935), no. 2, pp. 107–180. After Neyman's departure she headed the Group of Mathematical Statistics at SGGW in 1936/1937. In 1938 she moved back to Wilno and in 1938–1943 worked there at the *Centralne Biuro Statystyczne Zarządu Miejskiego* [Central Statistical Office of the City Executive Board].[144]

Geometry was represented in the research of mathematicians of the University of Warsaw (Tarski, Wundheiler — see the section on logic) and the Warsaw Polytechnic (Straszewicz, who in 1935 proved a theorem on approximation of extreme points of a compact convex subset of a Euclidean space by its exposed points), but there was no coherent research group.

One PhD in mathematics (besides Wundheiler's) related to this area was awarded at the University to Witold Sosnowski (1904–1941) in 1932 (thesis *Geometria rzutowa tworów urojonych płaskich* [Projective Geometry of Imaginary Planar Constructs], supervisor Żorawski).[145]

Significant geometric content was also present in a PhD in physics awarded at the University of Warsaw to Myron Mathisson (1897–1940) in 1930 (thesis *Ogólna teoria względności i dynamika elektronu* [General Relativity Theory and the Dynamics of an Electron], supervisor Czesław Białobrzeski (1878–1953)).[146] The results of the thesis were a product of scientific correspondence between Mathisson and Albert Einstein (1879–1955) and, with Einstein's help, appeared in three papers in *Zeitschrift für Physik*. In 1932 Mathisson obtained habilitation at the University of Warsaw and lectured there as a docent. Among other subjects, he taught applications of group theory to quantum theory and tensor calculus. With Białobrzeski

[144]From 1952 till 1959, K. Iwaszkiewicz worked as a senior lecturer in the Chair of Botany at the State Pedagogical Institute of Vilnius. She repatriated to Poland in 1959 and took a position as an adjunct (assistant professor) at the *Instytut Upraw, Nawożenia i Gleboznawstwa* [Institute of Soil Science and Plant Cultivation] (IUNG) in Puławy. In 1963 she was transferred to a branch of the Institute in Warsaw. She retired in 1964 and worked part-time at the Laboratory of Biometrics at IUNG (1964–1966) and at the *Instytut Warzywnictwa* [Research Institute of Vegetable Crops] in Skierniewice (1966–1967). See Krzyśko (2013); Krzyśko *et al.* (2018).

[145]A graduate in mathematics at the University of Warsaw, Sosnowski worked at the Warsaw Polytechnic and taught in state higher vocational schools. Since 1936 he headed the Laboratory of Descriptive Geometry at the Warsaw Polytechnic and since 1938 he was an instructor of mathematics at the Ministry of Religious Denominations and Public Education. Imprisoned by the Germans in the Auschwitz concentration camp, he died there.

[146]Białobrzeski was born to a Polish family in the Russian Empire. He studied at the University of Kiev and then at the Collège de France with Paul Langevin (1872–1946). In 1914 he became a professor at the University of Kiev. Repatriated to Poland in 1919, he worked first at the Jagiellonian University and since 1921 at the University of Warsaw. His work concerned thermodynamics, theory of relativity, quantum theory, theory of stellar evolution and structure, spectrography, astrophysics and philosophy of physics.

and Nikodym he also ran a seminar in theoretical physics. He also worked on partial differential equations of hyperbolic type and in 1935 visited Collège de France at the invitation of Hadamard and Langevin. From 1936–1937 he worked in Kazan (USSR) but managed to leave before the repressions against foreigners started. His candidacy for the chair of theoretical physics at the Hebrew University of Jerusalem fell through and in 1938 he went to Kraków. There he collaborated with the physicist Jan Weyssenhoff (1884–1972) and the physicist and mathematician Adam Bielecki (1910–2003).[147] (See, e.g., A. Bielecki, M. Mathisson and J.W. Weyssenhoff: *Sur un théorème concernant une transformation d'intégrales quadruples et intégrales curvilignes dans l'espace de Riemann*, Bulletin International de l'Academie Polonaise des Sciences et des Lettres. *Classe de Sciences Mathematiques et Naturelles, Série A*, 1939, pp. 22–28). In 1939 he went to Paris (where he obtained a solution of Hadamard's conjecture on the class of hyperbolic equations satisfying Huygens' principle) and to Cambridge. He died of tuberculosis in Cambridge.[148]

It is somewhat surprising that a research group on number theory did not arise in the interwar Warsaw. Sierpiński, who was interested in number theory in his early (and late) career, worked mainly on set theory in that period. However, two mathematicians active in Warsaw achieved significant results in this area. Arnold Walfisz (1892–1962), a native of Warsaw, studied in Munich, Berlin, Heidelberg and Göttingen. He got his PhD in Göttingen in 1921 under Edmund Landau (1877–1938) (thesis *Über die summatorischen Funktionen einiger Dirichletscher Reihen*).[149] In 1927 he returned to Warsaw. He worked for the insurance company *Europa* and lectured as a docent at the University (where he got habilitation in 1930). His interests included theory of lattice points in many-dimensional ellipsoids, additive theory of prime numbers, Diophantine approximations, representations of numbers by quadratic forms, and modular forms. Some of his publications in that period were co-authored by John Edensor Littlewood (1885–1977), Vojtěch Jarník (1897–1970), Sarvadaman Chowla (1907–1995) and Gábor Szegő (1895–1985).[150] In 1936 Walfisz went to Georgia (then a Soviet republic) at the invitation of Tbilisi Mathematical Institute. He remained in the USSR, working as a senior researcher at the Tbilisi Mathematical Institute (also serving as the head of the Department of Number Theory and Department of Algebra and Geometry) and teaching as a professor at the University of Tbilisi and at the Tbilisi Pedagogical Institute.[151]

Salomon Lubelski (1902–1941?) studied mathematics at the University of Warsaw and got his PhD in 1928 (thesis *O rozkładzie liczb na formy kwadratowe* [On

[147]For more information on Weyssenhoff and Bielecki as well as collaboration between Kraków physicists and mathematicians, see Chapter 4 in this volume and the references therein.
[148]See Sauer and Trautman (2008).
[149]For more information see Piotrowski and Kolankowski (2003).
[150]For more information see Schinzel (2013).
[151]See Duda (2007); Piotrowski (2013); Schinzel (2011); Turowicz (1964).

Figure 18. Third Congress of Mathematicians of USSR, 1956. From the right front: Andrzej Mostowski, Marceli Stark (1908–1974), Arnold Walfisz, Anna Walfisz (a mathematician, daughter of Arnold Walfisz), and others.[152]

Decomposition of Numbers into Quadratic Forms], supervisor Mazurkiewicz). He worked mainly on algebraic number theory. After the outbreak of war in 1939 he crossed to Białystok, then in the Soviet occupational zone, and became a professor at the State Pedagogical Institute. After the German invasion he was imprisoned in the Majdanek concentration camp and died there.[153]

Walfisz and Lubelski founded in 1935 another specialized journal, *Acta Arithmetica*. It was devoted to number theory but, according to the editorial statement, it also accepted works on geometry, algebra and (complex) function theory, if they used arithmetic methods, gave applications to number theory or concerned functions important for number theory, such as Riemann zeta function or Dirichlet L-functions. Thus in the interwar Poland there were several mathematical journals with international scope: besides *Fundamenta Mathematicae* and *Acta Arithmetica* there were also *Studia Mathematica* — a specialized journal published in Lwów, devoted mainly to functional analysis and related areas — and *Annales de la Société Polonaise de Mathématique*, a general mathematical journal published in Kraków. Out of these, *Acta Arithmetica* not only published works in "Congress" languages by mathematicians from different countries, but also had an international editorial

[152]Photograph from the collection of the Library of the Mathematical Institute of the Polish Academy of Sciences.

[153]Some of Lubelski's results were published posthumously, on the basis of his correspondence with Johannes Gualtherus van der Corput (1890–1975). See Piotrowski (2014) and Schinzel (1992).

board: Harald Bohr (1887–1951), van der Corput, Godfrey Harold Hardy (1877–1947), Jarník, Louis Joel Mordell (1888–1972), Alexander Markovich Ostrowski (1893–1986), Hans Rademacher (1892–1969), Teiji Takagi (1875–1960) and Nikolai Grigorievich Chebotaryov (1894–1947), with Walfisz and Lubelski as managing editors. The older journals *Wiadomości Matematyczne* and *Prace Matematyczno-Fizyczne* were still active, but published mainly the works of Polish scholars (although, for example, Karl Menger published a paper in Volume 37 of *Prace Matematyczno-Fizyczne*). *Prace Matematyczno-Fizyczne* were edited by Dickstein with the cooperation of Żorawski, Leon Lichtenstein (1878–1933), Władysław Natanson and Zaremba. Another venue for mathematical publications was *Mathesis Polska: miesięcznik poświęcony naukom ścisłym i ich metodologii* [Mathesis Polska: A Monthly Devoted to Exact Sciences and Their Methodology]. It was edited by Stanisław Warhaftman (1899–?), with the cooperation of Edward Stenz 1897–1956, a geophysicist) and Zarankiewicz, and published original and popularizing scientific works in Polish. Logical journals in Poland were much less successful: *Studia Logica*, founded in 1934, published just one paper, and the materials for the first issue of *Collectanea Logica*, to be launched in 1939, were destroyed in World War II.

As in the case of the community of set theory and topology, there were several promising junior people interested in analysis whose mathematical career was delayed or cut.

Zygmunt Zahorski (1914–1998) studied at the University of Warsaw from 1934 till 1938. His work *Über die Konstruktion einer differentierbaren, monotone, nicht konstanten Funktion, mit überall dichter Menge von Konstanzintervallen* was published in Sprawozdania z Posiedzeń Towarzystwa Naukowego Warszawskiego. Wydział III [*Comptes Rendus des Séances de la Sociétè des Sciences et des Lettres de Varsovie. Classe III*] **30**(1937), pp. 202–206). While teaching at Szkoła Podchorążych Lotnictwa [Air Forces Officer Cadets School], Zahorski was preparing his PhD thesis under the supervision of Mazurkiewicz (*O zbiorze punktów osobliwych funkcji mającej pochodne wszystkich rzędów* [On the Set of Singular Points of a Function Having Derivatives of All Orders]), but the outbreak of World War II interrupted the process. In 1939 Zahorski went to Lwów, where he became an assistant to Banach and an aspirant (an official PhD candidate in the Soviet academic system). He managed to publish a theorem characterizing the set of points of non-differentiability of a continuous real-valued function (*Punktmengen, in welchen eine stetige Funktion nicht differenzierbar ist*, Recueil Mathématique. Nouvelle Série (*Matematicheskii Sbornik. Novaya Seriya*) **9**(51)(1941), pp. 487–510), but again the PhD proceedings were interrupted by the German invasion on USSR. In 1941 he returned to Warsaw and worked in the Philips factory. After the Warsaw Uprising he was evacuated to

[154]From the collection of the Library of the Mathematical Institute of the Polish Academy of Sciences.

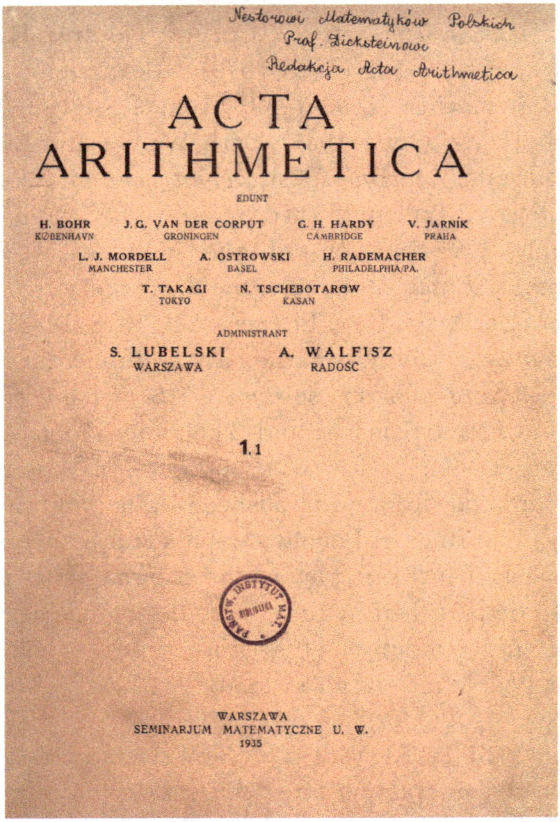

Figure 19. Cover of *Acta Arithmetica* (with a handwritten dedication to S. Dickstein from editors).[154]

Kraków and after the liberation he started working as an assistant at the Jagiellonian University. He finally obtained his PhD in 1946 under Ważewski. In 1947 he got his habilitation. He worked as a professor at the University of Łódź and the Silesian Polytechnic in Gliwice.[155]

Hanna Szmuszkowicz (1910–1993, also known in the feminine form Szmuszkowiczówna) got her master's degree in 1931 from the University of Warsaw. Working as a teacher she published a few papers, alone or jointly with Mazurkiewicz. In one of them (jointly with Mazurkiewicz) she gave a proof of the identity principle for quasianalytic functions. Since 1937 she lectured at the Free Polish University. During World War II she was involved in clandestine teaching. After the war she worked at the University of Łódź and at the Łódź Pedagogical College. She got her PhD in 1950 at the University of Warsaw for the thesis she submitted already in 1937, *O funkcjach quasi analitycznych* [On Quasi Analytic Functions]. From 1954 until 1962 she worked at the University of Warsaw and at IMPAN, and later she

[155]See Zahorski (1986).

headed the division of education of mathematics teachers at the *Instytut Kształcenia Nauczycieli* [Institute for Education of Teachers].[156]

Mojżesz Dawid Kirszbraun (1903–1942) obtained a notable result on extending a Lipschitz function from a subset of a Euclidean space to a Lipschitz function in the whole space with the same Lipschitz constant (*Über die zusammenziehende und Lipschitzsche Transformationen*, *Fundamenta Mathematicae* **22**(1934), pp. 77–108). He did not pursue further research and worked as an actuary. He perished in the Holocaust.

5. Conclusion

On the basis of the profiles of people and their activities described above it is not difficult to argue that the emergence and subsequent evolution of the Warsaw school was consistent with Janiszewski's vision. There was a group of senior researchers united by common mathematical interests. There was an influx of new adepts following in their masters' footsteps. Often identified already in their student years, they were able to enjoy camaraderie and collaboration between the senior and the junior mathematicians, sharing ideas, discussing, scrutinizing and developing them together, thus maintaining continuity of research (see Knaster, 1960). They traveled internationally and were visited by their foreign peers as their results became known and recognized abroad. In the egalitarian spirit of the Society for Scientific Courses, also embraced by Janiszewski, mathematical abilities counted more than gender, ethnicity or religion: women [Nikodymowa, Hosiasson-Lindenbaum, Milicer-Grużewska, Braunówna, Szmuszkowiczówna, K. Iwaszkiewicz, Wanda Szmielew (1918–1976)],[157] ethnic Jews (Rajchman, Knaster, Saks, Kuratowski, Felsztyn, Zalcwasser, Kerner, Tarski, Lindenbaum, Wundheiler, Wajsberg, Aronszajn, Eilenberg, Szpilrajn, Walfisz, Lubelski, Mathisson, Chojnacki, Perkal, Wojdysławski, Presburger and Kirszbraun)[158] and Poles with German (Wolibner) or Eastern-Slavic (Poprużenko) roots obtained their degrees, published in mainstream research journals, taught at academic institutions, sometimes becoming professors (Zalcwasser at the Free Polish University, Kuratowski at the University of Warsaw, Dickstein continued to lecture after his retirement, as a honorary professor of mathematics and history of science at the University of Warsaw) or managing research groups (K. Iwaszkiewicz).

---

[156]See Pawlikowska-Brożek (1994).

[157]Szmielew wrote her first paper, *On Choices from Finite Sets*, as a student under the direction of A. Lindenbaum in 1938. Because of the war it was published only in 1947. Szmielew obtained her PhD at Berkeley in 1950 with Tarski. She was affiliated all the time with the University of Warsaw, becoming an extraordinary professor there in 1967. She established a research group on the foundations of geometry.

[158]Nikodymowa, Hosiasson-Lindenbaum, Braunówna, Szmuszkowiczówna and Szmielew were also of Jewish descent.

Warsaw mathematicians interacted with their colleagues from other Polish academic centers, not only in scientific collaboration but also in organizational structures. This went somewhat against Janiszewski's own preferences: while seeking to create a "scholarly atmosphere" [159] he was reluctant to subject it to formal rules and regulations. Nevertheless, soon after the Mathematical Society was established in Kraków in 1919, he joined it along with Sierpiński, Mazurkiewicz and Dickstein. The Society became in 1920 the Polish Mathematical Society; its Warsaw branch was formed in 1923. The purposes of the Society were the universal cultivation of pure and applied mathematics through talks, publication of mathematical works, scientific collaboration with foreign mathematicians and organization of scientific congresses, and it fulfilled them well. Particularly important were the Congresses of Polish Mathematicians, organized in 1927 (Lwów), 1931 (Wilno), 1937 (Warsaw), as well as the Society's journal *Annales de la Société Polonaise de Mathématique*, which, besides original research work, also published records of meetings in local branches and thus preserved at least titles or abstracts of talks given there. In 1937 the seat of the main governing body of the Society was transferred to Warsaw, officially to better reflect the federational character of the Society (which was the even more highlighted in the new statute). However, this move more likely reflected the tendency to concentrate central institutions in the country's capital, since in 1919–1936 the Polish Mathematical Society was the only central institution representing Polish mathematics.[160]

Figure 20. Wanda Szmielew at the Eighth Congress of Polish Mathematicians, 1953.[161]

[159] During World War I, a group of Warsaw mathematicians and logicians gathered regularly for scientific meetings, often in Janiszewski's private apartment in Mariensztat street.

[160] A contemporaneous attempt to move the Polish Academy of Arts and Sciences from Kraków to Warsaw failed. See Duda (2016); Przeniosło (2011).

[161] Photograph from the collection of the Library of the Mathematical Institute of the Polish Academy of Sciences.

To what extent does it make sense to refer to "two Warsaws"? After all, they were two parts of the same mathematical community. But the two parts were not always perceived as equal. Criticism arose early on, pointing out the dominance of the set-theoretic research orientation; Rajchman went the farthest, supposedly saying that "set theory is not mathematics" or "in Warsaw we have set theory instead of mathematics".[162] According to the critics, the success brought by concentrating research on closely related topics came at a price, which was the marginalization of other disciplines. And indeed, the more traditional disciplines of mathematics were practiced not at the University, but at the Polytechnic, at the Central School of Rural Economy, at the Free Polish University or even outside academic structures. University students were taught differential equations and geometry by scholars who were primarily employed elsewhere. The apologists, mainly Sierpiński, argued that the set-theoretic orientation was broad enough to encompass mathematical analysis and that Polish mathematics could not afford having too many specialties with so few academic positions available.[163] The shortage of academic positions was real, even acute. With only 23 professorships and 27 assistantships[164] in mathematics in the whole country,[165] it was impossible to ensure full-time employment for all talented alumni. That is why so many mathematicians known for their research taught at high schools, continued their previous careers as active military officers or worked in institutions as varied as insurance companies, seed producers, surveying companies, and several state and municipal offices. Stipends for students, another thing Janiszewski had wished for in his program, were also scarce. Some students had to combine their studies with employment (Grużewski, Zalcwasser, Poprużenko, Pasenkiewicz), others sought postponement of tuition payments or defaulted on them (Presburger, against whom the University ultimately took legal action,[166] and

[162]See Borsuk (1978) and Mioduszewski (2014).

[163]This point of view Sierpiński presented when he summarized the state of Polish mathematics for *The Polish Review*, a multi-disciplinary American scholarly publication devoted to Polish topics, in 1959. He wrote (Sierpiński (1959, pp. 53–54)): *... the members of our school worked in such various fields as the theory of sets, topology, the theory of functions of a real variable and mathematical logic.* He placed only Żorawski and his students as well as Dickstein apart from the school. He further maintained that with too diverse mathematical interests *... there would have been no cooperation among mathematicians, there would have been no Polish mathematics, but only several individually working Polish mathematicians* (*ibid.*, p. 54).

[164]At the University of Warsaw there were four assistant positions: one in the Chair of Theoretical Mechanics was held initially by Bolesław Iwaszkiewicz (1902–1983, no relation to Karolina, later inspector of Polish schools in Berlin; after World War II professor at Wrocław Polytechnic, and author of geometry textbooks) and then by Wundheiler. Three other positions were in the three chairs of mathematics and were held by Rajchman, Szpilrajn and Borsuk (until his nomination for the III Chair in 1938).

[165]See Kuratowski (1980).

[166]See Zygmunt (1991).

Szmuszkowiczówna, whose material problems significantly delayed her PhD).[167] The newly reborn state of Poland could not spend significantly on its science and higher education, having to defend its borders and contain hyperinflation in the 1920s and deal with the consequences of a severe worldwide economic crisis in the 1930s.

Besides financial problems there were national tensions, manifesting themselves in discrimination against Jews in education and employment. In the late 1930s several academic schools, including the University of Warsaw, instituted restrictions on admissions (*numerus clausus*) and/or segregated seating for Jewish students ("ghetto benches").[168] Self-appointed groups of nationalist youth enforced these regulations, resorting to disruption of lectures (e.g., of Dickstein in 1937) and physical violence.[169] In the competitive academic environment it became even harder for Jews to get a position or promotion. When Wajsberg, Tarski and Lindenbaum applied in 1937 for a chair in philosophy at the University of Poznań, vacant after the moving of its holder, the philosopher and logician Zygmunt Zawirski (1882–1948), to the Jagiellonian University in Kraków, their applications were rejected and the chair remained unstaffed until after World War II.[170] Finally, the international situation grew precarious, especially for Poland located between Nazi Germany and the Soviet Union and conflicted with other neighbors, Czechoslovakia and Lithuania. It is understandable that under these circumstances many scholars (Mandelbrojt, Tarski, Poznański, Aronszajn, Eilenberg, Chojnacki, Walfisz, Mathisson and Neyman — a non-Jewish Pole) decided to emigrate from Poland.

Mathematicians actively sought solutions for problems affecting their whole community. In 1936 collaboration of scientific societies the Council of Exact and Applied Sciences was established, with the aim of coordinating scientific activities in particular disciplines. Within the Council a Mathematical Committee was formed (with Sierpiński as chairman, Banach as deputy, Kuratowski as secretary). At the Mathematical Congress in 1937 the Council presented a memorial "On the state and needs of mathematics in Poland". The Congress accepted it. The memorial highlighted the dangers of cultivating too narrow areas of mathematics and the stressed the necessity of development of classical analysis, which was neglected in the evolution of mathematical schools in Warsaw and Lwów. The Council expressed concern about the state of employment among research mathematicians. As a partial remedy, it postulated creation of a central institution overseeing mathematical research

[167] See Piotrowski (2014).
[168] These actions were explicitly permitted, but not mandated, by the Minister of Religious Denominations and Public Education.
[169] See Tomaszewski and Żbikowski (2001).
[170] See Woleński (1995).

and of two mathematical institutes, of pure sciences in Warsaw and of applied sciences in Lwów. The outbreak of World War II hindered implementation of these plans.

References

Z. Adamowicz, Wkład Wacława Sierpińskiego do ogólnej teorii mnogości [Wacław Sierpiński's Contribution to General Set Theory], *Wiadomości Matematyczne* **26**(1984), pp. 9–18. ISSN 0373-8302.

K. Bartnicka, Access to Higher Education and Study for Poles in the Second Half of the 19th Century, *Technical Transactions — Fundamental Sciences 1-NP* **7**(2014), pp. 13–35. ISSN 0011-4561.

K. Bartnicka, How to Study Mathematics — The Manual for Warsaw University 1st Year Students in the Interwar Period, *Technical Transactions — Fundamental Sciences 2-NP* **112**(2015), pp. 13–33. ISSN 0011-4561.

S. Bergman, R. Duda, B. Knaster, J. Mycielski and A. Schinzel, Kazimierz Zarankiewicz (2. V. 1902 – 5. IX. 1959), *Colloquium Mathematicum* **12**(1964), pp. 277–288. ISSN 0010-1354.

A. Betti, Polish Axiomatics and Its Truth: On Tarski's Leśniewskian Background and the Ajdukiewicz Connection, pp. 44–71, in D. Patterson (ed.), *New Essays on Tarski and Philosophy*, Oxford University Press, Oxford, 2008, viii + 434 pages. ISBN 978-0-19-929630-9.

R. Błędowski, S. Orłowski *et al.* (eds.), *Dziesięciolecie Wolnej Wszechnicy Polskiej TKN: sprawozdanie z działalności Towarzystwa Kursów Naukowych 1906–1916* [The Decade of the Free Polish University: A Report on the Activity of the Society for Scientific Courses 1906–1916], nakładem Zarządu TKN, skład główny w Księgarniach Gebethnera i Wolffa (Druk. K. Kowalewskiego), Warszawa, 1916, 265 pages.

P. Borowik and M. Kurkowski, Pewne osiągnięcia Kazimierza Kuratowskiego zastosowane w logice formalnej [Some Achievements of Kazimierz Kuratowski Applied in Formal Logic], X Szkoła Historii Matematyki [10th School of History of Mathematics] (Międzyzdroje, 1996), *Zeszyty Naukowe Uniwersytetu Opolskiego. Matematyka* **30**(1997), pp. 43–51. ISSN 1426-6881.

K. Borsuk, O osiągnięciach Prof. Dr Kazimierza Kuratowskiego w dziedzinie topologii [On the Achievements of Prof. Dr. Kazimierz Kuratowski in the Realm of Topology], *Wiadomości Matematyczne* **3**(1960), pp. 231–237. ISSN 0373-8302.

K. Borsuk, unpublished typescript, Polish Mathematical Society, Warszawa, 1978, 20 pages (including 6 handwritten) (photocopy handed in 1978 to the chair of Commission of History of Mathematics PTM).

A. Cała, *Asymilacja Żydów w Królestwie Polskim, 1864–1897: Postawy, konflikty, stereotypy* [Assimilation of Jews in the Polish Kingdom. Attitudes, Conflicts, Stereotypes], Państwowy Instytut Wydawniczy, Warszawa, 1989, 408 pages. ISBN 978-83-06-01789-2.

Z. Charzyński, A. Krzywicki and J. Zamorski, Witold Wolibner (9. IX. 1902–9. I. 1961), *Colloquium Mathematicum* **10**(1963), pp. 353–360. ISSN 0010-1354.

D. Ciesielska, On the 100 Anniversary of the Sierpiński Space-Filling Curve, *Wiadomości Matematyczne* **48**(2012), pp. 69–74. ISSN 2080-5519.

D. Ciesielska, Stanisława i Otton Marcin Nikodymowie w Polsce i Stanach Zjednoczonych [Stanisława and Otto Nikodym in Poland and United States of America — Childhood and School Years], pp. 143–156, in M. Kordas, K. Kucharski, I. Stefaniak (eds.), *Polacy w Ameryce: materiały z polsko-amerykańskiej konferencji naukowej z okazji 50-lecia Muzeum im. Kazimierza Pułaskiego w Warce, 27-28 czerwca 2017 roku* [Poles in America: Research Papers Presented at the Polish-American Conference on the 50th Anniversary of the Casimir Pulaski Museum in Warka, June 27–28, 2017], Volume 1, Publisher Muzeum imienia Kazimierza Pułaskiego w Warce, Warka, 2017, 517 pages. ISBN 978-83-61790-56-3. English translation: eadem, Stanisława and Otto Nikodym in Poland and United States of America — Childhood and School Years, *ibid.*, pp. 396–410.

D. Ciesielska, A Mathematician and a Painter Stanisława Nikodym and her Husband Otton Nikodym, *Mathematisches Forschungsinstitut Oberwolfach Report* **14**(2017), pp. 115–118. ISSN 1660-8933.

K. Ciesielski, Wykłady polskich matematyków na międzynarodowych kongresach matematycznych [Lectures of Polish Mathematicians at the International Congresses of Mathematics], *Wiadomości Matematyczne* **54**(2018), pp. 55–65. ISSN 2080-5519.

B. Cywiński, *Rodowody niepokornych* [The Lineages of the Defiant], Społeczny Instytut Wydawniczy "Znak", Kraków, 1971, 396 pages. ISBN 978-28-69-14003-5.

A. Derkowska, Otton Marcin Nikodým [sic] (1889–1974), *Wiadomości Matematyczne* **25**(1983), pp. 75–83. ISSN 0373-8302 (Polish).

A. Derkowska, Zenon Waraszkiewicz (1909–1946), *Wiadomości Matematyczne* **37**(2001), pp. 135–137. ISSN 0373-8302.

S. Domoradzki, Samuel Dickstein (1851–1939) w świetle korespondencji z Władysławem Natansonem i Marianem Smoluchowskim [Samuel Dickstein (1851–1939) in the Light of his Correspondence with Władysław Natanson and Marian Smoluchowski], pp. 84–103, in J. Mioduszewski (ed.), *Matematyka przełomu XIX i XX wieku. Nurt mnogościowy* [Mathematics of the Turn of the 19th and 20th Centuries. Set-Theoretical Trend], Prace Naukowe Uniwersytetu Śląskiego w Katowicach, Uniwersytet Śląski Katowice, 1992, no. 1253, 193 pages. ISSN 0208-6336.

S. Domoradzki, Czasopiśmiennictwo matematyczne polskie XIX i początku XX w., "Poradnik dla samouków [Polish Mathematical Periodicals of the 19th and the Beginning of the 20th Century, "A Guide for the Self-Study"], Periodyczne wydawnictwo miarą rozwoju potrzeb intelektualnych społeczeństwa [Periodical Publication as a Measure of the Development of Intellectual Needs of the Society], pp. 61–74, in J. Mioduszewski (ed.), *Matematyka przełomu XIX i XX wieku. Nurt mnogościowy* [Mathematics of the Turn of the 19th and 20th Centuries. Set-Theoretical Trend], Prace Naukowe Uniwersytetu Śląskiego w Katowicach, Uniwersytet Śląski Katowice, 1992, no. 1253, 193 pages. ISSN 0208-6336.

S. Domoradzki, *The Growth of Mathematical Culture in the Lvov Area in the Autonomy Period (1870–1920)*, History of Mathematics, Volume 47, Matfyzpress, Prague, 2012, 331 pages. ISBN 978-80-7378-178-1.

S. Domoradzki, Samuel Dickstein, pp. 324–326, in B. Orłowski (ed.), *Słownik polskich odkrywców, wynalazców oraz pionierów nauk matematyczno-przyrodniczych i techniki* [A Dictionary of Polish Discoverers, Inventors and Pioneers of Mathematical and Natural Sciences and Technology], Tom I A–G, Instytut Historii Nauki Polskiej Akademii Nauk, Oficyna Wydawnicza ASPRA-JR, Warszawa 2015, 521 pages. ISBN 978-83-86062-28-7.

S. Domoradzki, Józef Puzyna (1856–1919) — the Pioneer of Polish Mathematical School, pp. 11–22, in A. Lecko (ed.), *Current Research in Mathematical and Computer Sciences*, Uniwersytet Warmii i Mazur Press, Olsztyn, 2017, 318 pages. ISBN 978-83-8100-108-3.

S. Domoradzki and M. Stawiska, Lucjan Emil Böttcher and His Mathematical Legacy, pp. 127–161, in P. Pardalos, T. Rassias (eds.), *Mathematics Without Boundaries. Surveys in Pure Mathematics*, Springer, New York, 2014, xiv + 781 pages. ISBN 978-1-4939-1105-9.

S. Domoradzki and M. Stawiska, Polish Mathematics and Mathematicians in World War I, Part II, *Studia Historiae Scientiarum* **18**(2019), pp. 55–92. ISSN 2451-3202.

S. Domoradzki and M. Stawiska, *The Position of the Jagiellonian University and Its Mathematical Community Before and After WWI*, Chapter 4 in this book.

S. Domoradzki, M. Stawiska and M. Zarichnyi, Modern Mathematics in Lwów Before Banach, pp. 100–106, in Ch. Binder (ed.), *Neglected Parts of Mathematics and Their History*, XIV. Österreichisches Symposion zur Geschichte der Mathematik, Technische Universiät Wien, Wien, 2018, 236 pages.

S. Domoradzki and M. Zaricznyi, On Some Aspects of the Set Theory and Topology in J. Puzyna's Monumental Work, *Technical Transactions — Fundamental Sciences 1-NP* **7**(2014), pp. 85–97. ISSN 0011-4561.

R. Duda, Przemówienie na Walnym Zgromadzeniu PTM w Poznaniu 8 czerwca 1978 r [A Speech in the General Assembly of the Polish Mathematical Society in Poznań on June 8, 1978], *Wiadomości Matematyczne* **22**(1980), no. 2, pp. 296–302. ISSN 0373-8302.

R. Duda, Life and Work of Bronisław Knaster (1893–1980), *Colloquium Mathematicum* **51**(1987), pp. 85–102. ISSN 0010-1354.

R. Duda, Stulecie Wiadomości Matematycznych [The Centennary of Wiadomości Matematyczne] *Wiadomości Matematyczne* **32**(1997), pp. 111–135. ISSN 0373-8302.

R. Duda, On the Warsaw Interaction of Logic and Mathematics in the Years 1919–1939. Provinces of Logic Determined, *Annals of Pure and Applied Logic* **127**(2004), no. 1–3 pp. 289–301. ISSN 0168-0072.

R. Duda, *Matematycy XIX i XX wieku związani z Polską* [Mathematicians of the 19th and 20th Century Related to Poland], Wydawnictwo Uniwersytetu Wrocławskiego, Wrocław, 2007, 587 pages. ISSN 0239–6661.

R. Duda, *Pearls from a Lost City: The Lvov School of Mathematics*, Translated from the 2007 Polish original by Daniel Davies. History of Mathematics, Volume 40, American Mathematical Society, Providence, RI, 2014, xii + 231 pages. ISBN 978-1-4704-1076-6.

R. Duda, Matematyka [Mathematics], pp. 671–757, in A.K. Wróblewski (ed.), *Nauki przyrodnicze i ścisłe na Uniwersytecie Warszawskim* [Natural and Exact Sciences at the University of Warsaw], Monumenta Universitatis Varsoviensis, Wydawnictwa Uniwersytetu Warszawskiego, Warszawa, 2016, 863 pages. ISBN 978-83-235-1791-7.

R. Duda, Przedwojenni nauczyciele na powojennych uczelniach [Pre-War Teachers in the Post-War Academic Schools], *Antiquitates Mathematicae* **10**(2016), pp. 67–87. ISSN 1898-5203.

R. Duda, *Historia matematyki w Polsce na tle dziejów nauki i kultury* [History of Mathematics in Poland on the Background of the History of Science and Culture], Instytut Historii Nauki PAN im. L. i A. Birkenmajerów and Oficyna Wydawnicza ASPRA-JR, Warszawa, 2019, 626 pages. ISBN 978-8375458220.

R. Duda, *Roots of Polish Mathematics (1815–1915)*, in this book.

J. Dydak, Ideas and Influence of Karol Borsuk, *Wiadomości Matematyczne* **48**(2012), pp. 81–95. ISSN 2080-5519.

S. Eilenberg, Karol Borsuk — Personal Reminiscences, *Topological Methods in Nonlinear Analysis* **1**(1993), pp. 1–2. ISSN 1230-3429.

R. Engelking, Kazimierz Kuratowski (1896–1980). His Life and Work in Topology, pp. 431–452, in C.E. Aull, R. Lowen (eds.), *Handbook of the History of General Topology*, Volume 2, Papers from the Special Session of the American Mathematical Society Meeting held in San Antonio, TX, 1993, Kluwer Academic Publishers, Dordrecht, 1998, pp. i–x, 399–807 pages. ISBN 0-7923-5030-8.

A.B. Feferman and S. Feferman, *Alfred Tarski: Life and Logic*, Reprint of the 2004 original, Cambridge University Press, Cambridge, 2008, vi + 425 pages. ISBN 978-0-521-71401-3.

A. Garlicki (ed.), *Dzieje Uniwersytetu Warszawskiego 1915–1945* [History of the University of Warsaw 1915–1945], "Polska XIX i XX Wieku" [Poland of the 19th and 20th Centuries], Państwowe Wydawnictwo Naukowe, Warszawa, 1982, 359 pages. ISBN 8301028637.

Á. Garrido and U. Wybraniec-Skardowska (eds.), *The Lvov-Warsaw School. Past and Present. Studies in Universal Logic*, Birkhäuser, Springer, Cham, 2018, xviii + 815 pages. ISBN 978-3-319-65429-4.

B. Gleichgewicht, Wspomnienie o Samuelu Dicksteinie [A reminiscence of Samuel Dickstein], pp. 149–151, in S. Fudali (ed.): *Matematyka czasów Weierstrassa: materiały XV Ogólnopolskiej Szkoły Historii Matematyki, Kołobrzeg, 28 maja — 2 czerwca 2001 r.* [Mathematics of the Times of Weierstrass: Materials of the All-Poland School in History of Mathematics, Kołobrzeg May 28–June 2, 2001], Szczeciśki Oddział, Polskiego Towarzystwa Matematycznego, Instytut Matematyki Uniwersytetu Szczecińskiego, Szczecin, 2002, 212 pages.

M. Górny, *Concept of Mitteleuropa*, International Encyclopedia of the First World War, 2015. Online document. https://encyclopedia.1914-1918-online.net/article/concept_of_mitteleuropa. [2.10.2019].

M. Gruber, *Alfred Tarski and the "Concept of Truth in Formalized Languages"*, A running commentary with consideration of the Polish original and the German translation. Logic, Epistemology, and the Unity of Science, Volume 39, Springer, Cham, 2016, xii + 187 pages. ISBN 978-3-319-32614-6.

S. Hartman, W. Narkiewicz and C. Ryll-Nardzewski, Scientific work of Edward Marczewski, *Colloquium Mathematicum* **42**(1979), pp. 5–17. ISSN 0010-1354.

A. Indrzejczak, Stanisław Jaśkowski: Life and Work, pp. 457–464, in Á. Garrido and U. Wybraniec-Skardowska (2018).

J. Jadacki, Jan Łukasiewicz: A Creator of New Ideas in Logic and a Reinterpreter of Its History, pp. 33–46, in Á. Garrido and U. Wybraniec-Skardowska (2018).

Z. Janiszewski, O potrzebach matematyki w Polsce [On the Needs of Mathematics in Poland], *Nauka Polska, jej potrzeby, organizacja i rozwój* **1**(1918), pp. 11–18 (reprint *Wiadomości Matematyczne* **7**(1963), pp. 146–155.

I. Jóźwik and M. Terepeta, Matematyka na Politechnice Łódzkiej w latach 1945–70 [Mathematics in the Łódź Polytechnic in the Years 1945–70], *Antiquitates Mathematicae* **9**(2015), pp. 116–136. ISSN 1898-5203.

J. Kauffman, *Elusive Alliance: The German Occupation of Poland in World War I*, Harvard University Press, Cambridge, Massachusetts, 2015, 320 pages. ISBN 978-0-674-28601-6.

S. Kieniewicz and B. Grochulska (eds.), *Dzieje Uniwersytetu Warszawskiego 1807–1915* [History of the University of Warsaw 1807–1915], "Polska XIX i XX Wieku" [Poland of the 19th and 20th Centuries], Państwowe Wydawnictwo Naukowe, Warszawa 1981, 605 pages. ISBN 8301018321.

B. Knaster, Zygmunt Janiszewski (w 40-lecie śmierci) [Zygmunt Janiszewski (on the 40th Anniversary of His Death)], *Wiadomości Matematyczne* **4**(1960), pp. 1–9. ISSN 0373-8302.

S. Kolankowski and Z. Pawlikowska-Brożek, Samuel Dickstein (1851–1939), pp. 46–47, in S. Domoradzki, Z. Pawlikowska-Brożek and D. Węglowska (eds.), *Słownik Biograficzny Matematyków Polskich* [A Biographical Dictionary of Polish Mathematicians], Państwowa Wyższa Szkoła Zawodowa, Tarnobrzeg, 2003, 286 pages. ISBN 8391729338.

J. Kotas and A. Pieczkowski, Scientific Works of Stanisław Jaśkowski, *Studia Logica* **21**(1967), pp. 7–15. ISSN 0039-3215.

A. Koterski and Th. Uebel, Poznański and Wundheiler's 'The Concept of Truth in Physics': the Lvov-Warsaw School Contribution to Encyclopedism, pp. 291–307, in A. Brozek, F. Stadler and J. Woleński (eds.), *The Significance of the Lvov-Warsaw School in the European Culture*, Vienna Circle Institute Yearbook 21, Springer, Cham, 2017, xi + 353 pages. ISBN 9783319528694.

M. Krzyśko, Karolina Iwaszkiewicz-Gintowt (1902–1999), *Acta Universitatis Lodziensis, Folia Oeconomica* **286**(2013), pp. 31–35. ISSN 0208-6018.

M. Krzyśko (chair), W. Adamczewski, J. Berger, E. Gołata, K. Kruszka and B. Łazowska (eds.), *Statystycy polscy. Biogramy. Biblioteka Wiadomości Statystycznych* [Polish Statisticians. Biographical Entries. Library of "Wiadomości Statystyczne"], Zakład Wydawnictw Statystycznych, Warszawa, 2018, 504 pages. ISBN 978-83-7027-685-0.

C. Kuratowski, Stefan Mazurkiewicz et son oeuvre scientifique, *Fundamenta Mathematicae* **34**(1947), pp. 316–331. ISSN 0016-2736.

K. Kuratowski, Pięćdziesiąt tomów "Fundamenta Mathematicae". Wspomnienia i uwagi [Fifty Volumes of "Fundamenta Mathematicae". Reminiscences and Remarks], *Wiadomości Matematyczne* **7**(1963), pp. 9–17. ISSN 0373-8302.

K. Kuratowski, Wacław Sierpiński 1882–1969, *Nauka Polska* **6**(1969), pp. 169–172.

K. Kuratowski, Wacław Sierpiński (1882–1969) (1 plate, errata insert), *Acta Arithmetica* **21**(1972), pp. 1–5. ISSN 0065-1036.

K. Kuratowski, *Pół wieku matematyki polskiej 1920–1970*, Wiedza Powszechna, Warszawa 1973, 191 pages. English translation: *A Half Century of Polish Mathematics. Remembrances and Reflections*, Translated from the Polish by Andrzej Kirkor. With a preface by S. Ulam, International Series in Pure and Applied Mathematics, Volume 108, Pergamon Press, Inc., Elmsford, N.Y., PWN — Polish Scientific Publishers, Warsaw, 1980, viii + 200 pages. ISBN 0-08-023046-6.

K. Kuratowski, Zapiski do Autobiografii. Część pierwsza, Narodziny polskiej szkoły matematycznej [Notes for the Autobiography. Part One, the Birth of the Polish Mathematical School], *Kwartalnik Historii Nauki i Techniki* **24**(1979), no. 2, pp. 243–289. ISSN 0023-589X.

K. Kuratowski, *Notatki do autobiografii. Przedmowa: Ryszard Engelking* [Notes for the Autobiography. Foreword: Ryszard Engelking], Spółdzielnia Wydawnicza Czytelnik, Warszawa, 1981, 242 pages. ISBN 978-8-307-00489-1.

Sr. M. G. Kuzawa, *Modern Mathematics: the Genesis of a School in Poland*, College & University Press, New Haven, 1968, 143 pages.

Sr. M. G. Kuzawa, Fundamenta Mathematicae: An Examination of Its Founding and Significance, *American Mathematical Monthly* **77**(1970), pp. 485–492. ISSN 0002-9890.

G. G. Lorentz, Antoni Zygmund and His Work, *Journal of Approximation Theory* **75**(1993) pp. 1–7. ISSN 0021-9045.

L. Maligranda, Kazimierz Abramowicz (1888–1936), *Wiadomości Matematyczne* **52**(2016), pp. 251–288. ISSN 2080-5519 (Polish).

L. Maligranda and W. Piotrowski, Samuel Fogelson (1902–after 1941), *Antiquitates Mathematicae* **10**(2016), pp. 19–43. ISSN 1898-5203 (Polish).

L. Maligranda and W. Piotrowski, Aleksander Rajchman (1890–1940), *Wiadomości Matematyczne* **53**(2017), pp. 43–82. ISSN 2080-5519 (Polish).

P. Mancosu, R. Zach and C. Badesa, The Development of Mathematical Logic from Russell to Tarski, 1900–1935, pp. 318–470, in L. Haaparanta (ed.), *The Development of Modern Logic*, Oxford University Press, Oxford, 2009, x + 504 pages. ISBN 9780195137316.

S. Mandelbrojt, Souvenirs à bâtons rompus de Szolem Mandelbrojt [Fragments of Recollections of Szolem Mandelbrojt], Interview Recorded by Benoit Mandelbrot, Cahiers du séminaire d'histoire des mathématiques, Edited by René Taton and Pierre Dugac, *Institut Henri Poincaré, Paris* **6**(1985), pp. 1–46. ISSN 0767-7421.

E. Marczewski, *Rozwój matematyki w Polsce* [The Development of Mathematics in Poland], Polska Akademia Umiejętności Kraków, 1948, 47 pages.

E. Marczewski, Prace Kazimierza Kuratowskiego z teorii mnogości i teorii miary [Papers of Kazimierz Kuratowski in Set Theory and Measure Theory], *Wiadomości Matematyczne* **3**(1960), pp. 239–244. ISSN 0373-8302.

E. Marczewski, O pracach Wacława Sierpińskiego [On the Works of Wacław Sierpiński], *Wiadomości Matematyczne* **14**(1972), pp. 65–72. ISSN 0373-8302.

E. Marczewski, *Collected Mathematical Papers*, Edited and with a preface by Stanisław Hartman, Andrzej Hulanicki, Anzelm Iwanik, Zbigniew Lipecki, Czesław Ryll-Nardzewski and Kazimierz Urbanik. With a biography by Roman Duda and Hartman, Polish Academy of Sciences, Institute of Mathematics, Warszawa, 1996, xxxvii + 684 pages. ISBN 83-86806-00-1.

A. McFarland, J. McFarland and J. Smith (eds.), *Alfred Tarski: Early Work in Poland — Geometry and Teaching*, With a bibliographic supplement. Foreword by Ivor Grattan-Guinness. Birkhäuser, Springer, New York, 2014, xiii + 499 pages. ISBN 978-1-4939-1473-9.

J. Mioduszewski, Zenon Waraszkiewicz, *Wiadomości Matematyczne* **37**(2001), pp. 139-143. ISSN 0373-8302 (Polish).

J. Mioduszewski, *Cztery szkice z przeszłości matematyki. Euler, Cantor, Sierpiński w Moskwie, Dwie Warszawy* [Four Essays on the Past of Mathematics. Euler, Cantor, Sierpiński in Moscow, Two Warsaws], Oficyna Wydawnicza Impuls, Kraków, 2014, 186 pages. ISBN 9788378505273.

R. Murawski, Philosophical Reflection on Mathematics in Poland in the Interwar Period. Provinces of Logic Determined, *Annals of Pure and Applied Logic* **127**(2004), no. 1–3 pp. 325–337. ISSN 0168-0072.

R. Murawski, *Filozofia matematyki i logiki w Polsce międzywojennej*, Monografie Fundacji na Rzecz Nauki Polskiej, Wydawnictwo Naukowe Uniwersytetu Mikołaja Kopernika, Toruń, 2011, 268 pages. ISBN 978-83-231-2670-6. English translation: *The Philosophy of Mathematics and Logic in the 1920s and 1930s in Poland*, Translated from the 2011 Polish original by Maria Kantor, Science Networks. Historical Studies, Volume 48, Birkhäuser, Springer, Basel, 2014, xii + 228 pages. ISBN 978-3-0348-0830-9.

Yu.S. Nalbandjan, *On the Work of Mordukhai-Boltovskoi in Warsaw (1898–1916)*, Ministry of General and Professional Education of the Russian Federation and Rostov State University, Rostov-on-Don 1998, inprint 23 pages (Russian).

Z. Pawlikowska-Brożek, Zygmunt Janiszewski — organizator nauki i autor idei polskiej szkoły matematycznej [Zygmunt Janiszewski — Organizer of Science and Author of the

Idea of the Polish Mathematical School], pp. 45–52, in J. Mioduszewski (ed.), *Matematyka przełomu XIX i XX wieku. Nurt mnogościowy* [Mathematics of the Turn of the 19th and 20th Centuries. Set-Theoretical Trend], Prace Naukowe Uniwersytetu Śląskiego w Katowicach, Uniwersytet Śląski, Katowice, 1992, no. 1253, 193 pages. ISSN 0208-6336.

Z. Pawlikowska-Brożek, Matematyka polska w latach 1918–1951 [Polish Mathematics in the Years 1918–1951], *Studia i Materiały z Dziejów Nauki Polskiej, Seria II* **2**(1988), pp. 13–59.

Z. Pawlikowska-Brożek, "Poradnik dla samouków". Periodyczne wydawnictwo miarą rozwoju potrzeb intelektualnych społeczeństwa ["A Guide for the Self-Study". Periodical Publications as a Measure of the Development of Intellectual Needs of the Society], pp. 53–60, in J. Mioduszewski (ed.), *Matematyka przełomu XIX i XX wieku. Nurt mnogościowy* [Mathematics of the Turn of the 19th and 20th Centuries. Set-Theoretical Trend], Prace Naukowe Uniwersytetu Śląskiego w Katowicach, Uniwersytet Śląski, Katowice, 1992, no. 1253, 193 pages. ISSN 0208-6336.

Z. Pawlikowska-Brożek, Hanna Szmuszkowicz (1910–1993), *Wiadomości Matematyczne* **30**(1994), pp. 286–287. ISSN 0373-8302 (Polish).

W. Piotrowski, Ksiądz doktor Franciszek Sieczka [Father Franciszek Sieczka, PhD], Blog Ratowo klasztor czyli historia okolic Radzanowa nad Wkrą [Ratowo Monastery or the History of Surroundings of Radzanów on Wkra]. June 24, 2015. Online document. http://ratowoklasztor.pl/tag/ks-franciszek-sieczka/ [13.2.2021].

W. Piotrowski, Przyczynek do biografii Salomona Lubelskiego [A Contribution to the Biography of Salomon Lubelski], *Wiadomości Matematyczne* **49**(2013), pp. 61–63. ISSN 2080-5519.

W. Piotrowski, Jeszcze w sprawie biografii Ottona i Stanisławy Nikodymów [More on the Matter of Biography of Otto and Stanisława Nikodym], *Wiadomości Matematyczne* **50**(2014), pp. 69–74. ISSN 2080-5519.

W. Piotrowski, Doktoraty z matematyki i logiki na Uniwersytecie Warszawskim w latach 1915–1939 [Doctorates in Mathematics and Logic at the University of Warsaw in the Years 1915–1939], pp. 97–131, in W. Więsław (ed.), *Dzieje matematyki polskiej* [History of Polish Mathematics], Volume II, Instytut Matematyczny Uniwersytetu Wrocławskiego, Wrocław, 2013, 326 pages. ISBN 978-83-910055-8-3.

W. Piotrowski and S. Kolankowski, Arnold Walfisz (1892–1962), p. 253, in S. Domoradzki, Z. Pawlikowska-Brożek, D. Węglowska (eds.), *Słownik Biograficzny Matematyków Polskich* [A Biographical Dictionary of Polish Mathematicians], Państwowa Wyższa Szkoła Zawodowa, Tarnobrzeg 2003, 286 pages. ISBN 8391729338.

E. Poznański, Warszawska szkoła logiczno-filozoficzna (1918–1939) [The Warsaw Logico-Philosophical School (1918–1939)], edited and annotated by Jacek Jadacki from translation of the Hebrew original by Y. Becker and K. Kosakowska, *Filozofia Nauki* **18**(2010), no. 2(70), pp. 139–150. ISSN 1230-6894.

M. Przeniosło, Polskie Towarzystwo Matematyczne w okresie międzywojennym [Polish Mathematical Society in the Interwar Period], *Rocznik Lubelski* **36**(2010), pp. 65–76. ISSN 0080-3510.

M. Przeniosło, Powstanie i rozwój warszawskiej szkoły matematycznej w dwudziestoleciu międzywojennym [The Emergence and Development of Warsaw School of Mathematics in the Two Interwar Decades], *Przegląd Historyczny* **102**(2011), pp. 205–220. ISSN 0033-2186.

M. Przenioslo, International Mathematical Journals Published in Poland Between the Wars, *BSHM Bulletin — Journal of the British Society for the History of Mathematics* **24**(2009), pp. 20–26. ISSN 2637-5451.

R. Purdy and J. Zygmunt, Adolf Lindenbaum, Metric Spaces and Decompositions, pp. 505–550, in Á. Garrido and U. Wybraniec-Skardowska (2018).

R.D. Roman Duda, Z żałobnej karty: Haim Hanani (1912–1991) [From the Mourning Page: Haim Hanani (1912–1991)], *Wiadomości Matematyczne* **36**(2000), pp. 183–185. ISSN 2080-5519.

C. Reid, *Neyman*, Reprint of the 1982 original, Springer-Verlag, New York, 1998, viii + 298 pages. ISBN 0-387-98357-0.

T. Sauer and A. Trautman, Myron Mathisson, What Little We Know of His Life, *Acta Physica Polonica B, Proceedings Supplement* **1**(2008), pp. 7–26. ISSN 1899-2358.

A. Schinzel, Życiorys Wacława Sierpińskiego [A Biography of Wacław Sierpiński], *Wiadomości Matematyczne* **12**(1971), pp. 303–308.

A. Schinzel, Rola Wacława Sierpińskiego w historii matematyki polskiej [The Role of Wacław Sierpiński in the History of Polish Mathematics] *Wiadomości Matematyczne* **26**(1984), pp. 1–9.

A. Schinzel, Wacław Sierpiński, pp. 9–15, in J. Mioduszewski (ed.), *Matematyka przełomu XIX i XX wieku. Nurt mnogościowy* [Mathematics of the Turn of the 19th and 20th Centuries. Set-Theoretical Trend], Prace Naukowe Uniwersytetu Śląskiego w Katowicach, Uniwersytet Śląski, Katowice, 1992, no. 1253, 193 pages. ISSN 0208-6336.

A. Schinzel, Number Theory and Algebra in Salomon Lubelski's Papers, *Antiquitates Mathematicae* **5**(2011), pp. 237–248. ISSN 1898-5203.

A. Schinzel, Teoria liczb w pracach Arnolda Walfisza [Number Theory in the Works of Arnold Walfisz], pp. 157–165, in W. Więsław, *Dzieje matematyki polskiej* [History of Polish Mathematics], Volume II, Instytut Matematyczny Uniwersytetu Wrocławskiego, Wrocław, 2013, 326 pages. ISBN 978-83-910055-8-3.

W. Sierpiński, The Warsaw School of Mathematics and the Present State of Mathematics in Poland, *The Polish Review* **4**(1959), no. 1/2, pp. 51–63.

W. Sierpiński, *Oeuvres choisies. Tome I: Bibliographie, théorie des nombres et analyse mathématique*, publié par les soins de Stanisaw Hartman et Andrzej Schinzel. Contenant des articles sur la vie et les travaux de Sierpiński, par Kazimierz Kuratowski, Andrzej Schinzel et Stanisław Hartman. Comité de rédaction: Stanisław Hartman, Kazimierz Kuratowski, Edward Marczewski, Andrzej Mostowski, Andrzej Schinzel, Roman Sikorski, Marceli Stark. PWN — Éditions Scientifiques de Pologne, Warsaw, 1974, 300 pages (French).

P. Simons, Stanisław Leśniewski: Original and Uncompromising Logical Genius, pp. 209–221, in Á. Garrido and U. Wybraniec-Skardowska (2018).

G.I. Sinkevich, Wacław Sierpiński and the Founding of the Warsaw School of Set Theory and Measure Theory, *Voprosy Istorii Estestvoznaniia i Tekhniki* **1**(2011), pp. 71–82. ISSN 0205-9606 (Russian).

J. Splawa-Neyman, On the Application of Probability Theory to Agricultural Experiments. Essay on Principles. Section 9. Translated from the Polish and edited by D.M. Dąbrowska and T.P. Speed, *Statistical Science* **5**(1990), pp. 465–472. ISSN 0883-4237.

E.M. Stein, Calderón and Zygmund's Theory of Singular Integrals, pp. 1–26, in M. Christ, C.E. Kenig, C. Sadosky, *Harmonic Analysis and Partial Differential Equations* (Chicago, IL, 1996), Essays in Honor of Alberto P. Calderón, Chicago Lectures in Mathematics, University of Chicago Press, Chicago, IL, 1999, 360 pages. ISBN 9780226104560.

S. Surma, Mordchaj Wajsberg, Abstracts of the lectures delivered at the 19th Conference of the History of Mathematical Logic (Cracow, 1973), *Polish Academy of Sciences Institute of Philosophy and Sociology Bulletin of Section of Logic* **2**(1973), pp. 91–96.

H. Syta and R. van de Weygaert, *Life and Times of Georgy Voronoi.* Unpublished online document. https://arxiv.org/pdf/0912.3269.pdf [13.2.2021].

W. Szymański, Who Was Otto Nikodým? [sic], *Mathematical Intelligencer* **12**(1990), no. 2, pp. 27–31. ISSN 0343-6993.

W. Ślebodziński, Wspomnienia matematyka z lat 1903–1968 [Reminiscences of a Mathematician from the Years 1903–1968], *Wiadomości Matematyczne* **12**(1969), pp. 17–31. ISSN 0373-8302.

W. Ślebodziński, Kazimierz Żorawski, *Wiadomości Matematyczne* **11**(1969), pp. 49–64. ISSN 0373-8302 (Polish).

T. Świątkowski, Czy tylko przyczynki? [Contributions Only?], Nurt mnogościowy w teorii funkcji rzeczywistych [Set-Theoretical Trend in the Theory of Real Functions], Zapis odczytu na III Ogólnopolskiej szkole Historii Matematyki w Jaworzu [A written record of the talk at the III All-Poland school of history of mathematics in Jaworze], 1988, pp. 11–16.

K. Świętorzecka, Bolesław Sobociński: The Ace of the Second Generation of the LWS, pp. 599–613, in Á. Garrido and U. Wybraniec-Skardowska (2018).

K. Tatarkiewicz, Profesor Sobociński i kolega Bum [Professor Sobociński and Colleague Bum], *Wiadomości Matematyczne* **34**(1998), pp. 123–146. ISSN 0373-8302.

J. Tomaszewski and A. Żbikowski (eds.), *"Żydzi w Polsce. Dzieje i kultura. Leksykon"* [*Jews in Poland. History and Culture. A Lexicon*], Wydawnictwo Cyklady, Warszawa, 2001, x + 502 pages. ISBN 9788386859580.

A. Turowicz, Arnold Walfisz. A Biographical Note, *Acta Arithmetica* **10**(1964), p. 225. ISSN 0065-1036.

R. Urbaniak, Stanisław Leśniewski: Rethinking the Philosophy of Mathematics, *European Review* **23**(2015), no. 1, pp. 125–138. ISSN 1062-7987.

W. Wilczyński, Prace Wacława Sierpińskiego z funkcji rzeczywistych [Wacław Sierpiński's Works on Real Functions], pp. 299–306, in W. Więsław (ed.), *Dzieje matematyki polskiej* [History of Polish Mathematics], Volume I, Instytut Matematyczy Uniwersytetu Wrocławskiego, Wrocław, 2012, 311 pages. ISBN 978-83-910055-4-5.

P. Wojtaszczyk, O pracach S. Saksa z analizy funkcjonalnej [On the Papers of S. Saks in Functional Analysis], *Wiadomości Matematyczne* **24**(1982), pp. 156–158. ISSN 0373-8302.

Z. Wolak, Naukowa filozofia Koła Krakowskiego [Scientific Philosophy of the Kraków Circle], *Zagadnienia Filozoficzne w Nauce* **36**(2005), pp. 97–122. ISSN 0867-8286.

J. Woleński, Polish Logic, *Logic Journal of the Interest Group in Pure and Applied Logics (IGPL)* **12**(2004), no. 5 pp. 399–428. ISSN 1367-0751.

J. Woleński, Mathematical Logic in Poland 1900–1939: People, Circles, Institutions, Ideas, *Modern Logic* **5**(1995), pp. 363–405. ISSN 1047-5982.

J. Woleński, Philosophy of Exact Sciences (Logic and Mathematics) in Poland in 1918–1939 *Technical Transactions — Fundamental Sciences 2-NP* **112** (2015) pp. 255–265. ISSN 0011-4561.

J. Woleński, Alfred Tarski (1901–1983), pp. 361–371, in Á. Garrido and U. Wybraniec-Skardowska (2018).

J. Woleński, Jerzy Słupecki (1904–1987), pp. 567–573, in Á. Garrido and U. Wybraniec-Skardowska (2018).

J. Woleński, Mathematical Logic in Poland 1918–1939, *Antiquitates Mathematicae* **12**(2018), pp. 99–109. ISSN 1898-5203.

J. Woleński and J. Zygmunt, Jerzy Słupecki (1904–1987): Life and Work, *Studia Logica* **48**(1989), pp. 401–411. ISSN 0039-3215.

Z. Zahorski, Zarys biografii [A Biographical Sketch], *Zeszyty Naukowe Politechniki Śląskiej, Matematyka-Fizyka* **48**(1986), pp. 7–25. ISSN 0072-420X.

A. Zygmund, Stanisław Saks (1897–1942), *Wiadomości Matematyczne* **24**(1982), pp. 145–156 (Polish). An abbreviated English translation in A. Zygmund: *Stanisław Saks, 1897–1942*, Translated from the Polish by Christopher Atkin, Mathematical Intelligencer **9**(1987), no. 1, pp. 36–41. ISSN 0343-6993.

Anonymous, Antoni Szczepan Zygmund: December 26, 1900–May 30, 1992, *Studia Mathematica* **103**(1992), pp. 119–121. ISSN 0039-3223.

J. Zygmunt, Mojżesz Presburger: Life and Work, *History and Philosophy of Logic* **12**(1991), pp. 211–223. ISSN 0144-5340.

J. Zygmunt and R. Purdy, Adolf Lindenbaum: Notes on His Life, with Bibliography and Selected References, *Logica Universalis* **8**(2014), no. 3/4, pp. 285–320. ISSN 1661-8297.

https://doi.org/10.1142/9781786349316_0006

Chapter 6

Formation and Activity of the Lwów School of Mathematics

Stanisław Domoradzki, Margaret Stawiska-Friedland and Mykhailo Zarichnyi

1. A Brief History of Mathematics in Lwów Before 1914

On 20 January, 1661, the Polish king Jan Kazimierz signed a foundational act equating the Jesuit College in Lwów (existing since 1608) with the Kraków Academy in its rights and privileges. In 1772 the first Partition of Poland happened. Lwów fell outside of the range of reforms of *Komisja Edukacji Narodowej* [Commission of National Education], the first education ministry in Europe. Since 1795 Poland was no longer on the map of Europe. The country divided among three occupants could not function. The Academy in Lwów had two faculties — a philosophical and a theological one — but it did not grant degrees. It was closed in 1773, or rather demoted to the status of a high school.

In 1784, the emperor Joseph II (1741–1790) founded in Lwów a university with four faculties and Latin as the language of instruction. The Emperor intended Austria to have three big universities: in Vienna, Prague, and Galicia, whose capital was Lwów. Called the Josephine University after the ruler, the institution existed until 1804, when it was again demoted to the status of a high school (Lwów Liceum).

In 1817, the emperor Francis I (1768–1835) renewed the University with German as the language of instruction. The four faculties were reinstated. A beneficial change happened after the Spring of Nations (1848), thanks to the reforms of the Minister of Religion and Education, Leopold Graf von Thun und Hohenstein (1811–1888). The role of gymnasia crystallized: they were treated as "scientific institutions". Professionalization of teachers increased. Potential university students received increasingly better preparation. The philosophical faculty no longer had to remedy the students' educational shortcomings. The number of university students initially dropped, but soon kept increasing.

Figure 1. The old University building and St. Nicholas Church in Lwów.[1]

Already in 1744 a special Chair of Mathematics was established at the Lwów University. Below we list the names of mathematicians who lectured in Lwów before Polish was introduced as the language of instruction.

In Polish times, there were Faustyn Grodzicki (1709?–1787),[2] Michał Redziminowski (1709–1773?); Tomasz Siekierzyński (1720–1774),[3] Ludwik Hoszowski (1732–1802). After 1772, there were Ignacy baron Rain (1736–after 1807)[4] from 1774 until 1785, Piotr Lody (1764–1829) from 1794 until 1801, Franciszek Kodesch (1761–1831),[5] Joseph Johann von Littrow (1781–1840),[6] Jan Holfeld (1747–1814),

[1]A postcard ca. 1937 from the collection of Michael Zarichnyi.

[2]Grodzicki was an author of works on warfare and architecture.

[3]Siekierzyński published *Arytmetyka*, commonly used in Jesuit schools.

[4]Earlier, Rain was a professor of practical anatomy, surveyor and geographer in Lwów, professor of mathematics at Linz, professor of political learning and an adjunct of astronomy in Vienna. He did not publish. He vacated the chair after becoming a pastor of St. Magdalene Parish in Lwów.

[5]Kodesch was born in Nachod (Czechia). He worked in Lwów for 43 years, with a break as a professor in Kraków. He was offered a chair of pure mathematics in Vienna, but preferred to stay in Lwów.

[6]Littrow was appointed in Lwów since 1807, but soon was transferred to Kraków along with some other professors after the university's demotion. He taught in Kraków until 1810.

Leopold Schulz von Straschnitzki (1803–1852),[7] Ignacy Wojciech Lemoch (1802–1875).[8]

Another chair related to mathematics was the Chair of Practical Geometry, focused on geodesic surveying. It was held successively by the already mentioned Holfeld and Lemoch, who also lectured on elementary and advanced mathematics.

Mathematicians lectured, and published their lecture notes and textbooks. Until 1817 they lectured in Latin, afterwards in German. Except Straschnitzki, they did not produce original mathematical results. Some of them were highly praised, but it was for their lectures and the selection of the contents of their textbooks. Research activity, necessary for emergence of a scientific center, was absent.

In 1861 Galicia obtained autonomy with *Sejm Krajowy* [Domestic Parliament] and government in the capital city of Lwów. In its functional aspects the autonomy was taking shape until the early 1870s, and it was not an easy process. The Constitution from 1867 confirmed the functioning of the Domestic Parliament, which decided on the matters of education, economy and transportation. *Wysoka Rada Szkolna Krajowa* [High Domestic School Council] was also established. It granted significant autonomy to gymnasia. Polish was introduced as the language of instruction.

Before the autonomy, the level of the exact and natural sciences in Lwów was low. The University did not interact with gymnasia. After Thun's reforms and the Autonomy the situation changed. Gymnasia needed teachers of mathematics, who were getting their education mainly at the philosophical faculties of the universities. Mathematical talents could flourish and there were fairly paid positions for them.

In 1871 the Chair of Mathematics at the University was given to Wawrzyniec Żmurko (1824–1889), who arrived in 1851 at the Lwów Polytechnic School from a similar institution in Vienna. He mentored his successors: Placyd Dziwiński (1851–1936) at the Polytechnic and Józef Puzyna (1856–1919) at the University. Puzyna, who took the chair after Żmurko's death, contributed significantly to the growth of mathematical culture at the University and outside of it.[9]

Puzyna was known as a great specialist in analytic functions. He authored a two-volume monograph, *Teorya funkcyj analitycznych* [Theory of Analytic Functions].[10] It was Puzyna who invited Wacław Sierpiński to Lwów in 1908. This initiated the creation of a modern mathematical center. The fact that the dean of

---

[7]After Kodesch's death, a dozen or so candidates applied for his chair, including a professor of Prague University, Jakob Philipp Kulik (1793–1863). The position was given in 1834 to Straschnitzki, who in 1838 was transferred to the Vienna Polytechnic. The latter was the only one who had his own mathematical results.

[8]Lemoch was born in Netvořice (Czechia). In 1840 he was named a professor of practical geometry of the Lwów University. He served as the dean of the Faculty of Philosophy and as rector of the University. Since 1869 an ordinary professor. He was one of the organizers of the Technical Academy in Lwów; until 1870 he taught at both Lwów institutions of higher education.

[9]For more information, see Domoradzki (2017, 2018); Domoradzki and Zaricznyi (2014).

[10]Volume I was published in 1898, and Volume II was published in 1900.

the philosophical faculty was then Kazimierz Twardowski (1866–1938), an eminent philosopher, helped Puzyna in achieving his organizational successes.

The second scientific center in Lwów was the Polytechnic School, the only institution of higher technical education in Polish territories until 1915. It started its activity as an institution of higher education in 1844 (transformed from a former real high school) under the name of Imperial-Royal Technical Academy.

Figure 2. The main building of the Lwów Polytechnic.[11]

In 1852 the chair of mathematics was given to Żmurko, the first Pole among the professors of the Academy. Gradually more chairs were staffed by Poles. Since 1872/1873 all lectures were in Polish.

Since 1901 the Polytechnic School gained the right to grant a doctoral degree: *doctor rerum technicarum*. The licence to lecture (*venia legendi*) could be obtained without a doctoral degree; a track for specialist practitioners existed. Habilitations in mathematics and related sciences were obtained by Władysław Kretkowski (1840–1910) in advanced mathematics (1878), Gustaw Krammer in synthetic geometry (1881), Karol Skibiński (1849–1922) in mechanics of structures (1881), Mieczysław Łazarski (1852–1930) in descriptive geometry (1886), Kazimierz Żorawski in pure and applied mathematics (1893), and Antoni Pawłowski (1859–1942) in insurance mathematics (1913).

[11] A postcard ca. 1937 from the collection of Stanisław Domoradzki.

Below we list the names of those who lectured in mathematics at the Lwów Polytechnic since the mid-1870s. We write more, in the later sections, on those who were still active in Lwów in the interwar period.

Kazimierz Bartel (1882–1941) was an alumnus of the Polytechnic; he got all his academic degrees from it. He worked in descriptive and projective geometry. See below.

Figure 3. Kazimierz Bartel.[13]

Lucyan Böttcher (1872–1937) was a PhD student of Sophus Lie in Leipzig, and an author of important results in iteration theory and holomorphic dynamics, underappreciated throughout his professional life. See below.[12]

Wojciech Burtan (?–1917) was an assistant in the Chair of Mathematics in 1901–1906.

Placyd Dziwiński was named an extraordinary professor of the Second Chair of Mathematics in 1887; two years later he became an ordinary professor.

Stanisław Kępiński (1867–1908), a graduate of the Jagiellonian University, completed additional studies in Göttingen. Since 1899, he was a professor of the Polytechnic School. He was one of the most eminent, although lesser known, Polish mathematicians of the 19th century, and the author of valuable works in differential equations.[14]

---

[12]See also Domoradzki and Stawiska (2015); Stawiska (2014).

[13]From the collection of Zofia Pawlikowska-Brożek and Stanisław Domoradzki gathered for the *Słownik Biograficzny Matematyków Polskich* [Biographical Dictionary of Polish Mathematicians].

[14]See Koroński (2017); Pelczar (2001).

Gustaw Krammer (dates unknown) cooperated with the Chair of General Mechanics.

Jan Krassowski (1883–1947) was a founder of the chair of spherical astronomy and advanced geodesy from 1912 until 1914 and in 1916/1917.

Zdzisław Krygowski (1872–1955) got habilitation in 1907 at the Lwów University on the basis of the thesis *Sur le développement des fonctions hyperelliptiques en series trigonometriques*. After the establishment of the University of Poznań in 1919, he got the chair of mathematics there. At the request of the general staff of the Polish Army, he organized in Poznań a course in cryptology. Among its participants were Marian Rejewski (1905–1980), Jerzy Różycki (1909–1942) and Henryk Zygalski (1908–1978) — Polish mathematicians who contributed significantly to breaking the Enigma.

Mieczysław Łazarski (1852–1930) got a PhD in mathematics at the Lwów University in 1883. In 1885 he got habilitation in descriptive geometry at the Polytechnic School in Lwów and in 1887 became an extraordinary professor, and in 1889 an ordinary professor there.

Antoni Łomnicki (1881–1941) was a graduate of the Lwów University. In 1903 he got his PhD on the basis of the thesis *O odwzorowaniach cząsteczek funkcji hypergeometrycznych* [On Mappings of Particles of Hypergeometric Functions]. On a ministry scholarship he continued his studies in Göttingen (1906–1907) under the direction of Hermann Minkowski (1864–1909), David Hilbert (1862–1943), Felix Klein (1849–1925) and Gustav Herglotz (1881–1953). Since 1913/1914 he lectured as a Privatdozent at the Lwów Polytechnic; in 1919 he got the *venia legendi*. In August 1919 he took the position of a deputy mathematics professor after Krygowski. He had broad interests: analysis, probability, statistics, cartography and mathematical education. See below for an account of his activities and achievements in the interwar period.

Karol Maszkowski (1831–1886) was a precursor of the Lwów school of descriptive geometry.

Antoni Pawłowski (1859–1942) studied mathematics from 1878 until 1881 in Czernowitz (Chernivtsi) and economics in Vienna. From 1913 until 1922, he worked as a docent of actuarial mathematics at the Lwów Polytechnic. He is the author of numerous works in applied and trade mathematics. In his textbooks, concerning mainly political arithmetics and commercial accounts, he introduced many Polish terms in these areas.

Antoni Karol Plamitzer (1889–1954) got his doctorate in 1914 and habilitation in 1921, both at the Lwów Polytechnic. Since 1929, he was an ordinary professor in the Second Chair of Descriptive Geometry at the Mechanical Faculty.

Franciszek Ulkowski (1879–?) was an assistant of mathematics from 1906 until 1910, a teacher at the Higher Real School in Lwów (1910/1911), a teacher of mathematics and physics at the Imperial-Royal State Industrial School in Kraków (1912/1913).

Władysław Zajączkowski (1837–1898) lectured at the Lwów University. He was a fine mathematician and an excellent teacher, and the author of many works in differential equations and analytic geometry, including an extensive monograph *Wykład nauki o równaniach różniczkowych* [Exposition of the Science on Differential Equations], published in Paris in 1877.[15]

Wawrzyniec Żmurko, mentioned above.

Kazimierz Żorawski (1866–1953), after habilitation in 1892 at the Polytechnic School in Lwów, worked since 1893 in Kraków. In 1919, he moved to Warsaw. His scientific and organizational achievements are presented in the chapter on Kraków.[16]

2. Beginnings of the Mathematical School

2.1. *Introduction of Modern Mathematics to Lwów*

An outstanding role in the formation of a center on contemporary mathematics was played by Puzyna, Wacław Sierpiński (1882–1969), Zygmunt Janiszewski (1888–1920), and Stefan Mazurkiewicz (1888–1945). The last three arrived from Warsaw.[17]

Figure 4. Zygmunt Janiszewski.[18]

[15]See Ciesielska and Domoradzki (2014); Koroński (2017).
[16]See Domoradzki and Stawiska (Chapter 5 of this book).
[17]Georgy Feodosevich Voronoy (1868–1908) was one of Sierpiński's doctoral advisors. For more information see, e.g., Schinzel (1971, 1984).
[18]From the collection of Zofia Pawlikowska-Brożek and Stanisław Domoradzki gathered for the *Słownik Biograficzny Matematyków Polskich* [Biographical Dictionary of Polish Mathematicians].

The first world-class mathematical results in Lwów were related to the set-theoretic topology. Janiszewski[19] obtained important results in the continuum theory. In particular, he proved the following:

(1) The sum of two continua does not cut the plane provided that none of them cuts the plane and their intersection is either connected or empty.
(2) The sum of two continua such that their intersection is disconnected cuts the plane.

The work of Janiszewski titled *O rozcinaniu płaszczyzny przez kontinua* [On Cutting the Plane by Continua], *Prace Matematyczno-Fizyczne* **26**(1913), pp. 11–63, in Polish, was supplemented by an extended French abstract, which made the results more accessible to Western readers.[20]

Sierpiński was approved as a Privatdozent in mathematics at the University of Lwów in 1908.[21] One of his books written during the Lwów period [*Outline of Set Theory*] is closely connected to set-theoretic topology. Among Sierpiński's topological results of this period one should mention also the universality of Sierpiński's carpet in the sense that it contains a topological image of any planar curve.[22]

Mazurkiewicz's doctorate at the University of Lwów (1913) was supervised by Sierpiński. The results concerned the square-filling curves. One of his results consists of a complete proof (for $n = 2$) of statements announced by H. Lebesgue (1875–1941)[23] (the proof given by Lebesgue turned out to be incomplete):

---

[19]For more on Janiszewski's influence on Polish mathematics, see, e.g., Domoradzki and Stawiska (Chapter 5 of this book); Knaster (1960); Kuratowski (1980); Kuzawa (1968); Pawlikowska-Brożek (1992).

[20]An interesting discussion on the influence of Janiszewski's results as well as the results of other Polish topologists on the mathematical activity of some American mathematicians working approximately in the same period can be found in Th.L. Bartlow, D.E. Zitarelli, Who Was Miss Mullikin? American Mathematical Monthly **116**(2009), pp. 99–114.

[21]His Lwów period is considered to last until the end of March 1919; however, Sierpiński was interned in the years 1914–1917 and in the list of lectures delivered at the University of Lwów in 1916/1917 one can see the remark that *Professor Sierpiński does not have lectures in this semester.* Ultimately, he took the leave on demand in the Fall semester, 1918/1919.

[22]A three-dimensional generalization of Sierpiński's carpet, namely the Menger curve, is a universal object for all three-dimensional curves. Further generalizations of Sierpiński's carpet are the m-dimensional sets μ^n_m in the n-dimensional Euclidean space. The sets μ^{2m+1}_m are universal spaces for m-dimensional sets; they are also model spaces for the so-called Menger manifolds.

[23]H. Lebesgue, Sur la non applicabilité de deux domaines appartenant respectivement à des espaces à n et $n+p$ dimensions, *Mathematische Annalen* **70**(1911), pp. 166–168.

(1) every planar curve which fills a two-dimensional domain necessarily contains points of multiplicity 3;
(2) there is a curve that fills a two-dimensional domain such that the multiplicity of its points is at most 3.

Also, Mazurkiewicz proved in this paper that every planar nowhere dense continuum can be represented as the image of a Cantor set such that the preimage of every point consists of at most 2 points. Mazurkiewicz left Lwów for Warsaw in 1919, but still kept in contact with mathematicians from Lwów afterwards.

2.2. *Development of Functional Analysis*

After the turmoil of the WWI, Polish–Ukrainian and Polish–Bolshevik wars, the foundational role in Lwów mathematics was played by Hugo Steinhaus (1887–1972), who brought Stefan Banach (1892–1945) from Kraków to Lwów. Antoni Łomnicki should be credited for employing Banach as his assistant at the Polytechnic, despite Banach's incomplete academic credentials. This enabled Banach's later spectacular scientific career. Together, Banach and Steinhaus (soon joined by others) created the school of functional analysis, the focus of Lwów mathematics.[24]

Figure 5. Stefan Banach at the age of about 44.[25]

[24]For more general information see also Pawlikowska-Brożek (1982, 1988, 1992); Przeniosło (2007).
[25]Photograph from Jakimowicz and Miranowicz (2011a, p. 187). With the permission of the editors Emilia Jakimowicz and Adam Miranowicz.

The roots of functional analysis lie in the study of function spaces in connection with differential and integral equations and calculus of variations as well as the properties of transformations of functions such as the Fourier and Laplace transforms. The first notions and results of functional analysis can be found in the works of Vito Volterra (1860–1940), Salvatore Pincherle (1853–1936), Jacques Hadamard (1865–1963), David Hilbert, Erhard Schmidt (1876–1959), Frigyes Riesz (1880–1956), Hans Hahn (1879–1934), Eduard Helly (1884–1943), Eliakim Hastings Moore (1862–1932) and Norbert Wiener (1894–1964).[26]

Lwów mathematicians in the interwar period were responsible for the generalization of function spaces to more abstract ones, which could be treated with the use of linear algebra ("vector calculus") and topology. Consequently, they made giant advances in functional analysis, proving many foundational results and shaping the development of the discipline for many years to come.

The axiomatic definition of linear normed complete spaces was given by Stefan Banach in his PhD thesis. [27] Although complete normed spaces were known earlier, it was Banach who started to investigate them systematically. He proved some important properties, gave examples and applications of such spaces, nowadays called Banach spaces.[28] It is impossible to list all the results in functional analysis that are due to Banach. We will restrict ourselves to mentioning only some classical theorems associated with his name.[29]

The Hahn–Banach theorem, proved independently by Hahn and Banach,[30] allows the extension of bounded linear functionals defined on a subspace of a linear space to the whole space. As a corollary, it shows that for every (nontrivial) normed linear space there are (nontrivial) continuous linear functionals defined on this space. There are also geometric versions, known as hyperplane separation theorems, due to Stanisław Mazur and Meier Eidelheit (see below). The Hahn–Banach theorem is non-constructive, and in fact is equivalent to a weak form of the axiom of choice.[31]

---

[26]For more information see Bernkopf (1966); Birkhoff and Kreyszig (1984); Orlicz (1995); Pietsch (2007).

[27]Published as Banach (1922).

[28]Banach called these spaces the *type B spaces* in Banach (1922). The term *Banach spaces* was coined by Fréchet (1928). Note that independently such spaces were considered by Norbert Wiener. See also Pietsch (2007), where the birth of normed linear spaces is discussed in detail.

[29]See also Domoradzki (2006, 2007).

[30]See Domoradzki (2009); Narici and Beckenstein (1977) for an overview.

[31]The original proofs of Hahn and Banach concerned linear spaces over the field of real numbers. The extension to complex linear spaces was done by Andrew Sobczyk (1915–1981), an American mathematician of Polish origin, and his PhD advisor Henri Frederic Bohnenblust (1906–2000). See H.F. Bohnenblust and A. Sobczyk, Extensions of Functionals on Complex Linear Spaces, *Bulletin of the American Mathematical Society* **44**(1938), pp. 91–93.

The Banach–Steinhaus theorem (also known as the uniform boundedness principle) says that for a family of continuous linear operators (and thus bounded operators) whose domain is a Banach space, pointwise boundedness is equivalent to uniform boundedness in operator norm.[32]

The Banach fixed-point theorem (also known as the Banach contraction principle) states that every contraction in a complete metric space has a unique fixed point.[33]

The Banach–Alaoglu theorem[34] states that the closed unit ball of the dual space of a normed vector space is compact in the weak* topology. This theorem also requires the axiom of choice, unless the underlying space is assumed to be separable.

The open mapping theorem, also known as the Banach–Schauder theorem, states that if a continuous linear operator between Banach spaces is surjective then it is an open map.[35]

A Banach space has the Banach–Saks[36] property (respectively, the weak Banach–Saks property) if each bounded (respectively, weakly compact) sequence in it has a Cesàro convergent subsequence (in the norm topology). The Banach–Saks–Mazur theorem says that the spaces $L^p(\mathbb{R}, \mu)$, $1 < p < \infty$, for certain measures μ (e.g., the Lebesgue measure) have the Banach–Saks property. Józef Schreier (see below) gave examples demonstrating that not every Banach space has the weak Banach–Saks property.

The early achievements of Lwów mathematicians in functional analysis made their way into Banach's monograph *Teoria operacji liniowych* [Theory of Linear Operations].[38] In it, he treated sets and measurable operators in metric spaces, general vector spaces, normed spaces, various classes of normed spaces, weak convergence, biorthogonal systems, linear functionals, linear functional equations, topological groups and other topics. The concluding remarks, containing many open problems, were written in collaboration with Mazur (see below).[39]

---

[32]See S. Banach and H. Steinhaus, Sur le principe de la condensation de singularités, *Fundamenta Mathematicae* **9**(1927), pp. 50–61.

[33]See Banach (1922), This theorem is widely used in many different areas of mathematics; for example, in fractal geometry it is used for proving the existence of self-similar sets.

[34]The theorem was also obtained by the Greek–American mathematician Leonidas Alaoglu (1914–1981). See L. Alaoglu, Weak Topologies of Normed Linear Spaces, *Annals of Mathematics* **41**(1940), pp. 252–267.

[35]For Schauder's work, see Über die Umkehrung linearer, stetiger Funktionaloperationen, *Studia Mathematica* **2**(1930), pp. 1–6.

[36]Stanisław Saks (1897–1942), a Warsaw mathematician working on topology and analysis. For more information, see Chapter 5 of this book and references cited therein.

[37]From the collection of Stanisław Domoradzki.

[38]Published as Banach (1931).

[39]For more information see Jaëck (2016); Köthe (1960).

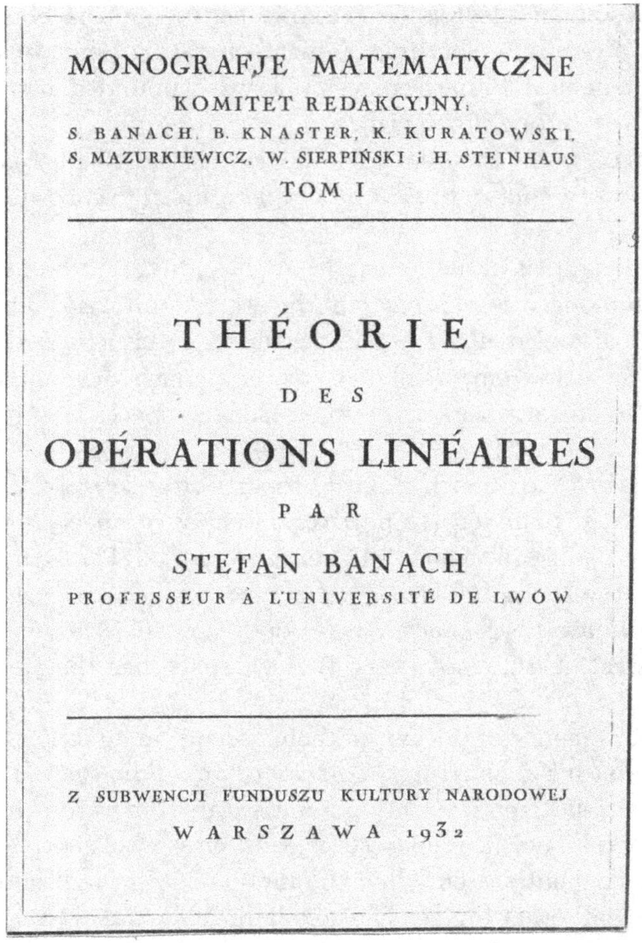

MONOGRAFJE MATEMATYCZNE
KOMITET REDAKCYJNY:
S. BANACH, B. KNASTER, K. KURATOWSKI,
S. MAZURKIEWICZ, W. SIERPIŃSKI i H. STEINHAUS
TOM I

THÉORIE
DES
OPÉRATIONS LINÉAIRES
PAR
STEFAN BANACH
PROFESSEUR À L'UNIVERSITÉ DE LWÓW

Z SUBWENCJI FUNDUSZU KULTURY NARODOWEJ
WARSZAWA 1932

Figure 6. The title page of Banach's monograph.[37]

Steinhaus used to joke that his greatest mathematical discovery was Banach. They published three mathematical articles jointly. But this should not overshadow Steinhaus's own multiple important mathematical contributions, including those in the field of functional analysis.

Steinhaus's interests were somewhat less abstract than those of his Lwów colleagues, Banach, Schauder and Mazur. Steinhaus worked on the series of trigonometric and other functions, their convergence and divergence, orthogonal expansions and complete systems. With Stefan Kaczmarz (1895–1939, see below) he co-authored a monograph *Theorie der Orthogonalreihen*.[40]

[40]Published in the series *Monografie Matematyczne*, Volume VI, Warszawa — Lwów, 1935, vi + 298 pages.

In fact, Steinhaus's work on functional analysis influenced his breakthrough results on probability. He used orthogonal systems of functions on the unit interval (starting with the Rademacher system) to model infinite sequences of random events. He initiated the study of random functional series.[41]

Kaczmarz graduated from the Jagiellonian University in 1922. In his PhD, which he obtained in Lwów in 1924 under Stanisław Ruziewicz (1889–1941). Kaczmarz considered topics close to the research of Kraków mathematicians. Namely, his PhD thesis title was *O związkach między pewnymi równaniami funkcyjnymi i różniczkowymi* [On Relations between Certain Functional and Differential Equations]. However, soon Kaczmarz picked up research topics more related to functional analysis. His habilitation in mathematics in 1929 at the Lwów University was titled *O zbieżności i sumowalności rozwinięć ortogonalnych* [On Convergence and Summability of Orthogonal Expansions].

Kaczmarz worked and published on many topics: orthogonal series (convergence and summability almost everywhere), functional equations, generalized convergence, existence of continuous functions with singularities of various kinds, integral transforms, number series, methods of summability, multipliers and Dini-type integrals, among others. He is best known for inventing a method of approximate (iterative) solutions of large systems of linear equations (called Kaczmarz algorithm or the cyclic projection method).[42] He also defined the Kaczmarz map, giving a homeomorphism of the Orlicz space with L^1. He delivered two lectures at International Congresses of Mathematicians: *Über eine Anwendung der Funktionalen an die Orthogonalreihen* in Bologna (1928) and *On the Orthogonal Series* in Oslo (1936).[43]

Stanisław Mazur (1905–1981), a graduate of the Lwów University, got his PhD in 1935 under Banach's supervision (thesis *O szeregach warunkowo sumowalnych* [On Conditionally Summable Series]). Afterwards he worked at the Lwów University and the Lwów Polytechnic.[44]

The Banach–Mazur theorem says that every real, separable Banach space $(X, \|\cdot\|)$ is isometrically isomorphic to a closed subspace of $C^0([0,1], \mathbb{R})$, the space of all continuous functions from the unit interval to the real line.

The Banach–Mazur distance is a distance on the set $Q(n)$ of n-dimensional normed spaces. With this distance, the set of isometry classes of n-dimensional

---

[41] See Woyczynski (2012).

[42] For information and references on Kaczmarz algorithm see Sznajder (2015).

[43] For more information see Maligranda (2007).

[44] In the years 1939–1941 and 1944–1946, Mazur worked at the Lwów University run by the Soviets. During the Nazi occupation he worked as a vendor. He repatriated to Poland in 1946 and became a professor of mathematics at the University of Łódź, University of Warsaw and the Mathematical Institute of the Polish Academy of Sciences. For a comprehensive overview of Mazur's contributions to functional analysis, see Köthe (1987).

normed spaces becomes a compact metric space, called the Banach–Mazur compactum.[45]

Mazur was one of the pioneers of game theory. He introduced the notion of a Banach–Mazur game, which is a topological game played by two players, trying to pin down elements in a set (space). Banach showed that for such a game there is a winning strategy.

The Mazur–Ulam theorem states that if V and W are normed spaces over $\mathbb{R}$ and the mapping $f\colon V \to W$ is a surjective isometry, then f is affine.[46]

Mazur's research also included normed algebras. One of his important results proved in his Lwów period is the Gelfand–Mazur theorem, which says that a Banach algebra with the unit over the complex numbers in which every nonzero element is invertible is isometrically isomorphic to the complex numbers. That is, the only complex Banach algebra that is a division algebra is the algebra of complex numbers $\mathbb{C}$ itself.[47]

In collaboration with Andrzej Turowicz (1904–1989), a graduate of the Jagiellonian University and a senior assistant at the Lwów Polytechnic, Mazur obtained a generalization of Weierstrass' theorem on approximation of continuous functions by polynomials, similar to what is now known as the Stone–Weierstrass theorem. Unfortunately, their joint results were never published.[48]

Juliusz Paweł Schauder (1899–1943) got his PhD at the Lwów University under the supervision of Eustachy Żyliński (see below). In his thesis he dealt with measure theory and generalized Stokes' theorem to some surface measures, but soon he got involved in functional analysis, to which he made some important contributions.[49]

The notion of an orthonormal basis in a Hilbert space is due to F. Riesz. Schauder generalized this notion by introducing the notion of a Schauder basis in a Banach space and giving examples of Schauder bases in the spaces C and L^p, $p \geq 1$.[50]

---

[45]Topological properties of the Banach–Mazur compacta are described in Ageev *et al.*, (2004) and Antonyan (2000).

[46]S. Mazur and S. Ulam, Sur les transformations isométriques d'espaces vectoriels normés, *Comptes Rendus Hebdomadaires des Séances de l'Académie des Sciences, Paris* **194**(1932), pp. 946–948.

[47]Israel Moiseevich Gelfand (1913–2009) proved a weaker version of this theorem independently of Mazur.

[48]For more information on Turowicz, a Benedictine monk and a professor of mathematics at the Polish Academy of Sciences after WWII, including his pre-war work in Lwów, see Domoradzki and Stawiska (2015); Pawlikowska-Brożek (2003); Ciesielski and Pogoda (1988); Turowicz (1989).

[49]See Derkowska (1990); Mawhin (2018).

[50]See Zur Theorie stetiger Abbildungen in Funktionalräumen, *Mathematische Zeitschrift* **26**(1927), pp. 317–320.

Schauder's theorem on adjoint operators says that a linear operator between Banach spaces is compact if and only if its adjoint operator is compact.[51]

Schauder's fixed point theorem[52] is a generalization of Brouwer's fixed point theorem to subsets of topological vector spaces. Some remarks on this result will be given later in the section that concerns topology.[53]

Herman Auerbach (1901–1942) proved for finite-dimensional normed spaces (not necessarily inner product spaces) the existence of a basis such that the vectors of this basis as well as of its dual have norms equal to 1.

Meier Eidelheit (1910–1943) got his PhD in 1938 under Banach's supervision (thesis *O rozwiązywaniu równań liniowych o nieskończenie wielu niewiadomych* [On Solving Linear Equations with Infinitely Many Unknowns]). In the paper Zur Theorie der konvexen Mengen in linearen normierten Räumen, *Studia Mathematica* **6**(1936), pp. 104–111, he proved the separation theorem for convex sets, i.e., a geometric version of the Hahn–Banach theorem.[54]

Władysław Orlicz (1903–1990) received his PhD in 1928 under the supervision of Żyliński (thesis *Z teorii szeregów ortogonalnych* [Some Problems in the Theory of Orthogonal Series]).[55] He is known for the introduction and systematic investigation of function spaces generalizing the L^p spaces which are now known as Orlicz spaces. The starting point for this research was Orlicz's collaboration with Zygmunt Wilhelm Birnbaum, also from Lwów, whom he met at Göttingen University in 1929/1930.[56] The early results are contained in the following papers: Z.W. Birnbaum, W. Orlicz: *Über die Verallgemeinerung des Begriffes der zueinander konjugierten Potenzen*, *Studia Mathematica* **3**(1931), pp. 1–67, and W. Orlicz, *Über eine gewisse Klasse von Räumen vom Typus B*, *Bulletin de l'Académie Polonaise des Sciences et des Lettres, Séries A*, **8/9**(1932), pp. 207–220.

---

[51]See J. Schauder, Über lineare, vollstetige Funktionaloperationen, *Studia Mathematica* **2**(1930), pp. 183–196.

[52]See J. Schauder, Der Fixpunktsatz in Funktionalräumen, *Studia Mathematica* **2**(1930), pp. 171–180.

[53]For more information about Banach's and Schauder's fixed point theorems, see, e.g., Kumar (2013).

[54]See Maligranda (2015).

[55]In 1931–1937 Orlicz worked at the Lwów Polytechnic; since 1937 he was a professor in Poznań. He spent the years of WWII in Lwów and returned to Poznań afterwards.

[56]Orlicz went to Göttingen on a scholarship in theoretical physics and Birnbaum was studying for his actuarial certificate. They wrote several papers together. For more information see Maligranda and Wnuk (2000a,b); Woyczyński (2001).

3. Diverse Faces of Modern Mathematics in Lwów

3.1. *Topology*

The Lwów School of Mathematics is known mostly because of its achievements in functional analysis. However, some significant results were obtained by members of the school also in other parts of mathematics.

The number of topological results and their importance demonstrate that Lwów was an important topological center also in the interwar period.[57] Many results of Lwów mathematicians deal with various kinds of convergence, and so are closely connected to set-theoretic topology (e.g., the open mapping theorem, Banach–Alaoglu theorem; see above).

Figure 7. The main building of the Lwów University, current view.[58]

In the theory of topological algebras, the Banach–Stone theorem characterizes compact Hausdorff spaces by their algebras of bounded continuous (real-valued)

[57]See also Borsuk (1995).
[58]Photographed by Stanisław Domoradzki.

functions (the algebras of scalars). Banach first proved it in the important case of metric spaces.[59]

During the period of 1928–1934, when Kazimierz Kuratowski was a professor at the General Faculty of the Lwów Polytechnic, he obtained some important topological results. He also published the first volume of his famous treatise: *Topologie. I: Espaces métrisables, espaces complets.*[60]

A criterion of planarity of graphs was established by Kuratowski while he worked in Lwów. [61] A graph is planar if and only if it does not contain topological copies of the graphs K_5 and $K_{3,3}$.

The Knaster–Kuratowski–Mazurkiewicz lemma (KKM-lemma) was published in 1929. Lwów is indicated as the city of work of all three authors. The lemma is equivalent to the Brouwer fixed point theorem, but in some applications (e.g., in the equilibrium theory) it seems to be more convenient than the latter.[62]

Also, Kuratowski obtained some important results in the dimension theory. He proved [63] that, under suitable conditions, n-dimensional sets are images of mappings with the cardinality of their preimages being less than or equal to $n + 1$.

Kuratowski and Ulam [64] obtained a result which is considered as a counterpart of the Fubini theorem for categories. It provides a characterization of meager sets in the topological products of second countable Baire spaces (in particular Polish spaces).

Stanisław Ulam (1909–1984) was the co-author of eight publications with Józef Schreier (1909–1943), who lived and worked in Drohobycz.[65] They worked both in the topological group and in topological semigroup theory. Ulam recognized the importance of investigation of topological-algebraic objects. Schreier obtained his PhD in 1934 under Banach. His thesis title was *O skończonej bazie w grupach topologicznych* [On Finite Base in Topological Groups].[66]

Schreier and Ulam proved[67] that every automorphism of the group S_ω of permutations with finite supports of a countable set is an inner automorphism, i.e., there exists an element $s \in S$ such that $x = sxs^{-1}$, for any $x \in S$. Here, S_ω is regarded

---

[59]The generalization is due to the American mathematician Marshall Harvey Stone (1903–1989). See M. Stone: Applications of the Theory of Boolean Rings to General Topology, *Transactions of the American Mathematical Society* **41**(1937), pp. 375–481.

[60]Published in the series *Monografie Matematyczne*, Volume 3, Seminarium Matematyczne Uniwersytetu Warszawskiego, Warszawa, Instytut Matematyczny PAN, 1933, x + 285 pages.

[61]In the paper Kuratowski (1930).

[62]See Knaster *et al.*, (1929). There are various generalizations of this result, e.g., the KKMS theorem proved by Lloyd Shapley (1923–2016).

[63]In Kuratowski (1932).

[64]In their joint paper Kuratowski and Ulam (1932).

[65]The results of Schreier and Ulam were reviewed and analyzed in modern terminology in Guran and Prytula (2013).

[66]See Maligranda (2013, 2014).

[67]See Schreier and Ulam (1934).

as a topological group with respect to the topology of pointwise convergence. In the same paper they also proved that the semigroup of continuous self-maps of the m-dimensional cube, $m \geq 1$, has a base consisting of 5 elements. (Recall that a base in a topological (semi)group is a subset that generates a dense sub(semi)group.)

Figure 8. Stanisław Ulam.[68]

The Borsuk–Ulam theorem states that for every continuous map from the n-sphere in $\mathbb{R}^{n+1}$ into the Euclidean n-space $\mathbb{R}^n$ there is some pair of antipodal points which are mapped to the same point. Here, two points on the unit sphere

$$S^n = \{x \in \mathbb{R}^{n+1} \mid \|x\| = 1\}$$

are called antipodal if they are in exactly opposite directions from the sphere's center. The first proof of this result was given by the Warsaw mathematician Karol Borsuk (1905–1982),[69] who attributed the formulation of the problem to Ulam.[70]

In their collaboration Borsuk and Ulam[71] defined the symmetric product $SP^n X$ of a topological space X as the orbit space of the natural action of the symmetric group on the nth power of X. The construction of symmetric product is often used in

[68]From the collection of Zofia Pawlikowska-Brożek and Stanisław Domoradzki gathered for the *Słownik Biograficzny Matematyków Polskich* [Biographical Dictionary of Polish Mathematicians].

[69]In Borsuk (1933). For more information on Borsuk and his achievements, see Chapter 5 of this book.

[70]Note that for $n = 1$ the theorem can be derived from the Mean Value Theorem. A combinatorial proof of the Borsuk–Ulam theorem is based on Tucker's lemma. The Borsuk–Ulam theorem has many applications; in particular, it implies the Ham Sandwich theorem.

[71]See Borsuk and Ulam (1931).

different areas of topology.[72] It also turned out that the infinite symmetric products can be used to introduce a topology in the set of persistence diagrams in Topological Data Analysis.

Schauder was the first to try to extend the Brouwer fixed point theorem to infinite-dimensional space. He formulated[73] the fixed-point theorem for compact convex subsets in infinite-dimensional topological linear spaces. However, it turned out soon that his proof worked smoothly only for the Banach spaces. Andrey Nikolayevich Tikhonov (Tychonoff, 1906–1993) extended the theorem to the locally convex case. Later the problem whether the result is true in its initial formulation was placed in *The Scottish Book* (Problem Number 54).

The representatives of the Lwów School of Mathematics frequently applied the Baire category theorem in order to prove the existence of objects with desired properties or to describe the behavior of "typical" objects in a given space. The method is based on the Baire category theorem proved by René-Louis Baire (1874–1932) in 1899: a (nonempty) complete metric space cannot be represented as the countable union of nowhere dense sets (i.e., it cannot comprise a first category set).[74] Since Banach spaces are complete, it is not strange that the Baire category theorem is of use in the Banach space theory. Actually, fundamental results of functional analysis (uniform boundedness principle, open mapping theorem, closed graph theorem) can be proved by applying Baire category arguments.

A classical example of the Baire category method is the Banach–Mazurkiewicz–Jarník[75] theorem, which says that among all continuous real-valued functions on a compact interval the functions that are nowhere differentiable form a residual set. The result was generalized by Herman Auerbach, both in the space of all continuous functions of a compact interval and in its subspace of Hölder continuous functions.[76]

---

[72] In particular, Albrecht Dold (1928–2011) and René Thom (1923–2002) used the infinite symmetric product $SP^\infty X = \varinjlim SP^n X$ to prove that its n-th homology group of X is isomorphic to the n-th homotopy group of $SP^\infty X$. A characterization of the G-symmetric power functors as open (equivalently, bicommutative) functors [in the framework of the theory of normal functors developed by Evgeny V. Shchepin (born 1951)] of finite degree is obtained in Zarichnyi (1992).

[73] See J. Schauder, Der Fixpunktsatz in Funktionalräumen, *Studia Mathematica* **2**(1930), pp. 171–180.

[74] By using this theorem, the existence of an object with prescribed properties among a universal class of objects can be proved as follows. First, the universal class is endowed with a complete metric. Second, one has to verify that the set of objects that do not satisfy the property is of the first category. An important class of spaces for which the Baire category method works is the class of topological spaces homeomorphic to separable complete metric spaces; the latter are called Polish spaces.

[75] Proved separately, and in somewhat different forms, by the three named mathematicians. See S. Banach, Über die Bairesche Kategorie gewisser Funktionenmengen, *Studia Mathematica* **3**(1931), pp. 174–179, S. Mazurkiewicz: Sur les fonctions non dérivables, *ibid.*, pp. 92–94, and V. Jarník, Über die Differenzierbarkeit stetiger Funktionen, *Fundamenta Mathematicae* **21**(1933), pp. 48–58.

[76] H. Auerbach, Démonstration nouvelle d'un théorème de M. Banach sur les fonctions dérivées des fonctions mesurables, *Fundamenta Mathematicae* **7**(1925), p. 263.

The method applied by Banach was further discussed in H. Auerbach, S. Banach: *Über die Höldersche Bedingung*, *Studia Mathematica* **3**(1931), pp. 180–184. In connection with these publications we also mention Kaczmarz's paper (S. Kaczmarz: *Integrale vom Dinischen Typus*, *ibid.*, pp. 189–199).[77]

3.2. *Set Theory*

The language of set theory was extensively used in Poland since Puzyna's treatise *Theory of Analytic Functions*. Later Sierpiński started to study set theory systematically. In 1909 he delivered a course in set theory and in the period before WWI he published a book on set theory.[78]

Figure 9. The building of the Mathematical Seminar, where the mathematicians of the Lwów School worked.[79]

Among the best known set-theoretic results of the Lwów mathematical school one should mention the Banach–Tarski paradox (also known as Banach–Tarski–Hausdorff paradox). In their joint work 1924, Banach and Tarski [80] proved a theorem about of the paradoxical decomposition of a ball in the three-dimensional Euclidean

[77]For more examples of the applications of Baire's theorem in analysis, see, e.g., Domoradzki (2005); Orlicz (1982); Jarnicki and Pflug (2015).
[78]See Domoradzki *et al.*, (2014) for more information.
[79]Photographed by Stanisław Domoradzki.
[80]See Banach and Tarski (1924).

space into finitely many disjoint subsets which can be reassembled (by translations and rotations) to give two balls identical to the original one. This result is a consequence of the Axiom of Choice. It should be noted that there are set theories in which the Banach–Tarski paradox is not possible.

Ulam introduced the notion of a measurable cardinal as an uncountable cardinal number κ such that there exists a κ-additive, nontrivial, 0-1-valued measure on the power set of κ.[81] He proved that the smallest cardinal κ that admits a nontrivial countably-additive two-valued measure on the power set of κ also admits a κ-additive measure. The notion of measurable cardinals is tightly connected with the notion of an inaccessible cardinal. Measurable cardinals were the first examples of large cardinals.

3.3. *Measure Theory*

Many achievements of the Lwów School of Mathematics deal with the notion of a measure. Measure theory rapidly developed after the invention of the Lebesgue measure and Lebesgue integral at the very beginning of the 20th century.

The following well-known theorem was proved by Steinhaus: if a subset A of reals is of positive measure, then the interior of the difference $A - A$ is nonempty.

As we saw above, some notions and results from set theory are related to the notion of a measure. Banach introduced the notion of a real-valued measurable cardinal.[82]

The notion of an additive measure is closely related to the notion of a Banach limit. Actually, the Banach limit on the space ℓ^∞ can be regarded as a (signed) finitely additive measure defined on all subsets of natural numbers.

As we have already mentioned, Banach and Tarski proved that there is a decomposition into five pieces of the unit ball in the three-dimensional space with the property that one can translate and rotate these pieces so that two unit balls can be obtained. The Banach–Tarski paradox implies that in the three-dimensional space there is no invariant Banach measure such that the measure of the unit cube is 1.

A Banach measure on a set X is a finite measure μ defined for every subset of X and such that $\mu(X) > 0$ and $\mu(\{x\}) = 0$ for every $x \in X$. (An Ulam measure is a 0-1-valued Banach measure.) Banach proved that there is a Banach measure in the plane which is consistent with the Lebesgue measure. This also proves that it is impossible to have an analog of the Banach–Tarski paradox in dimension 2.

Assuming the Continuum Hypothesis, Banach and Kuratowski proved that there is no Banach measure defined on all subsets of the segment $[0, 1]$.[83]

---

[81] See Ulam (1930).

[82] See Banach (1930).

[83] See Banach and Kuratowski (1929).

The following important problem of measure theory is named after Ruziewicz: is the Lebesgue measure on the n-sphere characterized, up to proportionality, by its properties of being finitely additive, invariant under rotations, and defined on all Lebesgue measurable sets? This problem is also known as the Banach–Ruziewicz problem. Also, the Banach–Tarski paradox seems to be a motivation for this problem. For $n = 1$, Banach found a negative solution to the Ruziewicz problem.[84]

Figure 10. A group of mathematicians in front of the building of the Lwów Polytechnic during the First Congress of the Polish Mathematical Society, 7–12, September 1927.[85]

Additionally, Z. Łomnicki and Ulam[86] considered the problem of existence of a probability measure on the product with given marginal measures. Their proof, however, turned out to be incorrect.

[84]For higher dimensions the situation is more complicated. That the answer is "yes" was proved independently by Gregory Margulis (born 1946) and Dennis Sullivan (born 1941) (for $n \geq 4$) and by Vladimir Drinfeld (born 1954) for $n = 2, 3$.

[85]From the collection of Stanisław Domoradzki. Donated by Mrs. Ewa Broszkiewicz.

[86]See Łomnicki and Ulam (1934).

3.4. *Differential Equations*

Functional analysis, the flagship of the Lwów mathematical school, had its roots in problems of calculus of variations as well as differential and integral equations. These disciplines were represented in the research of Lwów mathematicians.

Steinhaus's PhD thesis *Neue Anwendungen des Dirichlet'schen Prinzips*, written in Göttingen in 1911 under the supervision of David Hilbert, dealt with variational problems for second-order partial differential equations and their application to a packing problem. He did not revisit these topics later on.

Banach's PhD thesis, *O operacjach na zbiorach abstrakcyjnych z zastosowaniami do równań całkowych* [On Operations on Abstract Sets with Applications to Integral Equations] (1920, supervisor Kazimierz Twardowski),[87] not only introduced fundamental notions and properties of linear normed complete spaces, but also applied these notions to integral operators (defined by kernels). He dealt with a variational problem in his paper *Sur certains ensembles de fonctions conduisant aux équations partielles du second ordre*, *Mathematische Zeitschrift* **27**(1927), pp. 68–75.

Figure 11. Juliusz Paweł Schauder at a conference in Moscow in 1935.[88]

[87]Published as Banach (1922).
[88]From the collection of Michael Zarichnyi.

A breakthrough approach to partial differential equations was proposed by Schauder. He had already made several landmark contributions to functional analysis (Schauder bases, open mapping theorem, Schauder fixed point theorem) and thought of using topological properties of linear spaces to prove existence of solutions of differential equations. As recalled by Schauder's friend Henryk Schärf (see below),[89] Schauder turned to differential equations upon learning that in Göttingen they cared more about differential equations and variational problems than about the still-new functional analysis.[90]

Schauder's most important results in this new direction were obtained in collaboration with Jean Leray (1906–1998). In their joint work *Topologie et équations fonctionnelles, Annales Scientifiques de l'École Normale Supérieure, Serie* 3, **51**(1934), pp. 45–78, they introduced the topological degree of a mapping in an abstract setting and used it to prove the existence of solutions of certain partial differential equations. In 1938 they were jointly awarded the Malaxa prize for this achievement.[91]

Schauder continued to work on elliptic and hyperbolic equations of the second order. He published a paper *Quasilineare Differentialgleichungen zweiter Ordnung vom hyperbolischen Typus. Gemischte Randwertaufgaben*, *Studia Mathematica* **6**(1936), pp. 162–189 with Mirosław Krzyżański (1907–1965), then a voluntary assistant at the Stefan Batory University in Wilno.[92] Also in 1936, Schauder and Mazur gave a joint talk at the International Congress of Mathematicians in Oslo titled *Über ein Prinzip in der Variationsrechnung.*[93]

The classical approach to differential equations was also present in the research of Lwów mathematicians, particularly those educated at the Jagiellonian University in Kraków.

Włodzimierz Stożek (1883–1941) studied in Kraków (finished 1907) and Göttingen (finished 1909). He got his PhD with Stanisław Zaremba in 1922 (thesis *O wartościach charakterystycznych równań całkowych potencjału logarytmicznego* [On Characteristic Values of Integral Equations of the Logarithmic Potential]). Shortly afterwards he moved to Lwów and took the chair of mathematics in the Faculty of Civil Engineering at the Lwów Polytechnic. He wrote a few papers on potential theory — two of them with Władysław Nikliborc (see below), as well as

---

[89] In Schaerf (1993).

[90] It is true that abstract functional analysis was absent from the mathematical research in Göttingen in the 1920s and 1930s. Moreover, a new mathematics institute opened there in 1929, directed by Richard Courant (1888–1972), whose main interest was mathematical physics. However, at that time Hilbert worked on the foundations of mathematics and proof theory, moving away from his earlier interests in integral equations and mathematical physics.

[91] For more information see Mawhin (2018).

[92] After World War II, Krzyżański became a professor at the Jagiellonian University in Kraków. For more information see Domoradzki (2017).

[93] For more about talks of Polish mathematicians at International Congresses of Mathematics. See the *ICM Proceedings* (Date unknown) and Ciesielski (2018).

school textbooks with Sierpiński, Banach and Nikliborc. He was a regular in meetings of mathematicians in the Scottish Café.[94]

Władysław Nikliborc (1899–1948) finished his studies in mathematics at the Jagiellonian University (interrupted by World War I military service) in 1922. In the same year he moved to Lwów to become an assistant to Łomnicki in the Chair of Mathematics at the Faculty of Mechanics of the Lwów Polytechnic. In 1924 Nikliborc obtained his PhD degree at the Faculty of Philosophy of the Lwów University (thesis *Zastosowanie zasadniczego twierdzenia Cauchy'ego o istnieniu rozwiązań równań różniczkowych zwyczajnych do zagadnień na wartości brzegowe w równaniu* $y'' = f(x, y, y')$ [On Applications of the Fundamental Theorem of Cauchy on the Existence of Solutions of Ordinary Differential Equations to Boundary Value Problems for the Equation $y'' = f(x, y, y')$]). He went on to receive two habilitations, in mathematics and theoretical mechanics, respectively, at the Lwów University and the Lwów Polytechnic.

Nikliborc's habilitation at the Lwów University, in 1927, was based on a two-part paper *Sur les fonctions hyperharmoniques*, published in *Comptes Rendus* of the Paris Academy of Sciences in 1925 and 1926. In modern terminology, the functions he considered are called pluriharmonic. Nikliborc developed methods of several complex variables to deal with pluriharmonic functions and made significant advances in their theory.[95]

Nikliborc's interest in potential theory and differential equations extended in a natural way to problems of fluid mechanics and celestial mechanics. In 1930–1931 he visited Leon Lichtenstein (1878–1933) in Leipzig and reinforced his interested in fluid mechanics. Nikliborc got habilitation in theoretical mechanics in 1931 on the basis of a paper about a rotating fluid. He published papers in Polish, German and French journals. He also wrote academic and high school textbooks (some of them jointly with Steinhaus or Stożek). He took part in the Scottish Café meetings and even entered several problems in *The Scottish Book*. In 1937 Nikliborc moved to Warsaw as an extraordinary professor at the Warsaw Polytechnic.[96]

Banach did not explicitly work on mathematical problems of mechanics and there is no reliable information regarding his attending (or not) university lectures in mathematics and/or mechanics in Kraków. Nevertheless, Banach was the author of an extensive academic textbook in mechanics, *Mechanika w zakresie szkół akademickich* [Mechanics in the scope of academic schools],[97] also translated after World War II into English.

---

[94]Stożek and his two sons were among the Lwów professors and their family members shot by the Germans soon after their invasion on Lwów in 1941. See Duda (2007). For a personal reminiscence, see Dawidowicz (1981, 1989).

[95]See Domoradzki and Stawiska (2015) and references cited therein.

[96]See Ślebodziński (1948), Domoradzki and Stawiska (2015) and Duda (2007).

[97]Part I, *Monografie Matematyczne*, Volume 8, Warszawa–Lwów–Wilno, 1938, v + 234 pages, Part II, *Monografie Matematyczne*, Volume 9, Warszawa–Lwów–Wilno, 1938, pp. 235–255.

3.5. *Probability Theory*

At the turn of the 19th and 20th centuries, probability was gaining mathematical foundations and becoming a part of theoretical mathematics. Lwów scholars were at the forefront of this process.

Marian Smoluchowski (1872–1917) held the chair of theoretical physics at the Lwów University from 1900 to 1913. In 1906 he formulated the theory of (one-dimensional) Brownian motions, thus becoming one of the founders of the area of stochastic processes.[98]

Jan Łukasiewicz (1878–1956, see below) lectured on the foundations of probability in 1910/1911, developing calculus of probability from a few axioms all the way to conditional probability and Bayes' formula. These lectures were expanded into a monograph *Die logischen Grundlagen der Wahrscheinlichkeitsrechnung.*[99]

In Łukasiewicz's treatment, probability is a function of propositions. Another point of view — that probability is a function of sets — was represented by two Lwów mathematicians who also dealt with mathematical foundations of probability.

Antoni Łomnicki, a professor at the Lwów Polytechnic since 1921, was one of the pioneers of the use of measure theory in probability. His intuition was geometric, based on the graph of the density function. His theory appeared in the paper *Nouveaux fondements du calcul des probabilités* (*Définition de la probabilité fondée sur la théorie des ensembles*), *Fundamenta Mathematicae* **4**(1923), pp. 34–71. Subsequently, Łomnicki gave a proof of a version of the de Moivre–Laplace theorem,[100] and published several research papers on measurable and non-measurable functions as well as survey articles on probability and statistics.[101]

Steinhaus, a professor of the Lwów University since 1921, got interested in Lebesgue's theory of measure and integration during his studies in Göttingen (where he completed his PhD in mathematics in 1911). In the paper *Les probabilités dénombrables et leur rapport à la théorie de la mesure, Fundamenta Mathematicae* **4**(1923), pp. 286–310, he not only gave a measure-based system of axioms for probability, but also constructed a probability space suitable for dealing with infinite

---

[98]His publication Zur kinetischen Theorie der Brownschen Molekularbewegung und der Suspensionen, *Annalen der Physik*, *Series 4*, **14**(1906), pp. 756–780, was preceded by two short notes on the same topic in the Bulletin International de l'Académie des Sciences de Cracovie. Albert Einstein worked concurrently on Brownian motion, arriving at the same probability distribution in 1905. See Fuliński (1998).

[99]Kraków, Spółka Wydawnicza Polska, 1913. Translation: *Logical Foundations of Probability Theory*, in *Selected Works*, L. Borkowski (ed.), Amsterdam, North-Holland, 1970, pp. 16–63.

[100]Démonstration éleméntaire de la loi de Gauss dans le calcul des probabilités, *l'Enseignement Mathématique* **25**(1926), pp. 103–111.

[101]See Girlich (2005); Maligranda (2008); Pawlikowska-Brożek *et al.*, (1996).

Figure 12. Professor Antoni Łomnicki in uniform at the time of Lwów Defense.[102]

sequences of independent repeated Bernoulli trials and proved a version of the strong law of large numbers.[103]

Steinhaus went on to make major advances in probability, especially regarding the notion of independence and convergence of series with random terms. Among many papers on these subjects, he wrote *La courbe de Peano et les fonctions indépendantes, Comptes Rendus Hebdomadaires des Séances de l'Académie des Sciences* **202**(1936), pp. 1961–1963. In this paper he gave a construction of a square-filling curve in which the two coordinates, as functions of a parameter $t \in [0, 1]$, are independent random variables (the construction can be extended to higher dimensions). He also started applying statistics to various problems of science and everyday life (which became his lifelong pursuit). For example, he used statistics in his collaboration with the physician Ludwik Fleck (1896–1961) on the problem of counting leukocytes.[104]

[102]From the collection of Stanisław Domoradzki. Donated by Mrs. Ewa Broszkiewicz.

[103]See Girlich (2005); Martínez-Adame and Caballero (2016).

[104]Fleck got his doctorate in medicine at the Lwów University in 1920 and worked first in the department of internal medicine at Lwów General Hospital, then in his own private bacteriological laboratory. A survivor of the Lwów ghetto and the concentration camps Auschwitz and Buchenwald. After the World War II, Fleck worked at the Maria Skłodowska-Curie University in Lublin and the Instytut Matki i Dziecka [Mother and Child State Institute] in Warsaw before emigrating to Israel in 1961. He was known for his work in microbiology as well as for a very original approach to the sociology of science (theory of thought collective).

Steinhaus invented and patented several devices. The design of Steinhaus's longimeter is based on an idea akin to the Crofton formula in integral geometry, according to which the length of a curve equals the expected number of times it is crossed by a random line.

Steinhaus worked on the notion of independence with his PhD student Marek (Mark) Kac (1914–1984). They wrote three joint papers on the subject (published in *Studia Mathematica* in the years 1936–1938), dealing also with relations to convergence of series and Brownian motion.[105] Kac got his PhD in 1937 and in 1938 he went to the USA on a scholarship from Parnas Foundation. He remained there because of WWII, becoming a professor of mathematics at the Cornell University, the Rockefeller University and the University of Southern California. He was one of the most eminent probabilists of his time, working on the Ising model [with Theodore Berlin (1917–1962) and John Clive Ward (1924–2000)], rigorous theory of Feynman integrals [with Richard Feynman (1918–1988)] and probabilistic number theory [with Paul Erdős (1913–1996)].[106]

In 1938 Steinhaus contributed a chapter *La théorie et les applications des fonctions indépendantes au sens stochastique* to the volume *Les fonctions aléatoires ...*[107] The other invited contributors were two Russian (Soviet) mathematicians, Sergei Natanovich Bernstein (1880–1968) and Evgeny Evgenevich Slutsky (1880–1948).

Ulam was a student of the General Faculty at the Lwów Polytechnic, getting his master's degree in 1932 and doctorate in 1933 (both under the supervision of Kazimierz Kuratowski). He worked on a wide variety of problems, mostly in measure theory and topology. In 1935, at the invitation of John von Neumann (1903–1957), Ulam went to the USA to work at the Institute of Advanced Study in Princeton and at the Harvard University.[108]

Auerbach also published papers related to probability, e.g., *Über die Fehlerwahrscheinlichkeit einer Summe von Dezimalzahlen, Zeitschrift für angewandte Mathematik und Mechanik* **13**(1933), pp. 386–388.

Zbigniew Łomnicki (1904–1994), a nephew of Antoni Łomnicki, collaborated with Ulam on yet another treatment of probability based on measure theory.[109] Łomnicki worked as an actuary after finishing his studies in mathematics and physics at the

[105] The total number of papers in the series is 10. One was written by Steinhaus and Czesław Ryll-Nardzewski after WWII. Others were authored by Steinhaus alone.

[106] See Feigenbaum (1982); Kac (1985).

[107] Published in the series *Actualités Scientifiques et Industrielles*, no. 738, Hermann & Cie, Paris, 1938, 76 pages.

[108] During World War II, Ulam participated in the Manhattan Project developing nuclear weapons. This led to his subsequent work on multiplicative processes and inventing the Monte Carlo method of computation. He remained in the USA, working at the Los Alamos National Laboratory and lecturing at several universities. Ultimately he became a professor of mathematics at the University of Colorado at Boulder. For an autobiographical account see Ulam (1930).

[109] See their joint paper Łomnicki and Ulam (1934).

Lwów University. In 1933 he became the head of the Department of Mathematics in *Zakład Ubezpieczeń Społecznych* [Social Insurance Institution] in Warsaw.[110]

We should mention other graduates of the Lwów University in mathematics with experience in probability and actuarial mathematics.

Henryk Schärf (1907–2006, also Schaerf) finished his studies in Lwów in 1929. In 1931 he got an actuarial certificate at the University of Göttingen and afterwards worked as an actuary in Vienna, Lwów and Warsaw. During World War II he was able to get to Switzerland. In 1941 he published a paper on some problems in actuarial mathematics and in 1943 he got his PhD in mathematics at Eidgenössische Technische Hochschule (ETH), Zürich (thesis *Über links- und rechtsseitige Stieltjesintegrale und deren Anwendungen*).[111]

Ludwik Sternbach (1905–1942) graduated in mathematics in 1928, and from 1928 until 1934 taught mathematics and physics in Lwów gymnasia. Since 1929 he was a proofreader for *Studia Mathematica*. From 1934 until 1939 he worked in the Towarzystwo Ubezpieczeń na Życie Feniks [Life Insurance Society Feniks] and also as an editor of *Studia Mathematica*. He wrote two joint papers with Mazur.[112]

Zygmunt Wilhelm (later William) Birnbaum (1903–2000) first got a law degree and then, in 1929, a PhD under Steinhaus. He earned an actuarial certificate from Göttingen and worked as an actuary in Vienna and Lwów. He wrote several papers in his Lwów period, mostly on functions of a single complex variable, but also on functional analysis (jointly with Orlicz, see above), and one on probability (jointly with Schreier): *Eine Bemerkung zum starken Gesetz der großen Zahlen, Studia Mathematica* **4**(1933), pp. 85–89.[113]

Let us also mention another instance of introducing mathematical methods to statistics and its applications. Kazimierz Vetulani (1889–1941), a civil engineer and a professor in the chair of statics at the Lwów Polytechnic, published an article, *O zastosowaniu pewnego przedstawienia geometrycznego do biometrji i statystyki* [On the Application of Geometrie Approach to Biometrics and Statistics], *Roczniki Nauk Rolniczych i Leśnych* **28**(1932), pp. 389–400. This work was presented at the

---

[110]In World War II, Zbigniew Łomnicki fought in the Polish Army in Norway, Middle East and Italy. Afterwards he remained in emigration, first in Great Britain, and then in the USA. He worked as a mathematician in industry, lectured in the Polish University College (London) and the University of Washington (Seattle) and published papers on stochastic processes, statistics and insurance mathematics. See Pawlikowska-Brożek (1995).

[111]After the war Schärf moved to the USA, became a professor of mathematics at the Washington University in St. Louis, Missouri, and switched his interests to more abstract measure theory. See Schaerf (1993).

[112]For more information on him, see Maligranda and Prytuła (2013).

[113]In 1937 Birnbaum went as a reporter to New York and remained in the USA. He became a professor of mathematics at the University of Washington in Seattle, working on nonparametric testing and estimation, probability inequalities, survival distributions, competing risks, and reliability theory. See Łomnicki (1951); Woyczynski (1997).

Second Congress of Romanian Mathematicians in Turnu Severin in 1932 under the title *Über die Anwendung einer geometrischen Darstellung auf Biometrie und Statistik*.[114]

4. Mathematics Outside Lecture Halls: *Studia Mathematica* and *The Scottish Book*

In Poland in the first quarter of the 20th century, specialized mathematical journals were still a novelty. The first one was *Fundamenta Mathematicae* published in Warsaw since 1920 and concentrated on set theory and topology. Soon the need arose for a journal accommodating works in the growing area of functional analysis.[115]

In 1929, Banach and Steinhaus started to publish the journal *Studia Mathematica*, which accepted papers in foreign (Western) languages. The role of this journal for the Lwów Mathematical School was similar to that of *Fundamenta Mathematicae* for the Warsaw School. Up to 1940, nine volumes appeared. Besides Lwów school mathematicians, other Polish mathematicians also published there: Mirosław Krzyżański, Józef Marcinkiewicz (1910–1940),[116] Antoni Zygmund (1900–1992), Stefan Mazurkiewicz and Tadeusz Ważewski (1896–1972).

The foreign authors of papers published in *Studia Mathematica* from 1929 until 1940 were: Aurel Wintner (1903–1958), Jovan Karamata (1902–1967), Paul Lévy (1886–1971), Takehito Takahashi (Dates unknown), Francis Joseph Murray (1911–1996), Géza Grünwald (1910–1943), Herman Heine Goldstine (1913–2004), Paul Pierre Gillis (1910–2004), Cornelis Visser (1910–2001), Egbert Rudolf van Kampen (1908–1942), Leonid Vitaliyevich Kantorovich (1912–1986), Grigorii Mikhailovich Fichtenholz (1888–1959), Mark Grigorievich Krein (1907–1989), David Pinhusovich Milman (1912–1982),[117] Robert Fortet (1912–1998), Luigi Fantappié (1901–1956), Raymond Edward Alan Christopher Paley (1907–1933), Nikolai Maximovich Gunther (Gjunter) (1871–1941), Wilhelm Wirtinger (1865–1945), Henry Löwig (1904–1995), Reinhold Baer (1902–1979), Andrey Nikolayevich Kolmogorov (1903–1987), Felix Hausdorff (1868–1942), Simon Sidon (1892–1941), Wolfgang Doeblin

[114]For more information about Vetulani, see Maligranda and Prytuła (2013).

[115]Compare Duda (1995); Przenioslo (2009).

[116]Krzyżański and Marcinkiewicz spent a year in Lwów on scholarships from the National Cultural Fund. Marcinkiewicz was additionally employed as a junior assistant in Banach's chair in 1935/1936, substituting for Schauder. See Domoradzki (2017).

[117]The famous joint work by M.G. Krein and D. Milman, On Extreme Points of Regular Convex Sets, *Studia Mathematica* **9**(1940), pp. 133–138, was published in the last Lwów-based volume of *Studia Mathematica*.

Figure 13. The Scottish Café, view from about 1918.[118]

(1915–1940), Djuro Kurepa (1907–1993), Angus Ellis Taylor (1911–1999), and Paul Erdős (1913–1996).

The Lwów mathematical community was a very collaborative environment. Mathematicians used to gather in the Scottish Café[119] to discuss mathematical topics and formulate problems, often with prizes for solving them.[120] Since July 1935, the problems were collected in a notebook called *Księga Szkocka* [The Scottish Book]. The notebook was kept in the Café and problems were written down in it during sessions.

[118] A postcard ca. 1918 from the collection of Michael Zarichnyi.

[119] The Scottish Café, located in the Akademicka street, not far from the University, was long popular among various groups of scholars (e.g., philosophers) and others. Besides consumption and academic discussions, the patrons could enjoy other activities, such as playing chess. The mathematician Herman Auerbach was an accomplished an avid player. See Feigenbaum (1982); Zielinski (2019).

[120] Marcinkiewicz was very impressed by the mathematical atmosphere of the Scottish Café. After his return to Wilno (Vilnius) he tried to re-create the unique working environment in one of the local cafés, without much success. See Domoradzki (2017).

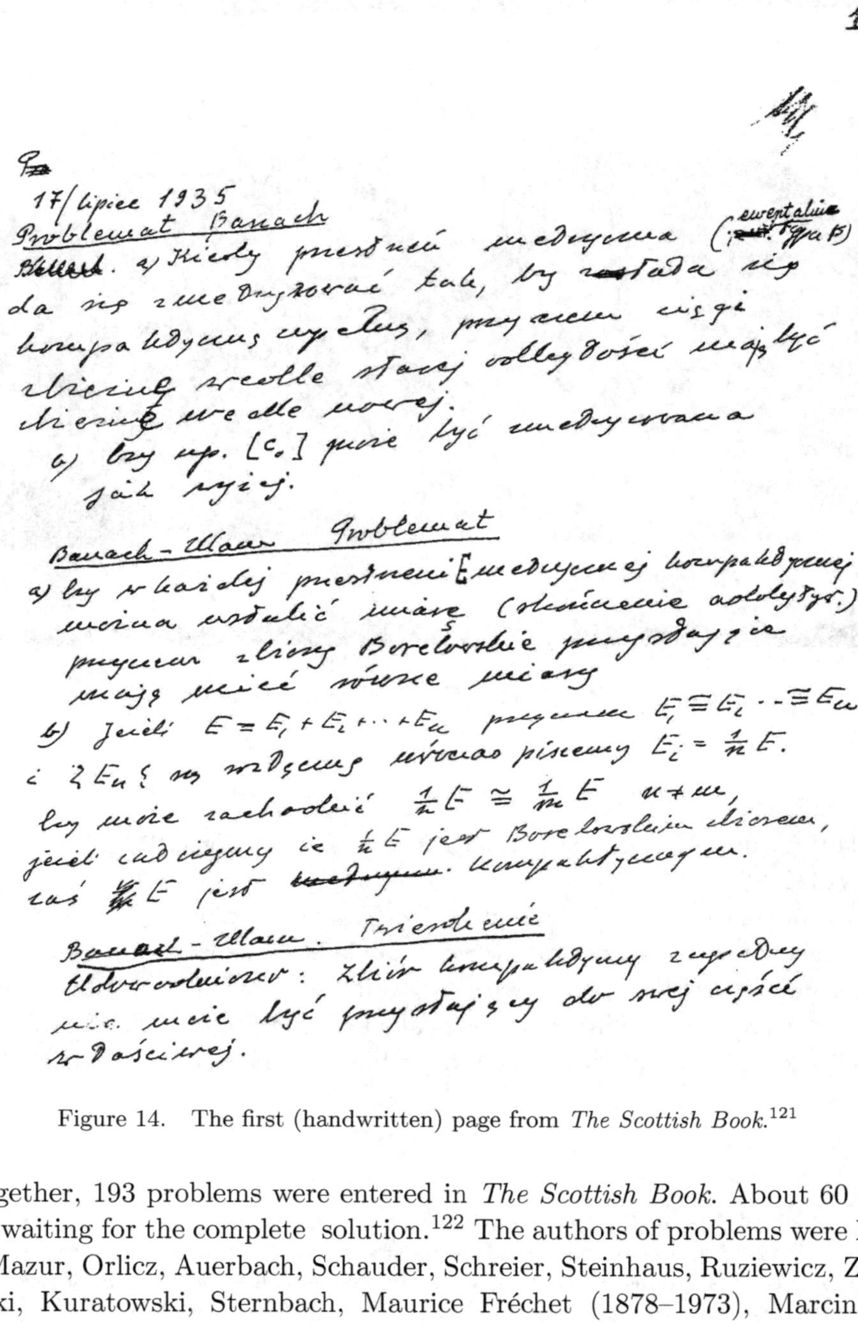
1

17/lipiec 1935
Problemat Banach
a) Kiedy przestrzeń metryczna (ewentualnie typu B) da się zmetryzować tak, by została się kompaktną zupełną, przyczem ciągi zbieżne według starej odległości mają być zbieżne według nowej.
b) czy np. $[c_0]$ może być metryzowana jak wyżej.

Banach - Ulam Problemat
a) czy w każdej przestrzeni E metrycznej kompaktnej można ustalić miarę (określenie addytywne) przyczem zbiory Borelowskie przystające mają mieć równe miary
b) Jeżeli $E = E_1 + E_2 + \cdots + E_n$ przyczem $E_1 \cong E_2 \cdots \cong E_n$ i $\{E_n\}$ są rozłączne wówczas piszemy $E_i = \frac{1}{n}E$. Czy może zachodzić $\frac{1}{n}E \cong \frac{1}{m}E$ $n \neq m$, jeżeli założymy że $\frac{1}{n}E$ jest Borelowskim zbiorem, zaś E jest kompaktycznym.

Banach - Ulam. Twierdzenie
Udowodniono: Zbiór kompaktyczny żaden nie może być przystający do swej części właściwej.

Figure 14. The first (handwritten) page from *The Scottish Book*.[121]

Altogether, 193 problems were entered in *The Scottish Book*. About 60 of them are still waiting for the complete solution.[122] The authors of problems were Banach, Ulam, Mazur, Orlicz, Auerbach, Schauder, Schreier, Steinhaus, Ruziewicz, Zbigniew Łomnicki, Kuratowski, Sternbach, Maurice Fréchet (1878–1973), Marcinkiewicz,

[121]From the home page of Stefan Banach. With permission from Emilia Jakimowicz and Adam Miranowicz, the authors of the Stefan Banach Wortal.

[122]One of the problems was solved very recently. See A. Eremenko, A Problem of Stanisław Saks, *Matematychni Studiï* **52**(2019), pp. 173–175.

Leopold Infeld (1898–1968), Kac, Nikliborc, Kaczmarz, Zygmund, Sierpiński, Samuel Eilenberg (1913–1998), Rolin Wavre (1896–1949), A.J. Ward (dates uknown), Simion Stoïlov (1887–1961), von Neumann, Edward Szpilrajn (1907–1976), Eidelheit, Borsuk, Cyril Offord (1906–2000), Joseph Kampé de Fériet (1893–1982), Bronisław Knaster (1893–1980), Saks, Nikolay Nikolayevich Bogolyubov (1909–1992), Pavel Sergeyevich Alexandrov (1896–1982), Sergei Lvovich Sobolev (1908–1989), Lazar Aronovich Lyusternik (1899–1981), Anisim Fedorovich Bermant (1904–1959).

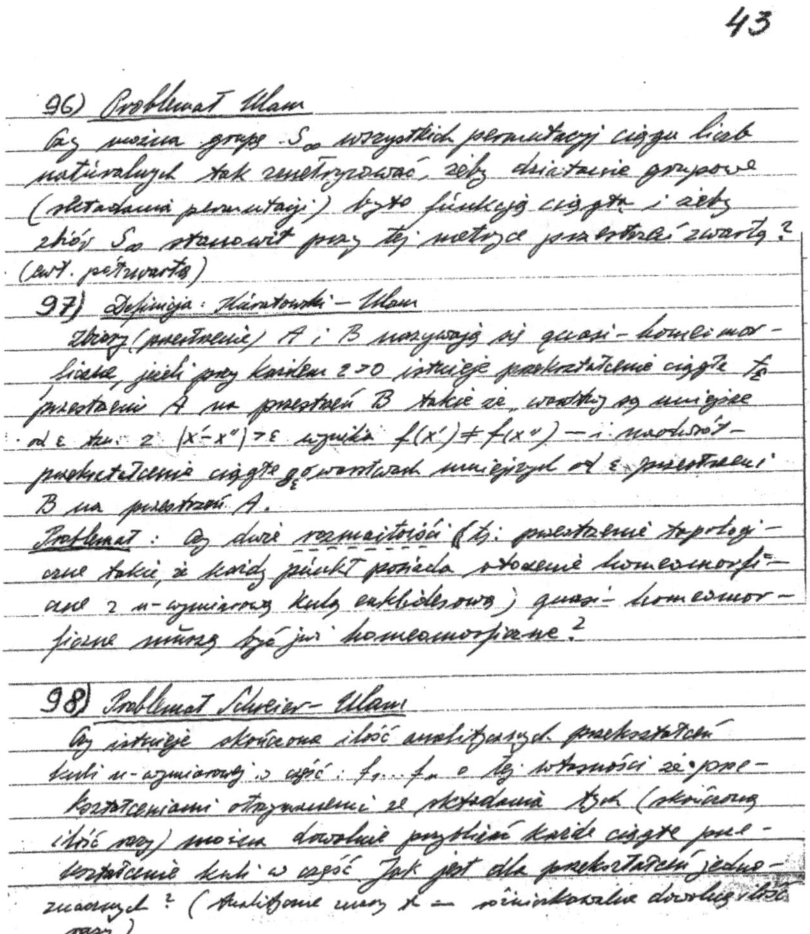
43

96) Problemat Ulam

97) Definicja: Kuratowski – Ulam

98) Problemat Schreier – Ulam

Figure 15. Page 43 from *The Scottish Book* Problem 96 (Ulam), 97 (Kuratowski–Ulam) and 98 (Schreier–Ulam).[123]

[123]From the home page of Stefan Banach. With permission from Emilia Jakimowicz and Adam Miranowicz, the authors of the Stefan Banach Wortal.

Some of the problems from *The Scottish Book* initiated completely new directions in mathematics. Random graphs were implicitly introduced by Ulam in Problem 38,[124] while infinite games were introduced by Mazur in Problem 43. Borsuk's Problem 175 whether the product of a triode (i.e., a curve homeomorphic to the letter T) and the Hilbert cube (i.e., the countable product of segments) is homeomorphic to the Hilbert cube led to the creation of the theory of Q-manifolds (i.e., manifolds modeled on the Hilbert cube).[125]

The book was hidden when the Germans invaded Lwów in 1941. After the WWII, a copy of *The Scottish Book* was sent to Ulam. He translated the problems into English and distributed them in the mathematical world.[126]

5. Logic in Lwów, and Relations Between Mathematics and Philosophy

Lwów was the place in which the first Polish exposition of symbolic logic appeared. In 1888 Stanisław Piątkiewicz (1849–1910), a graduate of the Lwów University in mathematics and a gymnasium teacher, published a 50-page article *Algebra w logice* [Algebra in Logic] presenting ideas of logical calculi and their application to syllogistics. Because it was printed in an annual school report of the Fourth Royal-Imperial Gymnasium in Lwów, it did not circulate widely, and so it had almost no impact.[127]

The presentation of mathematical logic in an academic setting had to wait until 1899/1900. In that academic year, Kazimierz Twardowski (1866–1938) offered the first course on logic at the Lwów University discussing some new developments in mathematical logic. A student of Franz Brentano (1838–1917), Twardowski got his PhD in philosophy 1891 at the University of Vienna and in 1895 became a professor of philosophy at the Lwów University. In his lectures (as well as in his writings) he strived for clarity and precision, took care of proper argumentation, supported rationalism and pursuit of truth. His teaching style appealed to many students. Twardowski supervised many PhDs in philosophy proper, logic and psychology (then regarded as branches of philosophy) and was a founder of the Lwow–Warsaw philosophical school.[128]

---

[124]That is, over 20 years before they were formally defined by Erdős and Alfred Rényi (1921–1970).

[125]See T.A. Chapman: *Lectures on Hilbert Cube Manifolds*, Expository lectures from the CBMS Regional Conference held at Guilford College, October 11–15, 1975, Regional Conference Series in Mathematics, no. 28, American Mathematical Society, Providence, R.I., 1976, x + 131 pages.

[126]In 1981, R.D. Mauldin published a book entitled *The Scottish Book. Mathematics in the Scottish Café*, which contains not only the list of problems and comments to them, but also biographical notes of their authors. In 2015 an updated edition of Mauldin's book (2015), was published. See also Kordos (2011); Stawiska (2015).

[127]For more details on Piątkiewicz and his report, see Batóg and Murawski (1996).

[128]See Woleński (1988); Brożek (2017); Jadacki (2017); Jadczak (1990); Wybraniec-Skardowska (1998).

Among the first students of Twardowski there was Łukasiewicz. He got his PhD in philosophy proper in 1902 (in his thesis he dealt with the distinction between inductive and deductive reasoning). He continued his studies in Berlin and Louvain and lectured at the Lwów University after getting his habilitation in 1906. Already in 1906 he gave lectures based on the material contained in the newly published *L'algebre de logique* by Louis Couturat (1868–1914). In 1910 Łukasiewicz published a monograph, *O zasadzie sprzeczności u Arystotelesa, Studium krytyczne* [On the Principle of Contradiction in Aristotle: A Critical Study],[129] in which he put into question the principle of contradiction (in three non-equivalent versions, namely an ontological, a logical and a psychological one), especially in connection with possibilities and future contingents. His philosophical considerations motivated him to develop an axiomatic system of three-valued logic (in 1917, published in 1920).[130]

Also in his Lwów period, Łukasiewicz published another monograph, *Die logischen Grundlagen der Wahrscheinlichkeitsrechnung*[131] in which he developed probability as a calculus of (finitely many) truth values of indefinite propositions. Later, as a professor at the University of Warsaw, Łukasiewicz worked on axiomatics of classical propositional logic, intuitionistic logic, modal logic, multi-valued logic as well as history of logic from the viewpoint of modern developments. He supervised several PhD theses on related topics. He also invented a parenthesis-free notation (the Polish notation) for logical propositions.[132]

Stanisław Leśniewski (1886–1939), born in the Russian Empire, was educated in philosophy and mathematics at the universities in Leipzig, Zurich, Heidelberg, Munich and St. Petersburg, and got his PhD in philosophy at the Lwów University with Twardowski in 1912. In his Lwów period he worked on existential propositions, criticized the principle of excluded middle, defended the principle of contradiction against Łukasiewicz's criticism, discussed the logical status of future contingents (polemicizing against Kotarbiński's view that they lack a definite truth value) and analyzed Russell's paradox. He spent the period of World War I in Moscow and afterwards became a professor of philosophy of mathematics at the University of Warsaw. He developed his original logical systems of mereology, ontology and protothetic, and initiated the theory of semantic categories. He was the PhD supervisor of Alfred Tarski (1901–1983), one of the greatest logicians of all times.[133]

Tadeusz Czeżowski (1889–1991) was a philosopher with a knowledge of and some interest in mathematics. He studied philosophy, mathematics and physics at the

---

[129]Kraków, Akademia Umiejętności, 1910, second edition: J. Woleński (ed.), PWN, Warsaw, 1987, 208 pages.
[130]See Jadacki (2018); Murawski (2014).
[131]Kraków, Spółka Wydawnicza Polska, 1913. Translation: *Logical Foundations of Probability Theory*, in Selected Works, L. Borkowski (ed.), Amsterdam, North-Holland, 1970, pp. 16–63.
[132]For more information, see Chapter 5 of this book, Jadacki (2018); Jordan (1967); Woleński (1988, 1995).
[133]For more information, see Woleński (1988); Betti (2008); Domoradzki and Stawiska (2019).

Lwów University and after his graduation, taught mathematics in Gymnasium VI in Lwów. His PhD, obtained in 1914 under Twardowski's supervision, concerned theory of classes. He was among the founders of the Polish Mathematical Society in Lwów, which first met on December 3, 1917.[134] In 1923 he became a professor of philosophy at the Stefan Batory University in Wilno (Vilnius).[135]

In the early 1920s Twardowski ran a seminar at the Lwów University devoted to reading Bertrand Russell. Among the participants were Steinhaus, Czeżowski, Kazimierz Ajdukiewicz (see below), Zygmunt Zawirski (1882–1948) and "the Gromskis".[136] Zawirski, another PhD student of Twardowski, later a professor in Poznań and Kraków, published on the relation between mathematics and logic and on the significance of multi-valued logic and its relation to probability. Daniela Gromska (neé Tenner, 1889–1973), a historian of philosophy, editor and translator, was then an assistant to Twardowski.

Writing about Twardowski, Zbigniew Jordan (1911–1977), a philosopher, editor and Polish diaspora political activist, pointed out the following: *The influence of his teaching was assuredly extended by the particular circumstances under which at that time the exact sciences were developing. After the analytic work of Cauchy, Weierstrass, Dedekind, and Cantor, mathematics had just reached the exactitude of method and the conceptual clearness for which it was prematurely famous in past centuries; the emerging Grundlagenproblem showed no limit to the control we can exercise over the methods adopted, every step forward being rewarded by ample and important results.*[137] However, with the exception of Łukasiewicz, Leśniewski and Miron Zarycki (1889–1961, see below), Twardowski's influence on Lwów mathematicians was rather indirect, if any.[138]

At any rate, the atmosphere at the Lwów University at the turn of the 19th and 20th centuries was favorable towards logic as well as the foundations and philosophy of mathematics. Let us recall that set theory also got introduced in Lwów around that time. First, professor of mathematics Puzyna used its elements in his monograph *Teoria funkcji analitycznych* [Theory of Analytic Functions].[139] Then Sierpiński, a professor at the Lwów University since 1910, gave courses and published (in 1912) a textbook on the subject.

---

[134]The other founders were Puzyna, Janiszewski, Steinhaus, A. Łomnicki and Dziwiński. See, e.g., Domoradzki (2012).

[135]See Czeżowski (1977).

[136]According to Steinhaus (2015, p. 135). Almost certainly, the Gromskis are Daniela Gromska and her husband.

[137]Jordan (1967, p. 347).

[138]For more on Polish logic and its interactions with mathematics see also Murawski (2004); (Woleński 1995, 1998, 2001, 2018).

[139]See Domoradzki and Zaricznyi (2014).

Along with Sierpiński, his junior colleagues in Lwów, Janiszewski (with a PhD in mathematics from Paris, 1911) and Mazurkiewicz, worked on problems in set theory and set-theoretic topology. All three spoke and wrote on the foundations and philosophy of mathematics on several occasions. They contributed to the series *Poradnik dla Samouków* [A Guide for the Self-Study].[140] Sierpiński and Mazurkiewicz wrote on set theory, and Janiszewski wrote on logic and philosophical problems of mathematics. The habilitation lectures of Sierpiński (1908) and Janiszewski (1913) in Lwów dealt with philosophical issues.[141] Some students of mathematics in that period went on to work on topics related to set theory, notably Ruziewicz (PhD in 1913 in Lwów) and Otto Nikodym (1887–1974; PhD in 1925 in Warsaw).[142]

Mathematicians who replaced Sierpiński, Janiszewski and Mazurkiewicz upon their move to the University of Warsaw after 1918 were familiar with set theory and frequently worked on its problems (see above), but their interest in logic, foundations and philosophy of mathematics varied.

Banach in his PhD thesis from 1920[143] gave axioms characterizing abstract spaces later called Banach spaces. He also worked with Mazur on a fragment of mathematical analysis based on numbers whose decimal expansions are represented by primitive recursive functions. They gave a joint talk on the subject at a meeting of the Lwów branch of the Polish Mathematical Society in 1937; the work was completed by Mazur after World War II.[144]

Banach participated in meetings of philosophers. In 1923 he gave a talk at a meeting of the Polish Philosophical Society in Lwów on paradoxes in mathematics. Among other things, he pointed out the axiom of choice as a source of paradoxes, but he did not consider the occurrence of paradoxes to cause problems for mathematical practice. This happened soon before his joint work with Tarski on a paradoxical decomposition of a ball (which we discussed above), and also a paradox of this kind.

Banach read the works of mathematically inclined philosophers and could interpret their writing in modern mathematical terms. He did so with the ideas of Józef Maria Hoene-Wroński (1776–1853) in the paper *Über das "Loi suprême" von J. Hoene-Wroński, Bulletin International de l'Academie Polonaise des Sciences et des Lettres, Classe des Sciences Mathématiques et Naturelles, Series A*, 1939, pp. 1–10.

---

[140]Łukasiewicz wrote the preface to the 1915 edition. For more information, see Pawlikowska-Brożek (1992b), Chapter 5 of this book and references cited therein.

[141]For more information, see Knaster (1960); Murawski (2014).

[142]This was not the case with Franciszek Leja (1885–1979), a graduate of mathematics in Lwów in 1910, later a professor of mathematics at the Jagiellonian University in Kraków. Leja's research concerned differential equations, complex analysis, potential theory, approximation theory and topological groups. For information about Ruziewicz, see below. For information about Nikodym, see Derkowska (1983).

[143]See Banach (1922).

[144]See Mazur (1963); Murawski (1998).

Steinhaus worked on some axiomatic systems. In 1912 he gave an axiomatic treatment of the notion of a limit.[145] In 1923 he proposed a system of axioms for probability, based on measure theory.[146] In 1925 he wrote an 82-page book *Podstawy geometrii* [Foundations of Geometry] (lecture notes published as an imprint by the Mathematical and Physical Circle of Students of the Jan Kazimierz University), in which he dealt with different systems of axioms for geometry, their consequences and consistency. One of his PhD students at the Lwów University, Sala Weinlös (1903–1941?) dealt with axiomatic systems in her PhD thesis defended in 1926, *O niezależności aksjomatów grup I, II i IV od układu aksjomatów geometrii euklidesowej 3-wymiarowej* [On Independence of Axioms of Group I, II and IV of the System of Axioms of Three-dimensional Euclidean Geometry].[147]

It is worth noting that Steinhaus formulated his views on mathematics, its object and method in a popularizing book *Czem jest a czem nie jest matematyka* [What Mathematics Is and Is Not].[148] He attended a philosophical seminar devoted to Russell's work, but was not enthusiastic about it. However, in his later career, together with Jan Mycielski (born 1932), Steinhaus proposed the axiom of determinacy for set theory, based on topological games and incompatible with the axiom of choice.

Kaczmarz proposed an axiomatics of arithmetics with only the relation '$<$' as a primitive term. It was identical with that given by Erhard Schmidt, but independent of it.[149]

Antoni Łomnicki wrote two notes on logic: *O układach zasad koniecznych i dostatecznych, służących do definicji pojęcia wielkości* [On Systems of Necessary and Sufficient Principles Used for the Definition of the Concept of Magnitude], *Wiadomości Matematyczne* **23**(1919), p. 37 and *O zasadzie dysjunkcji w logistyce i matematyce* [On the Disjunction Principle in Logistics and Mathematics], *Sonderabdruck aus dem Ruch filozoficzny* **6**(1922), no. 8, pp. 9–10. Most importantly, he was one of the originators of the idea that probability should be treated as a function of sets. He presented his fundamentals of probability in *Nouveaux fondements du calcul des probabilités* (*Définition de la probabilité fondée sur la théorie des ensembles*), *Fundamenta Mathematicae* **4**(1923), pp. 34–71. He used the axiomatic approach in his high-school geometry textbooks.

Eustachy Żyliński (1889–1954), a graduate of the Kiev University, was a professor at the Lwów University since 1919. During World War I he wrote a study *Mnogościowe ujęcie zasad logiki i matematyki* [Set-Theoretic Treatment of Principles of Logic and Mathematics], which remained unpublished. In 1925 he proved that there

[145] In Der Begriff der Grenze, *Mathematische Annalen* **71**(1912), pp. 88–96.
[146] Les probabilités dénombrables et leur rapport à la théorie de la mesure, *Fundamenta Mathematicae* **4**(1923), pp. 286–310.
[147] See Duda (2014); Prytuła (2012).
[148] Księgarnia Nakładowa, H. Altenberga, Lwów, 1923, 174 pages.
[149] See Maligranda (2007).

are exactly two binary functors (namely, binegation and the Sheffer stroke) each of which is sufficient for defining all other unary and binary functors of classical propositional logic.[150] His course on theoretical arithmetic for beginning students started with the system of Peano axioms. On several occasions (in public lectures or in writng) Żyliński remarked on the subject, method and role of mathematics.[151]

It is interesting that at Schauder's habilitation exam, in 1927, both Żyliński and Steinhaus asked the candidate both mathematical and philosophical questions. Żyliński's questions were:

(1) On the relation of the theory of infinite matrices with modern analysis;
(2) On "intuitionism", in particular on its position on the question of decidability.

Steinhaus' questions were:

(1) The argument between supporters of effective and ineffective definitions;
(2) Where do issues concerning the theory of orthogonal series in physics come from?

Józef Pepis (1910–1941) was Żyliński's student, but he was also strongly influenced by Leon Chwistek (see below). He studied mathematics at the Lwów University from 1928 until 1933 (master's thesis *Teorya niezmienników algebraicznych grupy rzutowej* [Theory of Algebraic Invariants of the Projective Group]). Working as a clerk and a private tutor, he published four papers in logic and one in functional equations and obtained his PhD in 1938 (thesis *O zagadnieniu rozstrzygalności w zakresie węższego rachunku funkcyjnego* [On the Issue of Decidability in Narrower Functional Calculus]).[152] He dealt with decidability of the first-order calculus of predicates. He also wrote some papers on recursive functions and intuitionistic mathematics, which were submitted to journals but did not appear because of World War II.[153]

The younger generation of mathematicians from the Lwów School received their university education mostly in the set-theoretic paradigm. The problems of abstract set theory appeared occasionally in their research, leading to new insights. For example, Ulam, while studying the problem of countably additive extension of Lebesgue measure to all subsets of $(0, 1)$, introduced non-measurable cardinals (also known as Ulam numbers).[154]

---

[150]Published as Some Remarks Concerning the Theory of Deduction, *Fundamenta Mathematicae* **7**(1925), pp. 203–209.
[151]See Kac (1985); Maligranda (2009); Murawski (2014).
[152]The thesis was defended at the Faculty of Mathematics and Natural Sciences. Żyliński was the supervisor, Banach and Chwistek were referees.
[153]See Murawski (1998); Mailgranda and Prytuła (2013).
[154]In the paper Ulam (1930).

In 1928 a chair in mathematical logic was created at the Lwów University. In a competition that polarized Lwów philosophers against mathematicians, Leon Chwistek (1884–1944) won against Tarski and moved to Lwów in 1930. Prior to his appointment, Chwistek formulated the theory of constructive types, an alternative to set theory and an improvement of the ramified type theory of Bertrand Russell (1872–1970) and Alfred North Whitehead (1861–1947).[155] In Lwów, Chwistek worked on a radically nominalistic system of foundations of mathematics, not containing existential axioms, such as the axiom of choice. He called this system *rational semantics.*

In 1928, Chwistek gave a talk, *Nouvelles recherches sur les fondements des mathématiques* at the International Congress of Mathematicians in Bologna. In 1935 he published a book, *Granice nauki. Zarys logiki i metodologii nauk ścisłych* [Limits of Science. An Outline of Logic and Methodology of Exact Sciences][156] in which he explicated his views on mathematics and logic as deductive sciences, the applicability of deductive method to particular sciences as well as on the plurality of realities.[157]

Chwistek brought with him from Kraków to Lwów a group of followers: Jan Herzberg (dates unknown), Jan Skarżeński (dates unknown) and Władysław Hetper (1909–1941). Hetper's master's thesis at the Jagiellonian University concerned integral equations, but in Lwów he worked solely on logic, getting his PhD in 1937 (thesis *Podstawy semantyki* [Foundations of Semantics]) and habilitation in 1939 (thesis *Relacje ancestralne w systemie semantyki*) [Ancestral Relations in the System of Semantics]).[158] Among his achievements there is a system of structural rules (in the style of sequent calculus of Gerhard Gentzen [(1909–1945), introduced in 1934] for a propositional calculus written in Polish notation, along with the proof of consistency and completeness of this system, as well as further development of Chwistek's rational semantics.[159]

Zarycki, an ethnic Ukrainian, studied mathematics and philosophy in Lwów. Since 1912 he was teaching in Ukrainian gymnasia in Bełżec, Zbaraż, Kołomyja and Tarnopol; since 1925 in Polish and Ukrainian gymnasia in Lwów. He got his PhD in mathematics in 1930 (supervisor Ruziewicz). In the thesis *Niektóre główne pojęcia teorii przestrzeni z punktu widzenia algebry logiki* [Some Main Notions of the Theory of Spaces from the Point of View of Algebra of Logic] he introduced axioms

---

[155] Russell's recommendation helped Chwistek to win the position. For a more detailed account of the competition and the surrounding controversy, see Woleński (1995). About Chwistek's theory, see Chwistek (2012); Świątkowski (1992).

[156] Published by Książnica-Atlas, Zjednoczone Zakłady Kartograficzne i Wydawnicze Towarzystwa Nauczycieli Szkół Średnich i Wyższych, Lwów–Warszawa, 1935, 264 pages.

[157] See Murawski (2014).

[158] Both theses were defended at the Faculty of Mathematics and Natural Sciences.

[159] See Domoradzki and Stawiska (2015).

for the boundary operation in topology. He maintained contact with Twardowski. In an article written in Ukrainian, *Pravda, krasa i matematyka* [Truth, Beauty and Mathematics],[160] Zarycki argued that mathematics had aesthetic value rather than the theoretical or practical one. [161]

Twardowski's disciples active in the interwar Lwów were philosophers. They rarely proposed axioms or proved theorems, but they nevertheless dealt with general topics of potential importance for mathematics, such as symbolic language and its semantics, theory of meaning, justification of knowledge and others, using explicative analysis and logical reconstruction in their work.

Kazimierz Ajdukiewicz (1890–1963) studied philosophy, mathematics and physics at the Lwów University, getting his PhD in 1912 with Twardowski. He continued his studies in 1913–1914 in Göttingen with Edmund Husserl (1859–1938), Leonard Nelson (1882–1927) and Hilbert. From 1925 until 1928, he worked at the University of Warsaw and in 1928 became an extraordinary professor at the University in Krakow. Ajdukiewicz defined modern notions of proof, theorem and logical consequence,[162] systematically studied definitions, proposed the classification of reasonings, introduced semantic categories and categorial grammar (where he introduced several functors on names and sentences behaving like fractions),[163] gave a criterion of syntactic coherence and initiated the logic of questions. He supervised a few PhDs and inspired many others through his teaching and writings.[164]

Izydora Dąmbska (1904–1983) got her PhD in 1927 with Twardowski. In 1930, traveling on a stipend from the National Cultural Fund, she visited Moritz Schlick (1882–1936) in Vienna and Hans Reichenbach (1891–1953) in Berlin. She dealt with the truth of scientific laws and classification of reasonings, and gave a logical analysis of analogy (as a homomorphism between structures).[165]

Maria Kokoszyńska-Lutmanowa (1905–1981) studied philosophy and mathematics in Lwów. After getting her PhD in philosophy in 1928 with Twardowski, she became an assistant to Ajdukiewicz. She dealt with the analysis of sentential functions, distinction between deductive and non-deductive sciences and logic of knowledge. She met Ludwig Wittgenstein (1889–1951) in Cambridge and took part in Vienna Circle meetings during her stay in Vienna in 1934/1935. Early on, she recognized the significance of Tarski's semantic theory of truth. Working together with

---

[160]Published in the volume *Matematyka sered nauk*, Lwów, 1927, pp. 25–37.
[161]See Zarichnyi and Ptashnyk (2009); Ivanyk (2013, 2017).
[162]Later Tarski characterized the logical consequence operation of the classical propositional calculus, proposing axioms similar to the closure axioms.
[163]The theory of semantic categories was influenced by some unpublished ideas of Leśniewski, so much so that Leśniewski accused Ajdukiewicz of plagiarism. Ajdukiewicz then acknowledged Leśniewski's priority and the conflict was attenuated. See Ingarden (1999).
[164]See Woleński (1988); Betti (2008); Buszkowski (2018); Olech (2018).
[165]See Brożek and Jadacki (2018); Woleński (2015).

Karl Popper (1902–1994) she helped translate Tarski's article on truth into German. She made further development of this theory.[166]

Seweryna Łuszczewska-Romahnowa (1904–1978) studied mathematics and philosophy in Lwów. After getting her PhD in philosophy in 1932 with Ajdukiewicz, she worked as an assistant in the First Chair of Philosophy and as a gymnasium teacher. She published after WWII (as a professor in Poznań) on classification systems, generalized Venn diagrams, induction and probability, and history of logic.[167]

Zygmunt Schmierer (1910–1943), a graduate of a rabbinical school and of the Lwów University, was an assistant to Ajdukiewicz. His PhD thesis from 1936 was titled *O funkcjach charakterystycznych w logikach wielowartościowych* [On Characteristic Functions in Multi-Valued Logics].[168]

Several individuals from Twardowski's circle outside of academia in Lwów also dealt with logic or philosophy of mathematics.

Milena Rudnicka (1892–1976), an Ukrainian political and social activist, teacher and journalist, got her PhD in 1918. Her thesis was titled *Rola matematyki w estetyce Albertiego. Przyczynek do zbadania matematycznych podstaw sztuki i estetyki wczesnego odrodzenia* [The Role of Mathematics in Alberti's Aesthetic. A Contribution to Mathematical Investigation of Foundations of Art and Aesthetic of Early Renaissance] (supervisor unknown).[169]

Pepi Spinnerówna (1905?–1969; later Josephine Mehlberg) submitted in 1927 a PhD thesis, *Rozumowanie matematyczne a logika tradycyjna* [Mathematical Reasoning and Traditional Logic]. There is no direct information of her being awarded a degree. In a letter to Roman Ingarden from February 11, 1928, Twardowski wrote that her promotion has taken place.[170]

Stanisław Kaczorowski (1888–1971), a teacher in the Fourth gymnasium in Lwów with a PhD in philosophy, published several articles in mathematical logic and high-school textbooks in logic.[171]

Kazimierz Sośnicki (1883–1976) studied philosophy, physics and mathematics at the Lwów University and later in Paris, Leipzig, Geneva and Zurich. He got his PhD in 1910 (thesis *Wyjaśnienie i opis w badaniach naukowych* [Explanation and Description in Scientific Research], supervisor Twardowski). He published articles, textbooks and collections of puzzles on logic.[172]

---

[166]See Brożek (2017); Kawczyński (2018).

[167]See Batóg (2001); Murawski and Pogonowski (2018).

[168]See Woleński (2015).

[169]See Ivanyk (2012).

[170]Spinnerówna married Henryk Mehlberg (1904–1979), a philosopher and logician from the Lwów-Warsaw school. The Mehlbergs survived WWII under false identities; Josephine took part in the resistance. After the war they emigrated from Poland. Henryk was a professor of philosophy at the University of Chicago, while Josephine was a professor of mathematics at the Illinois Institute of Technology in Chicago.

[171]See Woleński (1995).

[172]See Góra (2001).

Franciszek Kazimierz Smolka (1883–1947) studied law at the Lwów University, getting his JD in 1908. Then he studied philosophy with Twardowski, getting a PhD in 1925. Working as a gymnasium teacher, a lecturer in papyrology and a university librarian, he gave talks and published articles on Russell's paradoxes and on 3-valued logic in the 1920s.[173]

6. Other Mathematicians Active in Lwów Between the World Wars

In the interwar period there were other active mathematicians in Lwów. Some, but not all, of them can be fully considered members of the Lwów mathematical school.

We already mentioned some results of Auerbach on analysis and probability. However, his main interests were convex geometry, group theory and group representations. He got his PhD in 1930 (thesis *O polu krzywych wypukłych o średnicach sprzężonych* [On the Area of Convex Curves with Conjugate Diameters]) and habilitation in 1935 (thesis *Teorja grup linjowych ograniczonych* [Theory of Bounded Linear Groups], Książnica-Atlas, Lwów 1935, 16 pages) at the Lwów University.

In 1941 Auerbach characterized the ellipsoid as the only projectively homogeneous compact topological submanifold of codimension 1 in the $n + 1$-dimensional affine space over $\mathbb{R}$. Earlier, together with Mazur and Ulam he proposed another characterization of the ellipsoid. He also collaborated with Ulam on the paper *Sur le nombre de générateurs d'un groupe semisimple.*[174] Addressing Problem 19 from *The Scottish Book* posed by Ulam, Auerbach gave the necessary and sufficient conditions in a two-dimensional space for a figure with constant density ϱ to float freely on water. In particular, he gave an example of a cylinder of density $\varrho = 1/2$, with axis parallel to the surface of water, which floats freely on water and whose cross-section is not convex.[175]

Auerbach managed to write a mathematical paper while in the Lwów ghetto, using scraps of paper. He was killed in 1942, but his notes survived and were published in 1992.[176]

Maurycy (Mawrikij) Sperling (1910–1941?) studied mathematics from 1928 until 1932 at the University in Lwów. He was preparing for doctoral work under the direction of Mazur and Schauder. According to Banach's recollection, he was shot just after the German takeover. He published one paper on theory of relations (with electrostatic interpretation) and one in algebra (posthumously).[177]

---

[173]See Toczek (1999–2000).

[174]*Comptes Rendus de l'Academie des Sciences, Paris* **201**(1935), pp. 117–119.

[175]For more information on Auerbach's life and mathematics, see Derkowska and Mikosz (1992) and Plichko (1994).

[176]O geometrii trójkąta [Geometry of the Triangle], *Wiadomości Matematyczne* **29**(1992), pp. 234–251.

[177]See Maligranda and Prytuła (2013).

Several noted mathematicians worked at the Lwów Polytechnic. Bartel was a professor of descriptive geometry since 1917 and served as the rector of the Lwów Polytechnic in 1930/1931. He wrote an original monograph *Perspektywa malarska. Zasady, zarys historyczny, estetyka* [Painter's Perspective. Principles, Historical Outline, Aesthetics].[178]

Bartel was one of the top politicians of the Second Republic of Poland. From 1926 until 1929, he was five times the Prime Minister. He arranged for government subsidies of mathematical journals and mathematicians' scientific travels. At his initiative the General Faculty of the Lwów Polytechnic was formed in 1921.

The Chair of Descriptive Geometry B was held by Plamitzer. He published several papers on projective geometry. Since 1945 he worked in Kraków. He supervised the Chair of Descriptive Geometry at the Faculty of Hydraulic Engineering of the Kraków Polytechnic.

We have already mentioned several achievements of Antoni Łomnicki. Besides these, he also had deep understanding of the issues of mathematical cartography. He introduced a new method of projection and proposed a simple way of measurement by radiogoniometric signals based on a mesh he constructed. Thanks to his activity in cartography he was named an expert at the International Committee on Air Navigation under the direction of League of Nations (CINA). He was arrested and killed by the Germans soon after the Nazi invasion of Lwów in 1941 along with other Lwów professors and their family members.

Adam Maksymowicz (1880–1970) got his PhD in 1911 at the Lwów University and his habilitation in 1923 at the General Faculty of the Lwów Polytechnic. He published one article on summability of series and taught many courses. After 1945 he remained in Lwów.

Böttcher lectured at the Lwów Polytechnic since 1898. He made his last attempt to get habilitation at the Lwów University in 1918. Unsuccessful, he continued to teach at the Polytechnic until his retirement in 1935, but did not do any more research. His last mathematical publication was *Iteracye funkcyi kwadratowej* [Iterations of a Quadratic Function], *Wiadomości Matematyczne* **28**(1914), pp. 83–132.

Böttcher was one of pioneers of holomorphic dynamics. Remarkably, after World War I this topic saw major development in the works of Pierre Fatou (1878–1929), Gaston Julia (1893–1978) and Joseph Fels Ritt (1893–1951). The progress

[178] Volume 1 was published by Książnica-Atlas, Lwów-Warszawa, 1928, 312 pages.

happened mostly thanks to methods related to the notion of a normal family, introduced in 1912 by Paul Montel (1876–1975). Fatou gave credit to Böttcher for his theorem on the behavior of a holomorphic function near a superattracting fixed point, but did not reference any of Böttcher's publications.[179] Ritt cited a paper by Böttcher in Russian from 1904 while giving a detailed proof of the theorem mentioned.[180]

Böttcher seemed unaware of the recognition his contributions received and did not renew his mathematical interests. On the other hand, he was active in the Polish Mathematical Society and published two papers on philosophical issues of the foundations of mathematics: *O zasadzie sprzeczności* [On the Principle of Contradiction], *Przegląd Filozoficzny* **30**(1927), no. 4, p. 284, and *O antynomji Russela* [On Russell's Antinomy], *ibid.*, pp. 291–292.[181]

It should be noted here that Montel's results were known to some Polish mathematicians between the world wars, mainly to Mieczysław Biernacki (1891–1959), a former PhD student of Montel and a professor in Poznań since 1929, and to Leja. Montel published a paper in Annales de la Société Polonaise de Mathématique and gave a talk at a meeting of the Kraków branch of the Polish Mathematical Society in 1935.[182]

Since 1921, the Lwów Polytechnic comprised six faculties. Among them there was the General Faculty, which was to educate teachers of mathematics and physics, experimental technicians and graphic artists. Its curriculum in mathematics was deeper than the corresponding curricula of Polish universities. The Chair of mathematics was held first by Stożek, then by Kuratowski. The General Faculty existed until 1934. Its most prominent graduate was Ulam.[183]

Another institution of higher education in the interwar Lwów was the *Wyższa Szkoła Handlu Zagranicznego* [Higher School of Foreign Trade], since 1937 the Academy of Foreign Trade. It was established at the initiative of the mathematician Pawłowski, who also served as its first rector. From 1922 until 1937, he held the chair of Trade and Political Arithmetics at the Academy of Foreign Trade. Another mathematician working as a professor and serving as rector at this Academy was Ruziewicz, after losing his chair at the University.

[179] P. Fatou, Sur les équations fonctionelles, *Bulletin de la Société Mathématique de France* **47**(1919), pp. 161–271.

[180] J.F. Ritt, On the Iteration of Rational Functions, *Transactions of the American Mathematical Society* **21**(1920), pp. 348–356.

[181] See Domoradzki (2012); Domoradzki and Stawiska (2014); Duda (2014); Stawiska (2014).

[182] See Chapter 5.

[183] See Duda (2014a,b); Popławski (1992).

Figure 16. Stanisław Ruziewicz.[184]

Ruziewicz was one of the early members of the Lwów School of Mathematics, a student of Puzyna and Sierpiński.[185] He worked on the theory of real functions and set theory, and collaborated with Sierpiński and Banach. He co-authored textbooks with Żyliński: *Algebra* for high schools (1926) and Wstęp do matematyki. *Elementy algebry wyższej i teorii liczb* [Introduction to Mathematics. Elements of Higher Algebra and Number Theory] (1927). In 1934 his chair at the Lwów University was liquidated, ostensibly because of academic reforms, but more likely because of Ruziewicz's opposition to the ruling *Sanacja* block.

7. Mathematics in the Shevchenko Scientific Society

The Shevchenko Scientific Society, which is now considered to be the first Ukrainian national academy of sciences, was established in 1873 in Lwów. In 1892 the society

[184]From the collection of Zofia Pawlikowska-Brożek and Stanisław Domoradzki gathered for Słownik Biograficzny Matematyków Polskich [Biographical Dictionary of Polish Mathematicians].

[185]PhD 1913, thesis *O funkcji ciągłej, monotonicznej nie posiadającej pochodnej w nieprzeliczalnej mnogości punktów* [On a Continuous Monotone Function not Having Derivative in an Uncountable Set of Points]; habilitation 1918, thesis *O funkcjach ciągłych monotonicznych posiadających pantachiczne przedziały stałości* [On Continuous Monotone Functions Having Pantachic Intervals of Constancy].

was reorganized and its structure was determined by three sections: Historical-Philosophical, Philological and Mathematical-Natural-Scientific-Medical.

The mathematicians in the Shevchenko Scientific Society concentrated their activity in the following directions: conducting research and publication of results, writing textbooks for Ukrainian schools, the creation of Ukrainian scientific terminology in mathematics and natural sciences, reviewing the mathematical literature and writing popular science articles, teaching mathematics in schools, including Ukrainian ones. Some members of the Society were connected with the Ukrainian Underground University,[186] where they taught mathematical disciplines.

In this connection we mention Volodymyr Levytskyi (1872–1956), who was the author of the first mathematical paper in Ukrainian (published in 1894). Levytskyi graduated from a Polish Gymnasium in 1890 in Lwów and studied at the Faculty of Philosophy of the University. He taught at high schools after getting his PhD (in 1901).

Levytskyi published papers on number theory, functions of a complex variable, interpolation theory, geometry, history of mathematics and applications of mathematics to theoretical physics. He was the head of mathematical section of the Shevchenko Scientific Society. From 1931 to 1935 he served two terms as the President of the Society and the editor of its journal.

Publishing has been an important area of Society's activity. Since 1892, *Zapysky NTSh* [Notes of the Shevchenko Scientific Society], periodical section proceedings, and also 20 scientific series have been published. In the period 1897–1939, 32 volumes of *Zbirnyk matematychno-pryrodnycho-likars'koyi sektsiyi* [Collection of Mathematical, Natural Sciences and Medical Section] containing mathematical works were published. These editions fostered the development of Ukrainian mathematical terminology.

Till 1939, the activity of the mathematicians Volodymyr Levytskyi, Mykola Chaikovskyi (1887–1970), Myron Zarytskyi (in Polish Miron Zarycki, see above), Petro Ohonovskyi (1853–1917), Nykyfor Sadovskyi (1884–1935), Klym Hlibovytskyi (1875–1907), Vasyl Kalitsun (1878–1942), Vasyl Stasyuk (1887–1960) and others was especially fruitful.[187] Their investigations were devoted to many areas of mathematics: theory of analytic functions, algebra, topology, theory of differential and algebraic equations, descriptive and projective geometry, non-Euclidean geometry, actuarial mathematics and others.[188]

[186]Ukrainians who were not admitted to the Lwów University initiated the (Underground) Ukrainian University, in 1921. It was financed by private donations and existed till 1925.

[187]Sadovskyi and Hlibovytskyi also published in Polish. Hlibovytskyi (under the Polish name Klemens Hlibowicki) was an author of one of the first works on Galois theory in Polish, published in 1897/1898.

[188]The most complete survey of mathematical publications of members of the Shevchenko Scientific Society till 1939 can be found in Banakh and Prytula (2015).

Scientific interests of Zarytskyi were mostly focused on set-theoretic topology and theory of Boolean algebras with derivative. In the 25th volume of the Collection of Mathematical, Natural Sciences and Medical Section, he published his first mathematical paper *Une méthode d'introduction de la notion de bon ordre dans la Théorie des Ensembles*. We have already mentioned his results in general topology, but he also published papers in statistics as well as in the history of mathematics and teaching of mathematics (the latter mostly in Polish).

In 1927 at the anniversary celebration of the Mathematical, Natural History and Medical Section of the Shevchenko Scientific Society, Zarytskyi announced the report of the philosophical content *Truth, Beauty and Mathematics*, which we already discussed above. Note also that he translated the fundamental treatise of Banach[189] into Ukrainian. However, the name of the translator did not appear in the print.[190]

The list of foreign members of the Shevchenko Scientific Society included Albert Einstein, David Hilbert, Felix Klein, as well as mathematicians from the Soviet part of Ukraine: Kyiv mathematicians Dmytro Grave (1863–1939), Mykhaylo Kravchuk (1892–1942), and Mykola M. Krylov (1879–1955).

8. Conclusion

Undoubtedly, the Lwów School was a unique phenomenon in mathematics. Created almost from scratch, it reached the highest possible level in a relatively short time. Referring to the greatest mathematical achievements from 1909 to 1938,[191] R. Duda pointed out [192] that 48 of them are due to Polish mathematicians. A closer look shows that over 20 of them are due to mathematicians related to Lwów.[193] Functional analysis was complemented by topology, set theory, algebra,[194] topological algebra (theory of topological groups and semigroups), logic, game theory, differential equations, probability, and other disciplines. Mathematicians of various interests coexisted within the school or on its boundaries, more coherently than in Kraków and more harmoniously than in Warsaw.[195]

The person of Banach often serves as *pars pro toto* of the Lwów School of Mathematics. This is often a simplification, but in fact there are many good reasons to pick Banach to illustrate the character of the Lwów School. Of modest origin (and born out of wedlock), he became a world-class mathematician by combination of

[189]This means Banach (1931).

[190]On the other hand, some of Zaritskyi's results were cited in Kuratowski's *Topology* and Semadenis's monograph *Banach Spaces of Continuous Functions*.

[191]As listed year by year in Pier (1985).

[192]In his book Duda (1815–1915, pp. 295–296).

[193]The count may be different depending on whether Sierpiński, Łukasiewicz, Nikodym and Kuratowski are included.

[194]For more information see Domoradzki *et al.* (2016), Więsław (1995).

[195]See also Domoradzki (2006, 2007); Domoradzki *et al.*, (2018); Duda (2009, 2012, 2013, 2019).

his genius, continuous efforts and opportune timing.[196] Early on he developed a regular habit of discussing mathematics with like-minded companions (Wilkosz and Nikodym in his young days in Kraków), which he continued even in the conditions of World War I. One such discussion, in the Planty park, was overheard by Steinhaus, who soon started a collaboration with Banach and arranged for him a position of assistant at the Lwów Polytechnic.[197]

Banach lived for mathematics. He was open to many different mathematical (and not only mathematical) ideas, to interaction with other mathematicians, not only from Lwów. The mathematical discussions he valued so much included many people and took place in university classes, meetings of scientific societies and the popular Scottish Café. This intensified mathematical activity in Lwów and brought on a spectacular development, often by collaboration or building on each other's results. He took seriously his role as a teacher, contributing to later successes of his students.[198] As we were trying to demonstrate in this text, these characteristics were also representative of the Lwów School as a whole.

The topic of Lwów Mathematical School, its formation and activity is very broad and still requires investigation. For many reasons, systematic research in history of mathematics in Poland and Ukraine was taken up with significant delay after World War II. We are glad that we were able to contribute by outlining and presenting certain issues for the first time in this publication.

References

S.M. Ageev, S.A. Bogatyi, and D. Repovš, Banach-Mazur Compacta are Aleksandrov Compactifications of Q-manifolds, *Mathematical Notes* **76**(2004), pp. 3–9. ISSN 0001-4346.

S. Antonyan, The Topology of the Banach-Mazur Compactum, *Fundamenta Mathematicae* **166**(2000), pp. 209–232. ISSN 0016-2736.

S. Banach, Sur les opérations dans les ensembles abstracts et leur application aux équations intégrales, *Fundamenta Mathematicae* **3**(1922), pp. 133–181.

S. Banach, Über additive Massfunktionen in abstrakten Mengen, *Fundamenta Mathematicae* **15**(1930), pp. 97–101.

S. Banach, *Teoria operacyj. Tom I Operacje liniowe* [Theory of Operations, Volume I, Linear Operations] Kasa im. Mianowskiego Instytutu Popierania Nauki, Warszawa, 1931, viii + 236 pages.

S. Banach and K. Kuratowski, Sur une généralisation du problème de la mesure, *Fundamenta Mathematicae* **14**(1929), pp. 127–131.

S. Banach and A. Tarski, Sur la décomposition des ensembles de points en parties respectivement congruentes, *Fundamenta Mathematicae* **6**(1924), pp. 244–277.

[196] See Jakimowicz (2011); Domoradzki *et al.* (2011); Kałuża (1992).

[197] This famous encounter is marked by a commemorative bench at the place it likely happened, near the Wawel Hill, opposite Banach's home.

[198] See Barański (1993); Ciesielski and Pogoda (2011); Domoradzki (1999); Szałajko (1984, 1993); Ulam (1976).

T.O. Banakh and Ya.G. Prytula, Mathematical Publications of Shevchenko Scientific Society till 1939, *Mathematical Bulletin of Shevchenko Scientific Society* **12**(2015), pp. 130–146. ISSN 1812-6774.

F. Barański, Lwowskie wspomnienia o Stefanie Banachu [Lwów Reminiscences of Stefan Banach], *Opuscula Mathematica* **13**(1993), pp. 55–57. ISSN 0860-2727.

T. Batóg, Seweryna Łuszczewska-Romahnowa — Logic and Methodology of Science, pp. 113–119, in W. Krajewski (2001).

T. Batóg and R. Murawski, Stanisław Piątkiewicz and the Beginnings of Mathematical Logic in Poland, *Historia Mathematica* **25**(1996), pp. 68–73. ISSN 0315-0860.

M. Bernkopf, The Development of Function Spaces with Particular Reference to their Origins in Integral Equation Theory, *Archive for History of Exact Sciences* **3**(1966), pp. 1–96.

A. Betti, Polish Axiomatics and Its Truth: On Tarski's Leśniewskian Background and the Ajdukiewicz Connection, pp. 44–71, in D. Patterson (ed.), *New Essays on Tarski and Philosophy*, Oxford University Press, Oxford, 2008, viii + 434 pages. ISBN 978-0-19-929630-9.

G. Birkhoff and E. Kreyszig, The Establishment of Functional Analysis, *Historia Mathematica* **11**(1984), pp. 258–321. ISSN 0315-0860.

K. Borsuk, Drei Sätze über die n-dimensionale Euklidische Sphäre, *Fundamenta Mathematicae* **20**(1933), pp. 177–190.

K. Borsuk, *Polskie badania w zakresie topologii* [Polish Research in Topology], pp. 65–81, in Śródka (1995).

K. Borsuk and S. Ulam, On Symmetric Products of Topological Spaces, *Bulletin of the American Mathematical Society* **37**(1931), pp. 875–882.

A. Brożek and Maria Kokoszyńska, Between the Lvov-Warsaw school and the Vienna Circle, *Journal of the History of Analytic Philosophy* **5**(2017), no. 2, pp. 17–38. ISSN 2159-0303.

A. Brożek, Kazimierz Twardowski: A Great Teacher of Great Philosophers, pp. 15–32, in Á. Garrido and U. Wybraniec-Skardowska (2018).

A. Brożek and J. Jadacki, Izydora Dąmbska: The First Lady of the Twentieth-Century Polish Philosophy, pp. 223–233, in Á. Garrido and U. Wybraniec-Skardowska (2018).

A. Brożek, F. Stadler and J. Woleński (eds.), *The Significance of the Lvov-Warsaw School in the European Culture*, Vienna Circle Institute Yearbook, Volume 21, Springer, Cham, 2017, 353 pages. ISBN 9783319528694.

W. Buszkowski, Categorical Grammars and Their Logics, pp. 91–115, in Á. Garrido and U. Wybraniec-Skardowska (2018).

L. Chwistek, The Principles of the Pure Type Theory (1922), Translated by Adam Trybus with an Introductory Note by Bernard Linsky, *History and Philosophy of Logic* **33**(2012), pp. 329–352. ISSN 0144-5340.

D. Ciesielska and S. Domoradzki, On Mathematical Lectures at the Jagiellonian University in the Years 1860–1918. Essay Based on Manuscripts, *Technical Transactions — Fundamental Sciences* 1-NP **7**(2014), pp. 59–71. ISSN 0011-4561.

K. Ciesielski, Wykłady polskich matematyków na Międzynarodowych Kongresach Matematyków [Lectures of Polish Mathematicians at the International Congresses of Mathematics], *Wiadomości Matematyczne* **54**(2018), pp. 55–65. ISSN 2080-5519.

K. Ciesielski and Z. Pogoda, Conversation with Andrzej Turowicz, *The Mathematical Intelligencer* **10**(4) (1988), pp. 13–20. ISSN 0343-6993.

K. Ciesielski and Z. Pogoda, Stefan Banach and Lvov Mathematical School, pp. 129–142, in E. Jakimowicz (2011).

T. Czeżowski, Wspomnienia (Zapiski do autobiografii) [Reminiscences (Notes for an Autobiography)], *Kwartalnik Historii Nauki i Techniki* **22**(1977), no. 4, pp. 429–440.

A. Dawidowicz, Wspomnienia o Leonie Chwistku, Hugonie Steinhausie i Włodzimierzu Stożku [Reminiscences of Leon Chwistek, Hugo Steinhaus and Włodzimierz Stożek], *Wiadomości Matematyczne* **23**(1981), pp. 232–240. ISSN 0373-8302.

A. Dawidowiczowa, *Zeschnięte liście i kwiat* [Dried Leaves and Flower], Wydawnictwo Literackie, Kraków 1989, 215 pages. ISBN 978-83-08-02144-6.

A. Derkowska, Otton Marcin Nikodým [sic] (1889–1974), *Wiadomości Matematyczne* **25**(1983), pp. 75–83. ISSN 0373-8302 (Polish).

A. Derkowska, Juliusz Paweł Schauder (1899–1943), pp. 39–43, in S. Fudali (ed.), *Matematyka przełomu XIX i XX wieku. Materiały z IV Ogólnopolskiej Szkoły Historii Matematyki* [Mathematics of the turn of the 19th and 20th Centuries. Materials from the 4th All-Poland School on History of Mathematics], Szczecin, 1990, 192 pages (Polish).

A. Derkowska, M. Mikosz, and A. Neugebauer, Herman Auerbach (1901–1942), *Wiadomości Matematyczne* **29**(1992), pp. 225–231. ISSN 0373-8302 (Polish).

S. Domoradzki, Stefan Banach's Texbooks as an Example of a Mathematician's Approach to Teaching of Mathematics, pp. 387–389, in *Relationships Between Classroom Practice and Research in Mathematics Education, Proceedings of the CIEAEM 50 Neuchâtel, 2–7* août 1998, CIEAEM, Neuchâtel, 1999, 501 pages. ISBN 88-7967-036-0.

S. Domoradzki, Sozdanie i razvitie metoda kategorii Bĕra v rabotakh polskikh matematikov opublikovannykh v "Studia Mathematica" v mezhvoennyi period [Creation and Development of the Baire Category Method in the Works of Polish Mathematicians Published in Studia Mathematica during the Interwar Period], *Rossiĭskaya Akademiya Nauk. Institut Istorii Estestvoznaniya i Tekhniki. Voprosy Istorii Estestvoznaniya i Tekhniki* **3**(2005), pp. 39–46, 201. ISSN 0205-9606 (Russian).

S. Domoradzki, Remarks to the History of Banach-Tarski-Hausdorff Paradox, *Scientific Bulletin of Chełm, Section of Mathematics and Computer Science* **2**(2006) pp. 32–39.

S. Domoradzki, O metode summirovaniya Eulera v monografii S. Banakha [On Euler Summability Method in the Monograph of Stefan Banach], pp. 111–116, in L. Brylevskaia (ed.), *Leonard Euler i sovremennaya nauka* [Leonhard Euler and Modern Science], Russian Academy of Sciences, Saint Petersburg, 2007, 470 pages. ISBN 5-7187-0796-0.

S. Domoradzki, Uwagi do historii i zastosowań twierdzenia Hahna-Banacha [Remarks on History and Applications of the Hahn-Banach Theorem], *Studia Matematyczne Uniwersytetu Humanistyczno-Przyrodniczego Jana Kochanowskiego w Kielcach* **11**(2009), pp. 69–78. ISSN 1899-8267.

S. Domoradzki, *The Growth of Mathematical Culture in the Lvov Area in the Autonomy Period (1870–1920)*, History of Mathematics, Volume 47, Matfyzpress, Prague, 2012, 331 pages. ISBN 978-80-7378-178-1.

S. Domoradzki, Józef Puzyna (1856–1919) — The Pioneer of Polish Mathematical School, pp. 11–22, in A. Lecko (ed.), *Current Research in Mathematical and Computer Sciences*, Uniwersytet Warmii i Mazur Press, Olsztyn, 2017, 318 pages. ISBN 978-83-8100-108-3.

S. Domoradzki, Wileńskie środowisko matematyczne w czasach młodości profesora Krzyżańskiego [The Mathematical Community of Vilnius during the Young Years of Professor Krzyżański], pp. 57–66, in J. Koroński (ed.), *Wybrane zagadnienia dla równań różniczkowych* [Selected Issues for Differential Equations], Monografie Politechnika Krakowska, seria Nauki Podstawowe Matematyka, Kraków, 2017, 304 pages. ISBN 978-83-7242-954-4.

S. Domoradzki, Wzrost kultury matematycznej w obrębie oddziaływania Uniwersytetu Lwowskiego w okresie autonomii galicyjskiej [The Growth of Mathematical Culture in the Range of Influence of the Lwów University in the Period of Autonomy of Galicia], *Edukacja Technika-Informatyka* **150**(2018), no. 1, pp. 63–81. ISSN 2080-9069.

S. Domoradzki, Z. Pawlikowska-Brożek and M. Zarichny, Stefan Banach in the Light of Archives, pp. 103–128, in E. Jakimowicz and A. Miranowicz (2011).

S. Domoradzki and M. Stawiska, Lucjan Emil Böttcher and His Mathematical Legacy, pp. 127–161, in P. Pardalos, T. Rassias (eds.), *Mathematics without Boundaries. Surveys in Pure Mathematics*, Springer, New York, 2014, xiv + 781 pages. ISBN 978-1-4939-1105-9.

S. Domoradzki and M. Stawiska, Distinguished Graduates in Mathematics of Jagiellonian University in the Interwar Period. Part I: 1918–1926, *Technical Transactions — Fundamental Sciences* 2-NP **112**(2015), pp. 98–115. ISSN 0011-4561.

S. Domoradzki and M. Stawiska, Distinguished Graduates in Mathematics of Jagiellonian University in the Interwar Period. Part II: 1926–1939, *Technical Transactions — Fundamental Sciences*, 2-NP **112**(2015), pp. 117–141. ISSN 0011-4561.

S. Domoradzki and M. Stawiska, Polish Mathematics and Mathematicians in World War I, part I, *Studia Historiae Scientiarum* **17**(2018), pp. 23–49. ISSN 1731-6715.

S. Domoradzki and M. Stawiska, Polish Mathematics and Mathematicians in World War I, part II, *Studia Historiae Scientiarum* **18**(2019) pp. 55–92. ISSN 2451-3202.

S. Domoradzki and M. Stawiska, *Transformation of the Warsaw Mathematical Community in the Aftermath of World War I*, Chapter 5 in this book.

S. Domoradzki and M. Stawiska, *The Position of the Jagiellonian University and Its Mathematical Community Before and After WWI*, Chapter 4 in this book.

S. Domoradzki, M. Stawiska and M. Zarichnyi, Lucjan Böttcher: A Forgotten Polish Mathematician and his Influence on Modern Mathematics, pp. 222–226, in Ch. Binder (ed.), *XII. Österreichisches Symposion zur Geschichte der Mathematik, 'Mathematik — verschollen und gefunden'*, Österreichische Gesellschaft für Wissenschaftsgeschichte, Miesenbach, 2014, 276 pages.

S. Domoradzki, M. Stawiska and M. Zarichnyi, On Algebra in Lwów in the years 1870–1939, pp. 86–95, in Ch. Binder (ed.), *XIII. Österreichisches Symposion zur Geschichte der Mathematik*, Österreichische Gesellschaft für Wissenschaftsgeschichte, Miesenbach, 2016, 244 pages.

S. Domoradzki, M. Stawiska and M. Zarichnyi, Modern Mathematics in Lwów Before Banach, pp. 100–106, in Ch. Binder (ed.), *XIV. Österreichisches Symposion zur Geschichte der Mathematik — Neglected Parts of Mathematics and Their History*, Technische Universität Wien, Wien, 2018, 236 pages.

S. Domoradzki and M. Zaricznyi, On Some Aspects of the Set Theory and Topology in J. Puzyna's Monumental Work, *Technical Transactions — Fundamental Sciences* 1-NP **7**(2014) pp. 85–97. ISSN 0011-4561.

R. Duda, Fundamenta Mathematicae, Studia Mathematica, Acta Arithmetica — pierwsze trzy specjalistyczne czasopisma matematyczne [Fundamenta Mathematicae, Studia Mathematica, Acta Arithmetica — the First Three Specialized Mathematical Journals], *Zeszyty Naukowe Politechniki Śląskiej, Seria Matematyka-Fizyka* **76**(1995), pp. 47–80. ISSN 0072-470X.

R. Duda, *Matematycy XIX i XX wieku związani z Polską* [Mathematicians of the 19th and 20th Centuries with Links to Poland], Wydawnictwo Uniwersytetu Wrocławskiego, Wrocław, 2007, 587 pages. ISBN 978-83-229-3316-9.

R. Duda, Osiągnięcia i znaczenie lwowskiej szkoły matematycznej [Achievements and Significance of the Lwów School of Mathematics], Polska Akademia Umiejętności, *Prace Komisji Historii Nauki* **9**(2009), pp. 35–48. ISSN 1731-6715.

R. Duda, Matematyka na Uniwersytecie Lwowskim do usunięcia Polaków ze Lwowa [Mathematics at Lwów University Until the Removal of Poles from Lwów], pp. 291–323, in W. Lohman (ed.), *Universitati Leopoliensi. In memoriam*, Polska Akademia Umiejętności, Kraków, 2012, 534 pages. ISBN 978-83-7676-084-1.

R. Duda, Leaders of Polish Mathematics Between the Two World Wars, *Commentationes Mathematicae* **53**(2013), pp. 105–112. ISSN 0373-8299.

R. Duda, *Pearls from a Lost City: The Lvov School of Mathematics*, Translated from the 2007 Polish original by Daniel Davies. History of Mathematics, Volume 40, American Mathematical Society, Providence, RI, 2014a, xii + 231 pages. ISBN 978-1-4704-1076-6.

R. Duda, Katedry matematyczne i ich obsada na Politechnice Lwowskiej do 1945 r. [Mathematical Chairs and Their Staffing at the Lwów Polytechnic Until 1945], *Antiquitates Mathematicae* **8**(2014b), pp. 47–74. ISSN 1898-5203.

R. Duda, *Historia matematyki w Polsce na tle dziejów nauki i kultury* [History of Mathematics in Poland on the Background of the History of Science and Culture], Instytut Historii Nauki PAN im. L. i A. Birkenmajerów and Oficyna Wydawnicza ASPRA-JR, Warszawa, 2019, 626 pages. ISBN 978-8375458220.

R. Duda, *Roots of Polish Mathematics* (*1815–1915*), Chapter 3 in this book.

M. Feigenbaum, Reflections of the Polish Masters: An Interview with Stan Ulam and Mark Kac, *Los Alamos Science* **3**(3) (1982), pp. 54–65. ISSN 0273-7116.

M. Fréchet, *Les espaces abstraits et leur théorie considérée comme introduction à l'analyse générale, Collection de monographies sur la théorie des fonctions XII*, Gauthier-Villars, Paris, 1928, 296 pages.

A. Fuliński, On Marian Smoluchowski's Life and Contribution to Physics, *Acta Physica Polonica B* **29**(1998), pp. 1523–1536. ISSN 0587-4254.

Á. Garrido, U. Wybraniec-Skardowska (eds.), *The Lvov-Warsaw School. Past and Present. Studies in Universal Logic* Birkhäuser, Springer, Cham, 2018, xviii + 815 pages. ISBN 978-3-319-65429-4.

H.-J. Girlich, Łomnicki-Steinhaus-Kolmogorov: Steps to a Modern Probability Theory, pp. 47–56, in W. Więsław (ed.), *European Mathematics in the Last Centuries*, Proceedings of the Conference held in Będlewo, April 26–30, 2004, University of Wrocław, Mathematical Institute, Wrocław, 2005, 159 pages. ISBN 83-89366-38-X.

H. Góra, Sośnicki Kazimierz (1883–1976), pp. 612–615, in H. Markiewicz (ed.), *Polski Słownik Biograficzny* [Polish Biographical Dictionary], Volume 40, issue 4 (167), Polska Akademia Nauk — Instytut Historii im. Tadeusza Manteuffla, Warszawa — Kraków, 2001, 639 pages. ISBN 83-86301-01-5.

I. Guran and Ya. Prytula, Józef Schreier "On Finite Base in Topological Groups", *Matematychni Studii* [Proceedings of the Lviv Mathematical Society] **39**(2013), pp. 3–9. ISSN 1027-4634.

Unknown, *ICM Proceedings 1893–2018*, International Mathematical Union. Online document. https://www.mathunion.org/icm/proceedings [07.03.2021].

R. Ingarden, Dzieje mojej "kariery uniwersyteckiej" [A History of my "University Career"], *Kwartalnik Filozoficzny* **27**(1999), no. 2, pp. 183–201. ISSN 1230-4050.

S. Ivanyk, Mirona Zaryckiego koncepcja relacji matematyki i estetyki [Miron Zarycki's Concept of Relationship Between Mathematics and Aesthetics], *Semina Scientiarum* **11**(2012), pp. 92–106. ISSN 1644-3365.

S. Ivanyk, Mileny Rudnickiej próba rekonstrukcji systemu estetyki Leona Battisty Albertiego [Milena Rudnicka's Attempt to Reconstruct the System of Aesthetics of Leon Battista Alberti], *Semina Scientiarum* **12**(2013), pp. 21–33. ISSN 1644-3365.

S. Ivanyk, The Lvov-Warsaw School as a Multicultural Phenomenon. Ukrainian Aspect, pp. 29–44, in A. Brożek, F. Stadler and J. Woleński (2017).

J. Jadacki, The Lvov-Warsaw School from a Bird's Eye View, pp. 211–224, in A. Brożek, F. Stadler and J. Woleński (2017).

J. Jadacki, Jan Łukasiewicz: A Creator of New Ideas in Logic and a Reinterpreter of Its History, pp. 33–46, in Á. Garrido and U. Wybraniec-Skardowska (2018).

R. Jadczak, O tzw. szkole lwowsko-warszawskiej [On the so-called Lvov-Warsaw School], *Acta Universitatis Nicolai Copernici. Filozofia* **11**(197) (1990), pp. 19–37. ISSN 0208-564X.

F. Jaëck, Generality and Structures in Functional Analysis: The Influence of Stefan Banach, pp. 223–254, in K. Chemla, R. Chorlay and D. Rabouin (eds.), *The Oxford Handbook of Generality in Mathematics and the Sciences*, Oxford University Press, Oxford, 2016, xii + 507 pages. ISBN 978-0-19-877726-7.

E. Jakimowicz, A Remarkable Life, pp. 1–22, in E. Jakimowicz and A. Miranowicz (2011).

E. Jakimowicz, A. Miranowicz (eds.), *Stefan Banach. Remarkable Life, Brilliant Mathematics. Biographical Materials*, Third edition. Gdańsk University Press, Gdańsk, 2011, 186 pages. ISBN 978-83-7326-827-2.

M. Jarnicki and P. Pflug, *Continuous Nowhere Differentiable Functions. The Monsters of Analysis*, Springer Monographs in Mathematics, Springer, Cham, 2015, xii + 299 pages. ISBN 978-3-319-12669-2.

Z.A. Jordan, The Development of Mathematical Logic in Poland Between the Two Wars, pp. 346–397, in S. McCall (ed.), *Polish Logic 1920–1939*, Oxford University Press, Oxford 1967, viii + 406 pages. ISBN 978-0-19-824304-5.

M. Kac, *Enigmas of Chance. An Autobiography*, Alfred P. Sloan Foundation. Harper & Row, Publishers, New York, 1985, xxviii + 163 pages. ISBN 0-06-015433-0.

R. Kałuża, *Stefan Banach*, Wydawnictwo GZ, Warszawa, 1992, 167 pages. ISBN 83-900623-0-5. English translation: *The Life of Stefan Banach. Through a Reporter's Eyes.* Translated from the Polish, edited and with a preface by Ann Kostant and Wojbor Woyczyński, Birkhäuser Boston, Inc., 1996, x + 137 pages. ISBN 978-0-8176-3772-9.

F. Kawczyński, Maria Kokoszyńska-Lutmanowa: The Great Polemist, pp. 235–240, in Á. Garrido and U. Wybraniec-Skardowska (2018).

B. Knaster, Zygmunt Janiszewski (w 40-lecie śmierci) [Zygmunt Janiszewski (on the 40th Anniversary of His Death)], *Wiadomości Matematyczne* **4**(1960), pp. 1–9.

M. Kordos, *The Scottish Book*, pp. 143–152, in E. Jakimowicz and A. Miranowicz.

J. Koroński, *Równania różniczkowe zwyczajne i cząstkowe w publikacjach do I wojny światowej na tle rozwoju równań różniczkowych w świecie* [Ordinary and Partial Differential Equations in Publications until World War I against the Background of the Development of Differential Equations in the World], Monografie Politechniki Krakowskiej, Seria Nauki Podstawowe Matematyka, Kraków, 2017, 225 pages. ISBN 978-83-7242-927-8.

G. Köthe, Stanisław Mazur's Contributions to Functional Analysis, *Mathematische Annalen* **277**(1987), pp. 489–528. ISSN 0025-5831.

B. Knaster, C. Kuratowski and S. Mazurkiewicz, Ein Beweis des Fixpunktsatzes für n-dimensionale Simplexe, *Fundamenta Mathematicae* **14**(1929), pp. 132–137.

W. Krajewski (ed.), *Polish Philosophers of Science and Nature in the 20th Century*, Poznań Studies in the Philosophy of the Sciences and the Humanities, Volume 74, Rodopi, Amsterdam, 2001, 285 pages. ISBN 978-90-420-1497-8.

S. Kumar, A Short Survey of the Development of Fixed Point Theory, *Surveys in Mathematics and Its Applications* **8**(2013), pp. 91–101. ISSN 1843-7265.

C. Kuratowski, Sur le probléme des courbes gauches en topologie, *Fundamenta Mathematicae* **15**(1930) pp. 271–283.

C. Kuratowski, Sur l'application des espaces fonctionnels à la théorie de la dimension, *Fundamenta Mathematicae* **18**(1932), pp. 285–292.

K. Kuratowski, *Pół wieku matematyki polskiej 1920–1970*, Wiedza Powszechna, Warszawa 1973, 191 pages. English translation: *A Half Century of Polish Mathematics. Remembrances and Reflections.* Translated from the Polish by Andrzej Kirkor. With a preface by S. Ulam. International Series in Pure and Applied Mathematics, Volume 108, Pergamon Press, Inc., Elmsford, N.Y., PWN — Polish Scientific Publishers, Warsaw, 1980, viii + 200 pages. ISBN 978-0-08-023046-7.

C. Kuratowski and S. Ulam, Quelques propriétés topologiques du produit combinatoire, *Fundamenta Mathematicae* **19**(1932), pp. 247–251.

Sr. M. G. Kuzawa, *Modern Mathematics: The Genesis of a School in Poland*, College & University Press, New Haven, 1968, 143 pages.

Z. Łomnicki, *A Letter to Zygmunt Wilhelm (William) Birnbaum, 11/12/1951*, Zygmunt William Birnbaum Collection, University of Washington, University Libraries, Accession Number 5266-001. Online document. https://www.lib.washington.edu/specialcollections/collections/exhibits/spotlight-on-zygmunt-william-birnbaum-papers [02.10.2021].

Z. Łomnicki and S. Ulam, Sur la théorie de la mesure dans les espaces combinatoires et son application au calcul des probabilités. I. Variables indépendantes, *Fundamenta Mathematicae* **23**(1934), pp. 237–278.

L. Maligranda, Stefan Kaczmarz (1895–1939), *Antiquitates Mathematicae* **1**(2007), pp. 15–61. ISSN 1898-5203 (Polish).

L. Maligranda, Antoni Łomnicki (1881–1941), *Wiadomości Matematyczne* **44**(2008), pp. 61–112. ISSN 0373-8302 (Polish).

L. Maligranda, Eustachy Żyliński (1889–1954), *Antiquitates Mathematicae* **3**(2009) pp. 171–211. ISSN 1898-5203 (Polish).

L. Maligranda, Józef Schreier (1909–1943), Biography. *Wiadomości Matematyczne* **49**(2013), pp. 47–60. ISSN 2080-5519.

L. Maligranda, The Scientific Activity of Józef Schreier, *Wiadomości Matematyczne* **50**(2014), pp. 45–68. ISSN 2080-5519.

L. Maligranda, Meier (Maks) Eidelheit (1910–1943), *Wiadomości Matematyczne* **51**(2015), pp. 31–59. ISSN 2080-5519 (Polish).

L. Maligranda and J.G. Prytuła, Lwowscy uczeni wymienieni w przesłuchaniach Banacha z 1944 roku [Lvov Scholars mentioned in Banach's Interrogations from 1944], *Wiadomości Matematyczne* **49**(2013), pp. 29–66. ISSN 2080-5519.

L. Maligranda and W. Wnuk, Władysław Orlicz (1903–1990), *Wiadomości Matematyczne* **36**(2000a), pp. 85–147. ISSN 0373-8302 (Polish).

L. Maligranda and W. Wnuk, Władysław Orlicz: His Life and Contributions to Mathematics, pp. 23–29, in H. Hudzik and L. Skrzypczak (eds.), *Function Spaces*, *Proceedings of the*

5th International Conference held in Poznań, August 28–September 3, 1998, Lecture Notes in Pure and Applied Mathematics, Volume 213, Marcel Dekker, Inc., New York, 2000b, xvi + 511 pages. ISBN 978-0-8247-0419-3.

C. Martínez-Adame and J. Caballero, Borel y Steinhaus: dos piezas claves para comprender el surgimiento de la probabilidad moderna [Borel and Steinhaus: Two Key Players in Understanding the Origins of Modern Probability], *Lecturas Matemáticas* **37**(2016), pp. 37–61. ISSN 0120-1980.

R. D. Mauldin (ed.), *The Scottish Book. Mathematics from the Scottish Café. With Selected Problems from the New Scottish Book*, 2nd updated and enlarged edition, Birkhäuser – Springer, Cham, 2015, xviii + 322 pages. ISBN 978-3-319-22896-9.

J. Mawhin, A Tribute to Juliusz Schauder, *Antiquitates Mathematicae* **12**(2018), pp. 229–257. ISSN 1898-5203.

S. Mazur, Computable Analysis, *Rozprawy Matematyczne* **33**(1963), 110 pages.

J. Mioduszewski (ed.), *Matematyka przełomu XIX i XX wieku. Nurt mnogościowy* [Mathematics of the Turn of the 19th and 20th Centuries. Set-Theoretical Trend], Prace Naukowe Uniwersytetu Śląskiego w Katowicach, Uniwersytet Śląski Katowice, no. 1253, 1992, 193 pages. ISSN 0208-6336.

R. Murawski, The Contribution of Polish Logicians to Recursion Theory, pp. 265–282, in K. Kijania-Placek, J. Woleński (eds.), *The Lvov-Warsaw School and Contemporary Philosophy*, Papers from the School Held in Lvov, November 15–17, and Warsaw, November 19–21, 1995, Synthese Library, Volume 273, Kluwer Academic Publishers Group, Dordrecht, 1998, xii + 399 pages. ISBN 978-0-7923-5105-4.

R. Murawski, Philosophical Reflection on Mathematics in Poland in the Interwar Period. Provinces of Logic Determined, *Annals of Pure and Applied Logic* **127**(2004), pp. 325–337. ISSN 0168-0072.

R. Murawski, *Filozofia matematyki i logiki w Polsce międzywojennej*, Monografie Fundacji na Rzecz Nauki Polskiej, Wydawnictwo Naukowe Uniwersytetu Mikołaja Kopernika, Toruń, 2011, 268 pages. ISBN 978-83-231-2670-6. English translation: *The Philosophy of Mathematics and Logic in the 1920s and 1930s in Poland*, Translated from the 2011 Polish original by Maria Kantor. Science Networks, Historical Studies, Volume 48, Birkhäuser, Springer, Basel, 2014, xii + 228 pages. ISBN 978-3-0348-0830-9.

R. Murawski and J. Pogonowski, Seweryna Łuszczewska-Romahnowa, pp. 241–247, in Á. Garrido and U. Wybraniec-Skardowska.

L. Narici and E. Beckenstein, The Hahn–Banach Theorem: The Life and Times Proceedings of the First Ibero-American Conference on Topology and Its Applications (Benicassim, 1995), *Topology and Its Applications* **77**(1997), pp. 193–211. ISSN 0166-8641.

A. Olech, On Ajdukiewicz's Project of the Semantic Theory of Knowledge, pp. 65–90, in Á. Garrido and U. Wybraniec-Skardowska (2018).

W. Orlicz, Metoda kategorii Baire'a w zastosowaniu do pewnych zagadnień analizy matematycznej [Baire Cathegory Method in Application to Certain Problems of Mathematical Analysis], *Wiadomości Matematyczne* **24**(1982), pp. 1–15. ISSN 0373-8302.

W. Orlicz, Osiągnięcia polskiej matematyki w zakresie analizy funkcjonalnej [Achievements of Polish Mathematics in Functional Analysis], pp. 129–151, in A. Śródka (1995).

Z. Pawlikowska-Brożek, Wykaz profesorów i docentów matematyki pracujących w polskich uczelniach w latach 1919–1939 [A Roster of Professors and Docents of Mathematics Working in Polish Academic Schools in the Years 1919–1939], *Wiadomości Matematyczne* **24**(1982), pp. 219–223. ISSN 0373-8302.

Z. Pawlikowska-Brożek, Matematyka polska w latach 1918–1951 [Polish Mathematics in the Years 1918–1951], *Studia i Materiały z Dziejów Nauki Polskiej, Seria II,* **2**(1988), pp. 13–59. ISSN 0860-1011.

Z. Pawlikowska-Brożek, *Zygmunt Janiszewski — organizator nauki i autor idei polskiej szkoły matematycznej* [Zygmunt Janiszewski — Organizer of Science and Author of the Idea of the Polish Mathematical School], pp. 45–52, in J. Mioduszewski (1992).

Z. Pawlikowska-Brożek, *"Poradnik dla samouków". Periodyczne wydawnictwo miarą rozwoju potrzeb intelektualnych społeczeństwa* ["A Guide for the Self-Study". Periodical Publication as a Measure of the Development of Intellectual Needs of the Society], pp. 53–60, in J. Mioduszewski (1992).

Z. Pawlikowska-Brożek, *Dzieje nauk matematycznych* [History of Mathematical Sciences], pp. 7–28, in A. Śródka (1995).

Z. Pawlikowska-Brożek, Zbigniew Łomnicki (1904–1994), *Wiadomości Matematyczne* **31**(1995), pp. 187–190. ISSN 0373-8302 (Polish).

Z. Pawlikowska-Brożek, Andrzej Turowicz, pp. 246–247, in S. Domoradzki, Z. Pawlikowska-Brożek and D. Węglowska (eds.), *Słownik Biograficzny Matematyków Polskich* [A Biographical Dictionary of Polish Mathematicians], Państwowa Wyższa Szkoła Zawodowa, Tarnobrzeg, 2003, 286 pages. ISBN 978-83-917293-3-5.

Z. Pawlikowska-Brożek, D. Węglowska and Z. Węglowski, Pewne priorytety w rachunku prawdopodobieństwa [Some Priorities in Probability], pp. 175–183, in *Zeszyty Naukowe Politechnika Śląska*, no. 1287, Matematyka-Fizyka, zeszyt 76, Gliwice, 1996, 376 pages. ISSN 0072-470X.

A. Pelczar, Równania różniczkowe w Polsce. Zarys historii do połowy lat siedemdziesiątych XX wieku [Differential Equations in Poland. An Outline of History Until mid-1970s], *Wiadomości Matematyczne* **37**(2001), pp. 63–118. ISSN 0373-8302.

J.-P. Pier (ed.), *Development of Mathematics 1900–1950*, Birkhäuser, Basel, Boston, Berlin, 1985, pp. xvii + 729 pages. ISBN 978-3-7643-2821-4.

A. Pietsch, *History of Banach Spaces and Linear Operators*, Birkhäuser, Boston, MA, 2007, xxiv + 855 pages. ISBN 978-0-8176-4367-6.

A. Plichko, O Hermanie Auerbachu [On Herman Auerbach], *Wiadomości Matematyczne* **30**(1994), p. 273. ISSN 0373-8302.

Z. Popławski, *Dzieje Politechniki Lwowskiej 1844–1945* [History of Lwów Polytechnic], Zakład Narodowy imienia Ossolińskich, Wrocław–Warszawa–Kraków, 1992, 363 pages. ISBN 978-8304039322.

J. Prytuła, Doktoraty z matematyki i logiki na Uniwersytecie Jana Kazimierza we Lwowie w latach 1920–1938 [Doctorates in Mathematics and Logic at The Jan Kazimierz University in Lwów in the Years 1920–1938], pp. 137–161, in W. Więsław (ed.), *Dzieje matematyki polskiej* [History of Polish Mathematics], Instytut Matematyczny Uniwersytetu Wrocławskiego, Wrocław, 2012, 311 pages. ISBN 978-83-910055-4-5.

M. Przeniosło, Twórcy lwowskiej szkoły matematycznej (1920–1939) [Creators of the Lwów School of Mathematics (1920–1939)], *Dzieje Najnowsze* **39**(2007), no. 2, pp. 59–76. ISSN 0419-8824.

M. Przenioslo, International Mathematical Journals Published in Poland Between the Wars, *BSHM Bulletin — Journal of the British Society for the History of Mathematics* **24**(2009), pp. 20–26. ISSN 2637-5451.

A. Schinzel, Życiorys Wacława Sierpińskiego [A Biography of Wacław Sierpiński], *Wiadomości Matematyczne* **12**(1971), pp. 303–308. ISSN 0373-8302.

A. Schinzel, Rola Wacława Sierpińskiego w historii matematyki polskiej [The Role of Wacław Sierpiński in the History of Polish Mathematics], *Wiadomości Matematyczne* **26**(1984), pp. 1–9. ISSN 0373-8302.

The Scottish Book, Copies of the manuscript and the typescript in Polish and English. Online document. http://kielich.amu.edu.pl/Stefan_Banach/pdf/ks-szkocka/ks-szkocka1pol.pdf [02.18.2021].

H.M. Schaerf, My memories of Juliusz Schauder, *Topological Methods in Nonlinear Analysis* **2**(1993), pp. 15–19. ISSN 1230-3429.

J. Schreier and S. Ulam, Über topologische Abbildungen der euklidischen Sphären, *Fundamenta Mathematicae* **23**(1934), pp. 102–118.

M. Stawiska, Lucjan Emil Böttcher (1872–1937) — the Polish Pioneer of Holomorphic Dynamics, *Technical Transactions — Fundamental Sciences* 1-NP **7**(2014), pp. 233–243. ISSN 0011-4561.

M. Stawiska, Matematyka polska na Joint Mathematics Meetings w San Antonio (Texas), USA, styczeń 2015 [Polish mathematics at the JMM 2015 in San Antonio], *Anitquitates Mathematicae* **9**(2015), pp. 181–192. ISSN 1898-5203.

H. Steinhaus, *Wspomnienia i zapiski* [Reminiscences and Notes], Wydawnictwo Aneks, Londyn, 1992, 587 pages. ISBN 9780906601990. English translation: *Mathematician for All Seasons — Recollections and Notes*, **1**(1887–1945). With a foreword by Kazimierz Dziewanowski. Translated from the Polish by A. Shenitzer. Edited and with an introduction by R.G. Burns, I. Szymaniec and A. Weron, Series Vita Mathematica, Volume 18, Birkhäuser — Springer, Cham, 2015, xiii + 491 pages. ISBN 978-3-319-21983-7.

K. Szałajko, Wspomnienia o Kole Matematyczno-Fizycznym Studentów UJK we Lwowie [A Reminiscence of the Mathematical and Physical Circle of Students of UJK in Lwów], *Wiadomości Matematyczne* **26**(1984) pp. 85–96. ISSN 0373-8302.

K. Szałajko, Wspomnienie o Stefanie Banachu na tle Lwowa i Lwowskiej Szkoły Matematycznej [Reminiscences of Stefan Banach Against the Background of Lvov and the Lvovian School of Mathematics], *Opuscula Mathematica* **13**(1993), pp. 45–54. ISSN 0860-2727.

R. Sznajder, Kaczmarz Algorithm Revisited, *Technical Transactions — Fundamental Sciences* 2-NP **20**(2015), pp. 247–254. ISSN 0011-4561.

W. Ślebodziński, Władysław Nikliborc et son oeuvre scientifique, *Colloquium Mathematicum* **1**(1948), pp. 322–330. ISSN 0010-1354.

A. Śródka (ed.), *Historia nauki polskiej wiek XIX. Nauki ścisłe. Zeszyt pierwszy, matematyka. Fizyka. Chemia. Astronomia* [History of Polish science 19th Century. no. 1, Mathematics. Physics. Chemistry. Astronomy], PAN, Instytut Historii Nauki. Fundacja im. W. Świętosławskiego, Warszawa, 1995, 377 pages, xviii pages of photographs. ISBN 83-900891-9-X.

S. Świątkowski, Leon Chwistek (Polish), pp. 158–162, in J. Mioduszewski (1992).

A. Toczek, Franciszek Kazimierz Smolka, *Internetowy Polski Słownik Biograficzny*. Online document. https://www.ipsb.nina.gov.pl/a/biografia/franciszek-kazimierz-smolka [02.18.2021].

A. Turowicz, Unpublished Tape Recording, before 1989, Courtesy of Z. Pawlikowska-Brożek.

S.M. Ulam, Zur Maßtheorie in der allgemeinen Mengenlehre, *Fundamenta Mathematicae* **16**(1930), pp. 140–150.

S.M. Ulam, *Adventures of a Mathematician*, Charles Scribner's Sons, New York, 1976, xi + 317 pages (25 plates).

W. Więsław, *Algebra i teoria liczb w Polsce* [Algebra and Number Theory in Poland], pp. 153–164, in A. Śródka (1995).

J. Woleński, *Filozoficzna szkoła lwowsko-warszawska*, Państwowe Wydawnictwo Naukowe, Warszawa, 1985, 347 pages. ISBN 9788301053345. English translation: *Logic and Philosophy in the Lvov-Warsaw School*, Synthese Library, Volume 198, Kluwer Academic Publishers, Dordrecht, 1988, 378 pages. ISBN 978-94-010-7666-1.

J. Woleński, Mathematical Logic in Poland 1900–1939: People, Circles, Institutions, Ideas, *Modern Logic* **5**(1995), pp. 363–405. ISSN 1047-5982.

J. Woleński, *Filozoficzne problemy logiki* [Philosophical Problems of Logic], pp. 29–34, in A. Śródka (1995).

J. Woleński, *Logika Matematyczna* [Mathematical Logic], pp. 35–59, in A. Śródka (1995).

J. Woleński, The Reception of the Lvov-Warsaw School, pp. 3–19, in K. Kijania-Placek and J. Woleński (eds.), *The Lvov-Warsaw School and Contemporary Philosophy*, Synthese Library, Volume 273, Springer Science, 1998, 395 pages. ISBN 0792351053.

J. Woleński, Izydora Dąmbska — Between Conventionalism and Realism, pp. 107–111, in Krajewski (2001).

J. Woleński, Philosophy of Exact Sciences (Logic and Mathematics) in Poland in 1918–1939, *Technical Transactions — Fundamental Sciences* 2-NP **112**(2015), pp. 255–265. ISSN 0011-4561.

J. Woleński, Mathematical Logic in Poland 1918–1939, *Antiquitates Mathematicae* **12**(2018), pp. 99–109. ISSN 1898-5203.

J. Woleński, Zygmunt Schmierer (ok. 1910–1943). *Internetowy Polski Słownik Biograficzny.* Online document available at https://www.ipsb.nina.gov.pl/a/biografia/zygmunt-schmierer [02.18.2021].

W.A. Woyczyński, Seeking Birnbaum, or Nine lives of a Mathematician, *Mathematical Intelligencer* **23**(2) (2001), pp. 36–46. ISSN 0343-6993. English translation of idem, Szukając Birnbauma, *Wiadomości Matematyczne* **33** (1997), pp. 137–154. ISSN 0373-8302.

W. Woyczynski, The Legend of Hugo Steinhaus: Tales of Two Cities, up Close and Personal, *Wiadomości Matematyczne* **48**(2012), pp. 293–311. ISSN 2080-5519.

U. Wybraniec-Skardowska, Introduction. The School: Its Genesis, Development and Significance, pp. 3–14, in Á. Garrido and U. Wybraniec-Skardowska (2018).

M. Zakrzewski, Teoria Mnogości [Set Theory], pp. 83–127, in A. Śródka (1995).

M.M. Zarichnyi, Characterization of Functors of G-symmetric Power and Extension of Functors to the Kleisli Category, *Mathematical Notes* **52**(1992), pp. 1107–1111. ISSN 0001-4346.

M. Zarichnyi and B. Ptashnyk, Vydatnyi ukrayins'kyi matematyk i pedahoh Myron Zaryts'kyi (do 120-richchya vid narodzhennya) [Outstanding Ukrainian Mathematician and Teacher Myron Zaryts'kyj (to the 120-th anniversary)], *Visnyk L'vivs'kogo Universytetu. Seriya Mekhaniko-Matematychna* **70**(2009), pp. 191–207. ISSN 0320-6572.

K. Zielinski, Goście i gospodarze Kawiarni Szkockiej [Guests and Hosts of the Scottish Café], *Wiadomości Matematyczne* **55**(2019), pp. 197–215. ISSN 2080-5519.

https://doi.org/10.1142/9781786349316_0007

Chapter 7

Impact of World War I on the Mathematical Community in Hungary

Vojtech Bálint

1. Historical Background for the Period Before World War I

After the establishment of the Kingdom of Hungary, the country was in many periods essentially equal to the then European powers in terms of political importance. The same high standard also applies to the arts and culture, as evidenced by the findings, which are often more or less mere fragments of what remained after the many wars that took place in this area. The situation in the scientific field of culture is quite different, especially in mathematics. There was a huge decline in its development after the glorious era of Greek mathematics was over. In many parts of Europe, achievements in mathematics were revived as early as in the 16th–18th centuries. However, the situation was different in the Kingdom of Hungary. Attempts by major kings to establish a university in Hungary at that time lasted only a few years or decades. Thus, high-standard education could only be obtained at foreign universities and those who returned home did not find an environment suitable for their activities there. For this reason, the history of mathematics progress in Hungary started very late, at the beginning of the 19th century. This is well illustrated by the brief history of higher education in Hungary.

The first University of Hungary was established in Pécs by Louis I of Hungary in 1367. The university existed for a few decades only.[1]

The second University of Hungary was established by Sigismund of Luxembourg (1368–1437) in Óbuda in 1395. This university worked for no more than 15 years.[2]

[1]The modern University of Pécs was founded in 1912, and was originally located in Pozsony (now Bratislava, Slovakia). Following World War I Pozsony became part of the newly formed Czechoslovakia; therefore in 1921 the university moved to Pécs, where it has remained to this day. More detailed information can be found, e.g., in Filep (2012) and Font (2002).

[2]More detailed information can be found, e.g., in Szögi (1996) and also on the website http://uni-obuda.hu/en/university/history/universities-middle-ages [16.9.2019].

Figure 1. Nagyszombat University.[3]

The third University of Hungary was established in *Pressburg* (Istropolis by the Ancient Greek name; the name Bratislava originated in 1920) by the King Mátyás Hunyadi [Matthias Corvinus] (1443–1490) in 1467. This was made possible by the papal bull of 19 May 1465. The University under the name Academia Istropolitana was internally organized following the model of the University of Bologna. At Matthias Corvinus' invitation (and also under the influence of generous royal reward), several renowned teachers came to work there, e.g., Regiomontanus and Johannes Gattus. Moreover, Taddeo Ugoletto set up a university library as well. The Archbishop of Esztergom János Vitéz was a great supporter and organizer of the University. After his death in 1472, however, the University began to decline, famous teachers left, and teaching officially ended after Matthias Corvinus's death. After that, higher education did not exist in Pressburg for several centuries.[4]

The still working *Eötvös Loránd Tudományegyetem* in Budapest (official abbreviation ELTE) was established in 1635 by Esztergom Archbishop Péter Pázmány in *Nagyszombat* (now Trnava, Slovakia), then the seat of the Archbishopric, as a result of the Turkish occupation. Pázmány entrusted the Jesuit Order with the University management. György Dobronoki was its first Rector. The University was incomplete, it had only two faculties. However, the number of faculties gradually

[3]The picture is from the author's archive.
[4]More detailed information can be found, e.g., in Szögi (1996).

Figure 2. Royal University of Science in Budapest.[5]

increased: the Faculty of Law and the fourth Faculty of Medicine were established in 1667 and 1769, respectively, thus making the university complete — classical.

After the dissolution of the Jesuits in 1777, the University was moved to the center of the country, to the royal castle in the city of Buda. Many independent universities today have evolved from this institution. In 1784 the university moved to the left bank of the Danube to the city of Pest. The name Budapest was introduced in 1873, so until 1873 the University was named *Royal University of Pest*, and then the *University of Budapest* until 1921. The language of instruction was Latin until 1844 as a neutral language of the monarchy's nations. After the Austro-Hungarian Compromise of 1867, the era of the University's greatest development began. At the turn of the 19th and 20th centuries, it was among the 15 largest universities of the world according to the number of students, since there were only two classical universities in the then Hungary before 1914. Since 1895, women could also enroll in the University. In 1921–1950 the name of the university was *Pázmány Péter Tudományegyetem*, according to its founder.[6]

Magyar Királyi Ferenc József Tudományegyetem [Royal Hungarian Franz Joseph University] was the second modern university in the Hungarian realm of the Austro-Hungarian Empire. Founded in 1872, its seat was initially in *Kolozsvár* (today Cluj-Napoca, Romania). In the autumn of 1918, after the collapse of the Austro-Hungarian Empire and the proclamation of the union of Transylvania with Romania in *Alba-Iulia* on 1 December 1918, the Romanian forces began to occupy Hungarian territories. They took control of Kolozsvár on 24 December 1918. Therefore, after

[5]The picture is from the author's archive.

[6]More detailed information can be found, e.g., in Filep (2012); Szögi (1980, 1996).

Figure 3. Royal Hungarian Franz Joseph University at the end of the 19th century.[7]

Figure 4. University of Szeged.[8]

[7]The picture is from the author's archive.

[8]The picture is available on the website http://www.vasarhely24.com/iskolapad/orientaci/os-hetet-tartanak-a-szegedi-egyetemen [16.9.2019].

World War I, the University first moved from Kolozsvár to Budapest for a brief period (1919–1921), and later found its temporary home in Szeged (1921–1940). After a temporary return to Kolozsvár in 1940–1945 and a return to Hungary after WWII, it is now based in Szeged. This University was an important center of science and education in the Austro-Hungarian Monarchy. It was probably best known for its leading role in mathematics, earning the name 'Göttingen of the Monarchy'. The university attracted mathematicians such as Gyula Farkas, Lipót Fejér, Alfréd Haar, Frigyes Riesz, Ludwig Schlesinger, Béla Szőkefalvi-Nagy and Gyula Vályi.[9]

Emperor Joseph II established the *Institutum Geometrico-Hydrotechnicum* as part of the Faculty of Liberal Arts at the University of Buda in 1782. The Institutum is a direct predecessor of the present-day working *Budapesti Műszaki és Gazdaságtudományi Egyetem* (or shortly Műegyetem, official abbreviation BME) [Budapest University of Technology and Economics], and was the first in Europe to award engineering degrees to students of land surveying, river control, and road construction. BME is the most significant university of technology in Hungary and is considered the world's oldest institute of technology that has the university rank and structure. It was the first institute in Europe to train engineers at university level.

At the time, medium-level technical sciences were taught as preparation for university engineering studies at the Hungarian *Institutum Geometrico-Hydrotechnicum*.[10] The main subject at *Institutum Geometrico-Hydrotechnicum* was applied mathematics, which included teaching land surveying and hydraulic engineering. However, the rapid development of industry required increasingly better knowledge. The Hungarian Parliament (National Assembly) requested the establishment of an independent technical university in 1836 but did not receive a decisive answer. Therefore, those who decided to pursue an engineering career still studied in Vienna for some time. In 1856, teaching began to differentiate basically into three groups: general, mechanical and chemical engineering. In 1860, Hungarian (succeeding German) became the language of instruction, and the name *Királyi József Műegyetem* was used instead of *Polytechnikum*. On 10 July 1871, four years after the Compromise, Franz Joseph approved the organizational rules of the University.[11]

---

[9]More detailed information can be found, e.g., in Bisztray (1941); Filep (2012); Gazda (1997); Vekerdi (1996).

[10]It is also worth mentioning Georgikon Keszthely, established by Earl György Festetics in 1797.

[11]More detailed information can be found, e.g., in Császár (1998); Szögi (1980, 1996); Vekerdi (1996).

Figure 5. Royal Joseph University in 1909.[12]

Figure 6. Panorama BME today.[13]

The *Academy of Mining and Forestry* was established in Selmecbánya (today Banská Štiavnica, Slovakia) by the scientist Samuel Mikovíny (1700–1750) in 1735. On December 13, 1762, Maria Theresa established *Bergschule* ("Berg Schola") with

[12]The picture is from the author's archive.

[13]The picture is available on the website http://grammars.grlmc.com/SLSP2015/Travel.php [1.10.2019].

a formal academy status in order to train specialists in silver and gold mining and metallurgy for mines in the surrounding area and the whole country. It was the world's first institution of technology or technical university with tertiary technical education. Teaching started with the Department of Chemistry and Metallurgy and its first 40 students who were required to be over 18 years of age, have complete secondary education and one year of practical experience. Later, the Department of Mathematics, Mechanics, Hydraulics and Mining Engineering was established. The third, Department of Well Mining, was founded in 1770. The Department of Forestry was founded in 1807. The Department of Technical Drawing and Descriptive Geometry was established in 1839, followed in 1841 by the Department of Mineralogy, Geology and Palaeontology, the Department of Metallurgy and Testing and the Department of Chemistry and Physics in 1866, and finally the Department of Mining and Mineral Processing and the Department of Mining Surveying and Geodesy in 1872. Many members of the first teaching staff at BME arrived from Selmecbánya.[14]

2. The Two Bolyais

In the period before 1850 there were only two prominent Hungarian mathematicians — Farkas Bolyai and his son János.

The first Hungarian mathematician with independent achievements was Farkas Bolyai (1775–1856), who studied in Göttingen in 1796–1799 and met Gauss there. In 1832, Bolyai published the textbook *Tentamen*, into which he has *hidden* his own results. The verb *hid* is probably the most appropriate, because the results that had been unknown by the time, i.e., Farkas's own results, can only be found in that book by a real expert. Bolyai's best known assertion is that *if two planar polygons have the same area, both can be cut into polygons that are identical in pairs.* This seemingly simple assertion is not as trivial as it seems at first glance. Moreover, no one had for a long time been able to find out whether an analogous assertion also applies to the three-dimensional space. Therefore, Hilbert put this problem into his famous list as problem number 3 in 1900, and then it quite quickly became apparent that an analogous claim did not apply to the three-dimensional space.[15] Although Farkas Bolyai was a good mathematician for his time, his greatest merit was that he introduced his multiply talented son to mathematics.

János Bolyai (1802–1860), Farkas Bolyai's son, studied at the Military Academy in Vienna, since the family had no money to send him to study in Göttingen. During his studies, he was constantly thinking about Euclid's fifth postulate that many had

[14]More detailed information can be found, e.g., in Császár (1998); Szögi (1980, 1996); Vekerdi (1996).

[15]As early as in 1900, Max Dehn (1878–1952) proved that a regular tetrahedron and the cube with the same volume cannot be cut into identical pieces.

been unsuccessfully dealing with, including his father.[16] As basically everyone before him, János Bolyai also assumed that the negation of the fifth postulate holds — i.e., that at least two straight lines parallel to the straight line p can be drawn through the given point that does not lie on the straight line p — and tried to find some contradiction. However, he did not find any contradiction and, instead of a contradiction, he arrived at newer, unusual and seemingly nonsensical assertions. Therefore, János began (very rightly) to believe that the fifth postulate was independent of the others. In fact, it is one of the most revolutionary ideas in all mathematics, because it was to deny the general belief that Euclidean geometry was the only meaningful and correct one. As a 21-year-old, as early as in 1823, János wrote to his father in a letter that 'I have created another world, a new world from nothing', by which he meant non-Euclidean geometry. This was three years before Lobachevsky gave his first lecture on the issue. János Bolyai also developed the so-called absolute geometry, which does not assume either the validity of the fifth postulate or the validity of its negation. It is, therefore, a common (overlapping) part of both Euclidean and non-Euclidean geometries. The idea of a new, non-Euclidean geometry was so revolutionary that, like many others, even János's father Farkas did not believe in its correctness in 1825, because he could not understand that both geometries could be non-contradictory as they differ in one axiom. Therefore, a year later, János handed over a manuscript (in the German language) of his work to his former professor at the Academy of Military Engineers in Vienna, Captain Eckwehr, for assessment. However, the manuscript has not been found. So János published the results of his 1823 research no earlier than in 1832 as an Appendix in his father's book *Tentamen iuventutem studiosam in elementa matheseos ... introducendi* in Marosvásárhely. But this was already three years after Lobachevsky's (1792–1856) French article *Geometry imaginaire*. Moreover, Erdély (Transylvania), where the book *Tentamen* was published, was on the very edge of European civilization. That is why János's father Farkas wrote about his son's discovery to Gauss, who replied in a letter to Farkas that he, Gauss, had personally worked his way to that knowledge a long time ago. This denial of János's priority had crushed him personally so much that he never published anything else since.[17]

It is known (from a letter to Taurinus — see below) that Gauss had studied Euclidean geometry before 1799 without the fifth postulate, i.e. without parallelism. With his genius, there is no doubt that he had achieved some results. However, being 22-years old he was (rightly!) afraid of unkind reception and total misunderstanding, so he did not publish his results. Gauss wrote to his friend Franz Taurinus

[16] In today's modern understanding, this postulate says that just one straight line parallel to the straight line p can be drawn through the given point that does not lie on the straight line p.

[17] However, his left estate included 20,000 pages of mathematical notes that are still being studied. These mainly concern complex numbers, number theory and algebraic equations.

(1794–1874) sometime between 1824 and 1829 that he had solved the problem of parallel lines 30–35 years ago. Moreover, according to Felix Klein, who tried to shift the priority to Gauss, it is possible that Gauss mentioned the problem of parallel lines to also Martin Bartels (1769–1833), who taught Lobachevsky in Kazan, so maybe it was Bartels who directed his attention to this problem. Of course, this does not compromise Lobachevsky's or Bolyai's merits. However, it is not clear why Gauss did not inform even Lobachevsky in 1829 that the priority is his, Gauss's, and why he wrote this fact to János in 1832, although he called him a first-magnitude genius.

Of course, from the modern point of view of assessing the priority, the time of publishing the outcome (article on the issue) is crucial. But at that time, information spread much slower and most of the information was inaccessible to many for many years, and thus unheard of. In addition, János's letter to his father in 1823 contained the basic information three years before the first Lobachevsky's Russian-written article appeared in 1826, so perhaps the fairest thing to do is to attribute authorship to both Bolyai and Lobachevsky. *Congrès International de Bibliographie des Sciences Mathématiques* Secretariat led by Poincaré introduced the official title *Géométrie de Bolyai et Lobatschewsky.* In the book by J.Čižmár: *Dejiny Matematiky* [The History of Mathematics], Perfekt, Bratislava, 2017, the author (p. 639) concludes that "The Appendix written in Latin is a dense, formally mathematically more valuable interpretation, compared to Lobachevsky's first works, of non-Euclidean (hyperbolic) geometry in parallel to the synthetic geometry of the Euclidean space".

Richard Baltzer (1818–1887) in his two-volume book *Die Elemente der Mathematik* (1860, 1862) was the first scientist to address several mathematical achievements of the two Bolyais. It was from the Baltzer's book that a professor of the University of Bordeaux, G.J. Hoüel (1823–1886), learned about the geometric findings of Lobachevsky and Bolyai. In 1867, he published a French translation of Lobachevsky's book *Untersuchungen ...*, and also Bolyai's *Appendix* in the same year. Thus, Bolyai's study became much more accessible to mathematicians. Following the efforts of Boncompagni (1821–1894), an Italian translation of *Appendix* was published in 1868.

Bolyai-Lobachevsky's geometry had long struggled with very hostile reception. In 1868, Beltrami changed a little bit about this with a study that proved that the laws of hyperbolic geometry apply to a pseudosphere.[18] Johann Frischauf (1837–1924), Professor at the University of Graz, included lectures on non-Euclidean geometries in the geometry course as early as in the 1871/72 school year. These lectures were also published in a book (*Absolute Geometrie nach Johann Bolyai*, Leipzig, 1872), and it is clear from the title that it relied mainly on the work of János Bolyai. Frischauf's book *Elemente der absoluten Geometrie* (Leipzig, 1876) has a similar content. Due to his courageous attitude, Frischauf faced violent attacks

[18] A pseudosphere resembles the surface of a thin stream of water running out of the tap.

by some scientific circles and, in particular, by Austrian education authorities. The first (1891) and, incredibly, as much as the fourth (1896) editions of the English translation of the *Appendix* were published by Professor G.B. Halsted (1853–1922) of the Austin University of Texas. Professor Halsted even visited Bolyai's grave in Marosvásárhely in the summer of 1896. The Latin-written *Appendix* was translated into Hungarian as late as in 1897. The author of one translation was Ignác Rados, and the author of the second (the better) translation was József Suták. The German mathematician Paul Stäckel (1862–1919), an excellent expert in Gauss's geometrical works, studied (with the help of József Kürschák) Bolyai's left estate and published the results of his research in some articles written in the German language.

The Russian translation of *Appendix* penned by V.F. Kagan (Moscow–Leningrad, 1950) is of excellent standard. This translation also includes commentary, as well as a comparison of Gauss's, Bolyai's, and Lobachevsky's results. Even in the light of the latest research, only a few words could be changed in that study.[19]

3. Period from Two Bolyais to World War I

The two Bolyais do not fall into the reviewed interwar period. They are introduced and included because of the mathematical greatness of János Bolyai, and also for the purpose of completeness of this essay.[20]

The first Hungarian mathematician to gain recognition during his life was Jenő Hunyady (1838–1889). He studied at foreign universities for eight years — Vienna, Munich, Karlsruhe, Berlin, Paris and Göttingen, where he received his doctorate in 1864 for his work on algebraic curves. After returning home in 1865, Hunyady already had several own scientific achievements, mainly in the theory of determinants, and became Professor at the Technical University (today BME). The Hungarian Academy of Sciences[21] elected him a corresponding member in 1867 and then a full member in 1883, mainly on the basis of his work *A kúpszeleten fekvő hat pont föltételi egyenletének különböző alakjairól* [About the Different Shapes of the Equation for the Six Points Lying on the Conoid], Magyar Tudományos Akadémia,

[19] More detailed information can be found, e.g., in Dávid (1979); Filep (2012); Gazda (2000); Kiss (1999); Vekerdi (1996).

[20] It is clear that the criterion of completeness could not be applied to many national mathematics without too many pages, since mathematical research has been ongoing for many centuries. In the Kingdom of Hungary, however, there was no mathematical research before the Bolyais.

[21] In 1825, Count István Széchenyi offered the one-year return on all his property for the construction of the Hungarian Academy of Sciences building. The offer was joined by additional sponsoring from Ábrahám Vay, Count György Andrássy and Count György Károlyi. The designing and planning of the building began since the nation-wide money-raising campaign brought in quite a lot of money. In 1858, Baron Simon Sina donated a significant amount of 80,000 forints for construction work. The foundations of the building were laid in 1862 and its walls in 1863. The building was completed according to the design of the Berlin Professor Friedrich August Stüler on 29 August, 1863. In 1864, internal work took place and the first meeting in the Academy building was held on 24 April, 1865.

Figure 7. Hungarian Academy of Sciences.[22]

Budapest, 1876, pp. 1–23. Hunyady was one of the main organizers in establishing *Mathematikai és Physikai Társulat* [Mathematical and Physical Society, today Bolyai János Society] (see Section 4), in 1885–1891.[23]

Ágoston Scholtz (1844–1916) attended the schools of Igló, Rosenau and Lőcse (now Spišská Nová Ves, Rožňava and Levoča, Slovakia). After his secondary education he studied at the universities of Vienna and Berlin. After teaching several years at secondary level, he obtained the university habilitation in 1879 and began his teaching at the University of Budapest, as Professor since 1884. Scholtz's field of research was projective geometry and theory of determinants. He collaborated extensively with Jenő Hunyady; that is why both names are associated with their joint results: Hunyadi–Scholtz determinant theorem and Hunyadi–Scholtz matrix. Scholtz in Scholtz (1878) followed Hunyady's above-mentioned most important work *About the Different Shapes of the Equation for the Six Points Lying on the Conoid*, so his results also generated great interest abroad. Together with the Hunyady they initiated lectures (including dinner) for colleagues in 1885 (see Section 4).[24]

[22]The picture is available on the website https://welovebudapest.com/en/venue/hungarian-academy-of-sciences-mta [15.2.2021].
[23]Some of his works are included in References (Hunyady, 1882, 1876).
[24]His work (Scholtz, 1878) is included in References.

Mór Réthy (1846–1925) attended the technical universities of Vienna and Buda. (The name Budapest was introduced on November 17, 1873 after the unification of the two cities on both banks of the Danube.) After two years of teaching at the Modern Technical School of Körmöcbánya (today Kremnica in Slovakia), Réthy continued his studies in Göttingen and Heidelberg. In Heidelberg he obtained his doctoral degree under Leo Königsberger's supervision in 1874. In the same year he was appointed extraordinary professor of theoretical physics at the University of Kolozsvár (after WWI: Cluj-Napoca, Romania). Réthy stayed in Kolozsvár until 1886. Then he began teaching at the Technical University of Budapest in 1886, lecturing in geometry. However, then his interest changed to mathematical physics, lecturing in analytical mechanics and theoretical physics from 1892 until his retirement. His best-known student was Tódor Kármán (known as Theodore von Kármán). Réthy was a corresponding member of the Hungarian Academy of Sciences from 1878 and a full member from 1900. Réthy's literary heritage includes works in both mathematics and physics, and he was awarded a honorary degree by the University of Heidelberg in 1924. The last quarter of Réthy's life was shadowed by tremendous personal losses: the death of his second son as a victim of WWI, and the death of his four-year-old granddaughter. He published mainly in Hungarian and German.[25]

Lipót Klug (1854–1945) studied at the Technical University in 1870–1872 and then two years at the Pedagogical Institute. He taught in Pressburg until 1893. In the meantime he received a doctorate at the University of Budapest in 1882. Klug was Professor at the University of Kolozsvár from 1897 to 1917. He was mainly concerned with descriptive and synthetic geometry. For a short time in high school he also taught Edward Teller, who esteemed him especially, as well as (much later) the excellent geometrician Ferenc Kárteszi, who once wrote that without Klug, he would have certainly not become a geometer. Klug used his one-year retirement pension to establish a foundation to support the studies of gifted students. Unfortunately, the prize was awarded only in 1943 (won by Ferenc Zigány and László Fejes Tóth), because by then money lost value. Almost all historical sources state that the trace of Professor Klug was lost in 1944. But at that time the 90-year-old professor had left his house in Budapest to escape the Nyilas fascist rage and hid successfully. He reappeared in January 1945 and died in March 1945 at the age of 91.[26]

Gyula Kőnig (1849–1913) was an extraordinary personality. He studied medicine in Vienna for one year. Then he studied in Heidelberg, where he was taught by

[25]One of his works (Réthy, 1890) is included in References. More detailed information can be found, e.g., in Oláh-Gál (2013).

[26]His work (Klug, 1892) is included in References. More detailed information can be found, e.g., in Kárteszi (1973).

Helmholtz. His first publication in 1870 was about medical sciences, because on Helmholtz's impulse he measured the speed of functioning of the nervous system under external influences, e.g., after alcohol consumption. In Heidelberg, Kőnig received a doctorate in 1871 for his work *Zur Theorie der Modulargleichungen der elliptischen Functionen* under the leadership of Leo Königsberger. Kőnig completed his postdoctoral studies in Berlin, where he was taught by Kronecker and Weierstrass. After returning home in 1873 he became Professor at the Technical University of Budapest, where he worked for the rest of his life. In 1880 he became a corresponding member of the Hungarian Academy of Sciences, and a full member in 1889. In 1904 at the congress in Heidelberg, he showed a proof of inequality known as *Kőnig's inequality for cardinality of sets* and dedicated the second part of the lecture to the proof of incorrectness of *Cantor's continuum hypothesis.* Cantor himself was also present and did not protest against the proof. However, the next day Zermelo discovered a mistake in Kőnig's proof. The proof was based on a theorem due to F. Bernstein (1878–1956), which did not apply under such general conditions as Bernstein himself put it. At the establishment of the Mathematical and Physical Society, Kőnig became Vice-chairman, and he authored the very first article (*A gammafüggvények elemi tárgyalása* [Elementary Discussion of Gamma Functions]) in the first issue of *Mathematikai és Physikai Lapok* published by the Society. A year earlier, in 1883, Kőnig started to publish a journal entitled *Mathematikai és Természettudományi Értesítő* of the third Section of the Hungarian Academy of Sciences (it was a department of natural sciences). One of the most amazing achievements of G. Cantor as author of set theory was that *one- and multi-dimensional continuum* have the same cardinality. Gyula Kőnig found a much simpler proof for this assertion, which he presented on November 15, 1893 at a meeting of the Society. Although he had never published it himself, this proof made its way to the literature. Schoenflies published it in his book *Entwicklung der Mengenlehre und ihrer Anwendungen* (Teubner, 1913, with Hans Hahn), and also Felix Klein in his *Elementarmathematik vom höheren Standpunkte aus* (1908, 1911, 1924). In this way, the proof made its way into the set theory textbooks, always with the name of the author. Kőnig did an enormous amount of work in organizing mathematical life in Hungary, but he had to significantly slow down the pace due to health reasons at the turn of the 19th and 20th centuries. In 1917, his two sons established a foundation with 10,000-crowns initial capital at 6% interest. The interest yield on the principal deposited was supposed to be used every other year to award one original work in mathematics with 1,000 crowns. However, money lost value as a result of the war, and so the prize transformed into the Kőnig plaque after 1930. The moral value of this prize, however, remained very high. Let us recall the laureates of the Kőnig Prize until 1940: Mihály Bauer (1922), Gábor Szegő (1924), Gyula Szőkefalvi-Nagy (1926), Károly Jordan (1928), Ottó Szász (1930), Jenő Egerváry (1932), Pál Veress

(1934), László Kalmár (1936), István Lipka (1938), and László Rédei (1940). Kőnig published a great number of valuable articles in the Hungarian language.[27]

Gyula Vályi (1855–1913) studied at the University of Kolozsvár, and then in Berlin in 1877–1879 where he was taught by Kummer, Borchardt, Kronecker and Weierstrass. Vályi was most impressed by the latest. Upon his return into Kolozsvár, Vályi wrote an excellent thesis on second-order partial differential equations and had seven original works by the end of 1883. In 1884 he was appointed professor of theoretical physics and also professor of mathematics a year later. He was appointed a corresponding member of the Hungarian Academy of Sciences in 1891. His most renowned student was Frigyes Riesz.[28]

Manó Beke (1862–1946) received a teacher diploma in mathematics-physics from the University of Budapest in 1883. He spent the year 1884 in Göttingen, where he was taught by Felix Klein, and received a doctorate in the same year. Beke then taught at secondary grammar schools in Budapest until 1900, when he was granted professorship at the Technical University. His results on the irreducibility of linear differential equations were included by Picard in his work *Traité d'analyse*. Beke was a speaker in the International Congress of Mathematicians (ICM) in Rome in 1908. He also lectured at the conferences in Cambridge (1912) and in Paris (1914). He was elected a corresponding member of the Hungarian Academy of Sciences in 1914. Beke's best-known students include György Pólya, Gábor Szegő, and Dénes Kőnig. He was an excellent teacher and, like Gyula Kőnig, did an enormous amount of work in organising mathematical life in Hungary. He was particularly interested in differential equations.[29]

Gusztáv Rados (1862–1942) studied in Budapest at both the University and Technical University, and then in Leipzig. He focused mainly on algebra, but also on number theory. The criterion of his first scientific work on the number of different roots of a higher degree congruence became well-known very quickly. From 1893 he was Professor at the Technical University of Budapest and, like Kőnig and Beke, Rados did a lot of work in organizing mathematical life. He was a founding member of the *Mathematical and Physical Society* (see Section 4) and its chairman since 1931. From 1894 he was a corresponding member of the Hungarian Academy of Sciences and a full member from 1907. Rados lectured at the congresses of Zürich in 1897 and Rome in 1908. He published a number of valuable articles in the Hungarian language in the journal *Mathematikai és Természettudományi Értesítő* of the third Section of the Hungarian Academy of Sciences.[30]

[27]Some of his foreign language works are listed in References (Kőnig, 1872, 1905, 1914). More detailed information can be found, e.g., in Szénássy (1965).
[28]Some of Vályi's works are included in References (Vályi, 1889, 1906).
[29]His works (Beke, 1894, 1897) are included in References.
[30]One of his foreign-language works is included in References (Rados, 1987).

László Rátz (1863–1930) studied at the University of Budapest, and then one year in Berlin, studying philosophy, and another year in Strasburg, studying natural sciences. Rátz was a legendary secondary grammar school professor at the renowned *Fasori evangélikus gimnázium*. There he taught such personalities as Alfréd Haar, János Neumann or Jenő Wigner. According to Wigner, Nobel Prize winner in physics, Rátz was the best teacher he had ever known, and many of his students had similar statements about him. Rátz was the editor-in-chief of the journal *Középiskolai Matematikai Lapok* from 1896 to 1914 and this work has left an indelible mark on many other prominent personalities, e.g., Lipót Fejér, Tódor Kármán, Ede Teller, Ferenc Kárteszi, Pál Turán, Pál Erdős, etc. He also did a great deal of work in the curriculum commission and in 1910 was named *Officier d'Académie* by the Ministère de l'instruction publique et des Beaux-arts.[31]

Figure 8. László Rátz.[32]

Figure 9. A famous secondary grammar school, Fasori Evangélikus Gimnázium in Budapest.[32]

Dániel Arany (1863–1944) graduated from the University of Budapest, and then worked as an assistant at the Forestry Academy of Selmecbánya (today Banská Štiavnica, Slovakia). Later (1893–1896) he worked as a secondary grammar school teacher in Győr and then in Budapest. Following the role model of the journal *Mathematikai Lapok* published by the Mathematical and Physical Society (see Section 4), Arany as a 30-year-old secondary school teacher started to publish the KöMaL journal for secondary school students all by himself. The journal is still published today.

[31]More detailed information can be found, e.g., in Kántor (2016).

[32]The pictures are from the author's archive.

Nowadays, a prestigious competition for students of the first and second year of grammar schools is named after him. Arany wrote several high-profile secondary school textbooks, as well as articles on *insurance, Simson line, and probability theory*. He also gave a lecture at the Congress of Zürich in 1932. He died together with his wife in a ghetto in December 1944 during an onslaught on Budapest. Before leaving for the ghetto, Arany managed to donate his large and very valuable library to the Society.[33]

Lajos Schlesinger (1864–1933), born in Nagyszombat (today Trnava in Slovakia), was a German mathematician and geometer who perfectly mastered the Hungarian language. He studied in Heidelberg and Berlin, where he received a doctorate under the leadership of Lazarus Fuchs. Schlesinger taught at the University of Kolozsvár, where Klug had invited him. His extensive studies on linear differential equations and their systems are particularly valuable. In 1902 he became a corresponding member of the Hungarian Academy of Sciences. Later he taught for a very short time at the University of Budapest and then in Giessen, where he gradually supervised 24 PhD students. His most famous student in Hungary was Gyula Szőkefalvi-Nagy.[34]

József Kürschák (1864–1933) graduated from the Technical University in Budapest, where his teachers included, for example, Gyula Kőnig and Jenő Hunyady. He became a teacher at the Primary Grammar School in Rozsnyó (today Rožňava, Slovakia) and Debrecen. Kürschák got a doctoral degree in 1890. He taught at the Technical University from 1891. His students there included Dénes Kőnig, Lipót Fejér and János Neumann. He published more than 100 works and, *inter alia*, simplified Hilbert's proof of Waring's hypothesis. He was elected a corresponding member of the Hungarian Academy of Sciences in 1897 and a full member in 1914. For many years, Kürschák was Vice-chairman of the Mathematical and Physical Society (see Section 4) and did an enormous amount of work there. He was extremely popular because he was selflessly immersed in the scientific problems of his colleagues and helped them with valuable advice.[35]

Károly Jordán (1871–1959) studied at a secondary grammar school in Budapest and then at the universities in Paris, Zürich, Manchester and Geneva. He became a chemical engineer in Zürich and received his doctorate in chemistry in Geneva in 1896. After returning to Budapest in 1898, Jordán studied mathematics, earthquake science and astronomy. He worked as Director of the earthquake institute, and later as a meteorologist. Then he taught at BME. He devoted himself to mountaineering and caving, and especially to statistics regarding scientific work. Jordán wrote the first book on mathematical statistics in Hungary.[36]

---

[33] His works Arany (1905, 1912) are included in References.

[34] The Mathematical Library at the University of Kolozsvár still keeps more than 3400 pages of Schlesinger's manuscripts. Some of them are included in References (Schlesinger, 1912, 1899).

[35] Some of his most significant works are included in References (Kürschák, 1891, 1913, Kürschák and Hadamard (1910)).

[36] In References, Jordán (1927) is the French version of this book.

Zoárd Geőcze (1873–1916) graduated from the University of Budapest, but was also taught by professors from the Technical University. He had an extraordinarily hard-to-follow style (too many new terms defined), so it was difficult for him to be acknowledged. Geőcze taught at Ungvár Secondary Grammar School (today Uzhgorod, Ukraine) and there he suffered a great lack of professional literature. Nevertheless, he wrote a very interesting article, based on which he was granted a one-year scholarship from the Ministry for a study stay in Paris. There, Geőcze was influenced by Baire, Borel, Lebesgue and Poincaré, but due to his aforementioned hard-to-follow style Geőcze did not succeed, nor did he stay there when Marie Skłodowska tried to persuade him. He then spent one year in his homeland, but in 1910 he was back to Paris and received a doctorate at the Sorbonne. Geőcze died during World War I at the end of 1916 after a serious illness due to his exhaustion in the fulfilment of military duties.[37]

Győző Zemplén (1879–1916) was a sad and severe World War I victim among physicists. Despite his young age, Zemplén was internationally successful and was a corresponding member of the Hungarian Academy of Sciences. He died on the Italian front in the battle of Asiago on the Monte Dolore plateau, where he is buried. Another World War I victim was Ferenc Lukács (1891–1918) whose talent did not have time to develop since he was only 23 years old at the beginning of World War I.

4. Mathematikai és Physikai Társulat

Towards the end of the 19th century, especially after the Austro-Hungarian Compromise, there was a significant increase in economic growth, resulting in increased demand for solutions to problems that required a deep knowledge of mathematics and physics. In 1862, four students from Prague University founded *Spolek pro volné přednášky z mathematiky a fysiky* [The Association for Free Lectures in Mathematics and Physics], which transformed into *Jednota českých matematiků a fysiků* [The Union of Czech Mathematicians and Physicists] in 1869.[38] Foreign study stays for scientists became much easier and, at that time, there was a significant increase in the scientific standard at the Technical University of Budapest and also at the universities in both Kolozsvár and Budapest. This science-favorable state lasted basically until the beginning of World War I. In the autumn of 1885, at the initiative of Réthy and Scholtz, a more or less private association of mathematicians (*Mathematikai Társaság*) was established to meet twice a month at a Budapest restaurant and to inform each other about new scientific achievements: both about their own works and about experiences from foreign travels. At that time, information spread

[37]Some of his most significant works are included in References (Geőcze, 1906, 1907).

[38]A little earlier, the Moscow Society was founded in 1864, and the London Mathematical Society in 1865. After the founding of the Union of Czech Mathematicians and Physicists, other mathematical societies were established, e.g., the French Society in 1873, Tokyo Society in 1877, American Society in 1888 and successively many others.

very slowly, so the fastest and most reliable thing was when someone reported on their trip abroad, informing on 'what's new' in the mathematical world. Each such lecture and discussion was followed by a common dinner, so these meetings became very attractive. The association had no rules or a chairman, so they differed from ordinary 'dining societies' only by a black writing board. The association had about 25 members — professors of Budapest universities and secondary schools. Soon physicists also joined these events. Taking turns at the blackboard were most frequently Jenő Hunyady, Gyula Kőnig, Ágoston Scholtz, Kálmán Szily, Loránd Eötvös and Mór Réthy. The Kolozsvár group joined later, including Gyula Farkas, Lajos Schlesinger and Gyula Vályi. They summoned a meeting in Budapest with about 130 teachers at the initiative of the famous experimental physicist Loránd Eötvös in December 1890. Its motto was 'We want to learn from each other and teach each other'. Eötvös suggested establishing an association or society with rules that would bring together mathematicians and physicists, and also proposed the creation of a journal: to present the state of science advances at the meetings and to write in the journal everything that might be of interest to experts. Implementation of Eötvös's proposal was quick: the first double issue of the *Mathematikai és Physikai Lapok* journal was published in June 1891 — i.e., prior to the official establishment of the intended association (society). If we browse through the journal volumes from 1892 to 1943 (when the last issue of this journal was published), we find that its standard was basically very high all the time despite the fact that there were often problems with both lack of articles and financial issues with journal publication.

On 21 August 1891, after drawing up the statutes (i.e., 22 years after the establishment of the Union of Czech Mathematicians and Physicists), *Mathematikai és Physikai Társulat* [Mathematical and Physical Society] was established (hereinafter the Society), and its first Chairman was Loránd Eötvös, Vice-chairman Gyula Kőnig and Secretary Gusztáv Rados. The Society had 298 members in November 1891, and this number rose quite rapidly to 400, with the split between the capital city and the countryside being about 50:50. This number was more or less constant for a very long time. The Society did not accept every candidate; they were accepted by voting.

To compensate for the lack of a Nobel Prize for mathematics and to preserve Bolyai's message, the Hungarian Academy of Sciences founded the Bolyai Prize in 1902 on the occasion of the centenary of János Bolyai. According to the charter, the Prize was to be awarded every five years beginning in 1905 'for the best mathematical studies regardless of their language and place of publication.' The Prize was awarded to H. Poincaré in 1905, and to D. Hilbert in 1910. WWI interrupted the award for a long time. It was renewed as late as in 2000 when it was awarded to Saharon Shelah (Israel). The Society used the presence of foreign awardees as well as other foreign guests to organize their lectures.

The Second General Assembly of *Mathematikai és Physikai Társulat* was held in May 1894. The Secretariat of the *Congrès International de Bibliographie des Sciences Mathématiques*, under the leadership of Poincaré, intended to publish a bibliography of mathematical works and urged individual states to provide supporting documents. One of the chapters had a planned title *Géométrie de Lobatschewsky*, and the French Secretariat changed this title to *Géométrie de Bolyai et Lobatschewsky* based on the supporting documents from the Society. Note that in 1893 János Bolyai's grave almost disappeared, and there was only one person (Bolyai's faithful servant Júlia Szőts) who knew where the grave was located. The following year, the Society had a monument built on the grave of the creator of non-Euclidean and absolute geometry, and in 1911 the remains of János Bolyai were moved to a location next to his father's grave at the cemetery of the Reformed Church in Marosvásárhely.

Two years after founding the 'big' *Mathematikai és Physikai Lapok* journal, Arany Dániel, a young (30-year-old) secondary school teacher, founded its 'little brother', *Középiskolai Matematikai Lapok* (*KöMaL*), as a secondary school journal in 1894. The journal quickly became very popular and its editing in 1896 was taken over by the now legendary teacher Rátz László, who did this until 1914, when the journal ceased to exist due to war. It was renewed only in 1925 by Faragó Andor, but in 1939 the journal disappeared again due to lack of papers. Of course, today the journal exists and has a very high standard. It was restored in 1947 by Surányi János. It has included a part on physics since 1959, and a part on computer science since 2001. But it still works under the acronym KöMaL.

At the Second General Assembly of the Society in 1894, a decision was made to create a competition in mathematics (and later in physics). In order to ascertain the real state of mathematics (and physics) knowledge, the Society would organize a competition in Budapest and Kolozsvár every autumn for students who took their school-leaving exams that given year. It was an excellent decision, because many winners and also other participants later reached world fame. Here are some winners: Győző Zemplén (1896), Lipót Fejér (second place in 1897), Tódor Kármán (1898), Alfréd Haar (1903), Marcel Riesz (1904), Gábor Szegő (1912), Tibor Radó (1913), László Rédei (1918), and László Kalmár (1922). Among those who managed to win both competitions were, e.g., Ede Teller and Endre Makai. On the other hand, 'super-genius' János Neumann never won either of these competitions because he graduated in 1921 and the competition did not take place between 1919 and 1922 as a result of World War I. Tasks for these competitions were drawn up by the most renowned professors, e.g., Gyula Kőnig, Gusztáv Rados, László Rátz, József Kürschák, and later also by Lipót Fejér, because he was much younger than the others above. When Gaston Darboux (1842–1917) and Felix Klein (1849–1925) visited Budapest in 1905, they by all means demanded that they be present at the award ceremony of this competition to personally congratulate the winners. This is obvious, since names such as Zemplén, Fejér and Kármán had already been well-established by then.[39]

The 'big' *Mathematikai és Physikai Lapok* journal was edited by Gusztáv Rados until 1914, then by Lipót Fejér until 1932, and then by Dénes Kőnig until World War II.

The Society's work had to be reorganized after World War I — practically from scratch. In the spring of 1921, the Society's lectures were renewed thanks to J. Kürschák, and even the Society's General Assembly was held on 14 May. At this General Assembly, the memorable words of Sándor Mikola (professor of physics at the famous *Fasori gimnázium*) were delivered: 'In order for a larger number of individuals to work together, it is not necessary for them to be the same, not even to have the same views, but only to be able to honour each other and respect the views of others.' The wounds Hungary suffered as a result of peace treaties were felt not only economically, but also by the number of members of the Society. It was a loss of two thirds of the territory and with it a large population. There was severe disruption of infrastructure not only in transport. It was also an issue of paying reparations and completely ignorant isolation of scientists. The villagers did not have good opportunities to engage in the work of the Society, although some activities were resumed. It was already a success when the Ministry granted at least some high school professors leave to attend the congress of the Society.

Let us recall the words of Dénes Kőnig, Secretary of the Society, from his 1941 report Kőnig (1941), 'The unpleasant consequence of the extremely bad economic situation is that the Society had to stop paying royalties completely. It is true that scientists do not write papers for royalties, but the abolition of royalties is harmful in both social and moral terms and is particularly detrimental to journal editors. They cannot afford to ask someone to *work for free*, e.g., to write a paper on the development of a certain discipline. Such papers are time-consuming, usually require the best scientists and do not represent any scientific success for the author.'[40]

Before 1926, the organization of international conferences was an utopia, and there were very few mathematicians who attended them with the support of the Society in Hungary in those years. Even the conference in Szeged in 1928 was attended by only two foreign mathematicians: O.D. Kellogg and G.D. Birkhoff. But then the situation slightly changed. The list of mathematicians and physicists who gave lectures at the invitation of the Society is as follows: H. Petterson and F. Ehrenhaft,

[39]Today, the competition in mathematics bears Kürschák's name, and the one in physics bears Eötvös' name.

[40]In Slovakia, e.g., various ministries and their subordinate agencies routinely ask reputable experts to draw up reports — of course free of charge. This 'achievement of socialism' is still up to date 30 years after the change in the political system throughout the education system, and also in many other parts of the economy.

both from Wien, in 1927; K. Scheel (Berlin), E. Rupp (Berlin) and H. Rademacher (Wrocław) in 1929; A. Sommerfeld (Munich) and W. Lietzmann (Göttingen) in 1930; L.L. Silverman (Dartmouth College, USA) in 1931; J.A. Barnett (Cincinatti, USA) in 1932; O. Blumenthal (Aachen) in 1935; R. Wawre (Genève) in 1936; E. Finlay-Freundlich (Prague) and H. Mark (Vienna) in 1937; R.W. Pohl (Göttingen) and W. Sierpiński (Warsaw) in 1938; P. Debye (Berlin-Dahlem) in 1939; G. de Rham (Lausanne) in 1939. And, of course, there were lectures presented by Jenő Wigner and János Neumann from the USA as well. Béla Kerékjártó and Rudolf Ortvay were most active in organizing lectures by foreign guests. It was not easy because the Society could not afford to cover the travel costs of foreign or domestic speakers. In this light, the above list is of a very high profile.

And let us recall more words from the above-mentioned report Kőnig (1941). They concerned the issue whether mathematical publications in the Hungarian language were meaningful. His answer was a clear and unambiguous YES. If Hungarian mathematical books and papers do not appear, then both secondary and higher education would suffer greatly, because language must respond to the development of science by means of the printed word as well, of course. It is *a priori* clear that Hungarian-written scientific articles will have difficulties in finding their way into the world compared to articles written in 'world' languages. However, this essentially doesn't change the very essence, because science is part of the national culture. Of course, it is desirable to publish scientific results in world languages as well.

Kőnig's rationale is universal, so the answer to such a question seems to be YES in the case of every 'minor' language, such as Slovak, Czech, Romanian, etc.[41]

5. Historical Background for the Period after World War I

In addition to the deaths of the already established scientists Zoárd Geőcze and Győző Zemplén, as well as of younger talents like Ferenc Lukács, World War I outcomes had a very unpleasant impact on post-war Hungary in other respects as well. The victorious powers were not interested in the existence of a strong competitor in the form of a large state in Central-Eastern Europe, so they divided the Austro-Hungarian Monarchy into a number of successor states. Hungary, as a multinational part of the Austro-Hungarian Monarchy, lost 2/3 of its territory and a large part of its population. Disproportionately large war reparations had to be paid, resulting in enormous poverty in Hungary.

Note that the war reparations of Germany, Hungary and Austria resulted in the so-called 'Golden Twenties' in many other European countries. It is worth mentioning that it was the Austro-Hungarian Monarchy that began the war, not Hungary,

[41]Information from the period after WWII can be found, e.g., in Kántor (1996).

which, although it had some autonomy like the Croatian-Slovenian part within Hungary, externally acted as a whole together with Austria. *Unsere gemeinsamer Kaiser* declared the war, and all the nationalities of the monarchy fought for the Emperor (*Kaiser*) in the army. The Austrians and the Hungarians lost the World War I, but paradoxically, all the other nationalities of the monarchy came out of the World War I as victors, even though they fought for the Emperor, on the same side as the Austrians and the Hungarians. However, the other nationalities did not pay reparations, but received them.

In addition to the already mentioned huge poverty, the isolation of Hungarian scientists, including mathematicians, caused a great impact on this promisingly developing science field. Isolation was also supported by the ignorance of some leading Western mathematicians.[42] This was ignorance because World War I was not initiated by Hungarian mathematicians, just like Napoleonic conquest had not been started by French mathematicians...

The loss of a large part of territory caused Hungary to experience an 'increase in the density of educated people'. This was obvious, since a large proportion of scientists and teachers moved from the lost territories back to Hungary. Let us mention cities such as Kolozsvár, Pozsony, Selmecbánya, Kassa, Ungvár, Arad, Marosvásárhely, Nagyvárad, etc. The three factors — loss of much of the territory, poverty and isolation — caused that many young scientists and teachers were unable to find a job in Hungary. The then government showed profound ignorance when it wanted to address the percentage of education by the enactment of the *Numerus Clausus* Act of 26 September 1920. The discriminatory essence of that legal act was that the proportion of young people of a particular race or nationality in a given

---

[42] Émile Picard (1856–1941), Charles-Jean de la Vallée Poussin (1866–1962) and Gabriel Koenigs (1858–1931) played a key role in the excluding of German, Austrian, Hungarian and Bulgarian mathematicians from international cooperation.

In 1920, the ICM was held in Strasbourg. 'The two mathematicians in key administrative positions, Picard, the president of the *International Research Council* (IRC), and Koenigs, the Secretary General of the *International Mathematical Union* (IMU), were forceful advocates of the policy of excluding Germany from international cooperation.' (See page 30 in Olli Lehto, *Mathematics Without Borders: A History of International Mathematical Union*, Springer Verlag, New York, 1998, ISBN 0-387-98358-9.) On page 17 of Lehto's book we can read: 'Picard was a chief architect of the science policy that barred Germany and other Central Powers from the *IRC* and its *Union*'.

On page 225 of the book B. Bongiorno, G. Curbera, *Giovanni Battista Guccia: Pioneer of International Cooperation in Mathematics*, Springer, 2018, 301 pages, ISBN 978-3319786667 we can read: '...the Circolo still seemed, to be an isle of fraternal coexistence. However, several important mathematicians in the Circolo's Steering Council were also in charge of enforcing the exclusion policies in the new scientific-political structures of the time. Two notable examples of this were Émile Picard, who was president of the *IRC* from 1919 to 1931, and Charles de la Vallée Poussin, who was president of the *IMU* from 1920 to 1924. Picard and De la Vallée Poussin, among others, pushed for the expulsion of Germans from the Circolo's Steering Council.' Let us remark that Charles de la Vallée Poussin later distanced himself from the IMU's discriminatory policy.

institution should be equal to at least nine tenths of the nationwide proportion of the given race or nationality.[43]

It is clear that this Act was intended against the Jews, because it was they who accounted for a highly above-average percentage of educated people. Together, all these factors caused a great emigration wave of scientists, especially Jewish. It was a gigantic brain drain. A significant part of educated expatriates then formed the history of the 20th century, most of them in the USA. Fortunately, not everyone left. Between 1870 and 1914, a considerable community of intelligentsia formed in Hungary. For them, this post-war situation was not acceptable, but they did not leave because their ties to Hungarian national culture were strong.

6. From Riesz and Fejér to Erdős

With a few exceptions, all the above-mentioned mathematicians completed at least part of their studies at foreign universities. Now let us recall those whose mathematical work (both in teaching and in scientific research) did not begin until the early 20th century. It is a very large group comprising extremely high-quality individuals — and it is not a coincidence. It was already the result of the careful but uncompromising work of the afore-mentioned older generation that had brought up a whole range of truly excellent high school teachers and many university professors. So even if someone studied only in Hungary during this period, they could already have a good mathematics–physics education. The very first two in this list represent exceptional quality.[44]

Frigyes Riesz (1880–1956) is certainly one of the greatest personalities of Hungarian mathematics. He not only introduced and promoted various mathematical branches in Hungary, but also created them. Riesz studied at the universities of Zürich and Budapest for three years, and finally in Göttingen for two years. After a short high-school career he got to the University of Kolozsvár, which moved through Budapest to Szeged after WWI. There, Fr. Riesz, A. Haar and B. Kerékjártó created an extremely powerful mathematical workplace that became world-famous. Together they founded the outstanding *Acta Scientiarum Mathematicarum* journal, which continues to be published even today. Frigyes Riesz is one of the founders of functional analysis (together with Stefan Banach and Maurice Fréchet). Named after him are, e.g., *Riesz's lemma*, *Riesz representation theorem*, *Riesz space*,

[43]The *Numerus Clausus* Act was introduced in Tzarist Russia as early as in 1887. In addition to Hungary, this Act was introduced also in the Netherlands, Yugoslavia and Canada in 1920, Romania and Austria in 1926, Germany in 1933, and in Poland in 1937. During the Second Republic, it was also introduced in Czechoslovakia in 1938. From 1920, the Act was also applied in some universities in the USA (Harvard, Columbia, and Yale). One of the cruellest anti-Jewish laws was the *Jewish Code* of 9 September 1941 issued for the Slovak Republic.

[44]Note that the choice of names is subjective, since it cannot be different given the planned scope of the book.

Riesz-Fischer theorem, *Frigyes and Marcel Riesz theorem*, etc. His best-known students included János Aczél, Tibor Radó, Alfréd Rényi, and Béla Szőkefalvi-Nagy. In 1952, Frigyes Riesz and Béla Szőkefalvi-Nagy published the French-written textbook *Leçons d'analyse fonctionnelle*, which is still considered classical, has achieved many editions and reissues and has been translated into many languages. Frigyes Riesz was a corresponding member of the Hungarian Academy of Sciences from 1916 and a full member from 1936. In 1926, the Germans, Austrians, Hungarians and Bulgarians were allowed to renew their participation in the ICM, so Frigyes Riesz gave a plenary lecture at the 1932 Congress in Zürich (but he also lectured at other congresses).[45]

Lipót Fejér (1880–1959). After graduation he enrolled at the Faculty of Mechanical Engineering of the Budapest Technical University. Here he attended lectures by personalities such as Kőnig, Rados and Kürschák. However, for the second semester, he enrolled in the teacher program and the very next year at the University, where Beke and L. Eötvös lectured. Fejér spent his third university year (1899–1900) in Berlin, where he was taught by leading German mathematicians Frobenius, Fuchs and Schwarz. At one lecture, Schwarz proved one theorem with six reflections and immediately after the lecture Fejér showed him that two reflections were enough. No one has ever found a shorter proof since then. In another lecture, Schwarz noted that the problem of the Fourier series summation is closed. However, Fejér disagreed, replacing the partial sums with their arithmetic means and showing that the series of continuous functions would remain uniformly convergent. The article *Sur les fonctions intégrables et bornées* in the *Comptes Rendus* journal of the French Academy of Sciences made the 20-year-old Fejér immediately famous in 1900. With this work he also laid the foundations of modern theory of trigonometric series and it influenced the development of various areas of mathematics. He spent the winter semester of the 1902–1903 school year in Göttingen, where he attended lectures by D. Hilbert and H. Minkowski, and the summer semester in Paris, where he attended lectures by E. Picard and J. Hadamard. After returning home he led the exercises at the University of Budapest until 1905, and then accepted an invitation to the University of Kolozsvár, where he obtained habilitation and in the same year he got a position as an assistant to Professor Gyula Farkas. Fejér was elected a corresponding member of the Hungarian Academy of Sciences in 1908 and three years later he was appointed full professor in Kolozsvár. In spite of this newly acquired professorship, in the same year he accepted an invitation to the University of Budapest where he led one of the mathematical departments until his death. During this period, Fejér began to engage in interpolations and was awarded the Hungarian Academy of Sciences Grand Prize for his achievements in this area. In addition to developing and

[45]Some of his works are included in References (Riesz, 1904, 1944–1945, 1952; Riesz and Riesz, 1916). More detailed information can be found, e.g., in Császár (1998), Halmos (1981) and Vekerdi (1996).

expanding his work in the field of trigonometric series, he also achieved fundamental results in the theory of complex functions. Together with Riesz he produced the shortest proof of the basic theorem about conformal mappings.[46] Fejér was nominated as a full member of the Hungarian Academy of Sciences as early as in 1922, but as a result of the political situation in the post-war years he became a member of the Hungarian Academy of Sciences as late as in 1930. He published the results of his mathematical research in 109 articles.[47]

At the Fifth International Congress of Mathematicians in 1912, Fejér was asked to accept the post of Vice-chairman of the Sixth International Congress of Mathematicians, which was due to take place in Stockholm in 1916. However, this congress did not take place in 1916 due to WWI, and after this war, Hungarian mathematicians got isolated from the rest of the world. Nevertheless, Fejér was one of four European scientists invited to the 1933 Chicago World's Fair. On this occasion, Fejér and the Danish physicist Niels Bohr received an honorary doctorate at Brown University in Providence and, after the fair in Chicago, Fejér took a round-trip lecture tour across America at the invitation of 15 universities. In 1940, both Lipót Fejér and Frigyes Riesz were invited to deliver a plenary lecture and chair one section at the International Congress of Mathematicians in Cambridge. This time the congress was not held because of World War II, and it was postponed until 1950 (in Cambridge, USA). The organizers reiterated the invitation for Fejér and Riesz. However, it was a period of the so-called Cold War, Soviet mathematicians refused to participate,[48] and so even the mathematicians from the socialist countries (including Hungary) could not attend the congress.

Fejér was a senior (supervisor) to about 40 PhD students, many of whom became leading world mathematicians in various parts of the world. Let us mention just a few of them: Mihály Fekete (Jerusalem), György Pólya (Stanford), Kornél Lánczos (Dublin), Gábor Szegő (Stanford), Tibor Radó (Ohio), János Neumann (Princeton), Pál Turán (Budapest), Pál Erdős (a globetrotter, and one of the greatest figures in mathematics of the 20th century), and also the so-called *Big Five*, i.e., the PhD students born in 1924 who all became members of the Hungarian Academy of Sciences: János Aczél (Debrecen, Waterloo), Ákos Császár (Budapest), László Fuchs (Budapest, New Orleans), István Sándor Gaál (Minneapolis) and János Horváth (Maryland).

[46]They did not publish this proof, but T. Radó (1895–1967) published it with their permission on pages 241–242 of the work Über die Fundamentalabbildung schlichten Gebiete, *Szegedi Acta* **1**(1922/23), pp. 240–251.

[47]Leopold Fejér: *Gesammelte Arbeiten I, II.* (ed. Turán Pál), Akadémiai kiadó, Budapest, 1970, 1722 pages. In this bilingual work — written in German and Hugarian — Turán discusses in detail the scientific contribution and usability of Fejér's work.

[48]More precisely: The Soviet government did not allow them to participate. In an official letter from the president of the Soviet Academy of Sciences, they wrote that 'Soviet mathematicians are so busy with their daily work that they have no time to attend the Congress'.

Many Hungarian scientists left the country after WWI. Even Fejér received many offers to head a department abroad, but he refused all offers. These invitations were repeated before WWII, when the threat of fascism was increasingly topical. He refused all offers again, because (in his own words) he could not imagine living outside his homeland, outside his culture, which he knew so well and loved immensely. When the fascists (Nyilasi) took power in Hungary on 16 October 1944, Fejér was 'moved' to a Jewish ghetto on Tátrai Street. The ghetto was formally under the protection of the International Red Cross. Despite this 'protection", all residents of the house where he lived in the ghetto were expelled on Christmas 1944 to march to the Danube embankment where mass executions took place.[49] Fortunately and luckily, a Hungarian officer recognized Fejér and found enough courage to pull him out of the marching crowd despite the raging madness all around. Fejér survived and spent the rest of the war in his unheated faculty office, as his apartment was bombed and destroyed. He was saved from death by starvation by monetary donations, but mainly by food packages from friends and former students (e.g., Erdős and Pólya).[50]

Dénes Kőnig (1883–1944) studied at the Universities of Budapest and Göttingen. He received his doctorate in 1907 under Kürschák's supervision at the Technical University, where he also later taught. He was a specialist in the then emerging graph theory and also the author of the first textbook *Theorie der endlichen und unendlichen Graphen*, Leipzig, 1936, on this subject. This marked the beginning of graph theory and its own branch of mathematics. 'Kőnig's theorem' (in graph theory) and 'Kőnig's lemma' are named after him. He lectured at the Mathematical Congress in Cambridge (1912). From 1933 he served as Secretary of the Society. Kőnig got his professorship in 1935. His most famous student was Tibor Gallai, but Kőnig's activities and lectures played a key role in the growth of graph theory work by P. Erdős, T. Gallai, Gy. Hajós, T. Szele, P. Turán and many others. Dénes Kőnig committed suicide (by jumping from the height of his house) in 1944 when he was forced to move to a ghetto.[51]

Alfréd Haar (1885–1933) was taught by legendary László Rátz at secondary grammar school, and then he began to study at the Technical University (chemistry), but he also studied at the University of Budapest, and later in Göttingen, so he was taught by excellent teachers such as Eötvös, Kürschák, Carathéodory, Hilbert, Zermelo, and Klein. He received his doctorate under Hilbert's supervision in 1909. In the same year he become private docent at the university. In a 49-page thesis on systems of Sturm–Liouville functions and spherical functions, Haar introduced the

[49]Even today, the shoes belonging to the many murdered ones are still displayed on the riverbank in front of the Parliament.

[50]Some Fejér's works are included in References (Fejér, 1900, 1902). More detailed information can be found, e.g., in Bálint (2019), Császár (1998); Szász (1960); Vekerdi (1996).

[51]Some of his works are included in References (Kőnig, 1936, 1941). More detailed information can be found, e.g., in Császár (1998); Gallai, (1978); Vekerdi (1996).

widely used Haar orthogonal systems. From 1912 he was Professor at the University of Kolozsvár. He generalized the derivation of basic equations of calculus of variations. He is a co-founder of the journal *Acta Scientiarum Mathematicarum.* Haar was an excellent lecturer and lectured also at the Congress in Bologna (1928).[52]

Gyula Szőkefalvi-Nagy (1887–1953) graduated from the so-called 'Göttingen of the Monarchy', e.g., the University in Kolozsvár. He received his doctorate under Schlesinger's supervision in 1909, and then received support for his studies in Göttingen for 1911–1912. In particular, Szőkefalvi-Nagy contributed to the development of algebraic curves theory. In the interwar years he headed the Department of Geometry in Szeged. He was elected a corresponding member of the Hungarian Academy of Sciences in 1934 and a full member in 1946.[53]

Jenő Egerváry (1891–1958) received his doctorate in 1914 at the Pázmány Péter University in Budapest, where he studied under the supervision of Lipót Fejér. He dealt mainly with differential equations of mathematical physics and with technology in general. His article *Matrixok kombinatorikus tulajdonságairól* [About Combinatorial Properties of Matrices] written in 1931 was about the generalization of Dénes Kőnig's theorem. Twenty years later this article became the basis of the algorithm for the so-called *Hungarian method* in graph theory. In 1938, Egerváry was appointed *Privatdozent* at the Pázmány Péter University in Budapest. In 1941, he became full professor at the Technical University of Budapest, and in 1950 he was appointed Chairman of the Scientific Council of the Research Institute for Applied Mathematics of the Hungarian Academy of Sciences.[54]

Béla Kerékjártó (1898–1946) graduated from the University of Budapest, and then taught at the University of Szeged, and later in Budapest. He specialized in topology, where he discovered many basic theorems and, moreover, he simplified the proof of many other theorems. He lectured at the congresses in Bologna (1928) and in Zürich (1932). In 1929–1938 he lectured several times in Göttingen, Barcelona, Princeton and Sorbonne.[55]

László Rédei (1900–1981) taught for 20 years at secondary schools after graduating. From 1940 he was Professor at the University of Szeged, where he became a central figure in algebra research. His most important work is probably *Ein Satz über die endlichen einfachen Gruppen* (1950), which — in addition to its theoretical significance — also had consequences on coding theory.[56]

[52]His works Haar (1910, 1911) are included in References. More detailed information can be found, e.g., in Császár (1998); Vekerdi (1996) and in many others.

[53]His works Szőkefalvi-Nagy (1923, 1947) are included in References. More detailed information can be found, e.g., in Császár (1998); Vekerdi (1996).

[54]His work Egerváry (1931) is included in References. More detailed information can be found, e.g., in Császár (1998); Kántor (2012); Vekerdi (1996) and in many others.

[55]Some of his works are included in References (Kerékjártó 1923, 1944). More detailed information can be found, e.g., in Császár (1998); Vekerdi (1996).

[56]His work Rédei (1950) is included in References.

Pál Szász (1901–1978) graduated from the University of Budapest. He was Fejér's co-worker at the University of Budapest from 1933, and Professor from 1958. P. Szász wrote two significant books on differential and integral calculus in 1935. His most important achievements are those regarding Hilbert's axiomatics of hyperbolic geometry from 1973 (see Szász, 1958).

László Kalmár (1905–1976). After graduating from the University of Budapest he got into an excellent mathematical environment in Szeged. His advisor was Fejér. Upon completing his doctorate in Budapest, Kalmár took up a position at the University of Szeged. He began his career as a research assistant to Haar and Riesz. Kalmár was appointed a full professor at the University of Szeged in 1947. He founded Szeged's Cybernetic Laboratory and the Research Group for Mathematical Logic and Automata Theory. He got established in two new fields — game theory and algorithm theory. He wrote on theoretical computer science, including programming languages, automatic error correction and the connection between computer science and mathematical logic. Kalmár was elected to the Hungarian Academy of Sciences in 1949.[57]

Ferenc Kárteszi (1907–1989) graduated from the University of Budapest. As a result of the death of his older brother in 1926 and his father in 1928, he had to take care of his mother. In 1931–1940 Kárteszi taught at secondary school in Győr and obtained his doctorate in 1933 in Debrecen. With the support of Lipót Fejér, he spent the school year 1936/1937 in Bologna. He was a soldier in World War II. Kárteszi taught again after returning from the war prison at ELTE as Professor and head of the department from 1950. Kárteszi was interested mainly in geometry (under the influence of Professor Klug), and partly in combinatorics and graph theory.[58]

Ottó Varga (1909–1969) was born in Szepetnek. Since his father Imre was a Lutheran minister, the family moved first to Zagreb and soon after that to Poprad, where Imre Varga later worked as archdeacon of the Szepes region. Ottó attended the German secondary school in Kežmarok.[59] After his studies at the German University in Prague in 1928–1933 he obtained his doctorate in 1934 under the supervision of L. Berwald. In 1934–1935, Varga collaborated with W. Blaschke in Hamburg, obtained habilitation in Prague in 1937 and lectured there until 1941. At that time, Varga already had several outstanding results in integral geometry, which he applied later to Finsler geometry and geometrical optics. Since he worked closely with his teacher Berwald, the occupation of Prague caused him political inconvenience and he moved to Hungary. He accepted a post at the University of Kolozsvár, where he was

---

[57]Some of his works are included in References (Kalmár, 1926, 1928, 1950, 1957).

[58]The author of this essay had the honour of interpreting Professor Kárteszi's lecture to the Slovak language in Košice in 1968. Some Kárteszi's works are included in References (Kárteszi, 1957, 1966, 1973).

[59]These are the roots of his knowledge of languages — Hungarian, German and Slavic languages.

invited by Béla Szőkefalvi-Nagy, but after receiving his professorship Varga went to the University of Debrecen in 1942. In 1947 he gave lectures at the University of Zürich. He became a corresponding member of the Hungarian Academy of Sciences from 1950, and a full member from 1965. From 1967 he led the group of differential geometry at the Academy. Varga (with Rényi and Szele) founded the *Publicationes Mathematicae* journal in 1949.[60]

György Hajós (1912–1972). His great-grandfather, Adam Clark, was the famous Scottish engineer who built the Chain Bridge in Budapest. Hajós graduated from the University of Budapest in 1935, after which he taught at the Technical University. He completed his doctorate in 1938 under Fejér's supervision and then worked at ELTE from 1949. He was scientifically involved mainly in graph theory and geometry. His most important work was a proof of Minkowski's conjecture with the help of group theory in 1941. However, he also had important results in discrete geometry, graph theory, Bolyai-Lobachevsky geometry and numerical analysis. Hajós was President of the János Bolyai Mathematical Society from 1963 to 1972. He was a corresponding member of the Hungarian Academy of Sciences from 1948 and a full member from 1953.[61]

Tibor Gallai (1912–1992) obtained a doctorate under the supervision of Dénes Kőnig. Before the war he worked as a mathematician for an insurance company and later for a textile firm. Gallai was deployed in a labor camp in 1942–1945. After the war he taught first at a grammar school (his students included Vera Sós and Lajos Pósa, for example), and then at BME, and later he was a scientific associate of the Academy of Sciences. Gallai was scientifically involved in combinatorics and graph theory. He was an advisor to an extraordinary mathematical personality — László Lovász. His grammar school student Vera T. Sós, later a member of the Hungarian Academy of Sciences, said about him: 'Gallai was an extremely precious man, a role model of all role models, yet he was incredibly modest. He was the first to achieve many new results but did not publish them' (for example, theorems such as *Dilworth's theorem*, *Gallai-Sylvester theorem*, etc.) Also known is the Edmonds–Gallai decomposition theorem, proved independently by T. Gallai and Jack Edmonds. Gallai's friendship with Erdős started in the Anonymous group and lasted for his whole life.[62]

Barna Szénássy (1913–1995) received his secondary-school professor diploma in Debrecen in 1936. He first taught in Gyula, and then in 1939–1943 in Ungvár (today Uzhgorod in Ukraine). In 1947, Szénássy returned from a military prison in the Don region, and then taught at the University of Debrecen from 1950. He became one of

---

[60] Some of his 57 works are included in References (Varga, 1936a,b, 1951). More detailed information can be found, e.g., in Kántor (2010).
[61] Some of his works are included in References (Hajós, 1934, 1941).
[62] Some of his works are included in References (Gallai, 1961, 1978; Erdős, 1939; Erdős *et al.*, 1992, 1996).

the most important promoters of the history of Hungarian mathematics. Szénássy wrote 11 books and 44 scholarly studies.[63]

Béla Szőkefalvi-Nagy (1913–1998), son of the above-mentioned Gyula Szőkefalvi-Nagy, graduated from the excellent University of Szeged (relocated from Kolozsvár). He received his doctorate in 1937 and professorship in 1940. He was scientifically involved in functional analysis in a broader sense. Béla Szőkefalvi-Nagy with Frigyes Riesz wrote an extremely successful book on functional analysis. The book was translated into many languages. He further developed these ideas with Ciprian Foiaş. Béla Szőkefalvi-Nagy was a corresponding member of the Hungarian Academy of Sciences from 1945 and a full member from 1956.[64]

László Fejes-Tóth (1915–2005). After graduating from Pázmány Péter University in Budapest, he completed military service in 1939–1940. Then, until 1944, he was an assistant at the Geometric Institute of the University of Kolozsvár. After the war he taught at a grammar school until 1948. Fejes-Tóth obtained habilitation in 1946 and the title of Doctor of Science in 1952. He was a visiting professor at the University of Freiburg (1960–1961), then at the Universities of Wisconsin, Washington and Ohio in 1963–1964, and at the University of Salzburg from 1969 to 1990. He was elected a corresponding member of the Hungarian Academy of Sciences in 1962 and a full member in 1970. László Fejes-Tóth was Director of the *Mathematical Institute of the Hungarian Academy of Sciences* from 1970 to 1983. His most significant result in the field of analysis is about uniform convergence of Cauchy and Fourier series. L. Fejes-Tóth is considered a co-founder of the so-called Hungarian school in the field of discrete geometry, and it was these results that made him most famous: convex bodies, their extremal packings and coverings. He was a long-term chairman of the Bolyai János Mathematical Society.[65]

Alfréd Rényi (1921–1970) was taught by Fejér at the University of Budapest in 1940–1944. Rényi obtained doctorate under the supervision of Frigyes Riesz in 1947, and then taught in Debrecen. Rényi founded the *Mathematics Research Institute of the Hungarian Academy of Sciences* (today *Rényi Institute*) in 1950. This Institute was led by Rényi until his early death. He achieved results in combinatorics, graph theory and in probability theory. He wrote more than 350 papers in total, 32 of them in collaboration with Paul Erdős. The best-known of these papers are those introducing the Erdős–Rényi model of random graphs.[66]

---

[63]Some of his most significant works are included in References (Szénássy, 1965, 1970, 1975, 1995). More detailed information can be found, e.g., in Kántor (2014).

[64]Some of his works are included in References (Szőkefalvi-Nagy, 1967, 1953, 1959).

[65]Some of his most significant works are included in References (Fejes-Tóth, 1943, 1946, 1953, 1967).

[66]Some of his works are included in References (Rényi, 1970, 2004; Erdős, 1960), and a full list of his works can be found in *Matematikai Lapok* **21**(1970), no. 1–2, pp. 1–18. More detailed information can be found, e.g., in Turán (1970).

János Surányi (1918–2006) studied at the University of Szeged in 1937–1941, and then he was deployed for forced labor in 1942–1945. After the war Surányi taught in Szeged, and then at ELTE in 1951–1988. His main work ([ES] in co-authorship with Erdős) — a book entitled *Válogatott fejezetek a számelméletből* [Topics in the Number Theory] — was translated to many languages.

Pál Turán (1910–1976) graduated from the University of Budapest and received his doctorate under the supervision of Fejér in 1935. Until 1938 he made his living as a private teacher, and then as a high school professor.[67] Turán was deployed for forced labour during World War II. He was a visiting professor at the University of Copenhagen in 1947 and lectured at the Institute for Advanced Study (Princeton, USA) in 1948. Turán was Professor at ELTE from 1949, a corresponding member of the Academy of Sciences from 1948, and a regular member from 1953. Turán's scientific scope was very broad. His most significant scientific result is known as *the power sum method.* (This extremely profound result cannot be described to non-professionals without distortion.) Turán's studies on *interpolation* are a direct continuation of the Fejér School. One of his first articles on number theory concerned the Hardy–Ramanujan theorem, but due to Turán's approach, it grew into a mathematics branch called *probabilistic number theory.* He was the founder of functional algebra, which grew into the *analytic theory of algebraic equations.* Turán is also the founder of *extremal graph theory.* This originated as a generalization of the original question: What is the minimal number of edges of an n-vertices graph so that this graph contains a triangle? During 46 years of collaboration with Erdős they published 30 joint articles. Turán was undoubtedly one of the best Hungarian mathematicians of the 20th century.[68]

7. Pál Erdős

Sections 3–6 list those who spent their entire active lives in their homeland — Hungary. Section 8 introduces those who left Hungary, spent their entire active lives abroad, and eventually returned home at the end of their lives. In mathematical terms, Erdős represents the border between these two sets of people: he was always traveling, he was not home anywhere and at the same time he was home everywhere. Erdős spent a considerable part of his life outside Hungary (mainly for political reasons), but he was always happy to return home quite often — whenever he could, as his beloved *anyuka*[69] lived in Hungary. Erdős's scientific scope was very broad, fundamentally affecting large number of areas of mathematics: number

[67]However, his salary was so small that the following conversation took place. Businessman Weissfeld, Vázsonyi's father, said: 'Mister Turán, you have a monthly salary to cover one year!' Turán replied with a smile: 'One day I will have a yearly salary to cover one month.'

[68]Some of his works are included in References (Turán, 1950, 1951, 1970, 1984; Erdős, 1934).

[69]This is how he addressed his mother all his life.

theory, graph theory, combinatorics, discrete geometry,[70] probability, and analysis. Erdős was undoubtedly one of the greatest mathematicians of the 20th century. The number of his works — over 1500 — is only comparable to that of Leonhard Euler.

Pál Erdős (1913–1995). The 1920 *Numerus Clausus Act* was 'softened' in 1928, allowing Erdős to study in Budapest at the University as well as at the Technical University at the same time.[71] Thus, his teachers included such personalities as Lipót Fejér, József Kürschák and Dénes Kőnig. Erdős finished his PhD thesis under Fejér's supervision when he was just 19 years old, but he had to wait until graduation from the university. So, Erdős obtained his doctorate 'as late as' a 21-year-old in 1934.[72]

Erdős went to Manchester to L. Mordell in the same year. *Inter alia*, the famous combinatorics Erdős–Ko–Rado theorem was created in Manchester in 1938 but published as late as in 1961. Erdős spent the years 1938–1939 at the Institute for Advanced Study (Princeton). There, together with Aurel Wintner and the Polish mathematician Mark Kac, they laid the foundations of probabilistic number theory. Stan Ulam invited Erdős to be involved in the top secret Manhattan project. Erdős wrote an open postcard to Los Alamos, addressed to Péter Lax: 'Dear Péter, my spies report that Uncle Sam is working on an atomic bomb. Tell me if it's true.' János Kemény and Péter Lax, both very young, born in Budapest in 1926, invited Erdős to the renowned *La Fonda restaurant* in Santa Fe. The conversation was

[70]Discrete geometry investigates combinatorial as well as metric properties of geometric objects. Most questions and issues concern, in a sense, the extreme arrangement of finite sets of points, straight lines, hyperplanes, circles, spheres, angles and the like. In the 1930s, a very strong group of Hungarian promoters of this mathematics branch started to emerge: Erdős, Turán, L. Fejes-Tóth, etc. This group considerably expanded especially after WWII. Perhaps the best overview of discrete geometry problems can be found in Brass *et al.* (2005), but the books Bezdek (2003) and Matoušek (2002) are also useful.

[71]The *Anonymous* group formed around Erdős and Turán. It consisted of students, especially mathematicians, who were formed by teachers such as Lipót Fejér, József Suták, József Kürschák, Dénes Kőnig, Frigyes Riesz and Alfréd Haar. The youth met regularly in the city park of Városliget near the statue of *Anonymous* (hence the name of the group) and made trips to the Buda hills, talking about mathematics. Fejér and János Neumann often accompanied them. The life stories of the group members were significantly influenced by historical events, then WWII and its consequences. Dezső Lázár (1913–1943) and Géza Grünwald (1910–1942) did not survive the war. György Svéd and Márta Wachsberger, as well as György Szekeres (1911–2005) and Eszter Klein (1910–2005), got married (this second marriage initiated the name *Happy End problem*). Turán and Gallai stayed in Hungary and became the determinants of Hungarian mathematical life. Erdős emigrated to the USA and became a globetrotter. Instead of the then dominant continuity (in analysis), members of the *Anonymous* group, under the influence of Dénes Kőnig, turned their attention to discrete mathematics, graph theory and number theory. Instead of developing theories, they dealt with the issue called 'problem solving'.

[72]Later, Erdős, by the weight of his authority, achieved that László Lovász, one of his best students was awarded an academic title by the Academy as early as in the fourth year of university, i.e., a year before completing the study there.

in Hungarian, but Erdős did not hold up and asked *in English* pretty loudly: 'And how is work going with the A-bomb?' He wrote in a letter to Ed Teller that he wanted to return to Hungary after the war, so Teller did not involve him in the Manhattan project.

Erdős never had a lot of money, but if he heard that a student needed money for his studies, he opened not only his heart but also his wallet and sent him a check. Every time he was in Madras, he sent his salary to Srinivasa Ramanujan's poor wife. Although Erdős did not know him or her personally, Ramanujan's work impressed Erdős enormously. He attended conferences and as a grandmaster in a chess simultaneous game he moved from group to group, listening to their discussions, saying a few essential sentences, and then moved on. He was able to switch ideas in a split second and immediately find and use all the relevant information about every problem he had ever dealt with. Occasionally, after a 5-minute conversation at the conference break, he handed a young mathematician a piece of paper that indicated the main points of the proof for the problem the guy was dealing with. Of course, he had to have this prepared, but there were cases when he wrote something on a piece of paper out of the blue, right during such a short conversation, and the companion found the main idea of the proof for it so he could write an article.

Figure 10. Pál Erdős (1973).[73]

[73]The picture is from János Pach's archive.

Erdős's approach to mathematics was characterized by asking questions. He used to say that if we examine carefully selected trees, we would find the whole forest. In doing so, he suggested that well-asked questions were of the utmost importance, because if the answers were found to these questions, then a whole branch of mathematics would emerge. Not everyone agreed. For example, Saunders Mac Lane underestimated the approach to mathematics known as 'problem solving' and considered it secondary. Mac Lane believed that lectures on theory can span even more semesters. However, Erdős had the ability to raise extremely simple questions, which concerned the very essence of the problem exactly as a result of their almost trivial formulations. Mac Lane did not realize that if something really new was to emerge, it would certainly not emerge as a complete theory, but usually as a solo question. And then the answer, of course. And when there are many questions and many answers as well, a theory emerges.[74]

In 1948, Erdős finally got to Budapest to see his beloved *anyuka*. He found that four of his five uncles and aunts got lost in the Holocaust, just like Géza Grünwald, Dezső Lázár and many of his friends and acquaintances. Dénes Kőnig (author of the first textbook on graph theory) committed suicide when he was forced to move to a ghetto. Erdős also learned that Turán was stopped by a Russian patrol after the war, because on Stalin's order they sent people to gulags for *maleňkij rabot* (a little work). Turán lost his identification documents a few days earlier in those turbulent times. Back in 1935, however, Turán published a joint article with Erdős in Tomsk and, having found one copy in his bag, handed it to the patrol instead of his ID documents. The Russian soldier was so moved when he saw an article written in Russian that he released Turán. Turán then reported the incident to Erdős as a 'surprising application of number theory in practice'. But in order to avoid growing political pressures, Erdős quickly returned from Budapest to the USA. János Neumann, as chairman of the AMS, presented Erdős with the Cole Prize for results on number theory in 1951.

In 1953, it seemed that Erdős would stay in the USA for good. Arnold Ross, Head of the Department of Mathematics at Notre Dame University (South Bend, Indiana) gave Erdős a generous offer for one year: he was to teach only one lesson per week for upper grades, and assigned him an assistant to take over the job when

[74]With his questions — and of course answers — Erdős created many new and enormously useful areas of mathematics. As an example, we can ask the following question: 'What happens if we add one edge to the graph randomly?' As long as the number of edges does not reach half the number of vertices, basically nothing happens — the graph consists of several smaller components – the so-called connected subgraphs. But then a miracle happens: the *random addition* of a few edges makes the graph almost connected. It is very similar to the so-called phase transitions — formation of a traffic jam or freezing of water. After the introductory article by Erdős and Rényi, several hundred other works emerged, many books were written and conferences held, because it is about network analysis and, ultimately, about life structures analysis.

Erdős would be away. Erdős was a strict atheist but he especially enjoyed discussions with priests at the Catholic Notre Dame University. In his own words, he was only disturbed by the presence of too many 'plus' symbols. After one year, Ross gave him an offer that could not be refused: a job at Notre Dame University for an indefinite period under the same conditions. Erdős politely thanked and refused.

His mission as a 'travelling mathematician' continued. Essentially, even if he would have accepted the post, he would not have stayed in the USA. Those days, Joseph McCarthy intended to cleanse America of red danger. The 'red witch hunt' was so strong that a friend of Erdős in Los Angeles did not allow him to use his telephone to call *anyuka* in Budapest on her 73rd birthday, because he was afraid of a record on the telephone bill proving a call to a communist country. Even the British physicist Paul Dirac, Nobel Prize winner, was not allowed to enter the USA. Consequently, the American Psychological Society held its congress in Montreal, Canada, just to prevent foreigners from being subjected to McCarthyist humiliation. But even scientists living in America had trouble attending overseas conferences. Erdős wanted to go to Amsterdam to a congress in 1954 and, because he did not have US citizenship, applied for a visa with the possibility of returning. He didn't get the visa, but he left anyway. He wanted to go from the Netherlands to Britain but was not allowed to enter the country. So, after the congress, he went to Israel and got a job as permanent visiting professor at Technion in Haifa with a small but steady income. When he received the Wolf Prize in 1983, which, in addition to the (basically Nobel-like) fame, was $50,000 at that time, he kept only $720 out of the total amount. In honor of his parents and as thanks for 'giving him shelter' in 1954, Technion Haifa received $30,000 to set up a postdoctoral foundation. He distributed the remainder to competition prizes and study costs for young mathematicians.

Erdős became a corresponding member of the Hungarian Academy of Sciences in 1956, and a full member in 1962. Alfréd Rényi secured him a permanent position at the Academy, so he could receive some regular money. The Hungarian government granted him a consular passport so he could move freely. Since then he kept returning to Hungary several times a year and also took his *anyuka* on his travels from 1964.

Erdős ceaselessly sought out young talents. One such man — Béla Bollobás — highly emphasized Erdős's ability to adapt the scientific level to the audience capability. He discussed such combinatorics, geometric, and number theoretic problems that required no special knowledge, which extraordinarily raised interest by addressing open (i.e., unsolved) problems. And he was always ready to write a joint article with anybody, if that 'anybody' contributed at least a little bit to the effort. Being Erdős's co-author was an enormously powerful motivation for young people. For example, Béla Bollobás wrote his first article with Erdős as a 17-year-old. In 1956, Erdős travelled to Szeged to see a young student named András Hajnal. He has

Erdős number 1/56, which is the second smallest number (the smallest Erdős number is 1/62 and it belongs to András Sárközy).[75] Among the many young talents supported in this way are Vera T. Sós, János Pach, László Lovász, Zsolt Tuza, Zoltán Füredi, Carl Pomerance, Joel Spencer, Vojtěch Rödl, etc. However, one cannot forget the name that is much less known in the scientific community, but whom Erdős considered the greatest talent he had ever met: Lajos Pósa.

In 1959, Rózsa Péter[76] drew Erdős's attention to an 11-year-old boy who knew everything in mathematics that he could learn at secondary school. Of course, Erdős invited him to lunch with Madam Professor Péter. While Pósa was eating the soup, Erdős gave him this task: 'Prove that if we choose $n+1$ integers between 1 and $2n$, there will be two coprime between them.' Pósa's hand holding the spoon stopped in the air, and he pronounced a proof consisting of three words in Hungarian: '*Kettő közülük szomszédos.*' In English: 'Two of them are adjacent.'[77] But when Erdős once contemplated about this task as a boy, it took him almost 10 minutes to solve. From then on, Erdős did everything he could for Pósa, sending him tasks from abroad and seeking him while he was in Budapest. When Pósa was 13, Erdős explained to him the infinite case of the Ramsey theorem. Pósa understood what was going on in 15 minutes, went home and solved it before he went to bed in the evening. Pósa wrote his first joint article with Erdős as a 14-year-old. 'However, I never awoke an interest in geometry in Pósa,' said Erdős sadly. He was only concerned with what he was interested in. But he was really great in what he was interested in, i.e., the instruction process.[78]

Erdős took very good care of his *anyuka*. In 1971, she needed to go to hospital in Calgary. Essentially, Erdős did not want to believe that his 90-year-old *anyuka* could be seriously ill and that she was dying. After her death, Erdős began working 19 hours a day. Strong coffee was no longer enough, so he started taking pills such as amphetamine, benzedrine and the like, and combined them

[75] A person who wrote a joint article with Erdős has Erdős number 1. If he/she does not have Erdős number 1, but wrote an article with someone who has Erdős number 1, then he/she has Erdős number 2, etc. In order to differentiate between the owners of Erdős number 1, it was decided that if someone wrote k articles with Erdős, then he/she has Erdős number $1/k$.

[76] Author of the highly successful book Péter (1963), which has been translated into many languages, Professor at ELTE, and also lectured at the Congress in Zürich in 1932.

[77] Obviously, because adjacent numbers are coprime.

[78] Let us add that Pósa began teaching mathematics at Budapest's most prestigious mathematics gymnasium, *Fazekas gimnázium*, and after a few years at ELTE and the Academy of Sciences, Pósa returned to work as a high school teacher — fortunately. He taught students such as László Babai, György Elekes, Péter Komjáth, Imre Ruzsa, Rácz Béla András, etc. Perhaps not everyone is familiar with all these names, but these five of his students alone won eight gold medals (of which four times for full 42 points) at the IMO together with three silver and one bronze medals, and four special prizes — the most difficult to win. If you are able to teach such extraordinary student, then you must be a top-notch teacher.

with strong coffee... But he didn't take much and only took pills when he really worked very long hours. At that time, Ron Graham and his wife Fan Chung took over the main care of Erdős. 'I don't carry a lot of money with me, but I usually take about $1,000 when travelling,' Graham used to say. Erdős quickly found this out and used to borrow from him 300 to 400, most often for his Hungarian friends who needed money urgently. However, he always paid his debt. Graham wanted Erdős to quit taking pills and called him an addict. They bet $500 that Erdős would not last one month without pills. Erdős won the bet and said: 'Ron, it turns out I am not an addict. But you threw mathematics one month backwards because I did not have a single new idea for the whole month.' And he continued taking pills...

Erdős announced many prizes for problem solving: from $25 to several thousand dollars. The rewards still exist; Ronald Graham is the administrator of solutions. A solver can either get an original cheque signed by Erdős before his death (as a souvenir only, cannot be cashed) or a cashable cheque from Graham.

Erdős's heart began to worry him. On the one hand Erdős slowed the pace, but on the other hand he visited Atlanta, Memphis, three cities in Texas, Bell Laboratories, Rutgers University in New Jersey, New Haven, Baton Rouge, Colorado, France and Germany — all in the period from late 1995 to early 1996. During this long trip, he was taken to a hospital in Kalamazoo. When Erdős resumed consciousness at the intensive care unit, he turned to the mathematicians standing around his bed and said: 'Ralph (Faudree, with Erdős number 1/50), I was thinking about the problem we had talked about. Have you tried it from that side?'

Erdős died in Warsaw on 20 September 1996 after a heart attack.

If the Martians wanted to make contact with the Earthlings, Pál Erdős would be the most appropriate person on behalf of our planet's inhabitants. He could speak the language of the Universe, the language of natural numbers. That is what the *Economist* wrote in the obituary on 5 October 1996.

This obituary was nice. However, other obituaries and articles were also written that angered mathematicians with enormous injustice as they did not appreciate Erdős's goodness and generosity. In response to obituaries saying something like 'Died the man who loved numbers only', Ron Graham told a story at Erdős's funeral and stressed that there were dozens of such stories. 'I had a friend, Glen Whitney, a talented young mathematician who was admitted to Harvard, but could not afford to pay school fees. Whitney got some money but needed more. When I told Pali, he arranged a meeting with Whitney and lent him $1,000. It was a huge sum from a man who rarely had more than $30 with him and had almost no savings. Erdős made an agreement with Whitney that he would repay the loan once he would have enough money. A few years later, Whitney graduated from Harvard, taught in Michigan and asked me to tell Erdős that he was already able to repay the loan. I informed Erdős and he just said: 'Tell Whitney to do with the thousand dollars exactly what I have done.'

Of Erdős's over 1,500 works, it is too difficult to select 'some significant ones.' We have therefore included a very few of them in References, mainly because of his co-authors. A good overview of 'Erdős mathematics' can be found, for example, in Halász *et al.* (2002), but readers can also find a lot of information in Brass *et al.* (2015). There are numerous books and articles written about Erdős in which an interested person can find more detailed information. We can recommend, for example, Babai and Spencer (1998); Schechter (1999); Bálint (2013, 2019); Graham *et al.* (2013).[79]

8. Maybe the Martians?

As we have mentioned earlier, this section is about those mathematicians who spent their whole active lives abroad after leaving Hungary, and eventually returned home at the end of their lives. However, we will mention not only mathematicians but, marginally, other scientists as well[80] because their departure meant a considerable loss for Hungarian science due to their extremely high profile and total number. They shaped the global history of the 20th century outside Hungary. The reasons for this brain drain were the political set-up and economic situation after World War I.

Marcel Riesz (1886–1969) was the younger brother of Frigyes Riesz. He obtained his doctorate under Fejér's supervision. At the invitation of Gösta Mittag-Leffler (1846–1927), M. Riesz left Hungary and worked in Sweden at the University of Lund from 1911. There he raised the entire generation of the best Swedish mathematicians of the second half of the 20th century. M. Riesz is known for his work on summation methods, potential theory, and other parts of analysis, as well as number theory, partial differential equations, and Clifford algebras. He was elected a member of the *Royal Swedish Academy of Sciences* in 1936. M. Riesz raised a considerable amount of money in Sweden for the *Hungarian Mathematical Society*. The money was deposited with banks in Vienna and Paris, so that even after the devaluation of the Hungarian currency after World War I, mathematicians could at least have limited access to foreign books and journals.[81]

[79]Of course we also recommend the film titled *Paul Erdős: N is a number*, a 1993 biographical 57-minutes documentary about the life of the mathematician Paul Erdős, directed by George Paul Csicsery. The film was shot between 1988 and 1991, capturing Erdős in various countries along with some of his numerous collaborators. It covers his unusual career, personal life, and many of his recurring jokes and anecdotes, including that of Erdős numbers. The film won two Gold awards: Gold Apple Award (National Educational Film & Video Festival), and the Gold Plaque Award (Science/Nature, Documentary category, Chicago International Film Festival).

[80]Many of them deserve a separate chapter, but there is vast literature already written on each of them.

[81]Some of Riesz's articles are included in References (Riesz and Hilb, 1924; Riesz and Riesz, 1916). More detailed information can be found, e.g., in Horváth (1975) and also in Gårding (1970), which includes a chronological list of his 59 works as well.

Tódor Kármán (1881–1963) is better known as Theodore von Kármán. He received a degree in mechanical engineering from the Technical University of Budapest in 1902, then taught there for four years. From 1906 he continued his studies in Göttingen under the supervision of Klein, Hilbert and Prandtl. T. Kármán completed his doctorate under Prandtl's supervision in 1908 and became his assistant. Here he discovered the patterns in the water flow behind a ship, today called *von Kármán swirling flow.* From 1912, Kármán was Professor at the leading German university in Aachen, where he founded and led the Aeronautics Research Institute. During WWI he designed and constructed the so-called Kármán-Petróczy-Zurovecz helicopter at the Fischamend airport.

Figure 11. President Kennedy honors Theodore von Kármán.[82]

Since he could not get sufficient financial support for his experiments in Hungary, Kármán returned to Aachen in 1919. There he dealt with the impact of solid obstacles on the flow of liquids and gases, theory of turbulence and design of an aircraft wing profile. In 1926, the California Institute of Technology invited Kármán to organise the *Guggenheim Aeronautical Laboratory* in Pasadena (GALCIT). He left Germany for good in 1930 and became Director of GALCIT. Then he established the *Jet Propulsion Laboratory* (JPL) in 1944 to research rocket/missile technology.

[82]The picture is available on the website https://en.wikipedia.org/wiki/Theodore_von_Kármán [16.9.2019].

His research contributed significantly to the development of American aircraft and rocket technology.

In 1949 he resigned as honorary director of Caltech and settled in Paris. He was president of the NATO Advisory Group for Aeronautical Research and Development (AGARD), and in 1960 he succeeded in establishing the International Academy of Astronautics.

Kármán is known in the world of science, *inter alia*, for *von Kármán vortex street*, *von Kármán constant*, *von Kármán momentum integral*, *von Kármán ogive*, etc.[83] His most significant works have been published in all world languages.[84]

Kornél Lánczos (1893–1974) graduated from the University of Budapest in 1916 and became an assistant at the Technical University. His doctoral advisor was Rudolf Ortvay. He worked as Professor first in Frankfurt and then in Berlin. In 1921, Lánczos met Einstein and was his long-time collaborator afterwards. He was appointed Professor of Mathematics and Physics at Purdue University in the USA in 1931. Later he worked as a mathematician at the National Bureau of Standards, and then as an engineer-developer in Boeing. He discovered several very useful algorithms for numerical computations: Lanczos algorithm for finding eigenvalues of large symmetric matrices, Lanczos approximation for the gamma function, and the conjugate gradient method for solving systems of linear equations. These are of great importance also for today's computing technology. At the time of McCarthyism, Lánczos returned to Europe in 1952 and worked in Dublin. There he wrote several books, some of which are included in References.[85]

Mihály Fekete (1886–1957) studied at the Universities of Budapest and Göttingen and received his doctorate under Fejér's supervision in 1909. In 1914–1920 he taught at various schools but lost all his jobs as a result of the Numerus Clausus Act in 1920. In 1928, Fekete accepted the invitation to the Hebrew University in Jerusalem where he taught for a year. Then he was Director of the Mathematical Institute at the University until 1955. His most significant scientific achievement is the introduction of the *transfinite diameter of a point-set* and its use (see Fekete, 1930).

György Pólya (1887–1985) began studying law and then philosophy in Budapest. His teacher A. Bernát advised him to study mathematics as well, in which he received a doctorate under Fejér's supervision in 1912. From 1913, Pólya worked in Göttingen with such great personalities as Hilbert, Weyl and Landau. During a short study trip to Paris he was invited to work at the *Technische Hochschule* in Zürich.[86] Pólya

[83]He contributed to the theory of supersonic flying by, *inter alia*, simplifying the three-dimensional Navier-Stokes flow equations to a single equation. Kármán holds an honorary doctorate from a great number of universities and received the National Medal of Science (1963), Gauss, Kelvin, Guggenheim and Watt Medals, and Prandtl Commemorative Ring.

[84]We included in References only von Kármán and Biot (1940) and his comprehensive works von Kármán (1975).

[85]Lanczos (1956, 1986).

[86]Today ETH.

ignored the conscription order to join the army in 1914 and returned to Hungary as late as in 1967. His collaboration with Gábor Szegő began in 1923, which led to the publication in 1925 of their joint book in Berlin that forever linked the two names and lifted up the scientific reputation of both authors very high.[87] Pólya is considered the creator of modern mathematical heuristics and was an invited speaker in ICM 1928, 1936 and 1950. Although he had Swiss citizenship, he did not feel safe, so he emigrated to the USA in 1940, where he worked as Professor at Stanford University until retirement. Pólya was an honorary member of the Hungarian Academy of Sciences from 1976.[88]

The name Pólya is inseparable from the name Szegő. While Pólya fled WWI, Szegő belongs to a large set of those who fled fascism because of their Jewish origin.

Gábor Szegő (1895–1985) was the winner of the mathematical competition of secondary-school graduates in 1912, then studied at the University of Budapest, and attended lectures in Berlin in the summer of 1913 and Göttingen in the summer of 1914. Szegő fought in WWI in 1915–1918 and, still being a soldier, received a doctorate under Fejér's supervision in 1918. He taught at the Technical University in Budapest for two years. He published the book *Aufgaben und Lehrsätze aus der Analysis* with Pólya in 1925. He emigrated to the USA in 1934 after the rise of fascism and lived there for a long time only with the support of various foundations. In 1938, Szegő was appointed head of a department at Stanford and taught there until 1953. He was an honorary member of the Hungarian Academy of Sciences from 1965.[89]

Tibor Radó (1895–1967) joined the army in 1915 after two years of study at the Technical University of Budapest and was imprisoned in the Russian front in 1916. In the camp near Tobolsk, Radó learnt mathematics from his inmate Eduard Helly. After escaping and traveling thousands of miles through the Arctic land he managed to get home in 1920. He continued his studies in Szeged where Radó completed his doctorate under the leadership of A. Haar and F. Riesz in 1923. He taught for five years at the University of Szeged and worked as a researcher with the support of the Rockefeller Foundation in Germany in 1928. Since he did not get a job in Hungary, he went to Harward, USA in 1929 and then he was a professor at Ohio State University from 1930. Radó earned his greatest fame with solving the *problem of Plateau*.[90]

Ábrahám Wald (1902–1950) graduated from the Piarist Grammar School in Kolozsvár, which at that time belonged to Romania, so as a Hungarian Jew he had

[87]That book is still considered the best collection of tasks and a grand-master lesson in problem solving.

[88]Some of his works are included in References (Pólya and Szegő, 1964; Pólya *et al.*, 1934; Pólya, 2014; Pólya and Szegő, 1951). More detailed information can be found, e.g., in Alexanderson (1987).

[89]Some of his works are included in References (Pólya and Szegő, 1964, 1951; Szegő, 1939).

[90]Some of his works are included in References (Radó, 1925, 1930).

double trouble. Nevertheless, Wald graduated from the University of Cluj when he was 26 years old and completed his doctorate in Vienna under Karl Menger's supervision in 1931. He made his living on tutoring and extra classes and began to deal with economics. As a result of increasing anti-Semitism, Wald went to the USA in the summer of 1938 to deal with econometrics.[91] He was a member of the Statistical Research Group at Columbia University during WWII, where he performed statistical research for the US Army.[92] He is known for *Wald's equation*, *Wald test*, *sequential analysis*, etc. According to his students, Wald had a particularly clear way of explaining. He was invited to the Congress in Cambridge (USA) in 1950 where he delivered a plenary lecture. He died after a plane crash in India when he was on a lecture tour.[93]

Aurél Wintner (1903–1958) graduated from secondary school in Budapest in 1920, and then completed only four years at the university, probably for financial reasons. Until 1927, Wintner wrote several articles on quantum mechanics and astronomy, on the basis of which Lichtenstein invited him to the University of Leipzig, where he received a doctorate in astronomy in 1928. Wintner's book, *Spektraltheorie der Unendlichen Matrizen*, in which he laid the mathematical foundations of quantum mechanics, was published in 1929. He taught at Johns Hopkins University from 1930. Wintner was one of the founders of *probabilistic number theory* (together with Erdős and Kac). He wrote 437 articles and nine books.[94]

János Neumann (1903–1957), known as John von Neumann, is maybe the most influential of all Hungarian mathematicians. His work ranges across numerous fields of mathematics, as well as economics, physics and computer science.

As a 6-year-old, Neumann was fluent in Hungarian, German and Ancient Greek, but learnt English, French, Italian, Latin and Hebrew very quickly. As an 8-year-old he mastered differential and integral calculus.[95] Neumann graduated from the famous Evangelical Fasori Grammar School, where he was taught by the legendary László Rátz. As a 19-year-old, he had already completed two major works on ordinal numbers. He enrolled in mathematics at the University of Budapest in 1921, but also

[91]Except for one brother, all his relatives died in concentration camps.

[92]Wald was considered the smartest guy in the room where there were such personalities as Norbert Wiener and the future Nobel laureate Milton Friedman (who was considered the fourth smartest, by the way). The incident is well-known of hits on aircraft that have returned from combat. According to statistics, the engine had an average of 1.11 hits, the other parts of the aircraft had 1.55–1.8 hits. While others suggested that aircraft armor should be intensified (reinforced) in areas where there were many hits, Wald said immediately that it was similar to a hospital: there are many convalescents with a bullet in their arm or leg, but only a few convalescents with a bullet in their chest. So armor should be reinforced on the engine, because a hit in the engine very often results in the destruction of the aircraft.

[93]His works Wald (1939, 1945) are included in References.

[94] Two of Wintner's books are listed in References (Wintner, 1929, 2014).

[95]Fejér, Ortvay, Szegő, etc. were frequent visitors to the Neumanns.

Figure 12. John von Neumann.[96]

studied physics and chemistry there. In the same year he enrolled at the University of Berlin. Neumann also attended ETH, Zürich from 1924, where he became a chemical engineer in 1926, and in the same year he received a doctorate in Budapest under Fejér's supervision with his work on axiomatic set theory construction. At 23, he became the youngest professor at the University of Berlin.

Neumann was invited to Princeton University in 1930. Since then until 1933 he alternately taught in the USA and Germany. He co-founded the Institute for Advanced Study in Princeton in 1933, where he stayed all his life. Photographic memory and unimaginable speed and infallibility of thinking enabled him to fundamentally intervene in various areas of mathematics: game theory (1928), von Neumann paradox in set theory (1929), ergodic theory (1932), mathematical foundation of quantum mechanics (1932), measure theory (1936), lattice theory (1937), mathematical statistics (1941), and computer science.

In connection with theoretical work on the atomic bomb, he regularly went to Los Alamos during World War II, where everyone was curious about his opinion. After 1930, his attention was increasingly focused on computer technology. Today's computers are also built on the principles outlined by von Neumann.

In the opinion of many scientists, von Neumann truly deserved the Nobel Prize in physics, but he died too soon to be awarded it.[97] But if the Nobel Prize were given for mathematics and computer sciences, then he would have got at least three.

[96]The picture is available on the website https://en.wikipedia.org/wiki/John_von_Neumann [16.9.2019].

[97]He had cancer probably as a result of irradiation in atomic bomb experiments. A similar thing happened to Szilárd.

John von Neumann published 150 papers: 60 in pure mathematics, 30 in physics, and 60 in applied mathematics.[98]

This is probably the right place to disclose why the title of this section is what it is. In the early 1930s Hans Bethe[99] in Princeton asked jokingly whether von Neumann was a member of an extraterrestrial civilization, superior to humans. For the Americans, the Hungarian scientists were seemingly superhuman in intellect, spoke an incomprehensible native language, and came from a small obscure country. Moreover, they all spoke English with a strong accent. Let's present an original story from György Marx's book (Marx, 2000) *The Voice of the Martians.*

'The universe is vast, containing myriads of stars ... likely to have planets circling around them. The simplest living things will multiply, evolve by natural selection and become more complicated till eventually active, thinking creatures will emerge. Yearning for fresh worlds ... they should spread out all over the Galaxy. These highly exceptional and talented people could hardly overlook such a beautiful place as our Earth. And so Enrico Fermi came to his overwhelming question: if all this has been happening, they should have arrived here by now, so where are they?' It was Leo Szilárd, a man with an impish sense of humour, who supplied the perfect reply to the Fermi Paradox: 'They are among us, but they call themselves Hungarians.'

The joke was that Hungarian scientists are actually descendants of a Martian scout force which landed in Budapest around the year 1900, and later departed after the planet was found unsuitable, but leaving behind children by several Earth women, children who all became the famous scientists. John von Neumann used a number of facts as mock evidence to support this claim, such as the close geographic proximity of the Martian's birthplaces (Budapest); the well-traceable career path, which started with an interest in chemistry, led the individual in question to German universities where he moved towards physics, at which point the Martian left Europe for the United States.

When the question about Martians was put to Edward Teller — who was particularly proud of his monogram, E.T. (abbreviation for 'extraterrestrial') — he looked worried, and said: 'von Kármán must have been talking.'

The Martians was a term used to refer to a group of prominent Hungarian scientists (mostly, but not exclusively, physicists and mathematicians) who emigrated to the United States in the first half of the 20th century. The core of this group included von Neumann, von Kármán, Szilárd, Wigner, Teller, Erdős, Halmos, Pólya, Szegő, Wald, Kemény, Lax, Lánczos, Goldmark, and Orován, but some others are also included in the group sometimes.

[98]A subjective but representative selection (von Neumann, 1923, 1925, 1927, 1951, von Neumann and Morgenstern, 1944) of his work is included in References, and complete works can be found in Taub (1961–1963). More detailed information can be found, e.g., in Macrae (1992).
[99]Nobel Prize in physics in 1967.

Pál Halmos's (1916–2006) family moved from Hungary to the USA when he was 13 years old and he received his doctorate in 1938 under J.L. Doob's supervision with his work *The Mathematical Theory of Gambling Systems.* He then wrote his first book *Finite-Dimensional Vector Spaces* in Princeton, which immediately ensured him a high reputation and was published in many reissues. He taught successively at many universities: Syracuse University N.Y., University of Chicago 1946–1960, University of Michigan 1961–1967, University of Hawai 1967–1968, Indiana University 1969–1985, and from 1985 at Santa Clara University. Halmos made a significant contribution to several areas of mathematics: mathematical logic, probability theory, statistics, operator theory, ergodic theory, functional analysis (especially Hilbert spaces). He wrote altogether 16 books.[100]

János György Kemény (1926–1992) attended the excellent *Berzsenyi Dániel Gimnázium* in Budapest, but the family emigrated to the USA after the Second Jewish Law in 1940.[101] So János graduated from secondary school in New York with his name changed to John Georg Kemeny and after graduating from Princeton University in 1949 he earned a doctorate in logic. During the war he worked in Los Alamos in the Manhattan Project as a collaborator of Richard Feynman. Kemeny and Lax were probably the youngest scientists in the Manhattan Project and both belonged to the *Martians group.* J. G. Kemeny was Professor at Dartmouth Mathematics Department from 1953, and Head of this Department from 1955. He was a co-creator of Dartmouth Time-Sharing System and the BASIC programming language (both with Kurtz).[102] Kemeny visited Hungary only once, in 1964.

Péter Lax (born 1926) attended *Mintagimnázium* in Budapest[103] and because of his extraordinary talent, he was taught by such great teachers as Dénes Kőnig, Rózsa Péter and Pál Turán. The family emigrated via Lisbon to the USA in November 1941. He graduated from Stuyvesant High School in 1943.[104] Since he was only 17 years old, Lax did not have to go to war immediately, so he completed three semesters of university, where he was more a teacher than a student. In 1944, the U.S. Army sent him through Texas University and Oak Ridge National Laboratory to the Manhattan Project at Los Alamos, where he worked as a mathematician. Lax stayed there for one more year after the war, attending the university and studying for one semester at Stanford under Szegő and Pólya. With credits from the four universities at which he studied, Lax completed his studies in New York in 1947.

---

[100]A list of Halmos' books can be found, for example, at https://en.wikipedia.org/wiki/Paul_Halmos [16.9.2019].

[101]Kemény's grandfather, uncle and aunt did not emigrate and became victims of the Holocaust.

[102]Two of his books on this topic are included in References (Kemeny *et al.*, 1957; Kemeny and Kurtz, 1967).

[103]This grammar school was founded by Mór Kármán, father of the more famous Th. von Kármán.

[104]Already during high school Lax met von Neumann, R. Courant and P. Erdős, who introduced him to Einstein.

After receiving a doctorate in 1949 he got a job at the Department of Mathematics, Courant Institute of Mathematical Sciences in 1951, where he was Professor from 1958. Lax has made important contributions to various fields of pure and applied mathematics, including integrable systems, fluid dynamics and hyperbolic conservation laws. He is known for Lax-Wendroff method, Lax equivalence theorem, Babuška-Lax-Milgram theorem, and Lax pairs. Lax received the Wolf Prize in 1987 and the Abel Prize in 2005. He still visits Hungary regularly.[105]

The scientific work of Lax and Kemény dates back to the period after WWII, but both began their activity in the Manhattan Project and both belonged to the *Martians group. The Martians* also include a pharmacist, a doctor of philosophy and a Nobel Prize laureate in economics, who developed some of von Neumann's ideas about game theory.

János Harsányi (1920–2000) graduated from grammar school at *Fasori gimnázium* in 1938. He received a pharmacy diploma in Budapest in 1942, but as a Jew he was on forced labour from May to November 1944. At the beginning of 1945 he escaped from the railway station during the deportation and hid successfully. Harsányi studied philosophy from 1946 and completed his doctorate in 1947. He fled to Austria after nationalization of pharmacies in 1948. From there he emigrated to Australia in 1950, where he worked as a factory worker for three years. Harsányi taught economics at the University of Brisbane in 1954–1956. He then studied mathematics and statistics at Stanford University and received a doctorate with his thesis *A Bargaining Model for the Cooperative n-Person Game* in 1958. He was a researcher at the University of Canberra in 1958, and at the University of Detroit from 1961, and Professor at Berkley from 1964. In 1994, Harsányi received (together with John Nash and Reinhard Selten) the Nobel Prize in economic sciences.

The following three *Martians* are representatives of physics, so we will discuss them briefly, though each would deserve a separate section.[106]

Leó Szilárd (1898–1964) graduated from secondary school in 1916 and enrolled at the Technical University in Budapest, but a year later he was recruited in the army. Szilárd was one of the first men to get the Spanish flu, so he, fortunately, did not participate in the combat in which a substantial part of his regiment fell. He resumed studies at the Berlin Polytechnic at the end of 1919 to finally study physics and receive a doctorate under Max von Laue's supervision in 1922. He left Germany at the very last moment in 1933 — by the last train from Germany.

In Chicago in 1942, Szilárd and E. Fermi built the world's first functional nuclear reactor. Szilárd is known for *Nuclear Chain reaction, Szilárd petition, Einstein-Szilárd letter, Cobalt bomb, Absorption refrigerator, Szilárd engine, Szilárd-Chalmers effect, Einstein-Szilárd refrigerator.* At Szilárd's recommendation, the so-called

[105] His extensive work can be found in two volumes of selected papers (Lax *et al.*, 2005a, 2005b).
[106] Wigner also contributed to mathematics.

hotline between Moscow and Washington was created in 1962. He was diagnosed with cancer in 1953, with prospects of early death. But Szilárd was a versatile scientist. He studied biology as well and undertook self-administered treatment including irradiation, which he invented himself including dosing. When he died of a heart attack on 30 May 1964, the autopsy showed no sign of cancer. According to his own wish, a part of his ashes were scattered by balloons over the USA, and a part is placed in a cemetery in Budapest.[107]

Jenő Wigner (1902–1995) attended the famous *Fasori gimnázium*, where he met one-year-younger János Neumann. He was taught mathematics by the legendary László Rátz, who, in Wigner's words, was the best teacher he had ever known. He began studying chemistry at the Technical University in Budapest in 1920 but continued at the Technical University in Berlin from 1921. The *Wednesday afternoon colloquia of the German Physical Society* played an extraordinarily significant role in Wigner's life. Here he met scientists such as Max Planck, Max von Laue, Werner Heisenberg, Wolfgang Pauli, Albert Einstein and Leó Szilárd, who became his best friend. Wigner's doctoral work at the *Kaiser Wilhelm Institute for Physical Chemistry and Electrochemistry* was supervised by the physician and physical chemist Mihály Polányi, who became Wigner's second most important teacher after Rátz. Wigner was a Professor at Princeton from 1930. His work *Gruppentheorie und ihre Anwendungen auf die Quantenmechanik der Atomspektren*, Vieweg Verlag, Braunschweig 1931, 332 pages, was an essential contribution to theoretical physics as well as applied mathematics. He is known in mathematics, e.g., for Gabor–Wigner transform, Wigner quasi-probability distribution, and in theoretical physics for *Wigner D-matrix*, *Wigner crystal*, and many others. During the Manhattan Project, Wigner led a group responsible for the design of the production nuclear reactors that would convert uranium into weapons-grade plutonium. He received the Nobel Prize in physics in 1963. Wigner visited Hungary four times after 1976.

Edward Teller (1908–2003) graduated from secondary school in Budapest in 1925. He enrolled at the Technical University but left for Karlsruhe on 2 January 1926. Teller first studied mathematics and chemistry, and then began to study physics in 1927, in which he received a doctorate under Heisenberg's supervision in Leipzig in 1930. He is known for *Jahn-Teller effect, Ashkin-Teller model, and hydrogen bomb.* Teller left Germany in 1933 and after a short stay in Britain he moved to the USA. He then worked in Rome with Fermi, and later in Copenhagen under the leadership of Niels Bohr. Teller became a US citizen in 1941. At that time he taught in New York at Columbia University with his friend Szilárd. After a nuclear reactor was successfully launched in Chicago, Teller was invited to the Manhattan Project. E. Teller designed a more powerful version of the atomic bomb — a hydrogen bomb.

[107]More detailed information can be found, e.g., in Marx (1997).

Figure 13. Leo Szilárd with Albert Einstein writing a letter to President Roosevelt.[108]

Figure 14. Eugene Wigner.[108]

He worked on the development of the so-called Star Wars during President Reagan's era. He visited Hungary for the first time in 1990, and then repeatedly every year.

Péter Károly Goldmark (1906–1977). His family moved to Vienna in 1919, where Goldmark received a diploma in physics, and then a diploma in electrical engineering in Berlin. He lived in the USA from 1933. In 1940 he introduced the first functional color television and invented a way to jam German radars with interference during the war. Goldmark belonged to the *Martians.* He invented the LP, i.e. a long-playing phonograph disc in 1948, and became president of Columbia Records in 1956.

A strong impact on mathematics was also caused by the departure of excellent physicists, mechanical, electrical and chemical engineers, because new problems that they formulated were a driving force for mathematics as well. The following four emigrated to Britain and belong to the *Martians.*

Egon Orován (1902–1989) graduated from secondary school in Budapest, then studied chemistry in Vienna for one year, astronomy in the second year, and received a doctorate in solid-state physics in Berlin in 1932. His most important work (1934) is about plastic deformations, which he explained by dislocations.[109] For Tungsram, Orován together with Michael Polányi discovered a new process for the extraction of krypton from the air in 1936. During WWII he worked on the problems in producing and transporting ammunition, which deformed. Then in 1944, Orován showed that ships in the North Atlantic often broke due to poor welding. He worked at MIT in the USA from 1950.

[108]The pictures are from the author's archive.

[109]The crystal disorder was discovered by Vito Volterra in 1905.

György Hevesy (1885–1966) is known under the name George de Hevesy. He graduated from the Piarist Grammar School in Budapest, where he began his studies at the Technical University and then continued in Berlin and Freiburg, where he received a doctorate in physics. He worked in Copenhagen (under Niels Bohr) from 1920, and discovered hafnium there in 1922, and then *radioactive tracer* in 1923.[110] He was awarded the Nobel Prize for Chemistry in 1943.

Dénes Gábor (1900–1979) is better known as Dennis Gabor. After graduating from secondary school in Budapest he was in WWI on the Italian front for six months. He began his studies at the Technical University in Budapest and continued with electrical engineering at *Technische Hochschule Berlin* from 1920.[111] Gábor worked in the United Kingdom from 1934 and here he discovered holography in 1947, for which he was awarded the Nobel Prize in 1971.

Mihály Polányi (1891–1976) is better known as Michael Polanyi. He graduated from secondary school at *Mintagimnázium* and completed his university studies in medicine in 1914. Polányi served as a doctor in the army, and after his injury he earned a doctorate in physical chemistry on *adsorption* in 1917. He worked at the Kaiser Wilhelm Institute in Berlin from 1920 and became Professor in 1926.[112] He was a pioneer in the theory of diffraction analysis and (together with Orován) the dislocation theory of plastic deformation of ductile metals and other materials. Polányi emigrated to Manchester, UK, in 1933 where he later devoted himself to economics.[113]

9. Results of Previous Generations' Work

In Section 8, we described the huge brain drain from Hungary after World War I due to the economic and political situation.

Many excellent scientists and teachers left the country, but not all of them. There were quite a lot of secondary-school and university teachers who stayed in Hungary — all of them being basically the achievement of the work of previous generations. Their predecessors were excellent leaders and tutors. Those who did not leave their homeland then continued in the same kind of teaching they had learnt from the previous generations of scholars, and the instruction process kept a high standard.

Now we could include brief characteristics of the work of people born between 1925 and 1940, for example, from the very long (subjective and incomplete)

[110]Several atoms of the radioactive isotope are added to a substance and it is possible to trace the path of the substance in a living organism.

[111]At that time, the above-mentioned famous Einstein's seminar took place, involving Szilárd, Wigner, Neumann, and M. Polányi.

[112]He was Wigner's doctoral advisor.

[113]His son János as a Hungarian-Canadian scientist was awarded the Nobel Prize for chemistry in 1986.

Figure 15. Szilassi polyhedron in the yard of Fermat's birthplace.[114]

Figure 16. Raoul Bott, Endre Szemerédi, László Lovász, and János Kollár.

list, e.g., I. Reimann (1927–2012), Vera T. Sós (born 1930), A. Hajnal (1931–2016), A. Heppes (born 1933), P. Révész (born 1934), B. Uhrin (born 1938), etc. Or we could present the generation born in 1940–1950, e.g., E. Szemerédi (born 1940), A. Sárközi (born 1941), G. Halász (born 1941), G.O.H. Katona (born 1941),

---

[114]The picture is available on the website http://www.termeszetvilaga.hu/szamok/tv2003/tv0307/ogr.html [17.9.2019].

L. Szilassi (born 1942),[115] B. Bollobás (born 1943), E. Molnár (born 1943), M. Simonovits (born 1943), A. Gyárfás (born 1945), G. Fejes-Tóth (born 1947), I. Bárány (born 1947), J. Pelikán (born 1948), L. Lovász (born 1948), etc, or the 1950–1960 generation: G. Elekes (1949–2006), L. Babai (born 1950), M. Laczkovich (born 1948), J. Pintz (born 1950), J. Beck (born 1952), Z. Tuza (born 1953), P. Komjáth (born 1953), P. Frankl, Z. Füredi (both born 1954), J. Pach (born 1954), E. Győri (born 1954), K. Bezdek (born 1955), A. Bezdek, etc. And we are getting to mathematicians in their 60s now — just retired. Yet, we have not mentioned a single name from a wide range of really excellent (and extraordinarily significant for the next generations) secondary school teachers, for example J. Urbán, L. Pósa, F. Csorba, F. Pintér, J. Szoldatics, G. Kiss, S. Dobos, etc. Nor have we mentioned any well-established scientists who are now 40–60 years old. Instead, let us present just four names as representatives of generations:

Raoul Bott (1923–2005), Veblen Prize in 1964, Leroy P. Steele Prize in 2008, Wolf Prize in 2000[116];

Endre Szemerédi (born 1940), Leroy P. Steele Prize in 2008, Rolf Schock Prize in 2008, Abel Prize in 2012;[117]

László Lovász (born 1948), president of the International Mathematical Union 2007–2010, Wolf Prize and Knuth Prize in 1999, Gödel Prize in 2001, John von Neumann Theory Prize in 2006, Kyoto Prize in 2010;[118]

János Kollár (born 1956), Cole Prize in 2006, Nemmers Prize in 2016, Shaw Prize in 2017.[119]

Brain drain continues even today and not only from Hungary. However, the reason is usually no longer political but economic — considerably higher salaries and disproportionately better conditions, especially for those dealing with experimental sciences. But this is no longer a consequence of World War I.

---

[115]Szilassi earned world fame by designing a heptahedron whose every two faces share a common edge, so seven colors are necessary for the so-called regular coloring. Bronze sculptures of this so-called *Szilassi polyhedron* are displayed in the yard of Pierre de Fermat's birthplace in France, on the shore of Lake Michigan in the USA, in Canberra, Australia, and in the yard of the Evangelical Grammar School in Bonyhád. The author of this essay was pleased to support the habilitation of Lajos Szilassi at the University of Nitra as an opponent.

[116]The picture is available on the website https://link.springer.com/chapter/10.1007/978-3-319-51781-0_5 [17.9.2019].

[117]The picture is available on the website https://en.wikipedia.org/wiki/Endre_Szemerédi [17.9.2019].

[118]The picture is available on the website https://corvinlanc.kormany.hu/lovasz-laszlo [15.2.2021].

[119]The picture is available on the website https://www.math.princeton.edu/people/janos-kollar [17.9.2019].

References

G.L. Alexanderson, Obituary: George Pólya, *Bulletin of the London Mathematical Society* **19**(1987), pp. 559–608. ISSN 1469-2120.

D. Arany and L. Magyar, *Az élet- és járadékbiztosítás üzletelmélete* [The Business Theory of Life and Annuity Insurance], Korvin Testvérek Könyvnyomdája, Budapest, 1905, 64 pages.

D. Arany and A. Bánhegyi, *Unterricht an den höheren Gewerbeschulen und gewerblichen Fachschulen*, Budapest, 1912.

L. Babai, C. Pomerance and P. Vértesi, The Mathematics of Paul Erdős, *Notices of the American Mathematical Society* **45**(1998), no. 1, pp. 19–32. ISSN 0002-9920.

L. Babai and J. Spencer, Paul Erdős (1913–1996), *Notices of the American Mathematical Society* **45**(1998), no. 1, pp. 64–73. ISSN 0002-9920.

J. Bolyai, *Appendix: The Theory of Space* (Introduction by F. Kárteszi; supplement by B. Szénássy), Akadémiai Kiadó, Budapest, 1987, 238 pages. ISBN 978-9630544016.

V. Bálint, Z histórie kombinatorickej geometrie [From the History of Combinatorial Geometry], pp. 11–44, in J. Bečvář and M. Bečvářová (eds.), *34. Mezinárodní konference Historie matematiky,* Poděbrady, 23.8. až 27.8.2013, Matfyzpress, Praha, 2013, 201 pages. ISBN 978-80-7378-234-4.

V. Bálint, Erdős a ľudia okolo neho [Erdős and the People Around Him], pp. 35–42, in J. Bečvář and M. Bečvářová (eds.), *37. Mezinárodní konference Historie matematiky,* Poděbrady, 19.8. až 23.8.2016, Matfyzpress, Praha, 2016, 203 pages. ISBN 978-80-7378-317-4.

V. Bálint, Lipót Fejér (1880–1956), *Obzory matematiky, fyziky a informatiky* **48**(2019), no. 4, pp. 11–17. ISSN 1335-4981.

M. Beke, Die Irreducibilität der homogenen linearen Differentialgleichungen, *Mathematische Annalen* **45**(1894), no. 2, pp. 278–294.

M. Beke, Zur Gruppentheorie der homogenen linearen Differentialgleichungen, *Mathematische Annalen* **49**(1897), no. 3, pp. 573–580.

A. Bezdek (ed.), *Discrete Geometry,* In Honor of W. Kuperberg's 60th Birthday. Marcel Dekker Inc., New York, 2003, xvii + 461 pages. ISBN 0-8247-0968-3.

P. Brass, W.O.J. Moser and J. Pach, *Research Problems in Discrete Geometry,* Springer, Berlin, 2005, 499 pages. ISBN 0-387-23815-8.

Gy. Bisztray, T.A. Szabó and L. Tamás, *Erdély magyar egyeteme* [Hungarian University of Transylvania], Erdélyi Tudományos Intézet, Kolozsvár, 1941, 470 pages.

B. Csákány and A. Varga, A szegedi egyetemi matematikai intézetek hetvenöt éve [Seventy-five Years of the Mathematical Institutes of the University of Szeged], pp. 48–57, in L. Szentirmai (ed.), *A Szegedi Tudományegyetem múltja és jelene, 1921–1998* [The Past and Present of the University of Szeged, 1921–1998], József Attila Tudományegyetem, Szeged 1999, 517 pages.

Á. Császár, Magyar származású matematikusok hozzájárulása a matematika fejlődéséhez [Contribution of Hungarian-born Mathematicians to the Development of Mathematics], *Természet Világa* **129**(1998), III. extra number, pp. 3–10. ISSN 0040-3717.

J. Čižmár, *Dejiny Matematiky* [History of Mathematics], Perfekt, Bratislava, 2017, 896 pages. ISBN 978-80-8046-829-3.

L. Dávid, *A két Bolyai élete és munkássága* [Life and Work of the Two Bolyais], Gondolat, Budapest, 1979, 426 pages. ISBN 963-280-806-1.

J. Egerváry, Matrixok kombinatorikus tulajdonságairól [About Combinatorial Properties of Matrices], *Matematikai és Fizikai Lapok* **38**(1931), pp. 16–28.

P. Erdős and P. Turán, On a Problem in the Elementary Theory of Numbers, *American Mathematical Monthly* **41**(1934), pp. 608–611.

P. Erdős and T. Gallai, On Polynomials with Only Real Roots, *Annals of Mathematics* **40**(1939), no. 3, pp. 537–548.

P. Erdős, T. Gallai and Z. Tuza, Covering the Cliques of a Graph with Vertices, *Discrete Mathematics* **108**(1992), no. 1–3, pp. 279–289.

P. Erdős, T. Gallai and Z. Tuza, Covering and Independence in Triangle Structures, *Discrete Mathematics* **150**(1996), no. 1–3, pp. 89–101.

P. Erdős and J. Surányi, *Válogatott fejezetek a számelméletből* [Selected Chapters in the Number Theory], Tankönyvkiadó, Budapest, 1960, 250 pages.

P. Erdős, My Joint Work with Richard Rado, in *Surveys in Combinatorics 1987*, London Mathematical Society Lecture Note Series 123, Cambridge University Press, Cambridge, 1987, pp. 53–80. Also in Graham *et al.* (2013).

P. Erdős and A. Rényi, On the Evolution of Random Graphs, *Magyar Tudományos Akadémia Matematikai Kutatóintézet Közleményei* **5**(1960), pp. 17–61.

H. Eves, *An Introduction to the History of Mathematics*, 6th edition, Saunders College Publishing, Philadelphia, 1990, 775 pages. ISBN 978-0030295584.

L. Fejér, Sur les fonctions bornées et intégrables, *Comptes Rendus de l'Académie des Sciences, Paris* **131**(1900), pp. 984–987.

L. Fejér, Vizsgálatok a Fourier-féle sorok köréből 1–2 [Investigations on Fourier Series], *Mathematikai és Physikai Lapok* **11**(1902), pp. 49–68, pp. 97–123.

L. Fejes-Tóth, Einige Extremaleigenschaften des Kreisbogens bezüglich der Annäherung durch Polygone, *Acta Scientiarum Mathematicarum (Szeged)* **10**(1943), pp. 164–173.

L. Fejes-Tóth, Über die Fouriersche Reihe der Abkühlung, *Acta Scientiarum Mathematicarum (Szeged)* **11**(1946), no. 1–2, pp. 28–36.

L. Fejes-Tóth, *Lagerungen in der Ebene, auf der Kugel und im Raum*, Springer, Berlin, 1953, 197 pages.

L. Fejes-Tóth and A. Heppes, A Variant of the Problem of the Thirteen Spheres, *Canadian Journal of Mathematics* **19**(1967), no. 3–4, pp. 1092–1100.

M. Fekete, Über den transfiniten Durchmesser ebener Punktmengen, *Mathematische Zeitschrift* **32**(1930), pp. 108–114.

L. Filep, *A tudományok királynője* [The Queen of the Sciences], Typotex kiadó, Budapest, 2012, 510 pages. ISBN 978-963-2793-02-3.

M. Font, A középkori pécsi egyetem [Medieval University of Pécs], *Jelenkor* **45**(2002), no. 5, pp. 473–479. ISSN 0447-6425.

T. Gallai, Maximum–Minimum Sätze und verallgemeinerte Faktoren von Graphen, *Acta Mathematica Academiae Scientiarum Hungaricae* **12**(1961), pp. 131–173.

T. Gallai, The Life and Scientific Work of Dénes Kőnig, *Linear Algebra and Its Applications* **21**(1978), no. 3, pp. 189–205. ISSN 0024-3795.

L. Gårding, Marcel Riesz in Memoriam, *Acta Mathematica* **124**(1970), 11 pages. ISSN 0236-5294.

I. Gazda (ed.), *A magyar matematika történetéből* [From the History of Hungarian Mathematics], Magyar Tudománytörténeti Intézet, Piliscsaba, 2000, 319 pages. ISBN 963-9276-08-1.

I. Gazda (ed.), *Százhuszonöt éve nyílt meg a Kolozsvári Tudományegyetem* [One Hundred and Twenty Five Years Ago the Kolozsvár University Was Opened], Magyar Tudománytörténeti Intézet, Piliscsaba, 1997, 447 pages. ISBN 963-04-9192-3.

Z. Geőcze, A forgásfelület quadraturája [The Quadrature of the Rotation Surface], *Az Ungvári reáliskola 1905–1906 tanévi értesítője* [1905–1906 School Year Bulletin of the Real School in Ungvár], Ungvár, 1906, 12 pages.

Z. Geőcze, Quadrature des surfaces courbes, *Comptes Rendus de l'Académie des Sciences, Paris* **144**(1907), pp. 253–256.

R.L. Graham, J. Nešetřil and S. Butler (eds.), *The Mathematics of Paul Erdős I and II*, 2nd edition, Springer Verlag, New York, 2013, XIX + 563 pages, XIX + 607 pages. ISBN 978-1-4614-7257-5. ISBN 978-1-4614-7253-7.

A. Haar, Zur Theorie der orthogonalen Funktionensysteme I, *Mathematische Annalen* **69**(1910), no. 3, pp. 331–371.

A. Haar, Zur Theorie der orthogonalen Funktionensysteme II, *Mathematische Annalen* **71**(1911), no. 1, pp. 38–53.

A. Haar, Zur Variationsrechnung, *Abhandlungen aus dem Mathematischen Seminar der Universität Hamburg* **8**(1931), pp. 1–27.

G. Hajós, Ein neuer Beweis eines Satzes von Minkowski, *Acta Scientiarum Mathematicarum* (Szeged) **6**(1934), pp. 224–225.

G. Hajós, Über einfache und mehrfache Bedeckung des n-dimensionalen Raumes mit einem Würfelgitter, *Mathematische Zeitschrift* **47**(1941), pp. 427–467.

G. Halász, L. Lovász, M. Simonovits and V.T. Sós (eds.), *Paul Erdős and His Mathematics*, Springer, Berlin-Heidelberg, 2002, 728 pages. ISBN 978-3-540-42236-5.

P.R. Halmos, *Finite-Dimensional Vector Spaces*, Van Nostrand Reinhold Company, New York, 1958, 2nd edition, 212 pages.

P.R. Halmos, Riesz Frigyes munkássága [The Work of Riesz Frigyes], *Matematikai Lapok* **29**(1981), pp. 13–20.

J. Horváth, Riesz Marcel matematikai munkássága [The Work of Riesz Marcel], *Matematikai Lapok* **26**(1975), pp. 11–37; also in **28**(1980), pp. 65–100.

J. Hunyady, *A kúpszeleten fekvő hat pont föltételi egyenletének különböző alakjairól* [About the Different Shapes of the Equation for the Six Points Lying on the Conoid], Magyar Tudományos Akadémia, Budapest, 1876, pp. 1–23.

J. Hunyady, *Néhány determináns-egyenletről* [On Some Determinant-Equations], Magyar Tudományos Akadémia, Budapest, 1882, 24 pages.

K. Jordán, *Statistique mathématique*, Gauthier-Villars, Paris, 1927, 344 pages.

L. Kalmár, Zur Theorie der abstrakten Spiele, *Acta Scientiarum Mathematicarum (Szeged)* **4**(1928), pp. 65–85.

L. Kalmár, Az Interpolációról [On Interpolation], *Mathematikai és Physikai Lapok* **33**(1926), pp. 120–149.

L. Kalmár, Another Proof of the Gödel-Rosser Incompletability Theorem, *Acta Scientiarum Mathematicarum (Szeged)* **12**(1950), pp. 38–43.

L. Kalmár, Az úgy nevezett megoldhatatlan matematikai problémákra vonatkozó kutatások alapjául szolgáló Church-féle hipotézisről [On Church's-type Hypothesis, as the Base of Research of so-called Unsolvable Mathematical Problems], *A Magyar Tudományos Akadémia Matematikai és Fizikai Osztályának Közleményei* **7**(1957), pp. 19–38.

T. Kántor, Hungarian Mathematician in the 20th Century: Ottó Varga, *Teaching Mathematics and Computer Science, Debrecen* **8**(2010), no. 1, pp. 109–120. ISSN 1589-7389.

T. Kántor, Outstanding Mathematicians in the 20th Century: Egerváry Jenő (1891–1958), Establisher of the Hungarian Method, *Obzory matematiky, fyziky a informatiky* **41**(2012), no. 3, pp. 9–18. ISSN 1335-4981.

T. Kántor, A matematikatörténet szerény apostola: Szénássy Barna [A Humble Apostle of History of Mathematics: Szénássy Barna], *Egyetemtörténeti Közlemények* **5**(2014), no. 1–2, pp. 147–162. ISSN 2061-5132.

T. Kántor, A legendás hírű matematikatanár: Rátz László [A Legendary Math Teacher: Rátz László], *Természet Világa* **147**(2016), no. 9, pp. 161–164. ISSN 0040-3717.

T. Kántor, A Matematikai és Fizikai Társulat története (1945–1991) [History of the Mathematical and Physical Society (1945–1991)], *Debreceni Szemle* **4**, 2(1996), pp. 254–267. ISSN 1218-022X.

Th. von Kármán, *Collected Works of Theodore von Kármán 1952–1963*, Von Kármán Institute for Fluid Dynamics, Rhode-St-Genèse, Belgium, 1975, vii + 388 pages.

Th. von Kármán and M. A. Biot, *Mathematical Methods in Engineering: An Introduction to the Mathematical Treatment of Engineering Problems*, McGraw-Hill Book Company, Inc., New York, London, 1940, vii + 521 pages.

F. Kárteszi, *Ábrázoló geometria* [Descriptive Geometry], Tankönyvkiadó, Budapest 1957, 204 pages (Many editions, English and Italian translations, 1976, 1978).

F. Kárteszi, Találkozásaim Klug Lipóttal [My Meetings with Lipót Klug], *Matematikai Lapok* **24**(1973), pp. 219–223.

F. Kárteszi, *Szemléletes geometria* [Visual Geometry], Gondolat, Budapest 1966, 232 pages.

J.G. Kemeny, J.L. Snell and G. L. Thompson, *Introduction to Finite Mathematics*, Englewood Cliffs, New Jersey, Prentice-Hall, Inc., 1957, xi + 371 pages.

J.G. Kemeny and T. E. Kurtz, *BASIC Programming*, John Wiley & Sons, New York, 1967, 132 pages.

B. Kerékjártó, *Vorlesungen über Topologie*, Springer, Berlin, 1923, 270 pages.

B. Kerékjártó, *A geometria alapjairól* [On the Foundations of Geometry], Volume 1, Szegedi Városi Nyomda és Könyvkiadó, Szeged, 1937, 304 pages; Volume 2, Magyar Tudományos Akadémia, Budapest, 1944, 613 pages.

E. Kiss, *Matematikai kincsek Bolyai János kéziratos hagyatékából* [Mathematical Treasures from Bolyai János Legacy], Akadémiai Kiadó, Budapest, 1999, 224 pages. ISBN 9639132438.

L. Klug, *A projektív geometria elemei (Synthetikai-geometriai módszerrel)* [Elements of Projective Geometry (by Synthetic-Geometric Method)], Franklin-Társulat, Budapest, 1892, 283 pages.

G. Kőnig, Über eine reale Abbildung der s.g. Nicht-Euclidischen Geometrie, *Nachrichten von der Königlichen Gesellschaft der Wissenschaften und der Georg-August-Universität zu Göttingen* **9**(1872), pp. 157–164.

G. Kőnig, Zum Kontinuum-Problem, *Mathematische Annalen* **60**(1905), pp. 177–180.

G. Kőnig, *Neue Grundlagen der Logik, Arithmetik und Mengenlehre*, Veit, Leipzig, 1914, viii + 259 pages.

D. Kőnig, *Theorie der endlichen und unendlichen Graphen*, Akademische Verlagsgesellschaft m. b. h., Leipzig, 1936, 258 pages.

D. Kőnig, Az Eötvös Loránd Matematikai és Fizikai Társulat első ötven éve (Titkári beszámoló a Társulat 1941 jan. 30-án tartott jubiláris ülésén) [The First 50 Years of the Eötvös Loránd Mathematical and Physical Society (Secretary's Report to the Meeting of Society in Occassion of its 50th Anniversary on January 30, 1941)], *Mathematikai és Fizikai Lapok* **48**(1941), pp. 7–33.

J. Kürschák, A körmérés története és elmélete [The History and Theory of Measurement of Circle], *Mathematikai és Physikai Lapok* **1**(1891), no. 1–2, pp. 31–51.

J. Kürschák and J. Hadamard, Propriétés générales des corps des variétés algébriques, in *Encyclopédie des Sciences Mathématiques*, Paris, Leipzig, 1910, pp. 233–385.

J. Kürschák, Über Limesbildung und allgemeine Körpertheorie, *Journal für die reine und angewandte Mathematik* **142**(1913), pp. 211–253.

C. Lanczos, *The Variational Principles of Mechanics*, 4th revised edition, Dover Publications, 1986, 464 pages. ISBN 978-0486650678 (1st edition Toronto, 1949).

C. Lanczos, *Applied Analysis*, Prentice-Hall, New York, 1956, 539 pages.

P.D. Lax, P. Sarnak and A.J. Majda (eds.), *Selected Papers I*, Springer Verlag, Berlin, New York, 2005, 620 pages. ISBN 978-0-387-22925-6.

P.D. Lax, P. Sarnak and A.J. Majda (eds.), *Selected Papers II*, Springer Verlag, Berlin, New York, 2005b, 608 pages. ISBN 978-0-387-22926-3.

L. Lovász, I. Ruzsa and V.T. Sós (eds.), *Erdős Centennial*, Springer, Heidelberg, 2013, x + 720 pages. ISBN 978-3-642-39285-6.

N. Macrae, *John von Neumann*, Pantheon Books, New York, 1992, 405 pages. ISBN:0-679-41308-1.

J. Matoušek, *Lectures on Discrete Geometry*, Springer, Berlin, 2002, 462 pages. ISBN 0-387-95374-4.

G. Marx, *Szilárd Leó*, Akadémiai kiadó, Budapest, 1997, 166 pages. ISBN 9630574780 (Hungarian).

G. Marx, *A Marslakók érkezése* [The Voice of the Martians], Akadémiai kiadó, Budapest, 2000, 427 pages. ISBN 978-963-0577232.

J. von Neumann, Zur Einführung der transfiniten Zahlen, *Acta litterarum ac scientiarum Regiae Universitatis Hungaricae, Sectio Scientiarum Mathematicarum* **1**(1923), pp. 199–208.

J. von Neumann, Eine Axiomatisierung der Mengenlehre, *Journal für reine und angewandte Mathematik* **154**(1925), pp. 219–240.

J. von Neumann, D. Hilbert and L. Nordheim, Über die Grundlagen der Quantummechanik, *Mathematische Annalen* **98**(1927), pp. 1–30.

J. von Neumann and O. Morgenstern, *Theory of Games and Economic Behavior*, Princeton University Press, 1944, xviii + 625 pages (1st edition), ISBN 978-0-691-13061-3 (60th anniversary edition).

J. von Neumann, The General and Logical Theory of Automata, in L. A. Jeffress (ed.), *Cerebral Mechanisms in Behavior: The Hixon Symposium*, John Wiley & Sons, New York, 1951, xiii + 311 pages, pp. 1–31.

A.H. Taub (ed.), *Collected Works of John von Neumann*, Pergamon Press, Oxford, 1961–1963, 6 volumes, x + 654 pages; x + 568 pages; ix + 574 pages; x + 516 pages; ix + 784 pages; ix + 538 pages. ISBN 9780080095660.

R. Oláh-Gál, *Az értől az óceánig. Réthy Mór (1846–1925) akadémikus élete és munkássága* [From the Brook to the Ocean. The Academician Mór Réthy's (1846–1925) Life and Work], Magyar Tudománytörténeti Intézet, Budapest, 2013, 222 pages. ISBN 978-615-5365-04-1.

R. Péter, *Játék a végtelennel* [Playing with Infinity], 3rd edition, Gondolat, Budapest, 1963, 271 pages.

G. Pólya and G. Szegő, *Aufgaben und Lehrsätze aus der Analysis I, II*, 3. Berichtige Auflage, Springer Verlag, Berlin-Heidelberg-New York, 1964, xvi + 338 pages, x + 407 pages (1st edition 1925).

G. Pólya, G.H. Hardy and J.E. Littlewood, *Inequalities*, Cambridge University Press, Cambridge, 1934, xii + 314 pages.

G. Pólya, *How to Solve It, A New Aspect of Mathematical Method*, Princeton University Press, 2014, 288 pages. ISBN 9781400828678.

G. Pólya and G. Szegő, *Isoperimetric Inequalities in Mathematical Physics*, Princeton University Press, Princeton, 1951, 279 pages.

G. Rados, Zur Theorie der adjungierten Substitutionen, *Mathematische Annalen* **48**(1897), pp. 417–424.

T. Radó, Über den Begriff der Riemannschen Fläche, *Acta Scientarum Mathematicarum Universitatis Szegediensis* **2**(1) (1925), pp. 96–114.

T. Radó, The Problem of Least Area and the Problem of Plateau, *Mathematische Zeitschrift* **32**(1930), pp. 763–796.

L. Rédei, Ein Satz über die endlichen einfachen Gruppen, *Acta Mathematica* **84**(1950), pp. 129–153.

A. Rényi, *Ars Mathematica*, Typotex kiadó, Budapest, 2004, 326 pages. ISBN 978-963-7456-58-7.

A. Rényi, *Probability Theory*, North-Holland Publishing Company, Amsterdam, 1970, 670 pages. ISBN 0720423600.

M. Réthy, Végszerűen egyenlő területek [End-like Equal Areas], *Mathematikai és Temészettudományi értesítő* **8**(1890), no. 8–9, pp. 176–202.

F. Riesz, Über einen Satz der Analysis Situs, *Mathematische Annalen* **59**(1904), pp. 409–415.

F. Riesz, Sur la théorie ergodique, *Commentarii Mathematici Helvetici* **17**(1944–1945), pp. 221–239.

F. Riesz and B. Szőkefalvi-Nagy, *Leçons d'analyse fonctionnelle*, Akadémiai Kiadó, Budapest, 1952, 448 pages.

M. Riesz and E. Hilb, Neuere Untersuchungen über trigonometrische Reihen, in *Enzyklopädie der Mathematischen Wissenschaften*, II C 10, Leipzig, 1924, pp. 1189–1228.

M. Riesz and F. Riesz, Über die Randwerte einer analytischen Funktion, *Quatrième Congrès des Mathématiciens Scandinaves*, Stockholm, 1916, pp. 27–44. Uppsala, 1920.

B. Schechter, *Agyam nyitva áll* [My Brain Is Open], Vince, Budapest, 1999, 192 pages. ISBN 963-530-406-4.

L. Schlesinger, *Elliptikus függvények elmélete és alkalmazásai* [Theory and Applications of Elliptic Functions], Kalazantinum, 1898–1899, 290 pages.

L. Schlesinger, Über eine Klasse von Differentialsystemen beliebiger Ordnung mit festen kritischen Punkten, *Journal für die reine und angewandte Mathematik* **141**(1912), pp. 96–145.

Á. Scholtz, Sechs Punkte eines Kegelschnittes, *Archiv der Mathematik und Physik* **62**(1878), pp. 317–324.

P. Szász, Fejér Lipót, *A Magyar Tudományos Akadémia Matematikai és Fizikai Osztályának Közleményei* **10**(1960), no. 2, pp. 103–148.

P. Szász, A remark on Hilbert's Foundation of the hyperbolic plane geometry, *Acta Mathematica Academiae Scientiarum Hungaricae* **9**(1958), pp. 29–31.

G. Szegő, *Orthogonal Polynomials*, American Mathematical Society, Providence, Rhode Island, 1939, 444 pages.

B. Szénássy, *A Magyarországi Matematika története a 20. század elejéig* [History of Mathematics in Hungary Until the 20th Century], Akadémiai Kiadó, Budapest, 1970, 381 pages.

B. Szénássy, A Matematikai és Fizikai Társulat története [History of the Mathematical and Physical Society], *Debreceni Szemle* **3**(1995), no. 3, pp. 415–422. ISSN 1218-022X.

B. Szénássy, *Bolyai Farkas (1775–1856)*, Akadémiai Kiadó, Budapest, 1975, 157 pages. ISBN 963-05-0701-3.

B. Szénássy, *Kőnig Gyula (1849–1913)*, Akadémiai Kiadó, Budapest, 1965, 142 pages.

L. Szögi, *Mérnökképző Intézet a Bölcsészeti Karon, 1782–1850: Institutum Geometrico-Hydrotechnicum (Fejezetek az Eötvös Loránd Tudományegyetem történetéből)* [Institute of Engineering on the Faculty of Humanities, 1782–1850: Institutum Geometrico-Hydrotechnicum (Chapters from the History of the Eötvös Loránd University)], ELTE, Budapest, 1980, 263 pages. ELTE 80321.

L. Szögi, A magyar felsőoktatás kezdetei [The Beginnings of Hungarian Higher Education], *Természet Világa* **127**(1996), no. 1, pp. 2–5 and no. 2, pp. 2–6. ISSN 0040-3717.

G. Szőkefalvi-Nagy, Lage der Wurzeln von linearen Verknüpfungen algebraischer Gleichungen, *Acta Scientiarum Mathematicarum (Szeged)* **1**(1923), pp. 127–138. ISSN 0001-6969.

G. Szőkefalvi-Nagy, Generalisation of Certain Theorems of G. Szegő on the location of zeros of polynomials, *Bulletin American Mathematical Society* **53**(1947), no. 12, pp. 1164–1169. ISSN 0273-0979.

B. Szőkefalvi-Nagy, Riesz Frigyes tudományos munkásságának ismertetése [Presentation of Frigyes Riesz' Scientific Work], *Matematikai Lapok* **5**(1953), pp. 170–182.

B. Szőkefalvi-Nagy, *Haar Alfréd összegyűjtött munkái. Haar Alfréd Gesammelte Arbeiten*, Akadémia Kiadó, Budapest, 1959, 660 pages (Hungarian and German).

B. Szőkefalvi-Nagy and C. Foiaş, *Analyse harmonique des opérateurs de l'espace de Hilbert*, Akadémiai Kiadó, Budapest, 1967, 373 pages.

P. Turán, Fejér Lipót matematikai munkássága [Mathematical Work of Lipót Fejér], *Matematikai Lapok* **1**(1950), no. 3, pp. 160–170.

P. Turán (ed.), *Fejér Lipót összegyűjtött munkái. Leopold Fejér Gesammelte Arbeiten*, Volume 1 and 2, Akadémiai Kiadó, Birkhäuser Verlag, Budapest, Basel, Stuttgart, 1970, 872 + 850 pages.

P. Turán, *On a New Method of Analysis and Its Applications*, (Published posthumously, with the assistance of Gábor Halász and János Pintz), Wiley-Interscience, New York, 1984, xvi + 584 pages. ISBN 978-0471892557.

P. Turán, On Carlson's theorem in the theory of zeta function of Riemann, *Acta Mathematica Academiae Scientiarum Hungaricae* **2**(1951), pp. 39–72. ISSN 0236-5294.

P. Turán, Alfréd Rényi munkássága [The Work of Alfréd Rényi], *Matematikai Lapok* **21**(1970), pp. 199–210.

G. Vályi, A harmadrendű görbék elméletéhez [To the Theory of Third Order Curves], *Mathematikai és Temészettudományi értesítő* **8**(1889), no. 2, pp. 23–28.

G. Vályi, A másodrendű partialis differentialis egyenletek elméletéhez [To the Theory of Second Order Partial Differential Equations], *Mathematikai és Physikai Lapok* **15**(1906), no. 6–7, pp. 256–269.

O. Varga and W. Blaschke, Integralgeometrie 9: Über Mittelwerte an Eikörpern, *Mathematica (Cluj)* **12**(1936a), pp. 65–80.

O. Varga, Beiträge zur Theorie der Finslerschen Räume und der affinzusammenhängenden Räume von Linienelementen, *Lotos, Prag* **84**(1936b), pp. 1–4.

O. Varga, Az integrálgeometria alkalmazásai a geometriai optikában [Applications of Integral Geometry in Geometrical Optics], *Magyar Tudományos Akadémia III. Osztályának Közleményei* **1**(1951), pp. 192–201.

L. Vekerdi, *A Tudománynak háza vagyon* [The Science Has a House], Magyar Tudománytörténeti Intézet, Piliscsaba, Budapest, 1996, 227 pages. ISBN 963-04-6426-X.

A. Wald, Contributions to the theory of statistical estimation and testing hypotheses, *The Annals of Mathematical Statistics* **10**(4) (1939), pp. 299–326.

A. Wald, Sequential tests of statistical hypotheses, *The Annals of Mathematical Statistics* **16**(2)(1945), pp. 117–186.

A. Wintner, *Spektraltheorie der Unendlichen Matrizen*, S. Hirzel Verlag, Leipzig, 1929, xii + 280 pages.

A. Wintner, *The Analytical Foundations of Celestial Mechanics*, Dover Publications, 2014, 464 pages. ISBN 978-0486780603 (Unabridged republication of 1947 second printing of the work originally published in 1941 by Princeton University Press, Princeton, New Jersey).

https://doi.org/10.1142/9781786349316_0008

Chapter 8

Mathematical Community in Czechoslovakia Between World Wars

Jindřich Bečvář and Martina Bečvářová

1. Introduction

1.1. *General Historical Background*

From the 16th century until 1918, the Czech lands, Moravia, part of Silesia, Slovakia and West Ukraine (part of Galicia, from the 18th century) were under the rule of the Habsburg monarchy, i.e. they were a part of the Austrian Empire, later the Austro-Hungarian Empire, which was a multicultural and multinational country with many ethnic groups, with rich and poor parts, some social problems, etc. From 1914 until 1918, its countries suffered greatly from World War I.

In the middle of the 19th century, Bohemia (or the Czech lands) became the industrial backbone and one of the cultural and scientific centres of the Habsburg monarchy. Nevertheless, Czech political representation remained at the periphery of significance and maintained a marginal influence on Austrian politics. Therefore Czechs fought against Austrian authorities and struggled for the independence of the Czech nation. Fortunately, during WWI, Bohemia was out of the geographical range of the first-hand war battlefields!

On October 28, 1918, the *Czechoslovak Republic* was founded and became a democratic republic in the centre of Europe, connecting Czechs, Germans, Slovaks, Ukrainians and other inhabitants. Its official borders and independence were confirmed by the Versailles Agreement; later they were enlarged thanks to challenging discussions at the international peace conference (January–July 1919) and thanks to plebiscites in 1920 and 1921. Prague became the capital city of the new republic. The Czechoslovak Republic was destroyed in 1938–1939 by the Nazis.

Figure 1. Map of the Habsburg monarchy (1899).[1]

1.2. *Historical Background in Science and Education* (*Czech Lands from the 1860s to the 1910s*)

In the second half of the 19th century, due to the rise of nationalistic movements, the Czech and German communities, which had naturally coexisted in Bohemia for several centuries, were separated. This separation was also reflected in politics, culture, science and education. An important feature of the period was the process in which Czech science was "becoming independent". The first professional societies were bilingual, and Czechs and Germans collaborated. During the 1870s, however, the Czech community grew stronger and more active, and finally prevailed. This process was accompanied, on the one hand, by protracted national conflicts and, on the other, by the expensive constructions of independent schools and the establishment of new associations [for example, the Czech Technical University in Prague and the

[1]The map is taken from D.H. Lange: *Volksschul-Atlas*, Dreihundertste Auflage, George Westermann in Braunschweig, 1899.

Figure 2. Map of Central Europe (the 1920s).

German Technical University in Prague,[2] the Czech University in Prague and the German University in Prague,[3] Lotos (1848, an association of German scientists), the Union of Czech Mathematicians and Physicists (1862), the Czech Academy of the Emperor Franz Joseph for Sciences, Literature and Arts (1890), the Association

[2] The *Prague Technical University* was founded in 1707 as a local Polytechnic school. From the beginning, lectures were taught there in German. From 1864, Czech and German parallel regular lectures and seminars in mathematics, physics, descriptive geometry etc. were conducted. In 1869, the Prague Technical University was divided into two independent schools, the *Czech Technical University in Prague* and the *German Technical University in Prague*. For more information on the history of both schools, see Birk (1931); Jílek and Lomič (1973); Jílek *et al.* (1978); Stark *et al.* (1906); Tayerlová *et al.* (2004); Velflík (1906, 1909, 1910, 1925).

[3] *Charles University in Prague* is the oldest university in central Europe and was founded by the Czech king (later the Holy Roman Emperor) Charles IV in 1348. The law adopted in Vienna on 28 February 1882 led to the foundation of two independent universities in Prague, the *Czech University in Prague* and the *German University in Prague*. Let us note that from the end of the 18th century, the lectures were taught in German. From 1871, Czech and German parallel regular lectures and seminars in mathematics, physics, medicine, law, history, etc. were taught at Charles University in Prague. The Czech University in Prague and the German University in Prague were typical European universities of the 19th century. There were four faculties (Arts, Medicine, Law, and Theology) which specialised in the education and preparation of teachers, lawyers, physicians

Figure 3. Centre of German Mathematics in Prague (Viničná Street, now the Faculty of Natural Sciences, Charles University).[4]

for the Support of German Sciences, Fine Arts and Culture in Bohemia (1891) and the German Mathematical Circle in Prague (1913)].

We should note that in Moravia, the *Polytechnic in Brno* was founded in 1848 (according to the political proclamation, it should have been a typical bilingual school, but in reality, the lectures were taught exclusively in German). This school was known under the name of *German Polytechnic in Brno*.[5] In the 1890s in Vienna, Czech political representation opened long-term, difficult discussions about the foundation of a Czech university in Moravia. Due to numerous political, economic and nationalistic reasons, these efforts were fruitless. In 1899, the *Czech Polytechnic in Brno* was established.[6]

In the second half of the 19th century, no friendly connections existed between the two "classic" universities and the two technical universities in Prague. There were virtually no personal collaboration and no cooperation of professional and teaching activities. The German Technical University in Prague and the German University in

and surgeons, civil servants and officers, priests, etc. There were many obligatory lectures and seminars and few optional lectures and seminars. For more information, see Folta *et al.* (1967); Nový a kol. (1961); Svatoš *et al.* (1998).

[4]The picture is from the authors' archive.

[5]For more information on the development of that school, see Haussner (1924); see also Šišma (2002, 2009).

[6]For more information on the development of that school, see Froněk (1969); see also Bečvářová (2008); Bečvářová and Čižmár (2011); Kounovský (1926); Koštál (1968); Nový a kol. (1961).

Figure 4. The Centre of Czech Mathematics in Prague (Ke Karlovu Street in 1911, now the Faculty of Mathematics and Physics, Charles University).[7]

Prague were preferred by the Viennese government. They obtained better buildings, larger libraries, all seminars and more standardised places than the Czech ones. German mathematicians did not need to compete with Czech mathematicians in order to demonstrate their activities and professional significance.

For the sake of completeness, we should add that at the German University in Prague, there were two positions of full professor of mathematics from 1882 until 1920. At the German Technical University in Prague, there were two positions of full professor of mathematics and one position of a full professor of descriptive geometry. Few private docents lectured at these schools and an assistant position was assigned to each full professor position. Both schools offered lectures by private docents and assistants, or technicians, respectively. Prague also had a number of German high schools, with approximately 20 teachers of mathematics, physics and descriptive geometry.

[7]The picture is taken from Svatoš *et al.* (1998).

Figure 5. Building of the German Technical University in Prague (Husova Street in 1786).[8]

Figure 6. Building of the Czech Technical University in Prague (Karl's Place in 1874).[8]

[8]Both pictures are taken from Bečvář *et al.* (2006).

At the Czech University in Prague, there was only one position of full professor of mathematics from 1882 until 1903. From 1891 until 1903, there was one additional position for "supply" professor of "modern" geometry and no position for an assistant. For a limited time, one *Privatdozent* [associate professor or private docent] lectured there. The situation changed after 1903, resp. 1911. The Czech Technical University in Prague had two positions of full professor of mathematics and one position of full professor of descriptive geometry, i.e., there was the same situation as at the German school. One or two private docents offered special mathematical lectures and seminars at the Czech Technical University in Prague and an assistant position was assigned to each full professor position. Prague also had a number of Czech high schools with approximately 20 teachers of mathematics, physics and descriptive geometry, which were active in the *Union of Czech Mathematicians and Physicists.*[9]

At the German Technical University in Brno as well as the Czech Technical University in Brno, there were two positions of full professor of mathematics and one position of a full professor of descriptive geometry. Brno also had a number of German high schools with approximately 10 teachers of mathematics, physics and descriptive geometry. The same held for the Czech high schools in Brno, but their members were active in the *Union of Czech Mathematicians and Physicists.*

1.3. *Czech and German Mathematical Communities: General Information*

In the second half of the 19th century, the Czech mathematical community was growing thanks to the activities of the Czech University and the Czech Technical University plus some associations and societies. For more than three decades in the second half of the 19th century, Czech mathematicians tried to prove that they could compete with German mathematicians and even surpass them in many respects. These efforts required a lot of time and energy. Not until the last quarter of the 19th century, the works of Czech professional mathematicians reached a standard and they were able to keep up with the individual trends of science, comprehend them, and contribute to global scientific research. Since the 1880s, Czech mathematical works have specialised in individual branches and approached more critically the subjects of their research. Czech mathematicians paid attention to the latest results in descriptive and projective geometry, the theory of matrices and determinants, quadratic forms and analysis, and from the beginning of the 20th century,

---

[9]To show another difference between universities in Prague, we can compare the number of students and teachers at the Czech University, German University, Czech Technical University and German Technical University in the academic year 1899/1900: Czech University (2,805 students, 189 teachers), German University (1,162 students, 168 teachers), Czech Technical University (1,023 students, 80 teachers) and German Technical University (419 students, 57 teachers).

they started to contribute to the development of individual mathematical disciplines [for example, Matyáš Lerch (1860–1922), Karel Pelz (1845–1908), Jan Vilém Pexider (1874–1914), Jan Sobotka (1862–1931), František Josef Studnička (1836–1903), Eduard Weyr (1852–1903), Emil Weyr (1848–1894), Karel Zahradník (1849–1916)]. Some of them also published their significant results in foreign languages (German, French and even Italian) to make them accessible to the European mathematical community. On the other hand, they published Czech versions of their works that appeared in foreign languages, as well as informative, stimulating or methodological articles in local journals, textbooks and monographs for Czech students.[10]

At the end of the 19th century, the professional standard of the Czech mathematical lectures and scientific works became comparable to the German ones and even began to exceed them in terms of student enrolment [thanks to the activities of certain Czech professors from the Czech University and the Czech Technical University; for instance, older ones such as Gabriel Blažek, (1842–1910), Vincenc Jarolímek (1846–1921), Augustin Pánek (1843–1908), Karel Pelz, Bedřich Procházka (1855–1934), František Josef Studnička and Eduard Weyr, younger ones such as Bohumil Bydžovský (1880–1969), Václav Láska (1862–1943), František Nušl (1867–1951), Karel Petr (1868–1950), Jan Sobotka, Matěj Norbert Vaněček (1852–1922) and František Velísek (1877–1914)]. The importance of the Czech mathematical community was increasing due to the growth in the number of their teachers and students, the development of the *Union of Czech Mathematicians and Physicists* and the improvement and increase of publication facilities.

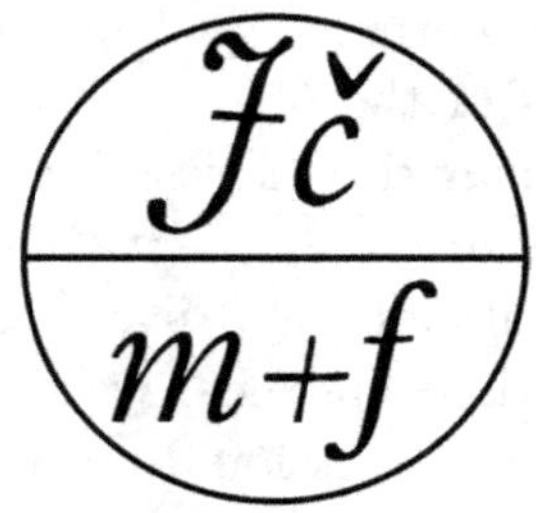

Figure 7. Logo of the *Union of Czech Mathematicians and Physicists* (1912).[11]

Jednota českých mathematiků a fysiků [The Union of Czech Mathematicians and Physicists] was founded in 1862 by four students from the Prague University as *Spolek pro volné přednášky z mathematiky a fysiky* [The Association for Free Lectures

[10]For more information on important Czech mathematicians and leaders of the Czech mathematical community in the 19th century, see Bečvář (1995, 1997); Bečvář *et al.* (2006); Bečvářová (2008, 2009); Bečvářová and Čižmár (2011); Folta *et al.* (1967); Kašparová and Nádeník (2010); Němcová (1998); Nový a kol. (1961).

[11]The picture is from the authors' archive.

of Mathematics and Physics]. It brought together Czech and German university students of mathematics and physics and Czech and German students from polytechnics. In the first years of its existence, it was a typical Prague bilingual association. Later the Czechs became dominant. In 1869, it was changed to *Jednota českých mathematiků* [The Union of Czech Mathematicians]. The association became a convenient center of mathematical (and physical) activities that were connected closely with those of the universities and professors, and bound together the university with the high school teachers and students, the teachers of the elementary schools, and recruited new people who were interested in mathematics and physics.

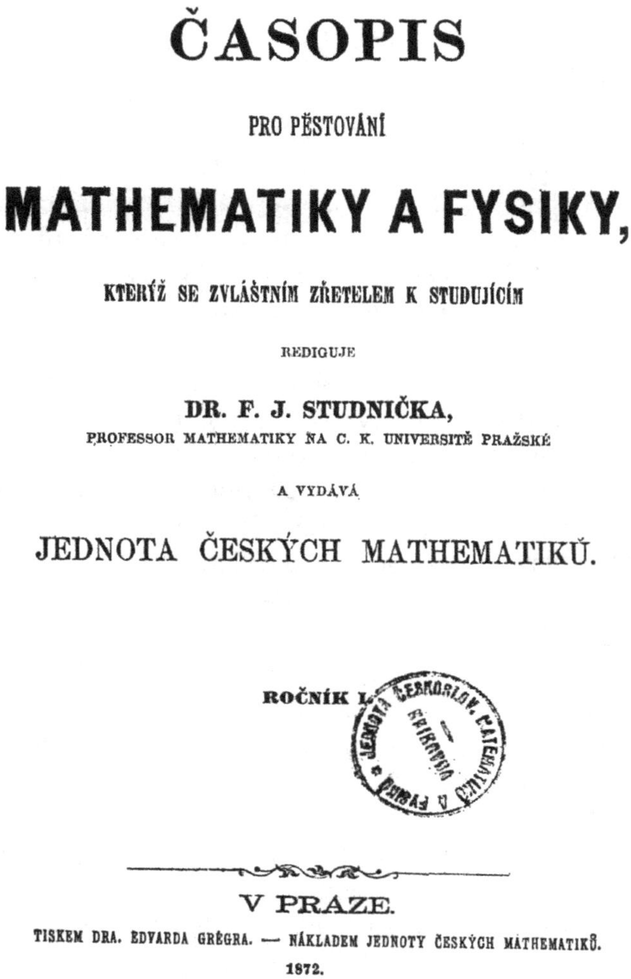

ČASOPIS

PRO PĚSTOVÁNÍ

MATHEMATIKY A FYSIKY,

KTERÝŽ SE ZVLÁŠTNÍM ZŘETELEM K STUDUJÍCÍM

REDIGUJE

DR. F. J. STUDNIČKA,

PROFESSOR MATHEMATIKY NA C. K. UNIVERSITĚ PRAŽSKÉ

A VYDÁVÁ

JEDNOTA ČESKÝCH MATHEMATIKŮ.

ROČNÍK I.

V PRAZE.

TISKEM DRA. EDVARDA GRÉGRA. — NÁKLADEM JEDNOTY ČESKÝCH MATHEMATIKŮ.

1872.

Figure 8. The title page of the *Journal for the Cultivation of Mathematics and Physics* (1872).[12]

[12]The picture is from the authors' archive.

In 1872, the Union started to publish its own journal entitled *Časopis pro pěstování mathematiky a fysiky* [The Journal for the Cultivation of Mathematics and Physics]. Only articles written in Czech could be published. The journal still exists, but under the name Mathematica Bohemica, and is published in English. From 1875 until 1878, Czech mathematicians tried to publish in the international journal *Archiv mathematiky a fysiky* [The Archive of Mathematics and Physics]. The articles could only be published in French, German or Italian. It was an unsuccessful attempt because only two volumes were published and then the journal ceased publication.

Prague became a center of the Czech mathematical community, which was scattered across the Austro-Hungarian Empire.[13]

On the other hand, in the two last decades of the 19th century, the number of German students decreased, because most of the German professors considered Prague to be merely a temporary stop on the way to Vienna or Germany [for example, professors Anton Josef Gmeiner (1862–1927), Wilhelm Weiß (1859–1904), Karl Ernst Zsigmondy (1867–1925)]. Some German mathematicians who connected their lives with Prague were not top-class productive mathematicians [for example, Karl Carda (1870–1943), Josef Grünwald (1876–1911), Anton Karl Grünwald (1838–1920), Anton Grünwald (younger, 1873–1932), Eduard Janisch (1868–1915)]. Gradually both German universities in Prague entered a position of personal and thematic isolation and acquired the character of provincial schools.

The same situation also occurred at the German Technical University in Brno in the second half of the 19th century and in the first two decades of the 20th century because some German mathematicians also considered Brno as a temporary stop on their way to Austria or Germany [for example, Emanuel Czuber (1851–1920), Oskar Peithner (1852–1923), Franz Hočevar (1853–1919), Heinrich Tietze (1880–1964)]. But there were also mathematicians who connected most of their productive lives with teaching in Brno [for example, Karl Prentner (1823–1904), Franz Unfendiger (1833–1890), August Leo Otto Biermann (1858–1909), Emil Waelsch (1863–1927)].[14]

Shortly before World War I the situation of the German mathematical community in Prague changed significantly because in a short period a group of young, energetic, talented professors educated at foreign universities, who belonged amidst the world's elite even at their youthful ages, arrived in Prague [for example, Wilhelm Blaschke (1885–1962), Philipp Frank (1884–1966), Paul Georg Funk (1886–1969), Gerhard Hermann Waldemar Kowalewski (1876–1950), Theodor Michael Friedrich Pöschl (1882–1955); during the war Karl Mack (1882–1943) and Roland Weitzenböck

[13]For more information on the Czech mathematical community and its activities, see Bečvář (1995); Bečvář and Bečvářová (2012); Houdek (1872); Košťál (1968); Němcová (1998); Nový a kol. (1961); Pátý (1987); Posejpal (1912); Řehořovský and Seydler (1887); Vetter (1952); Veselý (1962).
[14]For more information, see Šišma (2002).

(1885–1955), or, at the war's end, Ludwig Berwald (1883–1942)] and obtained regular positions there.

Young professors, private docents and assistants, and their students with a passion for mathematics, spiced up life in the Prague German mathematical community. They understood that regular professional meetings, which they were accustomed to from abroad, would be very useful and important for their future work. And so they created a special society called the *German Mathematical Circle in Prague*. From spring 1913 until spring 1934, 365 lectures, discussion contributions and reports on the newest discoveries, results and works published in German, French and Italian journals were given in their "mathematical circles". In lectures and reports, mathematicians paid attention to the following current mathematical topics: differentials, difference and integral equations, function theory and potential theory, functional analysis, geometric function theory, special functions, the axiomatic probability theory, classification of substitutions, fundamentals of statistics, fundamentals of mathematical logic, variation and adjustment calculus, mathematical theory of games, elementary, differential, affine and integral geometry, Minkowski geometry, Hilbert spaces, special geometric transformations, geometric extremal problems and geometric levelling, and the history of mathematics. It is not surprising that thanks to these activities the German mathematical community in Prague quickly became stronger, more active and well-known, not only in central Europe.[15]

A better situation for the development of German mathematics in Brno occurred from 1900 until 1912, due to the fact that at the German Technical University there taught as assistants of mathematics, as professor of mechanics or as professor of mathematics, Ernst Fischer (1875–1954), Georg Hamel (1877–1954), Richard von Mises (1883–1953), Johann Karl August Radon (1887–1956) and Heinrich Tietze, i.e., future well-known and top-class mathematicians who later obtained better positions in Germany. After the foundation of the Czechoslovak Republic, the same exceptional mathematicians did not obtain their regular positions in Brno. Unfortunately, they spent only a very short time in Brno.[16]

1.4. *WWI and the Situation in Prague and Brno from the Viewpoint of Mathematicians*

For both mathematical communities in Prague, World War I caused only some "small" difficulties (for example, an insufficiency of basic foods, problems with supplies and consumer goods, poor health and social care, lack of rooms for lectures and studies because buildings were turned into military hospitals, a shortage of

[15]For more information, see Bečvářová (2015, 2016, 2018); Kowalewski (1950); Nový a kol. (1961); Tobies (2006); Toepell (1991).

[16]For more information, see Šišma (2002).

international contacts, travel possibilities and foreign literature, a decrease in the number of students enrolled at high schools) because there were no major struggles and direct fights that took place there.

In Prague, only a relatively limited number of Czech and German mathematicians and professors of mathematics, but many young students of mathematics, had to interrupt their careers or studies to go to the battlefields. For example, professor František Velísek was executed on the Russian front in 1914,[17] Roland Weitzenböck was shortly on the Russian front as a high officer, Eduard Čech, as an older student, served as a lower officer in the Austrian war staff on the Italian front, Václav Hlavatý, as an older student, served as a lower officer in the Austrian war staff on the Italian front, Joseph Fuhrich, as a young student, fought as a soldier on the first lines of the Russian and Italian fronts.[18]

Prague mathematicians were "saved" because of their higher ages (for example, A.K. Grünwald, V. Láska, K. Petr, G.A. Pick, J. Sobotka) and because they were not Austro-Hungarian citizens (for example W. Blaschke, G. Kowalewski, T. Pöschl) or they were protected as people with high importance for special war tasks (standard counting for artillery, meteorological measurements for the navy or air force, arms research, ensuring connections and communications, ciphering and decoding, war planning and preparation of strategies). They usually served as officers or teachers at secondary military schools (Vienna, Prague, Olomouc, Luka, etc.).

During World War I, education in mathematics continued without "many obstacles" at high and secondary schools as well as universities in Prague. For example, in

---

[17]František Velísek (1877–1914) studied mathematics and physics at the Czech University in Prague from 1899 to 1901 and later from 1902 until 1904. In 1904, he successfully passed the state examinations to become a teacher of mathematics and physics at secondary schools. In 1905, he defended his PhD thesis *Příliv a odliv* [Flood and Tide]. The same year, he became an assistant of mathematics at the Czech Technical University in Prague. In 1909, he finished his habilitation procedure at the Czech Technical University in Prague and was appointed a *Privatdozent* [associate professor]. In 1909, i.e., at the age of 32, he obtained a post of extraordinary professor of mathematics. In August 1914, he had to start to serve as an officer on the Russian front. On 27 August 1914, he was shot in Krasno (Krásné) not far from Lvov. For more information, see Bečvářová (2008).

[18]In 1915, Joseph Fuhrich (1897–1945) matriculated at the German state secondary school, which, however, still qualified him to enroll at any institution of higher learning. He wished to study mathematics at the German University in Prague, but had to fight as a soldier on the first lines of the Russian front and later in the first line of the Italian front. In 1917, he was sent to a special military hospital in Vienna because he suffered from open TBC of lung. Later, he studied mathematics at the German University in Prague, at the Technical University in Leipzig, and at the University in Giessen, where he defended his PhD thesis. In the 1920s, he became a specialist in financial mathematics, problems of insurance, income tax and state budgets. In 1928, he was appointed a *Privatdozent* [associate professor] at the German Technical University in Prague. In 1937, he became an extraordinary professor of financial mathematics and mathematical statistics and in 1942 an ordinary professor. He probably taught in Prague until the spring of 1945. For more information, see Bečvářová (2016); Toepell (1991).

the war period, mathematics was taught at the Faculty of Arts of the German University by two ordinary professors and at the Faculty of Arts of the Czech University by two full professors and one extraordinary professor. Their minimal lecture duty was 5 lecture hours per week. The teachers usually listed course lectures in three to four year cycles in the scope of 3–5 hours per week, which they completed by facultative lectures and seminars in the scope of 2–4 hours per week. The low number of professors and associate professors, conjoined with the large scope of compulsory lectures, did not provide sufficient space for a larger thematic diversity of lectures compared to large universities abroad. Only two professors taught almost the entire spectrum of university mathematics, i.e., they did not specialise in a narrow circle of lectures and themes as nowadays. When it came to filling professor vacancies, versatility was preferred over narrow and deep specialisation. Teaching activities were stable from the perspective of their themes as well as the personal occupancy of mathematical fields. It is not possible, in Czech or foreign archives, to find any materials which provide detailed information on the content of the individual Prague mathematical lecture courses. Annotations or characteristics of subject content were not given in lecture lists. There were no compulsory syllabuses, nor lists of basic and secondary literature. The records of professors do not contain any data on proposals for new subjects, not even in facultative lectures. Professional content and the pedagogical level of teaching were fully in the control of individual professors and docents, or "professional sections" of the faculty, respectively. During WWI, we find only very few modifications of basic study conditions, of "partial and state" exams, modifications of conditions for awarding doctorates and habilitations, professional and processional demands of individual procedures, and structure and demands of teacher qualification exams. In the branch of mathematics, doctoral procedures, habilitation procedures, nominations of professors as well as teacher qualification examinations continued without any difficulties.[19]

[19]For example, during WWI, three PhD theses in mathematics were successfully defended at the Faculty of Arts of the German University in Prague [Emil Nohel (in 1914), Karl Löwner and Arthur Winternitz (both in 1917), all of them became assistants, later associate professors, extraordinary or ordinary professors at the German University in Prague] and six PhD theses in physics accompanied the main examinations in mathematics [Paul Torsch and Emma Becker (both in 1914), Emilie Heindl and Reinhold Fürth (both in 1916), Irene Schweizer and Elfriede Klepp (both in 1918)]. It should be emphasised that four PhD theses in physics were written by women and only two theses were written by men. At the Faculty of Arts of Czech University in Prague, four PhD theses in mathematics were successfully defended [Jindřich Kubíček (in 1916), Karel Šilháček, Miloš Kössler and Vincenc Nechvíle (all in 1918)] and 15 PhD theses in physics or astronomy accompanied the main examinations in mathematics [Jaroslav Šafránek and Alois Jemelka (both in 1915), Karel Teige, Josef Novák and Miloslav Valouch (all in 1916), František Pleskot, Emanuel Herolt, Rudolf Šimůnek (all in 1917), Božena Novotná, Vojtěch Štech, Marie Kuthanová, Ludmila Brázdilová, Slavko Rozgaj, Vincenc Zazvonil, František Ullrich (all in 1918)]. It should also be emphasised that three PhD theses in physics were written by women. For more information, see Bečvářová (2008, 2016); Tulachová (1965) and Výborná *et al.* (1965).

The *Union of Czech Mathematicians and Physicists* and the *German Mathematical Circle in Prague* operated regularly in Prague between 1914 and 1918. There were classical meetings with lectures and discussions which were closely connected with the professional work of individual pedagogues, doctoral candidates and students. The associations had high professional standards and they almost immediately responded to the newest facts and research trends.[20]

Let us add that during World War I, there was an absolutely different situation in Brno from that of Prague. Heinrich Tietze, the ordinary professor of mathematics, was sent to the Russian front, where he spent the entire time in the war. Lothar Schrutka (1881–1945), the extraordinary professor of mathematics, had to work from 1917 until 1918 as a teacher of mathematics at the military school in Vienna.[21]

Figure 9. Building of the German Technical University in Brno (Komenského náměstí — Commenius Square).[22]

[20]For more information, see Bečvář and Bečvářová (2012); Bečvářová (2008, 2015, 2016, 2018); Nový a kol. (1961); Vetter (1952) and Veselý (1962).
[21]For more information, see Šišma (2002).

Figure 10. Building of the Czech Technical University in Brno (Veveří).[22]

1.5. *Situation in Czechoslovakia from 1918 until the 1930s: General Information*

After the creation of the Czechoslovak Republic (1918), Czech high schools and the mathematical community developed rapidly. Let us note that from 1918 until 1921, numerous new state universities or faculties were established in Czechoslovakia: new Czechoslovak universities in Brno and Bratislava (now Slovakia), new faculties of technical universities in Prague (in the Czech and German independent versions), new business schools in Prague and Libverda (the first one for Czech students and the second one for German students), new faculties of technical universities in Příbram (the Czech mining school and the German mining school), new veterinary and agricultural schools in Brno (with Czech as the teaching language only). The last newly opened school was the Technical University in Košice (now Slovakia), which was founded in 1938.

[22]Both pictures are available on the website https://www.vutbr.cz [25.2.2018].

Figure 11. The Big Heraldic Emblem of the Czechoslovak Republic (1920).[23]

There were many national, religious, economic and social problems in Czechoslovakia. However, in contrast to other non-German parts of the former Austro-Hungarian Empire, German high and secondary schools, scientific and cultural communities and professional associations were not abolished in Czechoslovakia. On the contrary, for example, the German University in Prague became an equal, recognised and respected state university, not reduced, oppressed or financially suppressed by the new republic. Actually, it was the only official state minority university in post-war Europe divided into states based on a supranational principle. It was a relatively small but significant scientific and pedagogical workplace in Europe, and which maintained this position until the beginning of World War II.

The German University in Prague, the German Technical University in Prague, the German Technical University in Brno, the German Business School in Libverda and the German Mining School in Příbram flourished after World War I.

In the spring of 1939, after the Nazi occupation of the Czech lands, German universities were directly subordinated to the Reichsministry of Education in Berlin. Teaching at the German universities continued until the end of World War II. In May 1945, the German universities were abolished by Czechoslovak authorities with retroactive effect from 17 November 1939.

On 17 November 1939, all Czech universities and all other Czech higher-education institutions were initially closed by the Nazis for three years. However, they remained closed until the end of World War II.

[23]The picture is from the authors' archive.

2. Czech Mathematical Community at the Beginning of the 20th Century

2.1. *Introduction*

From the second half of the 19th century the Czech mathematicians had tried to prove that they were fully equal to the Germans and that they even surpassed them in many ways. These efforts cost a lot of time and strength. Only in the last quarter of the 19th century some of the Czech mathematicians reached such a level as to be able to follow, understand, and participate in global research. The number of people involved in teaching mathematics and attempting to solve mathematical problems independently, or seeking to apply mathematical results in practice (in industry, banking, government, etc.) had gradually grown.

An interesting feature of the 19th century was the establishment of scientific institutions, where lectures, scientific discussions, professional publications (journals, monographs and textbooks) and profesional reports were held, despite the initial lack of funds and a small number of experts. Scientific societies and associations, which originated in the Czech lands at the end of the 18th and the first half of the 19th century, associated Czech and German speaking and writing specialists from several fields. They were usually bilingual; their activities were not yet very much influenced by national conflicts. After the fall of Bach's absolutism (1859), a space for universal activities was opened up in the Czech society; various associations were formed and developed. From the middle of the 1860s, their professional and especially language profiling began.

The activities of *Jednota českých mathematiků* [The Union of the Czech Mathematicians] were important at the turn of the 19th and 20th centuries in terms of the development of Czech mathematical life. Over time, it became a natural center of national mathematical work, its activities were closely interconnected with universities, they united Czech university and secondary school teachers and students, teachers of lower schools and gained more people interested in mathematics and physics. However, only a few members worked for the Union with a great selfless commitment (e.g., František Houdek, Augustin Pánek, Jaroslav Sobička, Václav Starý, František Josef Studnička, Václav Řehořovský). These were the people who had sacrificed their time, professional careers, and financial resources for its further development, enabling it to expand and maintain its relatively extensive educational, publishing and popularisation activities at critical times.[24]

For completeness let us remind that mathematics was also cultivated in Bohemia in the mathematical and natural sciences section of the *Královská česká společnost*

[24]About the history of the Union, see Bečvářová (1999, 2008); Houdek (1872); Košt'ál (1968); Pátý (1987); Posejpal (1912); Řehořovský and Seydler (1887); Vetter (1952); Veselý (1962).

nauk [The Royal Bohemian Society of Sciences] founded in 1770 and in a similar section of the *Česká akademie císaře Františka Josefa pro vědy, slovesnost a umění* [The Czech Academy of the Emperor Franz Joseph for Sciences, Literature and Arts].[25]

Publishing, educational and popularising activities were also related to the activities of individual professional associations. Mathematical–Physical journals associated with the Union's work have influenced the development of Czech mathematical terminology and school mathematics for many years. They provided space for the publishing and educational activities of members and non-members of the Union, teachers and students. Especially the journal *Časopis pro pěstování mathematiky a fysiky* [The Journal for the Cultivation of Mathematics and Physics] tried to spread education. It encouraged both professional and educational activities, brought detailed information on the activities of the Union and supported an interest in active participation in the Union. It was a link between the Czech mathematical intelligence scattered throughout Austria-Hungary and beyond with the center in Prague.[26]

Since the 1880s, it is possible to observe increasing specialisation and a critical approach to focus on research in Czech mathematical production. Czech mathematicians responded to new results in descriptive and projective geometry, matrix theory, determinants, quadratic forms and analysis, and contributed to the development of several disciplines. Especially Matyáš Lerch, Karel Pelz, Jan Vilém Pexider, Jan Sobotka, František Josef Studnička, Emil Weyr, Eduard Weyr and Karel Zahradník had an important role in the Czech mathematical community. Significant results were published in foreign languages (German, French, or Italian) to be accessible to the European mathematical community.[27] Czech journals sometimes published Czech versions of foreign-language works and rather informative, popularising or methodological articles.[28]

---

[25]For more information, see Bečvářová (2008); Beran (1971); Kalousek (1885); Pokorný and Novotný (1993); Purš (1985).

[26]About the history of the *Journal for the Cultivation of Mathematics and Physics*, see Bečvářová (2008); esp. pp. 123–132. For more information, see Brdička and Schwabik (1987); Vetter (1922).

[27]For example: *Zprávy Královské české společnosti nauk*, *Rozpravy České akademie věd*, *Archiv der Mathematik und Physik*, *Sitzungsberichte der kaiserlichen Akademie der Wissenschaften in Wien*, *Anzeiger der kaiserlichen Akademie der Wissenschaften in Wien*, *Zeitschrift für Mathematik und Physik*, *Mathematische Annalen*, *Journal für die reine und angewandte Mathematik*, *Monatshefte für Mathematik und Physik*, *Nouvelles Annales de Mathématiques*, *Comptes Rendus hebdomadaires des séances de l'Académie des sciences*, *Mémoires de la Société des sciences physiques et natureles de Bordeaux*, *Bulletin de la Société Mathématique de France*, *Bulletin de l'Académie royale des sciences, des lettres et des beaux-arts de Belgique*, *Mémoires de la Société royale des sciences de Liège*, *Rendiconti del Reale Istituto Lombardo*, *Annali di matematica pura ed applicata*, *Giornale di matematiche*. For more see Bečvářová (2008), pp. 133–184.

[28]For more information see Bečvářová (2008), esp. pp. 133–184.

2.2. *Czech University in Prague*

Since its foundation in 1882, the Faculty of Arts of the Czech University in Prague, where mathematics was taught until 1920, had been focusing mainly on the education of high-quality and universally educated future high school professors. It was an important center of Czech mathematical life not only in Prague.

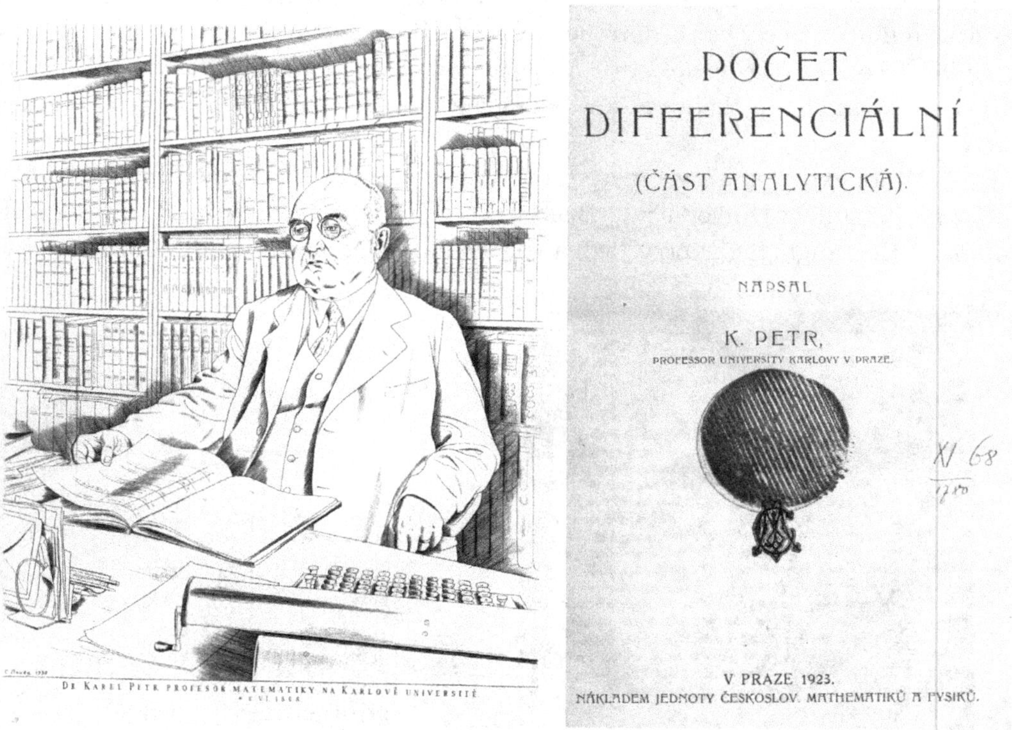

Figure 12. The picture of Karel Petr and the title page of his well-known textbook.[29]

In 1903, when two renowned professors of mathematics František Josef Studnička (1836–1903) and Eduard Weyr (1852–1903) died, it was necessary to immediately find new teachers of mathematics. In charge of the lectures was relatively unknown Karel Petr (1868–1950), who had been only a private associate professor at Czech Technical University in Brno since 1902. Already in 1903, he became an extraordinary and in 1908 a full professor at the Czech University in Prague. He devoted great effort to lectures in mathematical analysis and algebra, both seminar and proseminar. He raised a new generation of Czech teachers and mathematicians.

[29]The pictures are from the authors' archive. The picture of Karel Petr was made by Cyril Bouda (1901–1984), a Czech painter.

He was an expert in analytical number theory, theory of algebraic forms, mathematical analysis and numerical methods. He wrote a number of papers and quality textbooks on mathematical analysis — *Počet integrálný* [Integral Calculus, 1915, 1931] and *Počet differenciální* [Differential Calculus, 1923]. His lectures *O rovnicích differenciálních* [Differential Equations] were published already in 1911. For 15 years, Karel Petr was the editor of the mathematical part of the journal *Časopis pro pěstování mathematiky a fysiky* [The Journal for the Cultivation of Mathematics and Physics] and he actively worked in the Union.[30] His name is still known in mathematics today (see *Petrova-Douglasova-Napoleonova-Neumannova věta* [Petr–Douglas–Napoleon–Neumann's theorem] or the generalisation of the Napoleon triangle theorem).

In 1904, an experienced geometer Jan Sobotka (1862–1931) was transferred from the Czech Technical University in Brno to the Czech University in Prague as a full professor. He taught geometry, led a mathematical seminar and a mathematical

Figure 13. The picture of Jan Sobotka and the title page of his well-known textbook.[31]

[30] About the life and work of Karel Petr, see Crkalová (1992); Kořínek (1938, 1948); Nušl and Kössler (1928); Schwarz (1969).

[31] The picture of Jan Sobotka is taken from F. Kadeřávek, Jan Sobotka, profesor matematiky na universitě Karlově, šedesátníkem [Professor Jan Sobotka Sexagenarian], *Časopis pro pěstování matematiky a fysiky* **52**(1923), pp. 1–9. The picture of the title page is from the authors' archive.

proseminar with Karel Petr. He has greatly enhanced the preparation of high school professors and future mathematicians. During his stay in Brno, he wrote the monograph *Deskriptivní geometrie promítání parallelního* [Descriptive Geometry of Parallel Projection], which was published in 1906 and where he presented in a number of places an original interpretation of basic visualisation methods, which are based on parallel projection. He also published the three-part monograph *Diferenciální geometrie* [Differential Geometry, 1909, 1914, 1914]. His textbooks had become a basic study material for one generation of Czech students. Jan Sobotka focused on descriptive, differential, projective, kinematic and elementary geometry. His work was extensive. From 1885 to 1931, he published more than 100 papers. The most important works were in the field of visualisation methods and constructive geometry of curves and surfaces. Into the scientific literature entered the so-called *Sobotkovy konstrukce* [Sobotka's Constructions] that allowed the investigation of the relationships between axonometric and conjugated rectangular projections.[32]

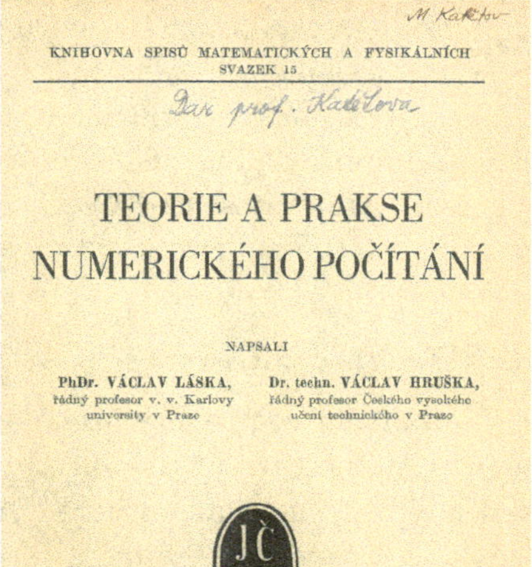

KNIHOVNA SPISŮ MATEMATICKÝCH A FYSIKÁLNÍCH
SVAZEK 15

TEORIE A PRAKSE
NUMERICKÉHO POČÍTÁNÍ

NAPSALI

PhDr. VÁCLAV LÁSKA, řádný profesor v. v. Karlovy university v Praze

Dr. techn. VÁCLAV HRUŠKA, řádný profesor Českého vysokého učení technického v Praze

JČMF

PRAHA 1934
NÁKLADEM JEDNOTY ČESKOSLOV. MATEMATIKŮ A FYSIKŮ

Figure 14. The picture of Václav Láska and the title page of the textbook written by V. Láska and V. Hruška.[33]

[32]About the life and work of Jan Sobotka, see Kašparová and Nádeník (2010); Košt'ál (1968).

[33]The picture of Václav Láska is taken from Q. Vetter: Profesor Dr. Václav Láska šedesátníkem [Professor Dr. Václav Láska Sexagenarian], *Časopis pro pěstování matematiky a fysiky* **53**(1924), pp. 1–19. The picture of the title page is from the authors' archive.

In 1911, Václav Láska (1862–1943) became a professor of applied mathematics at the Czech University in Prague. He came as an experienced teacher after many years of lecturing at the University in Lvov and at the Technical University in Lvov. From 1920, he was the director of the Department of Applied Mathematics. From 1927, he was the Director of the Geophysical Institute. He is the author of more than 300 works (function theory, probability, statistics, geodesy, etc.). He was also engaged in mathematical and physical geography, cartography, hydrology, hyetology, balneology, climatology, geophysics, geology, seismology, photogrammetry, astronomy and space physics, history and philosophy of exact and natural sciences, and popularisation. His contemporaries considered him the last polyhistorian of sciences. He was the editor of the text section of *Atlas Republiky Československé* [Atlas of the Republic of Czechoslovakia, 1935], he wrote the monographs *Počtářství geodetické t. j. návod ku počtům trigonometrickým a polygonálním pro účely katastrální* [Geodetic Surveying, i.e., Guidance on Trigonometric and Polygonal Calculations for Cadastral Purposes, 1894], *Vyšší geodesie* [Higher Geodesy, 1896], *Počet pravděpodobnosti* [Calculus of Probabilities, 1921] and *Úvod do geofyziky* [Introduction to Geophysics, 1927]. Together with Václav Hruška (1888–1964) he wrote two books — *Počet grafický a graficko-mechanický* [Graphical and Graphical-Mechanical Calculus, 1923] and *Theorie a prakse numerického počítání* [Theory and Practice of Numerical Calculation, 1934]. His name is still alive in science today (see *Láska's formula* for determining the approximate distance of the earthquake epicentre from the point of measurement). He represented Czech and Czechoslovak applied mathematics in a number of international societies.[34]

2.3. *Czech Technical University in Prague*

The second important centre of mathematical life in Prague was the Czech Technical University. At the turn of the 19th and 20th centuries, mathematics education experienced a significant change there. Due to the extension of the subject matter to professional subjects and the underestimation of the role of mathematics in the preparation of technicians, there was a great reduction in the number of compulsory mathematical lectures.

In 1908, the First Chair of Mathematics was occupied by Matěj Norbert Vaněček (1852–1922) the successor of Gabriel Blažek (1842–1910). Vaněček obtained habilitation for higher mathematics in 1906 and was appointed a full professor in 1908. He taught at the Czech Technical University in Prague until his death. Between 1880 and 1890, he wrote alone or with his brother Josef Sylvestr Vaněček (1848–1922) a series of papers on geometrical properties of curves and their constructions. After his arrival to the Technical University in Prague, he was mainly concerned with pedagogical issues. He published his views on teaching mathematics in 1914 in the

[34]About the life and work of Václav Láska, see Bečvářová (2011).

booklet called *K reformě studia středoškolského* [About the Reform of Secondary School Studies].[35]

After the death of Augustin Pánek (1843–1908), mathematics was taught in substitution by an assistant in the Second Chair of Mathematics, František Čuřík (1876–1944), who remained at the assistant post until the end of World War I. In 1920, he became a professor of mathematics and descriptive geometry at the Mining Academy in Příbram, where he taught until his death. He wrote several textbooks for technical college students (e.g., *Základy vyšší matematiky I* [Fundamentals of Higher Mathematics I, 1915, 1923], *Základy vyšší matematiky II* [Fundamentals of Higher Mathematics II, 1918, 1930], *Matematika. Technický průvodce 1* [Mathematics. Technical Guide 1, 1921, 1944]).[36] In November 1909, František Velísek (1877–1914) was appointed an extraordinary professor on the Second Chair of Mathematics. He was a talented mathematician who had obtained habilitation a month ago before his appointment. After the outbreak of World War I, he was mobilised as a reserve officer of the Austro-Hungarian army and sent to Galicia, where he was shot. His lectures were taken over by the assistant Karel Rychlík (1885–1968).

In 1908, the Czech Technical University in Prague acquired the Third Chair of Mathematics, because František Nušl (1867–1951) was appointed an extraordinary professor of mathematics by the Ministry in Vienna. He obtained habilitation in astronomy and geometric optics at the Czech University in Prague in 1905. In 1911, he was appointed a full professor. He remained at the Czech Technical University in Prague until the school year of 1920/1921, when he was given a vacation to manage the state observatory. In 1926, he was appointed a full professor of astronomy at the Faculty of Natural Sciences of Charles University, and his interest turned to astronomy. Nušl's main discipline was the practical astronomy and geometric optics; his works were published in both Czech (Czechoslovak) and foreign journals.[37]

After Velísek's death, the Czech Technical University in Prague had only two full professors of mathematics until the end of World War I. They were in charge of more than 600 students. Parallel lectures were substituted by assistants and private associate professors[38] until 1920, when Josef Klobouček (1875–1939) was appointed an extraordinary professor of mathematics.[39]

At the beginning of the 20th century, the interest in actuarial education, which was demanded by insurance companies and banks, began to increase. Josef Beneš

[35] About the life and work of Matěj Norbert Vaněček, see Folta (1963); Hruška (1923). About the life and work of Josef Sylvestr Vaněček, see Bečvářová (2013).
[36] About the life and work of František Čuřík, see Kotůlek (2014).
[37] About the life and work of František Nušl, see Libický (1951); Petr (1928).
[38] From 1910, some lectures were substituted by an associate professor Bohumil Bydžovský.
[39] About the life and work of Josef Klobouček, see Vojtěch (1935).

(1859–1927) was appointed its first full professor at the Czech Technical University in Prague. He worked at this school from 1904 to 1926.

Descriptive geometry had been an important field at the Czech Technical University in Prague since the middle of the 19th century. Bedřich Procházka (1855–1934) was appointed a professor of descriptive geometry in 1908. For him, the Czech Technical University in Prague was not an unknown place since he had worked there in the 1890s as an assistant. He stayed at this school until his retirement in 1925. He wrote a number of works on descriptive, kinematic and differential geometry, and devoted himself to stereographic and perspective projection and photogrammetry.[40] Especially his university textbooks on descriptive geometry, which he began to write during his time in Brno, became known. He published, together with Vincenc Jarolímek (1846–1921), the textbook *Deskriptivní geometrie pro vysoké školy technické* [Descriptive Geometry for Technical Universities, 1909, 1919, 1922] and the monograph *Vybrané stati z deskriptivní geometrie* [Selected Parts from Descriptive Geometry, 1912 to 1918]. Both titles gained considerable popularity.[41]

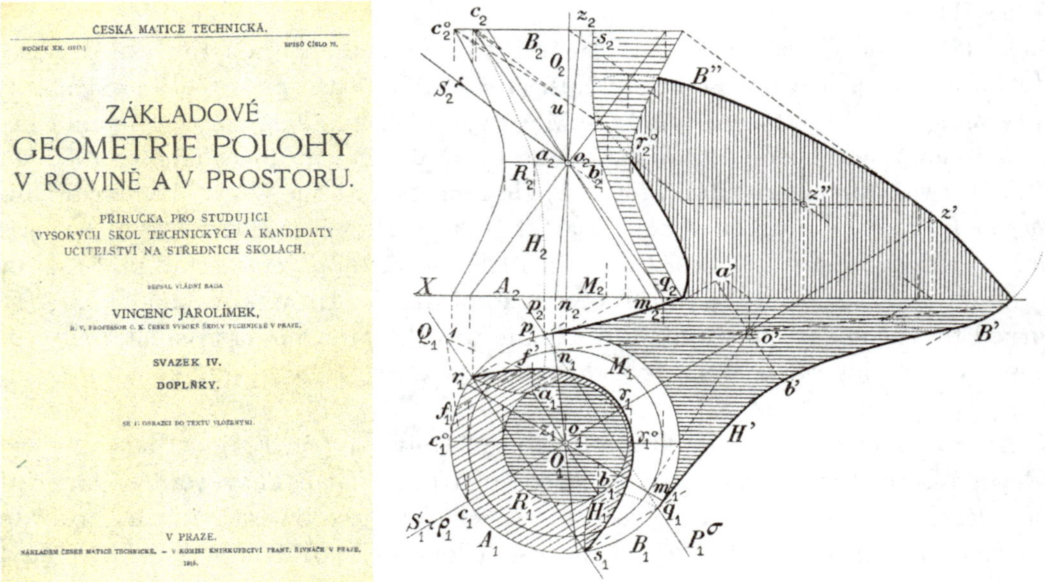

Figure 15. The title page of the textbook written by V. Jarolímek and the drawing of the lighting of a rotational hyperboloid.[42]

[40]Photogrammetry — processing information based on photographic images, reconstruction of shapes, measuring of dimensions, and positioning of objects that were displayed on a photo.

[41]About the life and work of Bedřich Procházka, see Kounovský (1926); Košťál (1968). About the life and work of Vincenc Jarolímek, see Folta (1958, 1959); Košťál (1968); Sobotka (1916).

[42]The pictures are from the authors' archive. The drawing is taken from the book V. Jarolímek, B. Procházka: *Deskriptivní geometrie pro vysoké školy technické* [Descriptive Geometry for Technical Universities], Praha, 1909.

In 1907, due to the large number of students at the Czech Technical University in Prague, there was a need to establish the Second Chair of Descriptive Geometry. The professors wanted to remove the imbalances from the early 20th century, when one parallel lecture was led by a full professor and the other by a mere substitute teacher. Vincenc Jarolímek was appointed the first professor of descriptive geometry of the Second Chair. He came from the Czech Technical University in Brno in 1906 and was giving lectures as a substitute teacher in Prague for one year. He stayed at the Czech Technical University in Prague until his retirement in 1915. He took over lectures for students of the construction department, but he became seriously ill in 1908 and from the following year he had a health holiday until 1915. Jarolímek published a number of articles on descriptive, projective, kinematic and differential geometry. As a high school professor, he wrote the quality textbook *Deskriptivní geometrie pro vyšší školy reálné* [Descriptive Geometry for Higher Grammar Schools, 1875], which gained considerable popularity, and several geometry textbooks for different classes of grammar schools.

From 1904, the assistant Bohumil Chalupníček (?–?) led parallel lectures on descriptive geometry.[43] He was appointed an adjunct after Jarolímek's arrival in Prague. At this point, the problem of running parallel lectures seemed to be solved. When Vincenc Jarolímek fell ill in 1908, the Czech Technical University in Prague was trying to get the third professor's post and it was just suggesting Chalupníček. However, the Ministry did not want to accept Jarolímek's retirement or to set up the third professorship. His lectures were substituted by Chalupníček and by František Kadeřávek (1885–1961), who obtained habilitation in the subject of synthetic geometry in space in 1912. In 1917, he was appointed an extraordinary professor and in 1920 he became a full professor and connected his whole active life with the Czech Technical University in Prague.[44]

2.4. *Czech Technical University in Brno*

After long efforts of the Czech political and cultural representation, the second Czech Technical University was established in Moravia. It became another center of Czech mathematical life. The decree on its establishment was signed by Emperor Franz Joseph I in Celovec (now Klagenfurt) in 1899.[45]

In 1899, the first institute of mathematics was founded at the Czech Technical University in Brno where Karel Zahradník (1849–1916) was appointed a full professor. He came to Brno after more than 20 years at the University of Zagreb

---

[43] About Bohumil Chalupníček, see http://bohumil-chalupnicek.cz/setlik.html [2.3.2019].

[44] About the life and work of František Kadeřávek, see Havlíček (1961); Kepr (1960); Menšík (1955).

[45] About the establishment and development of the Czech Technical University in Brno, see Froněk (1969). About teaching mathematics and Czech mathematical community in Brno, see Bečvářová (2008); Bečvářová and Čižmár (2011); Košťál (1968).

in Croatia as a routine mathematician with a number of publications, textbooks, didactic and popularising articles written in Czech, German, Croatian and French. In 1899, as a senior in the corps of professors, he became the first rector of the Czech Technical University in Brno and greatly contributed to its successful launching and further development. He remained in the rector's office even in the academic year 1900/1901. He could not focus on professional work at that time because he was overloaded with administrative duties. He was preparing the curricula of individual subjects, seeking new teachers and funds, leading correspondence, etc.

Karel Zahradník was an excellent teacher and educationist. He was a founding member of the *Jednota českých mathematiků* [The Union of the Czech Mathematicians]. In 1872 due to his contribution, the Union started to publish the journal *Časopis pro pěstování mathematiky a fysiky* [The Journal for the Cultivation of Mathematics and Physics]. He was actively involved in building Croatian mathematics in Zagreb, cooperating with the South Slavic Academy of Sciences and Arts, where he had been a member since 1882. Thanks to him, the best Czech mathematicians became the members of this society, and so successful cooperation was established. Karel Zahradník was elected a cross-member of the Royal Bohemian Society of Sciences in Prague, a corresponding member of the Czech Academy of the Emperor Franz Joseph for Sciences, Literature and Arts, and the Royal Serbian Academy of Science in Belgrade. Between 1900 and 1907, he was a member of the Provincial School Council for the Margraviate of Moravia. He is the author of almost a hundred professional papers and popularising articles, and university and secondary school textbooks. His work interferes with algebra (determinants, logarithms), mathematical analysis (basics of infinitesimal calculus, differential equations) and geometry (issues taught at secondary schools). The main topics of his professional work focused on geometry; he achieved the most significant results in the theory of rational curves of the third degree.[46]

After Zahradník's death, Václav Simandl (1887–1918) shortly taught mathematics at the Czech Technical University in Brno. Simandl was an assistant of descriptive geometry and a helping assistant of mathematics.[47] In the autumn of 1916, an experienced high school teacher Jan Vojtěch (1879–1953) took over the management of the institute, was appointed a titular in 1915, an extraordinary professor in 1918 and a full professor of mathematics two years later. However, he left Brno in 1923 and went to Prague to the Czech Technical University, where he worked until his retirement in 1949. He used his rich teaching experience from secondary schools and universities to create textbooks. He wrote high school geometry textbooks of great quality and together with Bohumil Bydžovský and František Teplý (1878–1929),

[46]About the work and life of Karel Zahradník, see Bečvářová and Čižmár (2011); Košťál (1968).
[47]About Václav Simandl, see Košťál (1968); Pelíšek (1919).

O PLOCHÁCH
DRUHÉHO STUPNĚ.

Z PŘEDNÁŠEK V ZIMNÍM POLOLETÍ 1910/1 NA C. K. ČESKÉ VYSOKÉ ŠKOLE TECHNICKÉ V BRNĚ

DRA. K. ZAHRADNÍKA.

V BRNĚ 1911.
NÁKLADEM VLASTNÍM. — TISKEM AKCIOVÉ MORAVSKÉ KNIHTISKÁRNY.

Figure 16. The picture of Karel Zahradník and the title page of his textbook.[48]

the textbook *Matematika pro nejvyšší třídy gymnázií a reálek* [Mathematics for the Highest Classes of Grammar Schools]. His very commonly used book *Základy matematiky ke studiu věd přírodních a technických* [Fundaments of Mathematics for Natural and Technical Sciences] was based on his lectures for students of chemistry and engineering at the Czech Technical University in Brno. Its popularity (despite the criticism of its little exactness) is evidenced by the fact that it was published in seven editions (from 1916 to 1946). In 1926, it was supplemented with the textbook *Přehled vyšší matematiky* [Higher Mathematics Overview].[49] Probably under Zahradník's influence, Jan Vojtěch devoted himself to the theory of transformations, the theory of planar curves of the sixth degree and projective geometry. In 1932, he published the comprehensive monograph *Geometrie projektivní* [Projective Geometry], which, among other things, contains a detailed explanation of synthetic and analytical methods enabling the study of projective relations.

[48]The pictures are from the authors' archive.
[49]About the work and life of Jan Vojtěch, see Drábek (1979a); Košt'ál (1968); Vojteková (2007); Vyčichlo (1953).

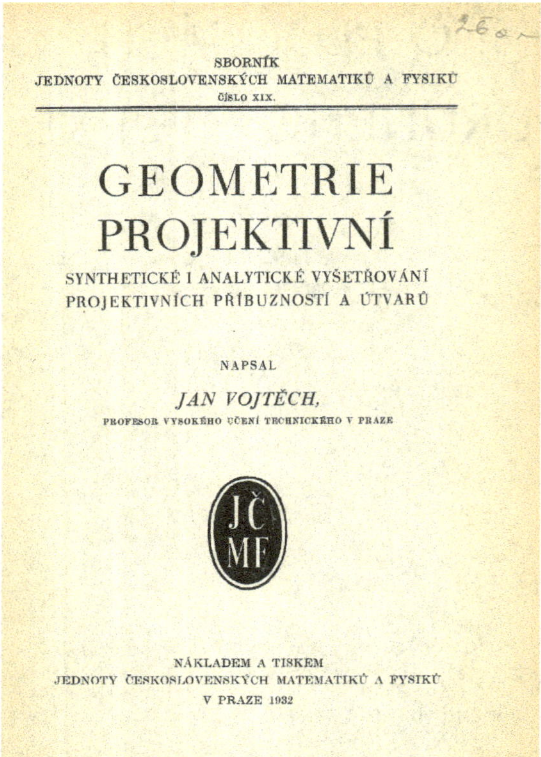
SBORNÍK
JEDNOTY ČESKOSLOVENSKÝCH MATEMATIKŮ A FYSIKŮ
ČÍSLO XIX.

GEOMETRIE
PROJEKTIVNÍ

SYNTHETICKÉ I ANALYTICKÉ VYŠETŘOVÁNÍ
PROJEKTIVNÍCH PŘÍBUZNOSTÍ A ÚTVARŮ

NAPSAL

JAN VOJTĚCH,
PROFESOR VYSOKÉHO UČENÍ TECHNICKÉHO V PRAZE

JČMF

NÁKLADEM A TISKEM
JEDNOTY ČESKOSLOVENSKÝCH MATEMATIKŮ A FYSIKŮ
V PRAZE 1932

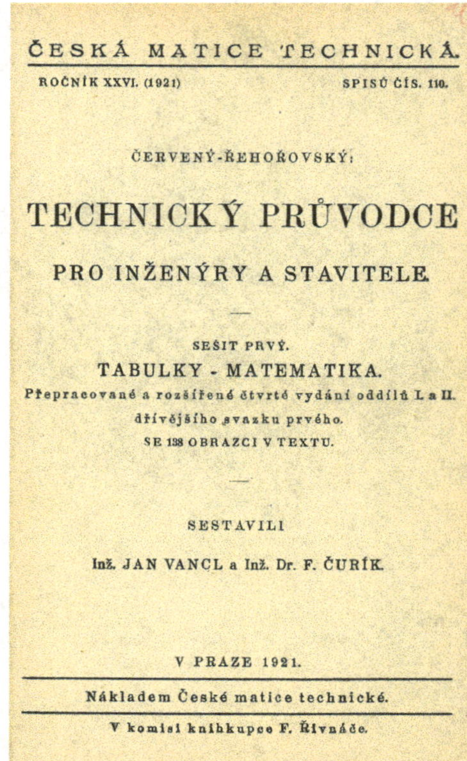
ČESKÁ MATICE TECHNICKÁ.
ROČNÍK XXVI. (1921) SPISŮ ČÍS. 110.

ČERVENÝ-ŘEHOŘOVSKÝ:

TECHNICKÝ PRŮVODCE
PRO INŽENÝRY A STAVITELE.

SEŠIT PRVÝ.
TABULKY - MATEMATIKA.
Přepracované a rozšířené čtvrté vydání oddílů I. a II.
dřívějšího svazku prvého.
SE 138 OBRAZCI V TEXTU.

SESTAVILI

Inž. JAN VANCL a Inž. Dr. F. ČUŘÍK.

V PRAZE 1921.

Nákladem České matice technické.
V komisi knihkupce F. Řivnáče.

Figure 17. The picture of the title page of the textbook written by J. Vojtěch and the picture of the title page of the textbook written by F. Čuřík.[50]

Due to the growing number of students as early as in 1900, the Second Institute of Mathematics was established at the Czech Technical University in Brno, where a descriptive geometer Antonín Sucharda (1854–1907) was called to. He became an extraordinary and in 1902 a full professor of mathematics. But he worked only briefly at the Czech Technical University in Brno; since 1904 he had not taught due to his poor health, he retired in 1906 and died a year later. Under the influence of the Weyr brothers, he focused on modern geometry. He published 45 works dedicated to different areas of geometry; he had the best results in the theory of algebraic curves and translational surfaces. He was also responsible for building a large collection of plaster models of third degree surfaces, for which he negotiated a special grant. However, he died before the installation of his collection.[51] Between 1905 and 1906, a high school teacher Jan Mayer (1861–1951) substituted lectures for the ill Antonín

[50]The pictures are from the authors' archive.
[51]About the life and work of Antonín Sucharda, see Košťál (1968); Sobotka (1908).

Sucharda. In 1907, Mayer was appointed a provincial school inspector of Moravia and worked at this post until his retirement in 1920.[52]

An excellent Czech mathematician Matyáš Lerch (1860–1922) was called to the Czech Technical University in Brno in 1906 from Fribourg in Switzerland; however, he left it for the newly founded Masaryk University in 1919. Since 1919, his health gradually deteriorated and he died of pneumonia in August 1922. At the Czech Technical University in Brno, he alternated in a two-year cycle with Karel Zahradník; most students attended the mathematics lectures only of one of them. Less talented students preferred the less demanding Karel Zahradník. Matyáš Lerch gave special lectures on number theory. In the first period of his working in Brno, he devoted himself to his favourite topics — calculations of simple and multiple integrals, study of special functions, infinite series, investigating properties of quadratic forms, etc. In 1912, he turned his attention to geometry.

Lerch has written about 240 works (in Czech, German and French), which he published in Czech and foreign journals. Probably the most valuable work published in the late 19th century was about number theory. He focused mainly on two issues — he studied the number of divisors of a given number and the number of binary quadratic forms of a given discriminant. In 1900, he received the so-called Grand Prix from the French Academy of Sciences for *Essais sur le calcul du nombre des classes de formes quadratiques binaires aux coefficients entiers*. In the study, he derived new formulas for determining the number of binary quadratic forms of special forms. To achieve the goal, he chose modern methods (e.g., the properties of the gamma function, Hermite's results, advanced parts of function theory). In the first decade of the 20th century, he wrote two excellent articles on Fermat's quotients and on the law of reciprocity. His works on number theory were world-class, and solved a number of special and very complex issues.

The second area in which he became well-known was function theory. In the 1880s, under the influence of Weierstrass's work, he was engaged in research into continuous functions that have no derivative at some points, resp. at no points, functions that have the derivative of all orders throughout the definition domain and cannot be developed at any point in the Taylor series. He also focused on functions that have no analytical continuation across the boundary of their domains of definition. The most valuable results of Lerch are contained in his article *Über Funktionen mit beschränkten Existenzbereiche* (1888), in which he generalised the Goursat-Poincaré theorem (1882), derived two new theorems allowing the construction of holomorphic funtions in a bounded domain having the boundary at the so-called natural boundary, and further constructed the functions that are holomorphic on the unit circle and have no analytical continuation over any point of the unit circle. At the beginning of the 20th century, he dealt with the theory of

[52]About the life and work of Jan Mayer, see Košťál (1968).

generating functions (see *Lerch's theorem*). His results also hit the theory of the Malmsten series, gamma function, theta function and Binet's function, theory of elliptic functions and integrals.

Tabulka důležitějších vzorců v textu odvozených.

Derivace funkcí *snu, cnu, dnu, amu.*

$$D_u snu = cnu\, dnu;\; D_u cnu = -snu\, dnu;\; D_u dnu = -k^2 snu\, cnu;\; D_u amu = dnu.$$

Zvláštní hodnoty funkcí *snu, cnu, dnu.*

	0	$\frac{K}{2}$	K	$2K$	$\frac{iK'}{2}$	iK'	$2iK'$
snu	0	$\frac{\sqrt{1-k'}}{k}$	1	0	$\frac{i}{\sqrt{k}}$	∞	0
cnu	1	$\frac{\sqrt{k'(1-k')}}{k}$	0	-1	$\sqrt{\frac{1+k}{k}}$	∞	-1
dnu	1	$\sqrt{k'}$	k'	1	$\sqrt{1+k}$	∞	-1

Periodické vlastnosti funkcí *snu, cnu, dnu.*

$$sn(u+2mK+2niK') = (-1)^m snu;\; cn(u+2mK+2niK') = (-1)^{m+n} cnu;\; dn(u+2mK+2niK') = (-1)^n dnu.$$

Součtové věty pro funkce *snu, dnu, cnu.*

$$sn(u+v) = \frac{snu\, cnv\, dnv + snv\, cnu\, dnu}{1-k^2 sn^2u\, sn^2v}.$$

$$cn(u+v) = \frac{cnu\, cnv - snu\, snv\, dnu\, dnv}{1-k^2 sn^2u\, sn^2v}.$$

$$dn(u+v) = \frac{dnu\, dnv - k^2 snu\, snv\, cnu\, cnv}{1-k^2 sn^2u\, sn^2v}.$$

Řady pro funkce *snu, cnu, dnu.*

$$snu = \frac{2\pi}{Kk} \sum_{\lambda=1,3,5,\ldots} \frac{q^{\frac{\lambda}{2}}}{1-q^\lambda} \sin\frac{\lambda u\pi}{2K};$$

Figure 18. The picture of the memorial plaque of Matyáš Lerch and one page from his textbook *Eliptické funkce* [Elliptical Function], Brno, without the date.[53]

Matyáš Lerch was probably the best, most well-known and world-renowned Czech mathematician of the last third of the 19th century and the first quarter of the 20th century.[54]

The Institute of Descriptive Geometry was one of the four institutes established at the foundation of the Czech Technical University in Brno. Its management was entrusted to Jan Sobotka, who came from the Technical University in Vienna, where he held the post of extraordinary professor. In Brno, he compiled a curriculum of descriptive geometry and merged it with projective and kinematic geometry. It was

[53]The pictures are from the authors' archive. The memorial tablet of M. Lerch is situated at the building of the Mathematical Institute of the Faculty of Natural Sciences, Masaryk University in Brno.

[54]About the life and work of Matyáš Lerch, see Bečvářová (2008); Borůvka (1957, 1972); Čupr (1923); Čupr and Rychlík (1925); Frank (1953, 1960); Fuchs and Lepka (2000); Košťál (1968); Lepka (1995); Škrášek (1953, 1960).

used in the main features until the end of World War I. Sobotka also founded a mathematical library complemented with a collection of models and tools. He worked at the Czech Technical University in Brno for only five years; in 1904 he left for the Czech University in Prague, where he was appointed a full professor.

After Sobotka's departure, Vincenc Jarolímek took over his lectures for a short time; in the same year he was replaced by Bedřich Procházka, a director of the secondary school in Náchod, who was then appointed a full professor. Procházka taught in Brno only for four years, because in 1908 he left for the Czech Technical University in Prague, where a full professor post was vacant.

Miloslav Pelíšek (1855–1940) was called to the Czech Technical University in Brno after Procházka's departure in 1908. He was immediately appointed a full professor and with this school he connected his whole life. He taught descriptive geometry six hours per week every semester and had six hours of exercises. For several years, he listed a one-hour facultative lecture on kinematic geometry, which he joined to regular lectures after World War I. The lectures were oriented in such a way that they would be most suitable for future engineers, because very few candidates of secondary school professors studied at the Czech Technical University in Brno until the beginning of World War I. Also, most of the exercises were devoted to technical applications. He retired in 1925, but continued to hold selective lectures.

He devoted his whole life to professional work and teaching, for 55 years he published articles on synthetic, kinematic, analytical and projective geometry. His most valuable works are on kinematic geometry. Modern Czech textbooks quote the so-called *Pelíškova konstrukce* [Pelíšek's Construction]. Miloslav Pelíšek was a great patron of science. Although he came from modest circumstances, he bequeathed in his will all his property to various scientific societies (his house in Brno, professional library and cash in the amount of one million Czech crowns).[55]

2.5. *Mining Academy in Příbram*

Příbram was not a big and important mathematical center. After long negotiations conducted since the 1820s, Emperor Franz Joseph I issued the decree (January 23, 1849 in Olomouc), by which he established two equal mining schools (academies) — for the northern area of the monarchy in Příbram and for the southern countries in Leoben (Styria). Thus the tradition of mining education was restored in Bohemia. The school was located in Příbram until 1939; during World War II it was closed and its building was used by the German army. The Czechoslovak presidential decree of 8 September 1945 transferred it to Ostrava.[56]

[55]About the life and work of Miroslav Pelíšek, see Hlavatý (1941); Košt'ál (1968); Piska (1956); Seifert (1926).

[56]For information on the history of the Mining Academy in Příbram, see Hrabák (1899); Majer (1984); Schenk (1974); Theurer (1924).

Josef Theurer (1862–1928) was the first professor of mathematics at the Mining Academy in Příbram. In 1895, he became a docent and was appointed to lecture mathematics and physics at a two-year preparatory course. In 1899, he was appointed an extraordinary and in 1902 a full professor of mathematics and physics. At this school he worked until his death. Between 1903 and 1905, he was the first rector of Czech origin (he was also rector in 1907–1909, 1909–1911, 1917–1918, 1918–1920, 1920–1921 and 1926–1927); he has significantly contributed to the reorganisation and modernisation of the school.

Theurer's activities were multifaceted. His professional work mainly concerned radiation, electrical oscillations, some problems of optics and thermodynamics. Together with František Reiss (1853–1918), a director of the Vinohrady Grammar School, he wrote the textbook *Fysika pro vyšší třídy středních škol* [Physics for Higher Secondary Schools, 1894], which was one of the first Czech modern textbooks of this subject and came out in several editions. He was a great promoter of electrical engineering; he put it in his college lectures.[57]

2.6. *Conclusion*

During the second half of the 19th century and the first two decades of the 20th century, mathematical work in the Czech language was born, and the Czech mathematical community was created and stabilised, and in some fields it began to progress at European level (theory of matrices, number theory, configuration theory, and mathematical analysis).[58]

The importance of Czech mathematical workplaces grew as the number of its professors and students increased. Their publishing, educational, cultural and popularising activities grew gradually. Czech mathematicians wrote Czech textbooks for almost all mathematical disciplines and all types of schools, and they stabilised Czech terminology, raised the level of Czech professional periodicals and also dared to translate classical and modern mathematical works.[59]

The Faculty of Arts of the Czech University in Prague, the Czech Technical University in Prague and the Czech Technical University in Brno were the centers of Czech mathematical life. Together with the Union, they played an important

[57] About the life and work of Josef Theurer, see Fryš (2006); Šebesta (1923, 1930).

[58] About the development of the Czech mathematical community until 1918, see Bečvářová (2008); Bečvář and Bečvářová (2019a); Nový a kol. (1961).

[59] About Czech textbooks of mathematics and translations of mathematical works, see Bečvářová (2002, 2007, 2008).

role in the development of Czech mathematical research in the first half of the 20th century.

3. The Czech Mathematical Community: Between the World Wars

3.1. *Introduction*

The foundation of the Czechoslovak Republic brought great impetus to the development of mathematics, and its study and application in technology, insurance and healthcare. The new state actively supported higher education and scientific work, built new Czech universities, created new professorships and assistant posts at Czech and German universities, and expanded the network of existing universities, and extended the network of existing general and specialised vocational secondary schools. It allowed young talented researchers to study abroad in prestigious foreign mathematics centers (Göttingen, Berlin, Hamburg, Munich, Paris, Padova, Rome, etc.), co-funded lecture stays of world mathematicians (e.g., Emile Borel (1871–1956), Élie Joseph Cartan (1869–1951), Guido Fubini (1879–1943), Henri Lebesgue (1875–1941), Vito Volterra (1860–1940)), and allowed the exchange of journals and the expansion of book collections. The activities of professional associations and societies were broadly supported.

The *Jednota československých matematiků a fysiků* [The Union of Czechoslovak Mathematicians and Physicists], the mathematical section of the *Česká akademie věd a umění* [The Czech Academy of Sciences and Arts] and a similar section of the *Královská česká společnost nauk* [The Royal Bohemian Society of Sciences] organised professional and educational lectures, seminars and conferences, funded the printing of professional monographs, textbooks for various types and grades of schools and professional and popularising journals, and played a significant role in the development of all round activities of the Czech mathematical community. In the interwar period, Czech mathematicians came to close the European level in a number of disciplines (number theory, differential geometry, probability theory, etc.).

The development of scientific and educational work in mathematics was forcibly interrupted in Czechoslovakia on 17 November, 1939, when Czech universities were closed down by Nazi occupants, nine student club officials were shot in Ruzyně barracks and approximately 1,200 students were deported to the Sachsenhausen concentration camp. In 1940, the Nazis began to significantly reduce the number of Czech secondary schools and set quotas for their students, graduates and high school graduates. Czech university professors first worked at home. Then by the ministerial decree of 11 July, 1940, they were *ex officio* on leave with a so-called "salary compensation for waiting", with effect from 31 July, 1940. They focused mainly on scientific work and writing textbooks. By the end of the war, many of them were totally deployed or involved in production. Some of them did not completely resign from their pedagogical activities. During the occupation, they met with former students, whom they helped with further education, possibly with the preparation of doctoral theses.[60]

3.2. *Charles University*

After long and complicated negotiations, the new Higher Education Act (the so-called *Lex Mareš*), which codified the relationship between both Prague universities came into effect on 19 February, 1920. The only follower of the ancient university remained the Czech University in Prague called *Universita Karlova* [Charles University]. It has been given the exclusive use of Karolinum, insignias, seals, old books, paintings, archives, etc. The German University in Prague received the name *Die Deutsche Universität in Prag* and obtained the appropriate portion of assets and equipment.

From the beginning of the 20th century, the development of natural sciences and the application of their results in practice forced the emergence of new, independent scientific disciplines, research institutes and educational institutions. It soon became apparent that the Faculty of Arts, in which mathematics and natural sciences were taught, could no longer provide sufficient space for further development. Therefore, in 1908, leading Czech naturalists submitted a proposal to create the Faculty of Natural Sciences, which was also supported by some leading representatives of the humanities. However, World War I delayed the implementation of the plans. Only after the establishment of the Czechoslovak Republic more negotiations were held, which resulted in the statutory government regulation No. 392/1920 Coll., which established the Faculty of Natural Sciences of Charles University and the Faculty of Natural Sciences of the German University in Prague. Teaching started from the school year 1920/1921. Its main task was to prepare high school professors of natural sciences and pharmacists.[61]

The Faculties of Natural Sciences quickly became respected centres of Prague mathematical research and teaching. These faculties developed classical mathematical disciplines, such as mathematical analysis, number theory, classical and differential geometry, as well as modern disciplines such as probability and statistics, actuarial mathematics, algebra, etc. The faculties prepared not only future high school teachers, but also mathematicians for higher education institutions, research institutes and the financial sector.

Karel Petr, Jan Sobotka and Václav Láska, professors at the Faculty of Arts, moved to the Faculty of Natural Sciences of Charles University. They continued

[60]On the development of the Czech mathematical community in the interwar period, see Bečvář and Bečvářová (2019b).

[61]About the history of the Faculty of Natural Sciences of Charles University, see Svatoš *et al.* 1998, Part IV, pp. 163–180.

their pedagogical and scientific work in mathematics and marked the direction of Czechoslovak mathematics. Petr's textbooks on mathematical analysis *Počet integrálný* [Integral Calculus, 1915, 1931] and *Počet differenciální* [Differential Calculus, 1923] played a major role in the development of Czech mathematics. It was Karel Petr who became the first dean of the Faculty of Natural Sciences (in 1920/1921). In the academic year 1925/1926, he was rector of Charles University. He retired in 1938.

Figure 19. The picture of Bohumil Bydžovský and the title page of his textbook.[62]

In the 1920s and 1930s, other mathematicians gradually came — professors, associate professors and assistants. A long-time high school professor Bohumil Bydžovský (1880–1969) was appointed a titular professor of mathematics in 1917. Two years later, he became an extraordinary and in 1920 a full professor. In the school year 1930/1931, he was the dean of the Faculty of Natural Sciences of Charles University.

After the forcible closure of universities by the German occupants, he left Prague and returned home to Veselí nad Lužnicí. In 1942, he and his wife were arrested and imprisoned by the Gestapo for suspicion that their son was in British emigration and that he was a military airman. The following year they were both released. After

[62]The picture of Bohumil Bydžovský is taken from Svatoš *et al.* (1998). The picture of the title page is from the authors' archive.

World War II, Bydžovský returned to the university, he was its rector in 1946/1947 and 1947/1948, and managed the 600th anniversary of its founding. He taught until he was 77, when he retired. He was the long-term chairman of the *Jednota československých matematiků a fysiků* [The Union of Czechoslovak Mathematicians and Physicists], between 1922 and 1935, and the editor-in-chief of the journal *Časopis pro pěstování matematiky a fysiky* [The Journal for the Cultivation of Mathematics and Physics]. He was also a member of the *Královská česká společnost nauk* [The Royal Bohemian Society of Sciences], the *Česká akademie věd a umění* [The Czech Academy of Sciences and Arts], the *Československá národní rada badatelská* [The Czechoslovak National Research Council] and later a member of the *Československá akademie věd* [Czechoslovak Academy of Sciences]. In 1920, he became a member of the *Národní komitét* [The National Committee] representing Czechoslovakia in the *International Mathematical Union*.[63] He participated in several congresses and conferences of mathematicians.

Bohumil Bydžovský focused on algebraic, analytical, synthetic and differential geometry, problems of geometric configurations and Cremona's transformations. He is the author of several university textbooks, for example, *Úvod do analytické geometrie* [Introduction to Analytical Geometry, 1923, 1946, 1956], *Základy teorie determinantů a matic a jich užití* [Fundaments of Determinant and Matrix Theory and Their Applications, 1930, 1947], *Úvod do algebraické geometrie* [Introduction to Algebraic Geometry, 1948], which were used by one generation of Czechoslovak mathematicians. He paid great attention to teaching mathematics. For many years, he worked at the Ministry of National Education, he was engaged in reform activities, and he is the author of the book *Naše středoškolská reforma* [Our High School Reform, 1932]. He wrote, alone or with co-authors, very popular and successful textbooks on arithmetic and a collection of problems for upper secondary school classes.[64]

Miloš Kössler (1884–1961) worked in mathematical analysis. He was a high school teacher for several years in Domažlice and Prague. In 1918, he defended his dissertation at the Faculty of Arts of the Czech University in Prague and two years later he obtained habilitation. He began lecturing at the Faculty of Natural Sciences in 1921 as an associate professor. At the end of 1922, he was appointed an extraordinary and five years later a full professor. In the school year 1935/1936, he served as the dean and in the following year as the vice-dean. He was an active member of the Union. From 1917 to 1947, he worked in its executive committee, between 1917 and 1926 he worked in a time-consuming position of a librarian and from 1939 to 1943 he even led the Union. Thanks to him, the Union's library was preserved during the war and at least a partial opportunity to publish in the field of mathematics and

[63]The Czechoslovak Republic was represented by Bohumil Bydžovský, Bohuslav Hostinský and Karel Petr.

[64]About the life and work of Bohumil Bydžovský, see Francová (2001); Metelka (1960); Nádeník (2011); Olejníčková (2005); Šindelář (1970).

physics was possible. He was a member of the *Královská česká společnost nauk* [The Royal Bohemian Society of Sciences], where he was a secretary for several years. In 1953, he was appointed a corresponding member of the *Československá akademie věd* [The Czechoslovak Academy of Sciences]. He was an active sportsman and he contributed to the building of university tennis courts in Albertov.

The center of Miloš Kössler's scientific interest was the theory of analytical functions. From the beginning of his university activities, he familiarised his students with the latest knowledge in this field and with the results of his own research. He led introductory lectures on mathematical analysis, and held a regular course on complex variable functions. In addition to the theory of analytical functions, he dealt with number theory and problems that combined both of the above-mentioned disciplines. He wrote more than 20 research papers and the excellent textbook *Úvod do počtu diferenciálního* [Introduction to Differential Calculus, 1926]. He was also interested in teaching problems and teaching at lower levels, and organising training courses that prepared teacher candidates for the proficiency tests they had to undergo.[65]

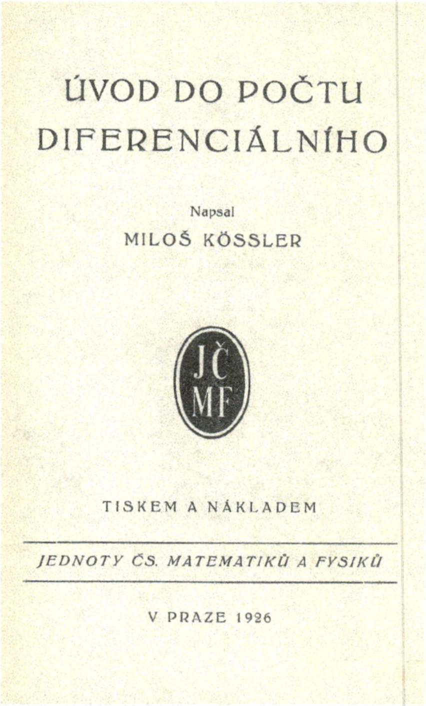

ÚVOD DO POČTU
DIFERENCIÁLNÍHO

Napsal
MILOŠ KÖSSLER

JČ
MF

TISKEM A NÁKLADEM
JEDNOTY ČS. MATEMATIKŮ A FYSIKŮ
V PRAZE 1926

Figure 20. The picture of Miloš Kössler and the title page of his textbook.[66]

[65]About the life and work of Miloš Kössler, see Jarník (1955); Kopřiva (1961); Nožička (1962); Pavlíková (2004).

[66]The picture of Miloš Kössler is taken from Nožička (1962). The picture of the title page is from the authors' archive.

Vojtěch Jarník (1897–1970) also lectured mathematical analysis. From 1919 to 1921, he worked at the Czech Technical University in Brno as an assistant to Jan Vojtěch. In 1921, he received his doctorate at the Faculty of Natural Sciences of Charles University. He became an assistant at the mathematical seminar at this faculty and worked under the leadership of Karel Petr. In the 1920s, he spent almost three school years at the University in Göttingen with Edmund Landau (1877–1938), Emmy Noether (1882–1935) and Bartel Leendert van der Waerden (1903–1996).[67] In 1925, he obtained habilitation at the Charles University. He was appointed an extraordinary professor in 1929 and a full professor of mathematics six years later. After World War II, he was involved in building a new education system. In 1947/1948, he was the dean of the Faculty of Natural Sciences of Charles University. From 1950 until 1953, he was the vice-rector. In the school year 1959/1960, he served as the dean of the Faculty of Mathematics and Physics of Charles University.

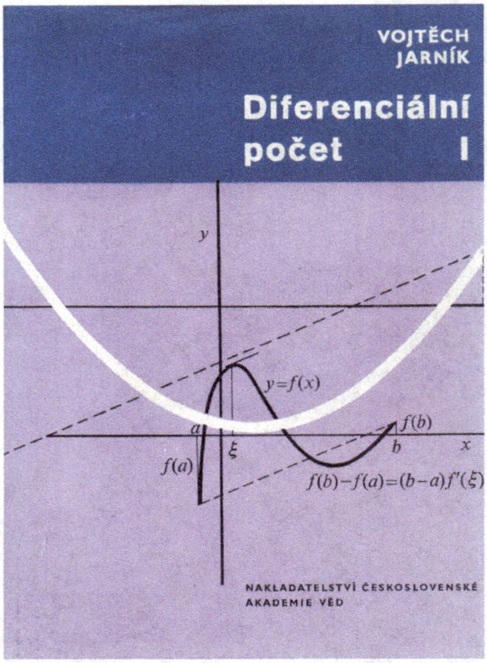

Figure 21. The picture of Vojtěch Jarník and the title page of his textbook.[68]

Vojtěch Jarník was a long-time member of the Union and its executive committee. For almost 15 years, he was the editor-in-chief of the mathematical section of the journal *Časopis pro pěstování matematiky a fysiky* [The Journal for the Cultivation

[67]About Jarník's stay in Göttingen, see Bečvářová and Netuka (2010).

[68]The picture of Vojtěch Jarník is taken from Novák (1999). The picture of the title page is from the authors' archive.

of Mathematics and Physics, from 1935 till 1950]; he managed to transform it into an internationally acclaimed journal. He worked in the editorial boards of the international journals *Czechoslovak Mathematical Journal* and *Acta Arithmetica.* He was a member of the governmental commission for the establishment of the *Československá akademie věd* [The Czechoslovak Academy of Sciences] and since its establishment, the first chairman of the Mathematics-Physics Section (from 1952 till 1955). After the establishment of the Scientific Collegium of Mathematics in 1962, he was a member of it and its chairman from 1964 to 1966. He participated in a number of international congresses and conferences; he was often invited to lectures at foreign institutions.

Jarník's 90 scientific papers interfere with several branches of mathematics. His main field of study was number theory, especially the theory of lattice points, Diophantic approximations and geometric representation of numbers, followed by mathematical analysis, especially the theory of functions of a real variable. His work *O jistém problému minimálním* [On a Certain Problem of Minimization, 1930], in which he presented an algorithm for finding the spanning tree of a simple connected weighted graph, is still widely cited. He was also interested in the history of mathematics, especially in the work of Bernard Bolzano. A great contribution to university mathematics teaching was his six textbooks of mathematical analysis: *Diferenciální počet I, II* [Differential Calculus I, II], *Integrální počet I, II* [Integral Calculus I, II, 1937, from 1946 to 1955, published repeatedly until 1984], *Diferenciální rovnice v reálném oboru* [Differential Equations in Real Domain, 1963] and *Diferenciální rovnice v komplexním oboru* [Differential Equations in Complex Domain, 1975]. Several generations of Czechoslovak mathematicians learned from them. His teaching activities were of great importance. He was not only a great mathematician but also an excellent teacher; he could clearly explain even the most demanding parts. His lectures were highly accurate.[69]

Václav Hlavatý (1894–1969) taught differential, algebraic and non-Euclidean geometry at the Faculty of Natural Sciences of Charles University. His university studies were disrupted by World War I. He spent the time on the Italian front. After 1918, he completed his studies and began working as a high school professor of mathematics and descriptive geometry. In the 1920s, he completed several foreign scholarships (the Netherlands, Italy, Great Britain and France). In 1925, he obtained habilitation at the Faculty of Natural Sciences of Charles University. In 1931, he was appointed an extraordinary and five years later a full professor of geometry and philosophy of mathematics.

In July 1948, he and his family went to the United States of America, where he was invited to a lecture stay. They did not return back. Hlavatý became a member of the Mathematics Institute at Indiana University in Bloomington. In 1967, he made a lecture trip around the world, during which he had 70 lectures in English, French,

[69]About the life and work of Vojtěch Jarník, see Novák (1999).

Italian and German. He was the first chairman of the *Společnost pro vědy a umění* [The Society for Sciences and Arts], which brought together Czech and Slovak exile scientists and artists.

Václav Hlavatý is the author of more than 150 scientific papers and university textbooks in the field of differential and algebraic geometry and general relativity. He was one of the most respected Czech modern geometers. He wrote textbooks: *Úvod do neuklidovské geometrie* [Introduction to non-Euclidean Geometry, 1926, 1949], *Diferenciální geometrie křivek a ploch a tensorový počet* [Differential Geometry of Curves and Surfaces and Tensor Calculus, 1937], *Diferenciální přímková geometrie* [Differential Straight-Line Geometry, 1941] and *Projektivní geometrie 1, 2* [Projective Geometry 1, 2, 1944, 1945]. The prestigious work of his emigration era is the monograph *Geometry of Einstein's Unified Field Theory* (1958).[70]

Václav Hlavatý was politically active after World War II; he was a member of the Czechoslovak National Socialist Party, for which he was a member of the Provisionel National Assembly after 1945.

Figure 22. The picture of Václav Hlavatý and the title page of his textbook.[71]

[70] About the life and work of Václav Hlavatý, see Durnová *et al.* (2017); Kowalski (1993); Nožička (1969).

[71] The picture of Václav Hlavatý is taken from Nožička (1969). The picture of the title page is from the authors' archive.

Vladimír Kořínek (1899–1981) contributed to the development of Czech algebra. He started to study mathematics and physics at the Faculty of Arts of the Czech University in Prague in 1918 and after five years he obtained a doctorate at the Faculty of Natural Sciences of Charles University. He spent the academic year 1923/1924 on a study stay at the Sorbonne in Paris. After his return, he worked as a voluntary assistant at the Mathematical Seminar of Charles University and at the same time he taught at secondary schools in Prague. From 1925 to 1931, he was an assistant at the Czech Technical University in Prague. There he taught physics for two years and mathematics for four years. During this time, he spent one year at the University in Hamburg where he met Emil Artin (1898–1962), who raised his interest in algebra. In 1931, he obtained habilitation in mathematics at the Faculty of Natural Sciences of Charles University. In the same year, he began working at the State Statistical Office, where he remained until he was appointed an extraordinary professor of mathematics (in 1935). After the war, he was appointed a full professor of mathematics (retroactively from 1940). From 1953 to 1955, he was the dean of the Faculty of Mathematics and Physics of Charles University. He lectured at numerous international congresses and conferences. He worked at the university until 1970.

Vladimír Kořínek was an active member of the Union. For more than one decade, he was its librarian. From 1969 to 1972, he was its chairman. He also became a member of the *Královská česká společnost nauk* [The Royal Bohemian Society of Sciences], the *Česká akademie věd a umění* [The Czech Academy of Sciences and Arts] and the *Československá národní rada badatelská* [The Czechoslovak National Research Council]. In 1952, he was appointed an academician at the Czechoslovak Academy of Sciences. For many years, he was a member of its Mathematical Section, and he worked on the scientific boards of the Institute of Mathematics of the Czechoslovak Academy of Sciences and the Mathematical Institute of Charles University. He chaired the Commission for the defence of candidate's dissertations and the Commission for the defence of doctoral dissertations in algebra and number theory. He was a member of the editorial boards of mathematical journals and worked in various commissions focusing on secondary school mathematics.

Vladimír Kořínek wrote several papers on statistics and actuarial issues. However, the focus of his professional activity was algebra. In the 1920s, he devoted himself to the arithmetic theory of quadratic forms, which was introduced him by his teacher, Karel Petr. After returning from Hamburg, he began dealing with the theory of associative algebra and its arithmetic, following the studies of Emil Artin and Heinrich Brandt (1886–1954). In the 1930s, he focused on abstract group theory and in the 1940s on the newly developing lattice theory. In the post-war period, he published only little, because he was involved in organisational and pedagogical work. His textbook *Základy algebry* [Fundaments of Algebra, 1953, 1956] comes from

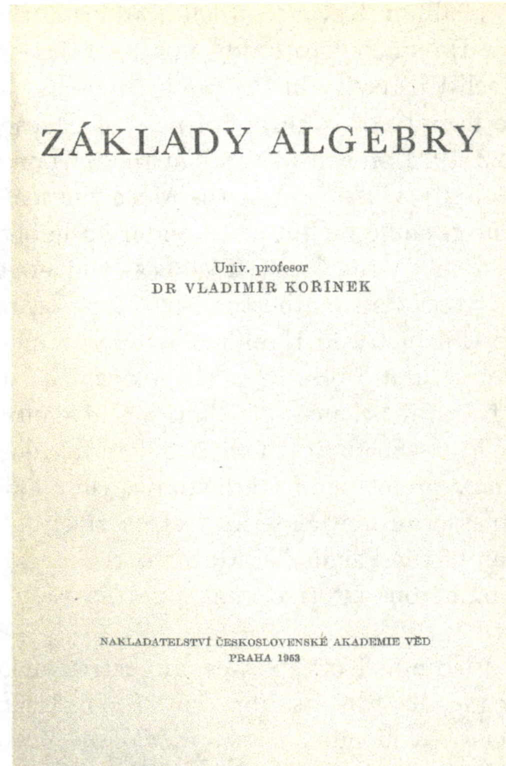

ZÁKLADY ALGEBRY

Univ. profesor
DR VLADIMÍR KOŘÍNEK

NAKLADATELSTVÍ ČESKOSLOVENSKÉ AKADEMIE VĚD
PRAHA 1953

Figure 23. The picture of Vladimír Kořínek and the title page of his textbook.[72]

this period. It was the first Czech university textbook of algebra, which influenced the teaching of mathematics at Czech universities for several decades.[73]

Emil Schoenbaum (1882–1967) was involved in actuarial and mathematical statistics. In 1906, after graduating from the Czech University in Prague, he left for a scholarship to Göttingen, where he listened to lectures of Felix Klein, David Hilbert and Hermann Minkowski (1864–1909) and where he devoted himself more deeply to actuarial mathematics, in particular social and private insurance. In 1919, he obtained habilitation at the Czech University in Prague in the field of actuarial mathematics and statistics. Four years later, he was appointed a full professor of actuarial mathematics. From 1930 to 1939, he edited the journal *Aktuárské vědy* [The Actuarial Sciences], which the Union began to issue on his initiative. As an employee of the *Všeobecný penzijní institut* [The General Pension Institute], he drafted several bills on pension and social insurance in Czechoslovakia. In 1930, he was commissioned

[72]The pictures are from the authors' archive.
[73]About the life and work of Vladimír Kořínek, see Bečvář and Kohoutová (2005).

to draft documents for social insurance laws in Greece. In 1939, he emigrated to America and applied his rich knowledge in the field of social insurance in Ecuador, Mexico, Bolivia, Chile, Paraguay, the United States of America and Canada. After the war, he returned briefly to Czechoslovakia. At the International Conference in Philadelphia in 1949, he was elected the Social Insurance Officer and then worked as an advisor in the International Labor Office. He died in Mexico City in 1967.[74]

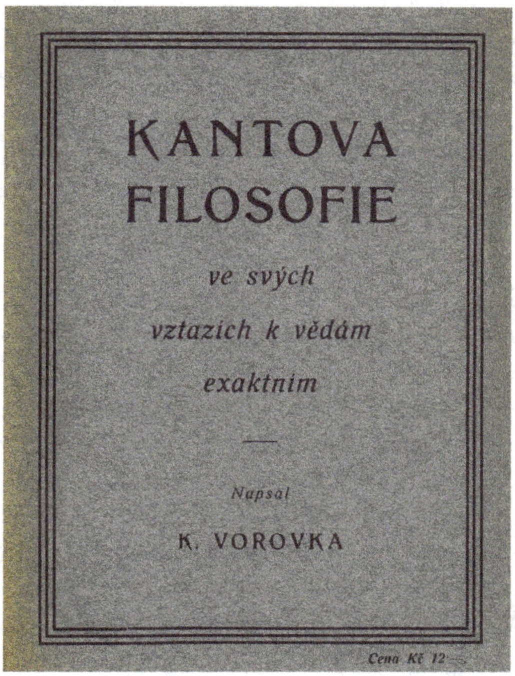

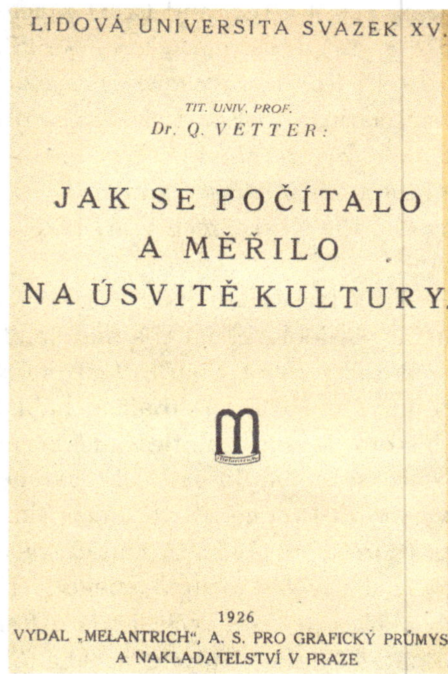

Figure 24. The picture of the title page of Karel Vorovka's book and the picture of the title page of Quido Vetter's book.[75]

The development of Czech philosophy of the exact sciences is credited to Karel Vorovka (1879–1929). In 1919, after several years at secondary schools, he obtained habilitation at the Czech University in Prague in philosophy and mathematics. In 1921, he was appointed an extraordinary and in 1924 a full professor of philosophy and exact sciences at the Faculty of Natural Sciences of Charles University. He worked in the seminar on methodology and history of natural and exact sciences and led the Department of Logic and History of Exact Sciences. For a long

[74]About the life and work of Emil Schoenbaum, see https://encyklopedie.soc.cas.cz/w/Schoenbaum-Emil [2.3.2019].
[75]The pictures are from the authors' archive.

time, he was involved in the Union and he was also a member of the *Královská česká společnost nauk* [The Royal Bohemian Society of Sciences] and a member of the standing committee for organising philosophical congresses. He was one of the leading Czech philosophers of the first third of the 20th century. He devoted himself to the philosophy of mathematics and physics, critically commenting on the state of Czech philosophy. His thoughts were significantly influenced by the works of Immanuel Kant (1724–1804), Tomáš Garrigue Masaryk (1850–1937), Henri Poincaré (1854–1912) and Ernst Mach (1838–1916). His most important works are *Skepse a gnóse: vyznání filosofické* [Skepticism and Gnosis: Philosophical Confession, 1921], *Kantova filosofie ve svých vztazích k vědám exaktním* [Kant's Philosophy in Its Relations to Exact Sciences, 1924], *Několik myšlenek o Masarykově filosofii a jeho Světové revoluci* [A Few Ideas About Masaryk's Philosophy and His World Revolution, 1926], *Česká mravní hegemonie a příští česká válka* [Czech Moral Hegemony and the Next Czech War, 1926] and *Americká filosofie* [American Philosophy, 1929].[76]

Quido Vetter (1881–1960) was originally a teacher of mathematics and physics at secondary schools. From the middle of the 1920s, he lectured mathematics history and methodology at Charles University and also at the Czech Technical University in Prague. In 1919, he obtained habilitation at the Czech University in Prague in the history of mathematics and became the first university teacher in this field in Czechoslovakia. In 1924, he extended his habilitation to the Czech Technical University in Prague. At the same time, he was appointed to Charles University as an extraordinary professor and charged with lectures on mathematics methodology at the Faculty of Natural Sciences. He devoted himself mainly to the history of mathematics in the Czech lands. After World War II, he returned to his teaching position at Charles University. The focus of his work was lectures on the history of mathematics at the Pedagogical University (later the Pedagogical Faculty of Charles University). He also worked at the Faculty of Natural Sciences of Charles University and later at the Faculty of Mathematics and Physics of Charles University.

Quido Vetter attended many international congresses and conferences dedicated to the history of exact sciences. At some of them he lectured on the history of mathematics and astronomy in the Czech lands. He was a member of Czech as well as foreign institutions (e.g., the *Královská česká společnost nauk* [The Royal Bohemian Society of Sciences], the *Národní komitét pro dějiny reálných věd při Národní radě badatelské v Praze* [The National Committee for the History of Exact Sciences at the National Research Council in Prague] and the *Mezinárodní akademie pro dějiny reálných věd* [The International Academy for the History of Exact Sciences]). His work contains more than 200 scientific studies, treatises and articles on the history

[76]About the life and work of Karel Vorovka, see Pavlincová (2010). See also Hykšová (2011, pp. 159–174).

of mathematics and the methodology of the history of sciences, and about 400 reviews of works on the history of mathematics. He dealt with the history of Babylonian, Egyptian, Arabic, Greek and also Czech mathematics. Among his best-known works are the book *Jak se počítalo a měřilo na úsvitě kultury* [How It Was Calculated and Measured at the Dawn of Culture, 1926] and articles *Úvod do dějin matematiky* [Introduction to the History of Mathematics, 1930], *Šest století matematického a astronomického učení na universitě Karlově v Praze* [Six Centuries of Mathematical and Astronomical Learning at Charles University in Prague, 1952], *Kratkij obzor razvitija matematiky v češskich zemljach do Bělogorskoj bitvy* [Short Survey on the Development of Mathematics in the Czech Lands until the Battle at the Bílá Hora, 1958], and *Kratkij obzor razvitija matematiky v češskich zemljach v period ot Bělogorskoj bitvy do konca XVII v.* [Short Survey on the Development of Mathematics in the Czech Lands from the Battle at the Bílá Hora until the 18th Century, 1961].[77]

In the second decade of the 20th century, several future significant mathematicians obtained habilitation at the Faculty of Arts or at the Faculty of Natural Sciences of Charles University. The most renowned of them were Bohuslav Hostinský (1884–1951), Karel Rychlík (1885–1968), Ladislav Seifert (1885–1956) and Eduard Čech (1893–1960). They became distinguished mathematicians, who influenced the development of Czechoslovak mathematics before World War II and especially after it.

In the following period, several associate professors obtained habilitation at Charles University, and some of them moved to Brno, Příbram or Slovakia after a short time in Prague. In 1923, after many years of study abroad, Jur Hronec (1881–1959) obtained habilitation in mathematical analysis. He switched to the Czech Technical University in Brno and later to Jan Ámos Komenský University in Bratislava, where he contributed to the development of Slovak higher education.

In 1926, Bohumil Machytka (1890–1928) became an associate professor of geometry, and then briefly lectured at Charles University and at the Czech Technical University in Prague.

Ladislav Truksa (1891–?) focused on applied mathematics in the mathematical department of the *Všeobecný pensijní ústav* [The General Pension Institute]. In 1931, he obtained habilitation in actuarial and mathematical statistics and lectured at Charles University as an external teacher until 1953 (with a war forced break). It was only in 1960 that he was appointed a full professor of mathematics. Due to the heavy workload in the pension institution, he did not write many works; however, his two-part textbook *Statistická dynamika* [Statistical Dynamics, 1956, 1958] is worth to mentioning.[78]

[77] About the life and work of Quido Vetter, see Bečvářová (2010a).

[78] About the life and work of Ladislav Truksa, see Bílý (1961).

Vladimír Knichal (1908–1974) who worked as an assistant of mathematics at Charles University from 1929 to 1939 focused on mathematical analysis. In 1936, he studied in Warsaw, where he was mainly interested in combinatorial topology. In the following year, he obtained habilitation at the Faculty of Natural Sciences of Charles University. He spent his war years in Prague as a high school teacher. In 1945, he was in charge of lectures at the Faculty of Natural Sciences of Masaryk University in Brno. Later he lectured at the Faculty of Education of Masaryk University, at the Faculty of Natural Sciences of Charles University and at the Faculty of Civil Engineering of the Czech Technical University in Prague. In 1950, he was appointed the head of the Department of Technical Mathematics at the *Ústřední matematický ústav v Praze* [The Central Mathematical Institute in Prague]. Four years later, he was appointed the director of the Institute of Mathematics of the Academy of Sciences of the Czech Republic, which he led until 1972. He professionally dealt with determinant theory, metric number theory, theory of real functions and applied mathematics.[79]

In 1939, František Vyčichlo (1905–1958) was appointed an associate professor at Charles University. Later he worked at the Czech Technical University in Prague and played there an important role in the post-war development of applied mathematics teaching.[80]

3.3. *Czech Technical University in Prague*

After the establishment of the Czechoslovak Republic, the Czech Technical University in Prague was reformed (1920). It followed the tradition of the Czech Polytechnic School. It brought together the universities (faculties) of civil engineering, architecture, mechanical and electrical engineering, chemical-technological engineering, agricultural and forestry engineering, a special education college and a business college, which had separate professorships. The common interests were represented by the rector's office and the senate. The Czech Technical University in Prague offered a wide and varied range of fields of study, providing new pedagogical and research posts at newly established institutes, research institutes and laboratories. Until the forcible closure of Czech universities by German occupants, it was the greatest university in Czechoslovakia providing education to Czech and Slovak future technical intelligence and was also a prominent center of applied Czechoslovak mathematics.[81]

Josef Klobouček (1875–1939), originally administering the Second Institute of Mathematics, took over the head office of the First Institute of Mathematics after the death of Matěj Norbert Vaněček (1852–1922). He managed the First Institute until his retirement in 1938. He originally taught mathematics and descriptive geometry

[79] About the life and work of Vladimír Knichal, see Kurzweil (1958, 1975).
[80] About the life and work of František Vyčichlo, see Bečvářová (2010b, 2010c); Nádeník (2011).
[81] See Kadeřávek and Pulkrábek (1956); Drábek (1979b); Tayerlová *et al.* (2004).

for several years at various secondary schools. From 1912, he was a private lecturer of selective lectures on synthetic and analytical geometry at the Czech Technical University in Prague. A study stay in Turin and Rome had a decisive influence on his professional work, because he met new methods of descriptive, analytical and differential geometry. In 1920, he was appointed an extraordinary and a year later a full professor of mathematics.

In 1923, after Klobouček's departure to the First Institute of Mathematics, Jan Vojtěch (1879–1953) took over the management of the Second Institute of Mathematics. He was managing this institute (except for a war-forced break) until his retirement in 1949.

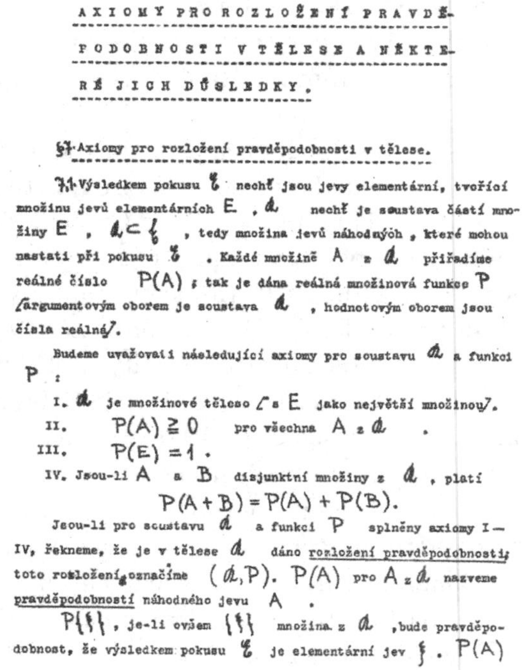

AXIOMY PRO ROZLOŽENÍ PRAVDĚPODOBNOSTI V TĚLESE A NĚKTERÉ JICH DŮSLEDKY.

§7. Axiomy pro rozložení pravděpodobnosti v tělese.

7.1. Výsledkem pokusu $\mathfrak{E}$ nechť jsou jevy elementární, tvořící množinu jevů elementárních E, $\mathfrak{A}$ nechť je soustava částí množiny E, $\mathfrak{A} \subset \mathfrak{E}$, tedy množina jevů náhodných, které mohou nastati při pokusu $\mathfrak{E}$. Každé množině A z $\mathfrak{A}$ přiřadíme reálné číslo $P(A)$; tak je dána reálná množinová funkce P [argumentovým oborem je soustava $\mathfrak{A}$, hodnotovým oborem jsou čísla reálná].

Budeme uvažovati následující axiomy pro soustavu $\mathfrak{A}$ a funkci P:

I. $\mathfrak{A}$ je množinové těleso [s E jako největší množinou].

II. $P(A) \geqq 0$ pro všechna A z $\mathfrak{A}$.

III. $P(E) = 1$.

IV. Jsou-li A a B disjunktní množiny z $\mathfrak{A}$, platí

$$P(A+B) = P(A) + P(B).$$

Jsou-li pro soustavu $\mathfrak{A}$ a funkci P splněny axiomy I—IV, řekneme, že je v tělese $\mathfrak{A}$ dáno rozložení pravděpodobnosti; toto rozložení označíme $(\mathfrak{A}, P)$. $P(A)$ pro A z $\mathfrak{A}$ nazveme pravděpodobností náhodného jevu A.

$P\{\xi\}$, je-li ovšem $\{\xi\}$ množina z $\mathfrak{A}$, bude pravděpodobnost, že výsledkem pokusu $\mathfrak{E}$ je elementární jev ξ. $P(A)$

Figure 25. The picture of František Vyčichlo and one page from Karel Rychlík's textbook *Introduction to Calculus of Probability*.[82]

One of the best mathematicians at the Czech Technical University in Prague in the first half of the 20th century was Karel Rychlík (1885–1968). He received his doctorate at the Faculty of Arts of the Czech University in Prague in 1909 and

---

[82]The picture of František Vyčichlo is taken from I. Babuška, K. Havlíček, F. Nožička: Památce prof. RNDr. Františka Vyčichla [In Memory of Prof. RNDr. František Vyčichlo], *Časopis pro pěstování matematiky* **83**(1958), pp. 374–388. The picture of the page from Karel Rychlík's textbook is taken from Hykšová (2003).

obtained habilitation there three years later. In 1920, he was appointed an extraordinary and three years later a full professor of mathematics at the Czech Technical University in Prague. He focused on algebra, number theory, probability theory and history of mathematics. He wrote three textbooks — *Úvod do elementární číselné theorie* [Introduction to Elementary Number Theory, 1931, 1950], *Úvod do počtu pravděpodobnosti* [Introduction to Calculus of Probability, 1938] and *Úvod do analytické theorie mnohočlenů s reálnými koeficienty* [Introduction to Analytical Theory of Polynomials with Real Coefficients, 1957]. He translated the works of Russian authors. He was deeply interested in the history of mathematics, especially in the mathematical works of Bernard Bolzano. He was the first Czech mathematician who understood Kolmogorov's probability theory and was able to give lectures on this subject at Charles University.[83]

In the 1920s, due to the large number of students who had to attend lectures on mathematics, František Rádl (1876–1956), the brother of the well-known philosopher and biologist Emanuel Rádl (1873–1942), was entrusted with substituting parallel lectures at the Czech Technical University in Prague. After years of teaching at high school, he spent the school year 1909/1910 on a study stay at the Sorbonne in Paris. He obtained habilitation at the Czech Technical University in Prague in 1917, but he was not appointed a full professor of mathematics until 1926. He taught here till his retirement in 1946. He dealt mainly with differential equations. Lack of exactness of his textbook *Učebnice matematiky pro vysoké učení technické* [Mathematics Textbook for Technical University, 1931, 1946] was sharply criticised.[84]

In 1919, Karel Dusl (1884–1948) obtained habilitation at the Czech Technical University in Prague. In 1931, he was appointed an extraordinary and five years later a full professor of mathematics at the Czech Technical University in Prague. After the war, he retired for health reasons. After his recovery, he left for the Faculty of Natural Sciences of Jan Ámos Komenský University in Bratislava (1946). He dealt mainly with vector calculus and mathematical analysis. He edited the mathematical part of *Technický slovník naučný* [Technical Encyclopaedia, 1927–1949] and he wrote the textbooks *Úvod do vektorového počtu* [Introduction to Vector Calculus, 1923, 1925], *O algebraických rovnicích a nižších funkcích transcendentních* [About Algebraic Equations and Lower Transcendent Functions, 1926], *Vybrané části z analyse* [Selected Parts of Analysis, 1934], *Diferenciální geometrie pro posluchače zeměměřičského inženýrství* [Differential Geometry for Students of Surveying Engineering, 1935], *Základy vyšší matematiky* [Fundaments of Higher Mathematics, 1937] and *Přehled teorie funkcí jedné komplexní proměnné* [An Overview of the Functions of One Complex Variable, 1941]. His mathematically most significant and interesting work is the more than 300 page book *Úvod do nauky o thetafunkcích*

[83] About the life and work of Karel Rychlík, see Hykšová (2003).
[84] About the life and work of František Rádl, see Rieger and Rychlík (1957).

[Introduction to Theta Functions, 1919]. Dusl was one of the few Czech mathematicians who regularly and actively participated in the international congresses of mathematicians.[85]

Descriptive geometry has been a classical discipline taught since the middle of the 19th century at all the technical universities in Europe. It was the same at the Czech Technical University in Prague. From 1920, František Kadeřávek (1885–1961) was the head of the Department of Descriptive Geometry and Stereometry. He received his education at the Czech Technical University in Prague and also at the Czech University in Prague. For many years he worked as an assistant of descriptive geometry at the Czech Technical University in Prague. From 1909, he lectured in place of the ill professor Vincenc Jarolímek. In 1917, he was appointed an extraordinary and in 1920 a full professor. At the same time, he began to teach geometrical methods at the Academy of Fine Arts in Prague. František Kadeřávek was one of the most important Czech geometers who elaborated some parts of projective geometry of conics and quadrics. He dealt with geometric affinities, warped surfaces, summing surfaces, special curves and geometrical theory of lighting. The two-part textbook *Deskriptivní geometrie* [Descriptive Geometry, First Part 1929, 1945, 1946, 1950, 1954; Second Part 1932, 1954] elaborated in cooperation with Josef Klíma and Josef Kounovský became well-known and it was used for many years in Czechoslovakia. His charming books *Perspektiva, příručka pro architekty, malíře a přátele umění* [Perspective, A Guide for Architects, Painters and Friends of Arts, 1922], *Relief, příručka pro sochaře a architekty* [Relief, A Guide for Sculptors and Architects, 1925] and *Geometrie a umění v dobách minulých* [Geometry and Arts in the Past, 1935, 1994, 1997] were also very popular among geometers and teachers of geometry.

In 1912, Josef Kounovský (1878–1949) obtained habilitation at the Czech Technical University in Prague in descriptive and synthetic geometry. He expanded his education in Zürich after his university studies in Prague. From 1922, he began to substitute lectures on descriptive geometry for the ill professor Bedřich Procházka. And after Procházka's retirement (1925), he was appointed to his post after long negotiations (1927). He also worked at the School of Higher Education Studies (the predecessor of the Pedagogical Faculty of Charles University). He lectured at the Czech Technical University in Prague until his death. In addition to the already mentioned textbook *Deskriptivní geometrie* [Descriptive Geometry] which he wrote together with F. Kadeřávek and J. Klíma, J. Kounovský wrote two books: *Zborcené plochy* [Warped Surfaces, 1947] and *Theoretické základy fotogrammetrie* [Theoretical Fundaments of Photogrammetry, 1948]. Together with František Vyčichlo, he wrote the popular textbook *Deskriptivní geometrie pro samouky* [Descriptive Geometry

[85]About the life and work of Karel Dusl, see the web pages https://sk.wikipedia.org/wiki/Karel_Dusl, http://biography.hiu.cas.cz/Personal/index.php/DUSL_Karel_29.5.1884-30.11.1948 [2.3.2019].

Figure 26. The picture of the title page of František Kadeřávek's book and one of his illustrations from the book.[86]

for Self-Taught Students, 1948, 1951, 1953], resp. its modified edition *Deskriptivní geometrie* [Descriptive Geometry, 1956, 1959].[87]

After World War I, applied mathematics was a new, rapidly developing field that was directly related to the development of technology. Its first full professor at the Czech Technical University was Václav Hruška (1888–1964), who firstly focused on theoretical mathematics (relations between periods of Abel functions), but later paid attention to numerical and graphical methods. In the 1920s, he worked intensively with the university professor Václav Láska. Their creative cooperation resulted in the publication of the books *Počet grafický a graficko-mechanický* [Graphical and Graphical-Mechanical Calculus, 1923] and *Teorie a prakse numerického počítání* [Theory and Practice of Numerical Computation, 1934]. In 1952, Václav Hruška published the monograph *Počet grafický a graficko-mechanický* [Graphical and Graphical-Mechanical Calculus], which included all known results of numerical and graphical methods. He was active in key negotiations and in setting up a new field — mathematical machines — and explained the necessity of opening the Laboratory of Mechanical Machines (later the Research Institute of Mathematical

[86]The pictures are from the authors' archive.

[87]About the development of teaching and scientific work on descriptive geometry, see, for example, Nádeník (2011). About the life and work of Josef Kounovský, see Kadeřávek (1950); Mikan (1939); Urban (1962).

Machines).[88] He was the first editor of the scientific edition *Stroje na zpracování informací* [Information Processing Machines, 1953], which opened the door to studying and constructing calculating machines (computers) in Czechoslovakia.

In 1930, Miloslav Hampl (1897–1974) obtained habilitation in applied mathematics and theoretical mechanics at the Czech Technical University in Prague. He graduated from the Faculty of Natural Sciences of Charles University and the Faculty of Mechanical Engineering and Electrical Engineering of the Czech Technical University in Prague. He worked as an assistant of the professors František Nušl, František Rádl and Václav Hruška. However, Miloslav Hampl soon left the Czech Technical University in Prague and entered the newly established Mathematical Department of Škoda Factory in Pilsen, where he worked until 1964.[89] Then, he moved to the Institute of Mathematics of the Czechoslovak Academy of Sciences and in 1966 to the Faculty of Mathematics and Physics of Charles University. He focused primarily on the theory of elasticity and strength.[90]

In the first quarter of the 20th century, the interest in actuarial mathematics began to increase because of its good application in the banking sector and public administration. In 1929, Jaroslav Janko (1893–1965) obtained habilitation in the field of actuarial mathematics and statistics at the Czech Technical University in Prague. He graduated from the Faculty of Arts of Charles University and did not take the usual path of a high school teacher. He obtained a position of mathematician at the Ministry of Social Welfare. In 1926, he graduated from a two-year course in actuarial and mathematical statistics at the Czech Technical University in Prague, where he was also appointed an extraordinary and five years later a full professor of actuarial mathematics. He stayed here until 1952, when he moved to the Department of Mathematical Statistics of the newly established Faculty of Mathematics and Physics of Charles University. He studied applied mathematics, actuarial mathematics and mathematical statistics.[91]

Otomar Pankraz (1903–1976) also studied actuarial theory and probability theory. Due to the difficult financial situation of his parents, he had very difficult conditions to study. Firstly, he was trained as a mechanical locksmith. Then he studied at an industrial college and attended university lectures as an extraordinary student. Only in 1929 at the age of 26, he passed a proper "matura" examination and applied for recognition of his university studies. In 1931, he received a doctorate

[88]The Research Institute of Mathematical Machines had focused on applied research since its foundation; associating mathematicians, physicists, technicians and designers.

[89]The Mathematical Department of Škoda Factory in Pilsen was one of the most important working places where applied mathematical research was carried out.

[90]Applied mathematics was also developed at the Mining Academy in Příbram where professor František Čuřík fostered it. About the life and work of Miloslav Hampl, see Polášek (1975); Špaček (1962).

[91]About the life and work of Jaroslav Janko, see Fabian (1964); Truksa (1954).

in natural sciences. Shortly thereafter, he completed actuarial and mathematical statistics studies at the Faculty of Natural Sciences of Charles University. Although he had a good position at the General Pension Institute in Prague, he was attracted by scientific work. Therefore, in 1931, he became an assistant professor of mathematics at the Czech Technical University in Prague. In this position, he remained until the closing of Czech universities by German occupiers. At the end of the 1930s, he was leading Czech expert in the axiomatic concept of probability theory. During the war, he managed to avoid all attempts to be deployed in war production. Shortly after the war, he was arrested, and charged with supporting Nazism and approving German persecution following the assassination of Reinhard Heydrich. The People's Court condemned him to five years in jail and loss of civil honor for ten years.[92]

3.4. *Masaryk University in Brno*

On 28 January, 1919, Masaryk University, the first Czech university in Moravia, was established by the Act No. 50/1919 Coll. Czech political and cultural representation had sought to establish it since the 1880s, but its voice was not heard in the Chamber of Deputies in Vienna and the Moravian Assembly in Brno. The requirement to create the second center of Czech national education and culture in Brno came under the resistance of the Germans in Brno, who feared that the establishment of the school would weaken their influence in the city.

In 1919, a university consisting of four faculties — law, medical, sciences and philosophical — was opened. Its professorships, which consisted of young professionals educated in Prague or in Brno, were appointed. It is therefore no wonder that new modern disciplines emerged in their work and teaching and that they took on new research methods. However, Masaryk University suffered from insufficient material security, since it was born in provisional spaces and due to the economic crisis, this situation failed to improve. In the 1930s, even the closing of the Faculty of Philosophy or the Faculty of Natural Sciences was considered. While the university happily overcame these problems, it was not able to resist the disaster that came with the onset of the German occupation. According to the Reich protector's internal instructions, the Nazis viewed the university as if it had never existed. The war rampage caused the death of many employees and students. The Faculty of Natural Sciences of Masaryk University lost a quarter of its professors during World War II.

At the newly established Faculty of Natural Sciences of Masaryk University, one institute of mathematics and two institutes of physics were established, where mathematics was also developed.[93] The foundation of Masaryk University in Brno

[92]About the life and work of Otomar Pankraz, see (Hykšová 2011, pp. 213–230).

[93]About Masaryk University, see Bednaříková a kol. (1969); Fasora and Hanuš (2009). About mathematics at Masaryk University, especially about its representatives and their activities, see Košťál (1968).

significantly strengthened the activities of the Czech mathematical community in Brno. The number of its members more than doubled.

In 1920, Matyáš Lerch (1860–1922) was appointed the first professor of mathematics at Masaryk University. A year later, Ladislav Seifert (1885–1956) became a professor of geometry. He graduated in Prague, where he also received a teaching qualification and worked as a high school teacher. The school year 1907/1908, which he spent at the University in Strasbourg and at the University in Göttingen, influenced his whole professional life. In 1914, he obtained his doctorate in technical sciences at the Czech Technical University in Prague. In 1920, he obtained habilitation in mathematics at Charles University and for descriptive, analytical and synthetic geometry at the Czech Technical University in Prague. In the following year, he was called to Masaryk University and remained faithful to it until his death. He was known as an excellent teacher and co-author of a number of textbooks of descriptive geometry for secondary and grammar schools. He wrote more than 30 scientific treatises on algebraic geometry (the theory of surfaces and straight lines in three-dimensional and four-dimensional space). His best-known publications include the books *Imaginární elementy v geometrii* [Imaginary Elements in Geometry, 1941, 1950], *Cyklografie* [Cyclography, 1949] and *Kubické a bikvadratické problémy* [Cubic and Biquadratic Problems, 1951].[94]

After Lerch's death, Eduard Čech was appointed an extraordinary professor of mathematics. His university studies in mathematics and descriptive geometry were interrupted by World War I, during which he served in the army in Italy. The state support which enabled him to spend the academic year 1921/1922 in Turin with the distinguished geometer Guido Fubini had a great influence on his scientific career. Based on their collaboration, two common books on projective differential geometry were created — *Geometria proiettiva differenziale I, II* (1926, 1927) and *Introduction à la géométrie projective différentielle des surfaces* (1931). After returning from Italy, Eduard Čech obtained habilitation in mathematics at the Faculty of Natural Sciences of Charles University (1922). In 1923, he was appointed an extraordinary professor and in 1928 a full professor of mathematics at Masaryk University. From 1931, he was engaged in general and combinatorial topology. His outstanding results attracted worldwide attention. In 1935, he was invited to lectures in Princeton at the *Institute for Advanced Study*, one of the best centers of mathematical research of that time. After returning to Brno, he founded the renowned topological seminar in which many respected publications were created by himself and his students. In 1936, he published the monograph *Bodové množiny* [Point Sets]. During the World War II, he worked on the creation of high school textbooks of mathematics. After the war, he moved to Charles University and influenced the development of Czechoslovak mathematics and mathematics teaching.

[94]About the life and work of Ladislav Seifert, see Hrdličková (2001); Nádeník (2011).

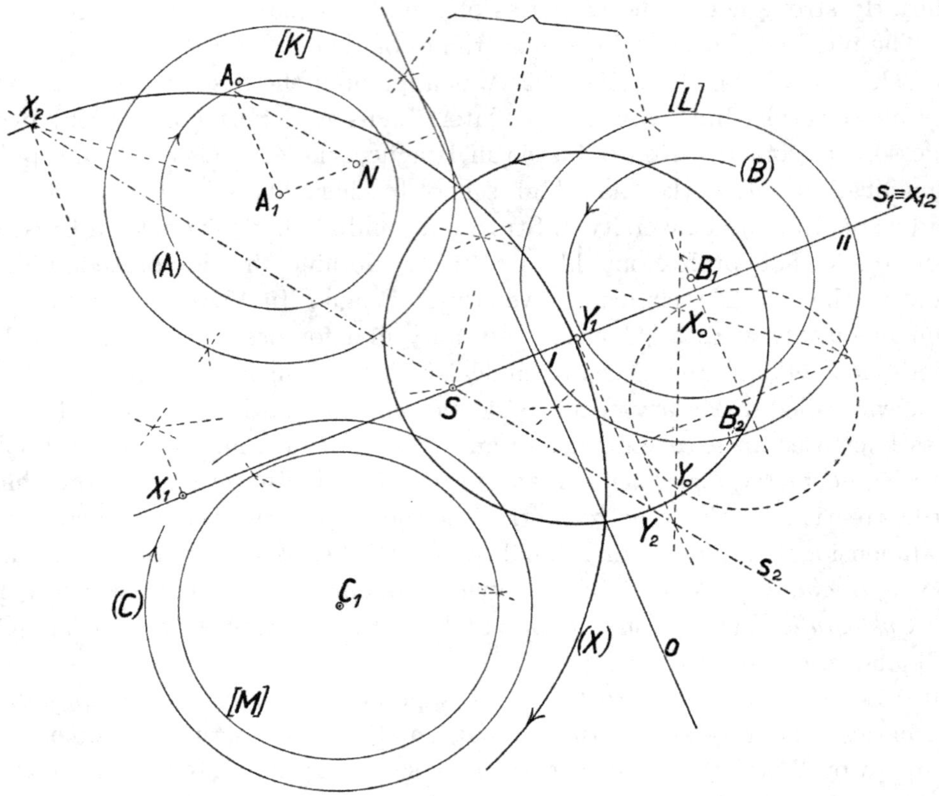

Figure 27. The geometrical construction from Ladislav Seifert's book *Cyclography.*[95]

Eduard Čech was one of the few world-renowned Czech mathematicians; his name is still alive in world mathematics (see Stone-Čech compactification, Čech-Brouwer dimension, Čech spaces, definition of homology and cohomology for general spaces). He published nearly a hundred papers on projective and differential geometry, topology, mathematical analysis and set theory. He was also deeply interested in the problems of teaching mathematics at primary and secondary schools as well as universities. He represented Czechoslovak mathematics at a number of international congresses and conferences, and lectured at many universities in the world. He was also a member of numerous foreign scientific societies. From 1953, he was the editor of the international mathematical journal entitled *Czechoslovak Mathematical Journal.* In 1960, he founded the international mathematical journal *Commentationes Mathematicae Universitatis Carolinae.*[96]

[95]The picture is from the authors' archive.

[96]About the life and work of Eduard Čech, see Katětov and Simon (1993); Nádeník (2011).

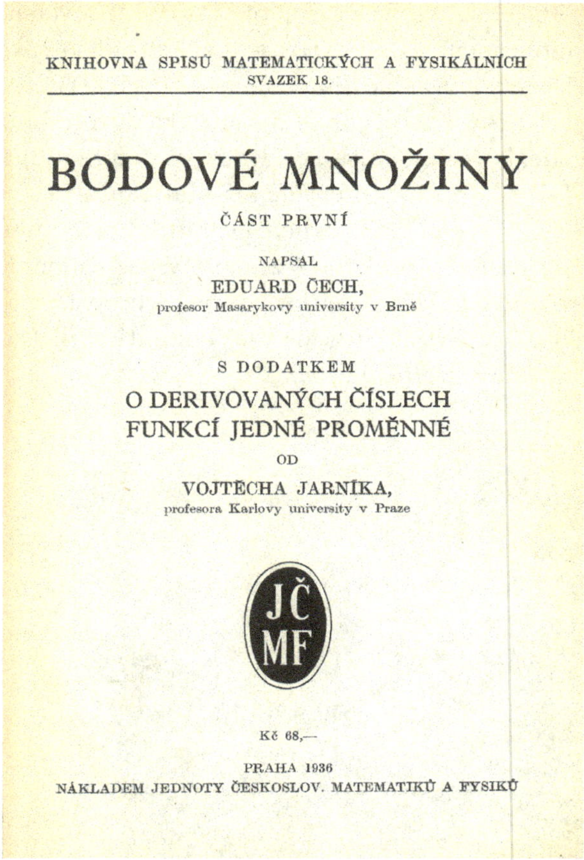
KNIHOVNA SPISŮ MATEMATICKÝCH A FYSIKÁLNÍCH
SVAZEK 18.

BODOVÉ MNOŽINY

ČÁST PRVNÍ

NAPSAL
EDUARD ČECH,
profesor Masarykovy university v Brně

S DODATKEM
O DERIVOVANÝCH ČÍSLECH
FUNKCÍ JEDNÉ PROMĚNNÉ
OD
VOJTĚCHA JARNÍKA,
profesora Karlovy university v Praze

JČMF

Kč 68,—

PRAHA 1936
NÁKLADEM JEDNOTY ČESKOSLOV. MATEMATIKŮ A FYSIKŮ

Figure 28. The picture of Eduard Čech and the title page of his book.[97]

In 1934, Otakar Borůvka (1899–1995) became an extraordinary professor of mathematics at Masaryk University. He was a student and assistant of Matyáš Lerch under whose leadership he began working on his dissertation which he defended (after Lerch's death) in 1923. In the academic year 1926/1927, he pursued a study stay at the Sorbonne in Paris, where he developed his talent under the leadership of Élie Joseph Cartan, a great mathematician who dealt with projective and differential geometry and the theory of invariants of transformation groups. At that time, Otakar Borůvka devoted himself mainly to differential geometry, algebra and graph theory. After returning in 1928, he obtained habilitation. He spent the academic year 1929/1930 on a second study stay in Paris. In the following year, he added another study stay with Wilhelm Blaschke at the University in Hamburg.

[97]The pictures are from the authors' archive.

In 1931, the professorship of Masaryk University submitted a proposal for Borůvka's appointment as an extraordinary professor. However, it came during the economic crisis, so it was not implemented until 1934. After the closure of Czech universities by the German occupants, Borůvka devoted himself to algebra. After World War II, he dealt mainly with the study of differential equations. In 1946, he was appointed a full professor of mathematics at Masaryk University (with effect from 1940). He lectured until 1970, when he retired. For many years, he worked at the Institute of Mathematics of the Czechoslovak Academy of Sciences in Brno (a branch of the Institute of Mathematics of the Czechoslovak Academy of Sciences in Prague). In 1965, he participated in the foundation of the international mathematical journal *Archivum Mathematicum.*[98]

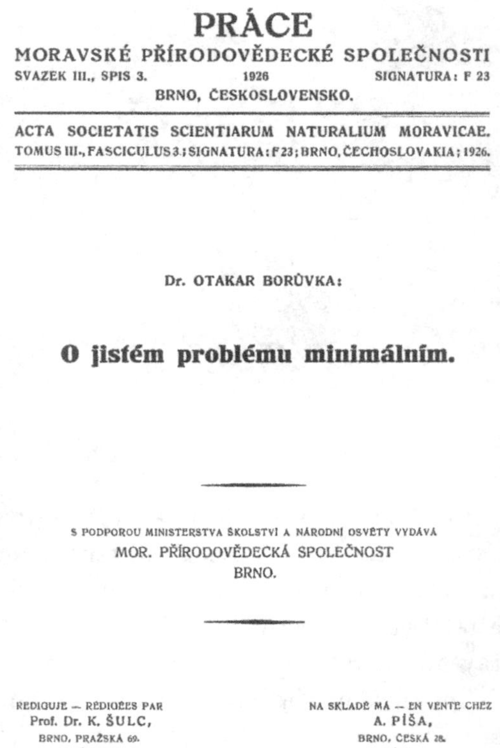

PRÁCE
MORAVSKÉ PŘÍRODOVĚDECKÉ SPOLEČNOSTI
SVAZEK III., SPIS 3. 1926 SIGNATURA: F 23
BRNO, ČESKOSLOVENSKO.

ACTA SOCIETATIS SCIENTIARUM NATURALIUM MORAVICAE.
TOMUS III., FASCICULUS 3; SIGNATURA: F 23; BRNO, ČECHOSLOVAKIA; 1926.

Dr. OTAKAR BORŮVKA:

O jistém problému minimálním.

S PODPOROU MINISTERSTVA ŠKOLSTVÍ A NÁRODNÍ OSVĚTY VYDÁVÁ
MOR. PŘÍRODOVĚDECKÁ SPOLEČNOST
BRNO.

REDIGUJE — RÉDIGÉES PAR
Prof. Dr. K. ŠULC,
BRNO, PRAŽSKÁ 69.
ČSR.

NA SKLADĚ MÁ — EN VENTE CHEZ
A. PÍŠA,
BRNO, ČESKÁ 28.
ČSR.

Figure 29. The picture of Otakar Borůvka and the title page of his well-known article *On a Certain Minimal Problem.*[99]

[98]About the life and work of Otakar Borůvka, see Třešňák *et al.* (1996).
[99]The picture of Otakar Borůvka is taken from Třešňák *et al.* (1996). The picture of the title page is from the authors' archive.

Borůvka's work *O jistém problému minimálním* [On a Certain Minimal Problem, 1926] became well-known. It described an algorithm for finding a minimum spanning tree. The algorithm bears his name in world literature until now. His important monographs are *Úvod do teorie grup* [Introduction to Group Theory, 1944, 1952, published in German, English and Romanian], *Základy teorie grupoidů a grup* [Fundaments of Groupoids and Groups, in German in 1960, in Czech in 1962, and in English in 1974, summarising the knowledge of groupoids and group theory], *Matice* [Matrices, 1947], *Základy teorie matic* [Fundamentals of Matrix Theory, 1971, providing introductory knowledge of matrix theory and their applications] and *Lineare Differentialtransformationen 2. Ordnung* (1967, later an extended English edition which contains, among other things, Borůvka's original results in the theory of linear second order differential transforms).

Josef Kaucký (1895–1982), who for many years worked as an assistant at the Institute of Theoretical Physics and taught at secondary schools, obtained habilitation in mathematics at the Faculty of Natural Sciences of Masaryk University in 1928. In 1937, he was appointed an unpaid extraordinary professor and a year later a professor of mathematics at the Technical University of Milan Rastislav Štefánik in Košice. After its relocation, he moved to Bratislava, where he worked until 1946. After World War II, he returned to the Technical University in Brno. After a short time, he began teaching at the Military Academy in Brno. He was interested in the theory of differential and difference equations, projective differential geometry and probability theory. His following books became well-known: *Úvod do počtu pravděpodobnosti a teorie statistiky* [Introduction to Probability and Statistics, 1934], *Elementární metody řešení obyčejných diferenciálních rovnic* [Elementary Methods for Solving Ordinary Differential Equations, 1953] and *Kombinatorické identity: úvod do studia metod kombinatorické analýzy* [Combinatorial Identities: Introduction to Combinatorial Analysis Methods, 1975].[100]

Let us recall that from 1920, Bohuslav Hostinský (1884–1951) worked at the Faculty of Natural Sciences of Masaryk University as a full professor of theoretical physics. In 1907, he received a doctorate at the Czech University in Prague. In the school year 1908/1909, he studied at the Sorbonne in Paris which influenced his further work. In 1912, he obtained habilitation in higher mathematics at the Czech University in Prague. Several times he was the dean of the Faculty of Natural Sciences of Masaryk University. In the academic year 1929/1930, he became the rector of Masaryk University. He was a member of Czech and foreign scientific societies. He was mainly active in the Czech Academy of Sciences and Arts and in Brno's branch of the Union. For many years he was the editor of the series *Spisy přírodovědecké fakulty Masarykovy university* [The Writings of the Faculty of Natural Sciences of Masaryk University].

[100]About the life and work of Josef Kaucký, see Čermák (1956); Novotný (1975).

Bohuslav Hostinský worked in differential geometry of curves and surfaces, probability theory (Markov chains), mathematical analysis and theoretical physics (kinetic theory of gases, vibration theory). He is the author of nearly 150 scientific papers and books, for example, *Diferenciální geometrie křivek a ploch* [Differential Geometry of Curves and Surfaces, 1915, 1935, 1942, 1950], *Mechanika tuhých těles* [Mechanics of Rigid Bodies, 1924], *Geometrické pravděpodobnosti* [Geometrical Probabilities, 1926], *Opérations infinitésimales linéaires* (co-author Vito Volterra, 1938), *Équations fonctionelles relatives aux probabilités continues en chaîne* (1939), *O mnohoúhelnících a mnohostěnech* [About Polygons and Polyhedrons, 1947] and *Počet pravděpodobnosti* [Calculus of Probability, 1950]. Let us note that Bohuslav Hostinský was a great opponent of relativity theory.[101]

3.5. *Technical University in Brno*

After the establishment of Czechoslovakia, there was great development of technical education also in Moravia. The Czech Technical University in Brno (founded in 1899) was transformed into a modern technical university. From the 1930s, it used the name *Vysoká škola technická Dra Edvarda Beneše* [The Technical University of Dr. Edvard Beneš]. Mathematics and descriptive geometry were taught here as a part of a general theoretical training. There were several top mathematicians at the posts of professors, private and honoured associate professors and assistants of the First or Second Institutes of Mathematics and the Institute of Descriptive Geometry.[102] The Technical University in Brno became an important center of both classical and applied mathematics. Members of the teaching staff played an important role in the establishment of the first universities in Slovakia.

From 1916, Jan Vojtěch taught mathematics at the First Institute. In 1923, he moved to Prague to the Czech Technical University, where he worked until his retirement in 1949. After his departure, František Nachtikal (1874–1939) substituted his lectures for a short time.[103] He also dealt with theoretical physics. In 1921, he published the first Czech-language book on relativity.[104]

In 1924, Jur Hronec was appointed a new extraordinary professor of mathematics at the First Institute. He was a student of Ludwig Schlesinger (1864–1933), professor at the University in Clausenberg (Cluj, Kolozsvár). After completing his university

[101]About Bohuslav Hostinský, see Beránek (1951); Košt'ál (1968). See also the information available at the web addresses: https://encyklopedie.brna.cz/home-mmb/?acc=profil_osobnosti&load=2922 [2.3.2019], https://cs.wikipedia.org/wiki/Bohuslav_Hostinsk%C3%BD [2.3.2019], https://edu.techmania.cz/cs/encyklopedie/vedec/2023/hostinsky [2.3.2019].

[102]For information on the origin and development of technology in Brno, see Froněk (1969); Košt'ál (1968).

[103]About František Nachtikal, see Košt'ál (1968); Valouch (1935).

[104]F. Nachtikal, *Princip Relativity. Názorný Výklad* [The Principle of Relativity. Illustrative Interpretation], A. Píša, Brno, 1921, 115 pages.

studies, Hronec received a post of secondary school teacher in Kežmarok, where he worked until 1923. At the same time, he went to the study stays at the University in Berlin (1910), Giessen (1911, 1912) and Paris (1914). In 1912, he defended his dissertation in Giessen under the supervision of L. Schlesinger. In the academic year 1923/1924, he studied at the universities in Prague, Göttingen and Giessen. On the impulse of Karel Petr and Jan Sobotka, he obtained habilitation at the Charles University in 1923 and began lecturing here as a private associate professor. In 1924, he was appointed to the Czech Technical University in Brno as an extraordinary professor of mathematics and four years later as a full professor. He worked there until 1938, when he left for Slovakia.

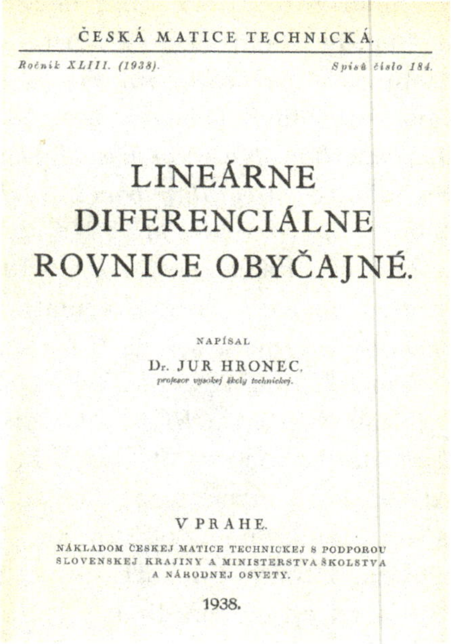
ČESKÁ MATICE TECHNICKÁ.

Ročník XLIII. (1938). *Spisů číslo 184.*

LINEÁRNE DIFERENCIÁLNE ROVNICE OBYČAJNÉ.

NAPÍSAL

Dr. JUR HRONEC,

profesor vysokej školy technickej.

V PRAHE.

NÁKLADOM ČESKEJ MATICE TECHNICKEJ S PODPOROU SLOVENSKEJ KRAJINY A MINISTERSTVA ŠKOLSTVA A NÁRODNEJ OSVETY.

1938.

Figure 30. The picture of Jur Hronec and the title page of his book.[105]

Jur Hronec was an excellent teacher. After leaving Brno, he devoted a great deal of energy to build higher education in Slovakia. When Milan Rastislav Štefánik Technical University in Košice was founded (1937), Jur Hronec was appointed its first rector and a professor of mathematics. When Košice fell to Hungary after the Munich Agreement, the school moved to Prešov, then to Martin and then to Bratislava. At that time, Hronec started to teach mathematics at Jan Ámos Komenský University in Bratislava. He participated in the establishment of the Faculty of Natural Sciences

[105]The pictures are from the authors' archive.

and from 1940 he taught there as a voluntary professor. After World War II, he contributed to the establishment of the Pedagogical Faculty and to the development of the Faculty of Natural Sciences of Milan Rastislav Štefánik University, the Technical University in Bratislava and the Agricultural and Forestry University in Košice. He was a member of the *Union*, the *Královská česká společnost nauk* [The Royal Bohemian Society of Sciences], the *Moravská přírodovědecká společnost* [The Moravian Society of Natural Sciences], the *Matica slovenská* [The Slovak Matica] and the *Slovenské múzeum* [The Slovak Museum]. In 1953, he was appointed an academician of the Slovak Academy of Sciences in Bratislava. He paid great attention to high school students; a mathematical competition of high school students (later the so-called Mathematical Olympiad) was organised in Slovakia on his initiative.

Under Schlesinger's influence, Hronec dealt mainly with the theory of linear differential equations. He wrote several original works and 10 publications, for example, *Algebraické rovnice a ich použitie na analytickú geometriu* [Algebraic Equations and Their Application to Analytical Geometry, 1932, 1949], *Lineárne diferenciálne rovnice obyčajné* [Linear Ordinary Differential Equations, 1938], *Diferenciálny a integrálny počet I, II* [Differential and Integral Calculus I, II, 1940, 1945, 1957; Second Part 1946], *Diferenciálne rovnice I, II* [Differential Equations I, II, 1956, 1960; Second Part 1958]. He summarised his views on the education reform and the position of mathematics in his works *Matematika ako prostriedok výchovy* [Mathematics as a Means of Education, 1906], *Vyučovanie a vyučovacia osobnost* [Teaching and Teaching Personality, 1923], *Učitel'ova osobnost* [Teacher's Personality, 1926], *Hlavné hl'adiská reformy rámového zariadenia nášho školstva* [The Mains Focuses of Our Education Reform, 1929], *O našich vysokých školách* [About Our Universities, 1930] and *Vysokoškolské požadavky Slovenska* [Higher Education Requirements of Slovakia, 1936] etc.[106]

After Hronec's departure in the autumn of 1939, Jiří Klapka (1900–1976) was appointed an extraordinary professor of mathematics at the Technical University in Košice. After the separation of Slovakia, he moved to Brno. During World War II, he was under German police supervision because his brother Otakar Klapka (1891–1941), a well-known lawyer, a long-standing deputy and a Prague mayor, was executed for resistance activity. During the occupation, Jiří Klapka worked as a mathematician at the Škoda Factory in Hradec Králové. In 1946, he was appointed a full professor of descriptive geometry at the Technical University in Brno (with retroactive effect from 1942). In 1952, he also took over the lectures on mathematics in place of the ill professor Karel Čupr.

Jiří Klapka focused on differential, projective, algebraic and descriptive geometry, graphical methods and vector calculus. He wrote several textbooks, lecture

[106] About the life and work of Jur Hronec, see Greguš (1981); Hronec *et al.* (1999); Žbirková (1975).

notes and small books — *Úvod do vektorového počtu a jeho užití v elektrotechnice* [Introduction to Vector Calculus and Its Use in Electrical Engineering, 1932], *Jak se studují útvary v prostoru?* [How to Study Objects in Space?, First Part, 1941; Second Part, 1942, 1947], *Deskriptivní geometrie, přednásky pro posluchače oboru strojního a elektrotechnického inženýrství* [Descriptive Geometry, Lectures for Students of Mechanical and Electrical Engineering, First Part 1949; Second Part 1953; other editions exist], *Deskriptivní geometrie pro směr stavební, zeměměřičský a architekturu* [Descriptive Geometry for Building, Surveying and Architecture, First Part and Second Part, 1951; other editions exist], *Analytická geometrie* [Analytical Geometry, 1960].

His works on projective differential geometry analysing the properties of straight line surfaces on the basis of Blaschke's and Čech's theory and his studies on the theory of Segre's W-congruencies have become recognised.[107]

Figure 31. The picture of Jiří Klapka and the title page of his book.[108]

[107]About the life and work of Jiří Klapka, see Borůvka and Svoboda (1960); Obůrka (1976); Košt'ál (1968).

[108]The picture of Jiří Klapka is taken from Obůrka (1976), and the picture of the title page is from the authors' archive.

Between 1906 and 1919, Matyáš Lerch headed the Second Institute of Mathematics of the Czech Technical University in Brno. Later, he moved to Masaryk University. Mathematical lectures were substituted by Karel Čupr (1883–1956). In 1923, Čupr was appointed an extraordinary and in 1928 a full professor of mathematics at the Czech Technical University in Brno. He became Lerch's direct successor and worked here (except for a war-forced break) until the early 1950s. After graduating from the Czech University in Prague, Karel Čupr taught at high schools. In 1914, he became Lerch's assistant and at the same time he taught at the Business Secondary School in Brno. He was a very prolific author, who wrote over 300 mathematical studies, textbooks, monographs, handbooks, popularising essays, commemorative and jubilee articles and reviews, etc. Let us mention some of his textbooks: *Přehled geometrie ke zkouškám* [Geometry Overview for Exams, 1941], *Přehled algebry ke zkouškám* [Algebra Overview for Exams, 1942], *Numerické řešení rovnic* [Numerical Solutions of Equations, 1945, 1947], *Matematika I. a II.* [Mathematics I and II, 1946], *Úvod do nomografie* [Introduction to Nomography, 1944, 1947], *Aritmetické hry a zábavy* [Arithmetic Games and Entertainment, 1942], *Geometrické hry a zábavy* [Geometric Games and Entertainment, 1949] and *Matematické zábavy a hry* [Mathematical Entertainment and Games, 1953]. His scientific work covers differential and difference equations of the nth order, special integrals and hyperbolic functions; their core is mainly in applications in electrical engineering, geodesy and statistics. He also dealt with the history of mathematics in Moravia, especially with the development of technical education. His lectures and essays *Z dějin matematiky v zemi Moravskoslezské* [From the History of Mathematics in the Moravian-Silesian Country], *Jak vznikalo inženýrství na Moravě* [How the Engineering in Moravia Was Born], *Inženýrské umění na Moravě* [Engineering Art in Moravia] and *Stanislav Vydra a jeho doba* [Stanislav Vydra and His Era] became known. Let us recall that he also dealt with three anonymous mathematical manuscripts for which he conducted a rigorous comparative study and showed that the author was Jan Ámos Komenský (1592–1670).[109]

From 1908 until 1925, Miloslav Pelíšek held the post of professor of descriptive geometry at the Czech Technical University in Brno. In 1927, Josef Klíma (1887–1943), who worked as a high school professor, assistant and private associate professor at the Czech Technical University for many years, was appointed an extraordinary professor and four years a full professor of descriptive geometry. He dealt with the theory of warped surfaces, conic sections and geometrical methods. Together with Václav Ingriš, he published the high school textbooks of drawing

[109]About the life and work of Karel Čupr, see Frank (1957); Košt'ál (1968); Lukášová (2002).

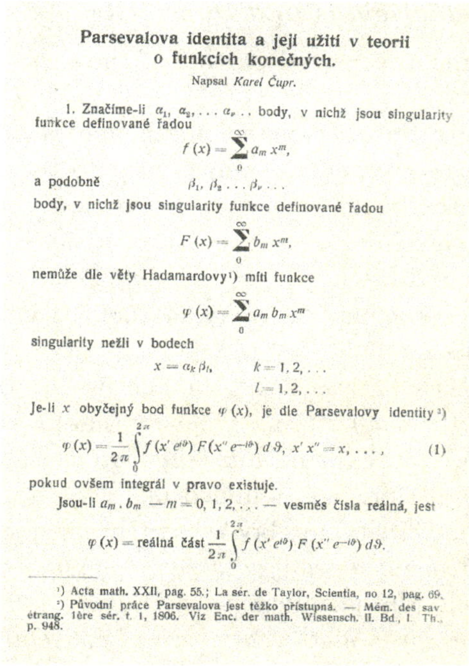

Parsevalova identita a její užití v teorii o funkcích konečných.

Napsal *Karel Čupr.*

1. Značíme-li $\alpha_1, \alpha_2, \ldots \alpha_\nu \ldots$ body, v nichž jsou singularity funkce definované řadou

$$f(x) = \sum_0^\infty a_m x^m,$$

a podobně $\beta_1, \beta_2 \ldots \beta_\nu \ldots$

body, v nichž jsou singularity funkce definované řadou

$$F(x) = \sum_0^\infty b_m x^m,$$

nemůže dle věty Hadamardovy[1]) míti funkce

$$\varphi(x) = \sum_0^\infty a_m b_m x^m$$

singularity nežli v bodech

$$x = \alpha_k \beta_l, \qquad k = 1, 2, \ldots \quad l = 1, 2, \ldots$$

Je-li x obyčejný bod funkce $\varphi(x)$, je dle Parsevalovy identity[2])

$$\varphi(x) = \frac{1}{2\pi} \int_0^{2\pi} f(x' e^{i\vartheta}) F(x'' e^{-i\vartheta}) \, d\vartheta, \; x' x'' = x, \ldots, \qquad (1)$$

pokud ovšem integrál v pravo existuje.

Jsou-li $a_m . b_m$ — $m = 0, 1, 2, \ldots$ — vesměs čísla reálná, jest

$$\varphi(x) = \text{reálná část} \, \frac{1}{2\pi} \int_0^{2\pi} f(x' e^{i\vartheta}) F(x'' e^{-i\vartheta}) \, d\vartheta.$$

[1]) Acta math. XXII, pag. 55.; La sér. de Taylor, Scientia, no 12, pag. 69.

[2]) Původní práce Parsevalova jest těžko přístupná. — Mém. des sav. étrang. 1ère sér. t. 1, 1806. Viz Enc. der math. Wissensch. II. Bd., I. Th., p. 948.

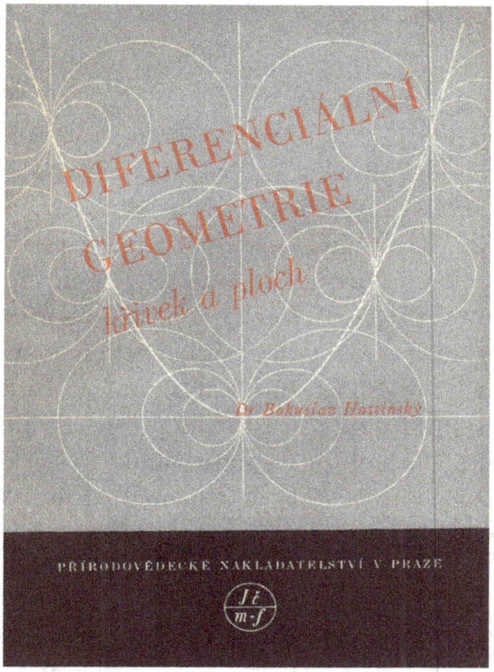

Figure 32. The first page of Karel Čupr's article *Parseval's Identity* ... and the title page of Bohuslav Hostinský's textbook.[110]

and descriptive geometry entitled *Rýsování pro III. a IV. třídu gymnasií a reálných gymnasií* [Drawing for 3rd and 4th Classes of Gymnasiums and Real Gymnasiums, 1934, Slovak version 1935], *Rýsování pro III. a IV. třídu reformních reálných gymnasií a reálek* [Drawing for 3rd and 4th Classes of Reformed Real Gymnasiums and Grammar Schools, 1934, Slovak version 1935], *Deskriptivní geometrie pro V. třídu reálek* [Descriptive Geometry for 5th Classes of Grammar Schools, 1934, 1946, 1947, 1948], *Deskriptivní geometrie pro VI. a VII. třídu reálek* [Descriptive Geometry for 6th and 7th Classes of Grammar Schools, 1935, 1946, 1947] and *Deskriptivní geometrie pro VII. a VIII. třídu reálných gymnasií a reformních reálných gymnasií* [Descriptive Geometry for 7th and 8th Classes of Real Gymnasiums and Reform Real Gymnasiums, 1935, 1938, Slovak edition 1937]. A well-known, popular and currently used two-part textbook *Deskriptivní geometrie* [Descriptive Geometry] was written by František Kadeřávek, Josef Klíma and Josef Kounovský.[111]

[110] The pictures are from the authors' archive.

[111] About the life and work of Josef Klíma, see Klapka (1949); Seifert (1946).

3.6. *The Union of Czechoslovak Mathematicians and Physicists*

After 1918, the *Jednota* [The Union][112] became a respected partner of the Czechoslovak government.

The Union started the work on reforming the teaching of mathematics and physics at all schools. The number of its members grew and its numerous professional activities developed (writing and reviewing textbooks, monographs, professional, educational and popularising articles, and organising lectures and competitions for talented students and emerging professionals and teachers). It maintained contacts with Czech and foreign scientific societies. Czech and German mathematicians, teachers, students and practitioners met on its ground.

In the 1920s and 1930s, its publishing activity was also rich because it was still publishing the *Časopis pro pěstování matematiky a fysiky* [The Journal for the Cultivation of Mathematics and Physics (found in 1872)] and *Rozhledy matematicko-přírodovědecké* [The Mathematical and Natural Horizons (founded in 1922)], textbooks for primary and secondary schools, university textbooks and professional monographs (a 20 volume series entitled *Sborník Jednoty českých mathematiků v Praze* [Proceedings of the Union of Czech Mathematicians in Prague, 1898–1937], a 56 volume series entitled *Knihovna spisů mathematických a fysikálních* [The Library of Mathematical and Physical Publications, 1915–1952], a 44 volume series entitled *Kruh* [The Circle, 1925–1953]), annual reports, and informative and educational prints. It became practically the only recognised publisher of literature in mathematics, physics and descriptive geometry in Czechoslovakia.[113]

In 1938, the Union completed the construction of its own house in the centre of Prague, which provided sufficient space for a large library, a textbook shop and teaching aids and lecture rooms. In the following year, the Czech lands were occupied by Germany, which began to restrict the activities of the Union. But the Union remained an important place where the sciences could be cultivated to a limited extent even during World War II. Moreover, the Union tried to spread information on the development of mathematics and physics, it took care of its library, organised training courses and lectures, and tried to publish educational texts and teaching aids.

[112]After the establishment of the Czechoslovak Republic, *Jednota* adopted the new name: *Jednota československých matematiků a fysiků* [The Union of Czechoslovak Mathematicians and Physicists].

[113]In 1919, the Union acquired a state license to print and sell textbooks. Shortly thereafter, it set up its own publishing house, which in 1933 merged with the publishing house of mining engineers and the *Prometheus* publishing house was established. Thus, a modern centre specialised in printing professional literature was founded in Prague.

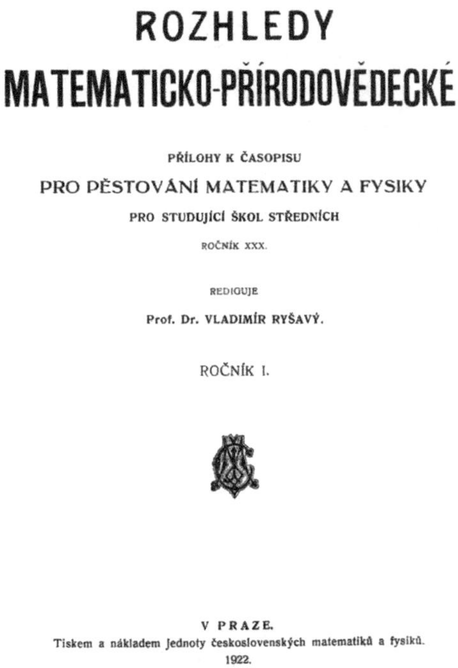

ROZHLEDY
MATEMATICKO-PŘÍRODOVĚDECKÉ

PŘÍLOHY K ČASOPISU
PRO PĚSTOVÁNÍ MATEMATIKY A FYSIKY
PRO STUDUJÍCÍ ŠKOL STŘEDNÍCH
ROČNÍK XXX.

REDIGUJE
Prof. Dr. VLADIMÍR RYŠAVÝ.

ROČNÍK I.

V PRAZE.
Tiskem a nákladem Jednoty československých matematiků a fysiků.
1922.

Figure 33. The title page of the journal *Mathematical and Natural Horizons* and the picture of the Union's building in the center of Prague in the 1930s.[114]

3.7. *Conclusion*

In the interwar period, Czechoslovak mathematicians have approached the European level in terms of both research and pedagogical work thanks to numerous contacts with foreign countries and international cooperation and thanks to the development of all types of schools, research institutes and laboratories and extensive and targeted support of publishing and association activities. At the end of the 1930s, the centers of higher mathematics in Slovakia began to be established.

4. German Mathematical Community in Bohemia and Moravia in the Early 20th Century

4.1. *Introduction*

In addition to the Czech mathematical community, there was a recognised German community in the Czech lands. Its members belonged to European mathematical

[114]The picture of the title page is from the authors' archive. The picture of the Union's building is taken from Veselý (1962).

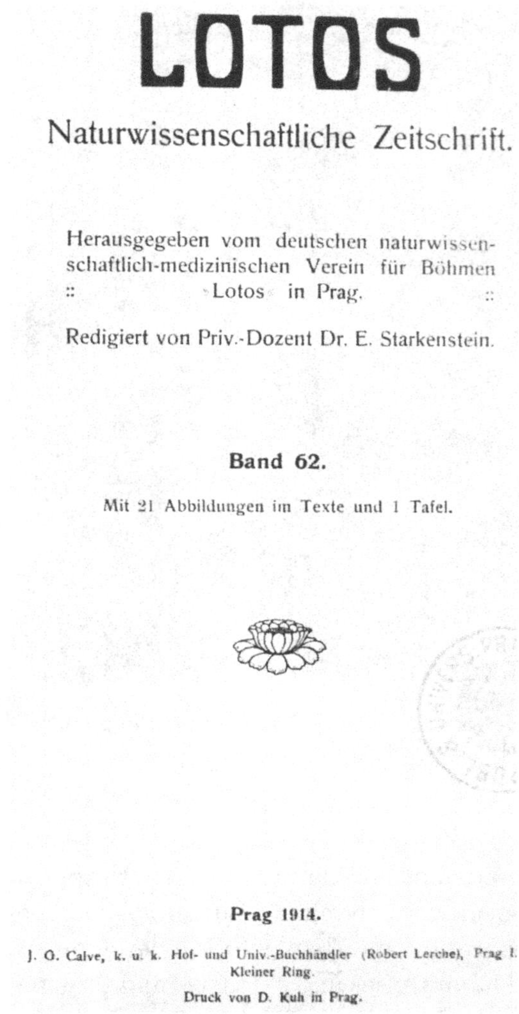

LOTOS

Naturwissenschaftliche Zeitschrift.

Herausgegeben vom deutschen naturwissenschaftlich-medizinischen Verein für Böhmen :: »Lotos« in Prag. ::

Redigiert von Priv.-Dozent Dr. E. Starkenstein.

Band 62.

Mit 21 Abbildungen im Texte und 1 Tafel.

Prag 1914.

J. G. Calve, k. u. k. Hof- und Univ.-Buchhändler (Robert Lerche), Prag I. Kleiner Ring.

Druck von D. Kuh in Prag.

Figure 34. The title page of the journal *Lotos* and the official stamp of the Mathematical Seminar of the German University in Prague.[115]

leaders. It had strong relations with the mathematicians of German-speaking countries and it was in contact with modern topics developed in world mathematics.

4.2. *Prague*

A strong German mathematical community existed in Prague from the middle of the 1860s. Its centers were the German Technical University in Prague (1869) and the German University in Prague (1882), schools preferred by the government in Vienna.

[115]The pictures are from the authors' archive.

They had better buildings, larger libraries, more specialised seminars and more systematised posts. The German mathematical community did not need to "compete" with the Czech community to demonstrate its activities and its professional significance and; to fight for its international position and recognition. However, it gradually got into isolation and gained a provincial character. Prague become the home for some German mathematicians, but for many it was only a transfer station on their way to more prestigious locations in the Austrian monarchy or in Germany.[116]

Outside the universities, the German mathematicians met in several professional associations and societies. The most important societies in the pre-war period and during World War I were *Lotos* (1848–1945), *Gesellschaft zur Forderung deutscher Wissenschaft, Kunst und Kultur in Böhmen* (1891–1945) and *Mathematisches Kränzchen in Prag* (1913–1934).

At the beginning of the 20th century, two regular chairs of mathematics were systematised at the German University in Prague, but there were no private associate professors and assistants. Few students studied mathematics.

In 1888, an extraordinary professor Georg Alexander Pick (1859–1942) joined the German University in Prague. After four years, he was appointed a full professor of mathematics at the First Chair. During his long career, he has raised a number of students in Prague. At the beginning of his career, he was influenced by Leo Königsberger (1837–1921), under whose leadership he defended his doctorate at the University in Vienna in 1880, by Ernst Mach (1838–1916), with whom he worked as an assistant at the German University in Prague, and by Felix Klein (1849–1925), with whom he stayed on a study stay in Leipzig from 1883 to 1885.

Pick wrote nearly 70 papers published in German and French in reports of several European academies and in prestigious journals. He dealt with linear algebra, invariant theory, integral calculus, complex analysis, differential and integral equations, potential theory, functional analysis, and differential geometry. His name is still alive in mathematics (Pick's formula, Schwarz-Pick lemma, Pick matrix, Pick-Nevanlinna interpolation).[117]

In 1912, Gerhard Hermann Waldemar Kowalewski (1876–1950) moved to the German University in Prague from the German Technical University in Prague. For eight years, he taught mathematics at the Second Chair. In 1920, he left Prague at his own request because he was not interested in working in a country where the German community is a recognised and equal national minority, but only a minority. He brought new modern topics to university lectures (theory of determinants and

[116]About mathematics at the German University in Prague, see Bečvářová (2016) and also Bečvářová (2008, esp. pp. 58–73).

[117]About the life and work of Georg Alexander Pick, see Bečvář and Bečvářová (2013b); Bečvářová (2015, 2016); Fritsch (2001a, 2001b, 2019); Pinl and Dick (1974); Toepell (1991).

Figure 35. The picture of Georg Pick and the illustration of Pick's formula.[118]

matrices, group theory, game theory, etc.). Note that as one of the few mathematicians of that time, he supported women's university studies.

Kowalewski is the author of a number of papers (group theory, transformation groups, differential geometry, approximation theory and interpolation, differential and integral equations) and of many textbooks — *Einführung in die Determinantentheorie einschließlich der Fredholmschen Determinanten* (1909, 1925, 1942, 1954), *Einführung in die analytische Geometrie* (1910, 1953), *Grundzüge der Differential- und Integralrechnung* (1909, 1919, 1923, 1932), *Das Integral und seine geometrische Anwendungen* (1910), *Die klassischen Probleme der Analysis des Unendlichen* (1910), *Die komplexen Veränderlichen und ihre Funktionen. Fortsetzung der Grundzüge der Differential- und Integralrechnung. Zugleich eine Einführung in die Funktionentheorie* (1911), *Mathematica delectans. Ausgewählte Kapitel aus der Mathematik der Spiele in gemeinverständlicher Darstellung* (1921), *Integralgleichungen* (1930, 2013), *Alte und neue mathematische Spiele. Eine Einführung in die Unterhaltungsmathematik* (1930), *Einführung in die Theorie der kontinuierlichen Gruppen* (1931), *Vorlesungen über Allgemeine natürliche*

[118]The pictures are taken from Bečvářová (2016).

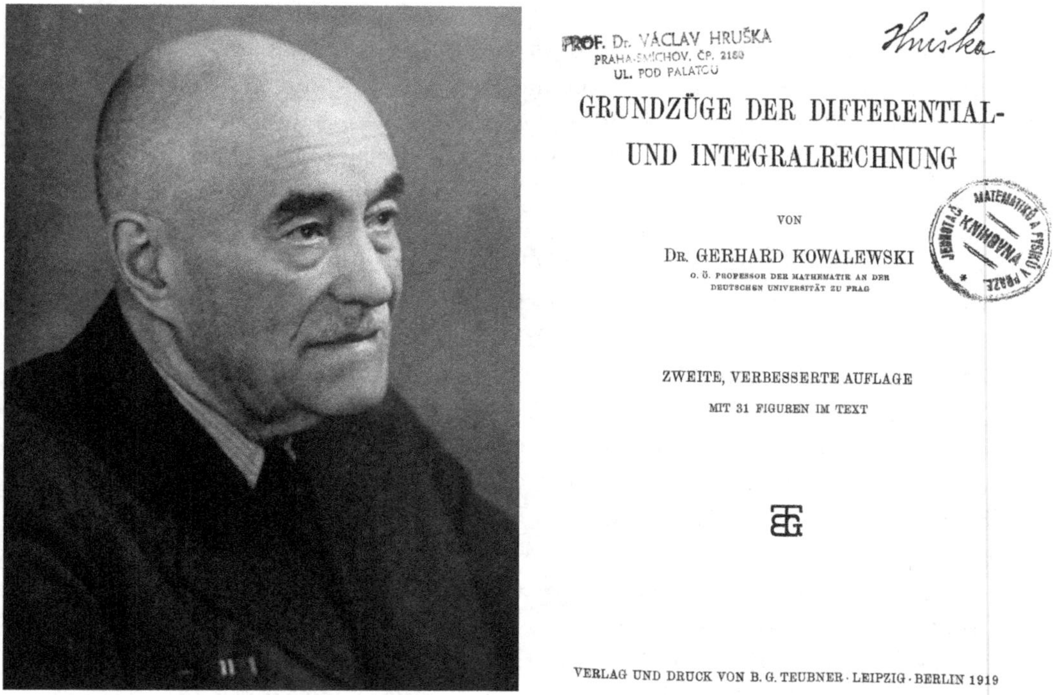

Figure 36. The picture of Gerhard Kowalewski and the title page of his book.[119]

Geometrie und Liesche Transformationsgruppen (1932), *Interpolation und genäherte Quadratur. Eine Ergänzung zu den Lehrbüchern der Differential- und Integralrechnung* (1932), *Lehrbuch der höheren Mathematik für Universitäten und Technische Hochschulen* (1933), *Magische Quadrate und magische Parkette* (1937), *Der Keplersche Körper und andere Bauspiele* (1938), *Grundbegriffe und Hauptsätze der höheren Mathematik insbesondere für Ingenieure und Naturforscher* (1938).[120]

Since the late 1860s, the German Technical University in Prague had been one of the leading German mathematical workplaces in the Czech lands, albeit it consisted only of a small community focused mainly on differential geometry, descriptive geometry and mathematics and applications in technical practice. At the beginning of the 20th century, there were three chairs of mathematics at schools that were

[119] The pictures are taken from Bečvářová (2016).

[120] Note that from 1920 to 1939 Kowalewski taught at the Technical University in Dresden. In 1939, he returned to Prague. Until 1945, he was a professor at the German University in Prague, resp. at the German Technical University in Prague. After 1946, he worked in Regensburg and Munich. About the life and work of Gerhard Kowalewski, see Bečvářová (2015, 2016, 2018); Kowalewski (1950); Toepell (1991).

filled with full or extraordinary professors, and one chair of descriptive geometry, some assistants and private associate professors were the teaching staff.

From 1907, the First Chair of Mathematics at the German Technical University in Prague was occupied by the full professor Karl Carda (1870–1943), who was until then an associate professor of mathematics at the Technical University in Vienna. Under the influence of older colleagues, he devoted himself to differential geometry at the beginning of his career, and later focused on modern topics — groups of transformations and invariant theory.[121] Gerhard Hermann Waldemar Kowalewski was appointed a full professor of mathematics at the Second Chair of Mathematics of the German Technical University in Prague in 1909. He came to Prague as an experienced mathematician from the University in Bonn, but he only lectured here until 1912, when he moved to the German University in Prague. Anton Grünwald (1873–1932) obtained habilitation at the German Technical University in Prague in 1909. In 1913, he was appointed an extraordinary professor of mathematics at the Third Chair. He was charged with teaching higher mathematics and selected geometry topics.[122]

In April 1913 (after Kowalewski's departure and thanks to a regular competition), Wilhelm Blaschke (1885–1962) was appointed a professor of mathematics. However, he worked at the school only until 1914, when he received a professorship at the University in Leipzig. He was an extraordinary mathematician, considered one of the founders of integral geometry. He brought new, prolific and modern geometric themes to Prague — conformal mappings, their properties and applications, convex theory, isoperimetric problems and affine differential geometry. After leaving Prague, he lectured at the Universities in Leipzig, Königsberg (Kaliningrad) and Tübingen. In 1919, he settled in the University in Hamburg. The German mathematical community in Prague maintained close professional contacts with him until the 1930s.[123]

After Blaschke's departure, the German Technical University in Prague was forced to search for a new professor of mathematics. They contacted Kowalewski, who agreed to return. In January 1915, the professorships made a proposal for his reappointment and sent it to the Ministry in Vienna. However, the Ministry disagreed with this non-standard procedure, requiring a proper competition for the post. In the summer of 1915, the professorships decided to insist on its original proposal. The Ministry after some time accepted the Prague proposal. When everything seemed to be resolved and the empty chair occupied, Kowalewski refused to accept the post.

[121] About the life and work of Karel Carda, see Bečvářová (2016); Birk (1931).

[122] About the life and work of Anton Grünwald, see Bečvářová (2016); Brik (1931).

[123] About the life and work of Wilhelm Blaschke, see Bečvářová (2015, 2016); Burau (1963); Reichardt (1966); Strubecker (1986); Toepell (1991).

A new competition for the post was put up in 1916 and led to the selection of the so-called "in professors" — Heinrich Tietze (1880–1964), Roland Weitzenböck (1885–1955) and Johann Radon (1887–1956). The appointment of a professor of mathematics did not take place until 1918; it could also have been due to the fact that there were only a few students during World War I, so the professors did not urge the appointment. Only in February 1918, when the number of students began to grow again, the rector asked for a quick solution. Due to the lack of funds, the Ministry in Vienna decided to appoint only an extraordinary professor, which was out of the question for Heinrich Tietze because he was already a full professor (at the German Technical University in Brno). That is why Ronald Weitzenböck was appointed an extraordinary professor in July 1918.

Weitzenböck served a few years in the army before his arrival to Bohemia. He taught at the Military Academy in Mödling, and then he studied at the University in Vienna and shortly also at the University in Bonn and at the University in Göttingen. As a private lecturer, he lectured at the Technical University in Graz. He left Prague in 1920; in 1920/1921 he taught again in Graz; between 1921 and 1945 at the University in Amsterdam. He wrote many works on invariant theory. He summarised his results in 1923 in his extensive monograph entitled *Invariantentheorie*.[124]

Descriptive geometry was an important field cultivated at the German Technical University in Prague. In 1898, Eduard Janisch (1868–1915) was appointed an extraordinary professor of descriptive geometry after long negotiations. He moved from the German State Industrial School in Brno, and did not have a doctorate, habilitation, or any major work. He quickly obtained habilitation in Prague and in 1904 he was appointed a full professor of descriptive geometry. He held this post until his death. As one of the few Prague mathematicians he regularly attended international congresses of mathematicians, and followed developments in mathematics, although he practically did not publish. After his death, a high school professor Karl Mack (1882–1943) was appointed an extraordinary professor of descriptive geometry (1916). Four years later, he became a full professor and connected his whole life with Prague.[125]

Gerhard Kowalewski, Wilhelm Blaschke and Roland Weitzenböck were among the top mathematicians whose results in mathematical analysis, integral geometry, number theory, differential equations, etc. are vivid and recognised.[126]

4.3. *Brno*

After long and demanding political negotiations and preparations, Emperor Franz Joseph I signed the decree establishing the Polytechnic in Brno on September 13,

[124] About the life and work of Roland Weitzenböck, see Bečvářová (2016); Toepell (1991).
[125] About the life and work of Eduard Janisch and Karel Mack, see Bečvářová (2016); Birk (1931).
[126] For more information on the German mathematical community in Prague, see Bečvářová (2016).

1849. It was created as a school with German and Czech language of instruction. The reality was different; the teaching was from the very beginning held only in German.[127] The German Technical University in Brno thus became a natural center of German mathematics in Moravia. At the beginning of the 20th century, there were two chairs of mathematics and one chair of descriptive geometry.

In 1912, Lothar Schrutka (1881–1945), an assistant at the Technical University in Vienna, was appointed a professor of the First Chair of Mathematics at the German Technical University in Brno. He received his education at the Universities in Vienna, Göttingen and Berlin. He taught in Brno until 1925, when he left for the Technical University in Vienna.[128]

Since 1895, Emil Waelsch (1863–1927) was on the Second Chair of Mathematics at the German Technical University in Brno. He influenced teaching there for over 30 years. He graduated from the German University in Prague and the German Technical University in Prague. From 1884 to 1885, he studied under Felix Klein in Leipzig. Between 1885 and 1886, he collaborated with Paul Gordan (1837–1912) and Max Noether (1844–1921) in Erlangen, where he obtained his doctorate. In 1890, he obtained habilitation in descriptive geometry at the German Technical University in Prague. He shortly worked as an assistant of professor Wilhelm Fiedler (1832–1912) at the Technical University in Zürich. In the school year 1892/1893, he studied under Sophus Lie (1842–1899) in Leipzig. In 1893, he was nominated the third professor of mathematics at the German University in Prague, but the post was not set up for financial reasons. He worked at the Second Chair of Mathematics of the German Technical University in Brno until 1910, when he was appointed a professor of descriptive geometry.[129]

In 1910, Emil Waelsch was replaced by Heinrich Tietze, a private associate professor at the Technical University in Vienna. Tietze studied at the universities in Vienna, Munich and Göttingen. He stayed in Brno until 1919. However, he spent World War I at the front. After the war, he left for the University in Erlangen, where from he moved to Munich in 1925. He dealt with topology, theory of transcendental numbers, theory of discrete groups, set theory, continued fractions, four color problem and mathematical questions of inheritance. He was probably the most important mathematician who worked at the German Technical University in Brno throughout its whole existence.[130]

[127] For more information on the German Technical University in Brno, see Haussner (1924); Hellmer (1899); Rohrer (1875).

[128] About the life and work of Lothar Schrutka, see Einhorn (1985); Ottowitz (1992); Šišma (2002).

[129] About the life and work of Emil Waelsch, see Bečvářová (2016); Šišma (2002); Toepell (1991).

[130] About the life and work of Heinrich Tietze, see Šišma (2002); Perron (1981); Toepell (1991).

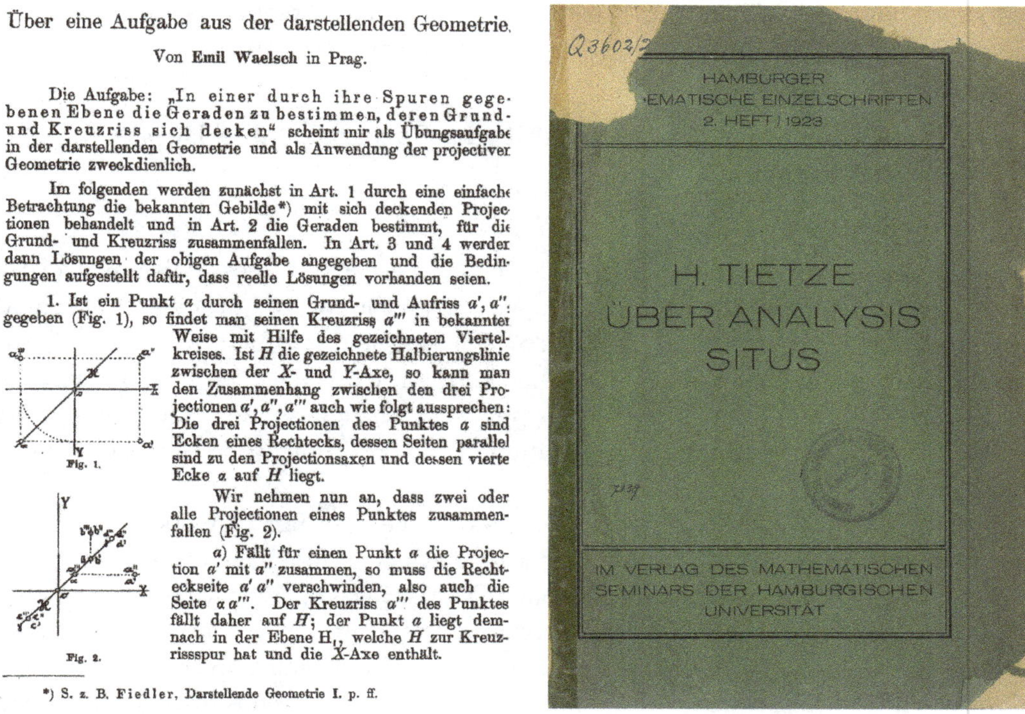
Über eine Aufgabe aus der darstellenden Geometrie.

Von **Emil Waelsch** in Prag.

Die Aufgabe: „In einer durch ihre Spuren gegebenen Ebene die Geraden zu bestimmen, deren Grund- und Kreuzriss sich decken" scheint mir als Übungsaufgabe in der darstellenden Geometrie und als Anwendung der projectiven Geometrie zweckdienlich.

Im folgenden werden zunächst in Art. 1 durch eine einfache Betrachtung die bekannten Gebilde*) mit sich deckenden Projectionen behandelt und in Art. 2 die Geraden bestimmt, für die Grund- und Kreuzriss zusammenfallen. In Art. 3 und 4 werden dann Lösungen der obigen Aufgabe angegeben und die Bedingungen aufgestellt dafür, dass reelle Lösungen vorhanden seien.

1. Ist ein Punkt a durch seinen Grund- und Aufriss a', a'' gegeben (Fig. 1), so findet man seinen Kreuzriss a''' in bekannter Weise mit Hilfe des gezeichneten Viertelkreises. Ist H die gezeichnete Halbierungslinie zwischen der X- und Y-Axe, so kann man den Zusammenhang zwischen den drei Projectionen a', a'', a''' auch wie folgt aussprechen: Die drei Projectionen des Punktes a sind Ecken eines Rechtecks, dessen Seiten parallel sind zu den Projectionsaxen und dessen vierte Ecke a auf H liegt.

Fig. 1.

Wir nehmen nun an, dass zwei oder alle Projectionen eines Punktes zusammenfallen (Fig. 2).

α) Fällt für einen Punkt a die Projection a' mit a'' zusammen, so muss die Rechteckseite $a'a''$ verschwinden, also auch die Seite $a\,a'''$. Der Kreuzriss a''' des Punktes fällt daher auf H; der Punkt a liegt demnach in der Ebene H_1, welche H zur Kreuzrissspur hat und die X-Axe enthält.

Fig. 2.

*) S. z. B. Fiedler, Darstellende Geometrie I. p. ff.

Figure 37. The first page of Emil Waelsch's article from 1892 and the title page of Heinrich Tietze's geometrical study from 1923.[131]

Until 1909, the Chair of Descriptive Geometry was not successfully occupied by a good candidate. In 1910, Emil Waelsch was transferred there from the Second Chair of Mathematics. He was fully qualified because he obtained habilitation in the field of descriptive geometry at the German Technical University in Prague in 1890. He taught in Brno until 1927, when he died.[132]

4.4. *Conclusion*

At the end of the 19th century and in the first two decades of the 20th century, the German mathematical community in the Czech lands suffered a repeated outflow of talented mathematicians to the surrounding German-speaking countries. Until the outbreak of World War I, the Czech lands were considered a transfer station on the way to a better career. After the establishment of Czechoslovakia, some nationalist German mathematicians refused to stay in the newly formed republic and left for

[131] The pictures are from the authors' archive.

[132] About mathematics and mathematicians as well as descriptive geometry and its teachers at the German Technical University in Brno, see Šišma (2002, 2004).

Austria or Germany. Those who stayed, but for some exceptions, never learned Czech and did not get used to the Czech environment very much.

5. German Mathematical Community in Czechoslovakia Between the World Wars

5.1. *Introduction*

German universities in Prague freely developed even after the establishment of the Czechoslovak Republic. They were not cancelled; on the contrary, they became equal, recognised and respected state schools, which the new republic had not reduced, reprimanded or financially limited. The German University in Prague was the only official state minority university in post-war Europe divided into states based on the principle of nationality. It maintained this position until the beginning of World War II.

Although the Faculty of Natural Sciences of the German University in Prague was relatively small, it was a significant European science and education center. It attracted German-speaking students from Czechoslovakia, who were still heading to Berlin, Dresden, Vienna and Budapest at the beginning of the 20th century, because the diplomas from German, Austrian and Hungarian universities had to be validated in Czechoslovakia, or supplemented by other Czechoslovak state exams. It also attracted foreign students of Jewish religion and of democratic beliefs from Lithuania, Latvia, Ukraine, Hungary, Poland and from the middle of the 1930s, also from Germany. The relatively low school fees and living costs in Prague and its good transport accessibility also played a positive role in its development. The reputation of some professors was also essential (from mathematics and physics there were L. Berwald, R. Carnap, Ph. Frank, K. Löwner, and G.A. Pick).

5.2. *Prague*

Professors, private associate professors and assistants who were educated in Prague, Austria or Germany and attended various study stays abroad where they had come into contact with modern mathematics worked at the German schools in Prague.

The German mathematical community in Prague, which consisted of teachers and students from the German University in Prague, the German Technical University in Prague, and the German secondary schools in Prague, was relatively cohesive. Its members regularly met at lectures held within the *Lotos*, the *Gesellschaft zur Forderung deutscher Wissenschaft, Kunst und Kultur in Böhmen*, the *Mathematisches Kränzchen in Prag* and the *Deutsche physikalisch-mathematische Gesellschaft in Prag* (1934–1939). They were also active members of the *Deutsche Mathematiker-Vereinigung.* The community played an important role in the development of Czechoslovak and European 20th century mathematics. It had a small but friendly relationship with Czechoslovak mathematicians, and a greater relationship with the mathematicians of German-speaking countries, and was in contact with modern topics developed in world mathematics (differential, projective and differential geometry, function theory, topology, logic, set theory, etc.).[133]

The interwar period was the time of the greatest boom of the German mathematical community in Prague. It reached its peak, and its results received world acceptance and recognition. In the 1920s and 1930s, it included were people of different nationalities in it (German-speaking citizens of Czechoslovakia, Austrians and Germans), of different religions (Catholics, Protestants, Jews and people without religion), of varied political affiliation (Democrats, Communists, SdP members, Zionists and non-political people), of diverse origins and with different relationships to Czech countries, or Czechoslovakia. It should be stressed that until the beginning of 1939, the German mathematical community in Prague was not significantly affected by the growing Nazi and anti-Semitic ideology and social problems. Most teachers were Jewish and democratic. They sought to help the colleagues who had to leave their posts in Germany, the students who could not properly complete their studies in their homes, and so on.

The boom of German schools and scientific work was interrupted in 1939, when Jewish professors had to leave their posts for racial reasons. In the first phase, they were replaced by assistants or quickly-called educators from Germany. Later, German professors, who had been thoroughly screened and who understood their transfer to Prague as an ideological task or a precursor to another career in Germany, came and received their posts.

German universities in Czechoslovakia were abolished in 1945, and their property was taken over by Czech schools.[134] After the expulsion of the German population, the German mathematical community in Czechoslovakia ceased to exist.

The Mathematical Institute of the Faculty of Natural Sciences of the German University in Prague was one of the top German university departments. The list of lectures held at this institute implies that the newly appointed (extraordinary) professors and newly established private or honoured associate professors usually offered new, probably more challenging and interesting lectures that followed the latest developments in world mathematics (for example, potential theory, topology, logic, group theory, groups of geometric transformations, vector and tensor calculus, spectral theory of linear operators). They did so only for the first two or four semesters after their arrival at the university. However, these efforts were not always met with an appropriate response and interest of students.

[133]On the relations between Czech and German mathematicians, see, e.g., Bečvář and Kohoutová (2005, pp. 221–233), about the German mathematical community and mathematical associations, see Bečvářová (2016). About *Lotos*, see Těšínská (1997), and also the annual reports of Lotos published between 1911 and 1942.

[134]See Svatoš *et al.* (1998, Part IV, pp. 181–212); Seibt (1984); Šišma (2002, 2004).

After 1918, the teaching of mathematics at the German University in Prague was significantly influenced by the Viennese native Georg Alexander Pick. After studying in Vienna, habilitating in Prague (1882) and studying under Felix Klein in Leipzig (1883–1885), he remained faithful to the German University in Prague. During the 30 years, he raised a new German-Czech generation of mathematicians. He prepared the way for the development of the Prague-German mathematical community which set the tone for European research in integral and differential geometry, functional analysis, matrix calculus applications, logic, etc. He lived in Vienna since his retirement in 1929 and returned to Prague after the Anschluss of Austria. In the summer of 1942, at the age of 83, he was deported to the ghetto in Terezín (Theresienstadt), where he died after a few days.

In 1919, Ludwig Berwald (1883–1942) obtained habilitation in modern mathematics at the German University in Prague. He studied at the University in Munich as a student of professor Edmund Aurel Voss (1845–1931), under whose supervision he defended his thesis *Über die Krümmungseigenschaften der Brennflächen eines geradlinigen Strahlsystems und der in ihm enthaltenen Regelflächen.* In 1922, he was appointed an extraordinary and in 1927 a full professor of mathematics at the German University in Prague. He lectured at the university until 1939 when he was deprived of his place for racial reasons. In 1941, he was deported to the ghetto in Łodz, where he died.

Ludwig Berwald wrote more than 50 mathematical papers published in German, Italian and English in Czechoslovak and foreign journals. He excelled especially in differential and integral geometry (non-cyclic surfaces, Monge and Serret surfaces, Finsler spaces, n-dimensional manifolds, theory of metric spaces, problems of convex bodies, etc.). He also dealt with algebraic and analytical geometry. He represented Prague German mathematicians at the interwar international congresses of mathematicians.[135]

Paul Funk (1886–1969) graduated from the universities in Tübingen, Vienna and Göttingen as a student of David Hilbert (1862–1943), who supervised his diploma thesis *Über Flächen mit lauter geschlossenen geodätischen Linien.* After the defense of the thesis in 1911, he was appointed a teacher of mathematics at a grammar school in Salzburg. Two years later, he was appointed an assistant of mathematics at the German Technical University in Prague where he worked under Wilhelm Blaschke and Karl Carda. In 1915, he obtained habilitation at the German University in Prague. In 1919, he also obtained habilitation at the German Technical University in Prague. In 1921, he was appointed there an extraordinary and in 1927 a full professor of mathematics. But he also taught at the German University in Prague.

[135]About the life and work of Ludwig Berwald, see Bečvář and Bečvářová (2013a); Bečvářová (2016, 2018); Fritsch (2001a, 2019); Pinl (1964, 1967); Pinl and Dick (1974); Toepell (1991).

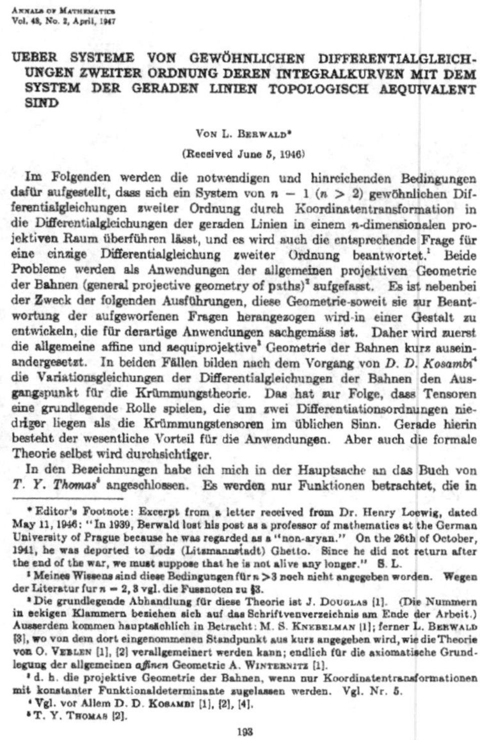

ANNALS OF MATHEMATICS
Vol. 48, No. 2, April, 1947

UEBER SYSTEME VON GEWÖHNLICHEN DIFFERENTIALGLEICHUNGEN ZWEITER ORDNUNG DEREN INTEGRALKURVEN MIT DEM SYSTEM DER GERADEN LINIEN TOPOLOGISCH AEQUIVALENT SIND

VON L. BERWALD*

(Received June 5, 1946)

Im Folgenden werden die notwendigen und hinreichenden Bedingungen dafür aufgestellt, dass sich ein System von $n - 1$ $(n > 2)$ gewöhnlichen Differentialgleichungen zweiter Ordnung durch Koordinatentransformation in die Differentialgleichungen der geraden Linien in einem n-dimensionalen projektiven Raum überführen lässt, und es wird auch die entsprechende Frage für eine einzige Differentialgleichung zweiter Ordnung beantwortet.[1] Beide Probleme werden als Anwendungen der allgemeinen projektiven Geometrie der Bahnen (general projective geometry of paths)[2] aufgefasst. Es ist nebenbei der Zweck der folgenden Ausführungen, diese Geometrie-soweit sie zur Beantwortung der aufgeworfenen Fragen herangezogen wird-in einer Gestalt zu entwickeln, die für derartige Anwendungen sachgemäss ist. Daher wird zuerst die allgemeine affine und aequiprojektive[3] Geometrie der Bahnen kurz auseinandergesetzt. In beiden Fällen bilden nach dem Vorgang von *D. D. Kosambi*[4] die Variationsgleichungen der Differentialgleichungen der Bahnen den Ausgangspunkt für die Krümmungstheorie. Das hat zur Folge, dass Tensoren eine grundlegende Rolle spielen, die um zwei Differentiationsordnungen niedriger liegen als die Krümmungstensoren im üblichen Sinn. Gerade hierin besteht der wesentliche Vorteil für die Anwendungen. Aber auch die formale Theorie selbst wird durchsichtiger.

In den Bezeichnungen habe ich mich in der Hauptsache an das Buch von *T. Y. Thomas*[5] angeschlossen. Es werden nur Funktionen betrachtet, die in

* Editor's Footnote: Excerpt from a letter received from Dr. Henry Loewig, dated May 11, 1946: "In 1939, Berwald lost his post as a professor of mathematics at the German University of Prague because he was regarded as a "non-aryan." On the 26th of October, 1941, he was deported to Lodz (Litzmannstadt) Ghetto. Since he did not return after the end of the war, we must suppose that he is not alive any longer." S. L.

[1] Meines Wissens sind diese Bedingungen für $n > 3$ noch nicht angegeben worden. Wegen der Literatur fur $n = 2, 3$ vgl. die Fussnoten zu §3.

[2] Die grundlegende Abhandlung für diese Theorie ist J. DOUGLAS [1]. (Die Nummern in eckigen Klammern beziehen sich auf das Schriftvenverzeichnis am Ende der Arbeit.) Ausserdem kommen hauptsächlich in Betracht: M. S. KNEBELMAN [1]; ferner L. BERWALD [3], wo von dem dort eingenommenen Standpunkt aus kurz angegeben wird, wie die Theorie von O. VEBLEN [1], [2] verallgemeinert werden kann; endlich für die axiomatische Grundlegung der allgemeinen *affinen* Geometrie A. WINTERNITZ [1].

[3] d. h. die projektive Geometrie der Bahnen, wenn nur Koordinatentransformationen mit konstanter Funktionaldeterminante zugelassen werden. Vgl. Nr. 5.

[4] Vgl. vor Allem D. D. KOSAMBI [1], [2], [4].

[5] T. Y. THOMAS [2].

193

Figure 38. The picture of Ludwig Berwald and the first page of his article from 1947.[136]

He had to leave his posts for racial reasons in 1939. At the end of the war, he was deported to the ghetto in Terezín (Theresienstadt).

After World War II, he and his family moved to Austria. In 1946, he was appointed a full professor of mathematics at the Technical University in Vienna and worked there until 1957. For the next four years, he helped with the lectures at the Technical University as well as the "classical" University in Vienna.

Paul Funk dealt mainly with applications of mathematics in technical practice. In 1962, he published his great book *Variationsrechnung und ihre Anwendung in Physik und Technik.* His name is still alive in world mathematics (Funk's transformation, or Minkowski-Funk transformation, Funk's solution of the fourth Hilbert's problem).[137]

An extraordinary lecture on the philosophical foundations of exact sciences was held at the German University in Prague by the neo-positivist Rudolf Carnap

[136] The picture of Ludwig Berwald is taken from Bečvářová (2016). The picture of the first page of Berwald's posthumously published article is from the authors' archive.

[137] About the life and work of Paul Funk, see Bečvářová (2016); Hornich (1969); Ottowitz (1992).

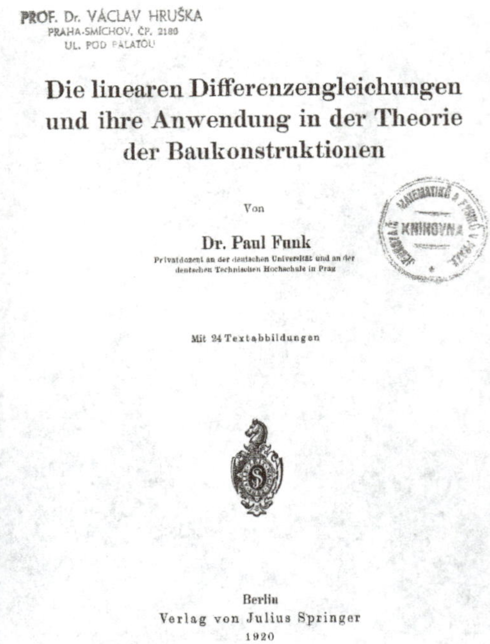

Die linearen Differenzengleichungen und ihre Anwendung in der Theorie der Baukonstruktionen

Von

Dr. Paul Funk

Privatdozent an der deutschen Universität und an der deutschen Technischen Hochschule in Prag

Mit 24 Textabbildungen

Berlin
Verlag von Julius Springer
1920

Figure 39. The picture of Paul Funk and the title page of his textbook.[138]

(1891–1970). He graduated from the Universities in Jena and Freiburg. In 1926, he obtained habilitation at the University in Vienna where he was appointed an extraordinary professor four years later. A year later, he accepted the offer of the post of extraordinary professor at the German University in Prague. Rudolf Carnap wrote significant works on metaphysics, philosophy, and logical syntax. In 1934, he published his most important and most quoted study *Logische Syntax der Sprache*. He also dealt with the fundamentals of mathematics, axiomatisation of mathematical theories, set theory and mathematical logic. In the 1930s, Carnap informed the German mathematical community in Prague about the fundamental results of David Hilbert and Kurt Gödel (1906–1978). He was well aware of the growing Nazi danger, so in 1936 he left for the USA. He worked at the University in Chicago (from 1936 to 1952), University in Princeton (from 1952 to 1954) and University in Los Angeles (from 1954 to 1961).[139]

[138]The picture of Paul Funk is taken from Bečvářová (2016). The picture of the title page is from the authors' archive.

[139]About the life and work of Rudolf Carnap, see Bečvářová (2015, 2016); Jaakko Hintikka (1975); Mormann (2000); Schillp (1999).

SCHRIFTEN ZUR WISSENSCHAFTLICHEN WELTAUFFASSUNG
HERAUSGEGEBEN VON
PHILIPP FRANK, O. Ö. PROFESSOR AN DER UNIVERSITÄT PRAG
UND
MORITZ SCHLICK, O. Ö. PROFESSOR AN DER UNIVERSITÄT WIEN
BAND 8

LOGISCHE SYNTAX DER SPRACHE

VON

RUDOLF CARNAP

WIEN · VERLAG VON JULIUS SPRINGER · 1934

Figure 40. The picture of Rudolf Carnap and the title page of his textbook.[140]

In the 1930s, Karl Löwner (1893–1968) worked at the German University in Prague. After his studies, he went to the University in Berlin where he obtained habilitation in 1923 and until 1928 he worked as a lecturer and private associate professor. From 1928 to 1930, he lectured as an extraordinary professor at the University in Cologne. In 1930, he became an extraordinary and in 1934 a full professor of mathematics at the German University in Prague. In 1939, he had to emigrate to the USA. He gradually went through a number of universities (Louisville, Brown University Providence, University of Syracuse, University of New York and Stanford University).

In Prague, Karl Löwner devoted himself mainly to geometric function theory, the theory of monotone matrix functions and measure theory on Hilbert spaces. In the USA, he dealt with partial differential equations, semigroups and applied mathematics. Löwner's absolutely essential work from the Prague period was

[140]The picture of Rudolf Carnap is taken from Bečvářová (2016). The picture of the title page is from the authors' archive.

Untersuchungen über schlichte konforme Abbildungen des Einheitskreises I, which gave not only the first non-trivial result of the well-known Bieberbach conjecture ($|a_3| \leq 3$), but also contributed to the geometric function theory a new method based on the so-called *Löwner's differential equation*. Surprisingly, this method was used not only by Louis de Branges in a complete solution of the Bieberbach conjecture (1984), but also after the year 2000, this method unexpectedly entered modern mathematics through the so-called ELS (stochastic Loewner evolution).[141]

SITZUNG VOM 26. FEBRUAR 1917.

Untersuchungen über die Verzerrung bei konformen Abbildungen des Einheitskreises $|z| < 1$, die durch Funktionen mit nicht verschwindender Ableitung geliefert werden.

Von

KARL LÖWNER (Prag).

Der Koebesche Verzerrungssatz in der von PICK und PLEMELJ gegebenen Verschärfung[1]) hat folgenden Inhalt: Wenn eine analytische Funktion $w = f(z)$ das Innere des Einheitskreises auf ein schlichtes, den Unendlichkeitspunkt nicht enthaltendes Gebiet abbildet, dann ist für je zwei Stellen z_1, z_2, falls $|z_1|$, $|z_2| \leqq \varrho < 1$ ist

$$\left| \frac{\left(\frac{dw}{dz}\right)_2}{\left(\frac{dw}{dz}\right)_1} \right| \leqq \left(\frac{1+\varrho}{1-\varrho}\right)^{2\varkappa}.$$

Die Bestätigung des von PICK und PLEMELJ vermuteten genauen Wertes von $\varkappa$, nämlich 2, ist durch eine nachher von FABER-BIEBERBACH[2]) entdeckte Reihenkoeffizientenabschätzung gegeben. Die endgültige Form des Verzerrungssatzes lautet also

$$\left| \frac{\left(\frac{dw}{dz}\right)_2}{\left(\frac{dw}{dz}\right)_1} \right| \leqq \left(\frac{1+\varrho}{1-\varrho}\right)^{4}.$$

Die Schranke wird von der Koebeschen Funktion

$$w = \frac{z}{(1+z)^2}$$

erreicht.

1) Vgl. PICK, Ber. d. Kgl. Sächs. Gesellschaft d. Wiss. 1916, sowie die auf diesen Aufsatz folgende Bemerkung über einen (ungedruckten) Vortrag von PLEMELJ.

2) BIEBERBACH, L., Berl. Ber. 1916 (XXXVIII) Seite 940—955. — FABER, G., Münch. Ber. 1916, Seite 39—42.

Figure 41. The picture of Karl Löwner and the first page of his well-known article.[142]

Arthur Winternitz (1893–1961) had a similar fate as Karl Löwner. He obtained habilitation in 1921 at the German University in Prague, and 10 years later he was appointed an extraordinary professor of mathematics. In 1939, he gained British citizenship, and he emigrated to Britain and began lecturing at the University of Oxford. He focused mainly on applied mathematics.[143]

[141] About the life and work of Karl Löwner, see Bečvářová (2016) and Bečvářová and Netuka (2015).

[142] The picture of Karl Löwner is taken from Bečvářová and Netuka (2015). The picture of the first page of his article is from the authors' archive.

[143] About the life and work of Arthur Winternitz, see Bečvářová (2016); Pinl and Dick (1974); Siegmund-Schultze (2012).

Heinrich Löwig (1904–1995) had an even more complicated life. He received a doctorate in mathematics at the German University in Prague in 1928 and obtained habilitation seven years later. For less than three years, he lectured as a private associate professor and at the same time worked as a professor of mathematics at various German secondary schools in Bohemia and Moravia. He was totally deployed during the war, and forced to work slavishly in various German labour camps from 1944 to 1945. In 1948, he left for the University of Hobart (Tasmania), where he taught until 1957. Then, he was appointed a professor at Edmonton University in Canada.

Heinrich Löwig specialised in differential equations, dimension theory, algebras and lattices. His remarkable interwar work was the article *Über die Dimension linearer Räume* (1932), in which he dealt with the issue of real and complex metric or normed linear spaces, a research topic started by Felix Hausdorff (1868–1942) and Stefan Banach (1892–1945). For real and complex linear spaces he defined two different notions of a dimension — the affine dimension and the metric dimension — and introduced the notion of a base. Without using Zorn lemma (it was not yet available) he proved, using transfinite induction, that every linear space has a base and that every two bases of the same space have the same cardinality. Then he studied the dimensions of linear spaces.[144]

Maximilian Pinl (1897–1978) also had a very complicated life. He studied at the Mining Academy in Leoben and the University in Vienna. From 1926 to 1935, he worked as the editor of the reference journal *Jahrbuch über die Fortschritte der Mathematik*. In 1936, he obtained habilitation at the German University in Prague and began to teach special topics in mathematics. In 1939, he was arrested by the Gestapo for disagreement with the removal of Jewish colleagues from the university, imprisoned for six months and then deployed at the Air Research Institute in Brunswick. In 1946, he was appointed an associate professor and two years later an extraordinary professor of mathematics at the University in Cologne. From 1949 to 1954, he taught at the University in Dacca (Pakistan). From 1954 to 1962, he lectured in the USA and from 1962 to 1967 he worked at the University in Münster. He dealt with differential geometry, partial differential equations, theory of relativity and gas dynamics.[145]

In the summer semester of 1938/1939, assistants and private associate professors Ernst Lammel (1908–probably 1961), Alfred Eduard Rössler (1903–?) and Otto

---

[144] About the life and work of Heinrich Löwig (Jindřich Löwig or Henry Loewig), see Bečvářová a kol. (2012); Bečvářová (2016).

[145] About the life and work of Maximilian Pinl, see Bečvářová (2016); Fritsch (2001c); Kraft (1981). A remarkable testimony to the events and sacrifices of World War II, see M. Pinl, Kollegen in einer dunklen Zeit I, II, III, *Jahresbericht der Deutschen Mathematiker-Vereinigung* **71**(1969), pp. 167–228, **72**(1971/72), pp. 165–189, **73**(1972), pp. 153–208, M. Pinl and A. Dick, Kollegen in einer dunklen Zeit, Schluß, *ibid.*, **75**(1974), pp. 166–208, **77**(1976), pp. 161–164.

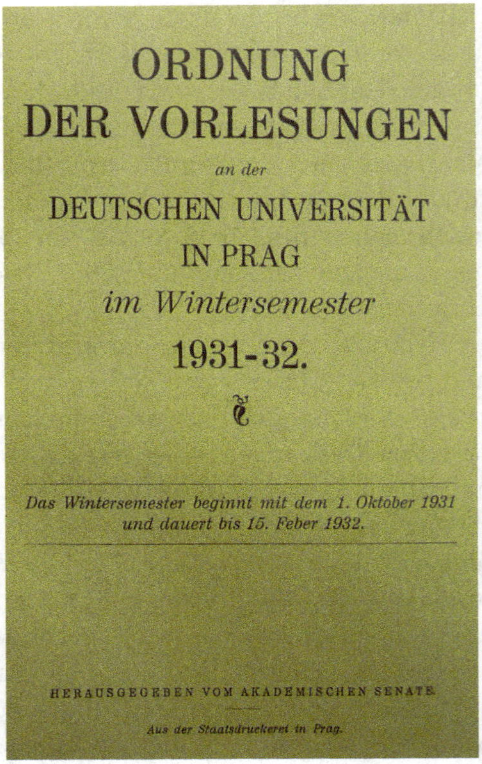

ORDNUNG
DER VORLESUNGEN
an der
DEUTSCHEN UNIVERSITÄT
IN PRAG
im Wintersemester
1931-32.

Das Wintersemester beginnt mit dem 1. Oktober 1931 und dauert bis 15. Feber 1932.

HERAUSGEGEBEN VOM AKADEMISCHEN SENATE.

Aus der Staatsdruckerei in Prag.

Figure 42. The picture of Heinrich Löwig and the title page of the *List of Lectures at the German University in Prague for the Winter Semester 1931/1932.*[146]

Varga (1909–1969) were entrusted with teaching mathematics at the German University in Prague. They were students, colleagues or co-workers of Ludwig Berwald, Karl Löwner, Arthur Winternitz and Paul Funk.

Ernst Lammel received his doctorate and habilitation in geometry at the German Technical University in Prague, where he lectured geometry between 1938 and 1945, firstly as a private associate professor, and later as an extraordinary professor. From 1950, he was a professor at the University in Tucumán (Argentina).[147] Alfred Eduard Rössler studied at the German University in Prague and the German Technical University in Prague (in the 1920s) where he became an assistant in 1926. He obtained habilitation at the German University in Prague in 1938 and worked as an associate professor at both Prague schools from 1938 to 1945. After World War II, he was appointed an extraordinary professor at the Technical University in Aachen.[148] Otto Varga graduated from the German University in Prague in 1934.

[146]The pictures are from the authors' archive.

[147]About the life and work of Ernst Lammel, see Bečvářová (2016); Toepell (1991).

[148]About the life and work of Alfred Eduard Rössler, see Bečvářová (2016); Toepell (1991).

In 1937, he obtained habilitation there and lectured there until 1942. After disagreements in Prague, he went to Hungary, where he was hiding for some time. After World War II, he worked at the University in Debrecen.[149]

After the establishment of Czechoslovakia, the German Technical University in Prague developed freely until 1939. In 1939, the school was transferred to the central Nazi administration. In 1945, it was abolished by the decree of the president of the Czechoslovak Republic. There were several professors who had been teaching there since the beginning of the 20th century. They spent more or less their whole active careers in Prague; however, they were neither the top of their field nor the top of the German mathematical community in Prague.[150]

Karl Carda connected a great part of his creative life with the Czech lands. After graduating from the Faculty of Arts of the University in Vienna, he received a doctorate in 1894 and a year later he received a teacher qualification in mathematics and physics. He also obtained a state scholarship that allowed him to attend Sophus Lie's lectures at the University in Leipzig (the winter semester of 1895/1896). In the summer semester, he went to Oslo and continued to visit Lie's lessons. From 1896, he was an assistant of mathematics at the German Technical University in Brno, from where he moved to the University in Vienna as a private associate professor and at the same time worked as an assistant at the Technical University in Vienna. In 1905, he was appointed an extraordinary professor at the Technical University in Vienna and two years later a full professor at the German Technical University in Prague, where he lectured until his retirement in 1939.

His publishing activity is not extensive. It consists of 19 original works. They are devoted to problems of hyperbolic functions, Bernoulli numbers, evaluation of certain integrals, fundamentals of differential geometry, theory of transformation groups and invariant theory.

Anton Grünwald, who was appointed a professor of mathematics in 1913, and taught at the German Technical University in Prague until his death in 1932. As a native of Prague, he loved his town and actually inherited the professor's post from his father Anton Karl Grünwald, who contributed to the development of German mathematical life in Prague in the 19th century. Although Anton Grünwald himself was not a creative mathematician, he educated a number of German-speaking technicians.

From 1921 until 1939, Paul Funk taught mathematics at the German Technical University in Prague. Then, he was forced to quit his teaching career because of his Jewish origins.

[149] About the life and work of Otto Varga, see Bečvářová (2016).

[150] For more information on the history of the German Technical University in Prague, see Birk (1931).

In the 1930s and 1940s, Ernst Lammel and Alfred Eduard Rössler taught at the German Technical University in Prague. They connected their scientific and pedagogical careers with the German mathematical community in the Czech lands.

Descriptive geometry was taught by a Viennese native Karl Mack, who graduated from the Technical University and also the "classical" University in Vienna. Firstly, he worked as a "volunteer" for the *Janus Life Insurance Office* in Vienna. In 1905, he was an assistant of descriptive geometry at the Technical University in Vienna. From 1908 to 1909, he was a professor at the State Secondary School in Vienna's 13th district. In 1911, he substituted lectures on descriptive geometry at the Technical University in Vienna and from 1913 to 1915 he taught mathematics at the *Exportakademie* in Vienna. In 1916, he was appointed an extraordinary and in 1920 a full professor of descriptive geometry at the German Technical University in Prague. Although he received various offers of professorships in Germany, he remained faithful to Prague's school until his death (1943).[151]

5.3. *Brno*

Even after the establishment of Czechoslovakia, the German Technical University in Brno was one of Czechoslovakia's excellent schools and scientific centers. World-well-known professors worked at two chairs of mathematics and one chair of descriptive geometry. In 1939, the school was transferred to the central Nazi administration. In 1945, it was abolished by the decree of the president of the Czechoslovak Republic.[152]

Since 1912, Lothar Schrutka held a professorship on the first chair of mathematics. He taught in Brno until 1925, when he left to Vienna. In 1927, Lothar Koschmieder (1890–1974), an experienced mathematician, came to his place. He received his education at the best German universities (Breslau, Freiburg and Göttingen). He was a student of Adolf Kneser (1862–1930), who has significantly influenced his scientific career. Even before World War I, Koschmieder was appointed an assistant at the University in Breslau. After the war, he was a private associate professor and finally an extraordinary professor at the same university. He worked at the German Technical University in Brno until 1940. Being its rector in 1938/1939 (and the supporter of Nazism, and a member of the NSDAP since 1939), he was involved in the transfer of the school to the German Nazi administration. At the beginning of 1940, he went to the Technical University in Graz, where he taught until 1946. After the war, he worked as a private mathematics teacher. In 1948, he accepted an offer of a post in Aleppo (Syria). A year later, he went to Argentina to the Faculty of Natural Sciences and Engineering of the University in Tucumán,

[151]For more information on the mathematicians at the German Technical University in Prague, see Bečvářová (2016).

[152]For more information on the German Technical University in Brno, see Šišma (2002, 2004).

where he stayed until 1953. From 1953 to 1958, he taught at the State University in Baghdad. From 1958 to 1973, he lectured mathematics at the University in Tübingen.

Lothar Koschmieder wrote more than 80 papers dealing with the calculus of variations, properties of special types of functions and methods of solving differential and integral equations. In 1933, his popular textbook *Variationsrechnung* was published in the well-known series *Sammlung Göschen.*[153]

From 1919 to 1923, the Second Chair of Mathematics of the German Technical University in Brno was not occupied. In 1923, Karl Mayr (1884–1940), a former private associate professor at the Technical University in Vienna, came to Brno. After a year of working in Brno, he left for the Technical University in Graz. In 1924, Rudolf Weyrich (1894–1971) was asked to substitute mathematical lectures in Brno. He studied at the German universities in Rostock and Breslau. After his studies, he became an assistant and a private associate professor at the University in Marburg. In 1925, he was appointed an extraordinary professor and in 1930 a full professor of mathematics at the German Technical University in Brno. He remained at the post until the end of World War II. Between 1945 and 1947, Rudolf Weyrich had no regular employment. From 1948 until 1950, he lectured at the Technical

Figure 43. The picture of Lothar Koschmieder and the title page of the textbook written by Josef Leopold Krames.[154]

---

[153]About the life and work of Lothar Koschmieder, see Šišma (2002); Volk (1980).

[154]The picture of Lothar Koschmieder (the conference in Oberwolfach in 1965) is taken for the *Archives of the Mathematisches Forschungsinstitut Oberwolfach*. The picture of the title page is from the authors' archive.

University in Braunschweig. From 1950 until 1958, he taught at the University in Istanbul (Turkey).

Rudolf Weyrich wrote a number of works in physics in which he pointed out various problems of electromagnetism. He was the owner of a patent for the special devices enabling the reception and transmission of electromagnetic waves. His mathematical work was directly related to mathematical physics. His monograph *Die Zylinderfunktionen und ihre Anwendungen* (1937) became recognised not only in German-speaking countries.[155]

The teaching of descriptive geometry was not neglected at the German Technical University in Brno. From 1910 until 1927, it was led by Emil Waelsch, and then for two years substituted by Rudolf Weyrich. It was not until 1929 that Josef Leopold Krames (1897–1986) was appointed a new professor. Krames was a student of Emil Müller (1861–1927), a distinguished geometer. After Müller's death, he was charged with substituting his lectures on descriptive and projective geometry in Vienna. In 1932, Krames left Brno for the Technical University in Graz. His teaching of descriptive geometry was again substituted by Rudolf Weyrich.

Krames is the author of more than a hundred treatises on descriptive geometry, most of which he finished after World War II. At the request of his teacher Emil Müller, Krames prepared for publication the second and third volumes of Müller's lectures *Vorlesungen über Darstellende Geometrie* that came out under the name *Die Zyklographie* (1929) and *Konstruktive Behandlung der Regelfächen* (1931).[156]

Let us note that the German mathematical community in Brno suffered from frequent departures of its members to Austria or Germany, where they obtained better professional opportunities. It is also interesting to note that the German mathematical communities in Prague and Brno maintained limited professional contacts. It was probably due to their different interests.

5.4. *Conclusion*

The German mathematical community developed freely also in the interwar Czechoslovakia. Its actors were well aware that their community in the Czech lands was not big enough and therefore they tried to maintain professional contacts with the mathematicians of other German-speaking countries. Many favourable circumstances have linked them. First of all, it was language, culture and education, and the fact that they were one "generation group" attuned to change, eager to accept new mathematical ideas and discoveries, longed to reorganise and reform the status of

[155] About the life and work of Rudolf Weyrich, see Šišma (2002). Note that Weyrich was sent to the Russian front in 1914, where he participated in first-line combat and suffered serious injury.
[156] About the life and work of Josef Leopold Krames, see Šišma (2002); Wunderlich (1987).

study and research. Experiences gained by individual mathematicians abroad during their studies, scholarships and first pedagogical activities also played an important role. It has to be emphasised that love for mathematics, physics and their applications was also a strong link. All this, together with the absence of "enclosure" of the small community and above all the desire to accomplish something, interact with each other and to show their results, led them to the interwar European mathematical leaders.

References

J. Bečvář a kol., *Eduard Weyr (1852–1903)*, Series Dějiny matematiky, Volume 2, Prometheus, Praha, 1995, 196 pages + 24 pictures. ISBN 80-7196-024-1.

J. Bečvář (ed.), *Jan Vilém Pexider (1874–1914)*, Series Dějiny matematiky, Volume 5, Prometheus, Praha, 1997, 110 pages. ISBN 80-7196-044-6. English version: J. Bečvář, A. Slavík (eds.), *Jan Vilém Pexider (1874–1914)*, Series History of Mathematics, Volume 38, Matfyzpress, Prague, 2009, 110 pages. ISBN 978-80-7378-058-6.

J. Bečvář and M. Bečvářová (eds.), *Matematika v proměnách věků VI* [Mathematics Throughout The Ages VI] Series Dějiny matematiky, Volume 45, Matfyzpress, Praha, 2010, 231 pages. ISBN 978-80-7378-146-0.

J. Bečvář and M. Bečvářová, 150 let Jednoty českých matematiků a fyziků [150 Years of the Union of Czech Mathematicians and Physicists], pp. 11–118, in J. Bečvář, M. Bečvářová (eds.), *33. Mezinárodní konference Historie matematiky, Velké Meziříčí, 24. 8. až 28. 8. 2012*, Matfyzpress, Praha, 2012, 303 pages. ISBN 978-80-7378-208-5.

J. Bečvář and M. Bečvářová, Ludwig Berwald, pp. 18–24, in M. V. Šimůnek, A. Kostlán (eds.), *Disappeared Science. Biographical Dictionary of Jewish Scholars from Bohemia and Moravia — Victims of Nazism, 1939–1945*, Series Studies in the History of Sciences and Humanities, Volume 29, Centre for the History of Sciences and Humanities, Institute of Contemporary History of the Academy of Sciences of the Czech Republic, Pavel Mervart — Kabinet dějin vědy ÚSD AV ČR, v. v. i., Červený Kostelec, Praha, 2013, 323 pages. ISBN 978-80-7465-041-3.

J. Bečvář and M. Bečvářová, Georg Pick, pp. 158–162, in M. V. Šimůnek, A. Kostlán (eds.), *Disappeared Science. Biographical Dictionary of Jewish Scholars from Bohemia and Moravia — Victims of Nazism, 1939–1945*, Series Studies in the History of Sciences and Humanities, Volume 29, Centre for the History of Sciences and Humanities, Institute of Contemporary History of the Academy of Sciences of the Czech Republic, Pavel Mervart — Kabinet dějin vědy ÚSD AV ČR, v. v. i., Červený Kostelec, Praha, 2013, 323 pages. ISBN 978-80-7465-041-3.

J. Bečvář and M. Bečvářová, Matematika v období 1850 až 1918 [Mathematics in the Period 1850–1918], pp. 427–465, in I. Kraus a kol., *Věda v českých zemích. Dějiny fyziky, geografie, geologie, chemie a matematiky* [Science in the Czech Lands. History of Physics, Geography, Geology, Chemistry and Mathematics], Česká technika — nakladatelství ČVUT, Praha, 2019, 553 pages. ISBN 978-80-01-06557-0.

J. Bečvář and M. Bečvářová, Matematika v období 1918 až 1950 [Mathematics in the Period 1918–1950], pp. 467–491, in I. Kraus a kol., *Věda v českých zemích. Dějiny fyziky, geografie, geologie, chemie a matematiky* [Science in the Czech Lands. History of Physics, Geography, Geology, Chemistry and Mathematics], Česká technika — nakladatelství ČVUT, Praha, 2019, 553 pages. ISBN 978-80-01-06557-0.

J. Bečvář, M. Bečvářová and J. Škoda, *Emil Weyr a jeho pobyt v Itálii v roce 1870/71* [Emil Weyr and his Study Tour in Italy in the Year 1870/71], Series Dějiny matematiky, Volume 28, Nakladatelství ČVUT, Praha, 2006, 166 pages + 50 pictures. ISBN 80-01-03438-0.

J. Bečvář and Z. Kohoutová, *Vladimír Kořínek (1899–1981)*, Series Dějiny matematiky, Volume 27, Ústav pro soudobé dějiny AV ČR, Praha, 2005, 329 pages + 40 pictures. ISBN 80-7285-050-4.

M. Bečvářová, *Z historie Jednoty 1862–1869* [From the History of Union 1862–1869], Series Dějiny matematiky, Volume 13, Prometheus, Praha, 1999, 138 pages + 17 pictures. ISBN 80-7196-141-8.

M. Bečvářová, *Eukleidovy Základy, jejich vydání a překlady* [Euclid Elements, its Editions and Translations], Series Dějiny matematiky, Volume 20, Prometheus, Praha, 2002, 297 pages. ISBN 80-7196-233-3.

M. Bečvářová, *Josef Smolík (1832–1915)*, Nakladatelství ČVUT, Praha, 2007, 254 pages. ISBN 978-80-01-03639-6.

M. Bečvářová, *Česká matematická komunita v letech 1848–1918* [Czech Mathematical Community from 1848 until 1918], Series Dějiny matematiky, Volume 34, Matfyzpress, Praha, 2008, 355 pages. ISBN 978-80-7378-028-9.

M. Bečvářová, *České kořeny bulharské matematiky* [Czech Roots of Bulgarian Mathematics], Series Dějiny matematiky, Volume 40, Matfyzpress, Praha, 2009, 482 pages. ISBN 978-80-7378-079-1.

M. Bečvářová, Quido Vetter (1881–1960) český historik exaktních věd světového jména [Quido Vetter (1881–1960) Czech Historian of Exact Sciences with World Renown], *Československý časopis pro fyziku* **60**(2010), pp. 382–386. ISSN 0009-0700.

M. Bečvářová, František Vyčichlo (1905–1958), pp. 209–220, in J. Bečvář and M. Bečvářová (eds.), *Matematika v proměnách věků VI* [Mathematics Throughout The Ages VI], Series Dějiny matematiky, Volume 45, Matfyzpress, Praha, 2010, 231 pages. ISBN 978-80-7378-146-0.

M. Bečvářová, Z korespondence Františka Vyčichla Vladimíru Kořínkovi [From the Correspondence of František Vyčichlo to Vladimír Kořínek], pp. 221–231, in J. Bečvář and M. Bečvářová (eds.), *Matematika v proměnách věků VI* [Mathematics Throughout The Ages VI], Series Dějiny matematiky, Volume 45, Matfyzpress, Praha, 2010, 231 pages. ISBN 978-80-7378-146-0.

M. Bečvářová, Václav Láska v Polsku [Václav Láska in Poland], pp. 149–158, in J. Bečvář, M. Bečvářová (eds.), *32. mezinárodní konference Historie matematiky, Jevíčko, 26. až 30. 8. 2011*, Matfyzpress, Praha, 2011, 301 pages. ISBN 978-80-7378-172-9.

M. Bečvářová a kol., *Zapomenutý matematik Henry Lowig (1904–1995)*, Series Dějiny matematiky, Volume 50, Matfyzpress, Praha, 2012, 294 pages. ISBN 978-80-7378-189-7. English version: M. Bečvářová *et al.*, *Forgotten Mathematician Henry Lowig (1904–1995)*, Series History of Mathematics, Volume 52, Matfyzpress, Prague, 2012, 279 pages. ISBN 978-80-7378-203-0.

M. Bečvářová, J. S. Vaněček a L. Cremona (nově objevená korespondence) [J.S. Vaněček and L. Cremona (Newly Discovered Correspondence)], pp. 63–80, in J. Bečvář, M. Bečvářová (eds.), *34. mezinárodní konference Historie matematiky, Poděbrady, 23. až 27. 8. 2013*, Matfyzpress, Praha, 2013, 201 pages. ISBN 978-80-7378-234-4.

M. Bečvářová, Mathematische Kränzchen in Prag — A Forgotten German Mathematical Society, *Technical Transactions, Fundamental Sciences — Czasopismo Techniczne, Nauki podstawowe, Cracow* **8**(2015), pp. 41–68. ISSN 0011-4561.

M. Bečvářová, *Matematika na Německé univerzitě v Praze v letech 1882–1945* [Mathematics at the German University in Prague from 1882 until 1945], Univerzita Karlova, Karolinum, Praha, 2016, 446 pages. ISBN 978-80-246-3182-0.

M. Bečvářová, Gerhard Hermann Waldemar Kowalewski and his Two Prague Periods, *Antiquitates Mathematicae* **12**(2018), pp. 111–159. ISSN 1898-5203.

M. Bečvářová and J. Čižmár, *Karel Zahradník (1848–1916). Praha–Záhřeb–Brno*, Series Dějiny matematiky, Volume 46, Matfyzpress, Praha, 2011, 410 pages. ISBN 978-80-7378-158-3.

M. Bečvářová and I. Netuka, *Jarník's Notes of the Lecture Course Punktmengen und reelle Funktionen by P.S. Aleksandrov (Göttingen 1928)*, Series History of Mathematics, Volume 43, Matfyzpress, Prague, 2010, 143 pages. ISBN 978-80-7378-113-2.

M. Bečvářová and I. Netuka, *Karl Löwner and His Student Lipman Bers — Pre-war Prague Mathematicians*, Heritage of European Mathematics, Volume 10, European Mathematical Society, Zürich, 2015, viii + 300 pages. ISBN 978-3-03719-144-6.

M. Bednaříková a kol., *Dějiny univerzity v Brně* [History of the University in Brno], Univerzita Jana Evangelisty Purkyně, Brno, 1969, 429 pages.

J. Beran, 200 let organisace české vědy, II. třída České akademie v letech 1891–1914 [200 Years of the Organization of Czech science, Second Class of Czech Academy in the Year 1891–1914], *Dějiny věd a techniky* **4**(1971), pp. 193–208.

J. Beránek, Bohuslav Hostinský, *Československý časopis pro fyziku* **1**(1951), p. 90.

A. Birk, *Die Deutsche Technische Hochschule in Prag 1806–1931*, Prag, 1931, 177 pages.

J. Bílý, K sedmdesátým narozeninám profesora RNDr. Ladislava Truksy [Professor Ladislav Truksa Septuagenarian], *Časopis pro pěstování matematiky* **86**(1961), pp. 492–496.

O. Borůvka, Dílo Matyáše Lercha v oboru matematické analysy [The Work of Matyáš Lerch in the Mathematical Analysis], *Práce Brněnské základny Československé akademie věd* **29**(1957), pp. 417–540.

O. Borůvka and K. Svoboda, Šedesátiny profesora Jiřího Klapky [Professor Jiří Klapka Sexagenarian], *Časopis pro pěstování matematiky* **85**(1960), pp. 377–384.

O. Borůvka, Vzpomínka na českého matematika Matyáše Lercha [A Remembrance of the Czech Mathematician Matyáš Lerch], *Pokroky matematiky, fyziky a astronomie* **17**(1972), pp. 130–134.

M. Brdička and Š. Schwabik, Časopis pro pěstování matematiky a fyziky a jeho pokračovatelé [Journal for Cultivation of Mathematics and Physics and Its Followers], in Pátý (1987), pp. 30–83.

W. Burau, Wilhelm Blaschkes Leben und Werk, *Mitteilungen der Mathematischen Gesellschaft in Hamburg* **9**(1963), pp. 24–40.

Z. Crkalová, *Život a dílo Karla Petra* [Life and Work of Karel Petr], diplomová práce, Matematicko-fyzikální fakulta, Univerzita Karlova, Praha, 1992, 121 pages.

J. Čermák, Šedesátiny profesora Kauckého [Professor Kaucký Sexagenarian], *Časopis pro pěstování matematiky* **81**(1956), pp. 126–127.

K. Čupr, Profesor Matyáš Lerch, *Časopis pro pěstování matematiky a fysiky* **52**(1923), pp. 301–313.

K. Čupr and K. Rychlík, Seznam vědeckých prací † prof. Matyáše Lercha [Scientific Works of † Matyáš Lerch], *Časopis pro pěstování matematiky a fysiky* **54**(1925), pp. 140–151.

K. Drábek, Sto let od narození prof. PhDr. Jana Vojtěcha [100 Years from the Birth of Professor Dr. Jan Vojtěch], *Pokroky matematiky, fyziky a astronomie* **24**(1979), pp. 223–225.

K. Drábek, 125 let katedry matematiky a deskriptivní geometrie stavební fakulty ČVUT [125 Years of the Department of Mathematics and Descriptive Geometry of the Faculty of Civil Engineering of the CTU], *Dějiny věd a techniky* **12**(1979), pp. 33–45.

H. Durnová, J. Kotůlek and V. Žádník, *Václav Hlavatý (1894–1969) — Cesta k jednotě* [Václav Hlavatý (1894–1969) — The Path to Unity], Masarykova univerzita, Brno, 2017, 187 pages. ISBN 978-80-210-8920-4.

R. Einhorn, *Vertreter der Mathematik und Geometrie an den Wiener Hochschulen 1900–1940*, Dissertationen an der Technischen Universität Wien, Verband der wissenschaftlichen Gesellschaften Österreichs, Wien, 1985, 741 pages.

F. Fabian, K sedmdesátým narozeninám prof. Dr. Jaroslava Janko DrSc. [Professor Jaroslav Janko Septuagenarian], *Časopis pro pěstování matematiky* **89**(1964), pp. 117–122.

L. Fasora and J. Hanuš, *Masarykova univerzita v Brně: Příběh vzdělání a vědy ve střední Evropě* [Masaryk University in Brno: A Story of Education and Science in Central Europe], Masarykova univerzita, Brno, 2009, 271 pages. ISBN 978-80-210-4850-8.

J. Folta, Historická poznámka k učebnicím deskriptivní geometrie Vincence Jarolímka [Historical Note to the Vincenc Jarolímek's Textbook on Descriptive Geometry], *Matematika ve škole* **8**(1958), pp. 165–172.

J. Folta, Přínos Vincence Jarolímka pro vyučování geometrie na českých středních školách [Vincenc Jarolímek's Contribution to Geometry Education in Czech Secondary Schools], *Pokroky matematiky, fyziky a astronomie* **4**(1959), pp. 618–627.

J. Folta, Čtyřicet let od smrti bratří Vaněčků [Forty Years from the Death of Vaněček Brothers], *Pokroky matematiky, fyziky a astronomie* **8**(1963), pp. 28–30.

J. Folta, J. Mandlerová and L. Nový, Matematika na pražské universitě v letech 1900–1918 [Mathematics at the University in Prague in the Period 1900–1918], *Acta Universitatis Carolinae — Historia Universitatis* **VIII-2**(1967), pp. 7–43.

L. Francová, *Bohumil Bydžovský*, doktorská disertační práce, Matematicko-fyzikální fakulta, Univerzita Karlova, Praha, 2001.

L. Frank, O životě profesora Matyáše Lercha [On the Life of Professor Matyáš Lerch], *Časopis pro pěstování matematiky* **78**(1953), pp. 119–137.

L. Frank, Vědecká a publikační činnost profesora Dr. Karla Čupra [Scientific and Publication Activity of Professor Dr. Karel Čupr], *Časopis pro pěstování matematiky* **82**(1957), pp. 117–120.

L. Frank, K 100. výročí narození Matyáše Lercha [On Centenary of Death of Matyáš Lerch], *Pokroky matematiky, fyziky a astronomie* **5**(1960), pp. 764–771.

R. Fritsch, Georg Pick und Ludwig Berwald — Zwei Mathematiker an der Deutschen Universität in Prag, *Schriften der Sudetendeutschen Akademie der Wissenschaften und Künste. Forschungsbeiträge der Naturwissenschaftlichen Klasse*, Volume 22, 2001, pp. 9–16. ISBN 978-3-936284-00-3.

R. Fritsch, Pick Georg, Neue Deutsche Biographie 20, Duncker & Humblot, Berlin, 2001, pp. 418–419. ISBN 3-428-00201-6.

R. Fritsch, Pinl Max, Neue Deutsche Biographie, Band 20, Duncker & Humblot, Berlin, 2001, pp. 452–453. ISBN 3-428-00201-6.

R. Fritsch, Georg Pick und Ludwig Berwald — Zwei Mathematiker an der Deutschen Universität Prag, Ansätze zu biographischen Würdigungen, pp. 1–5 (available on the page https://web.archive.org/web/20120131182944/http://www.mathematik.uni-muenchen.de/~fritsch/PICKBERW.pdf [1.3.2019]).

O. Froněk, *Dějiny české vysoké školy technické v Brně, 1. díl – do roku 1945* [History of Czech Technical University in Brno, 1st Part – until 1945], VUT, Brno, 1969, 414 pages.

J. Fryš, *Dvanáct osudů dvou staletí* [Twelve Fates on Two Centuries], J. Fryš, Příbram, 2006, 270 pages. ISBN 80-239-8119-6.

E. Fuchs and K. Lepka, Matyáš Lerch, *Pokroky matematiky, fyziky a astronomie* **48**(2000), pp. 50–62. ISSN 0032-2423.

M. Greguš (ed.), *Jur Hronec 1881–1981*, JČSMF, Praha, 1981, 107 pages.

A. Haussner, Geschichte der Deutschen Technischen Hochschule in Brünn 1849–1924, pp. 5–92, in *Festschrift der Deutschen Technischen Hochschule in Brünn zur Feier ihres fünfundsiebzigjährigen Bestandes im Mai 1924*, Brünn, 1924.

K. Havlíček, Prof. Dr. František Kadeřávek zemřel [Professor Dr. František Kadeřávek Died], *Pokroky matematiky, fyziky a astronomie* **6**(1961), pp. 231–234.

K. Hellmer, Geschichte der Deutschen Technischen Hochschule in Brünn, in *Festschrift der k. k. Technischen Hochschule in Brünn zur Feier ihres fünfzigjährigen Bestehens und der Vollendung des Erweiterungsbaues im Oktober 1899*, Brünn, 1899, pp. 1–97.

V. Hlavatý, Miloslav Pelíšek, *Česká akademie věd a umění*, Praha, 1941, 34 pages.

H. Hornich, Paul Funk, *Almanach der Österreichischen Akademie der Wissenschaften* **19**(1969), pp. 271–277.

F. Houdek, *Dějepis jednoty českých mathematiků v Praze* [History of the Union of Czech Mathematicians], Jednota českých mathematiků, Praha, 1872, 82 pages.

J. Hrabák, *Gedenkbuch zur Feier des fünfzigjährigen Bestandes der k.k. Bergakademie Příbram 1849 bis 1899*, Verlag der k.k. Bergakademie, Příbram, 1899, 265 pages.

J. Hrdličková, *Život a dílo Ladislava Seiferta (1883–1956)* [Life and Work of Ladislav Seifert (1883–1956)], disertační práce, Přírodovědecká fakulta, Masarykova univerzita, Brno, 2001, 150 pages + viii pictures.

O. Hronec, J. Sulaček and B. Riečan, *Starý pán. Kniha o Jurovi Hroncovi* [Old Man. Book on Jur Hronec], Veda, Bratislava, 1999, 218 pages. ISBN 80-224-0597-3.

V. Hruška, Posmrtná vzpomínka na Mat. Norb. Vaněčka [Matěj Norbert Vaněček (Obituary)], *Časopis pro pěstování mathematiky a fysiky* **52**(1923), pp. 313–319.

M. Hykšová, *Karel Rychlík (1885–1968)*, Series Dějiny matematiky, Volume 22, Prometheus, Praha, 2003, 286 pages. ISBN 80-7196-259-7.

M. Hykšová, *Filosofická pojetí pravděpodobnosti v pracích českých myslitelů* [Philosophical Concepts of Probability in the Works of Czech Thinkers], Series Dějiny matematiky, Volume 51, Matfyzpress, Praha, 2011, 274 pages. ISBN 978-80-7378-192-7.

Jaakko Hintikka (ed.), *Rudolf Carnap, Logical Empiricist. Materials and Perspectives*, Dordrecht, D. Reidel Publishing Company, 1975, lxviii + 406 pages. ISBN 978-94-010-1807-4.

V. Jarník, Vědecká práce M. Kösslera [Research Works of M. Kössler], *Časopis pro pěstování matematiky* **80**(1955), pp. 106–117.

F. Jílek and V. Lomič, *Dějiny ČVUT* [History of CTU], ČVUT, Praha, Volume 1, Part 1, 1973, 595 pages.

F. Jílek, V. Lomič and P. Horská, *Dějiny ČVUT* [History of CTU], ČVUT, Praha, Volume 1, Part 2, 1978, 447 pages.

F. Kadeřávek, In memoriam techn. dr. Josefa Kounovského [Techn. Dr. Josef Kounovský (Obituary)], *Časopis pro pěstování matematiky a fysiky* **75**(1950), pp. D345–D349.

F. Kadeřávek and J. Pulkrábek (eds.), *250 let technických škol v Praze 1707–1957* [250 Years of Technical Schools in Prague 1707–1957], Slavnostní sborník, České vysoké učení technické v Praze, ČSAV, Praha, 1956, 53 pages.

J. Kalousek, *Děje Král. České Společnosti Náuk spolu s kritickým přehledem publikací jejích z oboru filosofie, historie a jazykovědy ke stoletému jubilaeu této společnosti* ... [History of the Czech Royal Society of Sciences together with a Critical Overview of its Publications of Philosophy, History and Linguistics for the Centenary Anniversary of this Company ...], KČSN, Praha, 1885, vii + 256 pages.

M. Kašparová and Z. Nádeník, *Jan Sobotka (1862–1931)*, Series Dějiny matematiky, Volume 44, Matfyzpress, Praha, 2010, 250 pages. ISBN 978-80-7378-121-7.

M. Katětov and P. Simon (eds.), *The Mathematical Legacy of Eduard Čech*, Academia – Birkhäuser Verlag, Praha – Basel, 1993, 441 pages. ISBN 978-0348-7526-4.

B. Kepr, Sedmdesát pět let profesora Františka Kadeřávka [Seventy-Five Years of Professor František Kadeřávek], *Časopis pro pěstování matematiky* **80**(1960), pp. 384–385.

J. Klapka, Josef Klíma, *XXI. výroční zpráva Moravskoslezské akademie věd přírodních*, Brno, 1949, pp. 46–48.

J. Kopřiva, Prof. PhDr. Miloš Kössler zemřel [Prof. PhDr. Miloš Kössler Died], *Pokroky matematiky, fyziky a astronomie* **6**(1961), pp. 226–230.

V. Kořínek, Stručný přehled vědeckých prací profesora Karla Petra v desetiletí 1928–1938 [A Brief Overview of Scientific Works of Karel Petr in 1928–1938], *Časopis pro pěstování matematiky a fysiky* **67**(1938), pp. D245–D253.

V. Kořínek, Stručný přehled vědeckých prací profesora Karla Petra v desetiletí 1938–1948 [A Brief Overview of Scientific Works of Professor Karel Petr in 1938–1948], *Časopis pro pěstování matematiky a fysiky* **67**(1948), pp. D9–D18.

R. Košt'ál, *Vznik a vývoj pobočky JČMF v Brně* [Birth and Development of JČMF's Branch in Brno], JČMF, Praha, (1968), 192 pages.

J. Kotůlek, Hrdinou proti své vůli? Věnováno Františku Čuříkovi [Hero against his Will? Dedicated to František Čuřík], pp. 189–192, in J. Bečvář, M. Bečvářová (eds.), *Sborník 35. mezinárodní konference Historie matematiky, Velké Meziříčí 22. až 26. 8. 2014*, Matfyzpress, Praha, 2014, 273 pages. ISBN 978-80-7378-265-8.

J. Kounovský, Profesor Bedřich Procházka sedmdesátníkem [Professor Bedřich Procházka Septuagenarian], *Časopis pro pěstování matematiky a fysiky* **55**(1926), pp. 1–10.

G. Kowalewski, *Bestand und Wandel — Meine Lebenserinnerungen, zugleich ein Beitrag zur neueren Geschichte der Mathematik*, Oldenbourg, München, 1950, 309 pages.

O. Kowalski, Věnováno Václavu Hlavatému (Některé dokumenty o životě a díle) [Devoted to Václav Hlavatý (Some Documents about his Life and Work)], *Pokroky matematiky, fyziky a astronomie* **38**(1993), pp. 65–81. ISSN 0032-2423.

M. Kraft, Maximilian Pinl in memoriam, *Jahresbericht der Deutschen Mathematiker-Vereinigung* **83**(1981), pp. 119–124. ISSN 0012-0456.

J. Kurzweil, K padesátinám profesora Vladimíra Knichala [Fifty Years of Professor Vladimír Knichal], *Časopis pro pěstování matematiky* **83**(1958), pp. 127–128.

J. Kurzweil, O životě a díle člena korespondenta ČSAV prof. Vladimíra Knichala [On the Life and Work of Corresponding Member of the Czechoslovak Academy of Sciences Professor Vladimír Knichal (Obituary)], *Časopis pro pěstování matematiky* **100**(1975), pp. 314–324.

K. Lepka, *Matyáš Lerch's Work on Number Theory*, Faculty of Science, Masaryk University, Brno, 1995, 78 pages.

V. Libický, Univerzitní prof. PhDr. František Nušl zemřel [Professor PhDr. František Nušl Died], *Věstník Národního technického muzea* **32**(1951), pp. 153–154.

A. Lukášová, *Matematika na VUT. Život a dílo Karla Čupra* [Mathematics at the VUT. Life and Works of Karel Čupr], diplomová práce, Přírodovědecká fakulta, Masarykova univerzita, Brno, 2002, v + 76 pages + 14 appendices.

J. Majer, *Z dějin Vysoké školy báňské v Příbrami* [From the History of Mining University in Příbram], Symposium Hornická Příbram ve vědě a technice, Příbram, 1984, 252 pages.

M. Menšík, Prof. Ing. Dr. techn. František Kadeřávek sedmdesátníkem [Professor Ing. Dr. František Kadeřávek Septuagenarian], *Matematika ve škole* **5**(1955), pp. 498–500.

J. Metelka, K 80. narozeninám akademika Bohumila Bydžovského [To 80th Birthday of Academician Bohumil Bydžovský], *Pokroky matematiky, fyziky a astronomie* **5**(1960), pp. 603–612.

M. Mikan, Prof. Ing. Dr. techn. Josef Kounovský šedesátníkem [Ing. Dr. Josef Kounovský Sexagenarian], *Časopis pro pěstování matematiky a fysiky* **68**(1939), pp. D25–D27.

T. Mormann, *Rudolf Carnap*, Verlag C.H. Beck, München, 2000, 230 pages. ISBN 9783406419546.

Z. Nádeník, *Moji učitelé geometrie* [My Teachers of Geometry], Series Dějiny matematiky, Volume 48, Matfyzpress, Praha, 2011, 291 pages. ISBN 978-80-7378-159-0.

M. Němcová, *František Josef Studnička (1836–1903)*, Series Dějiny matematiky, Volume 10, Prometheus, Praha, 1998, 348 pages + 19 pictures. ISBN 80-7196-103-5.

B. Novák (ed.), *Life and Work of Vojtěch Jarník*, Prometheus, Praha, 1999, 197 pages. ISBN 80-7196-156-6.

M. Novotný, Osmdesát let profesora Josefa Kauckého [Eighty Years of Professor Josef Kaucký], *Pokroky matematiky, fyziky a astronomie* **20**(1975), pp. 286–288.

L. Nový a kol., *Dějiny exaktních věd v českých zemích* [History of Exact Sciences], ČSAV, Praha, 1961, 431 pages.

F. Nožička, Vzpomínka na profesora Miloše Kösslera [Recollection of Professor Miloš Kössler], *Časopis pro pěstování matematiky* **87**(1962), pp. 111–113.

F. Nožička, Profesor Václav Hlavatý, český matematik světového jména [Professor Václav Hlavatý, a Czech Mathematician of World Renown (Obituary)], *Časopis pro pěstování matematiky* **94**(1969), pp. 374–380.

F. Nušl and M. Kössler, Karel Petr, *Časopis pro pěstování matematiky a fysiky* **57**(1928), pp. 169–182.

O. Obůrka, Zemřel profesor Jiří Klapka [Prof. Jiří Klapka Died], *Časopis pro pěstování matematiky* **101**(1976), pp. 412–416.

J. Olejníčková, *Vědecké dílo Bohumila Bydžovského* [Scientific Work of Bohumil Bydžovský], doktorská disertační práce, Matematicko-fyzikální fakulta, Univerzita Karlova, Praha, 2005, 348 pages.

N. Ottowitz, *Mathematikunterricht an der Technischen Hochschule in Wien 1815–1918*, Dissertationen der Technischen Universität Wien, 1992, 2 Bande, 520 pages.

L. Pátý (ed.), *Jubilejní almanach. Jednota čs. matematiků a fyziků. 1862–1987* [Jubilee Almanac. Union of Czechoslovak Mathematicians and Physicists. 1862–1987], JČSMF, Praha, 1987, 229 pages.

H. Pavlincová, *Karel Vorovka. Cesta matematika k filosofii* [Karel Vorovka. The Path of Mathematicians to the Philosophy], Filosofia, Praha, 2010, 598. ISBN 978-80-7007-331-5.

M. Pavlíková, *Život a dílo Miloše Kösslera* [Life and work of Milos Kössler], doktorská disertační práce, Matematicko-fyzikální fakulta, Univerzita Karlova, Praha, 2004, 261 pages.

M. Pelíšek, Za zemřelým Dr. Václavem Simandlem [Dr. Václav Simandl (Obituary)], *Časopis pro pěstování matematiky a fysiky* **48**(1919), pp. 203–206.

O. Perron, Heinrich Tietze, 31. 8. 1880 — 17. 2. 1964, *Jahresbericht der Deutschen Mathematiker-Vereinigung* **83**(1981), nr. 4, pp. 182–191. ISSN 0012-0456.

K. Petr, František Nušl, *Časopis pro pěstování matematiky a fysiky* **57**(1928), pp. 73–80.

M. Pinl, In Memory of Ludwig Berwald, *Scripta Mathematica* **27**(1964), pp. 193–203. ISSN 0036-9713.

M. Pinl, Památce Ludwiga Berwalda [In Memory of Ludwig Berwald], *Časopis pro pěstování matematiky* **92**(1967), pp. 229–238.

M. Pinl and A. Dick, Kollegen in einer dunklen Zeit, Schluß, *Jahresbericht der Deutschen Mathematiker-Vereinigung* **75**(1974), pp. 166–208, 77(1976), pp. 161–164. ISSN 0012-0456.

R. Piska, K stému výročí narozenin prof. Miloslava Pelíška [For the 100th Anniversary of the Birth of Miloslav Pelíšek], *Matematika ve škole* **6**(1956), pp. 363–366.

V. Pleskot, Zemřel profesor Dr. Václav Hruška [Professor Dr. Václav Hruška Died], *Časopis pro pěstování matematiky* **79**(1954), pp. 375–378.

J. Pokorný and J. Novotný (eds.), *Česká akademie věd a umění 1891–1991: sborník příspěvků k 100. výročí zahájení činnosti* [Czech Academy of Sciences and Arts 1891–1991: Proceedings of the 100th Anniversary of the Establishment], Historický ústav AV ČR, Praha, 1993, 190 pages.

J. Polášek, Za profesorem Miloslavem Hamplem [Professor Miloslav Hampl], *Pokroky matematiky, fyziky a astronomie* **20**(1975), pp. 51–53.

V. Posejpal, *Dějepis Jednoty českých mathematiků* [History of the Union of Czech Mathematicians], JČM, Praha, 1912, 131 pages.

J. Purš, *200 let České společnosti nauk 1784–1984* [200 Years of Czech Society of Science 1784–1984], Ústav československých a světových dějin ČSAV, Praha, 1985, 573 pages.

H. Reichardt, Wilhelm Blaschke, *Jahresbericht der Deutschen Mathematiker-Vereinigung* **69**(1966), pp. 1–8. ISSN 0012-0456.

L. Rieger and K. Rychlík, Profesor Dr. František Rádl zemřel [Professor Dr. František Rádl Died], *Časopis pro pěstování matematiky* **82**(1957), pp. 378–382.

R.M. Rohrer, *Die k. k. Technische Hochschule in Brünn: Geschichtlich-statistische Skizze herausgegeben aus Anlass des 25-jährigen Bestandes der Lehr-Anstalt*, Brünn, 1875, 64 pages.

V. Řehořovský and A. Seydler, Na paměť 25letého trvání Jednoty českých mathematiků [In Memory of the 25 Years of the Union of Czech Mathematicians], *Časopis pro pěstování mathematiky a fysiky* **16**(1887), pp. 257–266.

J. Schenk, *125. výročí založení Vysoké školy báňské v Příbrami* [125th Anniversary of the Foundation of the Mining University in Příbram], Příbram, 1974, 16 pages.

P.A. Schillp (ed.), *The Philosophy of Rudolf Carnap*, Revised edition, Library of Living Philosophers, Book 11, Open Court, 1999, 1088 pages. ISBN 978-0812691535.

Š. Schwarz, Vedecká práca prof. K. Petra v oblasti teórie čísel [The Scientific Work of K. Petr in the Field of Number Theory], *Časopis pro pěstování matematiky* **94**(1969) pp. 358–361.

F. Seibt (ed.), *Die Teilung der Prager Universität 1882 und die intellektuelle Desintegration in den böhmischen Ländern*, Collegium Carolinum, Oldenbourgh, München, 1984, 220 pages.

L. Seifert, K sedmdesátinám prof. Miloslava Pelíška [Professor Miloslav Pelíšek Septuagenarian], *Časopis pro pěstování matematiky a fysiky* **55**(1926), pp. 329–332.

L. Seifert, Prof. Dr. techn. Josef Klíma, *Časopis pro pěstování matematiky a fysiky* **71**(1946), pp. D35–D42.

R. Siegmund-Schultze, German Speaking Migration of Mathematicians to and from Czechoslovakia, Caused by Nationalism in Germany, *Dějiny vědy a techniky* **45**(2012), pp. 141–166. ISSN 0300-4414.

J. Sobotka, Antonín Sucharda, *Časopis pro pěstování mathematiky a fysiky* **37**(1908), pp. 353–359.

J. Sobotka, Vincenc Jarolímek, *Časopis pro pěstování matematiky a fysiky* **45**(1916), pp. 439–449.

S. Stark, W. Gintl and A. Grünwald, *Die k. k. Deutsche Technische Hochschule in Prag 1806–1906*, Prag, 1906, 518 pages.

K. Strubecker, Wilhelm Blaschkes mathematisches Werk, *Jahresbericht der Deutschen Mathematiker-Vereinigung* **88**(1986), pp. 146–157. ISSN 0012-0456.

W. Süss, Berwald Ludwig, Neue Deutsche Biographie 2, Duncker & Humblot, Berlin, 1971, p. 173.

M. Svatoš, I. Čornejová, J. Havránek and Z. Pousta (eds.), *Dějiny Univerzity Karlovy 1918–1990* [History of Charles University, 1918–1990], Univerzita Karlova, Praha, 1998, 671 pages. ISBN 80-7184-539-6.

V. Šebesta, prof. Dr. Josef Theurer, *Časopis pro pěstování matematiky a fysiky* **52**(1923), pp. 283–286.

V. Šebesta, Za prof. Dr. Josefem Theurerem [Professor Dr. Jos. Theurer (Obituary)], *Časopis pro pěstování matematiky a fysiky* **59**(1930), pp. 1–2.

K. Šindelář, Památce akademika Bohumila Bydžovského [In Memoriam: Bohumil Bydžovský], *Časopis pro pěstování matematiky* **95** (1970), pp. 100–113.

P. Šišma, *Matematika na německé technice v Brně* [Mathematics at the German Technical School in Brno], Series Dějiny matematiky, Volume 21, Prometheus, Praha, 2002, 322 pages. ISBN 80-7196-256-2.

P. Šišma, *Učitelé na německé technice v Brně 1849–1945* [Teachers at the German Technical School in Brno 1849–1945], Společnost pro dějiny věd a techniky, Praha, 2004, 205 pages. ISBN 80-239-3279-9. German version: P. Šišma, *Zur Geschichte der Deutschen Technischen Hochschule Brünn. Professoren, Dozenten und Assistenten 1849–1945*, Schriftenreihe Geschichte der Naturwissenschaften und der Technik 13, Johannes Kepler Universität Linz, Trauner Verlag, Linz, 2009, 192 pages. ISBN 978-3-85499-451-0.

J. Škrášek, Seznam prací prof. Matyáše Lercha [List of Works of Prof. Matyáš Lerch], *Časopis pro pěstování matematiky* **78**(1953), pp. 139–148.

J. Škrášek, Život a dílo profesora Matyáše Lercha (K stému výročí jeho narození) [Life and Work of Professor Matyáš Lerch (On 100th Anniversary of his Birthday)], *Časopis pro pěstování matematiky* **85** (1960), pp. 228–240.

L. Špaček, Doc. Dr. Miloslav Hampl pětašedesátníkem [Sixty Five Years of Doc. Dr. Miloslav Hampl], *Časopis pro pěstování matematiky* **87**(1962), pp. 499–500.

M. Tayerlová, M. Efmertová, I. Jakubec and M. Josefovičová, *Česká technika* [Czech Technical University], České vysoké učení technické v Praze, Praha, 2004, 233 pages. ISBN 80-01-03165-9.

E. Těšínská, Fyzikální vědy v pražském německém přírodovědném spolku "Lotos" [The Physical Sciences in the Prague German Natural Sciences Club "Lotos"], *Pokroky matematiky, fyziky a astronomie* **42**(1997), pp. 35–47. ISSN 0032-2423.

J.A. Theurer, *Památník Vysoké školy báňské v Příbrami za léta od 1889 do 1924: k 75letému trvání vys. školy báňské* [Proceedings of the Mining University in Příbram for the Years 1889 to 1924: 75 Years of Its Duration], Vysoká škola báňská, Příbram, 1924, viii + 427 pages.

R. Tobies, *Biographisches Lexikon in Mathematik promovierter Personen an deutschen Universitäten und Technischen Hochschulen WS 1907/08 bis WS 1944/45*, Algorism, Studien zur Geschichte der Mathematik und der naturwissenschaftlichen herausgegeben von Menso Folkerts, Heft 58, Augsburg, 2006, 402 pages. ISBN 978-3-936905-21-2.

M. Toepell, *Mitgliedergesamtverzeichnis der Deutschen Mathematiker-Vereinigung 1890–1990*, München, 1991, 440 pages.

L. Truksa, Šedesát let profesora Dr. Jaroslava Janko [Professor Jaroslav Janko Sexagenarian], *Časopis pro pěstování matematiky* **79**(1954), pp. 181–185.

Z. Třešňák, P. Šarmanová and B. Půža, *Otakar Borůvka*, Series Osobnosti, Universitatis Masarykiana, Brno, 1996, 240 pages. ISBN 80-902004-0-0.

M. Tulachová, *Disertace pražské university 1882–1953* [Dissertations at the Prague University 1882–1953], Series Sbírka pramenů a příruček k dějinám University Karlovy, Volume 2, Universita Karlova, SPN, Praha, 1965, 479 pages.

A. Urban, Významní rakovničtí deskriptivní geometři [The Important Descriptive Geometers from the City of Rakovník], *Pokroky matematiky, fyziky a astronomie* **7**(1962), pp. 285–290.

M.A. Valouch, K šedesátce prof. Dr. Fr. Nachtikala [Professor Dr. Fr. Nachtikal Sexagenarian], *Časopis pro pěstování matematiky a fysiky* **64**(1935), pp. D1–D4.

A.V. Velflík, *Dějiny technického učení v Praze* [History of the Technical University in Prague], Praha, first volume, 1906 and 1909, ix + 632 + xxxxi pages, second volume, first part, 1910, 272 + xiv pages, second part, 1925, vi + 273–337 + xxv–xciv pages.

F. Veselý, *100 let Jednoty československých matematiků a fyziků* [One Hundred Years of the Union of Czechoslovak Mathematicians and Physicists], SPN, Praha, 1962, 129 pages + 26 pictures.

Q. Vetter, Půl století Časopisu pro pěstování matematiky a fysiky [Half Century of the "Journal for Cultivation of Mathematics and Physics"], *Časopis pro pěstování matematiky a fysiky* **51**(1922), pp. 53–56.

Q. Vetter, Devadesát let Jednoty československých matematiků a fysiků [Ninety Years of the Union of Czechoslovak Mathematicians and Physicists], *Časopis pro pěstování matematiky* **77**(1952), pp. 175–183.

L. Vojteková, Jan Vojtěch a jeho stredoškolské učebnice [Jan Vojtěch and his Textbooks for Secondary Schools], pp. 153–165, in J. Bečvář, M. Bečvářová (eds.), *Matematika v proměnách věků V* [Mathematics Throughout The Ages V], Series Dějiny matematiky, Volume 33, Matfyzpress, Praha, 2007, 331 pages ISBN 978-80-7378-017-3.

J. Vojtěch, K šedesátce prof. Dr. J. Klobоučka [Professor Dr. J. Klobouček Sexagenarian], *Časopis pro pěstování matematiky a fysiky* **64**(1935), pp. D101–D103.

O. Volk, Lothar Koschmieder, Neue Deutsche Biographie 12, Duncker & Humblot, Berlin, 1980, p. 610. ISBN 3-428-00193-1.

M. Výborná, J. Havránek and K. Kučera (eds.), *Disertace pražské university II (1882–1945)* [Dissertations at the Prague University II (1882–1945)], Series Sbírka pramenů a příruček k dějinám University Karlovy, Volume 3, Universita Karlova, SPN, Praha, 1965, 230 pages.

F. Vyčichlo, Prof. Dr. Jan Vojtěch zemřel [Professor Dr. Jan Vojtěch Died], *Časopis pro pěstování matematiky* **78**(1953), pp. 283–28.

W. Wunderlich, Josef Krames. Nachruf (mit Schriftverzeichnis), *Almanach der Österreichischen Akademie der Wissenschaften* **137**(1987), pp. 286–295.

V. Žbirková, *Juraj Hronec — pedagog* [Juraj Hronec — Educationalist], Bratislava, 1975, 143 pages.

https://doi.org/10.1142/9781786349316_0009

Chapter 9

The End of the 19th Century and the First Third of the 20th Century

Jindřich Bečvář

1. Scepticism in Natural Sciences at the Turn of the 20th Century

The turbulent progress in mathematics, physics, astronomy, and other natural as well as social sciences was a source of great satisfaction arising from the new results obtained during the 19th century. At the same time, these feelings were accompanied by a significant scepticism toward the future development of certain branches of science. Many researchers found it almost impossible that the new discoveries in physics, astronomy, chemistry, biology, geography and other disciplines might continue throughout the 20th century with undiminished strength.

On a side note, similar sceptical opinions on the future of mathematics appeared already at the turn of the 19th century. Let us quote an excerpt from a letter of Joseph Louis Lagrange (1736–1813), dated September 21, 1781, to none other than his friend and colleague Jean-Baptiste le Rond d'Alembert (1717–1783):

> *Il me semble aussi que la mine est presque déjà trop profonde, et qu'à moins qu'on ne découvre de nouveaux filons il faudra tôt ou tard l'abandonner. La Physique et la Chimie offrent maintenant des richesses plus brillantes et d'une exploitation plus facile; aussi le goût du siècle paraît-il entièrement tourné de ce côté-là, et il n'est pas impossible que les places de Géométrie dans les Académies ne deviennent un jour ce que sont actuellement les chaires d'arabe dans les Universités.*[1]

In 1870, James Clerk Maxwell (1831–1879), a Scottish physicist, whose most famous discoveries were in optics and electromagnetic field theory, and the author of the famous monograph *A Treatise on Electricity and Magnetism* (1873), expressed

[1]J.A. Serret, J.G. Darboux (eds.), *Œuvres de Lagrange*, Tome 13, Gauthier-Villars, Paris, 1882, 401 pages, see p. 368.

his admiration for the current state of knowledge in physics. He showed some scepticism about major progress in the future, and also his belief in the strength embodied in the unification of physics and mathematics:

> *It was a great step in science when men became convinced that, in order to understand the nature of things, they must begin by asking, not whether a thing is good or bad, noxious or beneficial, but of what kind is it? and how much is there of it? Quality and quantity were then first recognised as the primary features to be observed in scientific inquiry. As science has been developed, the domain of quantity has everywhere encroached on that of quality, till the process of scientific inquiry seems to have become simply the measurement and registration of quantities, combined with a mathematical discussion of the numbers thus obtained. It is this scientific method of directing our attention to those features of phenomena which may be regarded as quantities which brings physical research under the influence of mathematical reasoning.*[2]

In 1871, the world-famous Scottish physicist William Thomson — Lord Kelvin of Largs (1824–1907),[3] and the author of 18 books and 661 original treatises, holder of 70 patents, inventor of numerous electrical devices, who suggested the use of an absolute-temperature scale in 1848 and was engaged in laying down the transatlantic telegraph cable, outlined his view of the future development of physics by these words:

> *Accurate and minute measurement seems to the non-scientific imagination a less lofty and dignified work than looking for something new. But nearly all the grandest discoveries of science have been but the rewards of accurate measurement and patient long-continued labour in the minute sifting of numerical results.*[4]

Thomas Corwin Mendenhall (1841–1924), an American physicist and meteorologist, pioneer of the metric system in the USA, president of the American Association for the Advancement of Science in 1889, who also paid great attention to the works

---

[2]Sectional Proceedings. Section A. — Mathematical and Physical Science. — President, Prof. J. Clerk Maxwell, F.R.S., *Nature* **2**(1870), no. 47 (Sept. 22), pp. 419–428, see p. 419, ISSN 0028-0836.

[3]William Thomson was a member of the Royal Society of London since 1851, and presided the society in the years 1890–1895. He was a member of 88 academies and scientific societies, and received 21 honorary doctorates.

[4]Inaugural Address of Sir William Thomson, LL.D., F.R.S., President, *Nature* **4**(1871), no. 92 (Aug. 3), pp. 262–270, see p. 265, ISSN 0028-0836.
The following quote, which is however not authentic, is very often attributed to Lord Kelvin: *There is nothing new to be discovered in physics now. All that remains is more and more precise measurement.* See, e.g., C. Weiner, Who Said It First?, *Physics Today* **21**(1968), no. 8 (August), p. 9, ISSN 0031-9228, and L. Badash, The Completeness of Nineteenth-Century Science, *ISIS* **63**(1972), no. 1, pp. 48–58, ISSN 0021-1753.

of William Shakespeare (stylometry), declared his opinion on the future of scientific research in his 1887 book *A Century of Electricity*:

> *More than ever before in the history of science and invention, it is safe now to say what is possible and what is impossible. No one would claim for a moment that during the next five hundred years the accumulated stock of knowledge of geography will increase as it has during the last five hundred.... In the same way it may safely be affirmed that in electricity the past hundred years is not likely to be duplicated in the next, at least as to great, original, and far-reaching discoveries, or novel and almost revolutionary applications.*[5]

The outstanding American astronomer and mathematician Simon Newcomb (1835–1909), a leading expert in celestial mechanics,[6] author of the popular books *Popular Astronomy* (1878), *Astronomy for Everybody* (1902) and *The Reminiscences of an Astronomer* (1903), included in his 1888 paper *The Place of Astronomy Among the Sciences* the following words characterizing the future work of astronomers:

> *If the public are disappointed at not seeing brilliant discoveries coming from all the observatories, we must remember that what are called discoveries are not the main work of the scientific man of to-day. It would be too much to say with confidence that the age of great discoveries in any branch of science has passed by: yet, so far as astronomy is concerned, it must be confessed that we do appear to be fast reaching the limits of our knowledge.... At the same time, so far as we can see, one comet is so much like another that we cannot regard one as adding in any important degree to our knowledge. The result is that the work which really occupies the attention of the astronomer is less the discovery of new things than the elaboration of those already known, and the entire systemization of our knowledge.*[7]

The American physicist and metrologist Albert Abraham Michelson (1852–1931), known as the inventor of optical and spectral devices, who was obsessed with precise measurement and whose name is associated with an attempt to detect the relative motion of the Earth with respect to the aether (1881),[8] as well as with repeated measurements of the speed of light, the Nobel prize winner in 1907, expressed in

[5]T.C. Mendenhall, *A Century of Electricity*, Houghton, Mifflin, Boston, New York, 1887; revised 1890, 243 pages, see p. 223.

[6]S. Newcomb was also interested in statistics and economics, wrote textbooks in algebra, geometry, plane and spherical trigonometry, and was president of the American Association for the Advancement of Science in 1877.

[7]S. Newcomb, The Place of Astronomy Among the Sciences, *The Sidereal Messenger. A Monthly Review of Astronomy* **7**(1888), pp. 65–73, see pp. 69–70.

[8]In 1887, he repeated this experiment with the physicist and chemist Edward Morley (1838–1923). The experiment, which in fact ended in failure since it disproved the existence of the aether, became famous as the Michelson or Michelson-Morley experiment.

his paper *A Plea for Light Waves* the following opinion on the future of scientific research:

> *Still it is safe to say that at any rate the more important facts and laws as well as the more promising lines of the development of their applications, are now fairly well known, and that inducements to their further study and development are not wanting.*[9]

A similar opinion appears in his later treatise *Some of the Objects and Methods of Physical Science*:

> *It is never safe to affirm that the future of physical science has no marvels in store which may be even more astonishing than those of the past; but it seems probable that most of the grand underlying principles have now been firmly established and that further advances are to be sought chiefly in the rigorous application of these principles to all the phenomena which come under our notice...*
>
> *An eminent physicist has remarked that the future truths of physical science are to be looked for in the sixth place of decimals. In order to make such results possible the student and investigator must have at his disposal the methods and results of his predecessors, must know how to gauge their value, and to apply them to his own work; and especially must he have at his command all the modern appliances and instruments of precision which constitute a well-equipped physical laboratory — without which results of real value can be obtained only at immense sacrifice of time and labor.*[10]

The politician, historian and respected writer Charles Henry Pearson (1830–1894), born in London and active in Australia for many years, the author of the monographs *History of England During the Early and Middle Ages* (1868) and *History of England in the Fourteenth Century* (1876),[11] saw in his 1893 book *National Life and Character: A Forecast* the future of chemistry as follows:

> *Chemistry rests, when all abatements have been made, on the atomic theory, and even if future investigation enables us to forecast with absolute precision what the result of combinations hitherto unattempted will be, so that we can calculate in the study what is now worked out gropingly in the laboratory,*

[9] A.A. Michelson, A Plea for Light Waves, *Proceedings of the American Association for the Advancement of Science* **37**(1888), pp. 67–78, see pp. 69–70.

[10] *University of Chicago Quarterly Calendar* **3**(1894), no. 2 (August), pp. 12–15, see p. 15. See also R.S. Mulliken, Spectroscopy, Quantum Chemistry and Molecular Physics, *Physics Today* **21**(1968), no. 4 (April), pp. 52–57, see p. 56; C. Weiner, Who Said It First?, *ibid.* **21**(1968), no. 8 (August), p. 9, ISSN 0031-9228.

[11] We remark that Pearson published the *Biographical Sketch of Henry John Stephen Smith* in 1894, and recall that H.J.S. Smith (1826–1883) was an important Irish mathematician.

that discovery would hardly eclipse the merit of Dalton's contribution to science. So it is in every department of research.[12]

The American historian of mathematics Florian Cajori (1859–1930), famous for his books *A History of Mathematics* (1893), *A History of Elementary Mathematics* (1898), and the two-volume monograph *A History of Mathematical Notations* (1928–1929), expressed in his 1908 paper *The Age of the Sun and the Earth. Some Ancient and Modern Theories* the following opinion on the development of chemistry:

We are told that, about 1878, the president of a certain chemical society informed his hearers in an annual address that the age of discovery in chemistry was closed, and that henceforth we had better devote ourselves to a thorough classification of chemical phenomena.[13]

In 1909, the English experimental physicist Joseph John Thomson (1856–1940),[14] the discoverer of the electron, who was studying the motion of particles in electric and magnetic fields, and recipient of the 1906 Nobel Prize for his theoretical and experimental investigations on the conduction of electricity by gases, expressed himself on the future of research in physics:

Enthusiasm has been quickened, and there is a hopeful, youthful, perhaps exuberant, spirit abroad which leads men to make with confidence experiments which would have been thought fantastic twenty years ago. It has quite dispelled the pessimistic feeling, not uncommon at that time, that all the interesting things had been discovered, and all that was left was to alter a decimal or two in some physical constant. There never was any justification for this feeling, there never were any signs of an approach to finality in science. The sum of knowledge is at present, at any rate, a diverging not a converging series.[15]

The English mathematician and astronomer Ernest William Brown (1866–1938), expert on celestial mechanics (motion of the Moon, planets and asteroids), who spent most of his life in the USA (he was president of the American Mathematical Society in the years 1915–1916), wrote in his biographical article *George William Hill*:

For two hundred years mathematical astronomers, many of them of the first rank, had been devoting their energies to furnishing a complete demonstration of the power of the law of gravitation to account for the motions of all

[12] C.H. Pearson, *National Life and Character: A Forecast*, Macmillan, London, 1893, 357 pages, see p. 291; 2nd ed. 1894, reprint: 1913, 381 pages, see p. 308.
[13] F. Cajori, The Age of the Sun and the Earth. Some Ancient and Modern Theories, *Scientific American* **66**(1908), Supplement no. 1706, pp. 174–175, ISSN 0036-8733.
[14] Thomson was a member of the Royal Society of London since 1884 and presided the society in the years 1915–1920.
[15] Presidential Address to the British Association, *Report of the Seventy-Ninth Meeting of the British Association for the Advancement of Science* 1909, pp. 3–29, see p. 29.

> *the bodies in the solar system within the degree of accuracy of the observations. In the third quarter of the nineteenth century it was evident that this demonstration would soon be made.*[16]

In 1904, the British historian of science William Cecil Dampier Whetham (1867–1952) published a popular book *The Recent Development of Physical Sciences*[17] and many years later, he finished the book *A Shorter History of Science* (1944, 1945). He was a member of the Royal Society of London since 1901, and the first secretary of the Agricultural Research Council in the years 1931–1935. In his famous monograph *A History of Science and its Relations with Philosophy and Religion*, first published in 1929 and reprinted many times since then, he gave the following opinion on the development of science in the 20th century:

> *It seemed as though the main framework had been put together once for all, and that little remained to be done but to measure physical constants to the increased accuracy represented by another decimal place, and to carry farther those investigations which had seemed at intervals to be on the point of solving the problem of the structure of matter in terms of the luminiferous aether.*[18]

However, the views that the most important laws of physics had already been discovered and what remained was just a more precise determination of physical constants, soon turned out to be shortsighted. Since the turn of the 20th century, physics (and other disciplines) began to develop even more rapidly than before.[19]

The end of the 19th century already brought some fundamental discoveries in physics: the X-rays (1895), spontaneous radioactivity (1896), the electron (1897), polonium and radium (1898), and blackbody radiation (1900). The succession of discoveries in physics was perhaps even more intense throughout the 20th century. Among the most important ones during the first four decades were the Lorentz transformation (1904), the explanation of the Brownian motion (1905) and the photoelectric effect (1905), the photon (1905), special relativity (1905), superconductivity (1911), the atomic nucleus (1911), cosmic rays (1912), Bohr model of the atom (1913), general relativity (1915, 1916), the proton (1919), matrix mechanics, wave mechanics and quantum mechanics (1925, 1926), the prediction of the positron (1928), expansion of the universe (1929), the cosmological redshift (1929), white dwarf theory (1931), the prediction of the neutrino (1931), the neutron (1932), the

---

[16]E.W. Brown, George William Hill. 1838–1914, *Biographical Memoirs of the National Academy of Sciences* **8**(1916), pp. 275–309, see p. 281, ISSN 0077-2933.

[17]J. Murray, London, 1904, xii + 344 pages; 6th ed. 1927, later edition: HardPress, Miami, Fl., 2013, 396 pages, ISBN 978-1314392715.

[18]W.C. Dampier, *A History of Science and Its Relations with Philosophy and Religion*, Cambridge University Press, Cambridge, 1929, 1930 (xxii + 514 pages, see p. 382). Later editions: 1932, 1938, 1942, 1948 (527 pages, see p. 369), 1949, 1961, 1966; Literary Licensing, 2012 (536 pages, see p. 536), ISBN 978-1258239206.

[19]For more details, see L. Badash, The Completeness of Nineteenth-Century Science, *ISIS* **63**(1972), no. 1, pp. 48–58, ISSN 0021-1753.

positron (1932), neutron stars (1933), artificial radioactivity (1934), nuclear fission and the chain reaction (1939).[20]

We remark that a number of physical discoveries were directly linked to space exploration. The turn of the 20th century saw the beginning of a long and successful journey into the heart of the matter and an amazing development of particle physics.[21]

2. Mathematics at the Turn of the 20th Century

In comparison with physics, astronomy and other natural sciences, a considerable optimism was present in mathematics at the turn of the 20th century. In the second half of the 19th century, the efforts to provide a rigorous foundation for mathematical analysis culminated in the introduction of the ε-δ-language (arithmetization of analysis).[22] This made it possible to overcome the *second crisis on mathematics*,[23] i.e., problems related to ingenious but careless manipulation with infinitely small and infinitely large quantities, which was so typical for the 18th century.[24] Moreover, several rigorous theories of real numbers were put forward in the 1870s.[25]

---

[20]By a curious destiny, the first work on the chain reaction (N. Bohr, J.A. Wheeler, The Mechanism of Nuclear Fission, *Physical Review* **56**(1939), pp. 426–450, was published on September 1, 1939, the first day of World War II, which ended after dropping two atomic bombs.

[21]For a thorough introduction to the history of (modern) physics, we recommend the following titles: M. Jammer, *Concept of Space. The History of Theories on Space in Physics*, Harvard University Press, Cambridge, Mass., 1954, 196 pages; W.F. Magie, *A Source Book in Physics*, Harvard University Press, Cambridge, Mass., 1935, 1963, 1969, xiv + 620 pages, ISBN 978-0674823655; K. Simonyi, *A Cultural History of Physics*, Taylor & Francis, Boca Raton, London, New York, 2012, xiv + 622 pages, ISBN 978-1-56881-329-5; J.Z. Buchwald and R. Fox (eds.), *The Oxford Handbook of the History of Physics*, Oxford University Press, Oxford, New York, 2013, ix + 945 pages, ISBN 978-0-19-969625-3.

[22]The main persons involved in this process were Bernard Bolzano (1781–1848), Augustin-Louis Cauchy (1789–1857), Niels Henrik Abel (1802–1829), Peter Gustav Lejeune Dirichlet (1805–1859), Karl Theodor Weierstrass (1815–1897), and Richard Dedekind (1831–1916).

[23]The *first crisis in mathematics* refers to the presumed collapse of the foundations of mathematics after the discovery of incommensurability. It took place in ancient Greece during the 5th century B.C. It is assumed that the way out of this crisis consisted of the transition from the numerical conception of mathematics to a geometrical one, which is presented in Euclid's famous *Stoicheia* from the turn of the 3rd century B.C.

[24]The most influential representatives of the 18th century infinitesimal calculus were Johann Bernoulli (1667–1748), Colin MacLaurin (1698–1746), Leonhard Euler (1707–1783), Jean-Baptiste le Rond d'Alembert, and Joseph Louis Lagrange.

The contemporary handling of infinitely small quantities was criticized by the bishop George Berkeley (1685–1753) in his treatise *The Analyst, Or, A Discourse Addressed to an Infidel Mathematician* (1734), paragraph XXXV: *And what are these Fluxions? The Velocities of evanescent Increments? And what are these same evanescent Increments? They are neither finite Quantities nor Quantities infinitely small, nor yet nothing. May we not call them the Ghosts of departed Quantities?* See Ewald (1996, pp. 60–92), and in a broader context, pp. 11–92.

[25]Three most important theories of real numbers are due to Richard Dedekind, Georg Cantor (1845–1918), and Karl Theodor Weierstrass. However, it is also possible to remind the names of Charles Robert Méray (1835–1911) and Heinrich Eduard Heine (1821–1881).

Thus, in the end of the 19th century, mathematical analysis was placed on a firm foundation. It celebrated success also in the complex domain.[26] The beginning of the 20th century saw the birth of functional analysis,[27] Lebesgue's integration theory, topology, major progress in the area of integral equations, etc.

The 19th century brought many other important results on mathematics. Let us recall the discovery of non-Euclidean geometry[28] and its models, major advances in projective, algebraic, differential and n-dimensional geometry, the birth and spread of descriptive geometry, tensor calculus, etc.[29]

There appeared proofs of the fundamental theorem of algebra, the problem of solvability of algebraic equations in radicals was settled,[30] quaternions, octonions and other hypercomplex number system were discovered, the theory of algebras began to develop,[31] matrix theory[32] and vector calculus[33] were born, etc. A new algebra began to emerge — the algebra of structures.[34] No less successful was number

[26]Key contributors to the development of complex analysis in the 19th century were Augustin-Louis Cauchy, Bernhard Riemann (1826–1866), and Karl Theodor Weierstrass.

[27]Functional analysis emerged mainly from the 19th century works of Carl Gottfried Neumann (1832–1925), Salvatore Pincherle (1853–1936), Jules-Henri Poincaré, Vito Volterra (1860–1940), Jacques Hadamard (1865–1963), Ivar Fredholm (1866–1927), Niels Fabian Helge von Koch (1870–1924) and Frigyes Riesz (1880–1956).

[28]The pioneers of non-Euclidean geometries were Nikolai Ivanovich Lobachevsky (1792–1856), János Bolyai (1802–1860), Carl Friedrich Gauss (1777–1855), and later Bernhard Riemann.

[29]The development of geometry in the 19th century was influenced mainly by Gaspard Monge (1746–1818), Carl Friedrich Gauss, Jean-Victor Poncelet (1788–1867), August Ferdinand Möbius (1790–1868), Michel Chasles (1793–1880), Germinal Pierre Dandelin (1794–1847), Lambert Adolphe Jacques Quetelet (1796–1874), Jacob Steiner (1796–1863), Christian von Staudt (1798–1867), Julius Plücker (1801–1868), Hermann Günther Grassmann (1809–1877), Bernhard Riemann, Arthur Cayley (1821–1895) Eugenio Beltrami (1835–1900), William Kingdon Clifford (1845–1879), Jules-Henri Poincaré (1854–1912) Giuseppe Peano (1858–1932), Edouard Goursat (1858–1936), Christian Hugo Eduard Study (1862–1930), David Hilbert (1862–1943), and Élie Cartan (1869–1951).

[30]The main contributors to the proof of the fundamental theorem of algebra were Jean Robert Argand (1768–1822), Carl Friedrich Gauss, and later Leopold Kronecker (1823–1891). The solvability of algebraic equations in radicals was already studied by Joseph Louis Lagrange and Paolo Ruffini (1765–1822), important results were later put forward by Niels Henrik Abel and Évariste Galois (1811–1832).

[31]Various concrete algebras were discovered mainly by William Rowan Hamilton (1805–1865), John Thomas Graves (1806–1870), Arthur Cayley, William Kingdon Clifford, and Hermann Günther Grassmann. The general notion of a linear associative algebra is due to Benjamin Peirce (1809–1880).

[32]The theory of matrices and forms was developed in the 19th century by Arthur Cayley, James Joseph Sylvester (1814–1897), Georg Ferdinand Frobenius (1849–1917), Karl Theodor Weierstrass, Charles Hermite (1822–1901), and others.

[33]The vector calculus was elaborated by William Rowan Hamilton, Hermann Günther Grassmann, Josiah Willard Gibbs (1839–1903), James Clerk Maxwell, Oliver Heaviside (1850–1925), and others.

[34]The 19th century contributors to the birth of structural algebra were mainly Carl Friedrich Gauss, Ernst Eduard Kummer (1810–1893), Arthur Cayley, Leopold Kronecker, Augustin-Louis Cauchy, Richard Dedekind, Marie Ennemond Camille Jordan (1838–1922), Évariste Galois, Heinrich Weber (1842–1913), and Ludwig Otto Hölder (1859–1937).

theory: there was a major progress in the research on prime numbers,[35] and the numbers e and π were proved to be transcendental.[36]

The classical problems of ancient Greek mathematics were settled. After more than two millennia, the duplication of the cube, the angle trisection, the squaring of the circle, and the construction of regular n-gons were demonstrated to be unsolvable in general.[37]

Links between logic and algebra were found in the middle of the 19th century,[38] and an important relation between algebra and geometry was presented in the Erlangen program from 1872.[39]

In the 19th century, mathematics owed its success also to a number of applications, mainly in celestial mechanics. The 1846 discovery of a new planet, later called Neptune, was based on calculations involving the perturbations in the orbit of Uranus, and served as an excellent promotion of mathematics.[40] Mathematics was being used in geodesy, crystallography, cartography, as well as engineering.

In connection with the second International Congress of Mathematicians held in Paris, September 1900, David Hilbert, the most significant mathematician of the period, collected a list of 23 problems which he assumed to play a crucial role in the 20th century mathematics.[41] Many of them indeed had a strong influence on

---

[35]Important works were published by Carl Friedrich Gauss, Peter Gustav Lejeune Dirichlet, Ernst Eduard Kummer, Richard Dedekind, Jacques Hadamard, and Charles de La Vallée-Poussin (1866–1962).

[36]The transcendence of e and π was proved by Charles Hermite and Ferdinand Lindemann (1852–1939).

[37]These problems were solved by Pierre Laurent Wantzel (1814–1848), Ferdinand Lindemann, and Carl Friedrich Gauss.

[38]The main persons involved were Auguste de Morgan (1806–1871), George Boole (1815–1864), later Charles Sanders Peirce (1839–1914), John Venn (1834–1923), William Stanley Jevons (1835–1882), and Ernst Schröder (1841–1902).

[39]The term Erlangen program refers to Felix Klein's (1849–1925) inaugural address as a newly appointed professor at the University of Erlangen in 1872.

[40]The calculations were done independently by John Couch Adams (1819–1892) and Urbain Jean Joseph Le Verrier (1811–1877).

[41]Following the advice of Adolf Hurwitz (1859–1919) and Hermann Minkowski (1864–1909), he presented only ten problems during the congress; three were related to the foundations of mathematics, four to arithmetic and algebra, and three to complex function theory and calculus of variations.

A French translation of Hilbert's text entitled Sur les problèmes futurs des mathématiques was published in *Compte Rendu* (1902, pp. 58–114). A German version entitled Mathematische Probleme was published in *Nachrichten von der Gesellschaft der Wissenschaften zu Göttingen, Math.-phys. Klasse*, 1900, pp. 253–297, and in the journal *Archiv der Mathematik und Physik* **1**(1901), pp. 44–63 and 213–237. See also Hilbert (2001, III, pp. 290–329). An English translation entitled Mathematical Problems appeared in the *Bulletin of the American Mathematical Society* **8**(1902), pp. 437–479, ISSN 0273-0979, and in Ewald (1996, pp. 1096–1105).

For a broader discussion see F.E. Browder (ed.), *Mathematical Developments Arising from Hilbert Problems*, AMS, Providence, 1976, xii + 628 pages, ISBN 0-8218-1428-1; M. Hazewinkel, *David Hilbert, Paper on Mathematical Problems (1901)*, Grattan-Guinness (2005, pp. 732–747).

P.S. Aleksandrov (ed.), *Problemy Gil'berta*, Nauka, Moskva, 1969, 239 pages. German translation: *Die Hilbertschen Probleme. Erläutert von einem Autorenkollektiv unter der Redaktion von*

research in mathematics throughout the 20th century. At the same congress, Henri Poincaré (1854–1912) expressed a deep satisfaction in the current state of mathematics. In his contribution *Du rôle de l'intuition et de la logique en mathématiques*, he stated:

> *Il ne reste plus aujourd'hui en Analyse que des nombres entiers ou des systèmes finis ou infinis de nombres entiers, reliés entre eux par un réseau de relations d'égalité ou d'inégalité.*
>
> *Les Mathématiques, comme on l'a dit, se sont arithmétisées.*
>
> *... On peut dire qu'aujourd'hui la rigueur absolue est atteinte.*[42]

The "mathematical optimism" at the turn of the 20th century stemmed from the success of 19th century mathematics, as well as joyful expectations of important results of the 20th century.[43]

3. Set Theory

The 19th century saw the birth of set theory. Bernard Bolzano, a mathematician and philosopher of Italian-German origin, who spent his whole life in Bohemia, arrived at some set-theoretic ideas during the final years of his life. His book *Paradoxien des Unendlichen*[44] was published only posthumously in 1851.

Bernard Bolzano fully supported the idea of the *actual infinity*. Thus, he followed the idea of the German mathematician, physicist, philosopher, lawyer and diplomat Gottfried Wilhelm Leibniz (1646–1716), who was one of the few mathematicians

---

P.S. Aleksandrov, 4. Auflage, Harri Deutsch, Frankfurt am Main, 2007, 302 pages, ISBN 978-3-8171-3401-4.

[42]H. Poincaré, Du rôle de l'intuition et de la logique en mathématiques, *Compte Rendu* (1902, pp. 115–130), see p. 120 and 122.

[43]For more information about the development of mathematics, see, e.g., M. Kline, *Mathematical Thought from Ancient to Modern Times*, Oxford University Press, New York, 1972, xvii + 1238 pages; a three-volume reprint: 1990, xvii + 390 + xxii, xvii + 391–812 + xxii, xv + 813–1211 + xxii pages, ISBN 0-19-506135-7, 0-19-506136-5, 0-19-506137-3; U.C. Merzbach and C.B. Boyer, *A History of Mathematics*, 3rd ed. Wiley, Hoboken, NJ, 2011, xx + 688 pages, ISBN 978-0-470-52548-7; I. Grattan-Guinness (ed.), *Companion Encyclopedia of the History and Philosophy of the Mathematical Sciences I., II.*, Routledge, London, 1994, xiii + 842, xi + 843–1806 pages, ISBN 0-415-03785-9; later edition: Johns Hopkins University Press, Baltimore, London, 2003, ISBN 0-8018-7396-7, 0-8018-7397-5.

[44]B. Bolzano, *Paradoxien des Unendlichen*, C.H. Reclam sen., Leipzig, 1851, xiii + 134 pages (ed. F. Příhonský); 2nd ed. Mayer und Müller, Berlin, 1889. New ed. Meiner, Hamburg, 2012, iii + 228 pages, ISBN 978-3-7873-2161-2; English translation: *Paradoxes of the Infinite*, Routledge & Kegan Paul, London, 1950; Routledge Revivals, 2014, 190 pages, ISBN 9781317748571.

advocating the concept of the actual infinity. Bolzano used the following idea of Leibniz as a motto of his book:

> *Je suis tellement pour l'infini actuel, qu'au lieu d'admettre, que la nature l'abhorre, comme l'on dit vulgairement, je tiens qu'elle l'affecte par-tout, pour mieux marquer les perfections de son Auteur.*[45]

However, it seems that Bolzano did not realize that infinity can have "different sizes". On December 7, 1873, this fact was noticed by the German mathematician Georg Cantor while studying trigonometric series, and reported on the same day to Richard Dedekind:

> *So glaube ich schließlich zum Grund gekommen zu sein, weshalb sich der in meinen früheren Briefen mit (x) bezeichnete Inbegriff nicht dem mit (n) bezeichneten eindeutig zuordnen läßt.*[46]

As soon as December 1873, Cantor finished and submitted his work *Ueber eine Eigenschaft des Inbegriffes aller reellen algebraischen Zahlen*,[47] which is nowadays considered to be the first publication in set theory, although the term set (*Menge*) never appears there.

In the period 1873–1884, Cantor developed the theory of cardinal and ordinal numbers, which is now referred to as the *naive*, *intuitive* or *Cantorian* set theory.[48]

---

[45]G.W. Leibniz, *Opera omnia studio Ludovici Dutens*, Tom. II, Part I, 1768, p. 243.

[46]In H. Meschkowski, W. Nilson (eds.), *Georg Cantor. Briefe*, Springer, Berlin, 1991, xi + 535 pages, see pp. 35–37, ISBN 3-540-50621-7. In a letter dated November 29, 1873, Cantor introduced the symbols (n) a (x) as follows: *Man nehme den Inbegriff aller positiven ganzzahligen Individuen n und bezeichne ihn mit (n); ferner denke man sich etwa den Inbegriff aller positiven reellen Zahlgrössen x und bezeichne ihn mit (x); ...* (p. 31).

See also E. Noether and J. Cavaillès (eds.), *Briefwechsel Cantor-Dedekind*, Hermann, Paris, 1937, 60 pages; I. Grattan-Guinness: The Rediscovery of the Cantor-Dedekind Correspondence, *Jahresbericht der Deutschen Mathematiker-Vereinigung* **76**(1974/1975), pp. 104–139, ISSN 0012-0456. See also Cantor (1980, pp. 443–451), and Ewald (1996, pp. 843–878 and 923–940).

We remark that Cantor's diagonal method appeared only later in his work Über eine elementare Frage der Mannigfaltigkeitslehre, *Jahresbericht der Deutschen Mathematiker-Vereinigung, Ergänzungsband*, **1**(1891), pp. 75–78, ISSN 0012-0456; see also Cantor (1980, pp. 278–281).

[47]G. Cantor, Ueber eine Eigenschaft des Inbegriffes aller reellen algebraischen Zahlen, *Journal für die reine und angewandte Mathematik* **77**(1874), pp. 258–262, ISSN 0075-4102, see also Cantor (1980, pp. 115–118); English translation: *On a Property of the Set of Real Algebraic Numbers*, Ewald (1996, pp. 839–843); French translation: Sur une propriété du système de tous les nombres algébriques, *Acta Mathematica* **2**(1883), pp. 305–310, ISSN 0001-5962.

[48]G. Cantor, Über unendliche lineare Punktmannigfaltigkeiten, *Mathematische Annalen* 15(1879), pp. 1–7, **17**(1880), pp. 355–358, **20**(1882), pp. 113–121, **21**(1883), pp. 51–58 and 545–586, **23**(1884), pp. 453–488, ISSN 0025-5831; see also Cantor (1980, pp. 139–246).

G. Cantor, *Grundlagen einer allgemeinen Mannigfaltigkeitslehre. Ein mathematischphilosophischer Versuch in der Lehre des Unendlichen*, Teubner, Leipzig, 1883, 46 pages; see also Cantor (1980, pp. 165–209), or Ewald (1996, pp. 878–919).

For a long time, he was struggling with the problem whether there exists an uncountable set whose cardinality is smaller than the continuum, i.e., the cardinality of the set of real numbers; the negative answer is known as the *continuum hypothesis*. In the 1890s, he published a longer two-part work entitled *Beiträge zur Begründung der transfiniten Mengenlehre*,[49] in which he summarized the essential results of his longstanding research in set theory.

The great initial aversion toward set theory and its creator was gradually overcome. Already in the end of the 19th century, set theory was greeted with considerable sympathy and acknowledgment, and became widely accepted in the beginning of the 20th century.[50] Almost all mathematics began to be developed on the basis of set theory. In the first decade of the 20th century, a number of mathematical disciplines were already built on top of set theory.

4. The Third Crisis in Mathematics

The turn of the 20th century witnessed some important and alerting discoveries. These included a number of *antinomies* in set theory, which cast doubt on its correctness. The new situation began to be perceived as a gap in the foundations of mathematics, and is sometimes referred to as the *third crisis in mathematics*.

The first antinomy, already known to Cantor in 1895, was published in 1897 by the Italian mathematician Cesare Burali-Forti (1861–1931) in his paper *Una questione sui numeri transfiniti*.[51] It is concerned with the set of all ordinal numbers, and reads as follows:

The ordinal number of the set of all ordinal numbers is greater than every ordinal number.[52]

---

[49]G. Cantor, Beiträge zur Begründung der transfiniten Mengenlehre, *Mathematische Annalen* **46**(1895), pp. 481–512, **49**(1897), pp. 207–246, ISSN 0025-5831; also in Cantor (1980, pp. 282–356). Italian translation (only 1st part): Contribuzione al fondamento della teoria degli insiemi transfiniti, *Rivista di matematica* **5**(1895), pp. 129–162, ISSN 0035-6298; French translation: Sur les fondements de la théorie des ensambles transfinis, *Mémoires de la Société des sciences physiques et naturelles de Bordeaux* **3**(1899), pp. 343–437; English translation: *Contributions to the Founding of the Theory of Transfinite Numbers*, The Open Court, 1915, ix + 211 pages; reprint: Dover, New York, 1955, ix + 214 pages, also 2012, ISBN 0-486-60045-9.

[50]This is evidenced by translations of Cantor's works to other languages.

[51]C. Burali-Forti, Una questione sui numeri transfiniti, *Rendiconti del Circolo matematico di Palermo* **11**(1897), pp. 154–164. See also the short note Sulle classi ben ordinate, *ibid.* p. 260, ISSN 0009-725X; English translation: *A Question on Transitive Numbers*, van Heijenoort (1967, pp. 104–112).

[52]See I.M. Copi, The Burali-Forti Paradox, *Philosophy of Science* **25**(1958), pp. 281–286, ISSN 0031-8248; G.H. Moore and A.R. Garciadiego, Burali-Forti's Paradox: A Reappraisal of its Origins, *Historia Mathematica* **8**(1981), pp. 319–350, ISSN 0315-0860; Ch. Menzel, Cantor and the Burali-Forti Paradox, *The Monist* **67**(1984), pp. 92–107, ISSN 0026-9662.

Perhaps the best known antinomy is *Russell's paradox*. It was discovered by Bertrand Russell (1872–1970) in 1901, and formulated in his 1903 book *The Principles of Mathematics*.[53]

Consider the set of all sets that are not members of themselves:

$$M = \{X \mid X \notin X\}.$$

According to this definition, if $M \in M$, then $M \notin M$, and if $M \notin M$, then $M \in M$. Both possibilities lead to a contradiction.[54] In the same period, Georg Cantor was already struggling for some time with the idea of the set of all sets, i.e., a similar antinomy.

Several other antinomies appeared shortly, and all of them led to a serious conclusion: the concept of a set in Cantor's theory is quite vague.[55]

5. The Struggle for Exactness, Axiomatization

Tendencies leading to the axiomatic development of various disciplines and mathematics as a whole were growing stronger throughout the end of the 19th century.

To begin with, we point out that the methodological principles of building a scientific theory were given by Aristotle of Stageira (384–322) in a collection of six writings, known under the later title *Organon* (*Categories*, *On Interpretation*, *Prior Analytics*, *Posterior Analytics*, *Topics*, *On Sophistical Refutations*). Euclid's

---

[53] B. Russell, *The Principles of Mathematics*, Cambridge University Press, Cambridge, 1903, xxix + 534 pages, see p. 101; later editions: Allen & Unwin, London, 1938, ..., 1964, xxxix + 534 pages; Norton & Comp., 1996, 576 pages, ISBN 978-0393314045.

[54] See I. Grattan-Guinness, How Bertrand Russell Discovered his Paradox, *Historia Mathematica* **5**(1978), pp. 127–137, ISSN 0315-0860; A.R. Garciadiego, *Bertrand Russell and the Origins of the Set-Theoretic Paradoxes*, Birkhäuser, Basel, Boston, Berlin, 1992, xxix + 264 pages, ISBN 3-7643-2669-7; G. Link (ed.), *One Hundred Years of Russell's Paradox*, W. de Gruyter, Berlin, NY, 2004, ix + 662 pages, ISBN 3-11-017438-3.

[55] For more details about the birth and development of set theory, the antinomies, third crisis in mathematics and the efforts to overcome it, see F.A. Medvedev, *Razvitie teorii množestv v XIX veke*, Nauka, Moskva, 1965, 232 pages; H. Meschkowski, *Probleme des Unendlichen. Werk und Leben Georg Cantors*, Vieweg, Braunschweig, 1967, xii + 288 pages; P.E. Johnson, *A History of Set Theory*, Prindle, Weber & Schmidt, Boston, 1972, ix + 109 pages; A.A. Fraenkel, Y. Bar-Hillel and A. Levy (with the collaboration of D. van Dalen), *Foundations of Set Theory*, 2nd ed. North-Holland, Amsterdam, 1973, reprint 1984, x + 404 pages, ISBN 0-7204-2270-1 (1st ed. 1958, x + 415 pages); H. Meschkowski, *Hundert Jahre Mengenlehre*, Deutscher Taschenbuch, München, 1973, 160 pages, ISBN 978-3423041423; W. Purkert, Georg Cantor und die Antinomien der Mengenlehre, *Bulletin de la Société Mathématique de Belgique* **38**(1986), pp. 313–327, ISSN 0037-9476; W. Purkert and H.J. Ilgauds, *Georg Cantor. 1845–1918*, Birkhäuser, Basel, Boston, Stuttgart, 1987, 262 pages, ISBN 3-7643-1770-1 (1st ed. Teubner, Leipzig, 1985, 135 pages); J.W. Dauben, *Georg Cantor. His Mathematics and Philosophy of the Infinite*, Princeton University Press, Princeton, NJ, 1991, 424 pages, ISBN 9780691024479; J.W. Dauben, *Georg Cantor, Paper on the Foundations of a General Set Theory (1883)*, Grattan-Guinness (2005, pp. 600–612). See also Ewald (1996, pp. 838–940), Grattan-Guinness (2000) and van Heijenoort (2002).

Elements (Ancient Greek: *Stoicheia*) written around 300 B.C. followed these principles, and became perhaps the most famous mathematical text in history. It begins by introducing the basic notions, and proceeds with the formulation of postulates that describe the principles of straightedge and compass constructions (known as the Euclidean constructions). Only then follows a rigorously developed theory consisting of definitions, theorems, and their proofs. The whole work is written in a purely geometric spirit; even results belonging to arithmetic and number theory are presented in the language of geometry.

For more than two millenia, Euclid's approach served as a model for mathematical texts. Moreover, it has influenced philosophers, authors of non-mathematical writings, who respected mathematics, tried to mimic the axiomatic approach, and presented their ideas in a *geometric manner*, or using a *geometric method.* Perhaps the most important one was the Dutch philosopher Baruch (Benedictus) Spinoza (1632–1677), author of *Renati des Cartes Principia philosophiae more geometrico demonstrata* (1663) and *Ethica ordine geometrico demonstrata* (1677 in *Opera posthuma*). Their structure corresponds to a mathematical text, i.e., follows the axiomatic-deductive development. It begins with definitions and axioms, proceeds to theorems, proofs, and corollaries. Let us quote a very interesting introductory paragraph by Lodewijk Meyer (1629–1681) from the foreword to *Renati des Cartes Principia philosophiae*:

> *Mathematicorum in scientiis investigandis ac tradendis methodum, qua nempe ex* **definitionibus, postulatis** *atque* **axiomatis conclusiones** *demonstrantur, optimam esse tussimamque veritatis indagandae atque docendae viam, omnium, qui supra vulgus sapere volunt, uninamis est sententia. Et quidem iure merito. Etenim quum omnis rei ignotae certa ac firma cognitio non nisi ex certo praecognitis hauriri ac derivari queat, haec neccessario ab imo praestruenda erunt, tanquam stabile fundamentum, cui postmodum, ne sponte subsidat aut minimo impetu pessum eat, totum cognitionis humanae aedificium superimponatur. Istius autem, notae esse quae passim matheseos cultoribus nomine definitionum, postulatorum atque axiomatum venire solent, nemini dubium esse poterit qui nobilem istam disciplinam a limine tantum salutaverit.* **Definitiones** *enim nihil aliud sunt quam terminorum atque nominum, quibus res tractandae designantur, apertissimae explicationes.* **Postulata** *autem et* **axiomata** *seu communes animi notiones adeo clarae atque perspicuae sunt enuntiationes, ut iis omnes, qui ipsa vocabula solummodo recte intellexerint, assensum negare nequaquam possint.*[56]

[56]English translation: *It is admitted by all who have any claims to superior intelligence that the method of mathematics, viz., the method by which conclusions are demonstrated from definitions, postulates, and axioms is the best method of obtaining and imparting truth. And rightly so; for as certain knowledge of an unknown object can only be obtained through facts previously known, there must of necessity be certain premises on which the whole superstructure of human knowledge rests, provided it does not fall of its own weight, or succumb to some slight attack from without. No one*

It seems that even Jan Amos Comenius (1592–1670), called "the teacher of nations", was inspired by the brevity, clarity and structure of mathematical texts. In his charming and intellectually rich treatise *The Analytical Didactic* (1633), he formulated 186 pedagogical principles (theorems), each of which is followed by a paragraph with a concise and clear explanation (proof).[57]

From the 17th until the 19th century, the key figures who placed a major emphasis on the rigorous development of mathematical disciplines, mathematics as a whole, deductive principles and proofs of mathematical results included René Descartes (1596–1650), Gottfried Wilhelm Leibniz, Auguste de Morgan, George Boole, William Stanley Jevons, Charles Sanders Peirce, and Ernst Schröder.[58]

The successful process of developing rigorous foundations of mathematical analysis inspired thoughts on the exact development of other mathematical disciplines. The discovery of the non-Euclidean geometries, continuous nowhere differentiable functions, space-filling curves and other surprising phenomena led to deep reflection on the intuition, choice of axiom systems, and character of axioms. The view of axioms as simple assertions, which correspond to real-world observations and are self-evident and intuitively clear, has been strongly disputed. During the end of the 19th century, mathematicians were trying to obtain new characterizations of the notions of *axiom*, *axiomatic system*, *deductive reasoning*, *proof*, *truth*, *consistency*, etc. Logic began to play a major role in mathematics.

Already in the 1870s, under the influence of Dedekind's ideas, the German mathematician and philosopher Gottlob Frege (1848–1925) occupied himself intensely with the problem of logical foundations of mathematics. He was the first pioneer of *logicism*, a direction that considered mathematics as a branch of logic, and intended to present it exclusively by means and concepts of logic.

In 1879 he published the book *Begriffsschrift, eine der arithmetischen nachgebildete Formelsprache des reinen Denkens*,[59] in which he developed the axiomatic foundations of symbolic logic (predicate calculus) and outlined his intention — to build the arithmetic on top of logic, i.e., to prove arithmetic theorems by means of pure logic. He wrote in the foreword:

---

who has paid any attention to the noble study of mathematics can doubt its definitions or postulates or axioms. For definitions are but a very open explanation of the terms and names under which the subject is discussed, and the postulates and axioms of mathematics, or the general ideas of the mind, cannot be denied by any one who understands the use of his vocabulary. In B. Spinoza, *The Principles of Descartes' Philosophy*, Translated from the Latin, with an Introduction by Halbert Hains Britan, Open Court, La Salle, Illinois, 1961, lxxxi + 177 pages, see p. 1.

[57] *The Analytical Didactic of Comenius*, University of Chicago Press, Chicago, 1953, 239 pages.

[58] For instance, let us mention Descartes' treatises *Discours de la méthode*, *Regulae ad directionem ingenii*, and Leibniz's idea of *mathesis universalis*.

[59] G. Frege, *Begriffsschrift ...*, L. Nebert, Halle a. S., 1879, x + 88 pages; later editions: 1964, 1988; English translation: *Begriffsschrift, a Formula Language, Modeled upon that of Arithmetic, for Pure Thought*, van Heijenoort (2002, pp. 1–82).

> *Sie* [Begriffsschrift] *soll also zunächst dazu dienen, die Bündigkeit einer Schlusskette auf die sicherste Weise zu prüfen und jede Voraussetzung, die sich unbemerkt einschleichen will, anzuzeigen, damit letztere auf ihren Ursprung untersucht werden könne...*
>
> *Die Arithmetik ... ist der Ausgangspunkt des Gedankenganges gewesen, der mich zu meiner Begriffsschrift geleitet hat. Auf diese Wissenschaft denke ich sie daher auch zuerst anzuwenden, indem ich ihre Begriffe weiter zu zergliedern und ihre Sätze tiefer zu begründen suche.*[60]

Five years later, he published the book *Die Grundlagen der Arithmetik, eine logisch-mathematische Untersuchung über den Begriff der Zahl*,[61] in which he clarified the notions of a *number* and a *cardinal number* (*Zahl* and *Anzahl*), and outlined his program of reducing mathematics to logic. In doing so, he abandoned the symbolic notation introduced in his *Begriffsschrift*, expressed definitions and theorems in a natural language, and sketched only basic ideas of proofs.

In 1893 and 1903, he published the two-volume book *Grundgesetze der Arithmetik, begriffsschriftlich abgeleitet*.[62] Starting from a formal logic system, more advanced than in *Begriffsschrift*, he presented a finite system of axioms for the arithmetic and, in the language of his predicate calculus, derived the basic theorems of integer arithmetic, which he subsequently extended to real numbers and opened up a way to the logical foundations of mathematical analysis. He gave detailed proofs by means of his formal logic. The introduction begins with the following words (see p. 1):

> *In meinen Grundlagen der Arithmetik habe ich wahrscheinlich zu machen gesucht, dass die Arithmetik ein Zweig der Logik sei und weder der Erfahrung noch der Anschauung irgendeinen Beweisgrund zu entnehmen brauche.*

---

[60] G. Frege, *Begriffsschrift und andere Aufsätze*, Olms, Hildesheim, Zürich, New York, 2007, 124 pages, see pp. iv and viii, ISBN 978-3-487-00623-9.

[61] G. Frege, *Die Grundlagen der Arithmetik ...*, Koebner, Breslau, 1884, vii + xi + 118 pages; later edition: Meiner, Hamburg, 1988, xxii + 144 pages, ISBN 3-7873-0719-2.

English translation: *The Foundations of Arithmetic. A Logico-Mathematical Enquiry into the Concept of Number*, Basil Blackwell & Mott, Oxford, 1950, xxiv + xxii + 119 + 119 pages (German and English); Northwestern University Press, Evanston, Illinois, 1980, xi + 119 pages, ISBN 08101-0605-1. New English ed.: *The Foundations of Arithmetic. A Logical-Mathematical Investigation into the Concept of Number 1884. Translated with an Introduction and Critical Commentary by Dale Jacquette*, Pearson Education, New York, 2007, xxxii + 112 pages, ISBN 0-321-24189-4.

French translation: *Les fondements de l'arithmétique*, L'Ordre philosophique, Seuil, 1970, 240 pages, ISBN 978-2020027366.

[62] G. Frege, *Grundgesetze der Arithmetik ...*, Pohle, Jena, 1893, 1903, xxxii + 253, xv + 265 pages, reprint: Olms, Hildesheim, 1962; English translation: *The Basic Laws of Arithmetic*, Oxford University Press, 2016, 680 pages, ISBN 978-0198777304.

See M. Potter, *Reason's Nearest Kin, Philosophies of Arithmetic from Kant to Carnap*, Oxford University Press, Oxford, New York, 2000, x + 305 pages, ISBN 0-19-925261-0.

Shortly before the second volume went to press, Gottlob Frege learned from Bertrand Russell about his paradox.[63] His faith in the exactness of mathematics was shattered (he himself used the notion of the set of all sets). Thus, he wrote in the conclusion (p. 253) of his book:

> *Einem wissenschaftlichen Schriftsteller kann kaum etwas Unerwünschteres begegnen, als dass ihm nach Vollendung einer Arbeit eine der Grundlagen seines Baues erschüttert wird. In diese Lage wurde ich durch einen Brief des Herrn Bertrand Russell versetzt, als der Druck dieses Bandes sich seinem Ende näherte.*

During the end of the 19th century, Giuseppe Peano, a world-leading researcher in logic, was one of the first thinkers striving for a consistent axiomatization of mathematics, in a way that was more rigorous and abstract than before. In his 1888 book *Calcolo geometrico*,[64] he introduced the axioms of a vector space. The axioms of natural numbers (*Peano's axioms*) followed one year later in the treatise *Arithmetices principia, nova methodo exposita*;[65] on many pages, he presented the arithmetic only by means of formulas. In the years 1895–1908, he worked on a project called *Formulaire de mathématiques* (or *Formulario mathematico*), transcribing mathematics into his formal language, i.e., the language of formulas. His work *I principii di geometria logicamente esposti*[66] was written in a similar spirit.

We remark that a description of natural numbers similar to the one by Peano was given by Richard Dedekind in his book *Was sind und was sollen die Zahlen?*[67] which was published in 1888.[68]

---

[63]For the Russell-Frege correspondence see van Heijenoort (2002, pp. 124–128).

[64]G. Peano, *Calcolo geometrico secondo l'Ausdehnungslehre di H. Grassmann, preceduto dalle operazioni della logica deduttiva*, Fratelli Bocca Editori, Torino, 1888, xi + 171 pages; partially reprinted in Peano (1957–1959, II, pp. 3–19); English translation: *Geometric Calculus. According to the Ausdehnungslehre of H. Grassmann*, Birkhäuser, Boston, Basel, Berlin, 2000, xv + 150 pages, ISBN 0-8176-4126-2.

[65]G. Peano, *Arithmetices Principia ...*, Fratres Bocca, Torino, 1889, xvi + 20 pages, also in Peano (1957–1959, II, pp. 20–55; English translation: *The Principles of Arithmetic, Presented by a New Method*, van Heijenoort (2002, pp. 83–97). See also Peano's paper Sul concetto di numero, *Rivista di matematica* **1**(1891), pp. 87–102 and 256–267, ISSN 0035-6298.

[66]G. Peano, *I principii di geometria logicamente esposti*, Fratelli Bocca Editori, Torino, 1889, 44 pages; new ed. Kessinger, 2010, 42 pages, ISBN 978-1162494586. Also in Peano (1957–1959, II, pp. 56–91).

[67]R. Dedekind, *Was sind und was sollen die Zahlen?*, Vieweg, Braunschweig, 1888, xvii + 58 pages; 10th ed. 1965; English translation: *The Nature and Meaning of Numbers*, Open Court, Chicago, 1901; reprint: Dover, New York, 1963; also in Ewald (1996, pp. 787–833), more details in pp. 753–837.

[68]See J. Ferreirós, *Richard Dedekind (1888) and Giuseppe Peano (1889), Booklets on the Foundations of Arithmetic*, Grattan-Guinness (2000, pp. 611–626). See also H.C. Kennedy, The Origins of Modern Axiomatics: Pasch to Peano, *The American Mathematical Monthly* **79**(1972), pp. 133–136, ISSN 0002-9890; M. Segre, Peano's Axioms in their Historical Context, *Archive for History of Exact Sciences* **48**(1994), pp. 201–342, ISSN 0003-9519.

A major accomplishment in the axiomatic approach to mathematics was due to David Hilbert. In his 1899 book *Grundlagen der Geometrie*,[69] he carried out a rigorous axiomatization of elementary geometry. He followed the ideas outlined in 1882 by Moritz Pasch (1843–1930) in his book *Vorlesungen über neuere Geometrie*.[70] Hilbert's axiomatic system was structured as follows: eight axioms of incidence, four axioms of order, five axioms of congruence, two axioms of continuity, and one axiom of parallels. While Euclid's Elements relied on geometric imagination and many results had a constructive character, Hilbert abandoned the real world and developed his geometry in a purely formal and existential way.

He wrote in the introduction to *Grundlagen der Geometrie*:

> *Die vorliegende Untersuchung ist ein neuer Versuch, für die Geometrie ein vollständiges und möglichst einfaches System von Axiomen aufzustellen und aus denselben die wichtigsten geometrischen Sätze in der Weise abzuleiten, daßdabei die Bedeutung der verschiedenen Axiomgruppen und die Tragweite der aus den einzelnen Axiomen zu ziehenden Folgerungen möglichst klar zu Tage tritt.*

Lobachevsky, Bolyai, Gauss, Riemann, and Klein opened up a surprisingly rich world of geometries. In his book *Grundlagen der Geometrie*, David Hilbert no longer relied on the real world, although it provided an inspiration for his axiomatic system. In a strictly logical manner, he derived the consequences of axioms describing the relations between the primitive notions, proved one theorem after another, and developed the geometry corresponding to the chosen axiomatic system. He did not care about the relation between this geometry and the real world surrounding us — the

---

[69]D. Hilbert, *Grundlagen der Geometrie*, in Festschrift zur Feier der Enthüllung des Gauss-Weber Denkmals in Göttingen, Teubner, Leipzig, 1899, pp. 1–92; reprint: EOD Network, 2015, 94 pages, ISBN 978-3226016925. 2nd ed. 1903, v + 175 pages, 3rd. ed. 1909, vi + 279 pages, 7th ed. 1930, vii + 326 pages, 12th ed. Teubner, Stuttgart, 1977, vii + 271 pages, ISBN 3-519-22020-2.

English translation: *The Foundations of Geometry*, Open Court, Chicago, 1902, vii + 132 pages; later edition: BiblioLife, Charleston, SC, 2008, 144 pages, ISBN 9780559740640; French translations: Les principes fondamentaux de la géométrie, *Annales Scientifiques de l'École normale superieure* **17**(1900), pp. 103–209; *Les fondements de la géométrie*, 1902; Russian translation of the 7th ed.: *Osnovanija geometrii*, OGIZ, Moska, Leningrad, 1948, 491 pages.

See A. Schmidt, *Zu Hilberts Grundlegung der Geometrie*, in Hilbert (2001, II, pp. 404–414); J. Gray, *Hilbert and his Grundlagen der Geometrie*, in J. Gray, *Worlds Out of Nothing. A Course in the History of Geometry in the 19th Century*, Springer, London, 2007, xxiii + 376 pages, also 2010, xxiv + 400 pages, pp. 259–267, ISBN 978-0-85729-059-5; M.-M. Toepell, *Über die Entstehung von David Hilberts Grundlagen der Geometrie*, Vandenhoeck & Ruprecht, Göttingen, 1986, 293 pages, ISBN 978-3525403099; M.-M. Toepell, *David Hilbert, Grundlagen der Geometrie, 1st ed. (1899)*, Grattan-Guinness (2005, pp. 710–723).

[70]M. Pasch, *Vorlesungen über neuere Geometrie*, Teubner, Leipzig, 1882, iv + 201 pages. The second edition (Springer, Berlin, 1926, x + 275 pages) includes a treatise by Max Dehn (1878–1952) entitled *Die Grundlegung der Geometrie in historischer Entwicklung*, pp. 185–271; reprint: 1976, ISBN 3-540-06294-7. See also D. Tamari, *Moritz Pasch (1843–1930), Vater der modernen Axiomatik*, Shaker, Aachen, 2007, 360 pages, ISBN 978-3832262440.

only things essential were the consistency of his theory, and the independence of the axioms.[71]

David Hilbert emphasized the new interpretation of axioms and geometry even in his lectures. Occasionally he stated that the primitive notions can be interpreted in an arbitrary manner, and only the relations between them are essential:

One must be able to say at all times — instead of points, straight lines, and planes — tables, chairs, and beer mugs.[72]

Thus, the turn of the 20th century witnessed significant efforts toward a rigorous axiomatization of mathematics. They were successfully realized mainly in the arithmetic and geometry, as evidenced by the works of Frege, Peano, Dedekind, and Hilbert. Modern mathematical logic played and important role (Frege, Peano).

6. The Rise of Set Theory

During the first two decades of the 20th century, set theory became firmly embedded in mathematics, despite the controversies raised by the existence of the antinomies. This claim is easily supported by references to journal articles and books devoted to the new theory.

The German mathematician Gerhard Hessenberg (1874–1925) published in 1906 the work *Grundbegriffe der Mengenlehre*,[73] the married couple of William Henry Young (1863–1942) and Grace Chisholm Young (1868–1944) published in the same year the book *The theory of sets of points*,[74] and the Polish mathematician Wacław Sierpiński (1882–1969) wrote the book *Zarys teoryi mnogości*,[75] which was printed in 1912.

In 1900, Arthur Schoenflies (1853–1928) published a comprehensive treatise entitled *Die Entwickelung der Lehre von den Punktmannigfaltigkeiten*,[76] followed by a 1908 sequel bearing the same title,[77] and few years later by a modified version of the first volume entitled *Entwickelung der Mengenlehre und ihrer Anwendungen*,[78]

[71]The title of the second chapter of Hilbert's book is *Die Widerspruchslosigkeit und Gegenseitige Unabhängigkeit der Axiome.*

[72]See Reid (1996, p. 57).

[73]G. Hessenberg, Grundbegriffe der Mengenlehre, *Abhandlungen der Friesschen Schule*, Neue Folge 1, Heft IV, 1906, pp. 479–706; also in book form: Vandenhoeck und Ruprecht, Göttingen, 1906, viii + 220 pages.

[74]W.H. Young and G.C. Young, *The Theory of Sets of Points*, Cambridge University Press, Cambridge, 1906, xii + 316 pages; reprinted: 2009, xiv + 316 pages, ISBN 9780511694240.

[75]W. Sierpiński, Zarys teorji mnogości, *Biblioteka Matematyczno-fizycna* **9**, Warszawa, 1912, 158 pages.

[76]A. Schoenflies, Die Entwickelung der Lehre von den Punktmannigfaltigkeiten, *Jahresbericht der Deutschen Mathematiker-Vereinigung, Ergänzungsband*, **8**(1900), pp. 1–250, ISSN 0012-0456.

[77]A. Schoenflies, *Die Entwickelung ...*, Teubner, Leipzig, 1908, x + 331 pages.

[78]A. Schoenflies, *Entwickelung der Mengenlehre und ihrer Anwendungen*, Teubner, Leipzig, 1913, xi + 389 pages.

completed with the help of Hans Hahn (1879–1934).[79] Arthur Schoenflies also wrote the treatise *Mengenlehre* (1899) published in the Encyklopädie der mathematischen Wissenschaften,[80] as well as several works related to the axiomatization of set theory and the antinomies.[81]

One of the first textbooks on set theory, *Grundzüge der Mengenlehre*, was published in 1914 by the prominent German mathematician Felix Hausdorff (1868–1942),[82] who wrote important treatises on ordered sets already in the period 1906–1909. Another successful textbook was *Einleitung in die Mengenlehre. Eine gemeinverständliche Einführung in das Reich des unendlichen Größen*[83] written by Abraham Adolf Fraenkel (1891–1965) and published in 1919. We also recall the little book *Mengenlehre* written by Erich Kamke (1890–1961) for the popular series Sammlung Göschen.[84]

The textbooks and monographs provide evidence that throughout the first two decades of the 20th century, set theory secured a firm position in mathematics.

---

[79]Hans Hahn took great interest in set theory and the problems of infinity, as evidenced by some of his works. H. Hahn *et al.*, Diskussion zur Grundlegung der Mathematik, *Erkenntnis* **2**(1931), pp. 135–149. H. Hahn, *Die Krise der Anschauung*, in *Krise und Neuaufbau in der exakten Wissenschaften. Fünf Wiener Vorträge*, Deuticke, Wien, Leipzig, 1933, ii + 123 pages, pp. 41–64; H. Hahn, *Gibt es Unendliches?*, in *Alte Probleme — Neu Lösungen in den exakten Wissenschaften. Fünf Wiener Vorträge*, Deuticke, Wien, Leipzig, 1934, 122 pages, pp. 93–116. Both papers were reprinted in the book H. Hahn, *Empirismus, Logik, Mathematik*, B.F. McGuinness, ed., Suhrkamp, Frankfurt am Main, 1988, 179 pages, ISBN 351828245X; English version: *Empiricism, Logic and Mathematics. Philosophical Papers*, Reidel, Dordrecht, 1980, xx + 142 pages, pp. 73–102, 103–131, ISBN 978-90-277-1065-9 (*The Crisis in Intuition, Does the Infinite Exist?*).

[80]Bd. I-1. *Arithmetik und Algebra*, Teubner, Leipzig, 1900–1904, pp. 184–207.

[81]A. Schoenflies, Über die logischen Paradoxien der Mengenlehre, *Jahresbericht der Deutschen Mathematiker-Vereinigung* **15**(1906), pp. 19–25, ISSN 0012-0456. A. Schoenflies, Zur Grundlegung der Mengenlehre, *Mathematische Annalen* **72**(1912), pp. 551–561. A. Schoenflies, Zur Axiomatik der Mengenlehre, *ibid.* **83**(1921), pp. 173–200. A. Schoenflies, Bemerkung zur Axiomatik der Größen und Mengen, *ibid.* **85**(1922), pp. 60–64, ISSN 0025-5831. A. Schoenflies, Die Krisis in Cantor's mathematischen Schaffen, *Acta Mathematica* **50**(1927), pp. 1–23.

[82]F. Hausdorff, *Grundzüge der Mengenlehre*, Veit, Leipzig, 1914, viii + 476 pages. Page iii of this book contains the following dedication: *Dem Schöpfer der Mengenlehre Herrn Georg Cantor in dankbarer Verehrung gewidmet.* Reprints: Chelsea, New York, 1949, 1965 and 1978, ISBN 0-8284-0061-X. The second edition appeared in 1927 under the shorter title *Mengenlehre* (285 pages); 3rd ed. W. de Gruyter, Berlin, Leipzig, 1935, 307 pages, reprinted 1944 in New York (Dover); 4th ed. 1991, 352 pages, ISBN 0-8284-1119-0; English translation: *Set theory*, Chelsea, New York, 1957, 352 pages; later editions: 1962, 1978, 1991; AMS, Chelsea, 2005, ISBN 978-0821838358.

[83]A.A. Fraenkel, *Einleitung in die Mengenlehre. ...*, Springer, Berlin, 1919, vi + 156 pages, 2nd ed. 1923, viii + 251 pages, 3rd ed. 1928, xiii + 424 pages; reprint: Dover, New York, 1946.

Other Fraenkel's books related to set theory are: *Abstract Set Theory*, North-Holland, Amsterdam, 1953, xii + 479 pages, 2nd ed. 1961, viii + 295 pages, 3rd ed. 1966, viii + 297 pages, 4th ed. (ed. Azriel Levy), 1976, x + 295 pages, ISBN 0-444-11032-1; *Foundations of Set Theory* (with Y. Bar-Hillel), North-Holland, Amsterdam, 1958, x + 415 pages, 2nd ed. 1973, 1984, x + 404 pages, ISBN 0-7204-2270-1; *Set Theory and Logic*, Addison-Wesley, Reading, Mass., 1966, vii + 102 pages.

[84]Volume 999, W. de Gruyter, Leipzig, Berlin, 1928, 160 pages. The book saw its fifth edition in 1965, and was translated to English.

7. Way Out of the Crisis

During the first three decades of the 20th century, mathematicians, logicians and philosophers were bothered by the state of set theory after the discovery of the antinomies. Many of them were involved in solving problems of the rigorous foundations of mathematics, problems of the axiomatization of individual disciplines and mathematics as a whole, and took part in discussions on consistency and certainty in mathematics. They still felt the crisis in the foundations of mathematics, and therefore sought ways out of this uncomfortable situation.

Logicism was an important program created in response to the crisis in the foundations of mathematics. It followed the ideas presented in the works of Gottlob Frege. According to logicists, mathematics had to be regarded as a branch of logic, all mathematical notions had to be defined by means of logic, and all mathematical theorems had to be proved strictly by means of the inference rules from logic. Thus, the problem of the consistency of mathematics was reduced to the problem of the consistency of modern symbolic logic.

Bertrand Russell characterized "pure" mathematics already in the first short chapter of his 1903 book *The Principles of Mathematics*. The chapter *Definition of pure mathematics* begins with these words (see p. 3):

> *Pure Mathematics is the class of all propositions of the form "p implies q", where p and q are propositions containing one or more variables, the same in the two propositions, and neither p nor q contains any constants except logical constants...*

Later he stated (see p. 5):

> *The fact that all Mathematics is Symbolic Logic is one of the greatest discoveries of our age; and when this fact has been established, the remainder of the principles of mathematics consists in the analysis of Symbolic Logic itself.*

Bertrand Russell and Alfred North Whitehead (1861–1947) attempted a revision of set theory. In the period 1910–1913, they published the three-volume monograph *Principia Mathematica*,[85] which represents a basic cornerstone of logicism. They created a new axiomatic system and introduced the *theory of types*, which made it possible to avoid the known antinomies using a complicated hierarchy of types. They were able to build a major part of mathematics on top of logic in the spirit of

[85]B. Russell and A.N. Whitehead, *Principia Mathematica*, Cambridge University Press, Cambridge, UK, 1910, 1912, 1913, xiii + 666, xxxiv + 772, x + 491 pages; 2nd ed. 1925–1927, xlvi + 674, xxxi + 742, viii + 491 pages; reprints: 1935, ..., 1997, ISBN 0-521-62606-4.

Frege's ideas.[86] They supposed that their *Principia Mathematica* will become the foundation for all mathematics.[87]

In the book *My Philosophical Development*, Bertrand Russell expressed the visions that he shared with Whitehead while compiling their work by the following words:

> *The primary aim of Principia Mathematica was to show that all pure mathematics follows from purely logical premisses and uses only concepts definable in logical terms.*

Russell's and Whitehead's extensive work influenced the development of logic and found response among philosophers, but did not quite succeed among mathematicians. Both authors were quite disappointed. Bertrand Russell wrote:

> *Both Whitehead and I were disappointed that Principia Mathematica was only viewed from a philosophical standpoint. People were interested in what was said about the contradictions and in the question whether ordinary mathematics had been validly deduced from purely logical premisses, but they were not interested in the mathematical techniques developed in the course of the work. I used to know of only six people who had read the later parts of the book.*[88]

Other supporters of the logicist program included the logician and philosopher Rudolf Carnap (1891–1970)[89] and the American mathematician, logician and philosopher Willard van Orman Quine (1908–2000),[90] who simplified the theory of types.

---

[86]See, e.g., B. Russell, On Some Difficulties in the Theory of Transfinite Numbers and Order Types, *Proceedings of the London Mathematical Society* **4**(1906), pp. 29–53, ISSN 0024-6115. B. Russell, Mathematical Logic as Based on the Theory of Types, *American Journal of Mathematics* **30**(1908), pp. 222–262, ISSN 0002-9327, also in van Heijenoort (2002, pp. 150–182).

[87]See I. Grattan-Guinness, *A.N. Whitehead and Bertrand Russell, Principia Mathematica, First Edition (1910–1913)*, Grattan-Guinness (2005, pp. 784–794).

[88]B. Russell, *My Philosophical Development*, Allen & Unwin, London, 1959, 279 pages, see p. 74 and 86. In chapter seven Russell describes, among other things, the discovery of the paradox bearing his name.

[89]R. Carnap, Die logizistische Grundlegung der Mathematik, *Erkenntnis* **2**(1931), pp. 91–105. R. Carnap, *Logische Syntax der Sprache*, Springer, Wien, 1934, xi + 274 pages; later edition: 1968; English translation: *The Logical Syntax of Language*, Kegan Paul, Trench, Trubner, London, 1937, xvi + 352 pages; later edition: 2002, ISBN 0812695240. R. Carnap, *Logical Foundations of the Unity of Science*, 1938. R. Carnap, *Foundations of Logic and Mathematics*, Chicago University Press, Chicago, 1939, viii + 71 pages; later editions: 1947, 1950, 1970.

[90]W. Quine, New Foundations for Mathematical Logic, *The American Mathematical Monthly* **44**(1937), pp. 70–80, ISSN 0002-9890. W. Quine, On the Theory of Type, *The Journal of Symbolic Logic* **3**(1938), pp. 125–139, ISSN 0022-4812. W. Quine, On What There is, *The Review of Metaphysics* **2**(1948), no. 5, pp. 21–38, ISSN 0034-6632. W. Quine, *From a Logical Point of View. 9 Logico-Philosophical Essays*, Cambridge University Press, Cambridge, Mass., 1953, vi + 184 pages; later editions: 1961, 1964, 1971.

Another approach created in response to problems arising from the antinomies was the *intuitionism.* The intuitionists, whose leading representative was the Dutch mathematician Luitzen Egbertus Jan Brouwer (1881–1966),[91] criticized the foundations of classical mathematics, abandoned actual infinity, which has no counterpart in the real world, insisted on finitistic procedures, accepted only constructive proofs, and rejected existential ones. They criticized the set-theoretic, logicist as well as the developing formalist conceptions of mathematics, doubted the use of the law of excluded middle (*tertium non datur*) for infinite sets, and claimed that it is erroneous to deduce mathematics from logic, since mathematical logic is in fact a branch of mathematics. According to intuitionists, the existence of antinomies is a consequence of the concept of actual infinity, which has no intuitive basis.

Intuitionists were inspired by the ideas of Cantor's big opponent Leopold Kronecker,[92] whose goal was to develop all mathematics on top of the concept of

[91]The basic ideas of intuitionism were put forward already in Brouwer's dissertation *De onbetrouwbaarheid der logische principes* (*Tijdschrift voor Wijsbegeerte* **2**(1908), pp. 152–158), and later expanded in a number of works published mainly in the 1920s.

L.E. Brouwer, Intuitionism and Formalism, *Bulletin of the American Mathematical Society* **20**(1913), pp. 81–96, ISSN 0273-0979. L.E. Brouwer, Begründung der Mengenlehre unabhängig vom logischen Satz vom ausgeschlossenen Dritten I., II., *Verhandelingen der Kroninklijke Nederlandse Akademie van Wetenschappen te Amsterdam* **12**(1918/19), no. 5, pp. 1–43, no. 7, pp. 1–33. L.E. Brouwer, Intuitionistische Mengenlehre, *Jahresbericht der Deutschen Mathematiker-Vereinigung* **28**(1919), pp. 203–208, ISSN 0012-0456; English translation: *Intuitionist Set Theory,* Mancosu (1998, pp. 23–27). L.E. Brouwer, Besitzt jede reelle Zahl eine Dezimalbruchentwicklung?, *Mathematische Annalen* **83**(1921), pp. 201–210, ISSN 0025-5831; English translation: *Does Every Real Number Have a Decimal Expansion?*, Mancosu (1998, pp. 28–35). L.E. Brouwer, Zur Begründung der intuitionistischen Mathematik, I., II., III., *Mathematische Annalen* **93**(1925), pp. 244–257, **95**(1926), pp. 453–472, **96**(1927), pp. 451–488, ISSN 0025-5831. L.E. Brouwer, Intuitionistische Zerlegung mathematischer Grundbegriffe, *Jahresbericht der Deutschen Mathematiker-Vereinigung* **33**(1925), pp. 251–256, **36**(1927), pp. 127–129, ISSN 0012-0456; English translation: *Intuitionist Splitting of the Fundamental Notions of Mathematics*, Mancosu (1998, pp. 290–292, 293–295). L.E. Brouwer, Intuitionistische Betrachtungen über den Formalismus, *Sitzungsberichte der Preussischen Akademie der Wissenschaften, Physikalisch-mathematische Klasse*, 1928, pp. 48–52; English translation: *Intuitionistic Reflections on Formalism*, Mancosu (1998, pp. 40–44), or van Heijenoort (2002, pp. 490–492). L.E. Brouwer, Mathematik, Wissenschaft und Sprache, *Monatshefte für Mathematik und Physik* **36**(1929) pp. 153–164; English translation: *Mathematics, Science, and Language*, Mancosu (1998, pp. 45–53), or Ewald (1996, pp. 1170–1185). L.E. Brouwer, *Die Struktur des Kontinuums*, Vienna, 1930; English translation: *The Structure of the Continuum*, Mancosu (1998, pp. 54–63), or Ewald (1996, pp. 1186–1196). For a broader context see Ewald (1996, pp. 1166–1207). L.E. Brouwer, *Collected Works I., II.*, North-Holland, Amsterdam, 1975, 1976, xv + 628, xxvii + 706 pages, ISBN 0-444-10474-7, ISBN 0-444-10643-X.

[92]L. Kronecker, Grundzüge einer arithmetischen Theorie der algebraischen Grössen, *Journal für die reine und angewandte Mathematik* **92**(1882), pp. 1–122, also in Kronecker (1968, II, pp. 237–387). L. Kronecker, Ueber den Zahlbegriff, *ibid.* **101**(1887), pp. 337–355, ISSN 0075-4102, also in Kronecker (1968, III-1, pp. 251–274). See Ewald (1996, pp. 941–955).

a natural number.[93] They also followed some ideas of Henri Poincaré,[94] who, after the discovery of antinomies, rejected actual infinity, Cantor's set theory, Russell's logicism, and emphasized the major role of intuition.

> *Il n'y a pas d'infini actuel; ce que nous appelons l'infini, c'est uniquement la possibilité de créer sans cesse de nouveaux objets, quelque nombreux que soient les objets déjà créés.*[95]

Other scientists that might be ranked among intuitionists were the German mathematician Hermann Weyl (1885–1955), the French mathematicians Émile Borel (1871–1956) and Henri Lebesgue (1875–1941), the Russian mathematician Nikolai Nikolaevich Luzin (1883–1950), and especially the Dutch mathematician Arend Heyting (1898–1980).[96]

The intuitionist mathematics had to gave up many results of classical mathematics. This was perhaps the main reason why most mathematicians diverted from this direction. Even Brouwer and Weyl based a number of their results (topology, mathematical physics) on classical mathematics.

---

[93]Kronecker's statement from an 1886 meeting of the Berliner Naturforschender-Versammlung is well known and often quoted: *Die ganzen Zahlen hat der liebe Gott gemacht, alles andere ist Menschenwerk.* See H. Weber, Leopold Kronecker, *Mathematische Annalen* **43**(1893), pp. 1–25, see p. 15, ISSN 0025-5831.

[94]See H. Poincaré, Les mathématiques et la logique I., II., III., *Revue de métaphysique et de morale* **13**(1905), pp. 815–835, **14**(1906), pp. 17–34, 294–317; English translation in Ewald (1996, pp. 1021–1071).

Poincaré is the author of the treatises *La science et l'hypotèse*, *La valeur de la science* and *Science et méthode* (Paris, 1902, 1905, 1908), which were subsequently published in German under the titles *Wissenschaft und Hypothese*, *Der Wert der Wissenschaft*, *Wissenschaft und Methode* (Leipzig, 1904, 1906, 1914); English translation: *The Foundation of Science: Science and Hypothesis. The Value of Science. Science and Method*, The Science Pr., New York, 1929, xi + 553 pages. See Ewald (1996, pp. 972–1074).

[95]H. Poincaré, Réflexions sur les deux notes précédentes, *Acta Mathematica* **32** (1909), pp. 195–200, see p. 196. Poincaré's paper was a response to the earlier two articles: A. Schoenflies, Über eine Vermeintliche Antinomie der Mengenlehre, pp. 177–184, E. Zermelo, Sur les ensembles finis et le principe de l'induction complète, pp. 185–193.

[96]A. Heyting, Die formalen Regeln der intuitionistischen Logik, *Sitzungsberichte der Preussischen Akademie der Wissenschaften, Physikalisch-mathematische Klasse*, 1930, pp. 42–56; English translation: *The Formal Rules of Intuitionistic Logic*, Mancosu (1998, pp. 311–327). A. Heyting, Die formalen Regeln der intuitionistischen Mathematik, *ibid.* (1930, pp. 57–71, 158–169). A. Heyting, Sur la logique intuitionniste, *Académie Royale de Belgique, Bulletin de la Classe des Sciences* **16**(1930), pp. 957–963; English translation: *On Intuitionistic Logic*, Mancosu (1998, pp. 306–310). A. Heyting, Die intuitionistische Grundlegung der Mathematik, *Erkenntnis* **2**(1931), pp. 106–115. A. Heyting, *Mathematische Grundlagenforschung. Intuitionismus. Beweistheorie*, Springer, Berlin, 1934, iv + 73 pages, reprint: 1974, ISBN 3-540-06298-X; French translation: *Les fondements des mathématiques. Intuitionnisme. Théorie de la démonstration*, Gauthier-Villars, Paris, 1955. A. Heyting, *Intuitionism. An Introduction*, North-Holland, Amsterdam, 1956, viii + 132 pages; later editions: 1966, 1971, ix + 147 pages, 1976, ISBN 0-7204-2263-9.

David Hilbert was deeply preoccupied with the problems of rigorous foundations of mathematics. While working on his book *Grundlagen der Geometrie* in the end of 19th century, he was well aware of the important problem of consistency. After all, the second of his problems presented in Paris in 1900 was concerned with the consistency of the arithmetic. He presented his ideas on the construction of mathematics, consistency and completeness of axiomatic systems in his 1900 paper *Über den Zahlbegriff*[97]:

> *Es entsteht dann die notwendige Aufgabe, die Widerspruchslosigkeit und Vollständigkeit dieser Axiome zu zeigen, d.h. es muß bewiesen werden, daß die Anwendung der aufgestellten Axiome nie zu Widersprüchen führen kann, und ferner, daß das System der Axiome zum Nachweis aller geometrischen Sätze ausreicht. Wir wollen das hier eingeschlagene Untersuchungsverfahren die axiomatische Methode nennen.*[98]

At the Third International Congress of Mathematicians held in Heidelberg, 1904, he delivered a talk entitled *Über die Grundlagen der Logik und der Arithmetik*,[99] in which he again spoke about the rigorous foundations of mathematics and the consistency of the arithmetic:

> *Allein bei aufmerksamer Betrachtung werden wir gewahr, daß bei der hergebrachten Darstellung der Gesetze der Logik gewisse arithmetische Grundbegriffe, z.B. der Begriff der Menge, zum Teil auch der Begriff der Zahl, insbesondere als Anzahl, bereits zur Verwendung kommen. Wir geraten so in eine Zwickmühle, und zur Vermeidung von Paradoxien ist daher eine teilweise gleichzeitige Entwicklung der Gesetze der Logik und der Arithmetik erforderlich.*[100]

Hilbert came back to these questions in his 1918 paper *Axiomatisches Denken*,[101] which was based on a lecture delivered at the meeting of the Schweizerische mathematische Gesellschaft in Zürich, September 1917.

David Hilbert was the most important representative of *formalism*, a view that rejected any form of intuition, emphasized rigorous axiomatic foundations of

---

[97]D. Hilbert, Über den Zahlbegriff, *Jahresbericht der Deutschen Mathematiker-Vereinigung* **8**(1900), pp. 180–184, ISSN 0012-0456, also in Hilbert (1930), Anhang VI., pp. 241–246; English translation: *On the Concept of Number*, Ewald (1996, pp. 1089–1095).

[98]See Hilbert (1930, p. 242).

[99]D. Hilbert, *Über die Grundlagen der Logik und der Arithmetik*, Verhandlungen des dritten internationalen Mathematiker-Kongresses in Heidelberg vom 8. bis 13. August 1904, Teubner, Leipzig, 1905, pp. 174–185, also in Hilbert (1930, Anhang VII, pp. 247–261); English translation: *On the Foundations of Logic and Arithmetic*, van Heijenoort (2002, pp. 129–138).

[100]See Hilbert (1930, p. 250).

[101]D. Hilbert, Axiomatisches Denken, *Mathematische Annalen* **78**(1917), pp. 405–415, ISSN 0025-5831; also in Hilbert (2001, III, pp. 146–156); English translation: *Axiomatic Thought*, Ewald (1996, pp. 1105–1115).

mathematics, and insisted on the complete formalization of mathematics. A large number of mathematicians regarded this direction to be the best way out of the crisis in the foundations of mathematics.

A major success of formalism in the first two decades of the 20th century was the birth of the Zermelo-Fraenkel axiomatic system, the first rigorous axiomatic system of set theory. This system began to be developed by Ernst Zermelo (1871–1953) in the years 1904–1908, and was modified and completed by Abraham Adolf Fraenkel in the subsequent years (until 1921).[102]

The Zermelo-Fraenkel axiomatic system imposed more stringent requirements on the construction of sets, and thus eliminated all known antinomies. We recall that in 1904, Zermelo formulated the axiom of choice (*Auswahlsprinzip, Auswahlsaxiom*).[103]

At the same time, i.e., during the 1920s, David Hilbert successively formulated a program directed toward a formalized axiomatic development of mathematics, and

---

[102]E. Zermelo, Beweis, daß jede Menge wohlgeordnet werden kann, *Mathematische Annalen* **59**(1904), pp. 514–516, ISSN 0025-5831; English translation: *Proof that Every Set Can Be Well-ordered*, van Heijenoort (2002, pp. 139–141). E. Zermelo, Neuer Beweis für die Möglichkeit einer Wohlordnung, *ibid.* **65**(1908), pp. 107–128; English translation: *A New Proof of the Possibility of a Well-ordering*, van Heijenoort (2002, pp. 183–198). E. Zermelo, Untersuchungen über die Grundlagen der Mengenlehre I., *ibid.* **65**(1908), pp. 261–281; English translation: *Investigations in the Foundations of Set Theory I.*, van Heijenoort (2002, pp. 199–215). See also Ewald (1996, pp. 1208–1232).

A.A. Fraenkel, Zu den Grundlagen der Cantor-Zermeloschen Mengenlehre, *ibid.* **86**(1922), pp. 230–237, ISSN 0025-5831. A.A. Fraenkel, Der Begriff "definit" und die unabhängigkeit des Auswahlsaxioms, *Sitzungsberichte der Preussischen Akademie der Wissenschaften* 1922, pp. 253–257; English translation: *The Notion "Definite" and the Independence of the Axiom of Choice*, van Heijenoort (2002, pp. 284–289). A.A. Fraenkel, Untersuchungen über die Grundlagen der Mengenlehre I., II., *Mathematische Zeitschrift* **22**(1925), pp. 250–273, ISSN 0025-5874, *Journal für die reine und angewandte Mathematik* **155**(1926), pp. 129–158, ISSN 0075-4102.

We also mention the book A.A. Fraenkel, *Zehn Vorlesungen über die Grundlegung der Mengenlehre*, Teubner, Leipzig, 1927, x + 182 pages, based on Fraenkel's lectures delivered at the meetings of the Kant-Gesellschaft in Kiel, July 1927.

For more information, see, e.g., G.H. Moore, The Origins of Zermelo's Axiomatization of Set Theory, *Journal of Philosophical Logic* **7**(1978), pp. 307–329, ISSN 0022-3611; H.-D. Ebbinghaus and V. Peckhaus, *Ernst Zermelo. An Approach to His Life and Work*, Springer, Berlin, 2007, xiv + 356 pages, ISBN 978-3-540-49551-2, 2nd ed. 2015, xvi + 383 pages, ISBN 978-3-662-47996-4; S. Hayden and J.F. Kennison, *Zermelo-Fraenkel Set Theory*, Merrill, Columbus, Ohio, 1968, xi + 164 pages.

[103]G.H. Moore, *Zermelo's Axiom of Choice. Its Origins, Development, and Influence*, Springer, New York, Heidelberg, Berlin, 1982, xiv + 410 pages, ISBN 0-387-90670-3.

There are several possible formulations of the axiom of choice: *For an arbitrary set X, there exists a mapping that assigns to each nonempty subset of X one of its elements (selects a unique element from each nonempty subset of X.) – If X is a system of nonempty pairwise disjoint sets, then there exists a set that contains exactly one element in common with each of the sets in X. – For each system X of nonempty sets, there exists a mapping $f : X \to \cup_{x \in X} x$ such that $f(x) \in x$ for each $x \in X$. – The Cartesian product of a nonempty system of sets is a nonempty set.*

proof of its consistency (*Hilbert's program*). He delivered a number of lectures on this topic, whose contents were published in the articles

Neubegründung der Mathematik. Erste Mitteilung,[104]

Die logischen Grundlagen der Mathematik,[105]

Über das Unendliche,[106]

Die Grundlagen der Mathematik,[107]

Die Grundlegung der elementaren Zahlenlehre,[108]

Beweis des Tertium non datur.[109]

As early as 1922, Hilbert heavily criticized intuitionism and the ideas of its supporters. Among other things he stated:

> *Was Weyl und Brouwer tun, kommt im Prinzip darauf hinaus, daß sie die einstigen Pfade von Kronecker wandeln: sie suchen die Mathematik dadurch zu begründen, daß sie alles ihnen unbequem Erscheinende über Bord werfen und eine Verbotsdiktatur à la Kronecker errichten. Dies heißt aber, unsere Wissenschaft zerstückeln und verstümmeln, und wir laufen Gefahr, einen*

---

[104] D. Hilbert, Neubegründung der Mathematik, *Abhandlungen aus dem Mathematischen Seminar der Hamburgischen Universität* **1**(1922), pp. 157–177, also in van Heijenoort (2002, III, pp. 157–177); English translation: *The New Grounding of Mathematics*, Ewald (1996, pp. 1115–1134), or Mancosu (1998, pp. 198–214). This article was based on Hilbert's lectures for the Mathematische Gesellschaft in Kopenhagen in spring 1922, and for Mathematisches Seminar der Universität Hamburg in summer 1922.

[105] D. Hilbert, Die logischen Grundlagen der Mathematik, *Mathematische Annalen* **88**(1923), pp. 151–165, ISSN 0025-5831, also in Hilbert (2001, III, pp. 178–191); English translation: *The Logical Foundations of Mathematics*, Ewald (1996, pp. 1134–1148). This article was based on Hilbert's lecture for the Deutsche Naturforscher-Gesellschaft in September 1922.

[106] D. Hilbert, Über das Unendliche, *Mathematische Annalen* **95**(1926), pp. 161–190, ISSN 0025-5831; abridged version in *Jahresbericht der Deutschen Mathematiker-Vereinigung* **36**(1927), pp. 201–215, ISSN 0012-0456, or in Hilbert (1930, Anhang VII., pp. 262–288); French translation: Sur l'infini, *Acta Mathematica* **48**(1926), pp. 91–122, ISSN 0001-5962; English translation: *On the Infinite*, van Heijenoort (2002, pp. 367–392). This article was based on Hilbert's lecture delivered in Münster, June 1925.

[107] D. Hilbert, Die Grundlagen der Mathematik, *Abhandlungen aus dem mathematischen Seminar der Hamburgischen Universität* **6**(1928), pp. 65–85, also in Hilbert (1930, Anhang IX., pp. 289–312); English translation: *The Foundations of Mathematics*, van Heijenoort (2002, pp. 464–479). This article was based on Hilbert's lecture at the Mathematisches Seminar der Hamburgischen Universität. It was also published in the series *Hamburger Mathematische Einzelschriften*, **5**. Heft, Teubner, Leipzig, Berlin, 1928, 28 pages.

[108] D. Hilbert, Die Grundlegung der elementaren Zahlenlehre, *Mathematische Annalen* **104**(1931), pp. 485–494, ISSN 0025-5831, also in Hilbert (2001, III, pp. 192–195); English translation: *The Grounding of Elementary Number Theory*, Ewald (1996, pp. 1148–1157), or Mancosu (1998, pp. 266–273). This article was based on Hilbert's lecture for the Philosophische Gesellschaft in Hamburg, September 1930.

[109] D. Hilbert, Beweis des Tertium non datur, *Nachrichten von der Gesellschaft der Wissenschaften zu Göttingen, Math.-phys. Klasse*, 1931, pp. 120–125.

> *großen Teil unserer wertvollsten Schätze zu verlieren, wenn wir solchen Reformatoren folgen.*[110]

In the 1926 treatise *Über das Unendliche*, which is perhaps Hilbert's most important contribution dealing with the foundations of mathematics and Hilbert's program, he stated:

> *In der Geometrie und den physikalischen Theorien gelingt der Nachweis der Widerspruchsfreiheit durch Zurückführung auf die Widerspruchsfreiheit der arithmetischen Axiome. Diese Methode versagt offenbar bei der Arithmetik selbst. Indem unsere Beweistheorie auf Grund der Methode der idealen Elemente diesen letzten wichtigen Schritt ermöglicht, bildet sie den notwendigen Schlußstein in dem Lehrgebäude der Axiomatik. Und insbesondere, was wir zweimal erlebt haben, als es sich einmal um die Paradoxien der Infinitesimalrechnung und dann um die Paradoxien der Mengenlehre handelte, das ist künftig im Reiche der Mathematik unmöglich.*[111]

In the same paper, Hilbert responded to the attacks of the intuitionists, defended the concept of actual infinity (but did not suggest its existence in the real world) as well as set theory (the second sentence of the following excerpt is often quoted):

> *Fruchtbaren Begriffsbildungen und Schlußweisen wollen wir, wo immer nur die geringste Aussicht sich bietet, sorgfältig nachspüren und sie pflegen, stützen und gebrauchsfähig machen. Aus dem Paradies, das Cantor uns geschaffen, soll uns niemand vertreiben können.*[112]

In his 1928 work *Die Grundlagen der Mathematik*, Hilbert wrote:

> *Dieses Tertium non datur dem Mathematiker zu nehmen, wäre etwa, wie wenn man dem Astronomen das Fernrohr oder dem Boxer den Gebrauch der Fäuste untersagen wollte. Das Verbot der Existenzsätze und des Tertium non datur kommt ungefähr dem Verzicht auf die mathematische Wissenschaft überhaupt gleich...*
>
> *... die Mathematik ist eine voraussetzungslose Wissenschaft. Zu ihrer Begründung brauche ich weder den lieben Gott, wie Kronecker, noch die Annahme einer besonderen auf das Prinzip der vollständigen Induktion abgestimmten Fähigkeit unseres Verstandes, wie Poincaré, noch die Brouwersche Urintuition und endlich auch nicht, wie Russell und Whitehead, Axiome der Unendlichkeit, Reduzierbarkeit oder der Vollständigkeit, die ja wirkliche inhaltliche, überdies gar nicht plausible Voraussetzungen sind.*[113]

[110] D. Hilbert, *Neubegründung der Mathematik. Erste Mitteilung*, Hilbert (2001, III, p. 159).
[111] D. Hilbert, *Über das Unendliche*, Hilbert (1930, p. 287).
[112] D. Hilbert, *Über das Unendliche*, Hilbert (1930, p. 274).
[113] D. Hilbert, *Die Grundlagen der Mathematik*, Hilbert (1930, pp. 307, 312).

Hilbert's collaborators in the 1920s, working on Hilbert's program and on the questions of consistency, were Wilhelm Ackermann (1896–1962),[114] Paul Bernays (1888–1977),[115] John von Neumann (1903–1957),[116] and Jacques Herbrand (1908–1931).[117]

The aim of the formalists was to build individual mathematical disciplines on an axiomatic basis. Emphasis was placed on the choice of the axioms — they should not be taken over from logic (as in logicism) or the real world (as in intuitionism), but should describe the elements of the theory. Inference rules are provided by the classical predicate calculus, including the law of excluded middle.

Let us now briefly summarize the basic ideas of Hilbert's program of a complete formalization of mathematics.

To begin with, it is necessary to choose a finite number of symbols devoid of any meaning (alphabet) and provide precise rules for the construction of formulas (words) — abstract expressions free of all contentual meaning — out of these symbols. The next step consists in providing precise inference rules, which describe the construction of new formulas out of existing ones. The choice of a suitable collection

---

[114] W. Ackermann, Begründung des "tertium non datur" mittels der Hilbertschen Theorie der Widerspruchsfreiheit, *Mathematische Annalen* **93**(1925), pp. 1–36. W. Ackermann, Zur Hilbertschen Aufbau der reellen Zahlen, *ibid.* **99**(1928), pp. 118–133; English translation: *On Hilbert's Construction of the Real Numbers*, van Heijenoort (2002, 493–507). W. Ackermann, Über die Erfüllbarkeit gewisser Zählausdrücke, *ibid.* **100**(1928), pp. 638–649. W. Ackermann, Die Widerspruchsfreiheit der allgemeinen Mengenlehre, *ibid.* **114**(1937), pp. 305–315. W. Ackermann, Zur Widerspruchsfreiheit der reinen Zahlentheorie, *ibid.* **117**(1940), pp. 162–194, ISSN 0025-5831.

[115] P. Bernays, Über Hilberts Gedanken zur Grundlegung der Arithmetik, *Jahresbericht der Deutschen Mathematiker-Vereinigung* **31**(1922), pp. 10–19, ISSN 0012-0456; English translation: *On Hilbert's Thoughts Concerning the Grounding of Arithmetic*, Mancosu (1998, pp. 215–222). P. Bernays, Axiomatische Untersuchungen des Aussagen-kalküls der Principia Mathematica, *Mathematische Zeitschrift* **25**(1926), pp. 305–320, ISSN 0025-5874. P. Bernays, Zusatz zu Hilberts Vortrag über Die Grundlagen der Mathematik, *Abhandlungen aus dem mathematischen Seminar der Hamburgischen Universität* **6**(1928), pp. 89–92; English translation: *Appendix to Hilbert's Lecture "The Foundations of Mathematics"*, van Heijenoort (2002, pp. 485–489). P. Bernays, Die Philosophie der Mathematik und die Hilbertsche Beweistheorie, *Blätter für deutsche Philosophie* **4**(1930), pp. 326–367; English translation: *The Philosophy of Mathematics and Hilbert's Proof Theory*, Mancosu (1998, pp. 234–265).

[116] J. von Neumann, Zur Einführung der transitiven Zahlen, *Acta litterarum ac scientiarum Regiae Universitatis Hungaricae Francisco-Josephinae, Sectio scientiarum mathematicarum* **1**(1923), pp. 199–208; English translation: *On the Introduction of Transitive Numbers*, van Heijenoort (2002, pp. 346–354). J. von Neumann, Zur Hilbertschen Beweistheorie, *Mathematische Zeitschrift* **26**(1927), pp. 1–46, ISSN 0025-5874. J. von Neumann, Die formalistische Grundlegung der Mathematik, *Erkenntnis* **2**(1931), pp. 116–121.

[117] J. Herbrand, Recherches sur la théorie de la démonstration, *Prace Towarzystwa Naukowego Warszawskiego*, III, 1930, no. 33; English translation: *Investigations in Proof Theory: The Properties of True Propositions*, van Heijenoort (2002, pp. 525–581). J. Herbrand, Sur la non-contradiction de l'arithmétique, *Journal für die reine und angewandte Mathematik* **166**(1931), pp. 1–8, ISSN 0075-4102; English translation: *On the Consistency of Arithmetic*, van Heijenoort (2002, pp. 618–628).

of formulas (axiomatic system) gives rise to a theory, which can be interpreted as the collection of all formulas (theorems) deducible from the axiomatic system by means of the given inference rules. The proof of a theorem is a finite sequence of formulas, each of which is either an axiom, or follows from the previous formulas using the inference rules. A theory developed in this way is purely formal, and in general has no relation to the real world.[118] Formalism has completely eliminated intuition, which can be misleading.

The goal of the formalization of mathematical proofs by finitistic methods, as worked out by Hilbert and his collaborators, was to develop a proof theory (*Beweistheorie*), sometimes referred to as metamathematics.

Hilbert himself has wittily characterized his formalist approach to mathematics as follows:

> *Mathematics is a game played according to certain simple rules with meaningless marks on paper.*[119]

Hilbert imposed two essential requirements on the choice of the axiomatic system — completeness and consistency:

- Completeness: the truth of any assertion involving notions of the given theory can be decided on the basis of the axiomatic system: the assertion can be shown to be true or false.
- Consistency: it is impossible to reach a contradiction, i.e., impossible to prove two assertions that contradict each other (an assertion and its negation cannot be proved at the same time).

Hilbert spoke about decidability (*Entscheidbarkeit*, *Entscheidungsproblem*), i.e., the existence of an algorithm that can, in finitely many steps, decide whether a given assertion is provable or not. The problem of decidability is closely related to both completeness and consistency.

The requirement of independence, which was also sometimes imposed, was not essential, and played merely an aesthetic role: to show that no axiom follows from the remaining ones.

Hilbert's program called for the formalization of mathematics as a whole, starting with the arithmetic. It was necessary to

- arithmetize individual disciplines of mathematics,
- provide a full formalization of the arithmetic, and
- demonstrate the consistency of the arithmetic.

---

[118] On the other hand, it was the goal of physics and geometry to determine which geometry is valid in the real world (universe).

[119] In R. Goldstein, *Incompleteness: The Proof and Paradox of Kurt Gödel*, Atlas Books, Norton, New York, London, 2005, 296 pages, see p. 136, ISBN 0-393-05169-2.

If these steps had been carried out, mathematics would have been put on a firm foundation.

In 1928, Hilbert and his student Wilhelm Ackermann published the book *Grundzüge der theoretischen Logik.*[120] After choosing a suitable collection of axioms and suitable inference rules, an important question was whether their predicate calculus was complete, i.e., whether each tautology (a formula that is always true) is provable. They concluded the introduction of their book (see p. 2) as follows:

> *In jüngster Zeit hat Hilbert in einer Reihe von Abhandlungen und Universitätsvorlesungen den Logikkalkül dazu verwendet, um auf einem neuen Wege zu einem Aufbau der Mathematik zu gelangen, der die Widerspruchsfreiheit der zugrunde gelegten Annahmen erkennen läßt.*

During his talk *Probleme der Grundlegung der Mathematik* at the International Congress of Mathematicians in Bologna, September 1928, Hilbert again highlighted the importance of the completeness of the arithmetic and logical calculus:

> *Die Behauptung der Vollständigkeit des Axiomensystems der Zahlentheorie lässt sich auch so aussprechen: Wird eine dem Bereich der Zahlentheorie angehörige, aber nicht beweisbare Formel zu den Axiomen der Zahlentheorie hinzugenommen, so kann aus dem erweiterten Axiomensystem ein Widerspruch abgeleitet werden...*
>
> *Die Frage nach der Vollständigkeit des Systems der logischen Regeln, in allgemeiner Form gestellt, bildet ein Problem der theoretischen Logik.*[121]

An expanded version of Hilbert's talk contains the following paragraph:

> *Es entsteht nun die Frage, ob alle diese Formeln aus den Regeln des logischen Schließens unter Hinzunahme der genannten Gleichheitsaxiome beweisbar sind, mit anderen Worten, ob das System der üblichen logischen Regeln vollständig ist.*[122]

Hilbert's program was successfully developing throughout the end of the 1930s. At that time, the decidability and completeness of the predicate calculus of one-place predicates were already demonstrated by finitistic methods. In the years 1925 and 1927, Wilhelm Ackermann and John von Neumann proved the consistency of an arithmetic with restrictions on the axiom of induction, and Mojżesz Presburger

---

[120]D. Hilbert and W. Ackermann, *Grundzüge der theoretischen Logik*, Springer, Berlin, 1928, viii + 120 pages, 2nd ed. 1938, viii + 133 pages, 3rd ed. 1949, viii + 155 pages, 6th ed. 1972, viii + 188 pages, ISBN 3-540-05843-5; English translation of the 2nd ed.: *Principles of Mathematical Logic*, Chelsea, Providence, RI, 1950, reprint: 1999, xii + 172 pages, ISBN 978-0-8218-2024-9.

[121]D. Hilbert, *Probleme der Grundlegung der Mathematik*, Atti del Congresso internazionale dei matematici, Bologna 3–10 settembre 1928 (VI), Tomo I, Zanichelli, Bologna, 1929, pp. 135–141, see p. 140. Also in *Mathematische Annalen* **102**(1930), pp. 1–9, ISSN 0025-5831, also in Hilbert (1930, Anhang X., pp. 313–323), pp. 321–322; English translation: *Problems of the Grounding of Mathematics*, Mancosu (1998, pp. 227–233).

[122]See Hilbert (1930, p. 322).

proved by finitistic methods the consistency of a weaker arithmetic theory (containing addition but no multiplication).[123]

In 1930, David Hilbert concluded his Königsberg lecture *Naturerkennen und Logik* in a very optimistic manner:

> *Der wahre Grund, warum es Comte nicht gelang, ein unlösbares Problem zu finden, besteht meiner Meinung nach darin, daß es ein unlösbares Problem überhaupt nicht gibt. Statt des törichten Ignorabimus heiße im Gegenteil unsere Losung:*
>
> *Wir müssen wissen,*
>
> *Wir werden wissen.*[124]

In her 1970 book *Hilbert*, Constance Reid comments on this often-quoted statement by the following words:

> *We must know. We will know. It was in every respect a great last line. However, lives do not always end on great last lines. At almost the same time that Hilbert was making his speech at Königsberg, a piece of work was being brought to a conclusion which was to deal a death blow to the specific epistemological objective of the final program of Hilbert's career. On November 17, 1930, the Monatshefte für Mathematik und Physik received for publication a paper by 25-year-old mathematical logician named Kurt Gödel.*[125]

In his book *Philosophie der Mathematik und Naturwissenschaft*, Herman Weyl wrote:

> *Das Ziel von Hilberts "Beweistheorie" war, wie er erklärte, "die Grundlagenfragen ein für allemal aus der Welt zu schaffen". Im Jahre 1926 bestand Grund zu der optimistischen Erwartung, daß es ihm und seinen Mittarbeitern in ein paar Jahren beharrlicher Anstrengung gelingen würde, die Widerspruchslosigkeit für das formale Äquivalent unserer klassischen Mathematik nachzuweisen. Tatsächlich waren die ersten Schritte erfolgreich und vielversprechend.*[126]

Both mathematicians and philosophers took interest in the problems related to the foundations of mathematics. Let us recall only a few names. In 1924, Moritz Geiger (1880–1937), who was active in Munich and Göttingen, published a book

---

[123] M. Presburger, *Über die Vollständigkeit eines gewissen Systems der Arithmetik ganzer Zahlen, in welchem die Addition als einzige Operation hervortritt*, Sprawozdanie z I. kongresu matematyków krajów slowiańskich 1929, Waszawa, 1930, pp. 92–101, 395.

[124] D. Hilbert, Naturerkennen und Logik, *Die Naturwissenschaften* **18**(1930), pp. 959–963 and p. 963, ISSN 0028-1042; also in Hilbert (2001, III, pp. 378–387), see p. 387. We note that the inscription *Wir müssen wissen Wir werden wissen* was placed over Hilbert's grave in Göttingen.

[125] See Reid (1996, pp. 196–197).

[126] H. Weyl, *Philosophie der Mathematik und Naturwissenschaft*, 1966, see p. 279.

entitled *Systematische Axiomatik der Euklidischen Geometrie.*[127] In 1930, Felix Kaufmann (1895–1949), a Privatdozent of the University of Vienna, who attended meetings of the Vienna Circle (*Wiener Kreis*), published the treatise *Das Unendliche in der Mathematik und seine Ausschaltung. Eine Untersuchung über die Grundlagen der Mathematik.*[128] A prominent German mathematician Otto Hölder (1859–1937), a profesor at the Leipzig University who did important work in group theory, theory of complex functions and infinite series, also devoted himself to general questions in his lectures and writings.[129]

There is a wealth of monographs on logicism, intuitionism and formalism.[130]

---

[127] M. Geiger, *Systematische Axiomatik der Euklidischen Geometrie*, Filser, Augsburg, 1924, xxiii + 271 pages.

[128] F. Kaufmann, *Das Unendliche in der Mathematik und seine Ausschaltung ...*, Deuticke, Leipzig und Wien, 1930, x + 203 pages; a comprehensive list of references is included on pages 198–203. Later edition: Wissenschaftliche Buchgesellschaft, Darmstadt, 1968, x + 203 pages; English translation: *The Infinite in Mathematics: Logico-mathematical Writings*, Reidel, Dordrecht, 1978, xvii + 235 pages, pp. 1–164, ISBN 90-277-0847-9.

[129] O. Hölder, *Anschauung und Denken in der Geometrie*, Teubner, Leipzig, 1900, 75 pages, reprint: 1968. O. Hölder, *Die Arithmetik in strenger Begründung*, Teubner, Leipzig, 1914, iv + 74 pages, 2nd ed. Springer, Berlin, 1929, v + 73 pages. O. Hölder, *Die Mathematische Methode. Logisch erkenntnistheoretische Untersuchungen im Gebiete der Mathematik, Mechanik und Physik*, Springer, Berlin, 1924, x + 563 pages. O. Hölder, Der angebliche Circulus vitiosus und die sogennante Grundlagenkrise in der Analysis, *Sitzungsberichte der Leipziger Akademie* **78**(1926), pp. 243–250; English translation: *The Alleged "Circulus Vitiosus" and the So-Called Foundational Crisis in Analysis*, Mancosu (1998, pp. 143–148).

[130] S. Shapiro (ed.), *The Oxford Handbook of Philosophy of Mathematics and Logic*, Oxford University Press, Oxford, New York, 2005, xv + 833 pages, ISBN 978-0-19-514877-0; the following chapters are worth highlighting: W. Demopoulos and P. Clark, *The Logicism of Frege, Dedekind, and Russell*, pp. 129–165, M. Detlefsen, *Formalism*, pp. 236–317, D.C. McCarty, *Intuitionism in Mathematics*, pp. 356–386.

A.D. Irvine (ed.), *Philosophy of Mathematics*, Elsevier, Amsterdam, 2009, xvi + 717 pages, ISBN 978-0-444-51555-1; the following chapters are worth highlighting: J. Hintikka, *Logicism*, pp. 271–290, P. Simons, *Formalism*, pp. 291–310, A. Kanamori, *Set Theory from Cantor to Cohen*, pp. 395–459.

Excerpts from the works of important mathematicians–philosophers (Bernays, Brouwer, Carnap, Frege, Gödel, Heyting, Hilbert, Poincaré, Quine, Russell atd.) are available in the book P. Benacerraf and H. Putnam (eds.), *Philosophy of Mathematics. Selected Readings*, Prentice-Hall, Englewood Cliffs, NJ, Basil Blackwell, Oxford, 1964, v + 536 pages.

See also H.B. Curry, *Outlines of a Formalist Philosophy of Mathematics*, North-Holland, Amsterdam, 1951, viii + 75 pages, also 1958. H. Meschkowski, *Evolution of Mathematical Thought*, Holden-Day, San Francisco, London, Amsterdam, 1965, ix + 157 pages. V. Peckhaus, *Hilbertprogramm und kritische Philosophie. Das Götinger Modell interdisziplinärer Zusammenarbeit zwischen Mathematik und Philosophie*, Vandenhoeck und Ruprecht, Göttingen, 1990, x + 291 pages, ISBN 978-3525403143. W. Sieg, Hilbert's Programs: 1917–1922, *The Bulletin of Symbolic Logic* **5**(1999), pp. 1–44, ISSN 0022-4812. Ch. Tapp, *An den Grenzen des Endlichen. Das Hilbertprogramm im Kontext von Formalismus und Finitismus*, Springer, Berlin, Heidelberg, 2013, 553 pages, ISBN 978-3-642-29654-3.

For more on Hilbert's life and work, see H. Weyl, David Hilbert and his Mathematical Work, *Bulletin of the American Mathematical Society* **50**(1944), pp. 612–654, ISSN 0273-0979, also in Reid (1996, pp. 245–283), also in Weyl (1968, IV, pp. 130–172). O. Blumenthal, *Lebensgeschichte*, Hilbert

8. The End of Hope — Kurt Gödel

The Austrian mathematician, logician and philosopher Kurt Gödel (1906–1978) was deeply influenced by Hilbert's and Ackermann's book *Grundzüge der theoretischen Logik*, as well as by Russell's monograph *Introduction to Mathematical Philosophy*.[131] In 1930, he published the results of his doctoral dissertation supervised by Hans Hahn in the paper *Die Vollständigkeit der Axiome des logischen Funktionenkalküls*.[132] He demonstrated that the theory presented in Hilbert's and Ackermann's book is indeed complete, i.e., he proved the completeness of their first-order predicate calculus.

In September 1930, he participated in a conference on the epistemology of exact sciences (*Tagung für Erkenntnislehre der exakten Wissenschaften*) organized in Königsberg by the Gesellschaft für empirische Philosophie, and presented the results of his dissertation thesis.[133]

On the next day, during the final discussion chaired by Hans Hahn, Gödel briefly announced his new, highly important discovery, which he shortly afterwards published in the work *Über formal unentscheidbare Sätze der Principia mathematica und verwandter Systeme I*.[134] Gödel's result and its immediate corollary can be succinctly elucidated as follows (Gödel's first and second incompleteness theorems):

- Any consistent axiomatic theory containing the arithmetic is necessarily incomplete, i.e., there exist undecidable statements.

(2001, III, pp. 388–429); P. Bernays, *Hilberts Untersuchungenüber die Grundlagen der Arithmetik*, Hilbert (2001, III, pp. 196–216). G. Kreisel, Hilbert's Programme, *Dialectica* **12**(1958), pp. 346–372. See also Ewald (1996, pp. 1087–1165).

[131] B. Russell, *Introduction to Mathematical Philosophy*, Allen & Unwin, London, 1919, 2nd ed. 1920, vii + 208 pages. Later edition: Dover, 1993, 224 pages, ISBN 978-0486277240.

[132] K. Gödel, Die Vollständigkeit der Axiome des logischen Funktionenkalküls, *Monatshefte für Mathematik und Physik* **37**(1930), pp. 349–360; English translation: *The Completeness of the Axioms of the Functional Calculus of Logic*, Grattan-Guinness (2005, I, pp. 102–123), or van Heijenoort (2002, pp. 582–591).

[133] Among the participants of this conference were prominent representatives of intuitionism (A. Heyting), logicism (R. Carnap), as well as formalism (J. von Neumann). For some of their contributions see *Erkenntnis* **2**(1931).

[134] K. Gödel, Über formal unentscheidbare Sätze ..., *Monatshefte für Mathematik und Physik* **38**(1931), pp. 173–198; English translation: *On Formally Undecidable Propositions of Principia Mathematica and Related Systems I.*, Gödel (2001, I, pp. 144–195). Published also as a standalone text: Basic Books, New York, 1962, viii + 72 pages; reprint: Dover, New York, 1992. Also in Davis (1965, pp. 39–88), in van Heijenoort (2002, pp. 596–616), in a broader context van Heijenoort (2002, pp. 592–617).

See I. Grattan-Guinness, In Memoriam Kurt Gödel: His 1931 Correspondence with Zermelo on his Incompletability Theorem, *Historia Mathematica* **6**(1979), pp. 294–304, ISSN 0315-0860.

- The consistency of an axiomatic theory containing the arithmetic cannot be proved within this theory.[135]

The requirements of completeness and consistency of axiomatic systems were ruined. Hilbert's program turned out to be unrealisable. Still, it was a significant contribution to the debate on the rigorous foundations of mathematics. The third crisis in mathematics has not been and will not be overcome.

Kurt Gödel's result provided a negative answer to Hilbert's second problem: it is impossible to prove the consistency of the arithmetic. His results also imply that it is impossible to prove neither the consistency of Russell's and Whitehead's *Principia Mathematica*, nor the consistency of Zermelo-Fraenkel's set theory (within this theory).

Gödel's results were shocking. It took over 10 years before they were understood and accepted by the wider mathematical community.

A second axiomatic system (which distinguished sets and classes) for set theory was developed in the years 1937–1954 by Kurt Gödel and Paul Bernays,[136] but Adolf Abraham Fraenkel and John von Neumann also deserve credit for their earlier results.[137]

The Gödel-Bernays axiomatic system is essentially equivalent to the Zermelo-Fraenkel axiomatic system. A statement provable in the Zermelo-Fraenkel set theory is also provable in the Gödel-Bernays theory. A statement (involving sets) provable in the Gödel-Bernays theory is also provable in the Zermelo-Fraenkel theory.[138] Both

---

[135] We refer the reader to the following nice books: E. Nagel and J.R. Newman (D.R. Hofstadter, ed.), *Gödel's Proof.* Revised Edition, New York University Press, New York, 2008, 160 pages, ISBN 978-0814758373 (original edition: 1958, ix + 118 pages); R. Goldstein, *Incompleteness. The Proof and Paradox of Kurt Gödel*, Atlas Books, Norton, London, New York, 2005, 296 pages, ISBN 0-393-05169-2. See also R. Zach, *Kurt Gödel, Paper on the Incompleteness Theorems (1931)* in Grattan-Guinness (2005, pp. 917–925).

[136] P. Bernays, A System of Axiomatic Set Theory, *The Journal of Symbolic Logic* **2**(1937), pp. 65–77, **6**(1941), pp. 1–17, **7**(1942), pp. 65–89, 133–145, **8**(1943), pp. 89–106, **13**(1948), pp. 65–79, **19**(1954), pp. 81–96, ISSN 0022-4812.

[137] J. von Neumann, Eine Axiomatisierung der Mengenlehre, *Journal für die reine und angewandte Mathematik* **154**(1925), pp. 219–240, **155**(1926), p. 128, ISSN 0075-4102; English translation: *An Axiomatization of Set Theory*, van Heijenoort (2002, pp. 393–413). J. von Neumann, Die Axiomatisierung der Mengenlehre, *Mathematische Zeitschrift* **27**(1928), pp. 669–752, ISSN 0025-5874. J. von Neumann, Über die Widerspruchsfreiheitsfrage in der axiomatischen Mengenlehre, *Journal für die reine und angewandte Mathematik* **160**(1929), pp. 227–241, ISSN 0075-4102.

Furthermore, see P. Bernays, *Axiomatic Set Theory with a Historical Introduction by Abraham A. Fraenkel*, North-Holland, Amsterdam, 1958, viii + 226 pages, 2nd ed. 1968; later edition: Dover, 1991, 256 pages, ISBN 978-0486666372. See also J.R. Shoenfield: *Axioms of Set Theory*, in J.K. Barwise (ed.), *Handbook of Mathematical Logic*, North-Holland, Amsterdam, 1977, xi + 1165 pages, pp. 321–344, ISBN 072042285X.

[138] During the 1960s and 1970s, the Gödel-Bernays axiomatic system was popular in Poland, Czechoslovakia, and the Soviet Union.

systems are sufficiently rich in order to encompass the "whole" mathematics. Later axiomatic systems of set theory did not gain much popularity.

Another important result due to Gödel is the proof of the fact that if the Gödel-Bernays (equivalently, the Zermelo-Fraenkel) set theory is consistent, then it will remain consistent after adding the axiom of choice and the continuum hypothesis.[139] In 1963, Paul Joseph Cohen (1934–2007) proved that if the Zermelo-Fraenkel set theory is consistent, then it will remain consistent after adding the negation of the axiom of choice, or the negation of the continuum hypothesis.[140] Hence, the continuum hypothesis (as well as the axiom of choice) is logically independent of the Zermelo-Fraenkel theory, that is, it can be neither proved nor disproved. This settled the question that was bothering Cantor and was the subject of Hilbert's first problem.

In his book *Real analysis*, G.B. Folland expressed his opinion on the continuum hypothesis, which is probably shared by many mathematicians:

> *It is known, thanks to Gödel and Cohen, that the continuum hypothesis and its negation are both consistent with the standard axioms of set theory including the axiom of choice, assuming that those axioms are themselves consistent. ... Some mathematicians are willing to accept the continuum hypothesis as true, seemingly as a matter of convenience, but Gödel [..] and Cohen [..] have both expressed suspicions that it should be false, and as of this writing no one has found any really compelling evidence on one side or the other. My own feeling, subject to revision in the event of a major breakthrough in set theory, is that if the answer to one's question turns out to depend on the continuum hypothesis, one should give up and ask a different question.*[141]

Even after the publication of Gödel's results, David Hilbert and Paul Bernays continued to develop their ideas on the formalization of mathematics in the spirit

---

[139]K. Gödel, The Consistency of the Axiom of Choice and of the Generalized Continuum Hypothesis, *Proceedings of the National Academy of Sciences of the U.S.A.* **24**(1938), pp. 556–557, also in Gödel (2001, II, pp. 26–27).

K. Gödel, Consistency-Proof for the Generalized Continuum-Hypothesis, *ibid.* **25**(1939), pp. 220–224, ISSN 0027-8424. Also in Gödel (2001, II, pp. 28–32).

K. Gödel, The Consistency of the Axiom of Choice and of the Generalized Continuum Hypothesis with the Axioms of Set Theory, *Annals of Mathematics Studies*, Volume 3, Princeton University Press, Princeton, 1940, 66 pages; reprinted with further notes in 1951 and in 1966, and in Gödel (2001, II, pp. 33–101); see also Gödel (2001, II, pp. 1–32). Later edition: *The Consistency of the Continuum Hypothesis*, ISBN 9780691079271.

[140]P.J. Cohen, The Independence of the Continuum Hypothesis I., II., *Proceedings of the National Academy of Sciences of the U.S.A.* **50**(1963), pp. 1143–1148, **51**(1964), pp. 105–110, ISSN 0027-8424. P.J. Cohen, *Set Theory and the Continuum Hypothesis*, Benjamin, New York, 1966, 154 pages; reprint: 1977, 1980, ISBN 0-8053-2327-9.

[141]G.B. Folland, *Real Analysis. Modern Techniques and Their Applications*, Wiley, New York, 2nd ed. 1999, ix + 386 pages, p. 17, ISBN 978-0-471-31716-6.

of Hilbert's program. These were published in the years 1934 and 1939 in the two-volume monograph *Grundlagen der Mathematik I., II.*[142]

In the introduction to the first volume (*Zur Einführung*), David Hilbert responded to Gödel's results:

> *Dieser Ergebnisstand weist zugleich die Richtung für die weitere Forschung in der Beweistheorie auf das Endziel hin, unsere üblichen Methoden der Mathematik samt und sonders als widerspruchsfrei zu erkennen.*
>
> *Im Hinblick auf dieses Ziel möchte ich hervorheben, daß die zeitweilig aufgekommene Meinung, aus gewissen neueren Ergebnissen von Gödel folge die Undurchführbarkeit meiner Beweistheorie, als irrtümlich erwiesen ist. Jenes Ergebnis zeigt in der Tat auch nur, daß man für die weitergehenden Widerspruchsfreiheitsbeweise den finiten Standpunkt in einer schärferen Weise ausnutzen muß, als dieses bei der Betrachtung der elementaren Formalismen erforderlich ist.*

In the foreword to the second volume he wrote:

> *Der vorliegende Band schließt die Darstellung der Beweistheorie ab, die ich vor einigen Jahren zusammen mit P. Bernays begann.*

But in the meanwhile, Hilbert's program suffered an additional blow by the results of Alonzo Church (1903–1995)[143] and Alfred Tarski (1901–1983).[144]

Although Gerhard Gentzen (1909–1945) was able to prove the consistency of the arithmetic in 1936, he did so only at the cost of giving up Hilbert's finitistic methods. He used a transfinite procedure, rejected by Hilbert.[145]

Otto Blumenthal (1876–1944) commented on Hilbert's obstinate efforts to fulfill his program:

> *Aber bis zur Ausführung dieser weitreichenden Pläne wird es noch ein langer, schwerer Weg sein, besonders weil die bisherigen Veröffentlichungen nur die skizzenhafte Form von Vorträgen haben. Einen großen Schritt vorwärts bedeutet es deshalb, daß 1934 der erste Band des groß angelegten*

---

[142] D. Hilbert and P. Bernays, *Grundlagen der Mathematik I., II.*, Springer, Berlin, 1934, 1939, xii + 471, xii + 498 pages, 2nd ed. 1968, 1970, xv + 473, xiv + 561 pages. There exist English, French and Russian translations, German-English version (dual text version) is from 2011. See W. Sieg and M. Ravaglia, *David Hilbert and Paul Bernays, Grundlagen der Mathematik, 1st ed. (1934, 1939)*, Grattan-Guinness (2000, pp. 981–999).

[143] A. Church, An Unsolvable Problem of Elementary Number Theory, *American Journal of Mathematics* **58**(1936), pp. 345–363, ISSN 0002-9327. A. Church, A Note on the Entscheidungsproblem, *The Journal of Symbolic Logic* **1**(1936), pp. 40–41, 101–102, ISSN 0022-4812. See Davis (1965, pp. 88–107, 108–115).

[144] A. Tarski, Der Wahrheitsbegriff in der formalisierten Sprachen, *Studia Philosophica* **1**(1936), pp. 261–405.

[145] G. Gentzen, Die Widerspruchsfreiheit der reinen Zahlentheorie, *Mathematische Annalen* **112**(1936), pp. 493–565, ISSN 0025-5831. G. Gentzen, Die Widerspruchsfreiheit der Stufenlogik, *Mathematische Zeitschrift* **41**(1936), pp. 357–366, ISSN 0025-5874.

> *Werkes Hilbert-Bernays "Grundlagen der Mathematik" erschienen ist. Eine Darstellung der Hilbertschen Ausgestaltung des Logikkalküls ist schon 1928 herausgekommen: Hilbert-Ackermann "Grundzüge der theoretischen Logik". Die Fragen sind noch im Fluß, es sind, besonders von K. Gödel, gewichtige Einwände gegen die Beweistheorie erhoben worden. Aber Hilbert ist durchdrungen von der Überzeugung ihres endlichen Erfolgs, und jedenfalls ist sie eine Leistung von so hoher Originalität und Kraft, daß dieses Werk seines Alters zu seinen größten zählen wird.*[146]

Another attempt at a rigorous and comprehensive presentation of mathematics was initiated in the 1930s by a group of excellent French mathematicians publishing under the collective pseudonym Nicolas Bourbaki.[147] The problems of consistency were put aside, since the group did not consider them to be important. They expressed their point of view in 1954 in the final paragraph of the introduction to their book *Théorie des ensembles*:

> *En résumé, nous croyons que la mathématique est destinée à survivre, et qu'on ne verra jamais les parties essentielles de ce majestueux édifice s'écrouler du fait d'une contradiction soudain manifestée; mais nous ne prétendons pas que cette opinion repose sur autre chose que sur l'expérience. C'est peu, diront certains. Mais voilà vingt-cinq siècles que les mathématiciens ont l'habitude de corriger leurs erreurs et d'en voir leur science enrichie, non appauvrie; cela leur donne le driot d'envisager l'avenir avec sérénité.*[148]

9. Epilogue — A New View of Mathematics

At the first International Congress of Mathematicians held in Zürich, 1897, Henri Poincaré delivered a philosophically toned plenary lecture entitled *Sur les rapports de l'analyse pure et de la physique mathéématique*, which focused on the relation between mathematical analysis and mathematical physics. In the beginning he said:

> *On vous a sans doute souvent demandé à quoi servent les mathématiques et si ces délicates constructions que nous tirons tout entières de notre esprit ne sont pas artificielles et enfantées par notre caprice.*

---

[146] O. Blumenthal, *Lebensgeschichte*, Hilbert (2001, III, pp. 388–429), see p. 425.

[147] See N. Bourbaki, The Architecture of Mathematics, *The American Mathematical Monthly* **57**(1950), pp. 221–232, ISSN 0002-9890. N. Bourbaki, Foundations of Mathematics for the Working Mathematician, *The Journal of Symbolic Logic* **14**(1949), pp. 1–8, ISSN 0022-4812. J. Dieudonné, The Work of Nicholas Bourbaki, *The American Mathematical Monthly* **77**(1970), pp. 134–145, ISSN 0002-9890. L. Corry, *Bourbaki*, in N. Koertge (ed.), *New Dictionary of Scientific Biography I*, Charles Scribner's Sons, Gale, Centage Learning, Detroit, 2008, pp. 363–366. See also Ewald (1996, pp. 1263–1276).

[148] N. Bourbaki, *Théorie des ensembles I., II.*, Hermann, Paris, 1954, 136 pages, see p. 9.

He also spoke on the relation of pure and applied mathematics, emphasized a triple role of mathematics, as well as its detachment from the real world:

> *D'ailleurs une science uniquement faite en vue des applications est impossible; les vérités ne sont fécondes que si elles sont enchaînées les unes aux autres. Si l'on s'attache seulement à celles dont on attend un résultat immédiat, les anneaux intermédiaires manqueront, et il n'y aura plus de chaîne.*
>
> *Les mathématiques ont un triple but. Elles doivent fournir un instrument pour l'étude de la nature. Mais ce n'est pas tout: elles ont un but philosophique et, j'ose le dire, un but esthétique. Elles doivent aider le philosophe à approfondir les notions de nombre, d'espace, de temps.*
>
> *C'est pourquoi je n'hésite pas à dire que les mathématiques méritent d'être cultivées pour elles-mêmes et que les théories qui ne peuvent être appliquées à la physique doivent l'être comme les autres.*[149]

Already in the beginning of the 20th century, Bertrand Russell advocated a new conception of mathematics. He emphasized its abstract character – mathematics need not be related to the real world:

> *We start, in pure mathematics, from certain rules of inference, by which we can infer that if one proposition is true, then so is some other proposition. These rules of inference constitute the principles of formal logic. We then take any hypothesis that seems assuring, and deduce its consequences. If our hypothesis is about anything, and not about some one or more particular things, then our deductions constitute mathematics. Thus mathematics may be defined as the subject in which we never know what we are talking about, nor wherhet what we are saying is true...*
>
> *As one of the chief triumphs of modern mathematics consists in having discovered what mathematics really is, ...*[150]

Growing tendencies toward the rigourisation of mathematical disciplines, creation of solid foundations of mathematics, axiomatization, logicism and formalism have weakened the relations between mathematics and the real material world. Mathematics intensely began to develop theories detached from the real world. At the turn of the 20th century, the view of mathematics has radically changed. During the next decades, abstract mathematics became increasingly alienated from reality. These facts are corroborated by excerpts from the entries of Encyclopedia Britannica (Vol. 15, 1941, pp. 87, 83) written by Alfred North Whitehead and Frank Plumpton Ramsey (1903–1930).[151]

---

[149] H. Poincaré, *Sur les rapports de l'analyse pure et de la physique mathématique*, Verhandlungen des ersten Internationalen Mathematiker-Kongresses in Zürich vom 9. bis 11. August 1897, Teubner, Leipzig, 1898, viii + 307 pages, pp. 81–90, see pp. 81–82.

[150] B. Russell, Recent Work on the Principles of Mathematics, *International Monthly* **4**(1901), pp. 83–101, see p. 84.

[151] F.P. Ramsey, The Foundations of Mathematics, *Proceedings of the London Mathematical Society* **25**(1926), pp. 338–384, ISSN 0024-6115. F.P. Ramsey, *Foundations. Essays in Philosophy, Logic,*

In the encyclopedia's entry "*Mathematics, Nature of*", Alfred North Whitehead characterized mathematics by these words:

> *... science concerned with the logical deduction of consequences from the general premises of all real reasoning.*

In the entry "*Mathematics, Foundations of*", Frank Plumpton Ramsey wrote about geometry:

> *... as a branch of mathematics, geometry has no essential reference to physical space ...*
> *... regards geometry as simple tracing the consequences of certain axioms dealing with undefined terms, which are really variables in the ordinary mathematical sense, like x and y. And he demands of his axioms, not that they should be true on some particular physical interpretation of the undefined terms, but merely that should be consistent with one another.*

The crisis in the foundations of mathematics bothered mainly those researchers dealing with logic and problems of a rigorous development of mathematics.[152] Other mathematicians were aware of it, but it did not restrain themselves from intense work in their disciplines. Nevertheless, the acceptance of Gödel's results has led to the loss of the earlier absolute confidence in mathematical results, theories, and mathematics itself.[153]

We conclude with two quotations from the works of Hermann Weyl, one of the foremost mathematicians of the 20th century, author of the famous book *Raum. Zeit. Materie. Vorlesungen über allgemeine Relativitätstheorie* from 1918, who also took interest in physics and philosophy of mathematics and natural science.[154] In

---

Mathematics and Economics, ed. D.H. Mellor, Humanities Press, Atlantic Highlands, New Jersey, 1978, viii + 287 pages, ISBN 0-394-00814-5.

[152]For example, the inaugural lecture *Über Gewissheit in der Mathematik* (Berlin, 1930) given by Erhard Schmidt as a newly appointed rector went in this direction.

[153]Readers are encouraged to consult the book M. Kline, *Mathematics. The Loss of Certainty*, Oxford University Press, New York, 1980, 366 pages, ISBN 0-19-502754-X.

[154]For the monograph A. Bäumler and M. Schröter (eds.), *Handbuch der Philosophie* (Oldenbourg, München und Berlin, 1927–1934), Weyl wrote the treatise *Philosophie der Mathematik und Naturwissenschaft* (Abt. II.A, 1927, 162 pages), which was also published independently (Leibniz-Verlag, München, 1928, 162 pages). It was later substantially expanded (3rd ed. Oldenbourg, München, Wien, 1966, 406 pages), new edition: Oldenbourg, W. de Gruyter, 2008, 2014, 408 pages, ISBN 978-3-486-46794-9; English translation: *Philosophy of Mathematics and Natural Science*, Princeton University Press, Princeton, 1949, x + 311 pages; later edition: 2009, 336 pages, ISBN 9780691141206.

H. Weyl, *Das Kontinuum. Kritische Untersuchungen über die Grundlagen der Analysis*, Veit, Leipzig, 1918, v + 83 pages, reprint: 1932; English translation: *The Continuum*, Dover, New York, 1994.

H. Weyl, Die heutige Erkenntnislage in der Mathematik, *Symposion* **1**(1925), pp. 1–32, also in Weyl (1968, II, pp. 511–542); English translation: *The Current Epistemological Situation in Mathematics*, Mancosu (1998, pp. 123–142).

a very cogent way, they express not only his loss of mathematical certainty, but also his attitude toward the new situation.

In the introduction to his 1920 paper *Über die neue Grundlagenkrise der Mathematik*,[155] Weyl took the following view on the antinomies:

> *Die Antinomien der Mengenlehre werden gewöhnlich als Grenzstreitigkeiten betrachtet, die nur die entlegensten Provinzen des mathematischen Reichs angehen und in keiner Weise die innere Solidität und Sicherheit des Reiches selber, seiner eigentlichen Kerngebiete gefährden können.*

After a quarter century, in 1946, he concluded his treatise *Mathematics and Logic. A brief survey serving as preface to a review of The Philosophy of Bertrand Russell* with the following words:

> *... we are less certain than ever about the ultimate foundations of (logic and) mathematics. Like everybody and everything in the world today, we have our "crisis". We have had it for nearly fifty years. Outwardly it does not seem to hamper our daily work, and yet I for one confess that it has had a considerable practical influence on my mathematical life: it directed my interests to fields I considered relatively "safe", and has been a constant drain on the enthusiasm and determination with which I pursued my research work. This experience is probably shared by other mathematicians who are not indifferent to what their scientific endeavors mean in the context of man's whole caring and knowing, suffering and creative existence in the world.*[156]

References

G. Cantor, *Gesammelte Abhandlungen mathematischen und philosophischen Inhalts. Mit erläuternden Anmerkungen sowie mit Ergänzungen aus dem Briefwechsel Cantor-Dedekind, nebst Lebenslauf Cantor's von Adolf Fraenkel*, Hrsg. Ernst Zermelo, Springer, Berlin, 1932, vii + 486 pages. Reprint: Olms, Hildesheim, 1962. Reprint: Springer-Verlag, Berlin, Heidelberg, New York, 1980. ISBN 3-540-09849-6.

Compte Rendu du deuxième congrès international des mathématiciens tenu à Paris du 6 au 12 août 1900, Gauthier-Villars, Paris, 1902, 455 pages.

M. Davis (ed.), *The Undecidable. Basic Papers on Undecidable Propositions, Unsolvable Problems and Computable Functions*, Raven Press, Hewlett, New York, 1965, 440 pages.

---

[155] H. Weyl, Über die neue Grundlagenkrise der Mathematik, *Mathematische Zeitschrift* **10**(1921), pp. 39–79, see p. 39, ISSN 0025-5874. Also in Weyl (1968, II, pp. 143–180), Weyl (1956, pp. 211–248, p. 211); English translation: *On the New Foundational Crisis of Mathematics*, Mancosu (1998, pp. 86–118). Reprint: Wissenschaftliche Buchgesellschaft, Darmstadt, 1965, 41 pages. This article was based on Weyl's lecture at the Mathematische Kolloquium in Zürich.

[156] H. Weyl, Mathematics and Logic, *The American Mathematical Monthly* **53**(1946), pp. 2–13, see p. 13, ISSN 0002-9890. Also in Weyl (1968, IV, pp. 268–279).

Other interesting publications due to Weyl are the paper A Half-Century of Mathematics, *ibid.* **58**(1951), pp. 523–553, and the book *The Open World. Three Lectures on the Metaphysical Implications of Science*, Yale University Press, New Haven, 1932, v + 84 pages (*God and the Universe*, pp. 1–29; *Causality*, pp. 31–56; *Infinity*, pp. 57–84).

W. Ewald (ed.), *From Kant to Hilbert. A Source Book in the Foundations of Mathematics I., II.*, Clarendon Press, Oxford, 1996, xviii + 648 + xxviii, xv + 649–1340 pages. ISBN 0-19-850535-3, 0-19-850536-1.

K. Gödel, *Collected Works I., II., III., IV., V.*, S. Feferman (ed.), Oxford University Press, New York, 1986, 1990, 1995, 2003, 2003, xvi + 474, xv + 407, xvii + 532, xix + 662, xxiii + 664 pages. Reprint: 2001. ISBN 0-19-514720-0, 0-19-514721-9, 0-19-514722-7, 0-19-850073-4, 0-19-850075-0.

I. Grattan-Guinness, *The Search for Mathematical Roots, 1870–1940. Logic, Set Theories and the Foundations of Mathematics from Cantor through Russell to Gödel*, Princeton University Press, Princeton and Oxford, 2000, xiv + 690 pages. ISBN 0-691-05857-1.

I. Grattan-Guinness (ed.), *Landmark Writtings in Western Mathematics 1640–1940*, Elsevier, Amsterdam, 2005, xvii + 1022 pages. ISBN 0-444-50871-6.

D. Hilbert, *Grundlagen der Geometrie, Siebente umgearbeitete und vermehrte Auflage*, B.G. Teubner, Leipzig und Berlin, 1930, vii + 326 pages.

D. Hilbert, *Gesammelte Abhandlungen I., II., III.*, Springer-Verlag, Berlin, 1932, 1933, 1935, xiv + 539, viii + 453, vii + 435 pages. Reprint: Chelsea, Bronx, New York, 1965, 2nd ed., 1970. Reprint: 2001. ISBN 0-19-514720-0.

L. Kronecker, *Werke I., II., III.-1, III.-2, IV., V.*, 1895, 1897, 1899 + 1931, 1929, 1930. Reprint: Chelsea, New York, 1968, ix + 483, vii + 540, vii + 473 + iii + 215, x + 509, x + 527 pages.

P. Mancosu, From *Brouwer to Hilbert. The Debate on the Foundations of Mathematics in the 1920s*, Oxford University Press, New York, Oxford, 1998, x + 337 pages. ISBN 0-19-509632-0.

G. Peano, *Opere Scelte I., II., III.*, Cremonese, Roma, 1957–1959, vii + 530, vi + 518, vii + 470 pages.

C. Reid, *Hilbert*, Springer-Verlag, New York, Heidelberg, Berlin, 1970, xi + 290 pages. Later edition: Copernicus, New York, 1996. ISBN 0-387-94674-8.

J. van Heijenoort (ed.), *From Frege to Gödel. A Source Book in Mathematical Logic, 1879–1931*, Harvard University Press, Cambridge, MA, 1967, x + 660 pages; 4th ed., 2002, 680 pages. ISBN 9780674324497.

H. Weyl, *Selecta*, Birkhäuser, Basel, Stuttgart, 1956, 592 pages.

H. Weyl, *Gesammelte Abhandlungen, I.–IV.*, Hrsg. K. Chandrasekharan, Springer-Verlag, Berlin, Heidelberg, 1968, xx + 698, xx + 647, xviii + 791, x + 694 pages. ISBN 978-3-662-43804-6, 978-3-662-44236-4, 978-3-662-44251-7, 978-3-662-44289-0.

https://doi.org/10.1142/9781786349316_0010

Chapter 10

Set Theory — A Vehicle to Modern Mathematics in Poland

Roman Duda

1. Beginnings

In the theory of trigonometric series there was a problem[1]: if the sum of a trigonometric series is equal to 0 in all points except the points of a set Z, are all coefficients equal to 0? Heinrich Heine (1821–1881) has shown that the answer is positive if Z is finite and Georg Cantor (1845–1918) went further, showing that the answer is positive also for some infinite Z. Beginning in 1872, he was able to characterize such sets using new notions such as a *limit point* of a set, *derivative* of a set (the set of all limit points of the given set), *closed set* (a set which contains all its limit points), *perfect set* (identical with its derivative), *open set* (the complement of a closed set), etc. The sets initially considered were *linear* (i.e., subsets of the real line), but fairly soon Cantor extended the scope to sets lying in a Euclidean space of any dimension. In this way he began the construction of a whole new world[2] which in the next decades was to influence all mathematics. Cantor has called that world *set theory* (*Mengenlehre*). Since it initially dealt with sets in Euclidean spaces, the theory was for a time intermingled with the Euclidean geometry, thus leading to *geometric topology* and to the *theory of real functions* which were then emerging simultaneously,[3] being a strong stimulus for their development. The increasing generality of

[1]The problem was proposed to Cantor by his mentor, Professor Heinrich Heine, see, e.g., J. Dieudonné, *Geschichte der Mathematik 1700–1900*, Berlin, 1985, p. 401. And it is reminded in Cantor's first paper on the topic: G. Cantor, Ueber die Ausdehnung eines Satzes aus die Theorie der trigonometrischen Reihen, *Mathematische Annalen* **5**(1872), pp. 123–132; the paper is reprinted in Cantor (1984, pp. 9–18).
[2]Cantor's writings can be found in his collected works Cantor (1932). His set-theoretical works were also edited separately, with many useful comments, see Cantor (1984).
[3]See Kline (1972), Chapter 44, *The Theory of Functions of Real Variables*, pp. 1014–1051, and Chapter 50, *The Beginnings of Topology*, pp. 1158–1181. See also Mannheim (1964).

the theory seemed also to offer a set-theoretical basis for the whole of mathematics and, in consequence, to reshape it radically.

Among the first to understand the meaning of set theory for mathematics was Camille Jordan (1838–1922) whose textbook[4] presented the principles of mathematical analysis based on set theory *avec toute la précision et généralité* and its second edition of 1893 contained a special supplement devoted to that theory. The book has become influential. Generations of French (and not only French) mathematicians have learned from it. In particular, the book has opened a new phase of the French theory of functions.[5]

2. Lectures on Set Theory in Kraków and Lwów

In the first decade of the 20th century set theory was already an object of rising interest both for the considered problems and for the role it seemed to play in mathematics.

The main problem of the theory at that time seemed to be the *Continuum Hypothesis*. Stated by Cantor in 1878, it was saying that each infinite subset of reals is either countable or of the cardinality of all reals. In his famous lecture during the 2nd Congress of Mathematicians, held in 1900 in Paris, David Hilbert (1862–1943) began his list of open problems with that hypothesis. There were some unsuccessful attempts to prove it by Paul Tannery (1884), Beppo Levi (1900), Felix Bernstein (1904), Julius König (1904), and some others. The hypothesis turned out thus to be a difficult problem and the situation raised much excitement and interest.[6]

Another problem, also mentioned by Hilbert, was whether the set of all reals can be well ordered. To prove that assumption, Ernst Zermelo (1871–1953) has introduced the *Axiom of Choice* (1904). His proof was correct but the Axiom itself has become an object of heated debates involving many eminent mathematicians of the time: René Baire (1874–1932), Henri Lebesgue (1875–1941), Henri Poincaré (1854–1912), Jacques Hadamard (1865–1963), Bertrand Russell (1872–1970), Godfrey Hardy (1877–1947), Felix Klein (1849–1925), Giuseppe Peano (1858–1932) and others.[7]

[4]See Jordan (1893–1896).
[5]See Medvedev (1976). See also H. Gispert-Chambraz, *Camille Jordan et les fondements d'analyse. Comparaison de la 1ère edition 1882–1887 et du la 2ème 1893*, Publications Mathématiques d'Orsay 82-05 (Thèse du doctorat), 1982, 168 pages, esp. pp. 82–105.
[6]Later Sierpiński has devoted a monograph Sierpiński (1934) to it. Long efforts to solve the hypothesis were completed by Kurt Gödel in 1940 and Paul Cohen in 1963. Relying on commonly accepted axiomatics of set theory, they have shown that the hypothesis can be neither refuted (Gödel) nor proved (Cohen). It is independent.
[7]See Medvedev (1982).

Another source of interest were antinomies (paradoxes) found within set theory by Cantor, Russell, Cesare Burali-Forti (1861–1931) and others.[8]

Aside from intriguing problems, set theory offered a new language to deal with mathematics. Simple and precise, it could be used in all branches whatsoever, and even more, it seemed to offer a chance to base mathematics on set theory. The chance was pursued by Bertrand Russell (1872–1970) and Alfred North Whitehead (1861–1947) in their widely read *Principia Mathematica.*[9]

Chapters on set theory began to appear in some mathematical textbooks (cf. Jordan) and there came first special university lectures. The first textbook in Polish dealing with that theory was the monograph *The Theory of Analytic Functions* by Józef Puzyna (1856–1912).[10] The author presented that theory as based on set theory and included an extensive Chapter III called *On Set Theory*, presenting its basics.[11] Following Jordan's lead, the monograph turned out marvelous but, written in Polish, its influence was rather limited. However, it introduced Polish terminology of set theory, and Polish mathematicians of the next generation have regarded it highly: *The book is a genuine encyclopedia of Analysis. ... If it were published in a more common foreign language, it would soon enjoy further, still more perfect editions, having every reason to become a classic textbook.*[12]

Among the first university lectures on set theory were those given by Wacław Sierpiński (1882–1969) in Lwów and by Stanisław Zaremba (1863–1942) in Kraków.[13] Sierpiński lectured in 1910 and for him, as one can judge from his book which arose from those lectures and was published two years later (see Section 3), set theory was becoming a new field. The book starts with an emphatic motto, taken from Louis Couturat (1868–1914): *S'il y a une idée première et fondamentale en Mathématique, ce n'est donc pas l'idée de nombre, mais bien celle d'ensemble.* Consequently, in the eyes of Sierpiński, sets became fundamental objects of mathematics and set theory gained the status of an independent field worthy of interest also in itself. In contrast to Sierpiński, for Zaremba any considered set is "an entirety

[8]See Grattan-Guinness (1994); particularly Chapter 5.3 in volume I: A.R. Garciadiego, *The Set Theoretic Paradoxes*, pp. 629–634. See also van Heijenoort (1967).

[9]B. Russell, *The Principles of Mathematics*, Cambridge University Press, Cambridge, 1903, 534 pages. Extended and rewritten as Russell and Whitehead (1910).

[10]J. Puzyna, *Teorya funkcyj analitycznych* [Theory of Analytic Functions], 2 volumes, Published by the Author with the benefit of the Academy of Sciences in Kraków, Lwów, 1899–1900, 549 + 633 pages.

[11]See S. Domoradzki, Teoria mnogości w dziele Józefa Puzyny Teorya funkcyj analitycznych [Set Theory in the Józef Puzyna's Work Teorya Funkcyj Analitycznych], in Więsław (2012), pp. 45–58.

[12]See Saks and Zygmund (1938), p. viii. After World War II the book has been translated into English (1952, reedited 1968 and 1971) and French (1970).

[13]R. Duda, Pierwsze wykłady uniwersyteckie teorii mnogości na ziemiach polskich [The First University Lectures on Set Theory on Polish Soil], *Analecta* **23.1**(2014), pp. 163–178, ISSN 1230-1159.

of objects which satisfy conditions stated in its definition" which clearly shows that in his view set theory played a servant role only. In fact, his lectures were devoted to the theory of reals and the only sets considered by him were well defined sets of real numbers. Set theory was for him a method to make statements more precise and proofs shorter.

The difference of opinion on set theory between Sierpiński and Zaremba seems to be typical for mathematicians at the turn of the 20th century. Most were careful but there were also those who, like Sierpiński, bravely investigated the new vistas.

3. Sierpiński and His Lwów Seminar

A high respect of Sierpiński for set theory is seen not only in his lectures and the book to which they led, but also in his seminar which he started soon after his arrival to Lwów in 1908. In those times, set theory, still in its beginnings, began to take hold. In 1908 Zermelo provided an axiomatics for it, but difficult problems, including Continuum Hypothesis, were still laid open while the Axiom of Choice was treated suspiciously and by some strongly opposed. In addition, most sets in consideration were lying in Euclidean spaces that gave them some geometric flavor illustrated by the concept of a limit point. Their study has led to another rising discipline, initially called *Analysis Situs* but soon renamed to *Topology*. Before World War I and some time thereafter *set theory* and *topology* were hardly discernible. And surely they were not in Sierpiński's seminar. First original results obtained there could be classified more as topological than set-theoretical, and some as belonging rather to the theory of real functions.

The seminar was attended by the young mathematicians Stefan Mazurkiewicz (1888–1945), Stanisław Ruziewicz (1889–1941), and Zygmunt Janiszewski (1888–1920), all three of whom later became well known. Mazurkiewicz studied in Kraków, Munich, Göttingen and Lwów, where he obtained his PhD in 1913 for the thesis *Contributions to Set Theory*. Contributions were original and dealt with continuous images of a segment that later led to the characterization of continuous images of a segment (at present called Hahn–Mazurkiewicz–Moore characterization of Peano continua).[14] In 1913 also Ruziewicz, who had spent all his life in Lwów, got there his PhD for the thesis on a function which is monotone and continuous but does not possess a derivative in uncountably many points. Also his research was original but his main topics were real functions (from a set-theoretical standpoint).[15] And Janiszewski who came to Lwów after studies in Zurich, Munich, Göttingen and

[14]See Mazurkiewicz (1969). Contributions of Mazurkiewicz to topology are analyzed in: K. Kuratowski, Mazurkiewicz et son œuvre scientifique, *Fundamenta Mathematicae* **34**(1947), pp. 316–331; R. Pol, *The Works of Stefan Mazurkiewicz in Topology*, in Aull and Lowen (1998), pp. 415–430.

[15]See W. Wilczyński, Twórczość Stanisawa Ruziewicza [Work of Stefan Ruziewicz], in Więsław (2013), pp. 239–245.

Paris, where he got his PhD, was already known for his contributions to the theory of continua.[16] And Sierpiński was then already well known both for his original contributions to set theory and for the above-mentioned book *An Outline of Set Theory*, one of the very first books in the world on that topic.[17]

Although set theory has eventually led to several distinct fields such as point-set topology, theory of real functions, or measure theory, and in its pure core there became discernible parts like the theory of cardinals, theory of order, or descriptive set theory, Sierpiński has treated all that as one common discipline, vastly contributing to all of its parts.[18] And this general attitude of him has been adopted by the Polish school of mathematics of the interwar period. Before that, however, there came World War I.

4. World War I

World War I, which broke out in 1914, changed everything. At first it brought an end to the Lwów seminar: Sierpiński was interned in Russia; Janiszewski volunteered to the Polish Legiony and spent the winter of 1914/1915 in the Carpathian Mountains fighting against Russians; Mazurkiewicz returned to his native Warsaw; Ruziewicz was enrolled into the Austrian army. With no leader behind, the group ceased to exist.

Of importance, however, was another development. In July 1915 Russians left Warsaw, taking with them the Russian university, which existed there since 1869, together with its staff and belongings to Rostov upon Don. And Germans, who took over Central Poland, agreed to the petition to found the Polish university in Warsaw (analogous petition in Vilnius was rejected by the Germans). The founding was done in the autumn of 1915 and the first two mathematical chairs in that newly founded university were offered to Janiszewski and Mazurkiewicz. Polish intellectual life was entering a new phase and the newly founded journal *Nauka Polska*[19] invited Polish

[16]See Janiszewski (1962). The book contains an article evaluating his work.

[17]Originally published in Polish: W. Sierpiński, *Zarys teoryi mnogości* [An Outline of Set Theory], Biblioteka Matematyczno-Fizyczna, Warszawa, 1912, 158 pages (extended and reedited 1923 and 1928). Translated into English as: W. Sierpiński, *Introduction to General Topology*, University of Toronto Press, Toronto, 1934, 158 pages (Reedited 1952).

[18]See Sierpiński (1974–1975). Volumes 2 and 3 have the common subtitle *Théorie des ensembles et ses applications* and contain not only original articles by Sierpiński but also the evaluation of his contributions to set theory, topology, and theory of real functions. There are also other articles on him and his contributions. The case of topology in Poland has been treated in: R. Duda, *Początki topologii w Polsce* [Beginnings of Topology in Poland], in Strzałkowski (2001), pp. 101–137.

[19]The full title was *Nauka Polska. Jej potrzeby, organizacja i rozwój* [Polish Science. Its Needs, Organization and Development]. The journal was edited in the period 1918–1947 (with a break of World War II) by the Mianowski Fund. In the communist times the Fund was closed down while the journal was intercepted by the newly founded Polish Academy of Sciences, which edited it until 1991 under the shortened title *Nauka Polska*. After the collapse of communism the Fund restored it in 1991 and took the journal back, reintroducing the full title and continuing its editing.

scientists to report on the status and the needs of their disciplines. Response was wide, including three mathematicians: Zaremba, Janiszewski and Mazurkiewicz.[20]

In 1918 Poland regained its independence. In the opinion of Zaremba, the most important task of the reborn state was the "organization of public education" including "educating adequate number of teachers for secondary schools and teachers' seminars". For him, teachers were a proper circle for selecting future scientists. In other words, a talented student (in mathematics) should first become a teacher and then try to continue his scientific development by undertaking some research (naturally adopted from foreign centers).[21] The opinion was based on his own personal experience (serving for many years as a teacher in French lycées, Zaremba wrote his first mathematical papers) but it turned out inadequate to the situation.

In contrast to Zaremba, Janiszewski suggested a goal "to win an independent status for Polish mathematics" and proposed some steps to achieve it. His plan was: to select a single area in mathematics (a natural candidate was set theory) and to concentrate efforts of the whole young generation upon it; to foster a culture of mutual close collaboration; to establish a journal dedicated exclusively to the selected area and published in foreign languages only.[22] Although the project was also based upon his personal experience as a member of the Sierpiński seminar in Lwów with the addition of his own postulate of founding a new journal, it looked undoubtedly shocking. If Polish mathematicians were to focus most of their efforts exclusively on a single field, there was a real danger that other areas will be neglected, and if the choice was incorrect, the whole generation would be lost. Friendly collaboration is in sharp contrast with the culture of competition, considered by many an all important factor of progress. And in that time there was no single mathematical journal restricted to a particular area and there came several warnings that such a journal either will soon disappear for lack of material or will lose any significance by publishing inadequate papers. And Janiszewski's demand that this journal will be published in internationally recognized languages only raised protests against abolishing the Polish language.

Mazurkiewicz was the third author. He was supporting Janiszewski's plan and openly suggested "set theory, including axiomatics, logistics and applications" as the area to choose. And he turned his attention to some particular problems like libraries, printing shops, scholarships, necessity of decentralization.[23]

---

[20] All reports have filled in the first two volumes of *Nauka Polska*, thus yielding a wide panorama of the status of Polish science at the time.

[21] S. Zaremba, O najpilniejszych potrzebach nauki w Polsce ze szczególnym uwzględnieniem matematyki [On the Most Urgent Needs of Science in Poland with a Particular Attention to Mathematics], *Nauka Polska* **1**(1918), pp. 1–10.

[22] Z. Janiszewski, Stan i potrzeby matematyki w Polsce [State and Needs of Mathematics in Poland], *Nauka Polska* **1**(1918), pp. 11–18. An English translation appeared in the book Kuzawa (1968).

[23] S. Mazurkiewicz, O potrzebach matematyki w Polsce [On the Needs of Mathematics in Poland], *Nauka Polska* **2**(1919), pp. 1–5.

Poland had yet to fight for its very existence and borders, including Polish–Ukrainian war in 1919 and Polish–Soviet war in 1920, and so, for the time being, the project of Janiszewski could not be realized on a full scale. Nevertheless, Janiszewski began to collect papers for the first issue of *Fundamenta Mathematicae*, the title he invented for a "journal devoted to set theory and its applications" (quotation taken from the cover of the journal when it appeared).

In spite of turbulent times, still in 1919 there were founded (or refounded) five state universities (Warsaw, Kraków, Lwów, Vilnius, and Poznań) and three state polytechnics (Lwów, Warsaw, Kraków). When peace at last came, they could start normal work, in particular offering a modest basis to mathematics with around 80 academic positions. With the growing community; however, it was not enough and many mathematicians had to work as teachers. One may say that Zaremba's project was forcefully applied.

5. Moscow: Sierpiński and Lusin

When World War I broke out, Sierpiński was on holiday in Russia. Being an Austrian citizen, he was interned and sent to Wiatka. However, thanks to his Russian friends, he was allowed to move to Moscow, where he cooperated with Lusin until the end of 1917. There he was a witness of Souslin's discovery of analytic sets and thereafter he became one of the founders, together with Lusin, of the so-called *descriptive set theory* devoted to searching for a wider realm of sets than so far.

The basic realm for set theory was the real line and its subsets. For Cantor a fundamental class consisted of closed subsets, but Borel showed (1898) that taking their complements, countable unions and countable intersections one receives a great many new sets. All those sets are now called *Borel sets* and Lebesgue showed (1905) that they form a hierarchy of uncountably many levels with each level richer than all preceding ones. It was an incredible richness. Borel sets are sufficient to cultivate classical analysis including the theory of real functions and so the French mathematicians were fully satisfied with them. However, in 1919 Souslin showed that some continuous images of Borel sets, presently called *analytic sets*, were not Borel. It was the first glimpse into a new incredible realm and research on analytic sets has shown that they have some interesting properties: they satisfy the continuum hypothesis (Lusin), Borel sets are exactly those whose complements are analytic (Souslin, Sierpiński), analytic sets are Lebesgue measurable (Souslin), analytic sets enjoy the Baire property (Souslin, Sierpiński), etc.[24]

However, it turned out soon that also complements of analytic sets need not be Borel and these discoveries led to an analogous (to Borel sets) hierarchy of *projective sets*, still richer than that of Borel sets. But neither projective sets are all possible

[24]G. Sinkevich (2012), Chapter Teoria mnożestw [Theory of Sets], pp. 83–127.

subsets of the real line, since there exist subsets of the real line which are not projective. The so-called *descriptive set theory* is aimed at research in the area of all subsets of the real line with the use of all possible methods including the Axiom of Choice. The main early contributors to that theory were Lusin and Sierpiński (Souslin died in 1919), and main early centres were Moscow and Warsaw (after the return of Sierpiński to Warsaw in 1918).[25]

6. *Fundamenta Mathematicae* and the Warsaw School of Mathematics

Soon after announcing his project, Janiszewski began to collect papers for the proposed new journal with the intention "to present in the first volume of the journal, which I have invented, I have created and I am editing, possibly all Polish mathematicians working in the area of set theory to which the journal is devoted".[26] He managed to prepare the first volume with 25 papers by the following eight authors: Stefan Banach (1892–1945), Zygmunt Janiszewski, Kazimierz Kuratowski (1896–1980), Stefan Mazurkiewicz, Stanisław Ruziewicz, Wacław Sierpiński, Hugo Steinhaus (1887–1972), Stefan Wilkosz (1891–1941). It was also the time when Sierpiński returned to Warsaw.[27] As he later recalled, *When in 1919 all three of us — Janiszewski, Mazurkiewicz, and myself — found ourselves together as the first professors of mathematics in the revived Polish university in Warsaw, we decided to realize the idea of Janiszewski to publish in Warsaw, in foreign languages, a journal devoted to the theory of sets, topology, theory of functions of a real variable, and mathematical logic. In that way "Fundamenta Mathematicae" came into existence.*[28]

The first volume of *Fundamenta Mathematicae* appeared in 1920. Although Janiszewski prepared it, he didn't see it (he died in January 1920, a few months before the event). The chief editor became Sierpiński and his policy, not openly declared but clear to everyone interested, was to accept for publication any paper containing original results, without considering such aspects as "value", "depth", etc. Such a policy, obviously greatly encouraging young researchers, helped the group

[25]See Sinkewicz (2012). See also C. Phili, *La genèse et le dévelopment de la théorie descriptive des ensembles*, Istoriko-matematièeskije issledovanja, II series, Moscow, 2010, 62 pages, ISBN 978-5-8037-0474-4; J. Mioduszewski, Problemy mnogościowe w pracach Mikołaja Łuzina i ich związki z matematyką polską [Set-Theoretical Problems in the Works of Nicolai Lusin and Their Relations to Polish Mathematics], *Analecta* **23.1**(2014), pp. 163–178, ISSN 1230-1159 (Polish).

[26]It is a quotation from Janiszewski's letter, see Preprint C-1 (1980) of the Institute of Mathematics of the Polish Academy of Sciences.

[27]Sierpiński left Moscow after the February Revolution but he went first to Lwów and only thereafter to his native Warsaw.

[28]W. Sierpiński, Polish School of Mathematics, *Polish Perspectives* **6**(8)(1963), pp. 25–35.

to expand rapidly. This was the way in which the journal became instrumental in raising Polish mathematics.[29]

The beginnings, however, were not smooth. When Lebesgue saw the first volume, he responded by sending a friendly letter to Sierpiński in which he praised the content but expressed doubts as to the possibility of obtaining enough valuable papers to support its further existence.[30] Nevertheless, the work continued. Volume II, which appeared in 1921, consisted of 30 papers, including nine by Sierpiński and one by Lebesgue, and there appeared new Polish names: Stanisław Saks (1897–1942), Stefan Kempisty (1892–1940), Bronisław Knaster (1893–1980). The journal was noted and appraised. Raymond Clare Archibald (1875–1955), the editor of the *American Mathematical Monthly*, wrote in the 1921 issue: *Of the ten mathematical periodicals started since January 1919, none are of such notable importance for mathematical research as "Fundamenta Mathematicae" of which two volumes have appeared.* And in 1921 Lebesgue published a note,[31] in which he described the new journal and went on to evaluate its position and possible development. Of interest are also Lusin's comments on the Polish mathematics of that time.[32]

The journal has continued its existence involving more and more authors from the country and a number of good names from abroad including Norbert Wiener (FM 4), Louis Antoine (FM 5), Pavel Urysohn (FM 7), Nicolai Lusin (FM 10), John von Neumann (FM 13), Eduard Čech (FM 19), Salomon Bochner (FM 20), Albert Thoralf Skolem (FM 23), and others. And very important: published results were excellent. To quote only a few: the first construction of the pseudo-arc by Knaster (FM 3),[33] the Banach-Tarski Paradox (FM 6), dimension theory of Urysohn (FM 7 and 8), Leja's definition of a topological group (FM 9),[34] the theory of analytical sets by Lusin (FM 10), the theory of retracts by Borsuk (FM 17), the homology theory by Čech (FM 19), the famous "Antipodensatz" by Borsuk and his cohomotopy theory (FM 20), the Bochner integral (FM 20), and many others. No wonder then that the jubilee volume 25 (1935) comprised 47 papers from 12 countries by many

[29] R. Duda, *Fundamenta Mathematicae and the Warsaw School of Mathematics*, in Goldstein *et al.*, (1996), pp. 481–498.

[30] The letter was burnt by Germans in 1944 after the Warsaw Uprising together with the whole library of Sierpiński.

[31] H. Lebesgue, Á propos d'une nouvelle revue mathématique 'Fundamenta Mathematicae', *Bulletin de la Société mathématique de France* **46**(1922), pp. 35–48.

[32] P. Dugac and N. Lusin, Lettres à Arnaud Denjoy avec introduction et notes, *Archives Internationales d'histoire des sciences* **27**(1977), pp. 179–206.

[33] The construction was torturous. When E.E. Moise and R.H. Bing in 1948 rediscovered the pseudo-arc and learned that Knaster did it as early as 1922, they came with disbelief and each became a "pilgrim" to see Knaster in Wrocław.

[34] J.-P. Pier, L'apparition de la théorie des groupes topologiques, *Cahiers du séminaire d'histoire des mathématiques* **9**(1988), pp. 1–21.

first-class mathematicians. As Tamarkin has noted, *the twenty-fifth volume of FM represents a notable event in the mathematical life of the whole world.*[35]

In 1939, FM 32 appeared while FM 33 was in print. Then came World War II and Germans ordered to destroy the typesetting (some parts of it, however, were rescued by the printers). In such circumstances the volume FM 33, with some articles added, appeared only in December 1945, just a few months after the end of the war. It brought the first list of war victims from among Polish mathematicians. The journal still continues to exist.[36]

One of the characteristic features of the Warsaw school of mathematics was a free use of the so-called *ineffective methods*, i.e., methods allowing assertions on the existence of mathematical objects without providing their construction (see the next Section). Aside from the Axiom of Choice such was the Baire Category Theorem. The mastery of Polish mathematicians in using the category method was commonly recognized and found its expression in naming by Bourbaki the class of spaces to which the method applies — *Polish spaces.*[37]

7. Sierpiński and the Axiom of Choice

Intrigued by reading the Paris lecture of 1911 that Bertrand Russell gave on the Axiom of Choice, Sierpiński extended his work in uncovering implicit uses of the Axiom. Still in Lwów, he wrote two papers on the subject, published during World War I when he was in Russia.[38] In the first one he discovered (independently of Cipolla) that, for real functions, the equivalence of continuity at a point and sequential continuity at a point relies on the Axiom. And even more important was his recognition that the countable additivity of Lebesgue measure depended on a weaker Denumerable Axiom which Lebesgue opposed. These (and some other) results showed how deeply the Axiom of Choice resides in mathematics and how much depends on it. Still more impressive was his second paper, an extensive survey, in which he detailed the Axiom's contribution to set theory and analysis. And he assumed an attitude of educated neutrality toward the Axiom: since many important

[35] J.D. Tamarkin, 25 Volumes of Fundamenta Mathematicae, *Bulletin of the American Mathematical Society* **42**(1936), p. 300.

[36] Three volumes of *Fundamenta Mathematicae* are published each year. The publisher is the Institute of Mathematics, Polish Academy of Sciences, Warszawa, ISSN 6026-2736.

[37] See Bourbaki (1958), Chapitre. 9: *Utilisation des nombres réels en topologie générale.* See also R. Duda: Od twierdzenia Baire'a o kategorii do przestrzeni polskich [From the Baire Category Theorem to Polish Spaces], *Antiquitates Mathematicae* **2**(2008), pp. 163–187, ISSN 1898-5203(e).

[38] W. Sierpiński, Sur le rôle de l'axiome de M. Zermelo dans l'analyse moderne, *Comptes Rendus Paris* **163**(1916), pp. 688–691; W. Sierpiński, L'axiome de M. Zermelo et son rôle dans la théorie des ensembles et l'analyse, *Bulletin International de l'Académie des Sciences de Cracovie, Série A*, 1918, pp. 97–152. The latter article has been reprinted in Sierpiński (1974–1975), Volume III, pp. 208–255.

theorems depend on it and none consequence of it led to a contradiction, one may use it but it should be characterized precisely which proofs relied on the Axiom.

In the 1920s, Warsaw mathematicians discovered several further interconnections between the Axiom and some propositions in various branches of mathematics. Tarski gave definitions of a finite set whose equivalence required the Axiom and showed that several propositions in cardinal arithmetic are equivalent to the Axiom. Together with Banach he also demonstrated, via the Axiom, that any ball can be decomposed into finitely many pieces and reassembled into two balls of the same radius.[39] Kuratowski found a maximal principle,[40] later called the Kuratowski–Zorn lemma.[41] Relying on the lemma, Szpilrajn (Marczewski) proved the Order Extension Principle.[42] There were also contributions by Adolf Lindenbaum (1904–1941), Stanisław Ulam (1909–1984), and some others. While these results were set-theoretical, the Axiom of Choice assumed an increasingly vital role in algebra, logical and topology,[43] and set theory evolved into a fundamental discipline for the rest of mathematics with the basic concept of a set.

The history of the Axiom of Choice lies at the cross-roads where mathematics and philosophy meet and between the two World Wars there was a growing dichotomy in attitudes toward it. The principal distinction has already been stated by Hadamard in 1905: the Axiom posits the *existence* of a choice function, not its *construction*. Consequently, with the Axiom there appeared *ineffective* proofs of the existence of an object, with no underlying construction. It was a fundamental change in mathematics. So far, "to exist" meant "to be explicitly constructed" and such was the mood of the preceding millennia. The break of that tradition was abrupt and caused a justifiable resistance the more because there was no convincing philosophical analysis. There was, however, an immense body of mathematical results testifying to the Axiom's deductive strength, developed largely by the analysis of Sierpiński and the Warsaw school. The very weight of that body compelled great many mathematicians to accept a wider understanding of the existence of a mathematical object: its *consistency*. A great role in the dissemination of that view was played by Sierpiński who openly stated that he was a mathematician, not a philosopher, and who accepted the new understanding, with the only reservation he allowed was to denote proofs using the Axiom by a special sign. *He illustrated in detail how thoroughly the Axiom*

---

[39]S. Banach and A. Tarski, Sur la décomposition des ensembles de points en parties respectivement congruentes, *Fundamenta Mathematicae* **6**(1924), pp. 244–277; reprinted in Banach (1967), pp. 118–148, and in Tarski (1986), pp. 119–154. See also Wagon (1985).

[40]C. Kuratowski, Une méthode d'élimination des nombres transfinis des raisonnements mathématiques, *Fundamenta Mathematicae* **3**(1922), pp. 76–108; reprinted in Kuratowski (1988), pp. 62–94.

[41]P.J. Campbell, The Origins of Zorn's Lemma, *Historia Mathematica* **5**(1978), pp. 77–89.

[42]E. Szpilrajn (Marczewski), Sur l'extension de l'ordre partiel, *Fundamenta Mathematicae* **16**(1930), pp. 386–389; reprinted in Marczewski (1996), pp. 18–21.

[43]For an extensive survey, see Moore (1982), esp. Chapter 4, *The Warsaw School*, pp. 196–292.

was intertwined, not just with set theory, but with real analysis as well. He argued that the number of the Axiom's uses were unavoidable because they implied some special case of the Axiom. On the other hand, he granted that many mathematicians had opposed the Axiom and that their position might be justified. Nevertheless, he turned this opposition into a compelling reason to investigate the Axiom's role.[44]

And this view was later generally accepted, serving as a model for solving similar crises.

8. Further Development

Between the two World Wars there appeared around 1500 papers on set theory. Sierpiński himself wrote around 100, his Warsaw followers a further 100, to which logicians and Lwów mathematicians added around 75. Thus the Polish share in the development of set theory itself can be estimated at around 20%.[45] Many other Polish papers, however, were using set-theoretical methods (Sierpiński alone wrote 250 papers of this sort), thus strongly promoting the theory.

Concerning set theory itself, an important influence of Polish mathematicians can be seen in three specific areas:

(1) Axiom of Choice and Continuum Hypothesis;
(2) Descriptive set theory;
(3) Abstract measure problem and measurable cardinals.

Let us recall some main achievements. Concerning the Axiom of Choice, great was the role of Sierpiński because of both his results and the general attitude. Among his results was the theorem on measure–category duality[46] and the Lindenbaum–Tarski–Sierpiński theorem that the General Continuum Hypothesis implies the Axiom of Choice (Lindenbaum and Tarski formulated the theorem, and Sierpiński found its proof). The latter theorem is considered to be the finest Polish contribution to pure set theory. Sierpiński's monograph *Hypothèse du continu*[47] still retains its value. Among other results worth mentioning is the Hurewicz paradox: Continuum Hypothesis is equivalent to the assertion that the countable product of real lines contains

[44]See Moore (1982), esp. Chapter 4.1, *A Survey by Sierpiński*, pp. 197–209.
[45]Cf. M. Zakrzewski, Teoria mnogości [Theory of Sets] in *Historia Nauki Polskiej* ... (1995), p. 118.
[46]The theorem says: there exists a mapping f of the real line onto itself such that $f(E)$ is a set of measure 0 if E is a set of the first category, see W. Sierpiński, Sur la dualité entre la première catégorie et measure nulle, *Fundundamenta Mathematicae* **22**(1934), pp. 276–280; reprinted in Sierpiński (1934), Volume III, pp. 207–210.
[47]See Sierpiński (1934).

an infinite-dimensional subspace H such that each finite-dimensional subspace of H is countable and thus 0-dimensional.[48]

The first steps into the new realm called descriptive set theory were made by Souslin. Lusin and Sierpiński, who together with their followers, have built the theory of analytic sets which form the first level of the projective hierarchy. Kuratowski and Tarski discovered connections between Borel or projective hierarchy and logic, Hurewicz found the first example of a "natural" analytic set which is not a Borel one.

Collecting all those and some other results, the book *Topologie I* by Kuratowski[49] was long considered (at least until the 1980s) the most important synthesis of the descriptive set theory.

In 1914 Hausdorff showed that the general measure problem has no solution in $\mathbb{R}^n$ for $n > 2$ but a couple of years later Banach proved, applying the Axiom of Choice (as Hausdorff did), that for $n = 1$ and 2 the answer is affirmative.[50] And in the next year, he discovered, together with Tarski, the paradoxical decomposition of the unit ball (see Sections 6 and 7). Banach's continuing interest in the measure problem resulted in his reformulation of it in a more set-theoretic way: he asked if there is a function m assigning a nonnegative real number to every subset of the interval $[0, 1]$ which is not trivial (that is, $m(E) > 0$ for some E), is countable additive and vanishes for singleton sets (the latter condition replaced the requirement that congruent sets have the same measure). And in the joint paper Banach and Kuratowski showed, assuming the continuum hypothesis, that the answer was negative.[51] Their theorem completed in a sense the story of the classical measure problem. However, by breaking relations between measure and geometry Banach transferred the measure problem into a pure set-theoretic one by asking for what sets the abstract measure problem has a solution. In other words, for what cardinalities the answer is positive? And this problem of *measurable cardinalities* is among the most vivid ones in the modern pure set theory. First results in that area were obtained by Banach and Ulam.[52]

---

[48]W. Hurewicz, Über unendlich-dimensionale Punktmengen, *Koninklijke Akademie van Wetenschappen, Proceedings of the Section of Sciences* **31**(1928), pp. 916–922; reprinted in Hurewicz (1995), pp. 171–178.

[49]See Kuratowski (1933). This monograph had to consist of two volumes, but the second volume appeared only in 1950. However, it was devoted to metric topology. The monograph has been systematically extended and republished in the original French (1948, 1952, 1957, 1961) but also translated into English (1966, 1968) and Russian (1966, 1969).

[50]S. Banach, Sur le problème de mesure, *Fundamenta Mathematicae* **4**(1923), pp. 7–33; reprinted in Banach (1967), pp. 66–89.

[51]S. Banach and K. Kuratowski, Sur une généralisation du problème de la mesure, *Fundamenta Mathematicae* **14**(1929), pp. 127–131; reprinted in Banach (1967), pp. 182–186, and in Kuratowski (1988), pp. 327–331.

[52]S. Ulam, Zur Masstheorie in der allgemeinen Mengenlehre, *Fundamenta Mathematicae* **16**(1930), pp. 140–150.

In the later development, those results led to the so-called *infinite combinatorics* which included the classic construction of *Aronszajn trees*[53] and some concepts of Sierpiński and Szpilrajn (Marczewski), in particular the notion of *almost disjoint sets.*

9. Summary

The most astonishing fact, when we look back at the beginnings of Polish mathematics in the interwar period, was a strategic thinking.

The strategic thinking can be noticed already in an observation made by Sierpiński in 1911 after the 11th Congress of Polish Physicians and Natural Scientists in Kraków. Dissatisfied by the fact that all five lecturers in mathematics in that Congress were working in different mathematical fields and so there was no cooperation between them, he "came to the conclusion it would be much better if a larger number of our mathematicians focused on a single field of research".[54] Back in Lwów, where he was then a professor, he lectured on set theory and started a senior seminar on mathematics which attracted a group of active members.

One of them was Janiszewski who, responding to an invitation by *Nauka Polska*, has formulated a program with the goal "to win an independent status for Polish mathematics" (see Section 4). Notwithstanding the war misery and political instability, Janiszewski has invented the name *Fundamenta Mathematicae* for the yet nonexistent journal and began to collect papers to its first issue. He largely succeeded (in the first volume there appeared 25 papers by eight Polish mathematicians), but did not live to see it. He died in January 1920. Prior to that, however, Sierpiński returned from Moscow and joined Janiszewski and Mazurkiewicz in Warsaw. As he recalled later: *When the three of us, Janiszewski, Mazurkiewicz and I, found ourselves in Warsaw at the reborn university, we decided we will realize Janiszewski's plan* (see Section 6).

And so they did. And they were successful.

There was a bit of luck in their selection of set theory since the theory soon became important for all mathematics. There was a common conviction that the theory is assuming the role of a new, solid basis of the whole of mathematics. Although the conviction never fully materialized, it still lives and the set-theoretical way of thinking has permeated all branches of mathematics, sometimes even leading to new ones. Greatly influenced was the *theory of real functions*, completely reshaped in a set-theoretical way, while a new one was *topology*. In both of them Polish mathematicians played a vivid role, thus corroborating correctness of the initial choice.

---

[53]The construction was known although it was published by Kurepa in 1938 only (Aronszajn did not publish it).

[54]W. Sierpiński, O polskiej szkole matematycznej [On the Polish School of Mathematics], pp. 413–434, in J. Hurwic (ed.), *Wkład Polaków do nauki. Nauki ścisłe* [*The Scholarly Contributions of Poles. The Exact Sciences*], Państwowe Wydawnictwo Naukowe, Warszawa, 1967, 457 pages.

Mutual collaboration, proposed by Janiszewski, became legendary and much efficient. The most vivid legend is that of the *Scottish Café* in Lwów and the *Scottish Book*.[55] That sort of collaboration is in sharp contrast with the now prevailing domination of rivalry and competition.

Also successful was the journal *Fundamenta Mathematicae*. It was soon followed by *Studia Mathematica*, which was a Lwów journal supporting the Lwów school of functional analysis, and *Acta Arithmetica*, another Warsaw journal restricted to number theory. All three journals were the very first restricted mathematical journals in the world and this kind of journals is now prevailing.

Set theory became for Polish mathematics a trampoline to jump into world mathematics. The jump was brave and turned out efficient.

References

C.E. Aull and R. Lowen (eds.), *Handbook on the History of General Topology*, Volume II, Kluwer, Dordrecht, 1998, pp. 399–807. ISBN 0-7923-5030-8.

S. Banach, *Œuvres*, Volume I, PWN — Éditions Scientifiques de Pologne, Warszawa, 1967, 382 pages.

N. Bourbaki, *Topologie Générale: Deuxième Edition*, Hermann, Paris, 1958, 169 pages.

G. Cantor, *Gesammelte Abhandlungen Mathematischen und Philosophischen Inhalts*, Springer, Berlin, 1932, 489 pages (Reeditions: 1962, 1980).

G. Cantor, *Über unendliche, lineare Punktmannigfaltigkeiten. Arbeiten zur Mengenlehre aus den Jahren 1872–1884*, Teubner-Archiv zur Mathematik, Band 2, Leipzig, 1984, 180 pages.

R. Duda, *Pearls from a Lost City. The Lvov School of Mathematics*, History of Mathematics 40, American Mathematical Society, 2014, 231 pages. ISBN 978-1-4704-1076-6.

C. Goldstein, J. Gray and J. Ritter (eds.), *L'Europe Mathématique — Mythes, Histoires, Identités*, Éditions de la maison des sciences de l'homme, Paris, 1996, 575 pages. ISBN 2-7351-0685-3.

I. Grattan-Guinness (ed.), *Companion Encyclopedia of the History and Philosophy of the Mathematical Sciences*, 2 volumes, Routledge, London and New York, 1994, 1826 pages. ISBN 0-415-03785-9.

J. van Heijenoort (ed.), *From Frege to Gödel: Source Book on Mathematical Logic 1879–1931*, Harvard University Press, Cambridge MA, 1967, 660 pages.

Historia Nauki Polskiej. Wiek XX. Nauki ścisłe. Zeszyt Pierwszy [History of Polish Science. XX Century. Exact Sciences. Fascicle One], Polska Akademia Nauk. Instytut Historii Nauki. Fundacja im. W. Świętosławskiego, Warszawa, 1995, 377 pages. ISBN 83-900891-9-X.

W. Hurewicz, *Collected Works*, American Mathematical Society, Providence, Rhode Island, 1995, 598 pages. ISBN 0-8218-0011-6.

Z. Janiszewski, *Œvres Choisies*, Państwowe Wydawnictwo Naukowe, Warszawa, 1962, 319 pages.

C. Jordan, *Cours d'analyse de l'Ecole Polytechnique*, 3 volumes, Gauthiers-Villars, Paris, 1893–1896, 612 + 627 + 542 pages.

[55] See Duda (2014), Chapter 10, *Social Life*, pp. 55–62.

M. Kline, *Mathematical Thought from Ancient to Modern Times*, Oxford University Press, New York, 1972, 1238 pages.

C. Kuratowski, Topologie I, *Monografie Matematyczne 3*, Warszawa–Lwów, 1933, 285 pages.

K. Kuratowski, *Selected Papers*, PWN — Polish Scientific Publishers, Warszawa, 1988, 610 pages. ISBN 83-01-07169-9.

S.M.G. Kuzawa CSFN, Modern Mathematics, *The Genesis of a School in Poland*, College and University Press, New Haven, 1968, 143 pages.

J.H. Mannheim, *The Genesis of Point-Set Topology*, Pergamon Press, Oxford, 1964, 166 pages.

E. Marczewski, *Collected Mathematical Papers*, Polish Academy of Sciences, Institute of Mathematics, Warszawa, 1996, 684 pages. ISBN 83-86806-00-1.

S. Mazurkiewicz, *Travaux de Topologie et ses Applications*, PWN — Édition Scientifiques de Pologne, Warszawa, 1969, 380 pages.

F.A. Medvedev, *Francuzskaja škola teorii funkcii i množestv na rubeže XIX–XX v.* [The French School of the Theory of Functions and Sets at the Turn of XIX–XX C.], Izdatelstvo Nauka, Moskwa, 1976, 230 pages (Russian).

F.A. Medvedev, *Rannaya istoria aksjomy wybora* [The Early History of the Axiom of Choice], Izdatelstvo Nauka, Moskwa, 1982, 202 pages (Russian).

G.H. Moore, *Zermelo's Axiom of Choice. Its Origins, Development, and Influence*, Springer-Verlag, New York, Heidelberg, Berlin, 1982, 440 pages. ISBN 0-387-90670-3.

B. Russell and A.N. Whitehead, *Principia Mathematica*, 3 volumes, Cambridge University Press, Cambridge, 1910, 666 + 772 + 491 pages (reedited 1923).

S. Saks and A. Zygmund, *Funkcje analityczne* [Analytic Functions], Monografie Matematyczne 10, Warszawa, 1938, 431 pages.

W. Sierpiński, *Hypothèse du continu*, Monografie Matematyczne 4, Warszawa, 1934, 192 pages (extended edition 1956, USA).

W. Sierpiński, *Œuvres Choisies*, 3 volumes, PWN — Éditions Scientifiques de Pologne, Warszawa, 1974–1975, 300 + 780 + 685 pages.

G. Sinkewicz, Georg Cantor i polskaya škola teorii množestv [Georg Cantor and Polish School of Set Theory], Ministerstvo obrazovanya i nauki Rossijskoj Federacii & Sankt-Peterburskij arhitekturno-stroitelnyj universitet, Sankt Peterburg, 2012, 347 pages. ISBN 578-5-9227-0360-4 (Russian).

A. Strzałkowski (ed.), *Recepcja w Polsce nowych kierunków i teorii naukowych* [Reception in Poland of New Scientific Trends and Theories], Polska Akademia Umiejętności, Komisja Historii Nauki, Monografie 4, Kraków, 2001, 348 pages. ISBN 83-88857-18-5.

A. Tarski, *Collected Papers*, Volume I, 1921–1932, Birkhäuser, Basel, Boston, Stuttgart, 1986, 659 pages. ISBN 3-7643-3280-8.

S. Wagon, *The Banach-Tarski Paradox*, Cambridge University Press, Cambridge, 1985, 253 pages.

W. Więsław (ed.), *Dzieje matematyki polskiej* [History of Polish Mathematics], Instytut Matematyczny Uniwersytetu Wrocławakiego, Wrocław, 2012, 311 pages. ISBN 978-83-910055-4-5.

W. Więsław (ed.), *Dzieje matematyki polskiej II* [History of Polish Mathematics II], Instytut Matematyczny Uniwersytetu Wrocławskiego, Wrocław, 2013, 326 pages. ISBN 978-83-910055-8-3.

https://doi.org/10.1142/9781786349316_0011

Chapter 11

Transformation of Mathematics Between World Wars: The Case of Potential Theory

Ivan Netuka

Motto: *La théorie du potentiel est un véritable carrefour de la Mathématique.*[1]

1. Historical Background

An apt description of the development of potential theory is given in Simon (2015b), pp. 273–274:

> *Initially, because of its relevance to physics, and later, additionally, because of its importance to complex analysis (in part because of Riemann's use of it in his proof of his mapping theorem), potential theory has been a central object of mathematics for almost 300 years, and its greatest practitioners are almost a who's-who of analysts: Lagrange, Legendre, Laplace, Poisson, Gauss, Green and Thomson (aka Kelvin) in the period around 1800; Dirichlet, Riemann, Schwarz, and Neumann in the period centered sixty years after; Poincaré, Hilbert, Zaremba and Lebesgue in the period around 1905; Kellogg, Frostman, Evans, Perron, Wiener, F. Riesz, and Brelot in the period 1925–1940; and at mid-century, a quartet of French mathematicians: Brelot, Cartan, Choquet, and Deny. Since then, the field has blossomed into many branches: general elliptic theory, probability and potential theory, and some axiomatic approaches going back to Brelot and to Deny.*

The beginnings of mathematical potential theory can be found in the work of Green (1871) and especially the extensive paper of Gauss (1877).

Gauss discussed several important problems and solved them at the level of rigor of 19th century mathematics. We recall three of them which had a decisive and enduring impact on the development of mathematical analysis.

[1]Potential theory is a veritable crossroad in mathematics. (From the opening speech held by Pierre Jacquinot at the Colloquium on Potential Theory in Paris, Orsay, 1964; see du Plessis (1970, p. vii).)

The *problem of equilibrium* asks for a distribution of a given total mass (charge) on a closed surface S (conductor) which generates a potential that is constant on S.

The problem of *sweeping* (balayage) on S consists of finding, given a mass inside of S, a new mass distribution concentrated on S (an *equivalent layer*) inducing the same potential in the exterior of S. In the language of physics, this new distribution is the result of a phenomenon known as electrostatic induction.

In the present chapter, we mainly focus attention on the third problem, the so-called *Dirichlet problem.* This fundamental boundary value problem of mathematical physics appears, for instance, in electrostatics, elasticity theory, gravitation and heat conduction. Let us recall that the Dirichlet problem consists of determining a harmonic function on a domain D (for instance, bounded by a surface S) in the Euclidean space $\mathbb{R}^n$ that satisfies a prescribed boundary condition. More precisely, given a function f on the boundary ∂D of D, one seeks a harmonic function, i.e., a C^2-function h such that

$$\Delta h := \frac{\partial^2 h}{\partial x_1^2} + \cdots + \frac{\partial^2 h}{\partial x_n^2} = 0 \quad \text{on } D,$$

and h admits a continuous extension to ∂D which coincides there with f. Whereas the uniqueness of this *classical Dirichlet problem* is guaranteed by the maximum principle for harmonic functions, the question of existence occupied prominent mathematicians for more than 80 years. During this period, the work of establishing an existence proof is an instructive illustration of how the history of mathematics develops. On one hand, potential theory opened the door to new methods such as, for instance, variational methods and the method of integral equations, and notions, such as Schwartz's distributions, Wiener's capacity and Choquet's integral representation as well as theory of capacities. On the other hand, the development of potential theory was strongly influenced by emerging new mathematical tools, such as topology, measure and integration, functional analysis, partial differential equations and later on by probability theory. This interplay between various disciplines is a beautiful feature of the development of 20th century mathematics.

Let us note that one approach to establishing the existence of a solution to the Dirichlet problem that was used, for instance, by Riemann, was based on the so-called *Dirichlet principle.* It involves considering admissible functions, i.e., smooth extensions u of the given boundary condition f, and the *energy integral* $\int \operatorname{grad}^2 u \, dx$. If there is an admissible function u_0 such that the energy integral attains its minimum, then u_0 is a solution of the Dirichlet problem for f. This variational approach was sharply criticised by Weierstrass, who pointed out that the existence of a minimiser was not guaranteed.[2] Also, there are continuous functions

[2]For more information on the history of the Dirichlet principle, refer to Archibald (2016); Bottazzini (1986); Burkhardt and Meyer (1899–1916); Courant (1977); Dieudonné (1981); Gårding

that do not admit a smooth extension possessing a finite energy integral; cf. Maz'ya (1998, pp. 373–377). It may happen that admissible functions for f exist, and yet the Dirichlet principle does not give a solution to the Dirichlet problem; see Lebesgue (1913b). Hence the *Dirichlet principle* and the *Dirichlet problem* are not equivalent. In fact, the Dirichlet principle as a method for proving existence fell into disrepute after Weierstrass' criticism in 1869.

Let us mention that towards the end of the 19th century some attempts to revive Dirichlet's principle appeared. In 1900 and 1901, Hilbert used the so-called direct method for variational problems to provide a correct proof under restrictive conditions on the domain and boundary data; see Hilbert (1970, pp. 10–38). Another important contribution is due to Lebesgue (1907), see also Lebesgue (1972b, pp. 91–122) and Lebesgue (1972a, pp. 154–159). For almost 20 years Lebesgue retained his interest in the Dirichlet problem and influenced in a fundamental way the development of potential theory.

With appropriate assumptions imposed on the boundary, various new alternative proofs of existence were devised. These include Schwarz's alternating process, Carl Neumann's method of the arithmetic mean, and Poincaré's *méthode du balayage*, which turned out to be very influential in the development of potential theory.[3]

This chapter is organised as follows. Section 2 describes the idea of Poincaré's méthode du balayage. We first recall the Poisson integral, which provides a solution to the Dirichlet problem on a ball, and Harnack's theorem on monotone sequences of harmonic functions. In fact, these are the basic ingredients for Poincaré's méthode du balayage. Section 3 is devoted to an example of a domain where a solution of the classical Dirichlet problem fails to exist for certain continuous boundary conditions. Such an example (a domain with the so-called *Lebesgue spine*) was published in 1913. For historical reasons, we mention a similar example published by Urysohn in 1923. In the early 1920s, the existence of non-regular domains triggered extensive research on a reformulation of the classical Dirichlet problem. We discuss in Section 4 the idea of the generalised Dirichlet problem and properties which a reasonable generalisation should enjoy. Section 5 deals with Wiener's approach to the generalised Dirichlet problem. Wiener's papers, published in the period 1923–1925, had a profound effect on the development of potential theory. In an aside, we briefly comment on difficulties in the relationship between Wiener and Kellogg. In Section 6, Perron's method from 1923 of solving the generalised Dirichlet problem is analyzed. We also mention

(1979–1980); Gray (2015); Hilbert (1970); Jahnke (2003); Kline (1972); Lichtenstein (1909–1921); Monna (1975).

[3]We mention a few references dealing with the history of potential theory prior to, say, 1910: Brelot (1952, 1970, 1983, 1985); Burkhardt and Meyer (1899–1916); Dieudonné (1981); Gårding (1979–1980); Gray (2015); Hilbert (1970); Jahnke (2003); Kellogg (1967); Kline (1972); Mawhin (2010); Monna (1975); Netuka (2020); Shraer (1973); Sologub (1975).

the lesser known fact that, at the same time, a quite similar approach was independently developed by Remak. An important paper published by Wiener in 1925, devoted to a comparison of his and Perron's solution, is discussed. From a historical viewpoint, it is interesting to point out a mistake in this paper. Remarkably, this mistake held up the investigation of the Dirichlet problem for discontinuous boundary conditions till 1939. Logically, after Sections 4–6, an investigation of boundary behaviour of solutions to the Dirichlet problem should follow. However, we decided to insert Section 7, devoted to new mathematical tools which changed the landscape of potential theory significantly. The development of measure theory and integration in the first two decades of the 20th century enabled the definition of potentials in appropriate generality and facilitated the investigation of boundary values of harmonic functions (Fatou's theorem). A functional analysis perspective on integration, namely representation of linear functionals as measures and a compactness property of measures, found wide application in potential theory (for instance, to harmonic measure and the Riesz-Herglotz representation). Then we discuss subharmonic and superharmonic functions, notions which are fundamental for modern potential theory, and move on to the likewise important concept of capacity in the last part of Section 7. The crucial significance of capacity is illustrated by several results, which play a vital part in potential theory. In Section 8, the boundary behavior of a solution to the generalised Dirichlet problem is investigated. Several criteria for regularity of boundary points are recalled. However, we mainly focus on the way in which the understanding of the nature of irregular points developed. We discuss the notion of a barrier as well as the famous Wiener test of regularity. Here capacity plays a remarkable role, as it also does in results concerning the size of the set of irregular points. Looking beyond the mathematical details, we also touch on the issue of international cooperation, or more precisely the competition between French and American mathematicians.

Section 9 refers to a certain culmination of knowledge on the Dirichlet problem and balayage in the period up to the end of the second world war. New developments in mathematics influenced potential theory in an essential way. We discuss the final form of the Dirichlet problem (the Perron–Wiener–Brelot solution) and a uniqueness result proved by Keldysh.

As far as the theory of balayage is concerned, during the late 30s and the 40s, the focus shifted from potentials and mass distributions to superharmonic functions. Where measures appear, they tend to be a by-product at the end. This anticipates the future development of abstract potential theory.

We should mention that our exposition deliberately concentrates on $\mathbb{R}^n$ for $n \geq 3$ and we only rarely discuss potential theory in $\mathbb{R}^2$ because of its significant differences with higher-dimensional situation. Green's potentials are left aside. Here we refer to Ransford (1995), Armitage and Gardiner (2001), Helms (1969), Simon (2015a,b). Also, due to space limitation, we had to omit quadratic potential theory which relates to the notion of energy concept. We only mention several references: Cartan (1942, 1945); Helms (1969); Simon (2015b); Frostman (1935); Brelot (1952, 1965, 1970, 1972).

At the end of the chapter, a list of crucial results and/or notions having a decisive impact on the development of potential theory between the World Wars is appended, as well as thumb-nail biographies of mathematicians who significantly influenced potential theory during the period studied.

2. Poincaré's Méthode du Balayage

We first recall two very old results which continued to play a significant role even in modern potential theory, for instance, in the axiomatic potential theory created in the second half of the 20th century.

The first goes back to Poisson (1923) and shows that the Dirichlet problem is solvable for a ball by means of a formula called the *Poisson integral.* We present the result for $\mathbb{R}^n$, $n \geq 2$, cf. Armitage and Gardiner (2001, p. 6). The open ball of centre x_0 and radius r in $\mathbb{R}^n$ is denoted by $B(x_0, r)$, and $S(x_0, r)$ denotes the sphere $\partial B(x_0, r)$. Let σ stand for surface area measure on $S(x_0, r)$ and $\sigma_n := \sigma(S(0, 1))$. The *Poisson kernel* of $B(x_0, r)$ is the function

$$K_{B(x_0,r)}(x, y) := \frac{1}{\sigma_n r} \cdot \frac{r^2 - |x - x_0|^2}{|x - y|^n}, \quad y \in S(x_0, r), \quad x \in \mathbb{R}^n \setminus \{y\}. \tag{11.1}$$

For a σ-integrable function f on $S(x_0, r)$, we define

$$I_{B(x_0,r)}(x) := \int_{S(x_0,r)} f(y) \cdot K_{B(x_0,r)}(x, y)\, \mathrm{d}\sigma(y), \quad x \in B(x_0, r).$$

If f is a continuous function on $S(x_0, r)$, then $I_{B(x_0,r)}$ is a solution of the Dirichlet problem on $B(x_0, r)$ for the boundary condition f. Let us note that the first rigorous proof of this fact has been given by Schwarz (1872).

The second result has to do with sequences of harmonic functions on a domain D. The mean value property of harmonic functions can be used to prove that the limit of a locally uniformly convergent sequence of functions harmonic on D is harmonic. A deeper result, called *Harnack's theorem*, states that the limit of an increasing sequence of harmonic functions on D is either identically ∞ or harmonic. This result goes back to Harnack's book (1887, p. 67), and is stated there (for series) as follows:

> *Ist für eine beliebige F eine unbegrenzte Reihe von harmonischen Funktionen $u_1, u_2, \ldots, u_n, \ldots$ gegeben, die alle innerhalb der Fläche einerlei Zeichen haben, und konvergiert die Summe*
>
> $$u_1 + u_2 + u_3 + \cdots + u_n + \ldots,$$

> *an irgend einer Stelle im Innern der Fläche, so konvergiert sie auch für alle inneren Punkte der Fläche, und ist in derselben eine harmonische Funktion.*[4]

Now let us turn our attention to Poincaré's method announced in 1887 and developed in Poincaré (1890, 1899). We first explain what is meant by the *balayage of the unit point mass* ε_x at the point $x \in B = B(x_0, r) \subset \mathbb{R}^3$. We want to "sweep" ε_x on ∂B in such a way that the Newtonian potential of ε_x and of the swept-out mass distribution coincide in the complement of B. Let us define

$$V(z) := \int_{\partial B} \frac{1}{|z-y|} K_B(x,y)\,\mathrm{d}\sigma(y), \quad z \in \mathbb{R}^3.$$

This is the Newtonian potential of mass distribution σ_x on ∂B having the density $K_B(x,\cdot)$ with respect to σ. For every $z \in \mathbb{R}^3 \setminus (B \cup \partial B)$, the function $y \mapsto 1/|y-z|$, $y \in \mathbb{R}^3 \setminus \{z\}$, is harmonic. Hence $V(z) = 1/|z-x|$ on $\mathbb{R}^3 \setminus B$ and $V(z) < 1/|z-x|$ for $z \in B$. Considering now a mass distribution $\mu > 0$ on B having a continuous potential, integration of σ_x with respect to μ yields a new mass distribution μ' on ∂B with potential coinciding with that of μ on $\mathbb{R}^3 \setminus B$ and strictly smaller on B. The measure μ' is called a *swept-out measure* of μ (or *balayage* of μ). (For a mass distribution μ not necessarily living on B, we perform the balayage on the restriction of μ to B and simply add the potential of the part of μ living on $\mathbb{R}^3 \setminus B$.) Let us note that the balayage operation consists of replacing the values on B of the potential in question by the Poisson integral of its restriction to ∂B.

After this preparation, we present the idea of Poincaré's *méthode du balayage*.[5]

Let $D \subset \mathbb{R}^3$ be a bounded domain and f be a continuous function on the boundary ∂D of D. Uniform approximation makes it possible to reduce the problem of solving the Dirichlet problem to the case of polynomial boundary conditions. Moreover, taking U to be a ball containing the closure of D, one can, in addition, suppose that our boundary condition coincides on ∂D with a polynomial p such that $p > 0$ and $\Delta p < 0$ on U. (Let us note that here an important fact emerges, namely that p is in fact superharmonic on U; this notion explicitly appears in the 1920s, see Section 7.) Indeed, any polynomial can be written as a difference of two polynomials with such additional properties. Put $\varphi := -(4\pi)^{-1}\Delta p$. Then the volume potential

$$V(x) := \int_U \frac{\varphi(y)}{|x-y|}\,\mathrm{d}y, \quad x \in \mathbb{R}^3, \tag{11.2}$$

---

[4]For an arbitrary surface F, let an infinite series of harmonic functions $u_1, u_2, \ldots, u_n, \ldots$ be given which all are of the same sign in the interior of the surface and let the sum

$$u_1 + u_2 + u_3 + \cdots + u_n + \cdots$$

be convergent at some point of the interior of the surface. Then the series is convergent for all interior points of the surface and is a harmonic function there.

[5]Cf. Brelot (1985); Mawhin (2010); Shraer (1973); Sologub (1975).

is a C^2-function satisfying $\Delta V = \Delta p$ on U, whence V differs from p by a harmonic function on U. So it is sufficient to consider the Dirichlet problem on D for a boundary condition which is a restriction of potential of the form (11.2) to ∂D. Now an ingenious idea comes into the picture. Poincaré constructs a sequence of balls $B_1, B_2, \dots$ such that $\overline{B_j} \subset D$ and $D = \bigcup_{j=1}^{\infty} B_j$. To begin with, Poincaré sweeps B_1. This gives a new potential harmonic on B_1. Then B_2 is swept. The resulting potential is harmonic on B_2, but harmonicity may be destroyed on B_1. So next B_1 is swept once more and B_2 is swept again. Continuing, the balls are successively swept according to the scheme $B_1; B_2; B_1, B_2, B_3; B_1, B_2, B_3, B_4; \dots$. Hence each ball is swept infinitely often. The resulting sequence of potentials is decreasing, and for each ball B_j, there is a subsequence of potentials which are harmonic on B_j. By Harnack's theorem, the limit of the potentials is a harmonic function on $\bigcup_{j=1}^{\infty} B_j = D$, which we denote by h. It is not at all guaranteed that, for a given $y \in \partial D$, h tends to the prescribed value $V(y)$. Poincaré shows that this is the case provided that there exists a ball B such that $B \subset \mathbb{R}^3 \setminus \overline{D}$ and $y \in \partial B$. To this end, he constructs a function $\tilde{h}$ continuous at y, coinciding with V on ∂B, harmonic on D and satisfying $\tilde{h} \le V$ on D. Since every potential arising from successive sweeping is between $\tilde{h}$ and V, the limit h satisfies $\tilde{h} \le h \le V$ on D. Hence h tends to the common value of $\tilde{h}(y) = V(y)$.

Let us emphasise two important features of Poincaré's idea. The idea of balayage appears in a more or less explicit form of solutions of the generalised Dirichlet problem proposed in the 1920s and discussed in the following sections. The search for control of the boundary behavior of a candidate for a solution led to the development of the fruitful notion of a barrier several decades later. And finally, the idea of a balayage process gained new impetus through consideration of the balayage of superharmonic functions, and thus found a prominent role in modern potential theory in the second half of the 20th century.

3. The Lebesgue Spine[6]

In all approaches to existence proofs of the Dirichlet problem, restrictive assumptions on the geometry of the boundary of domains in question were imposed. Kellogg (1967, p. 285) writes: *So far, it was generally believed that the Dirichlet problem was solvable for any region, and that limitations of generality were inherent in methods, rather than in the problem itself.* Zaremba (1911) pointed out that a punctured ball provides an example of a domain where the Dirichlet problem is impossible, that is,

[6]The term *Lebesgue spine* appears in Exercise 10 in Kellogg (1967, p. 334): *Show that if q is the vertex of spine of Lebesgue, generated by rotating about the x-axis the curve*

$$y = e^{-\frac{1}{x}}, \quad 0 < x,$$

lying outside the spine and bounded by it in the neighbourhood of q, then q is an exceptional point.

a solution of the Dirichlet problem fails to exist for an arbitrary continuous boundary condition. Vasilesco (1937a, p. 8) remarks:

> *Mais l'example de M. Zaremba pouvait rester sans frapper beaucoup les esprits, car instinctivement on entendait encore à cette époque par frontière d'un domaine une surface plus ou moins bizarre, mais une surface tout de même.*
>
> *C'est pourquoi l'example d'impossibilité donné quelques années plus tard, en 1912, par M. Lebesgue, a été décisif pour l'orientation de recherches concernant le problème de Dirichlet.*[7]

Vasilesco refers here to a striking example of a simply connected bounded domain in $\mathbb{R}^3$ containing no isolated boundary points such that the Dirichlet problem does not admit a solution for some continuous boundary conditions. The one-page (!) note Lebesgue (1913a) can be considered as a real breakthrough in potential theory, changing its complexion for many decades to come. We quote again from Vasilesco (1937a, pp. 7–8): *L'example de M. Lebesgue marque, dans l'évolution des recherches concernant le problème de Dirichlet, une étape decisive comportant un élargissement d'horizon. Il ne fallait plus désormais vouloir à tout prix résoudre le problème classique de Dirichlet, puisque ce problème pouvait être impossible, même pour des surfaces analytiques...*[8]

Lebesgue considered a mass distribution on the line segment $I := \{(t,0,0) : 0 \leq t \leq 1\}$ which increases linearly from $(0,0,0)$ to $(1,0,0)$. The Newtonian potential of this distribution at a point $(x,y,z) \in \mathbb{R}^3$ is given by

$$V(x,y,z) = \int_0^1 \frac{t\,\mathrm{d}t}{\left((x-t)^2 + y^2 + y^2\right)^{\frac{1}{2}}}. \tag{11.3}$$

We have $V(0,0,0) = 1$ and $V = \infty$ on $I \setminus \{(0,0,0)\}$. The integral in (11.3) can be evaluated explicitly and can be written as

$$V(x,y,z) = v(x,y,z) - x\log\left(y^2 + z^2\right),$$

where $v(x,y,z) \to 1$ as $(x,y,z) \to (0,0,0)$; see Helms (1969, p. 175). For $k > 0$ put

$$S_k := \{(x,y,z) : x > 0, y^2 + z^2 = e^{-k/x}\}.$$

---

[7]However, the example of Mr. Zaremba had little impact on perceptions, since instinctively the boundary of a domain was at that stage understood as a surface, perhaps an unusual one, but still a surface. This is why the example of impossibility given a few years later, in 1912, by Mr. Lebesgue, was decisive for the direction of research concerning the Dirichlet problem.

[8]The example of Mr. Lebesgue marks a decisive step in broadening the horizon in the development of research concerning the Dirichlet problem. It was no longer necessary to seek exclusively a solution to the classical Dirichlet problem, since this problem might prove to be impossible, even for analytic surfaces.

M. H. Lebesgue : *Sur des cas d'impossibilité du problème de Dirichlet ordinaire.*

Soit un domaine D et une fonction f continue sur la frontière F de D, le problème de Dirichlet ordinaire consiste à trouver une fonction harmonique V continue dans D et sur F et se réduisant sur F à la fonction f donnée. Lorsque la fonction V existe quelle que soit la fonction continue f on dit que le problème de Dirichlet est possible pour le domaine D.

L'auteur rappelle qu'il a démontré dans les *Rendiconti* de Palerme que le problème de Dirichlet est possible pour tous les domaines plans qui sont des ensembles ouverts bornés et tels que tout point de leur frontière appartienne à un continu frontière. De sorte, par exemple, que tous les domaines limités par un nombre fini de courbes ordinaires rentrent dans cette catégorie. Le cas de l'espace conduit à des résultats tout différents; il existe dans l'espace des domaines simplement connexes limités par des surfaces de révolution dont la méridienne est une courbe analytique présentant un seul point singulier et pour lesquelles le problème de Dirichlet est impossible.

Considérons un segment OA porteur d'une masse attirante dont la densité, en chaque point P, est égale à la distance OP. Les surfaces équipotentielles $V = \alpha$ sont des surfaces de révolution entourant le segment attirant; pour $\alpha \geq 1$ elles passent par le point O, en lequel $V = 1$. Soit D le domaine limité par les deux surfaces équipotentielles $V = 2$, $V = \frac{1}{2}$; si l'on essaie de résoudre le problème de Dirichlet pour ce domaine avec, comme fonction continue f sur sa frontière, une fonction égale au potentiel, sauf en O où l'on prendra $f = 2$ pour la continuité, on voit facilement que, si ce problème avait une solution, ce serait la fonction potentielle dont on est parti, laquelle ne satisfait pas en O aux conditions de continuité du problème.

Le problème de Dirichlet est donc impossible pour ce domaine D. En modifiant légèrement l'exemple indiqué on arriverait à un domaine simplement connexe de la nature indiquée ci-dessus.

Figure 1. The Lebesgue spine paper.[9]

[9]H. Lebesgue, Sur des cas d'impossibilité du problème de Dirichlet ordinaire, *Comptes Rendus des séances de la Société Mathématique de France* **41**(1913), p. 17. The picture is from the author's archive.

Then

$$V(x, y, z) \to 1 + k \quad \text{as } (x, y, z) \to (0, 0, 0), (x, y, z) \in S_k. \tag{11.4}$$

If we remove from the open unit ball centred at 0 a sharp thorn defined as

$$\{(0, 0, 0)\} \cup \{(x, y, z) : V(x, y, z) \geq 2\},$$

we obtain a bounded simply connected open set D such that $(0, 0, 0) \in \partial D$. Defining $f = V$ on $\partial D \setminus \{(0, 0, 0)\}$ and $f((0, 0, 0)) = 2$, we obtain a continuous function on ∂D. Let us suppose that h is a solution of the classical Dirichlet problem for f on D. Then

$$h(x, y, z) \to 2 \quad \text{as } (x, y, z) \to (0, 0, 0) \tag{11.5}$$

and a simple argument shows that $h = V$ on D; cf. Brelot (1985). If $k \in (1, 2)$, then, by (11.4),

$$h(x, y, z) \to 1 + k, \quad (x, y, z) \to (0, 0, 0), (x, y, z) \in S_k \cap D,$$

which contradicts (11.5).

From the history of mathematics viewpoint, it is of interest to mention the article Urysohn (1925) submitted on 2 October, 1923 and published in Mathematische Zeitschrift in 1925. (Let us recall that P.S. Urysohn died on 17 August, 1924, at the age of 26 years.) His example is similar to the rotationally symmetric example of Lebesgue. The spine removed is described by means of an increasing continuous function φ on $[0, \frac{1}{2}]$, $\varphi(0) = 0$, $\varphi(x) < \exp(-8/x)$ for $x \in (0, \frac{1}{2}]$. Urysohn showed that the Dirichlet problem was not solvable for the boundary condition $(x, y, z) \mapsto 1 - (x^2 + y^2 + z^2)$. The editors remarked in a footnote: *Erst nach Drucklegung der Arbeit ist der Schriftleitung bekannt geworden daß Herr Lebesgue vor längeren Zeit schon ein sehr einfaches beispiel für die hier in Frage kommende Eigentümlichkeit gegeben hatte.*[10] The editors provided a reference to earlier contributions of Lebesgue, Bouligand and Wiener from the period 1913–1924.

4. The Generalised Dirichlet Problem

A bounded domain in $\mathbb{R}^n$, $n \geq 2$, is called *regular* if the classical Dirichlet problem has a solution for every continuous boundary function. Lebesgue's discovery of non-regular sets was a decisive turning point in potential theory investigations. Lebesgue realised that one should split the problem into two separate parts. We quote from Wiener's note (1924c):

> *Dans un Mémoire récent* Lebesgue (1924a) *M. Lebesgue a montré que le problème de Dirichlet se divise en deux parties dont la première est*

[10]Only after printing the work did the editors learn that Mr. Lebesgue had long ago given a very simple example of the peculiarity discussed.

> *la détermination d'une fonction harmonique correspondant à certaines conditions sur la frontière, et la seconde est la recherche des propriétés de cette fonction dans le voisinage de cette frontière. Dans un Mémoire* Wiener (1924a) *paru la même semaine, j'ai fait la même remarque indépendamment, et en plus j'ai donné une définition exacte du sens dans lequel une fonction harmonique dépend des conditions frontières. J'ai démontré aussi que ma méthode assigne une fonction bien déterminée à n'importe quelle condition frontière continue sur un esemble borné de points dans un nombre quelconque de dimensions.*[11]

As we know, for special boundary data, a separation of the problem into two parts appeared in *Poincaré's méthode du balayage*. Already in 1912, such an approach is indicated in Lebesgue (1912a). In the 1920s, several methods for attacking such a generalised Dirichlet problem were suggested; see Vasilesco (1936) and the extensive survey Vasilesco (1937a). We shall recall only two of the most important ones, Wiener's and Perron's methods, which turned out to be extremely fruitful for the further development of potential theory. Let us note that at an early stage of the study of the generalised Dirichlet problem, no requirements for the first step, i.e., assigning a harmonic function to the given boundary data, were specified.

Of course, it was understood that a solution of the generalised Dirichlet problem has to give the solution of the classical Dirichlet problem, if it exists. An outline of various known methods is given in Vasilesco (1937a). It was shown that any two of them lead to the same result. This does not imply, however, that some other method could not be devised providing a different reasonable solution of the generalised Dirichlet problem. Here we should be more precise about what the word *reasonable* could mean. One possibility is to postulate this in the following form.

Let $D \subset \mathbb{R}^n$ be a bounded domain. A *Keldysh operator*[12] $A : f \mapsto Af$ is a mapping sending continuous functions on ∂D to harmonic functions on D satisfying the following conditions:

(a) A is linear;
(b) A is increasing, i.e., if $f_1 \leq f_2$ on ∂D, then $Af_1 \leq Af_2$ on D;
(c) Af is a solution of the classical Dirichlet problem, if it exists for f.

---

[11]In a recent contribution Lebesgue (1924a), Mr. Lebesgue has shown that the Dirichlet problem splits into two parts, the first of which is the determination of a harmonic function corresponding to certain conditions on the boundary, and the second is the investigation of the properties of this function in the neighbourhood of this boundary. In a paper which appeared the same week, I made the same remark independently and, moreover, I gave a precise definition of the sense in which a harmonic function depends on boundary conditions. I also proved that my method assigns a well-determined function to an arbitrary continuous boundary condition for a bounded set in any number of dimensions.

[12]The reason for the terminology is explained in Section 9.

With this definition, the generalised Dirichlet problem can be understood as seeking a functional extension of the classical Dirichlet problem in the sense of the definition of a Keldysh operator. Let us mention here that the uniqueness question remained open until 1941; see Section 9.

5. Wiener's Solution of the Generalised Dirichlet Problem

Brelot (1966) says:

> *Wiener did not work in potential theory for very long (only about two years around 1924), but that was enough to bring about in this field (the so-called classical potential theory), as in many others, some fundamental contributions: a definitive form of the generalized solution of the Dirichlet problem for a continuous given boundary function, the notion of capacity for general compact sets and the famous Wiener criterion of regularity.*

Wiener published five papers and one Comptes Rendus note: Wiener (1923a,b, 1924a,b,c, 1925); see also Wiener (1976), where comments on Wiener's contributions to potential theory by L. Lumer are included.

Wiener's first potential theory paper Wiener (1923a), written jointly with his colleague H.B. Phillips from the Massachusetts Institute of Technology, reflects clearly the electrical origin of his interest. The approach here is based on using discrete nets and a natural mean value property characterizing discrete harmonic functions. The Dirichlet problem is solved for polycubic domains and, by a limiting procedure, the classical Dirichlet problem is solved on domains with boundaries possessing certain regularity properties. Already in Wiener (1923a), the idea of exhausting a general domain by regular ones appeared.[13] The idea was developed in Wiener (1924a).

In the footnote, Wiener writes: *The author wishes to express his thanks to Dr. J.L. Walsh and to Prof. O.D. Kellogg, both of Harvard University, with whom he has discussed matters in this paper, and to whom he owes much in the way of suggestion and stimulus.* In his autobiography *I am a Mathematician*, Wiener (1956) writes on pp. 79–85 about the development of his relations with Kellogg:

> *I used to talk occasionally with Professor O.D. Kellogg of Harvard concerning problems of possible interest on which I might do research. I did not realize at that time how carefully many professors conserve problems for their own graduate students and how sharply they regard proprietary rights in new problems. ... I was not officially a student of Kellogg's. He was very helpful to me, but I consumed too much of his time, and I think that he considered me as something of a nuisance. I learned from Kellogg that the old problem of potential distribution was attracting renewed interest. ... In this intermediate field Professor Kellogg had done vitally important work, and two of his young friends were writing doctoral dissertations on the subject at Princeton. When Kellogg informed me of the work that was being done in this*

[13]Bouligand used this idea earlier for introducing Green's function on general domains; see Bouligand (1919).

corner of potential theory, I immediately began to reflect how I myself should attack the problem.

I very soon found that I could make quick headway in this subject and that in fact, within a matter of days, I had got even further than the two Princeton doctoral candidates. When I showed Professor Kellogg my mathematical results, there was a sudden change in his attitude. At the beginning I had found him rather pleased with my interest in potential theory. Now, however, he was chiefly concerned with the effect that my pursuit of the subject might have on the acceptance of the doctoral dissertations of the other two young men. ...

I am afraid I did not accept with any grace Professor Kellogg's dictum that I was to erase from my mind the work I had already done on the potential problem and to clear the tracks for the two doctoral candidates. I had been aware through Dr. Kellogg's leak, and only through Dr. Kellogg's leak, that other people were working on the problem, but I had no information as to their methods and tools, and my result was genuinely original.

Furthermore, I did not accept with alacrity Kellogg's suggestion that I was now an established mathematician, who did not need these papers and who ought to give them up as a charity to youth and inexperience. Both of the candidates were older than myself, and both of them enjoyed the secure position being the pupils of men with influence in American mathematics. I had never received favors from those in power, and it was only when it was to my disadvantage that any Harvard man thought of treating me as an established mathematician. ...

I had visited her [Margaret Engemann, Wiener's future wife] *and her family at about the time of my run-in with Kellogg on a raw and bleak December day. In waiting for the streetcar home, I had got thoroughly soaked and had developed the beginnings of a bad cold. When I talked over the problem of the publication of my paper with Kellogg when we met at a local meeting of the American Mathematical Society, I was already sick and half-delirious with what later turned out to be pneumonia. Instead of acquiescing in his point of view, I felt that it was an unfair assertion of solidarity among the insiders against the outsider and asserted my intention of securing quick publication of my result in our new mathematical journal. This started a storm of antagonism against me, and both Kellogg and Birkhoff thundered at me from an exalted moral elevation.*

Let us return to mathematics, namely to the paper Wiener (1924a). In comments on this paper in Wiener (1976, p. 392), L. Lumer writes:

This is generally considered as Wiener's most fundamental paper on potential theory. The main object is the determination of a generalized solution of the Dirichlet problem for any continuous function on the boundary of an arbitrary Euclidean domain (Theorem I), which constitutes a major and definite breakthrough in the theory.

Wiener's procedure goes as follows: For a domain D and a continuous function f on ∂D, Wiener considers an increasing sequence of regular domains D_j such that $\overline{D_j} \subset D$ and $\bigcup_{j=1}^{\infty} D_j = D$, and a continuous extension F of f to D. Defining f_j to be the restriction of F to ∂D_j, we have a solution h_j of the classical Dirichlet problem for D_j and f_j. Wiener shows that the limit $h := \lim_{j\to\infty} h_j$ exists and gives a harmonic function on D. Moreover, the limit is independent of the particular choice of regular domains and continuous extension. The function h is called *Wiener's solution of the generalised Dirichlet problem* for D and f. Let us note that Wiener's proof is complicated and difficult to follow. However, the idea of harmonic measure to be discussed later in Sections 7 and 9 appears there. In fact, speaking in functional analysis language, Wiener's "outer weight", which corresponds to the harmonic measure, can be understood as a weak*-limit of the harmonic measures on the boundaries of regular sets D_j. The second very important point of the paper is the introduction of the notion of capacitary potential defined as a solution of the exterior Dirichlet problem for the boundary value 1 on the "conductor" (and zero at infinity). L. Lumer justly remarks:

Further researches done on capacity ... all suggest, almost 50 years later, that the notion of capacity is perhaps Wiener's most important and long-lasting contribution to potential theory; see Wiener (1976, pp. 392–393). Capacity turned out, as we shall see, to be the right tool for measuring the size, or massiveness/smallness of a set in potential theory.

6. Perron's Solution of the Generalised Dirichlet Problem

Perron's solution is based on the notion of a superharmonic function to be discussed more closely in Section 7. The final form of the so-called Perron–Wiener–Brelot method will be given in Section 9. Hence we only discuss here the idea of Perron's approach from Perron (1923).

Let us consider a bounded domain D in $\mathbb{R}^n$ (Perron takes $n = 2$, but the generalisation is straightforward), and a bounded function f on ∂D. Perron proposes to solve the following problem: *Find a harmonic function on D such that, for every $y \in \partial D$,*

$$\liminf_{z\to y} f(z) \leq \liminf_{x\to y} h(x) \leq \limsup_{x\to y} h(x) \leq \limsup_{z\to y} f(z). \quad (11.6)$$

For a bounded function u on $\overline{D}$ and a ball $B \subset \overline{D}$, Perron introduces the so-called *Poisson modification* $P_B u$, defined as u on $\overline{D} \setminus B$ and as the Poisson integral of u on B. (Integrability in the Lebesgue sense is assumed.)

Let f be a bounded function on ∂D. A function ψ on $\overline{D}$ is called an *upper function* for f, if ψ is continuous, $\psi \geq f$ on ∂D and $\psi \geq P_B\psi$ for every ball $B \subset \overline{D}$. (Let us remark that upper functions are in fact superharmonic; see Section 7.) The set of upper functions is min-stable and $P_B\psi$ is an upper function, whenever ψ is an upper function. Let h be the pointwise infimum of all upper functions. Using the Lebesgue monotone convergence theorem, Perron shows that $h = P_B h$ for every ball B, whence h is harmonic. The function h is, in later terminology, *Perron's upper solution of the generalised Dirichlet problem for f*. For a regular domain D, Perron proves that h solves the problem (11.6). One can define *Perron's lower solution* analogously.

However, Perron does not discuss whether these two solutions coincide, at least for continuous functions. Perron's method can be understood as an improvement of Poincaré's méthode du balayage. (Let us note that a polynomial p as in Section 2, satisfying $\Delta p \leq 0$, is an upper function for the restriction of p to ∂D.)

In the paper, Perron does not give any indication of the motivation for his approach. We only remark that an idea of "squeezing" a desired solution using approximation from above and below appears earlier in a different context, for instance, in the definition of Perron's integral (see Pier (1994, p. 536)), or in Peano's approach to the ordinary differential equation $y' = f(x, y)$ using upper and lower functions satisfying the corresponding differential inequalities.

At this point we mention a historical coincidence. The paper by Perron (1923) was received by *Mathematische Zeitschrift* on 18 September, 1922. The same journal received the article by R. Remak *Über potentialkonvexe Funktionen* on 15 December, 1922, which overlapped, as far as the method is concerned, with that of Perron. The editors of the journal asked Remak to adapt the notation to that of Perron. In fact, the paper by Remak (1924) offers proofs of key properties of upper functions. Unlike Perron, Remak avoids the use of the Lebesgue integral. The paper Remak (1927) contains an alternative proof of harmonicity of the infimum of the upper functions.

A simplified and elegant version of Perron's arguments can be found in Radó (1925).

Naturally, Wiener's solution and Perron's solution invited comparison. This was done in Wiener (1925), the last of Wiener's potential theoretic papers. Wiener begins by noting that Perron's formulation (11.6) has a flaw. We quote:

> *Now this form of putting the Dirichlet problem says at once too much and too little: too much, because it goes beyond the case of continuous functions f, and too little, because for all but the simplest discontinuous boundary conditions, I* [= the condition (11.6)] *is not strong enough to secure the uniqueness of u* [= h is our notation]. *We shall therefore confine our attention for the present to continuous boundary conditions f, and shall reserve for the end of the paper the more general case.*

A further criticism by Wiener points out that the linearity of Perron's solution was not established and, for general domains, the coincidence of the upper and lower Perron solutions was not resolved. Wiener's main result shows that, for every continuous boundary condition, the upper and lower solutions coincide (the so-called *Wiener's resolutivity theorem*) and their common value coincides with Wiener's solution. Consequently, Wiener's and Perron's solutions of the generalised Dirichlet problem define the same Keldysh operator in the terminology introduced above.

At the end of the paper, Wiener deals with discontinuous boundary conditions, the subject of his earlier paper Wiener (1923b). Brelot (1966) writes:

> *He* [Wiener] *thought of course of the identity of the Perron envelopes for larger classes of f. He surely also would have developed this point, except for a simple*

Eine neue Behandlung der ersten Randwertaufgabe für $\varDelta u = 0$.

Von

Oskar Perron in München.

§ 1.

Die Aufgabe. Unter- und Oberfunktionen.

Sei ein ebenes beschränktes Gebiet G gegeben (Gebiet = zusammenhängende Menge von lauter *inneren* Punkten). Es braucht nicht schlicht zu sein, sondern darf Teile der Ebene mehrfach bedecken (Fig. 1 und 2); doch soll G keine Windungspunkte enthalten. Über den Rand R des Gebietes G braucht vorläufig gar nichts vorausgesetzt zu werden; er darf z. B. auch Windungspunkte enthalten (Fig. 2). Nun stellen wir die folgende

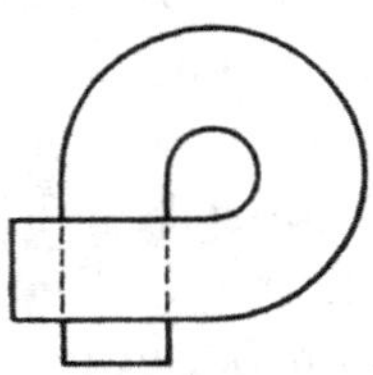

Fig. 1.

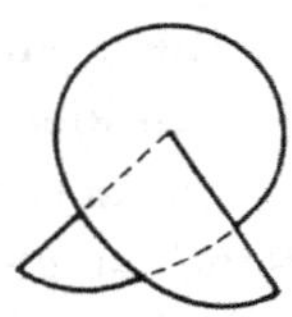

Fig. 2.

Randwertaufgabe. *Auf dem Rand R sei eine beschränkte Funktion f gegeben; ihre untere und obere Limesfunktion seien $\underline{f}$ und $\bar{f}$. Gesucht wird eine Funktion u im abgeschlossenen Bereich $G + R$, die den folgenden beiden Bedingungen genügt:*

I. *Die untere und obere Limesfunktion $\underline{u}$ und $\bar{u}$ erfüllen auf dem Rand R die Ungleichungen*[1])

$$\underline{f} \leqq \underline{u} \leqq \bar{u} \leqq \bar{f}.$$

II. *Im Gebiet G ist u harmonisch; d. h. in G ist u nebst den partiellen Ableitungen $\frac{\partial u}{\partial x}, \frac{\partial u}{\partial y}, \frac{\partial^2 u}{\partial x^2}, \frac{\partial^2 u}{\partial x \partial y}, \frac{\partial^2 u}{\partial y^2}$ stetig und genügt der partiellen Differentialgleichung*

$$\frac{\partial^2 u}{\partial x^2} + \frac{\partial^2 u}{\partial y^2} = 0.$$

[1]) Da u den Definitionsbereich $G + R$ hat, so sind die Limesfunktionen $\underline{u}$, $\bar{u}$ natürlich nicht wie $\underline{f}$, $\bar{f}$ in bezug auf R, sondern in bezug auf $G + R$ zu verstehen.

Figure 2. Perron's method paper.[14]

[14]O. Perron, Eine neue Behandlung der ersten Randwertaufgabe für $\Delta u = 0$, *Mathematische Zeitschrift* **18**(1923), no. 1, pp. 42–54. The picture is from the author's archive.

> *counterexample that the present writer* [Brelot] *discovered to be wrong 15 years later (then the identity of the suitable defined envelopes was proved to be equivalent to the summability of f with respect to the harmonic measure).*[15]

Here we quote the erroneous example of Wiener (1925, p. 29):

> *In the simplest case of discontinuous boundary values, the Perron generalization of the Dirichlet problem is still satisfactory. ...*
>
> *However, we do not need to go much further to find an instance where the Perron generalization breaks down. For example, let G be the interior of a circle with boundary R. Let us establish a system of polar coordinates with the center of G as pole. Let $f(P)$ be 1 for those points P of R with a Θ-coordinate rational in terms of 2π, and let $f(P)$ be 0 elsewhere on R. Clearly any upper function f is at least 1, and no lower function is greater than 0. Hence the Perron method yields us here no unique generalization of the Dirichlet problem.*

(The error is in *Clearly any upper function of f is at least* 1.)

To conclude this section, we summarise the following: Lebesgue's example of non-regular sets opened a new field of research. The classical Dirichlet problem had to be replaced by the generalised Dirichlet problem. The fact that at some boundary points (the so-called irregular points) the solution need not tend to the prescribed boundary value triggered the investigation of the character of irregular points and their distribution in the boundary. There remained open problems concerning discontinuous boundary data, and also, as we have already mentioned, whether all constructions of generalised Dirichlet problem lead to the same result.

New tools of mathematical analysis created in the first decades of the 20th century turned out to be useful for addressing these problems.

7. New Mathematical Tools and Potential Theory

Until the end of the 19th century, potential theory mostly used tools of classical analysis, that is, of differential and integral calculus. However, they turned out to be insufficient for the rigorous solution of fundamental problems, such as those, for instance, treated previously by Gauss. Only potentials of special distributions, typically of surface and volume distributions, were considered. The complexion of potential theory changed substantially during the first decades of the 20th century, mainly thanks to new theories of integration, measure theory, capacity theory and, later, the influence of functional analysis, harmonic analysis, topology, distribution theory and probability theory.

The link between potential theory and probability theory might seem somewhat unexpected, the former discipline being, so to say, deterministic in nature, while the latter is stochastic in character. In fact, this link, as the author personally heard in

[15]This is Brelot's resolutivity theorem, to be discussed in Section 9.

1973 from Professor Brelot, was, in a sense, in the air in Paris during the 30s. The first paper on Brownian motion and a probabilistic solution of the Dirichlet problem was published by Kakutani in 1944. Nowadays, probabilistic potential theory is a well-established mathematical discipline with a wide spectrum of applications. We refer to Doob (1984), Brelot (1972), Constantinescu and Cornea (1972), Bliedtner and Hansen (1986). By the way, from time to time specialists in potential theory informally discuss the astonishing fact that Wiener, who contributed essentially to potential theory and also, almost at the same time, significantly to the mathematical theory of Brownian motion, was apparently unaware of any link between these two fields.

However, let us return to the beginning of the 20th century, when mathematical analysis was given several important impulses; see von Renteln (1994), Pier (2001). One of the first applications illustrating the power and importance of the then newly created theory of Lebesgue integration has to do with potential theory. Fatou, in his dissertation Fatou (1906), proved several deep results on the boundary behaviour of analytic and harmonic functions on the unit circle B. He showed, in particular, that the Poisson integral of a bounded Lebesgue integrable function f possesses a radial limit $f(x)$ at almost every point $x \in \partial B$, that is, everywhere except for a set of Lebesgue measure zero. Fatou's results opened a new chapter of potential theory and harmonic analysis.

An enormous impact on the development of mathematical analysis was made by Radon's crucial work Radon (1913) dealing with a generalisation of the Lebesgue integral for functions of several variables. This so-called *Lebesgue–Stieltjes–Radon integral* provided an essential service to potential theory. In fact, this is the right tool for enabling the definition of potentials of an arbitrary mass distribution. Let us remark that Fréchet, in 1915, pointed out that such an integral can be considered in a very general setting as an abstract integral with respect to a positive measure (that is, a positive σ-additive set function) defined on a σ-algebra of sets. We cannot go here into the details, so we refer readers, as far as integration is concerned, to Bourbaki (1994), Choquet (2004), Hawkins (1970), Pier (1994, 2001), Pietsch (2007), Rudin (1987) and to Bauer's commentary on Radon's work in measure, integration and potential theory in *Gesammelte Abhandlungen* (pp. 29–44), quoted in the reference Radon (1913).

Thus attention was no longer restricted to potentials of special distributions, and the all-important mass distributions were represented by means of a general notion of measure. To be more specific, let us recall the classical Newtonian and logarithmic kernels of potential theory in $\mathbb{R}^n$:

$$\text{for } n \geq 3,\ G(x,y) := \frac{1}{(n-2)\sigma_n} \cdot \frac{1}{|x-y|^{n-2}}, \quad x, y \in \mathbb{R}^n,$$

$$\text{for } n = 2,\ G(x,y) := \frac{1}{2\pi} \log \frac{1}{|x-y|}, \quad x, y \in \mathbb{R}^2;$$

(we put $G(x,x) = \infty$; the reason for such a normalisation will be clear soon). Given a measure μ on $\mathbb{R}^n$ (with compact support if $n = 2$), we define the *potential of* μ as

$$G\mu(x) := \int G(x,y)\,\mathrm{d}\mu(y), \quad x \in \mathbb{R}^n.$$

We note that the function $G\mu$ is lower semicontinuous. (Recall that a function $f : \mathbb{R}^n \to (-\infty, \infty]$ is called *lower semicontinuous*, if $\{x \in \mathbb{R}^n : f(x) > c\}$ is an open set for every $c \in \mathbb{R}$.) Brelot (1952) writes that *G.C. Evans introduisit l'intégrale de Radon en théorie du potentiel* in 1920. However, already in 1919, we can read in Radon (1919, p. 1150):

> *Unter dem Potential der einfachen Schicht mit der Belegung $\Phi(E)$ — Φ ist eine absolut additive Mengenfunktion auf der Kurve C — verstehen wir das Integral:*
>
> $$V_P = \frac{1}{\pi}\int \lg \frac{1}{r_P}\,\mathrm{d}\Phi,$$
>
> *wo r die Entfernung des die Kurve durchlaufenden Punktes vom "Aufpunkte" P bedeutet.*[16]

Measures also appear as a tool enabling an integral representation of the solution to the generalised Dirichlet problem. More specifically, if $D \subset \mathbb{R}^n$ is a bounded domain and $x \in D$, then the mapping $f \mapsto Hf(x)$, sending a continuous function f on ∂D into the value at x of the solution of the Dirichlet problem, is linear and positive, that is, $f \geq 0$ implies $Hf(x) \geq 0$. Consequently (see Section 9), there exists a unique measure μ_x on ∂D (called the *harmonic measure*) such that, for every continuous function f on ∂D,

$$Hf(x) = \int f\,\mathrm{d}\mu_x.$$

Another extremely powerful tool is provided by the following compactness result for measures: *Let K be a bounded closed subset in $\mathbb{R}^n$ and (μ_n) is a sequence of measures on K such that* $\sup\{\mu_n(K) : n \in \mathbb{N}\} < \infty$. *Then there exists a subsequence (μ_{n_k}) of (μ_n) and a measure μ on K such that (μ_{n_k}) is weakly convergent to μ, that is*

$$\int f\,\mathrm{d}\mu_{n_k} \to \int f\,\mathrm{d}\mu$$

for every continuous function f on K. (See Rudin (1987, pp. 246 and 251).)

As an illustration, let us mention the *Riesz–Herglotz theorem* going back for $n = 2$ to Riesz (1911) and Herglotz (1911), which can easily be proved using the above

---

[16] Under a single layer potential with the mass distribution $\Phi(E)$ – Φ is an absolutely additive set function on the curve C — we understand the integral

$$V_P = \frac{1}{\pi}\int \lg \frac{1}{r_P}\,\mathrm{d}\Phi,$$

where r means the distance of a point running on the curve from the "reference point" P.

compactness result for measures; see Armitage and Gardiner (2001, p. 9), or Helms (1969, p. 29). *Let h be a positive harmonic function on $B = B(x_0, r)$. Then there exists a unique measure μ on ∂B such that*

$$h(x) = \frac{1}{\sigma_n r} \int \frac{r^2 - |x - x_0|^2}{|x - y|^n} \, \mathrm{d}\mu(y), \quad x \in B. \tag{11.7}$$

So h has the *Poisson integral representation* in terms of a measure living on the boundary of B.

It is natural to ask whether an analogous representation is available for domains other than a ball. The answer is yes, but for general domains one has to replace the Euclidean boundary by a different one originating in a pioneering work of R.S. Martin from 1941; see Armitage and Gardiner (2001, Chapter 8), and Brelot (1952, 1965, 1971, 1972). Here topology plays an essential role.

We return to Fatou's result mentioned above: *Let a positive harmonic function h on B have the representation* (11.7). *Then h has finite non-tangential limits almost everywhere on ∂B.* Here a *non-tangential limit* at $y \in \partial B$ means a limit with respect to every closed cone C with vertex at y such that $C \setminus \{y\} \subset B$; see Armitage and Gardiner (2001, p. 117).

In conclusion we note that measures were systematically used in the thesis by Frostman (1935). In particular, in his Chapter 7, some metric estimates of potential theoretic negligible sets in terms of Hausdorff measures are established using an ingenious application of the compactness property of measures.

In Section 2, we have already met superharmonic functions with continuous second partials in the discussion of Poincaré's *méthode du balayage* (recall the condition $\Delta p \leq 0$). As we know, Perron (as well as Remak) used continuous functions greater than or equal to their Poisson modification. In fact, intuitively, superharmonicity is a higher-dimensional analogue of one-dimensional concavity: if the graph of a concave function and a line intersect at two points, then the function is above the segment between those points. Analogously, if the values of a superharmonic function on a open set D are not smaller that the values of a harmonic function on the boundary of a ball contained in D, then the values of the superharmonic are not smaller that the harmonic function on the ball. Subharmonic functions are just the negative of superharmonic functions.

The terminology *fonction subharmonique* appears in F. Riesz's paper. Riesz (1926) mentions as his motivation the book by Poincaré (1899) and Hartogs' paper on functions of several complex variables from 1906, as well as a connection to Hardy's work on means of holomorphic functions. At this point we recall that, for a holomorphic function f, the function $\log |f|$ is subharmonic. This perhaps explains why the attention of Riesz and other authors was at first focused on subharmonic functions, whereas during later development preference has been given to superharmonic functions. So we recall Riesz's definition in the form of "super" version for $\mathbb{R}^n$.

Let D be a domain in $\mathbb{R}^n$. A function $u : D \to (-\infty, \infty]$ is called *superharmonic*, if

(a) u is lower semicontinuous on D and
(b) if D' is a domain such that $D' \cup \partial D' \subset D$ and h is a continuous function on $D' \cup \partial D'$, harmonic on D' and $h \leq u$ on $\partial D'$, then $h \leq u$ on D'.

We denote $\mathcal{S}(D)$ the set of all superharmonic functions on D and denote $\mathcal{H}^\star(D) := \mathcal{S}(D) \cup \{\infty\}$ (*hyperharmonic functions*). For various characterisations of $\mathcal{S}(D)$ we refer to Riesz (1926, 1930), Radó (1949), Brelot (1965), Armitage and Gardiner (2001), Helms (1969). We note that the set $\mathcal{H}^\star(D)$ is stable with respect to finite minima as well as limits of increasing sequences. Decreasing sequences of superharmonic functions need not have a superharmonic limit. However the limit may be corrected on a small set to become a superharmonic function; see Section 9.

Potentials of measures are examples of hyperharmonic functions. One of the main results of Riesz (1930) (see also Riesz (1960)) is that any superharmonic function may be written as a potential plus a harmonic function. We only state here a very special case for $\mathbb{R}^n$, $n \geq 3$: *Let s be a superharmonic function on $\mathbb{R}^n$ having a harmonic minorant. Then there exists a uniquely determined measure μ on $\mathbb{R}^n$ and a uniquely determined harmonic function h on $\mathbb{R}^n$ such that*

$$s = G\mu + h. \tag{11.8}$$

The usual formulation of the so-called *Riesz decomposition theorem* involves Green's potentials; see Armitage and Gardiner (2001); Brelot (1965); Helms (1969). From the point of view of modern mathematical analysis (Schwartz's distribution theory from the 40s), the *Riesz measure* μ of s, is (up to a sign) the *distributional Laplacian*. We only recall here that, with every locally integrable function u, we can associate a linear functional

$$L_u : \varphi \mapsto \int u \Delta \varphi \, \mathrm{d}\lambda,$$

where λ is the Lebesgue measure in $\mathbb{R}^n$ and L_u is defined on the space of infinitely differentiable functions with compact support. The point is that for s superharmonic, $-L_s$ can be represented by a measure μ. Taking $v := G\mu$, we have

$$L_v = -\mu,$$

whence $L_{s-v} = 0$, which leads to the harmonic function h in the decomposition (11.8). We refer here to Armitage and Gardiner (2001); Brelot (1965); Helms (1969); Simon (2015b).

The last crucial notion we wish to discuss here is capacity.

From an introductory course of physics we recall the formula $Q = CV$. Considering a conductor S in $\mathbb{R}^3$, we were taught that a charge of quantity Q could be distributed on S in such a way that the corresponding potential has a constant value V on S. If we vary the quantity Q, the ratio Q/V remains unchanged — this is the *capacity* C of the conductor S.

Inspired by electrostatics, Wiener (1924a) introduced the notion of capacity for a bounded closed set $F \subset \mathbb{R}^3$ as follows (a generalisation to $\mathbb{R}^n$, $n \geq 3$, is straightforward). Wiener's method allows us to solve the (exterior) Dirichlet problem for $D = \mathbb{R}^3 \setminus F$ with the boundary condition equal to 1 on ∂D and 0 at infinity. This yields a single layer potential u of a measure, the total mass of which can be evaluated by Gauss' integral

$$\frac{1}{4\pi} \int \frac{\partial u}{\partial n} \, \mathrm{d}\sigma;$$

here the integration is over a smooth closed surface containing F in the interior and n is the interior normal. This value is called the *capacity* of F. It was used in Wiener (1924b) for establishing Wiener's celebrated test of regularity of a boundary point; see also Evans (1947). In 1932, de la Valée Poussin (1932, p. 226), extended the notion of capacity to arbitrary sets as *la borne supérieure des charges qu'il peut soutenir sans que son potentiel surpasse l'unité* (the least upper bound of the charges supported [by the set] with potentials not exceeding 1). This capacity was renamed by Brelot (1939b) as *inner capacity* (notation $\mathrm{cap}_\star$). So, for an arbitrary E,

$$\mathrm{cap}_\star E := \sup \{\mu(K) : K \subset E \,\mathrm{compact},\ G\mu \leq 1\}.$$

If E is compact, then $\mathrm{cap}_\star E$ coincides with the capacity introduced by Wiener.

Evans (1933) showed that *a compact K is of capacity zero, if and only if there is a measure μ on K such that $G\mu = \infty$ on K.*

Brelot (1939b) and Monna (1940) introduced the notion of the *outer capacity* (notation $\mathrm{cap}^\star$) of an arbitrary set $E \subset \mathbb{R}^n$ by

$$\mathrm{cap}^\star E := \inf \{\mathrm{cap}_\star U : U \supset E \,\mathrm{open}\}.$$

Of course, inner and outer capacities depend on the Newtonian kernel. Brelot (1941) introduced the notion of *polar sets* in order to replace the notion of inner capacity. An arbitrary set E is called *polar*, if there exists a superharmonic function on $\mathbb{R}^n$ taking the value ∞ on E. (To be precise, Brelot's definition still used subharmonic functions.) Cartan (1945) showed that $\mathrm{cap}^\star E = 0$, if and only if the set E is polar. It turns out that polar sets play a role in potential theory similar to that of sets of measure zero in integration theory, that is, the role of negligible sets. For instance, the set of irregular points is polar; the limit of a decreasing lower bounded sequence of superharmonic functions equals a superharmonic functions except for a polar set. Also, they play a role in terms of removable singularities for superharmonic and harmonic functions. Here is a sample result; see Bouligand (1926): *Let D be an open set, E be a relatively closed polar subset of D, and h be a harmonic function on $D \setminus E$. If each point of E has a neighbourhood V such that h is bounded on $V \setminus E$, then h has a unique harmonic extension to D.*

During the 40s, an important question remained unanswered: for which sets E does one have $\text{cap}_\star E = \text{cap}^\star E$? Choquet's principal result from the early 50s says (for classical potential theory) that equality holds for an arbitrary Borel (or even analytic) set; see Armitage and Gardiner (2001); Helms (1969). In fact, Choquet created an abstract theory of capacities useful in disciplines far beyond potential theory; cf. Choquet (1953–1954) and a beautiful article Choquet (1986).

8. Boundary Behaviour of the Solution of the Generalised Dirichlet Problem

Lebesgue's discovery of non-regular sets totally altered the picture of classical potential theory. As we have already seen, the classical Dirichlet problem had to be replaced by the generalised one, and the investigation of its boundary behaviour kept many distinguished mathematicians occupied.

We consider again a bounded domain D in $\mathbb{R}^n$ and associate with every continuous function f on ∂D the solution Hf of the generalised Dirichlet problem in the Perron–Wiener sense. We know that, in general, for a boundary point y, either $Hf(x)$ does not have a limit for $x \to y$, or $Hf(x)$ has a limit, which may differ from $f(y)$.

The following important notion was introduced by Lebesgue (1924a): A point $y \in \partial D$ is called *regular* (we write $y \in \partial_{\text{reg}} D$), if

$$\lim_{x\to y} Hf(x) = f(y)$$

for every continuous function f on ∂D. Clearly, the set D is regular, if and only if $\partial D = \partial_{\text{reg}} D$. The points of $\partial D \setminus \partial_{\text{reg}} D$ are called *irregular.*

Early geometric criteria for regularity of a boundary point appeared before Lebesgue's discovery of irregular points. For instance, in our terminology, *a point $y \in \partial D$ is regular, if there is a ball B with $B \cap D = \emptyset$ and $y \in \partial B$* (Poincaré, 1890). Zaremba (1911) proved that *$y \in \partial D$ is regular, if there is a truncated closed solid cone in $\mathbb{R}^n \setminus D$ with vertex y.* Let us mention a criterion of Lebesgue (1924a): *A point $y \in \partial D$ is regular, if on every sphere centred at y having a sufficiently small radius r, there is a point whose distance from D is at least Kr, where K is a given positive constant.* (Other criteria are discussed in Bouligand (1926), Vasilesco (1937a, 1938).) We intend to reproduce only three more criteria going back to Wiener (reproduced here from Wiener (1924b)). In the joint paper Wiener and Phillips (1923a), the following theorem is stated for three-dimensional space:

The point O is regular whenever there are positive numbers a and b such that if $r < a$ and a sphere S of radius r is constructed with O as center, the content of the projection of the points in S and not in D on some plane exceeds br.

In footnote 3 of Wiener and Phillips (1923a) we read: *The use of the notion of projection in this paper is due to a suggestion of Prof. O.D. Kellogg.* In Wiener (1924b), Wiener recalls his results from Wiener (1924a). Part (b) below uses the notion of capacity, which turned out to be the right quantity to measure massiveness of sets in potential theory (see Wiener's test below):

In his paper on *Certain Notions in Potential Theory*, the author gives the following sufficient condition for the regularity of a point: *The point O of the boundary of a domain D is regular if there is a sequence of values of r tending to 0 such that either:*

(a) *The inner Lebesgue measure of the points on a sphere of radius r exterior to D in terms of the surface of the sphere itself exceeds a positive quantity independent of r; or*
(b) *The capacity of the part of the boundary of D interior to a sphere of radius r exceeds M times the capacity of the sphere, M being independent of r.*

All these criteria suggest an intuitive conclusion expressed by Lebesgue (1924a, p. 351), as follows (see also Vasilesco (1937a, p. 25)): *Ce qui semble être essentiel à la régularité d'un point de la frontière, c'est que le voisinage de ce point, extérieur au domaine, ne soit trop petit.*[17]

Geometric criteria were mostly deduced from suitably constructed *barriers*, which represent an effective instrument for studying boundary behaviour. Barriers were used by Poincaré (1890, 1899), and their importance was recognised by Lebesgue (1912a) who coined the term *barrier* (see also Lebesgue (1912b, 1924a)). The preliminary notion of a barrier required continuity up to the boundary and harmonicity on D (see Lebesgue (1912a,b)). Lebesgue (1924a) offers a generalisation and states a necessary and sufficient condition of regularity: *A point $y \in \partial D$ is regular, if and only if there exists a superharmonic function on D continuous at y attaining its lower bound at y and y only.*[18] He already emphasised the local character of regularity. Let us note that, for Lebesgue, superharmonic means a uniform limit of smooth functions s satisfying $\Delta s \leq 0$. The final form of the definition was given by Bouligand (1924). A function u is called a (*weak*) *barrier* (for D) at $y \in \partial D$, if u is positive and superharmonic on $D \cap U$ for some open neighbourhood U of y and $\lim_{x \to y} u(x) = 0$. If u is defined and bounded on all of D, and for each $\delta > 0$,

$$\inf\{u(x) : x \in D, \quad |x - y| \geq \delta\} > 0,$$

then u is called *a strong barrier* at y.

Bouligand (1926) proved that, for every regular point y, there exists a strong barrier at y.

[17] What seems to be essential for the regularity of a boundary point is that the neighbourhood of this point outside of the domain is not too small.

[18] As a consequence, Lebesgue states his *condition d'impossibilité*. If y is an irregular point and f is a continuous function on ∂D attaining its lower bound at y and y only, then the solution of the classical Dirichlet problem does not exist for f.

Concerning Lebesgue's barrier result, Wiener (1924c) says: *Quoique cette condition soit nécessaire et suffisante, comme elle manque cependant de charactérisation géometrique, on ne peut pas la considérer comme une solution définitive du problème de Dirichlet.*[19]

A similar opinion on the notion of a barrier is expressed in Wiener (1924b): *They all suffer from the defect of involving the geometrical character of the boundary only in a very indirect and devious manner. From the geometrical point of view, indeed, they are scarcely more than restatement of the regularity of* y.[20]

And then comes Wiener's famous test for regularity: *It is the purpose of this paper to develop a complete necessary and sufficient characterisation of regular points which shall be at least quasi-geometrical. Its statement reads as follows: Let* y *be a point of* ∂D*, the boundary of an open set of points* D*. Let* λ *be any positive quantity less than* 1*. Let* γ_n *be the capacity of the set of all points* x *not belonging to* D *such that*

$$\lambda^n \leq |x - y| \leq \lambda^{n-1}.$$

Then y *is regular or irregular according as*

$$\frac{\gamma_1}{\lambda} + \frac{\gamma_2}{\lambda^2} + \cdots + \frac{\gamma_n}{\lambda^n} + \ldots$$

diverges or converges.

This result is difficult and required deep insight into the subject as well as advanced technique. It shows how to express and to measure the non-smallness of the complement of D close to y, and confirms that regularity of a boundary point is a local notion. In Wiener's words from Wiener (1924b): *To return to Lebesgue: like the author, he points out that the solubility of the Dirichlet problem in the classical sense for a given region depends only on the* in Kleinem *properties of the boundary, and may be reduced to the investigation of what he terms the regularity of the individual points of the boundary.*

Despite of the significance of Wiener's test for classical potential theory, it was later overshadowed by other notions of smallness (such as thinness, see Section 9) and by a revival of the barrier criterion, that is, notions which did not depend on the Euclidean boundary or on a specific potential kernel, and thus were better adapted to the needs of abstract potential theory in the second half of the 20th century; see Constantinescu and Cornea (1972); Bliedtner and Hansen (1986); Bauer (1975); Brelot (1970, 1972).

---

[19] Although this is a necessary and sufficient condition, given its lack of geometric character one cannot consider it as a definitive solution of the Dirichlet problem.

[20] The notation is adapted here to that of our exposition.

We have seen that the period 1919–1925 was a very active one, as far as the Dirichlet problem is concerned, due, on one hand to European mathematicians (Lebesgue, Bouligand, Perron) and, on the other hand, American mathematicians (Wiener, Kellogg, Raynor, Gleason). Regarding the contribution of American researchers, Lebesgue (1924b), in response to Wiener's note (1924c), says:

> *Leurs théorèmes dépassent souvent les miens et ils ont si bien abordé les diverses questions qui m'ont occupé que, si j'avais connu leurs travaux j'aurais sans doute jugé inutile de revenir sur mes résultats de 1913. ...*
>
> *Les travaux de Savants américains avaient aussi échappé à M. Bouligand; il convient d'ajouter que, par contre, l'importante Note publiée par M. Bouligand, en 1919, semble être restée inconnue des Américains. Maintenant que ces efforts parallèles ne s'ignorent plus, on peut espérer qu'ils seront plus fructueux encore.*[21]

We find it interesting to quote from Wiener (1956, p. 92), on Wiener's reasons for submitting the note Wiener (1924c) to the *Comptes Rendus*:

> *In many scientific subjects there comes a time when the sharpness and definiteness of the new papers indicate that an important goal is about to be achieved. So it was with the work of Lebesgue and Bouligand. I knew that if I did not put forth a maximum effort, the whole topic was soon to be crossed off the account books of mathematics as one finally resolved. Accordingly, I put forth a maximum effort, employing those new tools which I had made my own, and I was delighted to find that I had achieved what was from the standpoint of research at that time a complete solution of the problem.*
>
> *I was aware that I must hurry. I went to a Mexican student of mine, Manuel Sandoval Vallarta, who was much more proficient in French than I was, and got him to help me to rewrite my ideas in acceptable French. I mailed a brief note to Lebesgue for publication in the Comptes Rendus and awaited the outcome.*
>
> *What followed is a coincidence of a sort much commoner in the history of discovery and invention than one might suppose. While my letter was crossing the ocean, Bouligand obtained some extremely important results which he had not yet had time to polish up. He took council with Lebesgue, who advised him to submit these results to the academy in a sealed envelope, after a custom sanctioned by centuries of academy tradition. The very next day, my paper came in and Bouligand's sealed envelope was opened. The two papers appeared side by side in the next number of Comptes Rendus, with a preface by Lebesgue covering both of them. While they were expressed in different mathematical language, the main idea of the two was*

[21]Their theorems often go beyond of mine and have dealt so well with various questions that have occupied me that, if I had known their work, I would undoubtedly have considered it pointless to return to my results of 1913...
The work of American researchers had also escaped the notice of Mr. Bouligand; it is worth mentioning, on the other hand that the important Note published by Mr. Bouligand in 1919 seems to have remained unknown to the Americans. Now, that these parallel efforts are no longer overlooked, we can hope for even greater fruitfulness in future. (The Note mentioned is Bouligand (1919).)

> *identical. However, the logic of Bouligand's paper was not as complete as mine, owing to the fact that what he had sent it was only a preliminary communication for purposes of record, and not a polished and finished job.*

The achievements of mathematicians during 1913–1925 marked a turning point in the development of potential theory. Nevertheless, a few fundamental problems remained open. One of them is related to the so-called *Kellogg lemma*, concerning the size and distribution of the set of irregular points in the boundary. Kellogg (1928, p. 336) writes: *It is natural to ask whether exceptional* [= irregular] *points can occur in sufficient frequency on the boundary to affect the solution of the Dirichlet problem. More precisely, can two different functions, harmonic and bounded in D, approach the same boundary values at all regular boundary points?*

If the answer were affirmative, Kellogg explains that there should exist a closed set of positive capacity consisting only of irregular points, and continues: *We should therefore had a contradiction if it were possible to establish the following lemma:* Every closed bounded set of positive capacity contains a regular point. *The corresponding lemma in two dimensions has been established...*[22] *In space of three or more dimensions, the lemma is still in doubt.*

Evans (1933) proved the result (usually referred to as the Kellogg–Evans theorem) for higher-dimensional spaces.[23] The proof depends on a detailed investigation of the character of discontinuities of the capacitary potential.

A remarkable result, the so-called *continuity principle*, can be understood as a logical outcome of research from Evans (1933). It was proved independently by Evans (1935) and Vasilesco (1935).

It reads as follows: *Let μ be a measure in $\mathbb{R}^n$, $n \geq 2$, of compact support S. Then $G\mu$ is continuous at $x \in S$ if and only if the restriction of $G\mu$ to S is continuous at x.*

Such a result was possible thanks to advanced techniques of analysis which were developed in the early decades of the 20th century. We quote from the introduction of Evans (1935): *A recent contribution to potential theory is characterised by the names of Lebesgue, Wiener, Kellogg, Vasilesco and Bouligand. Central features of this contribution are the notions of capacity and of regular boundary point, which are related to each other by Kelloggs's hypothetical lemma.*[24] *A recent memoir by de la Vallée Poussin reinterprets these theories in the light of potentials of positive mass and the Poincaré sweeping out process.*[25] *In the present memoir, the author's aim is*

---

[22]Kellogg refers to Kellogg (1928) and adds that the result was obtained on the basis of a theorem of Vasilesco (1930).
[23]For an elementary approach to the Kellogg–Evans theorem, see Arsove (1968).
[24]See Kellogg (1967, p. 337).
[25]The paper de la Valleé Poussin (1932) is meant.

to push on the development of these central problems of mass distribution, regularity, capacity and approximation, and to answer definitely some of the questions which have become important. Fortunately there is already at hand, in the analysis of the general integral or linear functional of Radon, Daniell and F. Riesz, the precise mathematical tool which is necessary.[26]

We close this section with a short summary. The third and the fourth decades of the 20th century turn out to be exceptionally fruitful and important for the development of classical potential theory. Not only were deep results obtained, but the picture of potential theory was transformed through the new horizons offered by other mathematical disciplines. These made it possible, for instance, to put old results of Gauss on a sound basis, as was done in Frostman's thesis and later in the important contributions by H. Cartan (see Frostman (1935), Cartan (1942, 1945)). These important aspects are, only for reasons of space limitations, omitted from the present contribution.

There is an extensive literature dealing with the development of potential theory. We refer to a few relevant historical studies, survey articles as well as historical comments in monographs and textbooks: Armitage and Gardiner (2001), Bliedtner and Hansen (1986), Bouligand (1926), Brelot (1933, 1952, 1965, 1970, 1971, 1972, 1983, 1985), Constantinescu and Cornea (1972), Doob (1984), du Plessis (1970), Gårding (1979–1980), Helms (1969), Jahnke (2003), Kellogg (1926a, 1926b, 1967), Kline (1972), Lichtenstein (1909–1921), Mawhin (2010), Monna (1975), Netuka (2020), Simon (2015b), Sologub (1975), Shraer (1973), Vasilesco (1936, 1937a,b).

9. The Dirichlet Problem and Balayage: The Ultimate Stage of the Development

We are now in the period around 1940. Thanks to the work of outstanding mathematicians, solutions of most of the key problems of classical potential theory in $\mathbb{R}^n$ had been achieved. Potential theory was ready for generalisation in various directions in the years to come (potentials of general kernels, relations to probability theory, semigroup theory, abstract or axiomatic theory, second order partial differential equations, quadratic potential theory, discrete potential theory, algebraic potential theory, infinite dimensional potential theory, fine potential theory, etc.).

A more detailed account of the further development is beyond the scope of the present contribution. Here we stay in the late 30s of the 20th century and deal mainly with the Dirichlet problem for general (discontinuous) boundary data. We also discuss the uniqueness of an operator associated with the generalised Dirichlet

[26] See Radon (1987); Daniell (1918, 1920); Riesz (1916).

problem and a certain ideological shift concerning balayage. We will use more modern notation than before.

For simplicity, we shall consider an arbitrary bounded open set D in $\mathbb{R}^n$, $n \geq 2$. We recall that a function u is called *hyperharmonic* on D if on each component of D either $u = \infty$ or u is superharmonic. Let f be an extended real-valued function on ∂D. A function u on D is said to be an *upper function* for f, if u is hyperharmonic, bounded below, and

$$\lim\ \inf_{x \to y} u(x) \geq f(y)$$

for each $y \in \partial U$. The infimum of upper functions is denoted by $\overline{H} f$ and called the *upper PWB solution*.[27] The *lower PWB solution* is defined in the obvious way (or one puts $\underline{H}\, f := -\overline{H}(-f)$). By the minimum principle, $\underline{H}\, f \leq \overline{H}\, f$ and $\underline{H}\, f$ and $\overline{H}\, f$ are ∞ or $-\infty$ or harmonic on each component of D.

An extended real-valued function is called *resolutive*, if $\overline{H} f$ and $\underline{H}\, f$ are equal and harmonic. The common value is the denoted Hf and called the PWB-*solution* of the Dirichlet problem. We have already mentioned Wiener's resolutivity theorem: *Every function continuous on ∂D is resolutive.* Wiener's original proof was difficult. However, one can first observe that f is resolutive, if it coincides on ∂D with a function continuous on $\overline{D}$ and superharmonic on D (because then $\overline{H} f$ is a *lower* function for f). Then one can use the fact that differences of functions of this type are uniformly dense in the space $C(\partial D)$ of continuous functions on ∂D. One can refer to the Stone–Weierstrass theorem (see Armitage and Gardiner (2001, p. 170)) or to the comment in Brelot (1939a, p. 140), where a similarity with an old idea of Poincaré (1890) is mentioned.

Now we come to one of the central notions of potential theory, the harmonic measure for a given point $x \in D$. Since the mapping

$$f \mapsto Hf(x), \quad f \in C(\partial D),$$

is linear and positive, there exists a uniquely determined measure μ_x on ∂D such that

$$Hf(x) = \int f \, \mathrm{d}\mu_x, \quad f \in C(\partial D). \tag{11.9}$$

We remark that this notion goes back in fact (under the name *outer weight*) to Wiener (1924a). In the two-dimensional case it appears as a tool in complex function theory in the contribution of R. Nevanlinna (see the reference in Simon (2015b), p. 721). The representation (11.9) is a consequence of the Riesz representation theorem.[28] It is worth mentioning that harmonic measure can be obtained as

---

[27]See Brelot (1939a); PWB refers to Perron, Wiener, Brelot.

[28]Cf. Armitage and Gardiner (2001, p. 306); or Rudin (1987, Chapter 2); for the history we refer to Bourbaki (1994, Chapter 11); and (Gray, 1984).

the balayage of the unit point mass at x on the complement of D; see de la Vallée Poussin (1932).

As we have already noted, Wiener (1923b) deals with the Dirichlet problem for discontinuous functions. His partial results rely on the fresh integration theory of Daniell (1918). However, the definitive result is due to Brelot (1939a) and is called the *Brelot resolutivity theorem*, which we state here, for simplicity, for D connected and for Borel extended real-valued functions f on ∂D: *The following are equivalent:*

(i) *f is resolutive;*
(ii) *f is μ_x-integrable for some $x \in D$;*
(iii) *f is μ_x-integrable for all $x \in D$.*

In a sense, this result completes the theory of the Perron–Wiener generalised Dirichlet problem.

Much of potential theory in the period between the World Wars dealt with finding a reasonable generalisation of the classical Dirichlet problem. As already said, reasonable surely means associating to a continuous boundary condition the solution of the classical Dirichlet problem, if it exists. It seems to be natural to ask linearity of such a mapping, because harmonic functions form a linear space. Finally, the minimum principle suggests that, to positive functions on ∂D, there should correspond positive harmonic functions. All methods for solving the generalised Dirichlet problem met these properties and any two methods proposed, as shown by Vasilesco (1937a), led to the same result. However, this does not guarantee any uniqueness statement concerning operators sending functions from $C(\partial D)$ into the space $\mathcal{H}(D)$ of harmonic functions on D with the reasonable properties discussed above; see the notion of the Keldysh operator introduced earlier in Section 4.

In 1941, Keldysh published in a short note Keldysh (1941b) the following result, which can be understood as a culmination of research on the generalised Dirichlet problem:

Let $A : C(\partial D) \to \mathcal{H}(D)$ be an operator enjoying the following properties:

(a) *A is linear;*
(b) *A is increasing, that is, if $f_1 \leq f_2$ on ∂D, then $Af_1 \leq Af_2$ on D;*
(c) *Af is the classical solution of the Dirichlet problem, if it exists for f.*
Then $Af = Hf$ holds for every $f \in C(\partial D)$.

The proof is based on the following very subtle result proved in Keldysh (1941a), Theorem XIX, using Wiener's test of regularity: *Let y be a regular point of D. Then there exists a function h continuous on $\overline{D}$ and harmonic on D such that $h(y) = 0$ and $h > 0$ on $\overline{D} \setminus \{y\}$.*

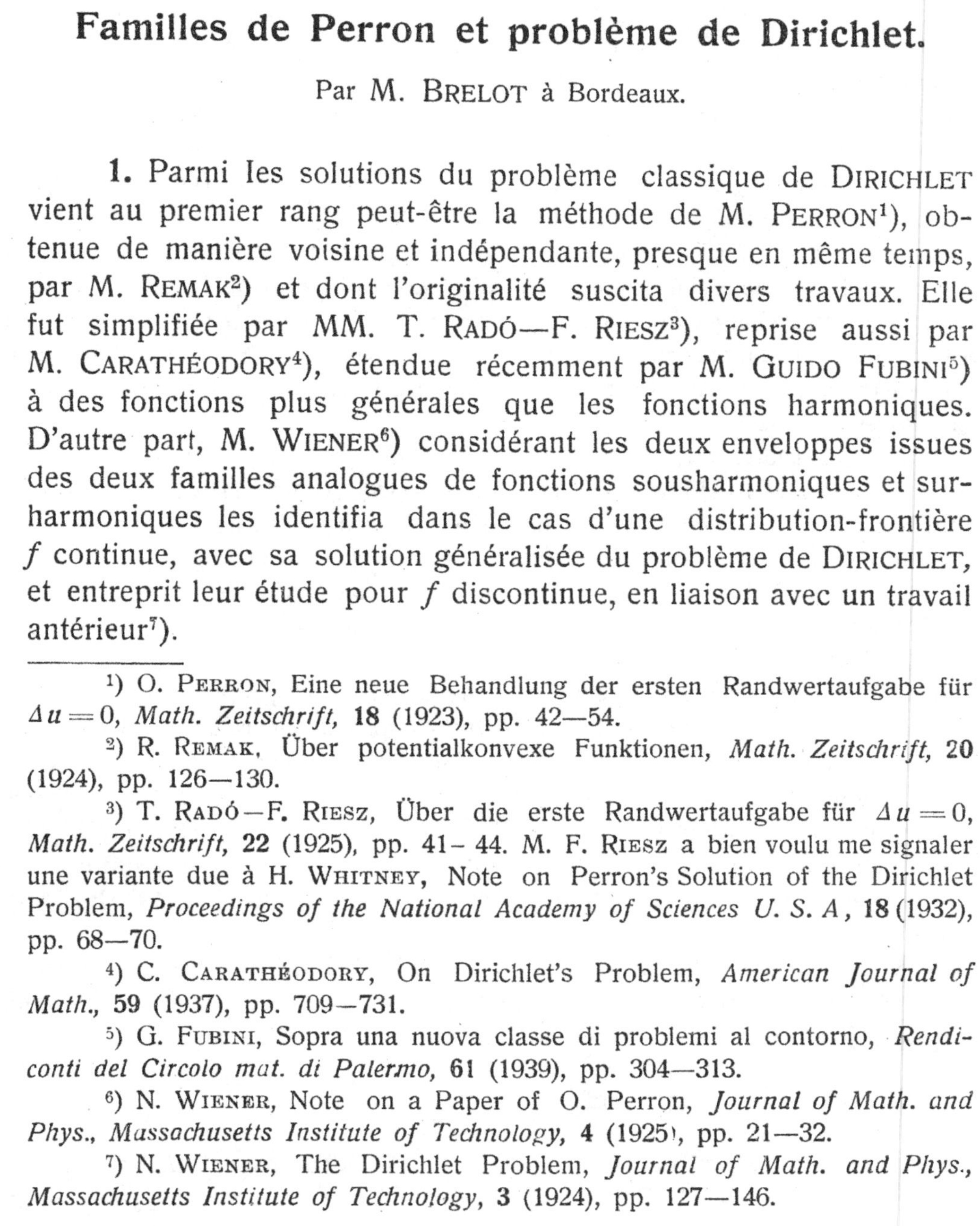

Familles de Perron et problème de Dirichlet.

Par M. BRELOT à Bordeaux.

1. Parmi les solutions du problème classique de DIRICHLET vient au premier rang peut-être la méthode de M. PERRON[1]), obtenue de manière voisine et indépendante, presque en même temps, par M. REMAK[2]) et dont l'originalité suscita divers travaux. Elle fut simplifiée par MM. T. RADÓ—F. RIESZ[3]), reprise aussi par M. CARATHÉODORY[4]), étendue récemment par M. GUIDO FUBINI[5]) à des fonctions plus générales que les fonctions harmoniques. D'autre part, M. WIENER[6]) considérant les deux enveloppes issues des deux familles analogues de fonctions sousharmoniques et surharmoniques les identifia dans le cas d'une distribution-frontière f continue, avec sa solution généralisée du problème de DIRICHLET, et entreprit leur étude pour f discontinue, en liaison avec un travail antérieur[7]).

[1]) O. PERRON, Eine neue Behandlung der ersten Randwertaufgabe für $\Delta u = 0$, *Math. Zeitschrift,* **18** (1923), pp. 42—54.

[2]) R. REMAK, Über potentialkonvexe Funktionen, *Math. Zeitschrift,* **20** (1924), pp. 126—130.

[3]) T. RADÓ—F. RIESZ, Über die erste Randwertaufgabe für $\Delta u = 0$, *Math. Zeitschrift,* **22** (1925), pp. 41- 44. M. F. RIESZ a bien voulu me signaler une variante due à H. WHITNEY, Note on Perron's Solution of the Dirichlet Problem, *Proceedings of the National Academy of Sciences U. S. A*, **18** (1932), pp. 68—70.

[4]) C. CARATHÉODORY, On Dirichlet's Problem, *American Journal of Math.,* **59** (1937), pp. 709—731.

[5]) G. FUBINI, Sopra una nuova classe di problemi al contorno, *Rendiconti del Circolo mat. di Palermo,* **61** (1939), pp. 304—313.

[6]) N. WIENER, Note on a Paper of O. Perron, *Journal of Math. and Phys., Massachusetts Institute of Technology,* **4** (1925), pp. 21—32.

[7]) N. WIENER, The Dirichlet Problem, *Journal of Math. and Phys., Massachusetts Institute of Technology,* **3** (1924), pp. 127—146.

Figure 3. The Brelot resolutivity theorem paper.[29]

[29]M. Brelot, Familles de Perron et problème de Dirichlet, *Acta Litterarum ac Scientiarum. Sectio Scientiarum Mathematicarum (Szeged)* **9**(1939), pp. 133–153. The picture is from the author's archive.

Having such a *high quality barrier* for $y \in \partial_{\mathrm{reg}} D$, one easily proves that $\lim_{x \to y} Af(x) = f(y)$. Therefore Af and Hf have the same limit at every regular point, hence $Af = Hf$.

Brelot (1960/1961) offered a simpler approach to the proof of Keldysh's result. An elementary proof can be found in Netuka (1980), where the problem is related to Choquet's integral representation theory developed in the 50s of the 20th century. A detailed exposition is given in Lukeš *et al.* (2010).[30]

Now we briefly return to balayage, which originated in the works of Gauss and Poincaré. Balayage modelled by electrostatic induction is an operation associated with transfer of a charge. So, originally, properties of potentials were involved. In the 30s, important papers were published still primarily operating with potentials of a mass distribution, mainly written by de la Vallée Poussin (see, for instance, de la Vallée Poussin (1932)).

Let us recall that a minimum of two potentials $G\mu_1$ and $G\mu_2$ is a potential of some other measure μ. However, there is no easy way to express μ in terms of μ_1, μ_2. Such a question does not arise when one deals with superharmonic functions, which form a min-stable cone. So the focus of balayage theory began gradually to shift to an operation on superharmonic functions rather than on potentials of a mass distribution. Balayage of measures then arises as a by-product of balayage of functions.

There is also another point: potentials are linked to a specific kernel, whereas the notion of superharmonic functions admits a broad generalisation to various abstract settings, as documented by the axiomatic approach to potential theory which developed in the second half of the 20th century; see, for instance, Bauer (1975); Constantinescu and Cornea (1972); Bliedtner and Hansen (1986). We shall still focus on potential theory in $\mathbb{R}^n$ for $n \geq 3$ here. The plane case requires non-trivial modifications; see Armitage and Gardiner (2001); Ransford (1995); Simon (2015b).

The modern balayage theory is based on the following idea.[31] Given an arbitrary set $E \subset \mathbb{R}^n$ and a positive superharmonic function u (we shall write $u \in \mathcal{S}^+$), the *réduite* is defined by

$$R_u^E := \inf\{v \in \mathcal{S}^+ : v = u \,\text{on}\, E\}.$$

For instance, taking in our special context a bounded open set D with a nice boundary, then for $E := \mathbb{R}^n \setminus D$ and u a potential of the unit point mass ε_x at a point

[30] We note that, in fact, linearity is not needed; uniqueness of A already follows from conditions (b) and (c).

[31] Here we closely follow the exposition of Bauer (1975). Let us emphasise that Brelot was a key figure in the development of potential theory between 1938–1945, as well as later on; see Brelot (1933, 1938, 1939a,b,c, 1940, 1941, 1944, 1945) and also Brelot (1965, 1966, 1970).

$x \in D$, R_u^E is the potential of the balayaged measure distributed on E (in fact, on ∂D). In general, R_u^E is not lower semicontinuous, and thus not superharmonic, so R_u^E has to be regularised, that is, replaced by the greatest lower semicontinuous function $\leq R_u^E$ denoted by $\hat{R}_u^E$ and called the *sweep* or *balayée function* of u relative to E. Then $\hat{R}_u^E \in \mathcal{S}^+$ and

$$0 \leq \hat{R}_u^E \leq R_u^E \leq u.$$

If E is a "nice" closed set and $x \in E$, then $\hat{R}_u^E(x) = u(x)$.

In general we define the *base* $b(E)$ of E by

$$b(E) := \{x \in \mathbb{R}^n : \hat{R}_u^E(x) = u(x) \quad \text{for every}\, u \in \mathcal{S}^+\}.$$

The base of E contains the interior of E and is contained in the closure of E. It turns out that $E \setminus b(E)$ is a small set, more precisely $E \setminus b(E)$ is polar.[32]

The sweeping of masses turns out to be a dual operation to sweeping of superharmonic functions. The richness of the system of superharmonic functions leads to the following theorem: *There exists a unique measure μ^E on $\mathbb{R}^n$ such that*

$$\int u \,\mathrm{d}\mu^E = \int \hat{R}_u^E \,\mathrm{d}\mu \quad \textit{for every}\, u \in \mathcal{S}^+.$$

The measure μ^E is called the *swept-out measure* of μ on E. It lives on $b(E)$ and $b(E) = \{x \in \mathbb{R}^n : \varepsilon_x^E = \varepsilon_x\}$. If μ has compact support then the swept-out measure is characterised as follows: *μ^E is the only measure living on $b(E)$ such that $G\mu^E = G\mu$ on E except for polar set.* Moreover,

$$\hat{R}_{G\mu}^E = G\mu^E.$$

There is a close link between swept-out measures and harmonic measures (see de la Vallée Poussin (1932), Frostman (1935)): *If D is a bounded open set, $x \in D$ and $D^c := \mathbb{R}^n \setminus D$, then*

$$\mu_x^D = \varepsilon_x^{D^c}.$$

If $y \in \partial D$, then y is a regular point of D if and only if $\varepsilon_y^{D^c} = \varepsilon_y$, that is $\partial_{\mathrm{reg}} D = b(D^c)$; see Frostman (1938/1940).

So the condition $\varepsilon_y^{D^c} \neq \varepsilon_y$ expresses the fact that D^c is, in a way, sufficiently thin close to y.

We have known that such a thinness of D^c close to y can be characterised by means of Wiener's test (Wiener's series converges). Brelot (1939b) introduced

---

[32]See Section 7. This follows from the so-called *fundamental convergence theorem: Let $\mathcal{F}$ be family of positive hyperharmonic functions, let $u := \inf \mathcal{F}$ and $\hat{u}$ be the lower semicontinuous regularisation defined by $\hat{u}(y) := \liminf_{x \to y} u(x)$, $y \in \mathbb{R}^n$. Then the set $\{\hat{u} < u\}$ is polar.* The result goes back to Cartan (1942, 1945); see also Brelot (1938, 1957, 1944, 1955/1956).

the notion of a thin set: *A set E is called thin* (*effilé* in French) *at a point y, if either $y \notin \overline{E}$ or there exists a superharmonic function u on a neighbourhood of y such that*

$$\liminf_{x \to y, x \in E \setminus \{y\}} u(x) > u(y);$$

see Brelot (1944, 1971). Hence *E is thin at y, if and only if $y \notin b(E)$, that is, if $\varepsilon_y^E \neq \varepsilon_y$* (see Brelot (1945)), and for *an open set D*, as proved in Brelot (1939b), *$y \in \partial D$ is regular, if and only if $\mathbb{R}^n \setminus D$ is not thin at y.*

Let us note that in a letter of December 30, 1940 addressed to M. Brelot, H. Cartan pointed out that, for a point $y \in \mathbb{R}^n$, the system of all complements of sets of the form $E \setminus \{y\}$ where E is thin at y, forms a local base for a topology; see Brelot (1944, 1971). Cartan later coined the term *fine topology.* In fact, the fine topology is the coarsest topology on $\mathbb{R}^n$ for which all superharmonic functions on open subsets of $\mathbb{R}^n$ are continuous. Brelot's notion of thinness and Cartan's observation opened a wide-ranging mathematical discipline, where a fine topology analogue of balayage theory plays an indispensable role (the *fine potential theory*).

There were numerous results from potential theory of the interwar period which could not be included here in view of the space limitations. We refer to some relevant sources only: Bouligand (1926), Brelot (1933, 1952, 1970, 1971, 1972), Gårding (1979/1980), Kellogg (1926a,b), Lichtenstein (1909–1921), Vasilesco (1936, 1937a,b, 1938).

10. Conclusion

The motto at the beginning of this article compares potential theory with a crossroads in mathematics. In his contribution to the International Congress of Mathematicians (Vancouver, 1974), Bauer (1975) says: *This was ten years ago. Meanwhile traffic has increased and crossroads had to be converted into interchanges of highways — also in potential theory.*

Needless to say the 45 years since 1974 have yielded a further progress in potential theory. For instance, MathSciNet contains, for this period up to August 1, 2019, a total of 6 846 items under MSC Primary 31 — Potential theory.

We close with a quotation from the historical survey by Brelot (1972):

> *La théorie du potentiel qui n'était d'abord qu'un chapitre de physique mathématique, a posé depuis cent cinquante ans des problèmes mathématiques difficiles et délicats qui ont attiré les mathématiciens les plus célèbres comme Gauss, Hilbert ou Poincaré. Elle a suscité l'introduction ou le développements de méthodes et d'outils nouveaux de grande portée. ... Elle s'est renouvelée par la topologie, approfondie et surtout élargie en axiomatiques diverses dont certaines s'appliquent à de vastes classes d'équations aux dérivées partielles du second ordre et elle s'est soudée aux*

> *probabilités par les processus de Markov; elle devient aussi en partie un chapitre d'analyse fonctionelle.*[33]

As expressed by J.-P. Pier (2001, p. 209):

> *The search for unity in mathematics may well be traced all through the first part of the century. Potential theory appears as an illustration of this permanent phenomenon.*

This is exactly what we intended to show in this contribution.

11. Selected Highlights in the Development of Potential Theory[34]

- *Dirichlet Problem on a Ball*: S.D.Poisson 1823; H.A. Schwarz 1870, 1872.
- *Potential Theory as a Part of Mathematics*: G. Green 1828; C.F. Gauss 1839.
- *Convergence Properties of Sequences of Harmonic Functions*: A. Harnack 1887.
- *Méthode du Balayage*: H. Poincaré 1887, 1890, 1899.
- *Boundary Limits of the Poisson Integral*: P. Fatou 1906.
- *Representation of Positive Harmonic Functions*: G. Herglotz 1911; F. Riesz 1911; R.S. Martin 1941.
- *Barriers*: H. Poincaré 1890; H. Lebesgue 1912, 1924; G. Bouligand 1924, 1926.
- *Existence of Non-regular Domains for the Dirichlet Problem*: S. Zaremba 1911; H. Lebesgue 1913.
- *Potentials of Measures*: J. Radon 1919; G.L. Evans; F. Riesz 1924.
- *Generalized Dirichlet Problem*: H. Lebesgue 1912, 1924; N. Wiener 1924.
- *Regular and Irregular Points*: H. Lebesgue 1924.
- *Perron's Solution of the Generalized Dirichlet Problem*: O. Perron 1923; R. Remak 1924.
- *Wiener's Solution of the Generalized Dirichlet Problem*: N. Wiener 1924.
- *Capacity*: N. Wiener 1924; Ch. de la Vallée Poussin 1932; M. Brelot 1939; A.F. Monna 1940; H. Cartan 1945; G. Choquet 1952, 1954.
- *Criteria of Regularity*: H. Poincaré 1890; S. Zaremba 1909; N. Wiener and H.B. Phillips 1923; N. Wiener 1924; H. Lebesgue 1924.

---

[33]Potential theory, which was at first merely a chapter of mathematical physics, has for 150 years posed difficult and delicate mathematical problems that have attracted the most celebrated mathematicians, such as Gauss, Hilbert and Poincaré. It gave rise to the introduction or development of new and far reaching methods and tools.... It has been renewed by topology, deepened and above all enlarged in various axiomatics, some of them applicable to large classes of partial differential equations of second order, and has been fused to probability by Markov processes; it has also become a chapter of functional analysis.

[34]Only highlights related to the concepts discussed in detail in this chapter are mentioned here.

- *Resolutivity of Continuous Functions*: N. Wiener 1925.
- *Set of Irregular Points*: O.D. Kellogg 1928; G.C. Evans 1933.
- *Superharmonic Functions and Potentials*: F. Riesz 1925, 1926, 1930.
- *Continuity Principle*: G.C. Evans 1935; F.A. Vasilesco 1935.
- *Negligible Sets in Potential Theory*: H.A. Schwarz 1872; H. Lebesgue 1923, G. Bouligand 1926; O.D. Kellogg 1928; F.A. Vasilesco 1930; G.C. Evans 1933; M. Brelot 1939, 1941; H. Cartan 1942.
- *Harmonic Measure*: Ch. de la Vallée Poussin 1931, 1932; O. Frostman 1935.
- *Balayage of Measures*: Ch. de la Vallée Poussin 1932.
- *Resolutivity of General Boundary Conditions*: M. Brelot 1939.
- *Balayage of Superharmonic Functions*: M. Brelot 1944, 1945.
- *Fundamental Convergence Theorem*: M. Brelot 1938; H. Cartan 1942, 1945; M. Brelot and G. Choquet 1957.
- *Uniqueness of the Operator of the Generalized Dirichlet Problem*: M.V. Keldysh 1941; M. Brelot 1960.
- *Thin Sets*: M. Brelot 1939, 1940, 1944.
- *Fine Topology*: H. Cartan 1940; M. Brelot 1944.
- *Brownian Motion and Dirichlet Problem*: S. Kakutani 1944.

12. Thumb-nail Biographies[35]

Georges Louis Bouligand (1889–1979) studied at the École Normale Supérieure. His thesis under the supervision of Picard dealt with Green's and Neumann's functions for a cylinder. At the Lycée de Rennes he taught the *classe de mathématiques spéciales*. In 1928, he became a professor at the University in Poitiers, and in 1938 he was appointed professor at the Sorbonne. He published papers and textbooks in mathematical analysis, analytic and differential geometry, topology, fractal geometry (the Minkowski-Bouligand dimension) as well as in physics (rational mechanics, mathematical physics, relativity theory). Bouligand had a deep interest in the didactics of mathematics and the development of mathematics. In the 1920s, he published important results in potential theory concerning regular points and the boundary behaviour of Green's function, the barrier theorem, and the idea of the generalised Dirichlet problem.

Marcel Brelot (1903–1987) graduated from the École Normale Supérieure in 1924. He was awarded the Arconati Visconti scholarship for the period 1924–1927 to start his research career and then the Rockefeller scholarship for the period

[35]Many of these biographical facts are drawn from the MacTutor History of Mathematics archive; see https://www-history.mcs.st-and.ac.uk/BiogIndex.html [28.8.2019].

1929–1931, which enabled him to stay a year in Rome with V. Volterra and a year in Berlin with E. Schmidt. In 1931, he defended his thesis, dealing with the application of subharmonic functions to a particular type of partial differential equation, under the supervision of Picard. After two years at C.N.R.S. as Chargé de Recherches, he moved to Algiers where he worked as a maître de conférence until 1938. He spent four years in Bordeaux as a professor of Mathématique générales. Then he became a professor in Grenoble, and in 1953 was appointed as maître de conférence and later as professor at the Sorbonne. In 1957, in collaboration with Choquet and Deny, Brelot founded the Séminaire de Théorie du Potentiel. Paris became a world centre for potential theory, and Brelot's scientific achievements made him a key international figure in the subject. In 1974, he was elected a Corresponding member of the Academy. Sadly, in the same year, his son died tragically in an air crash. Brelot's list of publications includes more than 150 items, the majority of them concerning potential theory. His work significantly changed the nature of modern potential theory. He obtained definitive results on the Dirichlet problem, introduced the notions of polar sets and thinness, as well as further notions related to balayage of superharmonic functions, and made an extensive study of the Martin boundary representation of harmonic functions. His name is associated with the convergence theorem for superharmonic functions and, in the 50s, with opening up a new field in potential theory, namely axiomatic potential theory (Brelot's harmonic spaces).

Henri Paul Cartan (1904–2008) was the son of the distinguished mathematician Élie Cartan (who worked on differential geometry, Lie groups and algebras). Both brothers of H. Cartan died tragically. Jean, a composer, died of tuberculosis when he was 25, and Louis, a physicist, served with the French resistance in WWII and was executed by the Germans in 1943. H. Cartan studied mathematics at the École Normale Supérieure and at the Sorbonne. He was a student of Montel. He briefly taught at the Lycée Caen and at the University of Lille. In 1931, he was appointed professor at the University of Strasbourg, which was relocated to Clermont-Ferrand in 1939 due to WWII. In 1940, H. Cartan was appointed professor at the Sorbonne. After the war, he spent two years in Strasbourg again and then returned to the Sorbonne. Five years before his retirement in 1975, he taught at the Université Paris-Sud at Orsay. In 1935, Cartan became a founding member of the group of mathematicians called Bourbaki. Between 1930 and 1970, Cartan was a central figure in several complex variables. He also worked in the theory of sheaves, homological theory and algebraic topology. Cartan made an important contribution to potential theory in the 40s, mainly through a systematic study of measures of finite energy from the perspective of Hilbert spaces. This led him to a crucial convergence theorem for superharmonic functions as well as to a new perspective on balayage as a projection in a pre-Hilbert space of measures.

Charles Jean Gustav Nicolas Baron de la Vallée Poussin (1866–1962) was the son of Charles Louis de la Vallée Poussin, professor of mineralogy and geology at the Catholic University of Leuven. He studied engineering, but mathematics soon became his main interest. In 1891, he was appointed as Gilbert's assistant. Gilbert died in 1892 and de la Vallée Poussin, at the age of 26, was elected to Gilbert's chair. In August 1914, de la Vallée Poussin escaped from Leuven at the time of its destruction by the German Army in WWI. He taught at Harvard University and returned to Europe in 1918. After a short stay in Paris (the Sorbonne and Collège de France) he came back to Belgium, where he spent his entire career. De la Vallée Poussin was awarded the Prix Poncelet in 1916, conferred with the title of Doctor Honoris Causa by many universities, served as President of the International Mathematical Union, and was made a baron by King Albert I in 1928. In 1896, de la Vallée Poussin and Hadamard independently proved the prime number theorem. De la Vallée Poussin also proved important results on differential equations, approximation theory, measure theory, the Riemann zeta function, complex analysis, and conformal mappings. His book on Lebesgue integration, published in 1916, influenced generations of mathematicians. In potential theory, de la Vallée Poussin's contributions concerned the notion of capacity and the theory of balayage of measures.

Griffith Conrad Evans (1887–1973) completed his studies at the English and Latin High School in Boston in 1903, and then entered Harvard University. Among his teachers were Osgood, Bôcher and Coolidge. Evans' thesis, supervised by Bôcher, dealt with Volterra's integral equation and appeared in the Transactions of the American Mathematical Society in 1910 and 1911. A travelling fellowship from Harvard made it possible for Evans to spend time in Europe, with Volterra in Rome and briefly with Planck in Berlin. He returned in 1912. After declining appointments at the Massachusetts Institute of Technology and the University of California at Berkeley he decided to accept a post at the Rice Institute in Houston, Texas. He was promoted to full professor in 1916. During WWI, he took part in scientific assignments related to bomb trajectories and anti-aircraft defense. He stayed at Rice and turned down further offers from Harvard and University of California at Berkeley. Finally, he received an offer from Berkeley in 1933 that was too good to refuse, and served as chairman of the mathematics department there for 15 years. During WWII Evans served on the Applied Mathematics Panel of the National Defense Research Council. His main mathematical contributions ranged across subjects including integral equations, functional analysis, minimal surfaces, and mathematical economics. In potential theory, he was one of the first mathematicians to use Lebesgue–Stieltjes–Radon integration. He proved several important results, such as the continuity principle (the Evans–Vasileco theorem) and a uniqueness result for the generalised Dirichlet problem (the Kellogg–Evans theorem), and also contributed to the theory of harmonic and superharmonic functions, balayage and capacity.

Oliver Dimon Kellogg (1878–1932) studied at Princeton University. He received a John S. Kennedy Fellowship and spent two years in Europe. After a year in Berlin he moved to Göttingen, where he worked on his doctorate under the supervision of Hilbert. After his return to the United States he taught at the University of Missouri from 1905, and in 1910 he was promoted to professor there. During WWI Kellogg was involved in the design of devices to detect submarines. In 1919, he was appointed to Harvard, first as a lecturer, later as associate professor and then as full professor in 1927. He published, jointly with G.D. Birkhoff, an important result generalising the Brouwer fixed point theorem, and worked on orthogonal sets of functions, Bernstein polynomials, and on problems in applied mathematics. In 1929, his book Foundations of Potential Theory appeared. It became a classic that was used for several decades. Kellogg published a series of important papers in potential theory, and is known for the Kellogg–Evans theorem on the negligibility of the set of irregular points in the Dirichlet problem.

Henri Léon Lebesgue (1875–1941) studied at the École Normale Supérieure. He was influenced by Baire, Borel, and Jordan. He taught at the Lycée Central at Nancy from 1899 until 1902. Then he became maître de conference at the University of Rennes. He moved to the Sorbonne in 1910, where he was appointed professor in 1919, and then to the Collège de France in 1921. He also taught at the École Normale Supérieure. His discoveries on the integral go back to 1901, but his work was opposed by some. Lebesgue fell out with both Baire and Borel, and this led to a persistent rivalry with them. However, the Lebesgue measure and integral, and their far-reaching generalisations, turned out to be efficient tools in many areas of mathematics, such as trigonometric series, differentiation theory, harmonic analysis, functional analysis, and real function theory. Lebesgue also made major contributions to other fields of mathematics such as topology, dimension theory, descriptive set theory, calculus of variations and potential theory. After 1922, his activity focused on pedagogy, elementary geometry and the history of mathematics. In potential theory, he published important results on the Dirichlet principle, discovered non-regular sets for the Dirichlet problem, contributed to the formulation of the generalised Dirichlet problem, introduced the notion of a barrier and developed criteria for the regularity of boundary points. His results fundamentally changed the complexion of potential theory.

Oskar Perron (1880–1975) studied at the University of Munich and the University of Berlin. In 1902, he received his doctorate in Munich under the supervision of von Lindemann. He continued his studies in Tübingen and then in Göttingen, where he worked with Hilbert. He returned to Munich where he was appointed a lecturer in 1906. In 1910, Perron was appointed an extraordinary professor in Tübingen, and from 1914 he taught as an ordinary professor in Heidelberg. Between 1915 and 1918 he undertook war work, and later returned to Heidelberg. In 1922 he was appointed to a chair in Munich in place of A. Pringsheim, who had retired. He became a member of the so-called *Munich mathematical triumvirate* together with C. Carathéodory and H. Tietze. The period from 1933 to 1945 was difficult for Perron for obvious reasons and, in view of his conflicts with Bieberbach and other pro-Nazis colleagues, his academic career was in danger. Perron retired in 1951, but continued to teach at Munich until 1960. His remarkable publication record includes more than 200 items from various fields. Perron is known for his notion of the integral, and for results on differential equations and continued fractions. However, he also worked in number theory, geometry, matrix theory and algebra. His books, *Die Lehre von den Kettenbrüchen*, *Irationalzahlen*, and *Algebra*, became classics. His final book, on non-Euclidean geometry, was published when he was 82. In potential theory, Perron is known for his method of solving the generalised Dirichlet problem.

Jules Henri Poincaré (1854–1912), a French mathematician and physicist, is usually regarded as the greatest mathematician of the late 19th century. As an aside, his cousin Raymond Poincaré served several terms as prime minister, and was president of the French Republic during WWI. Henri Poincaré's first degree was from the École Polytechnique. He then continued his studies at the École des Mines. For a short time he worked as a mining engineer, while completing his thesis under the supervision of Hermite to receive a doctorate in mathematics in 1879. After two years of teaching at the University of Caen he returned to Paris, and in 1886 was appointed to a chair at the Sorbonne as well as at the École Polytechnique. He held both chairs until his untimely death. His scientific achievements cover a wide range of mathematics, physics, celestial mechanics, philosophy and psychology. In mathematics, he contributed to topology, complex functions of several variables, algebraic geometry, group theory, differential equations and other fields. In potential theory, his méthode du balayage became famous. His work on C. Neumann's method of the arithmetic mean provided an impetus for the development of the theory of integral equations, spectral theory and later functional analysis.

Frigyes Riesz (1880–1956) is a key figure in 20th century analysis. His contributions cover a large part of abstract analysis. Riesz studied in Budapest, and then Göttingen and Zürich, and was awarded a doctorate from Budapest in 1902. He was appointed to a chair at the University of Kolozsvár in 1911. After WWI, Kolozsvár (now Cluj) became part of Romania, and the Hungarian university was moved to Szeged, where Riesz spent the next 25 years as professor. In 1922, together with Haar, Riesz founded the János Bolyai Mathematical Institute as well as the journal Acta Scientiarum Mathematicarum. He was appointed to the chair of mathematics in the University of Budapest in 1945. Marcel Riesz, a younger brother of F. Riesz, was also a mathematician. He mostly worked at the Swedish University of Lund, where he supervised the famous thesis of Frostman. F. Riesz was a pioneer in functional analysis (he paved the way to normed linear spaces, introduced L^p spaces (named after Lebesgue) and identified their duals as well as the dual of the space of continuous functions. He introduced the notion of weak convergence, proved the celebrated Riesz-Fischer theorem on orthogonal expansions, established important results on Hardy spaces, and made key contributions to operator theory (Riesz-Schauder theory) and ergodic theory. He was the first to define general subharmonic functions, and proved the Riesz decomposition theorem, which made it possible (locally) to express a subharmonic function as a sum of a potential and a harmonic function.

Florin Vasilesco (1897–1958) studied at the University of Bucharest and University of Iaşi where he graduated in 1918. A scholarship enabled him to spend research stays in Germany, Switzerland and France. He defended his thesis in Paris in 1925, and returned for a short time to Bucharest. Then he moved to Paris and later on to Harvard, where he met O.D. Kellogg. Later, he spent some time at the Rice Institute, where G.C. Evans worked. His return to Bucharest in 1930 was again brief. By 1932 he was back in Paris, from where he moved to Montpellier, and then in 1938 he was appointed to the chair of rational mechanics in Lille. He was interested in real function theory, fluid mechanics and elasticity theory. In the 30s he worked in potential theory. His results on singularities of harmonic functions, and his proofs of Kellogg's lemma and the continuity principle (the theorem of Evans-Vasilesco), became classical. He also published extensive surveys on the generalised Dirichlet problem, and on the notions of capacity and irregular points for the Dirichlet problem.

Norbert Wiener (1894–1964) was a child prodigy, who was essentially home schooled by his father, a professor of Slavic languages at Harvard. Norbert Wiener entered Harvard at the age of 14 to begin graduate studies in 1909. He began in zoology, then studied philosophy and mathematics at Cornell for a year and returned to Harvard to continue with philosophy. He received his PhD from Harvard at the age of 18 for a thesis on mathematical logic. In Cambridge, England, he studied under Russell and attended courses delivered by G.H. Hardy. His interests were wide. In 1914, he went to Göttingen to study differential equations under Hilbert, and also attended lectures by Landau on group theory. In 1915 he continued in Cambridge with Russell, returned to the United States, worked for a short period for the General Electric Company, and joined a group of scientists working on ballistics during WWI. At the end of the war, he became an instructor in mathematics at Massachusetts Institute of Technology, where he spent the rest of his career. Wiener's scientific achievements are vast, and include the mathematical theory of Brownian motion, the axiomatics of normed linear spaces, harmonic analysis, stochastic processes, ergodic theory, cybernetics, and the mathematical theory of communication. He worked in potential theory for only three years, from 1923 to 1925. Within this short period he produced work that had a major influence on the subject: it dealt with the generalised Dirichlet problem, the notion of capacity, Wiener's famous test of regularity, and Wiener's resolutivity theorem.

References

T. Archibald, Counterexamples in Weierstraß's Work, pp. 269–285 in W. König and J. Sprekels (eds.), *Karl Weierstraß (1815–1897), Aspects of His Life and Work*, Springer Spektrum, Wiesbaden, 2016, xv + 289 pages. ISBN 978-3-658-10619-5.

D.H. Armitage and S.J. Gardiner, *Classical Potential Theory*, Springer Monographs in Mathematics, Springer Verlag, London, 2001, viii + 333 pages. ISBN 1-85233-618-8.

M.G. Arsove, The Wiener-Dirichlet Problem and the Theorem of Evans, *Mathematische Zeitschrift* **103**(1968), pp. 184–194. ISSN 0025-5874.

H. Bauer, Aspects of Modern Potential Theory, pp. 41–51, in R.D. James (ed.), *Proceedings of the International Congress of Mathematicians (Vancouver, B.C., 1974)*, Volume 1, Canadian Mathematical Congress, Montreal, Quebec, 1975, viii + 600 pages. ISBN 0-919558-04-6.

J. Bliedtner and W. Hansen, *Potential Theory. An Analytic and Probabilistic Approach to Balayage*, Universitext, Springer Verlag, Berlin, 1986, xiv + 435 pages. ISBN 3-540-16396-4.

U. Bottazzini, *The Higher Calculus: A History of Real and Complex Analysis from Euler to Weierstrass*, Springer Verlag, New York, 1986, 332 pages. ISBN 0-387-96302-2.

G. Bouligand, Sur les fonctions bornées et harmoniques dans un domaine infini, nulles sur sa frontière, *Comptes Rendus Mathématique, Académie des Sciences Paris* **169**(1919), pp. 763–766.

G. Bouligand, Domaines infinis et le cas d'exception du problème de Dirichlet, *Comptes Rendus Mathématique, Académie des Sciences Paris* **178**(1924), pp. 1054–1057.

G. Bouligand, Sur le problème de Dirichlet, *Annales de la Société Mathématique Polonaise* **4**(1926), pp. 59–112.

N. Bourbaki, *Elements of the History of Mathematics*, Springer Verlag, Berlin, 1994, xviii + 301 pages. ISBN 3-540-19376-6.

M. Brelot, Le problème de Dirichlet sous sa forme moderne, *Mathematica* **7**(1933), pp. 147–166.

M. Brelot, Sur le potentiel et les suites de fonctions sous-harmoniques, *Comptes Rendus Mathématique, Académie des Sciences Paris* **207**(1938), pp. 836–838.

M. Brelot, Familles de Perron et problème de Dirichlet, *Acta Litterarum ac Scientiarum. Sectio Scientiarum Mathematicarum (Szeged)* **9**(1939a), pp. 133–153.

M. Brelot, Sur la théorie moderne du potentiel, *Comptes Rendus Mathématique, Académie des Sciences Paris* **209**(1939b), pp. 828–830.

M. Brelot, Points irréguliers et transformations continues en théorie du potentiel, *Journal de Mathématiques Pures et Appliquées, Neuvième Série* **9**(1940), pp. 319–337.

M. Brelot, Sur la théorie autonome des fonctions sous-harmoniques, *Bulletin des Sciences Mathématiques* (2) **65**(1941), pp. 72–98.

M. Brelot, Sur les ensembles effilés, *Bulletin des Sciences Mathématiques* **68**(1944), pp. 12–36.

M. Brelot, Minorantes sous-harmoniques, extrémales et capacités, *Journal de Mathématiques Pures et Appliquées, Neuvième Série* **24**(1945), pp. 1–32.

M. Brelot, La théorie moderne du potentiel, *Annales de l'Institut Fourier (Grenoble)* **4**(1952), pp. 113–140.

M. Brelot, Nouvelle démonstration du théorème fondamental sur la convergence des potentiels, *Annales de l'Institut Fourier (Grenoble)* **6**(1955/1956), pp. 361–368.

M. Brelot, Sur un théorème de prolongement fonctionnel de Keldych concernant le problème de Dirichlet, *Journal d'Analyse Mathématique* **8**(1960/1961), pp. 273–288.

M. Brelot, *Éléments de la Théorie Classique du Potentiel*, 3e édition, Les cours de Sorbonne. 3e cycle, Centre de Documentation Universitaire, Paris, 1965, 209 pages.

M. Brelot, Norbert Wiener and Potential Theory, *Bulletin of the American Mathematical Society* **72**(1966), pp. 39–41.

M. Brelot, Historical Introduction, pp. 1–21 in *Potential Theory*, Centro Internazionale Matematico Estivo, I Ciclo, Stresa, 1969, Edizioni Cremonese, Rome, 1970, 248 pages.

M. Brelot, *On Topologies and Boundaries in Potential Theory*, Enlarged edition of a course of lectures delivered in 1966, Lecture Notes in Mathematics, Volume 175, Springer Verlag, Berlin, New York, 1971, vi + 176 pages. ISBN 3-540-05327-1.

M. Brelot, Les étapes et les aspects multiples de la théorie du potentiel, *L'Enseignement Mathématique* (2) **18**(1972), no. 1, pp. 1–36. ISSN 0013-8584.

M. Brelot, La Fonction de Green, pp. 9–23, in *Actes du 108e Congrès National des Sociétés Savantes* (Grenoble, 1983), Section des Sciences, Fasc. III, Comité des Travaux Historiques et Scientifiques, Paris, 1983, 294 pages. ISBN 2-7355-0011-X.

M. Brelot, Le Balayage de Poincaré et l'épine de Lebesgue, pp. 141–151, in *Actes du 110e Congrès National des Sociétés Savantes* (Montpellier, 1985), Section des Sciences, Fasc. III, Comité des Travaux Historiques et Scientifiques, Paris, 1985, 264 pages. ISBN 2-7355-0092-6.

M. Brelot and G. Choquet, Le théorème de convergence en théorie du potentiel, *Journal of Madras University, Section B* **27**(1957), pp. 277–286.

H. Burkhardt and F. Meyer, Potentialtheorie, pp. 464–503, in *Encyklopädie der Mathematischen Wissenschaften II A 7 b*, B. G. Teubner, Leipzig, 1899–1916.

H. Cartan, Capacité extérieure et suites convergentes de potentiels, *Comptes Rendus Mathématique, Académie des Sciences Paris* **214**(1942), pp. 944–946.

H. Cartan, Théorie du potentiel newtonien : énergie, capacité, suites de potentiels, *Bulletin de la Société Mathématique de France* **73**(1945), pp. 74–106.

G. Choquet, Theory of Capacities, *Annales de l'Institut Fourier (Grenoble)* **5**(1953/1954), pp. 131–295.

G. Choquet, La naissance de la théorie des capacités: réflexion sur une expérience personnelle, *Comptes Rendus de l'Académie des Sciences, Série Générale, La Vie des Sciences* **3**(1986), no. 4, 385–397. ISSN 0762-0969.

G. Choquet, T. De Pauw, P. de la Harpe, J.-P. Kahane, H. Pajot and B. Sévennec, Autour du centenaire Lebesgue, Papers from the Meeting "*The Lebesgue Measure is 100 Years Old!*" held at the École Normale Supérieure, Lyon, April 27–28, 2001, Panoramas et Synthèses 18, Société Mathématique de France, Paris, 2004, xii + 156 pages. ISBN 2-85629-170-8.

C. Constantinescu and A. Cornea, Potential Theory on Harmonic Spaces, *Die Grundlehren der mathematischen Wissenschaften*, Band 158, Springer Verlag, New York, Heidelberg, 1972, viii + 355 pages. ISBN 0-387-05916-4.

R. Courant, *Dirichlet's Principle, Conformal Mapping, and Minimal Surfaces*, Interscience Publishers, Inc., New York, 1950, xi + 332 pages. Reprint 1977. ISBN 0-387-90246-5.

P.J. Daniell, A General Form of Integral, *Annals of Mathematics* (2) **19**(1918), no. 4, pp. 279–294.

P.J. Daniell, Further Properties of the General Integral, *Annals of Mathematics* (2) **21**(1920), no. 3, pp. 203–220.

J. Dieudonné, *History of Functional Analysis*, Notas de Matemática 77, North-Holland Publishing Company, 1981, 312 pages. ISBN 0-444-86148-3.

J.L. Doob, *Classical Potential Theory and Its Probabilistic Counterpart*, Grundlehren der Mathematischen Wissenschaften 262, Springer Verlag, New York, 1984, xxv + 846 pages. ISBN 0-387-90881-1.

G.C. Evans, Application of Poincaré's Sweeping out Process, *Proceedings of the National Academy of Sciences of the United States of America* **19**(1933), pp. 457–463.

G.C. Evans, On Potentials of Positive Mass I, *Transactions of the American Mathematical Society* **37**(1935), no. 2, pp. 226–253.

G.C. Evans, A Necessary and Sufficient Condition of Wiener, *American Mathematical Monthly* **54**(1947), pp. 151–155.

P. Fatou, Séries trigonométriques et séries de Taylor, *Acta Mathematica* **30**(1906), no. 1, pp. 335–400.

O. Frostman, Potentiel d'équilibre et capacité des ensembles avec quelques applications à la théorie des fonctions, *Meddelanden fran Lunds Universitats Matematiska Seminarium* **3**(1935), pp. 1–118.

O. Frostman, Sur le balayage des mesures, *Acta Litterarum ac Scientiarum, Sectio Scientiarum Mathematicarum (Szeged)* **9**(1938/1940), pp. 43–51.

L. Gårding, The Dirichlet Problem, *The Mathematical Intelligencer* **2**(1979/1980), no. 1, pp. 43-53. ISSN 0343-6993.

C.F. Gauss, *Algemeine Lehrsätze in Beziehung auf die im verkehrten Verhältnisse des Quadrats der Entfernung wirkenden Anziehungs-und Abstossungs-Kräfte*, Resultate aus den Beobachtungen des magnetischen Vereins, 1839, Werke, 5. Band, 1877, 51 pages.

J. Gray, The Shaping of the Riesz Representation Theorem: A Chapter in the History of Analysis, *Archive for History of Exact Sciences* **31**(1984), no. 2, pp. 127–187. ISSN 0003-9519.

J. Gray, *The Real and the Complex: A History of Analysis in the 19th Century*, Springer Undergraduate Mathematics Series, Springer Verlag, Cham, 2015, xvi + 350 pages. ISBN 978-3-319-23714-5.

G. Green, *An Essay on the Application of Mathematical Analysis to the Theories of Electricity and Magnetism*, printed for the author, by T. Wheelhouse, Nottingham, 1828, viii + 72 pages (also in *Mathematical Papers*, Macmillan, London, 1871, pp. 9–41).

A. Harnack, *Die Grundlagen der Theorie des logarithmischen Potentiales und der eindeutigen Potentialfunktion in der Ebene*, B.G. Teubner, Leipzig, 1887, 158 pages.

T. Hawkins, *Lebesgue's Theory of Integration. Its Origins and Development*, The University of Wisconsin Press, Madison, Wisconsin–London, 1970, xv + 227 pages. ISBN 0-299-05550-7.

L.L. Helms, *Introduction to Potential Theory*, Pure and Applied Mathematics, Volume 22, Wiley-Interscience, A Division of John Wiley & Sons, New York–London–Sydney, 1969, ix + 282 pages.

G. Herglotz, Über Potenzreihen mit positvem reelen Teil im Einhetskreis, *Sitzungsberichte der Sächsischen Akademie der Wissenschaften, Leipzig, Mathematische-Naturwissenschaftliche Klasse* **63**(1911), pp. 501–511.

D. Hilbert, *Gesammelte Abhandlungen, Band III, Analysis, Grundlagen der Mathematik, Physik, Verschiedenes, Lebensgeschichte*, Springer Verlag, Berlin–New York, 1970, viii + 435 pages.

H.N. Jahnke (ed.), *A History of Analysis*, History of Mathematics 24, American Mathematical Society, Providence, RI, London Mathematical Society, London, 2003, ix + 422 pages. ISBN 0-8218-2623-9.

M.V. Keldysh, On the Solvability and Stability of the Dirichlet Problem (Russian), *Uspekhi Matematicheskikh Nauk* **8**(1941a), pp. 171–231.

M.V. Keldysh, On the Dirichlet Problem (Russian), *Doklady Akademii Nauk SSSR* **32**(1941b), no. 5, pp. 308–309.

O.D. Kellogg, Recent Progress with the Dirichlet Problem, *Bulletin of the American Mathematical Society* **32**(1926a), no. 6, pp. 601–625.

O.D. Kellogg, On the Classical Dirichlet Problem for General Domains, *Proceedings of the National Academy of Sciences of the United States of America* **12**(1926b), pp. 397–406.

O.D. Kellogg, Unicité des fonctions harmoniques, *Comptes Rendus Mathématique, Académie des Sciences Paris* **187**(1928), pp. 526–527.

O.D. Kellogg, *Foundations of Potential Theory*, Reprint from the 1st edition of 1929, Die Grundlehren der Mathematischen Wissenschaften, Band 31, Springer Verlag, Berlin–New York, 1967, ix + 384 pages.

M. Kline, *Mathematical Thought from Ancient to Modern Times*, Oxford University Press, New York, 1972, xvii + 1238 pages.

H. Lebesgue, Sur le problème de Dirichlet, *Rendiconti del Circolo Matematico Palermo* **24**(1907), pp. 1–32.

H. Lebesgue, Sur le problème de Dirichlet, *Comptes Rendus Mathématique, Académie des Sciences Paris* **154**(1912a), pp. 335–337.

H. Lebesgue, Sur le principe de Dirichlet, *Comptes Rendus Mathématique, Académie des Sciences Paris* **155**(1912b), pp. 699–701.

H. Lebesgue, Sur des cas d'impossibilité du problème de Dirichlet ordinaire, *Comptes Rendus des séances de la Société Mathématique de France* **41**(1913a), p. 17.

H. Lebesgue, Sur l'équivalence du problème de Dirichlet et du problème du calcul des variations considéré par Riemann, *Comptes Rendus des séances de la Société Mathématique de France* **41**(1913b), pp. 48–50.

H. Lebesgue, Conditions de régularité, conditions d'irrégularité, conditions d'impossibilité dans le problème de Dirichlet, *Comptes Rendus Mathématique, Académie des Sciences Paris* **178**(1924a), pp. 449–354.

H. Lebesgue, Observation au sujet de la Note de M.N. Wiener, *Comptes Rendus Mathématique, Académie des Sciences Paris* **178**(1924b), pp. 1053–1054.

H. Lebesgue, *Œuvres Scientifiques*, Volume I., Institut de Mathématiques de l'Université de Genève, Geneva, 1972a, 339 pages.

H. Lebesgue, *Œuvres Scientifiques*, Volume IV., Institut de Mathématiques de l'Université de Genève, Geneva, 1972b, 391 pages.

L. Lichtenstein, Neuere Entwicklung der Potentialtheorie. Konforme Abbildung, pp. 177–377, in *Encyklopädie der Mathematischen Wissenschaften II C3*, B.G. Teubner, Leipzig, 1909–1921.

J. Lukeš, J. Malý, I. Netuka and J. Spurný, *Integral Representation Theory. Applications to Convexity, Banach Spaces and Potential Theory*, De Gruyter Studies in Mathematics 35, Walter de Gruyter & Co., Berlin, 2010, xvi + 715 pages. ISBN 978-3-11-020320-2.

J. Mawhin, Henri Poincaré and the Partial Differential Equations of Mathematical Physics, pp. 257–277, in É. Charpentier, É. Ghys and A. Lesne (eds.), *The Scientific Legacy of Poincaré*, History of Mathematics 36, American Mathematical Society, Providence, RI, 2010, xiv + 392 pages. ISBN 978-0-8218-4718-3.

V. Maz'ya and T. Shaposhnikova, *Jacques Hadamard, a Universal Mathematician*, History of Mathematics 14, American Mathematical Society, Providence, RI, London Mathematical Society, London, 1998, xxv + 574 pages. ISBN 0-8218-0841-9.

A.F. Monna, Sur la capacité des ensembles, *Proceedings Akadamie van Wetenschappen Amsterdam* **43**(1940), pp. 81–86.

A.F. Monna, *Dirichlet's Principle. A Mathematical Comedy of Errors and Its Influence on the Development of Analysis*, Oesthoek, Scheltema & Holkema, Utrecht, 1975, vii + 138 pages. ISBN 90-313-0175-2.

I. Netuka, The Dirichlet Problem for Harmonic Functions, *American Mathematical Monthly* **87**(1980), no. 8, pp. 621–628. ISSN 0002-9890.

I. Netuka, Lebesgue's Criticism of Carl Neumann's Method in Potential Theory, *Archive for History of Exact Sciences* **74**(2020), no. 1, pp. 77–108. ISSN 0003-9519.

O. Perron, Eine neue Behandlung der ersten Randwertaufgabe für $\Delta u = 0$, *Mathematische Zeitschrift* **18**(1923), no. 1, pp. 42–54.

J.-P. Pier, Intégration et Mesure 1900–1950, pp. 517–564, in J.-P. Pier (ed.), *Development of Mathematics 1900–1950, Luxembourg*, 1992, Birkhäuser, Basel, 1994, 729 pages. ISBN 3-7643-2821-5.

J.-P. Pier, *Mathematical Analysis During the 20th Century*, Oxford University Press, New York, 2001, x + 428 pages. ISBN 0-19-850394-6.

A. Pietsch, *History of Banach Spaces and Linear Operators*, Birkhäuser, Boston, Inc., Boston, MA, 2007, xxiii + 855 pages. ISBN 0-8176-4367-2.

N. du Plessis, *An Introduction to Potential Theory*, University Mathematical Monographs, No. 7, Hafner Publishing Co., Darien, Conn., Oliver and Boyd, Edinburgh, 1970, viii + 177 pages. ISBN 0-05-002054-4.

H. Poincaré, Sur les équations aux derivées partielles de la physique mathématique, *American Journal of Mathematics* **12**(1890), no. 3, pp. 211–294.

H. Poincaré, *Théorie du potentiel Newtonien*, Georges Carré et C. Naud, Paris, 1899, 366 pages.

S.D. Poisson, Addition au mémoire précédent et au mémoire sur la manière d'exprimer les fonctions par les séries de quantités périodiques, *Journal de l'École Royale Polytechnique, 9 cahier*, **12**(1923), pp. 145–162.

T. Radó, *Subharmonic Functions*, Chelsea Publishing Company, New York, 1949, 56 pages.

T. Radó and F. Riesz, Über die erste Randwertaufgabe für $\Delta u = 0$, *Mathematische Zeitschrift* **22**(1925), no. 1, pp. 41–44.

J. Radon, Theorie und Anwendungen der absolut additiven Mengenfunktionen, *Sitzungsberichte der Akademie der Wissenschaften zu Wien* **122**(1913), pp. 1295–1438 (also in P.M. Gruber, E. Hlawka, W. Nörbauer and L. Schmetterer (eds.), *Gesammelte Abhandlungen*, Band 1, pp. 45–144, Verlag der Österreichischen Akademie der Wissenschaften, Vienna, Birkhäuser Verlag, Basel, 1987. ISBN 3-7001-1102-9).

J. Radon, Über Randwertaufgaben beim logarithmischen Potential, *Sitzungsberichte der Akademie der Wissenschaften zu Wien* **128**(1919), pp. 1123–1167 (also in P.M. Gruber, E. Hlawka, W. Nörbauer and L. Schmetterer (eds.), *Gesammelte Abhandlungen*, Band 1, pp. 228–272, Verlag der Österreichischen Akademie der Wissenschaften, Vienna, Birkhäuser Verlag, Basel, 1987. ISBN 3-7001-1102-9).

T. Ransford, *Potential Theory in the Complex Plane*, London Mathematical Society Student Text 28, Cambridge University Press, Cambridge, 1995, x + 232 pages. ISBN 0-521-46120-0.

R. Remak, Über Potentialkonvexe Funktionen, *Mathematische Zeitschrift* **20**(1924), no. 1, pp. 126–130.

R. Remak, Über die erste Randwertaufgabe der Potentialtheorie, *Journal für reine und angewandte Mathematik* **156**(1927), pp. 227–230.

M. von Renteln, Zur Situation der Analysis um die Jahrhundertwende, pp. 107–130, in E. Eichhorn and E.-J. Thiele (eds.), *Vorlesungen zur Gedenken an Felix Hausdorff*, Berliner Studienreihe zur Mathematik 5, Heldermann, Berlin, 1994, viii + 250 pages. ISBN 3-8853-105-2.

F. Riesz, Sur certains systèmes singuliers d'équations intégrales, *Annales Scientifiques de l'École Normale Supérieure, Troisième Série* **28**(1911), pp. 33–62.

F. Riesz, Über lineare Funktionalgleichungen, *Acta Mathematica* **41**(1916), no. 1, pp. 71–98.

F. Riesz, Sur les fonctions subharmoniques et leur rapport à la théorie du potentiel I, *Acta Mathematica* **48**(1926), no. 3–4, pp. 329–343.

F. Riesz, Sur les fonctions subharmoniques et leur rapport à la théorie du potentiel II, *Acta Mathematica* **54**(1930), no. 1, pp. 321–360.

F. Riesz, *Œuvres Complètes*, Volume I., (ed.), Á. Császár, Akadémiai Kiadó, Budapest, 1960, 758 pages.

W. Rudin, *Real and Complex Analysis*, Third edition, McGraw-Hill Book Co., New York, 1987, xiv + 416 pages. ISBN 0-07-054234-1.

H.A. Schwarz, Zur Integration der partiellen Differentialgleichung $\partial^2 u/\partial x^2 + \partial^2 u/\partial y^2 = 0$, *Journal für reine und angewandte Mathematik* **74**(1872), pp. 218–253.

M.G. Shraer, H. Poincaré's Methods in Potential Theory (Russian), pp. 203–217, 338, in A.F. Lapko (ed.), *Istoriko-matematiceskije Issledovanija*, vypusk XVIII, Nauka, Moscow, 1973, 339 pages.

B. Simon, *Basic Complex Analysis. A Comprehensive Course in Analysis, Part 2A*, American Mathematical Society, Providence, RI, 2015a, xviii + 639 pages. ISBN 978-1-4704-1100-8.

B. Simon, *Harmonic Analysis. A Comprehensive Course in Analysis, Part 3*, American Mathematical Society, Providence, RI, 2015b, xviii + 759 pages. ISBN 978-1-4704-1102-2.

V.S. Sologub, *The Development of the Theory of Elliptic Equations in the Eighteenth and Nineteenth Centuries* (Russian), Naukova Dumka, Kiev, 1975, 280 pages.

P. Urysohn, Zur ersten Randwertaufgabe der Potentialtheorie. Ein Fall der Unlösbarkeit, *Mathematische Zeitschrift* **23**(1925), no. 1, pp. 155–158.

C. de la Vallée Poussin, Extension de la méthode du balayage de Poincaré et le problème de Dirichlet, *Annales de l'Institut Henri Poincaré* **2**(1932), pp. 169–232.

F.A. Vasilesco, Sur les singularité des fonctions harmoniques, *Journal de Mathématiques Pures et Appliquées* **9**(1930), pp. 81–111.

F.A. Vasilesco, Sur la continuité du potentiel à travers les masses, et la démonstration d'un lemme de Kellogg, *Comptes Rendus Mathématique, Académie des Sciences Paris* **200**(1935), pp. 1173–1174.

F.A. Vasilesco, Le problème de Dirichlet dans le cas le plus général, *L'Enseignement Mathématique* **35**(1936), pp. 88–106.

F.A. Vasilesco, Le problème généralisé de Dirichlet, *Mémoires de l'Académie Royale Belgique* **16**(1937a), pp. 3–55.

F.A. Vasilesco, *La notion de capacité*, Actualité scientifiques et industrielles, no. 571, Hermann, Paris, 1937b, 49 pages.

F.A. Vasilesco, *La notion de point irrégulier dans le problème de Dirichlet*, Actualité scientifiques et industrielles, no. 660, Hermann, Paris, 1938, 59 pages.

N. Wiener and H.B. Phillips, Nets and the Dirichlet Problem, *Journal of Mathematics and Physics, Massachusetts Institute of Technology* **2**(1923a), pp. 105–124.

N. Wiener, Discontinuous Boundary Conditions and the Dirichlet Problem, *Transactions of the American Mathematical Society* **25**(1923b), no. 3, pp. 307–314. Errata ibid. **26**(1924), no. 4, p. 494.

N. Wiener, Certain Notions in Potential Theory, *Journal of Mathematics and Physics, Massachusetts Institute of Technology* **3**(1924a), pp. 24–51.

N. Wiener, The Dirichlet Problem, *Journal of Mathematics and Physics, Massachusetts Institute of Technology* **3**(1924b), pp. 127–146.

N. Wiener, Une condition nécessaire et suffisante de possibilité pour le problème de Dirichlet, *Comptes Rendus Mathématique, Académie des Sciences Paris* **178**(1924c), pp. 1050–1054.

N. Wiener, Note on a Paper of O. Perron, *Journal of Mathematics and Physics, Massachusetts Institute of Technology* **4**(1925), pp. 21–32.

N. Wiener, *I am a Mathematician. The Later Life of a Prodigy*, Doubleday and Co., Garden City, New York, 1956, 380 pages.

N. Wiener, *Collected Works, Volume I., Mathematical Philosophy and Foundations; Potential Theory; Brownian Movement, Wiener Integrals, Ergodic and Chaos Theories, Turbulence and Statistical Mechanics, with Commentaries*, edited by P. Masani, Mathematicians of Our Time 10, MIT Press, Cambridge, Massachusetts, London, 1976, x + 761 pages. ISBN 0-262-23070-4.

S. Zaremba, Sur le principe du minimum, *Bulletin International de l'Académie des Sciences de Cracovie, Classe des Sciences Mathématiques et Naturelles* **7**(1909), pp. 197–264.

S. Zaremba, Sur le principe de Dirichlet, *Acta Mathematica* **34**(1911), pp. 293–316.

https://doi.org/10.1142/9781786349316_0012

Chapter 12

Transformation of Mathematics Between World Wars: The Case of Combinatorics

Antonín Slavík

1. Introduction

This chapter provides an overview of the major discoveries in combinatorics between the world wars. At that time, combinatorics was already a well-established mathematical discipline, with new results being published in respected journals, including the *Acta Mathematica*, *American Journal of Mathematics*, and *Annals of Mathematics*. The motivation for studying combinatorial problems frequently came from other disciplines, such as set theory, group theory, linear algebra, logic, and number theory. Recreational mathematics was an important source of inspiration for research in graph theory, and this fact contributed to the bad reputation of graph theory as a science of trivial problems. However, the publication of the first textbook of graph theory in 1936 (see Section 5), which presented it as a completely rigorous discipline with applications in linear algebra and set theory, helped to change this attitude. Meanwhile, some important results that are now classified as belonging to graph theory were formulated in the language of topological structures. Finally, one should not forget that numerous developments in combinatorics grew out of purely practical problems, such as the design of experiments, the construction of electricity networks, or the enumeration of chemical compounds.

Standard references on the history of combinatorics are Biggs, Lloyd and Wilson (1995) and Wilson and Watkins (2013). This chapter focuses on a very short historical period with the goal of providing additional details as well as some interesting excerpts in order to illustrate the spirit of the original works.

2. Latin Squares, Finite Projective Planes, and Block Designs

There is a wealth of literature on the history of Latin squares, orthogonal Latin squares, and Euler's conjecture (see, e.g., Andersen (2013), Frisinger (1981),

Klyve and Stemkoski (2006), Mullen (1995), Richardson (2019), Styan, Boyer and Chu (2009), Ullrich (2002), and references cited therein). Before turning our attention to the interwar period, we recall some basic notions and earlier results.

A Latin square of order n is an array consisting of n rows and n columns, in which each cell contains a number from the set $\{1, \ldots, n\}$ (or, more generally, one of n distinct symbols). Moreover, each number occurs exactly once in each row and exactly once in each column. It is easy to construct a Latin square of an arbitrary order; however, there is no efficient formula giving the total number of different Latin squares of order n. The number grows quickly: as of today, $n = 11$ is the largest order for which we know the total number of Latin squares (approximately 8×10^{47}).[1]

Given a pair of Latin squares of order n, suppose they are superimposed to give a square array containing ordered pairs (i, j) of integers from the set $\{1, \ldots, n\}$. If each ordered pair occurs exactly once throughout the array, the two Latin squares are said to be orthogonal. Alternatively, such a pair is sometimes referred to as a Graeco-Latin square. This name stems from the fact that the elements need not be integers: Some authors have considered squares filled with letters from the Latin and Greek alphabets, respectively. A Graeco-Latin square then arises as a combination of two such squares.

A systematic investigation of orthogonal Latin squares was initiated by Leonhard Euler (1707–1783), who is the author of the famous 36 officers problem[2]: *Is it possible to arrange six regiments consisting of six officers each of different ranks in a 6×6 square such that no row or column duplicates a rank or a regiment?* Denote the regiments by the Latin letters a, b, c, d, e, f and the ranks by the Greek letters α, β, γ, δ, ϵ, ζ. Clearly, the arrangement of the 36 officers corresponds to a 6×6 array consisting of pairs of letters. Each ordered pair (representing a unique officer) appears exactly once. The requirement that neither ranks nor regiments are repeated in any row or column implies that each Latin letter and each Greek letter occur exactly once in each row and exactly once in each column. Hence, the problem is equivalent to the construction of a pair of orthogonal Latin squares of order 6.

Euler showed that if n is odd or divisible by four, then there exist orthogonal Latin squares of order n. He observed that there is no such pair for $n = 2$ and was unable to find any pair corresponding to $n = 6$ or, more generally, $n = 4k + 2$ with $k \in \mathbb{N} \cup \{0\}$. Thus, he conjectured that no such pairs exist, and this conjecture

[1] See item A002860 in *The On-Line Encyclopedia of Integer Sequences*, available online at https://oeis.org/A002860 [30.8.2019].

[2] Euler's results on Latin squares, including the 36 officers problem, were presented to the St. Petersburg Academy of Sciences in 1779. They were subsequently published in a Dutch journal: L. Euler, Recherches sur une nouvelle espèce de quarrés magiques, *Verhandelingen uitgegeven door het zeeuwsch Genootschap der Wetenschappen te Vlissingen* **9**(1782), pp. 85–239. See item E530 in The Euler Archive, available online at http://eulerarchive.maa.org/pages/E530.html [30.8.2019]. Some earlier appearances of (orthogonal) Latin squares are documented in Andersen (2013), Richardson (2019), Styan, Boyer and Chu (2009) and Ullrich (2002).

remained an open problem for many years. The 36 officers problem was finally settled in 1900, when the French amateur mathematician Gaston Tarry (1843–1913) confirmed[3] by a clever but still quite laborious case-by-case analysis that there are no orthogonal Latin squares of order 6.

Instead of looking for pairs of orthogonal squares, one can consider the following generalization: Given an integer $n \in \mathbb{N}$, find the value $N(n)$ corresponding to the maximum number of mutually orthogonal Latin squares of order n, i.e., a collection of Latin squares where every two are orthogonal. Euler's conjecture is then equivalent to the statement that $N(4k+2) = 1$ for each $k \in \mathbb{N} \cup \{0\}$. It is not difficult to observe that $N(n) \leq n-1$ for each $n \in \mathbb{N}$: Without loss of generality, we can assume that all of the $N(n)$ Latin squares have the vector $(1, \ldots, n)$ in their first row. Then the entries in the first column of the second row have to be distinct values from the set $\{2, \ldots, n\}$, which gives at most $n-1$ possibilities.

In 1922, the American mathematician Harris Franklin MacNeish (1879–1953) published the paper *Euler Squares*[4] concerned with the numbers $N(n)$, $n \in \mathbb{N}$. He began by recalling Euler's 36 officers problem, mentioned the solution given by Tarry, and then proceeded to his own results, which can be summarized as follows:

- If n is a prime power, then $N(n) = n - 1$.
- Given k mutually orthogonal Latin squares of order m and k mutually orthogonal squares of order n, it is possible to construct k mutually orthogonal Latin squares of order mn. Thus,

$$N(nm) \geq \min(N(n), N(m)).$$

Consequently, if $n = p_1^{r_1} \cdots p_k^{r_k}$ is the prime factorization of an integer n, then $N(n) \geq \min(p_1^{r_1}, \ldots, p_k^{r_k}) - 1$. In fact, MacNeish believed that the statement holds with inequality replaced by equality, but this conjecture turned out to be incorrect. Note that if $n = 4k + 2$, then $n = 2^1 \cdot p_2^{r_2} \cdots p_k^{r_k}$, and hence MacNeish's conjecture implies $N(n) = 1$, i.e., the validity of Euler's conjecture.

Let us continue by recalling the concept of a finite projective plane, which is a pair of finite sets X (consisting of "points") and $P \subset \mathcal{P}(X)$ (consisting of "lines") with the following properties:

1. Every two lines $p_1, p_2 \in P$ intersect in a unique point ($|p_1 \cap p_2| = 1$).
2. Every two distinct points $x_1, x_2 \in X$ determine a unique line p ($x_1, x_2 \in p$).
3. There exist four points no three of which lie on a single line.

It turns out that for every finite projective plane, there exists a number $n \in \mathbb{N}$, called the order of the plane, such that each line contains exactly $n+1$ points, and each point is incident with exactly $n+1$ lines. Moreover, the total number of points

[3]G. Tarry, Le problème des 36 officiers, *Comptes Rendus de l'Association Française pour l'Avancement des Sciences* **29**(1900), pp. 170–203.
[4]H.F. MacNeish, Euler Squares, *Annals of Mathematics* **23**(1922), pp. 221–227.

as well as lines in such a plane is $n^2 + n + 1$. The smallest finite projective plane, called the Fano plane, has order $n = 2$, and therefore consists of $2^2 + 2 + 1 = 7$ points and 7 lines.

The relation between finite projective planes and orthogonal Latin squares has been fully clarified by the Indian mathematician (later active in the United States) Raj Chandra Bose (1901–1987). Let us quote from the introduction to his 1938 paper *On the Application of the Properties of Galois Fields to the Problem of Construction of Hyper-Græco-Latin Squares*[5]:

> *When $p-1$ mutually orthogonal squares of side p exist, then by their superposition we get what may be called a completely orthogonalised or Hyper-Græco-Latin square. The work of Fisher and Yates has shown that such squares are of fundamental importance in experimental design. It is easy to see that for a prime number p, a p-sided Hyper-Græco-Latin square exists. Recently Yates has shown that Hyper-Græco-Latin squares exist also for the cases $p = 4, 8, 9$. Professor Fisher, during his recent visit to India, in a Seminar held under the auspices of the Indian Statistical Institute, made the surmise that it should be possible to construct a Hyper-Græco-Latin square for every value of p, which is a prime or a power of a prime. It is the object of this paper to prove that this surmise is correct, by using the properties of Galois Fields. It is hoped that the properties of Galois Fields, and the finite geometries connected with them, will prove useful in many problems of experimental design and the author hopes to pursue this matter in subsequent papers.*

When dealing with finite projective planes, Bose chose an arbitrary line, and called it the line at infinity. Thus, his plane consisted of n^2 finite points, and $n+1$ points at infinity. He then provided a constructive proof of the following statement: *A finite projective plane of order n exists if and only if there exist $n - 1$ mutually orthogonal Latin squares of order n, i.e., if and only if $N(n) = n - 1$.* Bose's next goal was to show that if n is a prime power, then there exists a finite projective plane of order n, or equivalently, a set of $n-1$ mutually orthogonal Latin squares of order n. Thus, Bose (as well as Fisher) was unaware of MacNeish's 1922 paper containing the same result. Nevertheless, his construction based on finite fields (Galois fields) is quite elegant.

It might be helpful to pause for a moment and make the following observation: Let G be a finite quasigroup, that is, a finite set equipped with a binary operation $\circ$ such that for each $a, b \in G$, the equations $a \circ x = b$ and $y \circ a = b$ have unique solutions $x, y \in G$. Then it is easy to check that the quasigroup's multiplication table (Cayley table) is a Latin square. Conversely, every Latin square represents the multiplication table of a quasigroup.

Returning to Bose, he recalled that for every prime power $n = p^k$, there is a finite field $GF(p^k)$ containing exactly n elements (and conversely, the number of elements in each finite field is a prime power); denote them by $g_1, \ldots, g_n$, where $g_1 = 0$ is the

[5]R.C. Bose, On the Application of the Properties of Galois Fields to the Problem of Construction of Hyper-GræCo-Latin Squares, *Sankhyā: The Indian Journal of Statistics* **3**(1938), pp. 323–338.

zero element. Then for each $k \in \{2, \dots, n\}$, the matrix having the elements

$$a_{ij} = g_k \cdot g_i + g_j, \quad i, j \in \{1, \dots, n\},$$

is a Latin square of order n (since the operation $\circ$ given by $x \circ y = g_k \cdot x + y$ defines a quasigroup).[6] Repeating this construction for all possible k, we get $n - 1$ Latin squares, which can be shown to be mutually orthogonal. Bose concluded his paper with numerous examples, pointing out that if n is a prime, then $GF(n)$ is simply the set of integers modulo n, while higher prime powers correspond to finite fields whose elements are polynomials.

Bose correctly observed that his procedure does not give all possible finite projective planes. In fact, the problem of determining all possible orders for which there exists a finite projective plane is still an open problem. Bose's result together with Tarry's verification of the fact that $N(6) = 1$ implies that there is no finite projective plane of order 6. The existence of a plane of order 10 has been ruled out by an extensive computer search (for the history of this result see Lam, 1991). As of today, it is unknown whether there exists a finite projective plane of order 12.

The interest in Latin squares, finite projective planes and related combinatorial structures in the interwar period was fuelled by applications in the design of agricultural experiments. The pioneer of the use of combinatorial and statistical methods in experiments was the British mathematician Ronald Aylmer Fisher (1890–1962). Fisher, who is often referred to as the father of modern statistics, spent the years 1919–1933 at the Rothamsted Experimental Station, one of the oldest agricultural research institutions (founded in 1843), where he worked on the design and analysis of field experiments.

Fisher's work on experimental design was summarized in three influential books, all of which contain passages devoted to Latin squares: *Statistical Methods for Research Workers*,[7] *The Design of Experiments*,[8] and *Statistical Tables for Biological, Agricultural and Medical Research*[9] written together with Frank Yates (1902–1994), who became Assistant Statistician at the Rothamsted Station in 1931 and took over Fisher's position in 1933.[10]

---

[6]In fact, Bose defined a_{ij} to be the unique $\ell \in \{1, \dots, n\}$ such that $g_\ell = g_k \cdot g_i + g_j$ in order to obtain a Latin square whose entries are integers.

[7]First publication: R.A. Fisher, *Statistical Methods for Research Workers*, Oliver and Boyd, Edinburgh, London, 1925, ix + 239 pages. The book has 14 editions, several reprints and translations. Interesting details about this book, including complete bibliographic information, can be found in Edwards and Fisher (2005).

[8]First publication: R.A. Fisher, *The Design of Experiments*, Oliver and Boyd, Edinburgh, London, 1935, xi + 260 pages. This book has nine editions.

[9]First publication: R.A. Fisher and F. Yates, *Statistical Tables for Biological, Agricultural and Medical Research*, Oliver and Boyd, London, Edinburgh, 1938, viii + 90 pages. This book has six editions and four reprints.

[10]In 1964, the journal *Biometrics* published a special issue dedicated to the memory of Fisher. Yates, still working at the Rothamsted Station, contributed a nice survey Yates (1964) of Fisher's achievements in experimental design.

The goal of the agricultural experiments was to verify the effectivity of various fertilizers, herbicides, insecticides, etc. One possible situation, described in Chapter 5 of *The Design of Experiments*, corresponds to an experiment carried out in potatoes at Ely in 1932. There are six different types of treatments of potato plants, denoted by the letters A, B, C, D, E, F, consisting of different quantities of nitrogenous and phosphatic fertilizers. In order to minimize the influence of the soil heterogeneity, each treatment is repeated six times: The experiment is performed on a 6×6 array, and a single treatment is never repeated in the same row or column. Thus, the arrangement corresponds to a Latin square of order 6, whose entries are the letters A–F.

In 1926, Fisher wrote thus[11]:

> *Systematic arrangements in a square ... have been used previously for variety trials in, for example, Ireland and Denmark; but the term "Latin square" should not be applied to any such systematic arrangements. The problem of the Latin Square, from which the name was borrowed, as formulated by Euler, consists in the enumeration of every possible arrangement, subject to the conditions that each row and each column shall contain one plot of each variety. Consequently, the term Latin Square should only be applied to a process of randomization by which one is selected at random out of the total number of Latin Squares possible ...*

To give another example of the usefulness of Latin squares, consider the situation when we have n sorts of crops, n different herbicides, and n different insecticides. Testing each crop against each herbicide and each insecticide would require n^3 experiments, which might be too expensive. To reduce the costs and still get a representative sample, we can take a Latin square of order n whose rows correspond to herbicides and columns to insecticides, while the entries of the Latin square correspond to crops. Such an arrangement ensures that each herbicide is tested in combination with each insecticide. At the same time, one type of crop is never tested twice with the same herbicide nor with the same insecticide, and the total number of experiments is only n^2.

The concept of orthogonal Latin squares is also useful in the design of experiments. Consider again the previous example, but suppose that we also have n fertilizers. We may choose a second Latin square of order n whose rows again correspond to herbicides and columns to insecticides, while the entries represent fertilizers. Then we perform n^2 experiments such that for all $i, j \in \{1, \ldots, n\}$, the i-th herbicide and j-th insecticide are combined with the crop and the fertilizer given by the entries in the i-th row and j-th column of the two Latin squares. If possible, it is advantageous to choose the two squares to be orthogonal; then each type of crop is tested with

[11]R.A. Fisher, The Arrangement of Field Experiments, *Journal of the Ministry of Agriculture* **33**(1926), pp. 503–513. See also the book by Kotz and Johnson (1992), which contains a reprint of Fisher's 1926 paper together with a foreword by Speed.

each fertilizer exactly once. This procedure reduces the total number of experiments from n^4 to n^2 in a systematic way.

In their 1934 paper *The* 6×6 *Latin Squares*,[12] Fisher and Yates have successfully calculated the number of all Latin squares of order 6. It was already known that it suffices to determine the number of reduced Latin squares, i.e., squares whose first row and first column is the vector $(1, \ldots, n)$. The total number of Latin squares of order n is then obtained by multiplication by $n!(n-1)!$. Fisher and Yates found that there are 9408 reduced squares of order 6 (and therefore the total number of squares of order 6 is 812,851,200), which they divided into 17 classes. Their joint book *Statistical Tables for Biological, Agricultural and Medical Research* gives a list of all reduced Latin squares of orders 4 and 5, one member for each of the 17 classes of order 6, and examples of higher-order squares. It also provides examples of $n-1$ mutually orthogonal Latin squares of order n for $n = 3$, 4, 5, 7, 8, and 9.

Yates also investigated a different type of a combinatorial structure, called by him a "symmetrical incomplete randomized block arrangement", now known as a 2-design, or a balanced incomplete block design. The motivation came again from the design of experiments, namely from a situation where a certain plant is amenable to t different treatments, whose effectiveness is to be tested in a series of experiments. Each of them is called a block and involves exactly k distinct treatments. One possibility is to perform all $\binom{t}{k}$ experiments, but this might be too expensive. To reduce the costs, we only insist on the fact that each pair of treatments occurs in the same number of blocks. Let us quote an excerpt from the summary of Yates' 1936 paper *Incomplete Randomized Blocks*[13]:

> *The paper describes a general method of arranging replicated experiments in randomized blocks when the number of treatments to be compared is greater than the number of experimental units in a block. This new type of arrangement, for which the name of symmetrical incomplete randomized blocks is proposed, is such that every two treatments occur together in a block the same number of times. This restriction enables estimates of the treatment effects and of the experimental error to be obtained expeditiously by the ordinary procedure of the analysis of variance. Estimates of block differences can also be obtained if required. The special case in which the blocks are formed of pairs of experimental units is capable of specially simple treatment. The method of symmetrical incomplete randomized blocks is likely to be of most use in cases in which the experimental material naturally divides itself into groups, such as litters of experimental animals, containing numbers less than the number of treatments that it is desired to test, especially if the differences between these natural groups are of interest.*

[12] R.A. Fisher and F. Yates, The 6×6 Latin Squares, *Proceedings of the Cambridge Philosophical Society* **30**(1934), pp. 492–507.

[13] F. Yates, Incomplete Randomized Blocks, *Annals of Eugenics* **7**(1936), pp. 121–140.

The paper of Yates provided many real-world examples of experiments that can be described using 2-designs, including a statistical analysis of their results. He used the following notation: t is the number of available treatments, k is the number of experimental units (treatments) per block, b is the number of blocks. It turns out that each treatment necessarily occurs in the same number of blocks, which Yates denoted by r and referred to it as the number of replications of each treatment. It is clear that $tr = bk$ gives the total number of experimental units. Furthermore, if λ denotes the number of times any two treatments occur together in a block, one can show that $\lambda = r(k-1)/(t-1)$.

Special cases of 2-designs were already investigated in the 19th century. The most famous ones were Steiner triple systems, in which $k = 3$ and $\lambda = 1$. In 1846, the British mathematician Thomas Penyngton Kirkman (1806–1895) showed[14] that such triple systems exist for each t of the form $6\ell + 1$ or $6\ell + 3$ (it is easy to see that no other triple systems can exist). The interest in such systems was related to a famous problem in recreational mathematics — the Kirkman's schoolgirl problem: *Fifteen young ladies in a school walk out three abreast for seven days in succession: it is required to arrange them daily, so that no two shall walk twice abreast.* More information about this problem and the early history of block designs is available in Biggs and Wilson (2013).

But let us return to Yates' 1936 paper, in which he seems to be unaware of these earlier results. He considered the following problem: Given a pair of numbers t, k, is it possible to find the value of r (or, equivalently, λ) such that there exists a 2-design with these parameters? He provided a table corresponding to the values $t \leq 25$ and $k \leq 12$, whose entries give a lower bound on r for which a corresponding 2-design might exist; he also included an example of such a design, or remarked that he was unable to find it. For example, given $t = 8$ and $k = 4$, he found that the smallest possible r is 7, and provided an example of a 2-design consisting of $b = tr/k = 14$ blocks.

Yates also noted that 2-designs with $\lambda = 1$ and $r = k$ can be obtained from $r-2$ mutually orthogonal Latin squares of order $r-1$. He was not aware of the fact that such a 2-design is in fact a finite projective plane, whose points correspond to available treatments, and lines correspond to blocks; the relation $\lambda = 1$ implies that two points determine a unique line, while $r = k$ implies that the number of points k on each line coincides with the number of lines r incident with an arbitrarily chosen point. The order of such a projective plane is $n = r - 1 = k - 1$. Thus, Yates has in fact discovered one part of Bose's 1938 theorem on the relation between finite projective planes and orthogonal Latin squares. It is also worth pointing out that

[14] T.P. Kirkman, On a Problem in Combinations, *Cambridge and Dublin Mathematical Journal* **2**(1847), pp. 191–204.

Yates wrote: "It is known that completely orthogonalized squares exist when the side is a prime number and also for sides 4, 8 and 9. It is also known that no such square of side 6 exists. Higher non-primes have not been investigated". Thus, he was not aware of MacNeish's 1922 result, which says that $N(n) = n - 1$ whenever n is a prime power.

The ideas of Yates were quickly adopted by Fisher and Bose. In 1938, the previously mentioned *Statistical Tables for Biological, Agricultural and Medical Research*, a joint work of Fisher and Yates, already contained tables of balanced incomplete block designs. In 1939, Bose published a long paper entitled *On the Construction of Balanced Incomplete Block Designs.*[15] The next landmark paper *An Examination of the Different Possible Solutions of a Problem in Incomplete Blocks*[16] by Fisher, which came out in 1940, was primarily devoted to the enumeration of 2-designs with $t = 15$, $b = 35$, $k = 3$, $r = 7$. However, this paper is now best remembered for a result called Fisher's inequality. It says that each 2-design with $k < t$ (called a proper 2-design) necessarily satisfies $r \geq k$. Recalling that $bk = tr$, we may conclude that $b \geq t$. This form of Fisher's inequality, which is now standard, says that an experimental design with t treatments requires at least t blocks.[17] This condition is necessary, but not sufficient. As of today, it is still an open problem to find out necessary and sufficient conditions on the parameters t, k, and λ or r guaranteeing the existence of a corresponding 2-design.

Let us finally return to Euler's and MacNeish's conjectures, which have been resolved only in the second half of the 20th century. Nevertheless, their solutions relied heavily on the knowledge of relations between Latin squares and balanced incomplete block designs discovered in the interwar period. In 1959, the American mathematician Ernest Tilden Parker (1926–1991) showed[18] that if $m + 1$ is a Fermat prime greater than 3, then there exists a set of m mutually orthogonal Latin squares of order $m^2 + m + 1$. The proof used the fact that both m and $m + 1$ are prime powers, and hence there exist finite projective planes of orders m and $m + 1$. For $m = 4$, Parker's result implies the existence of 4 mutually orthogonal Latin squares of order 21, disproving MacNeish's conjecture. In the same year, Bose and the Indian mathematician Sharadchandra Shankar Shrikhande (1917–2020) found another general procedure for constructing orthogonal Latin squares from balanced incomplete block designs; as a special case, they obtained two orthogonal Latin

[15] R.C. Bose, On the Construction of Balanced Incomplete Block Designs, *Annals of Eugenics* **9**(1939), pp. 353–399.

[16] R.A. Fisher, An Examination of the Different Possible Solutions of a Problem in Incomplete Blocks, *Annals of Eugenics* **10**(1940), pp. 52–75.

[17] The number of treatments is now commonly denoted by the letter v, standing for "varieties".

[18] E.T. Parker, Construction of Some Sets of Mutually Orthogonal Latin Squares, *Proceedings of the American Mathematical Society* **10**(1959), pp. 946–949.

squares of order 22, disproving Euler's conjecture.[19] Finally, in 1960, Bose, Parker and Shrikhande showed[20] that there exists a pair of orthogonal Latin squares for each order except 2 and 6; their proof again involved balanced incomplete block designs.

3. Hall's Theorem and Sperner's Theorem

The interwar period saw the birth of two cornerstone results in combinatorial set theory (see also Anderson, 2013): Hall's theorem on systems of distinct representatives, and Sperner's theorem on the maximum size of an antichain in the family of all subsets of a finite set.

The standard formulation of Hall's theorem in terms of systems of distinct representatives appeared in a 1935 paper by the British mathematician Philip Hall (1904–1982); see later in this section. However, some results that are related or even equivalent to Hall's theorem, including sufficient conditions for the existence of perfect matchings in bipartite graphs, were discovered earlier.[21]

A noteworthy moment in the rich history of Hall's theorem is a 1916 paper, *Über Graphen und ihre Anwendung auf Determinantentheorie und Mengenlehre*[22] by the Hungarian mathematician Dénes Kőnig (1884–1944). He introduced a bipartite graph as a graph containing no cycle of odd length, and showed that this happens if and only if the vertices can be decomposed in two parts such that all edges connect vertices from different parts. Then he proved the following result, sometimes referred to as Kőnig's line coloring theorem: *Consider a bipartite graph with all vertices*

---

[19] R.C. Bose and S.S. Shrikhande, On the Falsity of Euler's Conjecture About the Non-Existence of Two Orthogonal Latin Squares of Order $4t+2$, *Proceedings of the National Academy of Sciences of the USA* **45**(1959), pp. 734–737.

[20] R.C. Bose, S.S. Shrikhande and E.T. Parker, Further Results on the Construction of Mutually Orthogonal Latin Squares and the Falsity of Euler's Conjecture, *Canadian Journal of Mathematics* **12**(1960), pp. 189–203.

[21] Hall's theorem is sometimes referred to as the "marriage theorem", and is formulated in terms of matchings in bipartite graphs. This interpretation goes back to the paper by H. Weyl, Almost Periodic Invariant Vector Sets in a Metric Vector Space, *American Journal of Mathematics* **71**(1949), pp. 178–205. In this article, Weyl wrote: *Given n distinct objects* $a_1, \dots, a_n$ *(boys) and n distinct objects* $b_1, \dots, b_n$ *(girls); moreover a scheme of linkage* Q_n *according to which an* a_i *and a* b_k *are either linked (friends) or not linked. ... What is the necessary and sufficient condition that the boys can be paired with the girls in such a fashion that in each of the n pairs the partners are friends?*

[22] D. Kőnig, Über Graphen und ihre Anwendung auf Determinantentheorie und Mengenlehre, *Mathematische Annalen* **77**(1916), pp. 453–465. This paper is a German translation of the Hungarian paper D. Kőnig, Graphok és alkalmazásuk a determinánsok és a halmazok elméletére, *Mathematikai és Természettudományi Értesítő* **34**(1916), pp. 104–119. An English translation of a part of this paper is available in Biggs, Lloyd and Wilson (1998, pp. 203–205).

having degree smaller than or equal to d. Then it is possible to color the edges using d colors in such a way that edges of the same color never share a vertex.[23]

In other words, the edge chromatic number of a bipartite graph never exceeds the largest vertex degree. An immediate consequence is the following result: *The edges of each d-regular graph might be decomposed into d disjoint perfect matchings.* In particular, every regular bipartite graph has a perfect matching. Kőnig's paper continues with a section entitled *Anwendung auf Determinanten*, which contains the following theorem: *Let A be a nonzero square matrix of order n consisting of nonnegative integers and such that all row sums as well as all column sums coincide. Then it is possible to choose n nonzero elements of A such that no two lie in the same row or column, i.e., when calculating the determinant of A, at least one of the terms* $a_{1\pi(1)}, \ldots, a_{n\pi(n)}$ *(where* π *is a permutation of* $\{1, \ldots, n\}$*) is nonzero.*

The proof is simple: Construct a bipartite graph with vertex set $V = V_1 \cup V_2$ such that vertices in the first part V_1 correspond to the numbers of rows, and vertices in the second part V_2 correspond to numbers of columns of the matrix A. Each pair of vertices $i \in V_1$ and $j \in V_2$ is connected by a_{ij} distinct edges. (Note that Kőnig dealt with multigraphs, where two vertices can be connected by more than one edge.) The resulting graph is bipartite and d-regular, where d is the common value of all row and column sums of A. Thus, the graph has a perfect matching, and the edges of this matching correspond to nonzero elements of A lying in different rows and columns.

The rest of Kőnig's paper deals with the validity of the previous results in the case of infinite graphs, having in mind potential applications in set theory. In fact, it seems that Kőnig's main goal was to prove the line coloring theorem for graphs with infinitely many vertices; this would have opened the way to a new and perhaps simpler proof of a certain set-theoretic result by Felix Bernstein (1878–1956). More details can be found in Franchella (1997) and Gallai (1978, pp. 195–197).

In 1917, Ferdinand Georg Frobenius (1849–1917) proved[24] the following theorem, sometimes referred to as the Frobenius-Kőnig theorem: *Let A be a square matrix of order n. Then all terms in the definition of the determinant of A vanish if and only if A contains a* $k \times \ell$ *zero submatrix,*[25] *where* $k + \ell = n + 1$. This result is known to be equivalent to Hall's theorem, see, e.g., Section 7.2 in Strang (1986). The abovementioned result of Kőnig from 1916 is an easy corollary of the Frobenius–Kőnig theorem; see, e.g., Zhan (2013, p. 59). In this context, Frobenius expressed a critical opinion on Kőnig's result, found it to be only of little worth, and remarked

[23] Kőnig did not refer to coloring, but considered assigning d different indexes to the edges.

[24] G. Frobenius, Über zerlegbare Determinanten, *Sitzungsberichte der Kőniglich Preussischen Akademie der Wissenschaften* **18**(1917), pp. 274–277.

[25] A $k \times l$ submatrix is the intersection of k rows and l columns, which need not be adjacent.

that graph theory is a method of little value for the development of the theory of determinants. More details of the dispute between Frobenius and Kőnig are available in Schneider (1977).

The history of Hall's theorem is also related to group theory. For an arbitrary group G, its subgroup H and an element $g \in G$, we define the left coset $gH = \{gh : h \in H\}$, and the right coset $Hg = \{hg : h \in H\}$. Every two left/right cosets are either disjoint, or coincide. If H is finite, then the cardinality of each coset is $|H|$. In this case, the group G can be expressed as the finite disjoint union of all the left/right cosets. A set consisting of one representative from each of the left/right cosets is called a transversal. In 1910, the American mathematician George Abram Miller (1863–1951) proved[26] that one may always find a common transversal for both the left and right cosets. However, in 1927, the Dutch mathematician Bartel Leendert van der Waerden (1903–1996) noted[27] that this result belongs to set theory rather than to group theory. He observed that the existence of a common transversal follows immediately from the following theorem: *Let X be a finite set, which is decomposed into n disjoint subsets of size ℓ in two different ways: $X = A_1 \cup \cdots \cup A_n = B_1 \cup \cdots \cup B_n$. Then there exist distinct elements $x_1, \ldots, x_n \in X$, no two of which belong to the same set A_i or the same set B_j.*

Van der Waerden's proof has less than two pages, and the paper ends with an interesting note. Van der Waerden observed that his result also follows from Kőnig's result on the existence of a perfect matching in a regular bipartite graph. Indeed, it suffices to consider the bipartite graph with parts $V_1 = \{A_1, \ldots, A_n\}$ and $V_2 = \{B_1, \ldots, B_n\}$, in which each edge corresponds to an element $x \in X$, and connects the unique two vertices A_i and B_j containing x. Thus, in fact, we get a regular multigraph (with vertex degrees equal to ℓ, the common size of the sets $|A_i|$ and $|B_j|$), for which Kőnig's result guarantees the existence of a perfect matching, and the edges of this matching correspond to the desired system of representatives $x_1, \ldots, x_n \in X$. Van der Waerden noted that it is also possible to go in the reverse direction, i.e., to use his own result to deduce Kőnig's theorem: Given a regular bipartite graph with parts V_1 and V_2, take A_i to be the set of all edges incident with the i-th vertex of V_1, and B_i the set of all edges incident with the i-th vertex of V_2. It is easy to prove that for a regular bipartite graph, the parts V_1 and V_2 necessarily have the same size. Thus, the number of sets A_i is the same as the number of sets B_i, and van der Waerden's theorem guarantees the existence of a common system of distinct representatives, which in turn gives rise to a perfect matching.

[26] G.A. Miller, On a Method Due to Galois, *The Quarterly Journal of Pure and Applied Mathematics* **41**(1910), pp. 382–384.

[27] B.L. van der Waerden, Ein Satz über Klasseneinteilungen von endlichen Mengen, *Abhandlungen aus dem Mathematischen Seminar der Universität Hamburg* **5**(1927), pp. 185–187.

We are finally able to explain the motivation behind the 1935 paper *On Representatives of Subsets*[28] by Philip Hall, a leading expert in group theory. This paper, which contains the first published statement of Hall's theorem as we know it today, begins as follows:

> *Let a set S of mn things be divided into m classes of n things each in two distinct ways, (a) and (b); so that there are m (a)-classes and m (b)-classes. Then it is always possible to find a set R of m things of S which is at one and the same time a C.S.R. (= complete system of representatives) for the (a)-classes, and also a C.S.R. for the (b)-classes. This remarkable result was originally obtained (in the form of a theorem about graphs) by D. Kőnig. In the present note we are concerned with a slightly different problem, viz. with the problem of the existence of a C.D.R. (= complete system of distinct representatives) for a finite collection of (arbitrarily overlapping) subsets of any given set of things. The solution, Theorem 1, is very simple. From it may be deduced a general criterion, viz. Theorem 3, for the existence of a common C.S.R. for two distinct classifications of a given set; where it is not assumed, as in Kőnig's theorem, that all the classes have the same number of terms. Kőnig's theorem follows as an immediate corollary.*

Hall introduced the notion of a complete system of distinct representatives as follows: Given any set S and any finite system of nonempty subsets of S denoted by $T_1, \ldots, T_m$, a complete system of distinct representatives is a set of m distinct elements of $a_1, \ldots, a_m \in S$ such that $a_i \in T_i$ for each $i = 1, \ldots, m$. He noted that if such a system of representatives exists, then the union of any k of the sets $T_1, \ldots, T_m$ must contain at least k elements. Finally, in Theorem 1, he proved that this condition is not only necessary but also sufficient. His proof was based on induction with respect to the number of the sets. Given m sets $T_1, \ldots, T_m$, the induction hypothesis implies the existence of the system of distinct representatives for the sets $T_1, \ldots, T_{m-1}$. If we consider all such systems, it suffices to prove that their intersection does not contain the set T_m. Hall proved this by contradiction, but this part of proof was not completely straightforward.

Finally, Hall showed how his theorem leads to the following more general version of van der Waerden's theorem, which seems to be the main result of his paper: *Let X be a finite set, which is decomposed into pairwise disjoint subsets in two different ways: $X = T_1 \cup \cdots \cup T_n = S_1 \cup \cdots \cup S_m$. Suppose that the union of any k sets from $T_1, \ldots, T_n$ contains elements from at least k sets from $S_1, \ldots, S_m$. Then there exist distinct elements $x_i \in T_i$, $i \in \{1, \ldots, n\}$, no two of which belong to the same set S_j.*

Hall's theorem soon found many other applications. Among the earliest ones were two remarkable results on Latin squares, both proved by the American

[28]P. Hall, On Representatives of Subsets, *Journal of the London Mathematical Society* **10**(1935), pp. 26–30.

mathematician Marshall Hall Jr (1910–1990).[29] The first theorem published in 1945 says that each Latin rectangle with r rows and n columns, where $r \leq n$, can be extended to a Latin square of order n.[30] In a second paper[31] published in 1948, M. Hall provided a stronger version of P. Hall's theorem with a lower bound for the number of systems of distinct representatives, and used it to prove that there exist at least $n!(n-1)!\cdots 2!1!$ Latin squares of order n.

Let us proceed to the next fundamental result of combinatorial set theory, Sperner's theorem, which finally led to the creation of a new branch of mathematics called the Sperner theory. Two basic references on this subject are Anderson (2002) and Engel (1997). The preface to Anderson's book contains the following passage:

> *The origins of the theory can perhaps be traced back to 1928, when Sperner published a simple theorem which has had repercussions far beyond his wildest dreams. Sperner's theorem simply asserts that if you want to find as many subsets of an n-element set as possible, subject to the condition that no subset is contained in another, then you cannot do better than to choose all the subsets of size $[\frac{n}{2}]$. This theorem has been reproved and generalized to such an extent that it has given rise to a whole branch of the theory of partially ordered sets (or posets) called Sperner theory.*

In 1927, the German mathematician Emanuel Sperner (1905–1980) was still a student of Wilhelm Blaschke (1885–1962) at the University of Hamburg when he finished the short paper *Ein Satz über Untermengen einer endlichen Menge*[32] containing his famous result. The paper, which was published in 1928,[33] has no references to earlier literature, but Sperner acknowledged in a footnote that the problem was suggested to him by Otto Schreier (1901–1929) in Hamburg. He went straight to the core of the matter — the paper begins as follows:

> *Gegeben sei eine endliche Menge $\mathfrak{M}$ von n Elementen. Von zwei Untermengen $\mathfrak{A}$ und $\mathfrak{B}$ der Menge $\mathfrak{M}$ heißt $\mathfrak{A}$ Teilmenge von $\mathfrak{B}$ und $\mathfrak{B}$ Obermenge von $\mathfrak{A}$, wenn alle Elemente von $\mathfrak{A}$ in $\mathfrak{B}$ vorkommen. Ordnung einer Untermenge von $\mathfrak{M}$ heißt die Anzahl ihrer Elemente.*
>
> *Ein System Σ von Untermengen heißt ausgezeichnet, wenn keine der in Σ vorkommenden Untermengen Teilmenge einer anderen solchen ist. Die Anzahl der in Σ vorkommenden Untermengen heißt Grad von Σ.*

---

[29]There is no family relationship between Philip Hall and Marshall Hall Jr, but they collaborated with each other.

[30]M. Hall, An Existence Theorem for Latin Squares, *Bulletin of the American Mathematical Society* **51**(1945), pp. 387–388.

[31]M. Hall, Distinct Representatives of Subsets, *Bulletin of the American Mathematical Society* **54**(1948), pp. 922–926.

[32]E. Sperner, Ein Satz über Untermengen einer endlichen Menge, *Mathematische Zeitschrift* **27**(1928), pp. 544–548.

[33]In the same year, Sperner published a second landmark paper containing the result known as Sperner's lemma, which deals with the coloring of triangulations of n-dimensional simplices.

Es gilt nun der Satz:

Ist Σ ein ausgezeichneten System von Untermengen der Menge $\mathfrak{M}$, so ist der Grad von Σ immer $\leq \binom{n}{[\frac{n}{2}]}$, wenn n die Ordnung von $\mathfrak{M}$ ist; und zwar gilt das Gleichheitszeichen

1. *bei geradem n nur einem Falle, nämlich wenn Σ besteht aus sämtlichen Untermengen der Ordnung $\frac{n}{2}$,*
2. *bei ungeradem n in genau zwei Fällen, nämlich*
   (a) *wenn Σ besteht aus sämtlichen Untermengen der Ordnung $\frac{n+1}{2}$,*
   (b) *wenn Σ besteht aus sämtlichen Untermengen der Ordnung $\frac{n-1}{2}$.*

A possible motivation for Schreier and Sperner might have been the following question from number theory: *Given a square-free integer n, what is the maximum number of its positive divisors, no one of which divides any other?* Each positive divisor can be identified with the set of all primes occuring in its factorization. Thus, if n is a product of ℓ primes, it follows immediately from Sperner's theorem that there are at most $\binom{\ell}{[\ell/2]}$ divisors having the desired property. Not long after the publication of Sperner's paper, in 1934, the German mathematician Felix Behrend (1911–1962) used this observation to prove that each sequence of positive integers, none of which divides any other, has lower density zero.[34] In the same year, this result was obtained by a different method by the Hungarian mathematician Paul Erdős (1913–1996).[35] The following comment by Erdős may be found in Elliot (1979, p. 214):

> *In 1934, end of September, I left Hungary for the first time; my mother accompanied me until Vienna. I was a bit despondent on the train to Paris, but soon proved that if $a_1 < \cdots < a_k \leq x$, $a_i \nmid a_j$, $i \neq j$, then $\sum_{a_i<x} 1/a_i < \frac{c\log x}{\sqrt{\log\log x}}$, i.e. I rediscovered Behrend's proof which he found a few months earlier. Davenport and Rado met me at Cambridge on October 1st. I told them my proof and learned that I was anticipated. (I was rather lucky during my long life this did not happen often to me.)*

Erdős used Sperner's theorem to solve other important problems. For example, his work on the Littlewood–Offord problem[36] is nicely described in Bollobás (1997):

> *Erdős was still a young undergraduate, when he became interested in extremal problems concerning set systems. It all started with his fascination with Sperner's theorem on the maximal number of subsets of a finite set with no subset contained in another. Sperner proved in 1928 that if the ground set has n elements then the maximum is*

---

[34]F. Behrend, On Sequences of Numbers Not Divisible One by Another, *Journal of the London Mathematical Society* **10**(1935), pp. 42–44. On the other hand, there exist sequences of positive integers, none of which divides any other, whose upper density is positive; see A.S. Besicovitch, On the Density of Certain Sequences of Integers, *Mathematische Annalen* **110**(1935), pp. 336–341.

[35]P. Erdős, Note on Sequences of Integers no One of Which is Divisible by Any Other, *Journal of the London Mathematical Society* **10**(1935), pp. 126–128.

[36]P. Erdős, On a Lemma of Littlewood and Offord, *Bulletin of the American Mathematical Society* **51**(1945), pp. 898–902.

attained by the system of all $\lfloor n/2 \rfloor$-subsets. Erdős was quick to appreciate the beauty and importance of this result, and throughout his career frequently returned to problems in this vein.

In 1939, Littlewood and Offord gave estimates of the number of real roots of a random polynomial of degree n for various probability spaces of polynomials. In the course of their work, they proved that for some constant $c > 0$, if $z_1, z_2, \ldots, z_n$ are complex numbers with $|z_i| \geq 1$ for each i, then of the 2^n sums of the form $\pm z_1 \pm z_2 \pm \ldots \pm z_n$, no more than $cr2^n(\log n)n^{-1/2}$ fall into a circle of radius r.

On seeing the result, Erdős noticed immediately the connection with Sperner's theorem, especially in the real case. In fact, Sperner's theorem implies the following best possible assertion. If $x_1, \ldots, x_n$ are real numbers of modulus at least 1, then no more than $\binom{n}{\lfloor n/2 \rfloor}$ of the sums $\pm x_1 \pm x_2 \pm \ldots \pm x_n$ fall in an open interval of length 2. From here it was but a short step to show that the maximal number of sums that can fall in an open interval of length $2r$ is precisely the sum of the r largest binomial coefficients $\binom{n}{k}$.

4. Ramsey Theory

The first famous result of Ramsey theory[37] goes back to the 1916 paper *Über die Kongruenz $x^m + y^m \equiv z^m \pmod{p}$*[38] by Issai Schur (1875–1941), who was born in the Russian Empire but spent most of his life in Germany. As the title of the three-page paper suggests, it deals with the congruence $x^m + y^m \equiv z^m \pmod{p}$. Schur recalls a result due to the American mathematician Leonard Eugene Dickson (1874–1954),[39] who showed in 1909 that for each prime number $m \in \mathbb{N}$, there exists a lower bound $M \in \mathbb{N}$ such that for all $p \geq M$, the above-mentioned congruence is satisfied by three relatively prime numbers x, y, z. The motivation for considering this congruence came from an attempt to prove Fermat's last theorem by showing that the congruence has no solution, but this possibility was ruled out by Dickson. Schur explained that his goal was to give a new, short proof of Dickson's theorem:

> *Im folgenden will ich zeigen, daß der Dicksonsche Satz sich fast unmittelbar aus einem sehr einfachen Hilfssatz ergibt, der mehr der Kombinatorik als der Zahlentheorie angehört:*
>
> **Hilfssatz.** *Verteilt man die Zahlen $1, 2, \ldots, N$ irgendwie auf m Zeilen, so müssen, sobald $N > m!e$ wird, in mindestens einer Zeile zwei Zahlen vorkommen, deren Differenz in derselben Zeile enthalten ist.*

In other words, for each $m \in \mathbb{N}$, there exists a smallest number $S(m)$ such that if the numbers $1, \ldots, S(m)$ are distributed into m classes, there exist x, y, z from

---

[37] An even older but relatively unknown Ramsey-type result was obtained by David Hilbert (1862–1943) in 1892, see Chapter 31 of Soifer (2009).

[38] I. Schur, Über die Kongruenz $x^m + y^m \equiv z^m \pmod{p}$, *Jahresbericht der Deutschen Mathematiker-Vereinigung* **25**(1916), pp. 114–117.

[39] L.E. Dickson, On the Congruence $x^n + y^n + z^n \equiv 0 \pmod{p}$, *Journal für die Reine und Angewandte Mathematik* **135**(1909), pp. 134–141.

the same class such that $x - y = z$. Moreover, Schur provided the upper bound $S(m) \leq m!e + 1$. The numbers $S(m)$, $m \in \mathbb{N}$, are now called Schur's numbers; as of today, their precise values are known only for $m \in \{1, \ldots, 5\}$.[40]

The next landmark result is due to van der Waerden, who in 1927 published the paper *Beweis einer Baudetschen Vermutung*[41] containing the following theorem: *For each pair $k, \ell \in \mathbb{N}$, there exists a number $W(k, \ell)$ such that if the numbers $1, \ldots, W(k, \ell)$ are distributed into k classes, then at least one class contains an arithmetic progression of length ℓ.* The numbers $W(k, \ell)$ are now called van der Waerden numbers; apart from the trivial cases $W(1, \ell) = \ell$ and $W(k, 2) = k + 1$, only a few exact values are known.[42]

The title of van der Waerden's paper refers to the Dutch mathematician Pierre Joseph Henry Baudet (1891–1921), who made the conjecture corresponding to $k = 2$. However, the book Soifer (2009), which contains a wealth of information on the history of Ramsey theory, provides evidence for the fact that the same conjecture was also posed by Schur. Soifer also explains how van der Waerden's result became popular despite being published in a relatively unknown Dutch journal (see Soifer, 2009, p. 309):

> *B.L. van der Waerden proved this pioneering result while at Hamburg University and presented it the following year at the meeting of D.M.V., Deutsche Mathematiker Vereinigung (German Mathematical Society) in Berlin. The result became popular in Göttingen, as the 1928 Russian visitor of Göttingen A.Y. Khinchin noticed and later reported, but its publication in an obscure Dutch journal hardly helped its popularity. Only Issai Schur and his two students Alfred Brauer and Richard Rado learned about and improved upon Van der Waerden's result almost immediately; and somewhat later, in 1936, Paul Erdős and Paul Turán commenced density considerations related to Van der Waerden's result. Only after the World War II, when Khinchin's book Three Pearls of Number Theory came out in Russian in 1947 and again in 1948, in German in 1951, and in English in 1952, the result became a classic, and has remained one of the most striking "pearls" of mathematics. ... Now that the success of Khinchin's booklet had made the result classic, the latter merited a special attention and commentary by its solver. Van der Waerden obligated, and in 1954 published an essay Der Beweis der Vermutung von Baudet with a more expressive English title How the Proof of the Baudet's Conjecture Was Found in a later published translation. This essay has appeared four times in German: twice in 1954, in 1965, posthumously in 1998; and once in English in 1971. It is not just invaluable as a historical document. The essay delivers a vibrant portrait of mathematical invention in the making. Van der Waerden presents all critical ideas of the proof in the most clear and engaging way.*

[40] See item A030126 in *The Online Encyclopedia of Integer Sequences*, available online at https://oeis.org/A030126 [30.8.2019].

[41] B.L. van der Waerden, Beweis einer Baudetschen Vermutung, *Nieuw Archief voor Wiskunde* **15**(1927), pp. 212–216.

[42] See https://en.wikipedia.org/wiki/Van_der_Waerden_number [30.8.2019].

The previously mentioned German mathematician Alfred Theodor Brauer (1894–1985) was indeed well acquainted with van der Waerden's theorem and its proof, since he acted as a reviewer of the paper for the *Jahrbuch über die Fortschritte der Mathematik*.[43] Already in the next year, he published the paper *Über Sequenzen von Potenzresten*[44] containing the following stronger version of van der Waerden's theorem: *For each pair* $k, \ell \in \mathbb{N}$, *there exists a number* $N(k, \ell)$ *such that if the numbers* $1, \ldots, N(k, \ell)$ *are distributed into* k *classes, then at least one class contains an arithmetic progression of length* ℓ *together with its difference.* Note that this result also generalizes Schur's theorem, which corresponds to the case $\ell = 2$. And, as Brauer himself admitted, it was his advisor Schur who discovered this more general theorem! But the main topic of Brauer's paper was the existence of arbitrarily long sequences of consecutive k-th power residues or nonresidues modulo p, for sufficiently large primes p, and Schur preferred not be listed as a co-author despite his important contribution.

The name "Ramsey theory" originated from a deep result by the British mathematician Frank Plumpton Ramsey (1903–1930) that appeared in his 1930 paper *On a Problem of Formal Logic*.[45] The introduction of the paper provides a clear description of its contents:

> *This paper is primarily concerned with a special case of one of the leading problems of mathematical logic, the problem of finding a regular procedure to determine the truth or falsity of any given logical formula.*[46] *But in the course of this investigation it is necessary to use certain theorems on combinations which have an independent interest and are most conveniently set out by themselves beforehand.*
>
> *The theorems which we actually require concern finite classes only, but we shall begin with a similar theorem about infinite classes which is easier to prove and gives a simple example of the method of argument.*

In the 1930s, Alonzo Church (1903–1995) and Alan Mathison Turing (1912–1954) proved that the *Entscheidungsproblem* is unsolvable in general; on the other hand, the goal of Ramsey's paper was to show that there exists a decision procedure for a certain restricted class of sentences. His first "theorem on combinations" reads as follows:

> *Let* Γ *be an infinite class, and* μ *and* r *positive integers; and let all those sub-classes of* Γ *which have exactly* r *members, or, as we may say, let all* r*-combinations of the members of* Γ *be divided in any manner into* μ *mutually exclusive classes* C_i

[43]The review is available online at https://zbmath.org/?q=an%3A53.0073.12 [30.8.2019].

[44]A. Brauer, Über Sequenzen von Potenzresten, *Sitzungsberichte der Preußischen Akademie der Wissenschaften* (1928), pp. 9–16.

[45]F.P. Ramsey, On a Problem of Formal Logic, *Proceedings of the London Mathematical Society* **30**(1930), pp. 264–286.

[46]Called in German the *Entscheidungsproblem*; see Hilbert und Ackermann, *Grundzüge der Theoretischen Logik*, 72–81 (Ramsey's footnote).

> $(i = 1, 2, \ldots, \mu)$, *so that every* r*-combination is a member of one and only one* C_i*; then, assuming the axiom of selections* [axiom of choice], Γ *must contain an infinite sub-class* Δ *such that all the* r*-combinations of the members of* Δ *belong to the same* C_i.

Ramsey noted that the statement is trivial for $\mu = 1$, or $\mu = 2$ and $r = 1$. The general proof is by double induction with respect to r and μ. The result is in fact an infinite version of the well-known Ramsey theorem, which came next:

> *Given any* r, n, *and* μ *we can find an* m_0 *such that, if* $m \geq m_0$ *and the* r*-combinations of any* Γ_m [any set with m elements] *are divided in any manner into* μ *mutually exclusive classes* C_i $(i = 1, 2, \ldots, \mu)$, *then* Γ_m *must contain a sub-class* Δ_n [a subset with n elements] *such that all the* r*-combinations of members of* Δ_n *belong to the same* C_i.

Note that if $r = 1$, then the Ramsey theorem reduces to the well-known pigeonhole principle, and it suffices to take $m_0 = (n-1)\mu + 1$. The case $r = 2$ is also easy to visualize: We can think of the 2-combinations as the edges of a complete graph on m vertices, and the theorem says that if we color them by μ colors, we will be able to find a monochromatic complete subgraph on n vertices. Ramsey remarked that in general, he was unable to calculate the optimal value of m_0:

> *This is the theorem which we require in our logical investigations, and we should at the same time like to have information as to how large* m_0 *must be taken for any given* r, n, *and* μ. *This problem I do not know how to solve, and I have little doubt that the values for* m_0 *obtained below are far larger than is necessary.*

The calculation of the Ramsey numbers m_0 still remains a difficult problem, and exact values are known only for a small fraction of the parameters r, n, and μ.

The Ramsey theorem soon found a striking application in a completely different problem, when Erdős together with George Szekeres (1911–2005) used it to prove the following result: *For every* $n \in \mathbb{N}$, *there exists an* $N \in \mathbb{N}$ *such that any set of at least* N *points in the plane in general position has a subset of* n *points forming a convex polygon.* Their paper *A Combinatorial Problem in Geometry*[47] begins by recalling the solution corresponding to $n = 4$ and $N = 5$ due to Esther Klein (future wife of Szekeres), and continues as follows:

> *Miss Klein suggested the following more general problem. Can we find for a given* n *a number* $N(n)$ *such that from any set containing at least* N *points it is possible to select* n *points forming a convex polygon?*
>
> *There are two particular questions:* (1) *does the number* N *corresponding to* n *exist?* (2) *If so, how is the least* $N(n)$ *determined as a function of* n*?* (*We denote the least* N *by* $N_0(n)$.)

---

[47] P. Erdős and G. Szekeres, A Combinatorial Problem in Geometry, *Compositio Mathematica* **2**(1935), pp. 463–470.

We give two proofs that the first question is to be answered in the affirmative. Both of them will give definite values for $N(n)$ and the first one can be generalised to any number of dimensions. Thus we obtain a certain preliminary answer to the second question. But the answer is not final for we generally get in this way a number N which is too large. Mr. E. Makai proved that $N_0(5) = 9$, and from our second demonstration, we obtain $N(5) = 21$ (from the first a number of the order 2^{10000}).

Thus it is to be seen, that our estimate lies pretty far from the true limit $N_0(n)$. It is notable that $N_0(3) = 3 = 2 + 1$, $N_0(4) = 5 = 2^2 + 1$, $N_0(5) = 9 = 2^3 + 1$. We might conjecture therefore that $N_0(n) = 2^{n-2} + 1$, but the limits given by our proofs are much larger.

It is still unknown whether the conjecture mentioned in the last sentence holds;[48] at least, in 1961, Erdős and Szekeres proved[49] that $N_0(n) \geq 2^{n-2} + 1$ for all $n \geq 3$. But let us return to their 1935 paper and give an outline of the first proof: It begins by observing that a finite set of points determines a convex polygon if and only if any four of them form a convex quadrilateral. The next step is to consider four-element subsets of the given N points and divide them into two classes, depending on whether the corresponding four points determine a convex quadrilateral or not. Finally, the Ramsey theorem with $r = 4$ and $\mu = 2$ implies that if N is sufficiently large, then there exist n points whose four-element subsets are all in the same class. This class necessarily corresponds to convex four-element subsets, since we already know that among at least five points, one can always choose four determining a convex quadrilateral.

In fact, Erdős and Szekeres were originally not aware of Ramsey's paper, and gave their own proof of the Ramsey theorem, providing more precise bounds on the Ramsey numbers. Their second proof is not based on the Ramsey theorem, but is equally interesting. First, it established the estimate $N_0(n) \leq \binom{2n-4}{n-2} + 1$, which remained the best known upper bound for over 60 years. Second, an inspection reveals that the core of the alternative proof is the following result: *An arbitrary sequence of $n^2 + 1$ real numbers contains a monotone subsequence of length $n + 1$.* Moreover, somewhat hidden in the text is the following generalization: *An arbitrary sequence of $(r - 1)(s - 1) + 1$ real numbers contains a nondecreasing subsequence of length r or a nonincreasing subsequence of length s.* This result is of independent interest and is sometimes called the Erdős–Szekeres theorem, while the main result of the 1935 paper is referred to as the "happy ending theorem".

[48]We now know that $N_0(6) = 17 = 2^4 + 1$, see G. Szekeres and L. Peters, Computer Solution to the 17-point Erdős-Szekeres Problem, *The ANZIAM Journal* **48**(2006), pp. 151–164. ISSN 1446-1811. The values of $N_0(n)$ for $n \geq 7$ remain unknown.

[49]P. Erdős and G. Szekeres, On Some Extremum Problems in Elementary Geometry, *Annales Universitatis Scientiarum Budapestinensis de Rolando Eötvös Nominatae. Sectio Mathematica* **3–4**(1961), pp. 53–62.

5. Graph Theory

Another major area of combinatorics is graph theory. Two excellent sources on the history of this discipline which are not limited to the interwar period are Beineke and Wilson (2013); Biggs, Lloyd and Wilson (1998).

In the 1930s, there were still many mathematicians who did not regard graph theory as a serious discipline, owing partly to the fact that many results in this field were originally inspired by problems of recreational mathematics.[50] A major impetus that helped to change this attitude was the publication, in 1936, of the first textbook of graph theory, *Theorie der endlichen und unendlichen Graphen*[51] by Dénes Kőnig, whose role in the development of this discipline is hard to overemphasize. The situation was succinctly characterized by William Thomas Tutte (1917–2002) in the introduction to the English edition of Kőnig's textbook[52]:

> *But the honour of presenting Graph Theory to the mathematical world as a subject in its own right, with its own textbook, belongs to Dénes Kőnig.*
>
> *Low was the prestige of Graph Theory in the Dirty Thirties. It is still remembered, with resentment now shading into amusement, how one mathematician scorned it as "The slums of Topology". It was the so-called science of trivial and amusing problems for children, problems about drawing a geometrical figure in a single sweep of the pencil, problems about threading mazes, and problems about colouring maps and cubes in cute and crazy ways. It was too hastily assumed that the mathematics of amusing problems must be trivial, and that if noticed at all it need not be rigorously established. Students tempted by Graph Theory would be advised by their supervisors to turn to something respectable or even useful, like differential equations. I am reminded that my own most recent research in Graph Theory has involved differential equations. Mathematics is One, after all.*

On a similar note, Paul Erdős wrote in 1977[53]:

> *I am very glad that the new Journal of Graph Theory is born – but I cannot help feeling sorry that Dénes Kőnig did not live to see the present flowering of graph theory to which he contributed so much.*
>
> *I myself got interested in graph theory when I was in high school and saw a paper by Kőnig published in the mathematical magazine for high school students. When I was a freshman, T. Gallai, G. Grünwald and I took Kőnig's course on graph theory. All three of us very rapidly began to "prove and conjecture" independently.*

---

[50] See also the interesting paper Wate-Mizuno (2014), which describes the interplay of graph theory and recreational mathematics in the work of Kőnig.

[51] D. Kőnig, *Theorie der endlichen und unendlichen Graphen*, Akademische Verlagsgesellschaft, Leipzig, 1936, xi + 258 pages.

[52] D. Kőnig, *Theory of Finite and Infinite Graphs.* Translated from the German by Richard McCoart. With a Commentary by W.T. Tutte, Birkhäuser Verlag, Boston, MA, 1990, 426 pages. ISBN 978-1-4684-8973-6.

[53] P. Erdős, A Note of Welcome, *Journal of Graph Theory* **1**(1977), p. 3. ISSN 0364-9024.

> *It is curious how little graph theory and combinatorial analysis was appreciated in those early dark ages. A friend of my parents, a statistician, once said about Dénes Kőnig: "He is great in his art but his art is so small". Ten years later J.H.C. Whitehead, the great English topologist, said about a graph theorist: "He works in the slums of topology". When I first got to Princeton in 1938, I was surprised how many of the topologists looked down upon the four colour problem and considered it an unimportant side issue.*

A nice overview of Kőnig's life and work, including detailed comments on his textbook, is available in the biography Gallai (1978), where Gallai wrote:

> *Kőnig collected in his book almost all the significant results of combinatorial graph theory, making many hard-to-find problems accessible. His monograph was also distinguished by the careful planning of the material, the precision of its language, and the completeness of its bibliography. The informal and comprehensible style and many interesting problems of the book evoked the interest of many young mathematicians in graph theory.*

Even today, Kőnig's textbook remains a pleasant reading, and contains material that is missing in more recent treatises. For example, Kőnig devoted a whole chapter to the "labyrinth problem", where he presented three algorithms for finding a way through a maze represented by a graph. A more important feature of the book is that it considers not only finite simple graphs, but also multigraphs and infinite graphs. Dénes Kőnig was clearly influenced by his father Gyula Kőnig (1849–1913), an expert in set theory, and realized that some set-theoretic problems and results might be more transparent when reformulated in the language of finite or infinite graphs. To provide an example, consider the Cantor–Bernstein–Schröder theorem: *If there exist injective functions* $f : A \to B$ *and* $g : B \to A$*, then there is a bijection* $h : A \to B$. Kőnig's graph-theoretic proof is as follows: Without loss of generality, assume that A and B are disjoint, and consider a bipartite graph G with parts A and B. Two vertices $x \in A$, $y \in B$ are joined by an edge if either $f(x) = y$, or $g(y) = x$; if both statements hold, there will be two edges joining x and y. Now, it suffices to show the existence of a perfect matching in G. This is accomplished by noticing that each component of G is either a cycle of even length, a singly infinite path, or a doubly infinite path; each such component obviously has a perfect matching. Kőnig wrote: *It cannot be asserted that this proof was substantially simplified by the graph theoretical formulation. It was only made clearer.* More details of Kőnig's graph-theoretic approach to set-theoretic problems can be found in Franchella (1997).

Despite the low prestige of graph theory, the interwar period saw the emergence of many significant results. An important source of progress was the ongoing work on the resolution of the four-color conjecture proposed in 1852 by Francis Guthrie (1831–1899), which asks whether every map can be colored using four colors in such a way that neighboring countries always have different colors. Equivalently, one may ask whether the vertices of each planar graph can be colored using four colors in such a way that neighboring vertices never share the same color. Although the 1879 proof by the British mathematician Alfred Bray Kempe (1849–1922) turned out to be incorrect,[54] his idea of reducible configurations (corresponding to situations where the problem of coloring can be reduced to a smaller graph) and unavoidable sets of configurations (collections of configurations one of which has to appear in each planar graph) proved to be fruitful, and finally led to the computer-assisted proof of the conjecture in 1976 by Kenneth Ira Appel (1932–2013) and Wolfgang Haken (born 1928).

Meanwhile, some important results were obtained by the American mathematician Philip Franklin (1898–1965), whose 1922 paper *The Four Color Problem*[55] dealt with reducible and irreducible configurations. He discovered some new reducible configurations, and showed that every irreducible map must contain more than 25 regions. Thus, he verified the four-color conjecture for each map containing at most 25 regions. The problem of coloring maps makes sense not only in the plane, but on other surfaces as well. In his 1934 paper *A Six Color Problem*,[56] Franklin improved a result by Percy John Heawood (1861–1955) by showing that for a one-sided (nonorientable) surface of characteristic zero, i.e., the Klein bottle, six colors are always sufficient.

Prior to Franklin, some results on reducible configurations and unavoidable sets of configurations were already obtained by the American mathematician George David Birkhoff (1884–1944) in his 1913 paper *The Reducibility of Maps.*[57] But Birkhoff's first paper related to the four-color conjecture, entitled *A Determinant Formula for the Number of Ways of Coloring a Map*,[58] appeared already in 1912, and is now famous for introducing the concept of a chromatic polynomial. Given a map M with n regions, Birkhoff denoted by $P(\lambda)$ the number of ways of coloring the map in λ colors ($\lambda \in \mathbb{N}$); the four-color conjecture claims that $P(4) > 0$ for each map.

[54]Another erroneous proof of the four-color theorem is mentioned in the biography of D. Kőnig (Gallai, 1978): ... *in the academic year 1904/5 he attended in Göttingen Minkowski's "Analysis Situs" lectures. In these, in addition to the characterization of the topological properties of two-dimensional surfaces and the generation of various normal types, Minkowski intended to present a proof, given by Wernicke, of the four-color conjecture. Only during the presentation of preliminary steps of the proof did it turn out that the proof was erroneous, and so only the proof of the five-color theorem was presented. The lectures of the distinguished professor at Göttingen must have had a great influence on Kőnig's subject selection* ...

[55]P. Franklin, The Four Color Problem, *American Journal of Mathematics* **44**(1922), pp. 225–236.

[56]P. Franklin, A Six Color Problem, *Journal of Mathematics and Physics* **13**(1934), pp. 363–369.

[57]G.D. Birkhoff, The Reducibility of Maps, *American Journal of Mathematics* **35**(1913), pp. 115–128.

[58]G.D. Birkhoff, A Determinant Formula for the Number of Ways of Coloring a Map, *Annals of Mathematics* **14**(1912), pp. 42–46.

Birkhoff gave a clear explanation of the fact that P is a polynomial function of order n, with the leading coefficient equal to 1:

> *In fact let m_i ($i = 1, 2, \ldots, n$) be the number of ways of coloring the map by using exactly i colors when mere permutations of the colors are disregarded. With this definition it is clear that*
>
> $$m_i \cdot \lambda \cdot (\lambda - 1) \cdots (\lambda - i + 1)$$
>
> *represents the number of ways of coloring the given map in exactly i of the λ colors counting two colorings as distinct when they are obtained by a permutation one from the other; for, of the i colors used, the first may be chosen in λ ways, the second in $\lambda - 1$ ways, and so on. If λ is less than i the above term reduces to zero.*
>
> *But the total number of ways of coloring the given map in λ colors is the sum of the number of ways of coloring it with $1, 2, \ldots, n$ of these colors, since no more than n colors can be used. Accordingly the total number of ways is represented by*
>
> $$P(\lambda) = m_1\lambda + m_2\lambda(\lambda - 1) + \cdots + m_n\lambda(\lambda - 1)\cdots(\lambda - n + 1)$$
>
> *for all values of λ. It is clear that in general $m_1 = 0$ inasmuch as for $n > 1$ no map can be colored in a single color, and that $m_n = 1$ since there is only one way of coloring M in n colors if permutation of the colors be disregarded.*

The rest of Birkhoff's paper contains the derivation of a formula for the coefficients of P, which is unfortunately not too practical since it requires checking all possible submaps of the given map.

Birkhoff kept returning to the four-color conjecture and chromatic polynomials throughout his life; for example, in 1929 he proved the estimate $P(\lambda) \geq \lambda(\lambda - 1)(\lambda - 2)(\lambda - 3)^{n-3}$ for each map with $n \geq 3$ regions and all $\lambda \neq 4$.[59] His obsession with the four-color conjecture was nicely characterized in Biggs, Lloyd and Wilson (1998, p. 169):

> *Birkhoff was intrigued by the four-colour problem, and in later years he regretted that he had wasted so much time on it. But he also declared that every great mathematician had at some time attacked the problem, and had, for a while, believed himself successful. From his son, Garrett, we learn that he would ask his wife to prepare suitably complicated maps for him to colour: Mrs. Birkhoff's opinion of this task has not been recorded.*

The method for calculating the coefficients of the chromatic polynomial was greatly simplified by Hassler Whitney (1907–1989), who was working on his doctoral dissertation entitled *The Coloring of Graphs* under the supervision of Birkhoff. Whitney did not restrict himself to planar graphs (maps), but considered the chromatic polynomial P for an arbitrary graph G having n vertices. He supposed that the edges of G are arranged in a definite order, considered all possible cycles in G, and removed from each the last edge; this procedure leads to a collection of

[59] G.D. Birkhoff, On the Number of Ways of Colouring a Map, *Proceedings of the Edinburgh Mathematical Society* **2**(1930), pp. 83–91.

broken cycles. Then he showed that the chromatic polynomial of G is given by $P(\lambda) = \lambda^n + \sum_{i=1}^{n-1}(-1)^i k_i \lambda^{n-i}$, where k_i is the number of ways of choosing a set of i edges of G that does not contain any of the broken cycles. Hence, Whitney was not only able to simplify Birkhoff's procedure for calculating the coefficients, but also discovered their combinatorial meaning.[60] He derived other properties of the chromatic polynomial, including the fact that $k_i > 0$ for all i between 0 and $n - c$, where c is the number of connected components of G, and $k_i = 0$ for $i > n - c$. Hence, the coefficients of $\lambda^0, \ldots, \lambda^{c-1}$ in P are zero, while the coefficients of $\lambda^c, \ldots, \lambda^n$ are nonzero and alternate in signs.

The four-color theorem deals with planar graphs, which can be drawn on the plane in such a way that the edges do not cross each other. The simplest examples of non-planar graphs are K_5, the complete graph on five vertices, and $K_{3,3}$, the complete bipartite graph on six vertices (also known as the utility graph). A result due to the Polish mathematician Kazimierz Kuratowski (1896–1980) essentially says that each non-planar graph has to contain one of the two previously mentioned graphs. The formulation that is now standard is as follows: *A graph is planar if and only if it does not contain a subgraph that is a subdivision of K_5 or $K_{3,3}$*. This result goes back to Kuratowski's 1930 paper *Sur le problème des courbes gauches en topologie*,[61] where it was discussed in a rather abstract topological context (the word "graph" never appears in the paper). Kuratowski recalled the problem of characterizing topological spaces that are homeomorphic to subsets of the Euclidean plane. He restricted himself to Peano continua that contain only finitely many simple closed curves, and proved that if such a continuum cannot be embedded in the plane, then it necessarily contains one of the two types of configurations:

> *Je prouve qu'un continu gauche* [non-planar] *de ce genre contient nécessairement une courbe homéomorphe à l'une des deux courbes suivantes:*
>
> (1) *courbe composée des arêtes d'un tétraèdre et d'un segment unissant deux arêtes disjointes* (*cette courbe contient* 6 *points d'ordre* 3)*:*

---

[60]The abovementioned result was originally announced in the short paper H. Whitney, The Coloring of Graphs, *Proceedings of the National Academy of Sciences of the USA* **17**(1931), pp. 122–125. A detailed derivation is given in H. Whitney, A Logical Expansion in Mathematics, *Bulletin of the American Mathematical Society* **38**(1932), pp. 572–579. This paper is written in an expository style, and the main topic is the inclusion–exclusion principle. Whitney noted that the formula *... is well known to logicians. It should be better known to mathematicians also; we give in this paper several applications which show its usefulness.* He gave three applications: the calculation of the number of integers not exceeding a given number that are not divisible by any of the given numbers, the number of derangements of a finite set, and finally the result on the coefficients of the chromatic polynomial. A much deeper and extensive treatment of Whitney's results on the chromatic polynomial followed in the paper H. Whitney, The Coloring of Graphs, *Annals of Mathematics* **33**(1932), pp. 688–718. A readable account of Whitney's mathematical results, including his work on chromatic polynomials, dual graphs and matroids, is available in Kendig (2018).

[61]C. Kuratowski, Sur le Problème des Courbes Gauches en Topologie, *Fundamenta Mathematicae* **15**(1930), pp. 271–283.

(2) *courbe composée des arêtes d'un tétraèdre et des* 4 *segments qui unissent le centre de gravité du tétraèdre à ses sommets* (*cette courbe contient* 5 *points d'ordre* 4).

Kuratowski noted that these "curves", which we recognize to be $K_{3,3}$ and K_5, obviously cannot be embedded in the plane, and therefore his result provides a necessary and sufficient condition. Interestingly, the characterization of planar graphs was also obtained independently by the American mathematicians Orrin Frink (1901–1988) and Paul Althaus Smith (1900–1980). The one-sentence abstract of their paper, submitted for presentation at a meeting of the American Mathematical Society, was published in 1930: "One of the results of this paper is a simple necessary and sufficient condition that an arbitrary linear graph be mappable on a plane".[62] However, the paper itself was never published, because Kuratowski's paper had already appeared. More details of the history of this result, including the role of Lev Semenovich Pontryagin (1908–1988), are available in Kennedy, Quintas and Sysło (1985).

Another famous result in graph theory that was originally discovered in the context of topology is Menger's theorem, which can be formulated as follows: *If P, Q are disjoint sets of vertices in a graph G such that no $n-1$ vertices disconnect P from Q, then there exist n internally disjoint paths from P to Q*. The corresponding topological result published in the 1927 paper *Zur allgemeinen Kurventheorie*[63] by Karl Menger (1902–1985) reads as follows: *Ist K ein kompakter regulär eindimensionaler Raum, welcher zwischen den beiden endlichen Mengen P und Q n-punktig zusammenhängend ist, dann enthält K n paarweise fremde Bögen, von denen jeder einen Punkt von P und einen Punkt von Q verbindet.* The graph-theoretical version appeared in the previously mentioned Kőnig's 1936 textbook. Much later, Menger published a treatise on the origins of his theorem, and explained how it made its way to Kőnig's textbook (see Menger, 1981):

> *Some graph theorists may be surprised to learn that this graph theoretical assertion first came up in 1926 as a lemma in proving an extremely general theorem of set theoretical curve theory. ... In the spring of 1930, I came through Budapest and met there a galaxy of Hungarian mathematicians. In particular, I enjoyed making the acquaintance of Dénes Kőnig, for I greatly admired the work on set theory of his father, the late Julius Kőnig — to this day one of the most significant contributions to the continuum problem — and I had read with interest some of Dénes' papers. Kőnig told me that he was about to finish a book that would include all that was known about graphs. I assured him that such a book would fill a great need; and I brought up my n-Arc Theorem which, having been published as a lemma in a curve-theoretical paper, had not yet come to his attention. Kőnig was greatly interested, but did not believe that the theorem was correct. "This evening," he said to*

[62] O. Frink and P.A. Smith, Irreducible Non-Planar Graphs, *Bulletin of the American Mathematical Society* **36**(1930), p. 214.

[63] K. Menger, Zur allgemeinen Kurventheorie, *Fundamenta Mathematicae* **10**(1927), pp. 96–115.

me in parting, "I won't go to sleep before having constructed a counterexample". When we met again the next day he greeted me with the words, "A sleepless night!" and asked me to sketch my proof for him. He then said that he would add to his book a final section devoted to my theorem. This he did; and it is largely thanks to Kőnig's valuable book that the n-Arc Theorem has become widely known among graph theorists.

But let us return to planar graphs. A purely combinatorial characterization of this class was discovered by Whitney, and is based on the notion of a combinatorial dual graph. The usual geometric definition of the dual graph G^* to a planar graph G depends on the choice of planar drawing. In the interior of each face of G, we place a single vertex of G^*. Both G and G^* have the same number of edges. Each edge e^* in G^* crosses a unique edge e in G and joins two vertices of G^* corresponding to faces separated by e. In his 1932 paper *Non-Separable and Planar Graphs*,[64] Whitney introduced a new definition of a dual graph, now called the combinatorial dual. His original definition was formulated in terms of rank and nullity, but a more transparent condition[65] can be found in the subsequent 1933 paper *Planar Graphs*[66]: Two connected graphs are combinatorial duals of each other if there is a one-to-one correspondence between their edges such that any circuit in one graph corresponds to a cut-set (i.e., a set of edges whose removal leaves the graph disconnected) in the other graph, and vice versa.[67] Whitney's planarity criterion then reads as follows: *A connected graph is planar if and only if it has a combinatorial dual.* After proving this theorem, Whitney deduced the following result corresponding to Kuratowski's criterion:

> *A necessary and sufficient condition that a graph have a dual is that it contain neither of the following types of graphs as a subgraph:*
>
> K_1 — *This graph is formed by taking five vertices, and joining each two by an arc or suspended chains.*
>
> K_2 — *This graph is formed by taking two sets of three vertices each, and joining each vertex in one set to each vertex in the other set by an arc or suspended chain.*

Trees represent an important class of planar graphs. It is a basic fact that each tree on n vertices has exactly $n-1$ edges; what is the total number of such trees? More precisely, we are interested in the number of trees whose vertices are labeled by n different symbols, say $1, \dots, n$. Arthur Cayley (1821–1895) discovered that

---

[64]H. Whitney, Non-Separable and Planar Graphs, *Transactions of the American Mathematical Society* **34**(1932), pp. 339–362.

[65]A modern and accessible treatment of Whitney's results, including both approaches to the definition of a combinatorial dual, is available in Thulasiraman and Swamy (1992); Deo (1974).

[66]H. Whitney, Planar Graphs, *Fundamenta Mathematicae* **21**(1933), pp. 73–84.

[67]Equivalently, any cycle in one graph corresponds to a minimal cut-set in the other graph, and vice versa.

the result is given by the simple formula n^{n-2}, but his 1889 paper *A Theorem on Trees*[68] lacked a complete proof and discussed only the cases $n = 4$ and $n = 6$. The first rigorous proof was given by Heinz Prüfer (1896–1934) in the 1918 paper *Neuer Beweis eines Satzes über Permutationen.*[69] Many other proofs of Cayley's formula have been discovered since then, but Prüfer's combinatorial proof still ranks among the most elegant ones.

Prüfer mentioned neither Cayley nor graph theory, but his goal was to give a simpler proof of a certain result on permutations due to the German mathematician Otto Dziobek (1856–1919) published in the previous year. Prüfer remarked that the problem can be reformulated in an intuitive geometrical form, as observed by Schur: If n cities have to be connected by a railway network of $n-1$ lines, the total number of such networks is n^{n-2}. The idea of the proof is to construct a bijection between trees with vertices $1, \ldots, n$, and sequences of such vertices of length $n - 2$. The procedure is simple: Find the leaf with the smallest label, remove it, and write down the number of its neighbors. The same step is repeated until we have a sequence of length $n-2$, which is now known as the Prüfer sequence or code. Prüfer gave some examples and showed that the whole process can be reversed; an English translation of his paper is available in Biggs, Lloyd and Wilson (1998), pp. 52–54.

Cayley's formula might be interpreted as giving the total number of spanning trees of the complete graph on n vertices. Another important problem related to spanning trees is the following: Given a connected weighted graph, find its minimum spanning tree, i.e., a spanning tree the sum of whose edge weights is as small as possible. The first algorithm for solving this problem was proposed by the Czech mathematician Otakar Borůvka (1899–1995), who considered the problem of designing an electricity network in Moravia. His 1926 paper *O jistém problému minimálním* [On a Certain Minimal Problem][70] was written in Czech with a German summary, and the whole problem was formulated in the language of matrices, whose elements correspond to edge weights:

[68]A. Cayley, A Theorem on Trees, *The Quarterly Journal of Pure and Applied Mathematics* **23**(1889), pp. 376–378. Note that Cayley refers to a related problem involving determinants that was considered in the paper Borchardt, Ueber eine der Interpolation entsprechende Darstellung der Eliminations-Resultante, *Journal für die reine und angewandte Mathematik* **57**(1860), pp. 111–121.

[69]H. Prüfer, Neuer Beweis eines Satzes über Permutationen, *Archiv der Mathematik und Physik* **27**(1918), pp. 142–144.

[70]O. Borůvka, O jistém problému minimálním, *Práce Moravské přírodovědecké společnosti* **3**(1926), pp. 37–58. Borůvka's biography in O'Connor and Robertson (2019) contains the following note: *In fact the problem had been suggested to Borůvka before he became a university student. He had a friend, Jindřich Saxel, who worked for the firm West-Moravian Powerplants and he suggested the problem which he stated in terms of cities and the distances between them. At the time that Saxel suggested the problem to Borůvka, World War I was still happening and Czech universities were closed. Borůvka was offered a job with West-Moravian Powerplants at this time but declined.*

Es möge eine Matrix der bis auf die Bedingungen $r_{\alpha\alpha} = 0$, $r_{\alpha\beta} = r_{\beta\alpha}$ positiven und von einander verschiedenen[71] *Zahlen $r_{\alpha\beta}$ $(\alpha, \beta = 1, 2, \ldots, n;\ n \geq 2)$ gegeben sein.*

Aus dieser ist eine Gruppe von einander und von Null verschiedener Zahlen auszuwählen, so dass

1° *in ihr zu zwei willkürlich gewählten natürlichen Zahlen p_1, p_2 $(\leq n)$ eine Teilgruppe von der Gestalt $r_{p_1c_2}, r_{c_2c_3}, r_{c_3c_4}, \ldots, r_{c_{q-2}c_{q-1}}, r_{c_{q-1}p_2}$ existiere,*
2° *die Summe ihrer Glieder kleiner sei als die Summe der Glieder irgendeiner anderen, der Bedingung 1° genügenden Gruppe von einander und von Null verschiedenen Zahlen.*

However, the final paragraph of the Czech text contains the following geometric interpretation: Consider n points in the r-dimensional space, whose distances are pairwise distinct. The problem is to connect them by a network whose total length is as small as possible.

Borůvka's description of the algorithm for finding the optimal solution is long and complicated, but the procedure becomes much more transparent when reformulated in graph-theoretical language: Begin by joining each vertex with its nearest neighbor, obtaining a certain forest. For each component, add the shortest edge joining it to a different component, and repeat this step until obtaining a connected graph. Essentially the same explanation, but without using graph-theoretical terminology, was given in Borůvka's subsequent two-page paper *Příspěvek k řešení otázky ekonomické stavby elektrovodných sítí* [Contribution to the Solution of a Problem of Economical Construction of Electrical Networks],[72] published in the same year in a Czech journal aimed at electrical engineers. Borůvka referred to his earlier theoretical paper, noted that the problem is of importance in the design of electrical networks, and explained the algorithm by means of an example with 40 vertices. The paper Nešetřil, Milková and Nešetřilová (2001) contains English translations of both papers by Borůvka, as well as a comparison with other algorithms for finding minimum spanning trees. Borůvka's algorithm turns out to be rather efficient, and is the basis of even faster algorithms.

Another algorithm for solving the same problem was described by the Czech mathematician Vojtěch Jarník (1897–1970). His 1930 paper *O jistém problému minimálním.* (*Z dopisu panu O. Borůvkovi*) [On a Certain Minimal Problem. (From a Letter to O. Borůvka)][73] is formulated as a response to Borůvka's paper, whose goal

[71]This condition means that the edge weights have to be all distinct, but the restriction is not a serious one.

[72]O. Borůvka, Příspěvek k řešení otázky ekonomické stavby elektrovodných sítí, *Elektrotechnický obzor* **15**(1926), pp. 153–154.

[73]V. Jarník, O jistém problému minimálním. (Z dopisu panu O. Borůvkovi), *Práce Moravské přírodovědecké společnosti* **6**(1930), pp. 57–63.

is to propose an alternative and perhaps simpler procedure. Jarník's style is similar to the one by Borůvka – the problem is formulated in an algebraic way, and no graph terminology is used. Only the end of the paper provides an intuitive interpretation: There are n balls connected by $\frac{1}{2}n(n-1)$ rods of pairwise distinct weights. The goal is to remove some of these rods in such a way that the balls still hold together, and the mass of the remaining rods is as small as possible. When reformulated in graph-theoretical language, the new algorithm suggested by Jarník is simple: Begin with an arbitrary vertex, and find the shortest edge incident with this vertex, giving rise to a tree with two vertices. Add the shortest edge joining the tree to a vertex that is not included in the tree. Repeat this step until obtaining a tree containing all vertices. Jarník's paper is followed by a two-page summary in German; an English translation of all important passages is available in Korte and Nešetřil (2001). The algorithm is often called Prim's algorithm after the American mathematician and computer scientist Robert Clay Prim (born 1921), who discovered it independently in 1957.

Let us conclude this section by returning once more to the work of Dénes Kőnig and his 1936 textbook, which contains the following famous theorem: *For a graph G let m be the (finite) minimal number of vertices $P_1, P_2, \ldots, P_m$ with the property that every edge of G ends in one of these vertices, and let n be the maximal number of edges $k_1, k_2, \ldots, k_n$ which pairwise have no common endpoint. Then if G is a bipartite graph, $m = n$.* In modern terminology, the number of edges in a maximum matching in any bipartite graph equals the number of vertices in a minimum vertex cover. This result appeared already in 1931 in Kőnig's Hungarian paper *Graphok és matrixok* [Graphs and Matrices].[74] In the 1936 textbook, he showed that his theorem implies Hall's theorem on systems of distinct representatives: He constructed a bipartite graph in which one part corresponds to n sets and the other part to the elements of their union; edges indicate that a certain element is contained in a particular set. Then he showed the existence of a matching consisting of n edges, whose vertices lying in the second part correspond to the desired system of distinct representatives.

Kőnig also provided a matrix formulation if his theorem, which is obtained by identifying nonzero elements of a matrix with edges in a bipartite graph (whose parts corresponds to rows and columns of the matrix, respectively): *The minimal number of rows and columns which contain in their totality every nonzero element of a matrix is equal to the maximal number of nonzero elements which pairwise belong to different rows and columns.* Also in 1931, another Hungarian mathematician Jenő Egerváry (1891–1958) published the paper *Matrixok kombinatorius tulajdonságairól* [Concerning Combinatorial Properties of

[74]D. Kőnig, Graphok és matrixok, *Matematikai és Fizikai Lapok* **38**(1931), pp. 116–119.

Matrices][75] containing an even more general result concerned with maximizing the sum $\sum_{i=1}^{n} a_{i,\nu(i)}$ over all permutations ν of $\{1, \ldots, n\}$; this problem corresponds to finding the maximum matching in a weighted bipartite graph represented by the matrix A. In 1953, the ideas of Kőnig and Egerváry became the basis of the "Hungarian method" developed by the American mathematician Harold William Kuhn (1925–2014) for solving the assignment problem.[76]

6. Additional Results

The survey of important results on combinatorics obtained throughout the interwar period would be incomplete without mentioning Whitney's pioneering work on matroids. His 1935 paper *On the Abstract Properties of Linear Dependence*[77] grew out of his earlier work on graph theory, and his observation of some surprising similarities with linear algebra. The paper provided four equivalent definitions of a matroid, including one that is perhaps the most familiar today:

> *Let M be a set of elements. Let any subset N of M be either "independent" or "dependent". Let the two following postulates be satisfied:*
>
> (I_1) *Any subset of an independent set is independent.*
> (I_2) *If $N = e_1 + \cdots + e_p$ and $N' = e'_1 + \cdots + e'_{p+1}$ are independent, then for some i such that e'_i is not in N, $N + e'_i$ is independent.*

Then the set M together with the collection of its independent subsets is called a matroid. The two main examples that gave rise to the notion of a matroid are: (1) M is the set of all columns of a certain matrix, a set $N \subset M$ being independent if and only if the corresponding set of vectors is linearly independent. (2) M is the set of all edges of a certain graph, a set $N \subset M$ being independent if and only if the corresponding subgraph of G contains no circuit.

The remaining three definitions proposed by Whitney characterize matroids in terms of rank functions, bases and circuits. After showing the equivalence of all four definitions, Whitney proceeded to separable matroids and dual matroids. He showed that each matroid has a dual, in contrast to graphs, which have a dual

---

[75] J. Egerváry, Matrixok kombinatorius tulajdonságairól, *Matematikai és Fizikai Lapok* **38**(1931), pp. 16–28.

[76] See H.W. Kuhn, The Hungarian Method for the Assignment Problem, *Naval Research Logistics Quarterly* **2**(1955), pp. 83–97. The story behind the discovery of the Hungarian method was described by Kuhn (2012), see also Martello (2010).

[77] H. Whitney, On the Abstract Properties of Linear Dependence, *American Journal of Mathematics* **57**(1935), pp. 509–533. This is the last paper from the interwar period included in Gessel's and Rota's *Classic Papers in Combinatorics*, Gessel and Rota (2009). The remaining ones are Ramsey's *On a Problem of Formal Logic*, Whitney's *Non-Separable and Planar Graphs*, Erdős's and Szekeres's *A Combinatorial Problem in Geometry*, and Hall's *On Representatives of Subsets*; see the references earlier in this text.

if and only if they are planar. A readable account of these ideas may be found in Kendig (2018), Chapter 8; see also the expository paper Wilson (1973). Interestingly, some ideas similar to those of Whitney were developed independently and almost simultaneously by the Japanese mathematician Takeo Nakasawa (1913–1946), see Nishimura and Kuroda (2009). More information on the history of matroid theory is available in Kung (1986).

Finally, let us mention two landmark papers in enumeration theory, *The Theory of Group-Reduced Distributions*[78] by John Howard Redfield (1879–1944), and *Kombinatorische Anzahlbestimmungen für Gruppen, Graphen und chemische Verbindungen*[79] by George Pólya (1887–1985). Both mathematicians worked independently on similar problems related to calculating the number of configurations while taking into account symmetries represented by groups (e.g., coloring the vertices of polyhedra, enumeration of trees, chemical compounds, etc.). Even a basic overview of their results would involve technical details that are beyond the scope of this text; we refer the reader to Lloyd (2013) for a nice historical overview of this subject.

7. Biographical Notes

It comes as no surprise that the lives of mathematicians mentioned in the previous text were affected by World War I, the rise of antisemitism, and World War II. Without any claim to completeness, let us mention some examples; the following biographical information relies mainly on O'Connor and Robertson (2019).

World War I interrupted university studies of several young mathematicians. Two of them were Heinz Prüfer and Alfred Brauer, students of the University of Berlin. Prüfer had more luck, and spent his military service period from 1917 till 1918 working for the Königlich Preussische Landesaufnahme, whose goal was to contribute to the war efforts by dealing with problems in surveying, cartography, and topography. On the other hand, Brauer had to serve for four years in the German army. He was seriously wounded during the war, and lost seven years of his professional career. His military service initially helped him to keep his lecturing post in Berlin in 1933; despite being a Jew, he was exempted from the 1933 law according to which all civil servants of non-Aryan descent were to be retired. Nevertheless, Brauer was dismissed in 1935. In 1938, Brauer's brother and mathematician Richard was already in the United States, but Alfred failed to obtain the "professor visa" as he did not satisfy the condition of having an academic position during the last two

[78]J.H. Redfield, The Theory of Group-reduced Distributions, *American Journal of Mathematics* **49**(1927), pp. 433–455.

[79]G. Pólya, Kombinatorische Anzahlbestimmungen für Gruppen, Graphen und chemische Verbindungen, *Acta Mathematica* **68**(1937), pp. 145–254. An English translation is included in G. Pólya and R.C. Read, *Combinatorial Enumeration of Groups, Graphs, and Chemical Compounds*, Springer, New York, 1987, viii + 148 pages. ISBN 0-387-96413-4.

years. He finally managed to escape to the United States in 1939, unlike his sister Alice, who died in a concentration camp.

The German mathematician Issai Schur, also of Jewish origin, was dismissed already in 1933, despite being a pre-war official. The decision was later revoked, but Schur was dismissed again in 1935. He received invitations from the United States and Great Britain, but did not accept them since he probably felt too old to emigrate. In 1938, the antisemitic mathematician Ludwig Bieberbach (1886–1982) expressed astonishment that his former co-author Schur is still a member of the Prussian Academy of Sciences despite being a Jew. Schur was forced to resign from the Academy, and finally decided to leave Germany for Palestine in 1939, where he died two years later at the age of 66.

Although his parents were of Jewish origin, Dénes Kőnig was brought up as a Christian, and hence was initially unaffected by the persection of Hungarian Jews. The situation changed in 1944, when Germany invaded Hungary and installed a new government composed of members of the Hungarian National Socialist Party. This government launched a reign of terror against the Jews of Budapest, and thousands of them were murdered. To avoid this fate, Kőnig committed suicide in 1944.

Paul Erdős, another Hungarian mathematician of Jewish origin, saved his life by leaving his native land shortly after receiving the doctorate in 1934, when he obtained a post-doctoral fellowship at Manchester. In 1938, he relocated to the United States, working subsequently in Princeton, Madison, and Purdue University. He had no contact with his family in Hungary from 1941 to 1945, when he learned that several of his relatives had been murdered.

From 1933 till 1939, George Szekeres worked as an analytical chemist in Budapest. Being endangered by the persecution of Jews, he and his wife Esther escaped to Shanghai, where George continued to work as a chemist for some time. Until 1939, no visas were required for entering Shanghai. Although China faced Japanese invasion and was far from being a safe place, many thousands of European Jews, including the Szekeres family, survived the war years in Shanghai.

Two young Hungarian mathematicians, Jenő Egerváry and George Pólya, have managed to bypass military service in World War I. Egerváry was found unfit to serve, despite the fact that during 1910–1920 he developed a strong reputation as a mountaineer, climbing the highest peaks of the Tatra mountains. In the beginning of the war, Pólya was found unfit to serve because of an old injury. Later, when the Hungarian army needed more soldiers, Pólya was urged to return from Switzerland. He refused, and subsequently feared to come back to Hungary after the war. Being afraid that he might be arrested for failing to undertake war service, he returned to Hungary no sooner than in 1967.

The events of World War I did not always have a detrimental effect on the mathematics community. In 1913, Poland did not exist as an independent state, and Warsaw was a part of the Russian Empire. Instruction in Polish was prohibited, and the young Kazimierz Kuratowski decided to take an engineering course at the University of Glasgow. After one year, the outbreak of World War I prevented him from returning to Scotland. In 1915, the Russian army withdrew from Warsaw, which fell under German occupation. Fortunately, the new German governor general reopened the University of Warsaw with Polish as the language of instruction. Kuratowski enrolled as a student of mathematics, and graduated in 1919 when Warsaw was already a capital of the Polish Republic. It came under German occupation again in 1939, and many academicians were murdered or sent to concentration camps. During the war, Kuratowski risked his life and taught at the illegal underground university in Warsaw.

The Austrian mathematician Karl Menger, whose mother was Jewish, left Vienna in 1937, one year before Germany's annexation of Austria, and accepted a chair at the University of Notre Dame, Indiana. From 1942 to 1944, Menger was teaching calculus to naval cadets as part of the Navy College Training Program. Although he was planning to return to Vienna after the war, the university found no position for him. In 1948, Menger moved to the Illinois Institute of Technology in Chicago, where he remained for the rest of his life.

Antisemitism was present also in the United States, where it reached its peak in the interwar period. New legislation adopted in the 1920s restricted immigration from Eastern European countries with large Jewish populations. In 1922, Harvard University considered introducing quotas on the number of Jewish students. Although the plan was abandoned, a number of universities adopted numerus clausus policies limiting admissions of students based on their religion or race, which effectively led to the reduction of the number of Jewish students. George Birkhoff, who became full professor at Harvard in 1919 and remained there throughout the rest of his life, was known for his antisemitic views. The following quote is from Chapter 3 of Nadis and Yau (2013):

> *Although Birkhoff went out of his way to help some Jewish scholars like Ulam, none was hired for a permanent math faculty position during his three-plus decades at Harvard (from 1912 to 1944).*
>
> *...it is undeniable that in 1934 Birkhoff opposed Solomon Lefschetz's election as president of the American Mathematical Society at least in part because he was Jewish. In a letter to his friend Roland Richardson, who was then secretary of the society, Birkhoff wrote: "I have a feeling that Lefschetz will be likely to be less pleasant even than he had been, in that from now on he will try to work strongly and positively for his own race. They are exceedingly confident of their own power and influence in the good old USA .... He will be very cocky, very racial, and use the Annals [of Mathematics, which Lefschetz edited from 1928 to 1958] as a good deal of racial perquisite. The racial interests will get deeper as Einstein's and all of them do".*
>
> *Although Birkhoff was not without company, statements of this sort would suggest that he had a far more jaundiced view of his Jewish colleagues than most of his contemporaries. Albert Einstein, to whom Birkhoff alluded above, called Birkhoff "one of the world's great anti-Semites".*

Another controversial person was the Dutch mathematician Bartel Leendert van der Waerden. In 1931, he became professor at the University of Leipzig, and kept this position until the end of World War II. In 1935, he protested against the dismissal of five Jewish professors. As one of the associate editors of the *Mathematische Annalen*, he came under pressure from the Nazis not to publish papers by Jewish authors; instead of a possible resignation, he chose to keep his post to avoid being replaced by a Nazi mathematician. Surprisingly, van der Waerden declined repeated invitations to leave Germany for Princeton or Utrecht. In the end of the war, when Leipzig was suffering from heavy bombing, van der Waerden with his wife and children moved to the countryside of Austria, wherefrom they were transported by the American army to the Netherlands. Because he had worked in Germany throughout the war, he faced great difficulties when trying to find a position in the Dutch academic system. A campaign by the influential newspaper *Het Parool* caused his appointment at the University of Amsterdam to be postponed from 1946 till 1948. English translations of the corresponding newspaper articles as well as van der Waerden's responses are available in Soifer (2009). Let us quote some excerpts:

> *The proposal to appoint Dr B C van der Waerden as professor in the faculty of mathematics and physics at the University of Amsterdam should surprise all those who know that Mr. Van der Waerden served the enemy throughout the entire war... For five years the Netherlands fought Germany and for all those five years Mr. Van der Waerden kept the light of science shine in Leipzig. He raised Hitler-followers... Not because Mr. Van der Waerden had been gang-pressed to the forced Arbeitseinsatz, not because it was impossible for Mr. Van der Waerden to go into hiding; no, Mr. Van der Waerden served the enemy, because he liked it at Leipzig; he was completely voluntary a helper of the enemy... Mr. Van der Waerden from the first until the last day of the war served science in the land of the enemy and this was compensated by the enemy's money. He who has voluntarily served the enemy from May '40 to May '45 is a bad Dutchman.*

During World War II, a number of mathematicians were participating in various military projects. In 1941, Philip Hall joined the Code and Cypher School at Bletchley Park to work on Italian and later Japanese ciphers. Marshall Hall Jr worked for the American Naval Intelligence and was also involved in deciphering Japanese coded messages, as well as German Enigma codes. During 1944, he spent six months at Bletchley Park. In 1943, Hassler Whitney became a member of the Applied Mathematics Panel of the National Defense Research Committee, where he specialized in air-to-air rocketry problems, studied the corresponding differential equations, and suggested improvements in aiming at moving targets; see McCleary (2006). On the other side of the conflict, Emanuel Sperner worked as an assistant in the German Navy's Weather Service from 1942 to 1944.

References

L.D. Andersen, Latin Squares, pp. 251–283, in R. Wilson, J.J. Watkins (eds.), *Combinatorics: Ancient and Modern*, Oxford University Press, Oxford, 2013, x + 381 pages. ISBN 978-0-19-965659-2.

I. Anderson, Combinatorial Set Theory, pp. 309–328, in R. Wilson, J.J. Watkins (eds.), *Combinatorics: Ancient and Modern*, Oxford University Press, Oxford, 2013, x + 381 pages. ISBN 978-0-19-965659-2.

I. Anderson, *Combinatorics of Finite Sets*, Dover Publications, Mineola, NY, 2002, xvi + 250 pages. ISBN 0-486-42257-7.

L. Beineke and R. Wilson, Modern Graph Theory, pp. 331–352, in R. Wilson, J.J. Watkins (eds.), *Combinatorics: Ancient and Modern*, Oxford University Press, Oxford, 2013, x + 381 pages. ISBN 978-0-19-965659-2.

N.L. Biggs, E.K. Lloyd and R.J. Wilson, The History of Combinatorics, pp. 2163–2198, in R.L. Graham, M. Grötschel, L. Lovász (eds.), *Handbook of Combinatorics*, Elsevier, Amsterdam, 1995, cii + 2198 pages. ISBN 0-444-88002-X.

N.L. Biggs, E.K. Lloyd and R.J. Wilson, *Graph Theory 1736–1936*, Clarendon Press, Oxford, 1998, x + 239 pages. ISBN 0-19-853916-9.

N. Biggs and R. Wilson, Block Designs, pp. 231–249, in R. Wilson, J.J. Watkins (eds.), *Combinatorics: Ancient and Modern*, Oxford University Press, Oxford, 2013, x + 381 pages. ISBN 978-0-19-965659-2.

B. Bollobás, Paul Erdős — Life and Work, pp. 1–41, in R.L. Graham, J. Nešetřil (eds.), *The Mathematics of Paul Erdős*. Volume I, Springer, Berlin, 1997, xiv + 399 pages. ISBN 3-540-61032-4.

N. Deo, *Graph Theory with Applications to Engineering and Computer Science*, Prentice-Hall, Englewood Cliffs, NJ, 1974, xvii + 478 pages. ISBN 0-13-363473-6.

A.W.F. Edwards and R.A. Fisher, Statistical Methods For Research Workers, First Edition (1925), pp. 856–870, in I. Grattan-Guinness (ed.), *Landmark Writings in Western Mathematics 1640–1940*, Elsevier, 2005, xvii + 1022 pages. ISBN 0-444-50871-6.

P.D.T.A. Elliot, *Probabilistic Number Theory I. Mean-Value Theorems*, Springer-Verlag, New York, 1979, xxii + 393 pages. ISBN 0-387-90437-9.

K. Engel, *Sperner Theory*, Cambridge University Press, Cambridge, 1997, ix + 417 pages. ISBN 0-521-45206-6.

M. Franchella, On the Origins of Dénes Kőnig's Infinity Lemma, *Archive for History of Exact Sciences* **51**(1997), pp. 3–27. ISSN 0003-9519.

H.H. Frisinger, The Solution of a Famous Two-Centuries-Old Problem. The Leonhard Euler-Latin Square Conjecture, *Historia Mathematica* **8**(1981), pp. 56–60. ISSN 0315-0860.

T. Gallai, The Life and Scientific Work of Dénes Kőnig (1884–1944), *Linear Algebra and Its Applications* **21**(1978), pp. 189–209. ISSN 0024-3795.

I. Gessel and G.-C. Rota (eds.), *Classic Papers in Combinatorics*, Birkhäuser, Basel, 2009, x + 489 pages. ISBN 978-0-8176-4841-1.

K. Kendig, *Never a Dull Moment. Hassler Whitney, Mathematics Pioneer*, MAA Press, Providence, RI, 2018, xix + 385 pages. ISBN 978-1-4704-4828-8.

J.W. Kennedy, L.V. Quintas and M.M. Sysło, The Theorem on Planar Graphs, *Historia Mathematica* **12**(1985), pp. 356–368. ISSN 0315-0860.

D. Klyve and L. Stemkoski, Graeco-Latin Squares and a Mistaken Conjecture of Euler, *The College Mathematics Journal* **37**(2006), pp. 2–15. ISSN 0746-8342.

B. Korte and J. Nešetřil, Vojtěch Jarník's Work in Combinatorial Optimization, *Discrete Mathematics* **235**(2001), pp. 1–17. ISSN 0012-365X.

S. Kotz and N.L. Johnson (eds.), *Breakthroughs in Statistics. Volume II. Methodology and Distribution*, Springer, New York, 1992, xxi + 600 pages. ISBN 0-387-94039-1.

H.W. Kuhn, A Tale of Three Eras: The Discovery and Rediscovery of the Hungarian Method, *European Journal of Operational Research* **219**(2012), pp. 641–651. ISSN 0377-2217.

J. P. S. Kung, *A Source Book in Matroid Theory*, Birkhäuser, Boston, 1986, 413 pages. ISBN 0-8176-3173-9.

C.W.H. Lam, The Search for a Finite Projective Plane of Order 10, *American Mathematical Monthly* **98**(1991), pp. 305–318. ISSN 0002-9890.

E.K. Lloyd, Enumeration (18th–20th Centuries), pp. 285–307, in R. Wilson and J.J. Watkins (eds.), *Combinatorics: Ancient and Modern*, Oxford University Press, Oxford, 2013, x + 381 pages. ISBN 978-0-19-965659-2.

S. Martello, Jenő Egerváry: From the Origins of the Hungarian Algorithm to Satellite Communication, *Central European Journal of Operations Research* **18**(2010), pp. 47–58. ISSN 1435-246X.

J. McCleary, Airborne Weapons Accuracy: Topologists and the Applied Mathematics Panel, *The Mathematical Intelligencer* **28**(2006), pp. 17–21. ISSN 0343-6993.

K. Menger, On the Origin of the n-arc Theorem, *Journal of Graph Theory* **5**(1981), pp. 341–350. ISSN 0364-9024.

G.L. Mullen, A Candidate for the "Next Fermat Problem", *The Mathematical Intelligencer* **17**(1995), pp. 18–22. ISSN 0343-6993.

S. Nadis and S.-T. Yau, *150 Years of Mathematics at Harvard (1825–1975)*, Harvard University Press, Cambridge, MA, 2013, xx + 249 pages. ISBN 978-0-674-72500-3.

J. Nešetřil, E. Milková and H. Nešetřilová, Otakar Borůvka on Minimum Spanning Tree Problem. Translation of Both the 1926 Papers, Comments, History, *Discrete Mathematics* **233**(2001), pp. 3–36. ISSN 0012-365X.

H. Nishimura and S. Kuroda, *A Lost Mathematician, Takeo Nakasawa. The Forgotten Father of Matroid Theory*, Birkhäuser, Basel, 2009, xii+234 pages. ISBN 978-3-7643-8572-9.

J.J. O'Connor and E.F. Robertson, *MacTutor History of Mathematics*, available online at https://www-history.mcs.st-andrews.ac.uk/ [30.8.2019].

J.T.E. Richardson, Who Introduced Western Mathematicians to Latin Squares?, *British Journal for the History of Mathematics* **34**(2019), pp. 95–103. ISSN 2637-5451.

H. Schneider, The Concepts of Irreducibility and Full Indecomposability of a Matrix in the Works of Frobenius, Kőnig and Markov, *Linear Algebra and Its Applications* **18**(1977), pp. 139–162. ISSN 0024-3795.

A. Soifer, *The Mathematical Coloring Book: Mathematics of Coloring and the Colorful Life of Its Creators*, Springer, New York, NY, 2009, xxx + 607 pages. ISBN 978-0-387-74640-1.

G. Strang, *Introduction to Applied Mathematics*, Wellesley-Cambridge Press, Wellesley, MA, 1986, xii + 758 pages. ISBN 0-9614088-0-4.

G.P.H. Styan, C. Boyer and K.L. Chu, Some Comments on Latin Squares and on Graeco-Latin Squares, Illustrated with Postage Stamps and Old Playing Cards, *Statistical Papers* **50**(2009), pp. 917–941. ISSN 0932-5026.

K. Thulasiraman and M.N.S. Swamy, *Graphs: Theory and Algorithms*, John Wiley & Sons, New York, 1992, xv + 460 pages. ISBN 0-471-51356-3.

P. Ullrich, Officers, Playing Cards, and Sheep. On the History of Eulerian Squares and of the Design of Experiments, *Metrika* **56**(2002), pp. 189–204. ISSN 0026-1335.

M. Wate-Mizuno, Mathematical Recreations of Dénes Kőnig and His Work on Graph Theory, *Historia Mathematica* **41**(2014), pp. 377–399. ISSN 0315-0860.

R. Wilson and J.J. Watkins (eds.), *Combinatorics: Ancient and Modern*, Oxford University Press, Oxford, 2013, x + 381 pages. ISBN 978-0-19-965659-2.

R. Wilson, An Introduction to Matroid Theory, *American Mathematical Monthly* **80**(1973), pp. 500–525. ISSN 0002-9890.

F. Yates, Sir Ronald Fisher and the Design of Experiments, *Biometrics* **20**(1964), pp. 307–321. ISSN 0006-341X.

X. Zhan, *Matrix Theory*, American Mathematical Society, Providence, RI, 2013, x + 253 pages. ISBN 978-0-8218-9491-0.

https://doi.org/10.1142/9781786349316_bmatter

Index

D

E

F

G

M

N

CPSIA information can be obtained
at www.ICGtesting.com
Printed in the USA
JSHW020424110621
15772JS00003B/18

9 781786 349309